Handbook of Experimental Pharmacology

Volume 141

Springer
Berlin
Heidelberg
New York
Barcelona
Hong Kong
London
Milan
Paris
Singapore
Tokyo

Ionotropic Glutamate Receptors in the CNS

Contributors

G. Adelmann, R. Bähring, P. Béhé, D. Bowie, D.W. Choi, D. Colquhoun, M. Frotscher, R. Ganel, J.R.P. Geiger, R. Gill, M. Hollmann, R.L. Huganir, P. Jonas, M.P. Kavanaugh, J.A. Kemp, M.B. Kennedy, G.A. Kerchner, J.N.C. Kew, A.H. Kim, H.-C. Kornau, T. Kuner, H.-K. Lee, J. Lerma, M.L. Mayer, K. Mews, H. Monyer, T.S. Otis, R.S. Petralia, J. Rossier, A. Roth, J.D. Rothstein, M.E. Rubio, B. Sakmann, P.H. Seeburg, B. Taskin, L.O. Trussell, R.J. Wenthold, L.P. Wollmuth, D.J.A. Wyllie

Editors

Peter Jonas and Hannah Monyer

Springer

Professor Dr. Peter JONAS
Albert-Ludwigs-Universität Freiburg
Physiologisches Institut
Hermann-Herder-Str. 7
D-79104 Freiburg
Germany
e-mail: jonasp@ruf.uni.freiburg.de

Professor Dr. Hannah MONYER
ZMBH, Im Neuenheimer Feld 282
D-69120 Heidelberg
Germany
e-mail: monyer@mpimf-heidelberg.mpg.de

With 70 Figures and 38 Tables

ISBN 3-540-66120-4 Springer-Verlag Berlin Heidelberg New York

Library of Congress Cataloging-in-Publication Data applied for

Die Deutsche Bibliothek – CIP-Einheitsaufnahme

Ionotropic glutamate receptors in the CNS / contributors G. Adelmann . . . Ed. Peter Jonas and Hannah Monyer. – Berlin ; Heidelberg ; New York ; Barcelona ; Hong Kong ; London ; Milan ; Paris ; Singapore ; Tokyo : Springer, 1999
(Handbook of experimental pharmacology ; Vol. 141)
ISBN 3-540-66120-4

Cover design: *design & production* GmbH, Heidelberg

Typesetting: Best-set Typesetter Ltd., Hong Kong

SPIN: 10674788 27/3020 – 5 4 3 2 1 0 – Printed on acid-free paper

Preface

It is now widely accepted that glutamate is the major excitatory neurotransmitter in the mammalian central nervous system. The main criteria for accepting a molecule as a chemical transmitter appear to be fulfilled at several synapses: Glutamate mimics the action of the natural transmitter in the postsynaptic neuron (CURTIS et al. 1959), glutamate is present in presynaptic elements (OTTERSEN and STORM-MATHISEN 1984), and glutamate is released from central neurons in an activity-dependent manner (BRADFORD 1970).

The postsynaptic receptors that mediate the effects of glutamate are markedly diverse. Based on their activation by agonists that act more selectively than the natural transmitter itself, α-amino-3-hydroxy-5-methyl-4-isoxazolepropionate (AMPA) receptors, kainate receptors, and *N*-methyl-D-aspartate (NMDA) receptors can be distinguished. Molecular cloning has revealed additional structural diversity. To date, almost 20 glutamate receptor subunit genes have been identified, and an even larger number of splice variants and edited versions are present in the mammalian brain.

Analysis of synaptic transmission revealed that "the" excitatory synapse does not exist. Glutamatergic synapses in different circuitries differ substantially in their signaling properties, although they use the same transmitter. We have learned that cellular, subcellular, and molecular factors determine synaptic function, and that glutamate receptor subunit diversity is of direct relevance in shaping the unique signaling properties of a glutamatergic synapse.

Whereas glutamate is an essential signaling molecule at central synapses, excessive glutamate release may have negative effects. Glutamate receptor overactivation mediates Ca^{2+} influx into neurons, either directly through the Ca^{2+} permeability of the receptor channels, or indirectly by depolarization of the postsynaptic cell. If the Ca^{2+} influx is sufficiently large, this may result in cell death.

In this book, we decided to concentrate on these three main topics (glutamate receptor structure and function, mechanisms of glutamatergic synaptic transmission, and excitotoxic glutamate effects). Even with these restrictions, however, it is impossible to cover all aspects exhaustively. We have thus concentrated on facets that have emerged more recently in these areas of research, while putting less weight on those that were reviewed extensively in the past.

The book is the first one on glutamate receptors in the *Handbook of Experimental Pharmacology* series, although certain aspects have been reviewed in previous volumes, for example in Vol. 102 *Selective Neurotoxicity* (1992; edited by H. Herken and F. Hucho). The emphasis in the present book is on physiology rather than pharmacology, but this, we believe, is justified by the close relationship between the two disciplines, which one day, according to Oswald Schmiedberg's famous dictum of 1883, may "merge into a unifying theory of life".

We would like to thank all of our colleagues who contributed to this volume. We also thank the Springer Verlag staff, particularly Mrs. Doris Walker, for support.

A final remark should be made regarding nomenclature. A unified glutamate receptor subunit nomenclature does not exist yet; different nomenclatures for the same subunits are in use in different laboratories. We agree with Wilhelm Feldberg that certain types of scientists, perhaps including ourselves, would "rather use his colleague's toothbrush than his terminology." Accordingly, we have left it to the authors of the chapters which nomenclature they prefer.

P. Jonas and H. Monyer
September 1999

References

Bradford HF (1970) Metabolic response of synaptosomes to electrical stimulation: Release of amino acids. Brain Res 19:239–247
Curtis DR, Phillis JW, Watkins JC (1959) Chemical excitation of spinal neurones. Nature 183:611–612
Ottersen OP, Storm-Mathisen J (1984) Glutamate- and GABA-containing neurons in the mouse and rat brain, as demonstrated with a new immunocytochemical technique. J Comp Neurol 229:374–392

List of Contributors

ADELMANN, G., Anatomisches Institut, Universität Freiburg, Postfach 111, D-79001 Freiburg, Germany

BÄHRING, R., Institut für Neurale Signalverarbeitung, Zentrum für Molekulare Neurobiologie, Universitäts-Krankenhaus Eppendorf, Martinistr. 52, D-20246 Hamburg, Germany

BÉHÉ, P., Department of Pharmacology, University College London, Gower Street, London WC1E 6BT, United Kingdom

BOWIE, D., Department of Pharmacology, Emory University School of Medicine, Rollins Research Center, 1510 Clifton Road, Atlanta, GA 30322, USA
e-mail: dbowie@pharm.emory.edu

CHOI, D.W., Center for the Study of Nervous System Injury and Department of Neurology, Washington University School of Medicine, Campus Box 8111, 660 S. Euclid Ave., St. Louis, MO 63110, USA
e-mail: choid@neuro.wustl.edu

COLQUHOUN, D., Department of Pharmacology, University College London, Gower Street, London WC1E 6BT, United Kingdom
e-mail: d.colquhoun@ucl.ac.uk

FROTSCHER, M., Anatomisches Institut, Universität Freiburg, Postfach 111, D-79001 Freiburg, Germany
e-mail: frotsch@uni-freiburg.de

GANEL, R., Johns Hopkins University, Dept. of Neurology and Neuroscience, Meyer 6-109, 600 N.Wolfe Str., Baltimore, MD 21287, USA

GEIGER, J.R.P., Physiologisches Institut, Universität Freiburg, Hermann-Herder-Str. 7, D-79104 Freiburg, Germany

GILL, R., F. Hoffmann-La Roche, Preclinical CNS Research CH-4070 Basel, Switzerland

HOLLMANN, M., Glutamate Receptor Laboratory, Max-Planck-Institute for Experimental Medicine, Hermann-Rein-Str. 3, D-37075 Göttingen, Germany
e-mail: Hollmann@mail.mpiem.gwdg.de

HUGANIR, R.L., Howard Hughes Medical Institute, Department of Neuroscience, Johns Hopkins University, School of Medicine, 725 N. Wolfe Str., 904 PCTB, Baltimore, MD 21205, USA
e-mail: rick.huganir@qmail.bs.jhu.edu

JONAS, P., Physiologisches Institut, Universität Freiburg, Hermann-Herder-Str. 7, D-79104 Freiburg, Germany
e-mail: jonasp@ruf.uni-freiburg.de

KAVANAUGH, M.P., Vollum Institute, L-474, Oregon Health Sciences University, 3181 SW Sam Jackson Park Rd., Portland, OR 97201, USA

KEMP, J.A., F. Hoffmann-La Roche, Preclinical CNS Research CH-4070 Basel, Switzerland
e-mail: john.kemp@roche.com

KENNEDY, M.B., Div. Biol. 216-76, Caltech, Pasadena, CA 91125, USA

KERCHNER, G.A., Center for the Study of Nervous System Injury and Department of Neurology, Washington University School of Medicine, 660 S. Euclid Ave., St. Louis, MO 63110, USA

KEW, J.N.C., F. Hoffmann-La Roche, Preclinical CNS Research CH-4070 Basel, Switzerland

KIM, A.H., Center for the Study of Nervous System Injury and Department of Neurology, Washington University School of Medicine, 660 S. Euclid Ave., St. Louis, MO 63110, USA

KORNAU, H.-C., BASF-LYNX Bioscience AG, Department of Neuroscience, Im Neuenheimer Feld 515, D-69120 Heidelberg, Germany
e-mail: kornau@basf-lynx.de

Kuner, T., Abt. Molekulare Neurobiologie,
Max-Planck-Institut für medizinische Forschung, Jahnstr. 29,
D-69120 Heidelberg, Germany – Present address:
Department of Neurobiology, Duke University Medical Center,
101 Bryan Research Bldg., Research Drive, P.O. Box 3209, Durham,
NC 27710, USA
e-mail: kuner@neuro.duke.edu

Lee, H.-K., Howard Hughes Medical Institute, Department of Neuroscience,
Johns Hopkins University, School of Medicine, 725 N. Wolfe Str.,
904 PCTB, Baltimore, MD 21205, USA

Lerma, J., Instituto Cajal, Consejo Superior de Investigaciones Cientificas,
Av. Doctor Arce 37, E-28002 Madrid, Spain
e-mail: lerma@cajal.csic.es

Mayer, M.L., Laboratory of Cellular and Molecular Neurophysiology,
National Institute of Child Health and Human Development, NIH,
49 Convent Drive, Bethesda, MD 20892, USA

Mews, K., Anatomisches Institut, Universität Freiburg, Postfach 111,
D-79001 Freiburg, Germany

Monyer, H., ZMBH, Im Neuenheimer Feld 282,
D-69120 Heidelberg, Germany
e-mail: monyer@mpimf-heidelberg.mpg.de

Otis, T.S., Department of Neurobiology, UCLA Medical Center,
10833 Le Conte Ave., Box 951763, Los Angeles, CA 90095-1763, USA
e-mail: otist@ucla.edu

Petralia, R.S., NIDCD/NIH, 36/5D08, 36 Convent Drive, MSC 4162,
Bethesda, MD 20892-4162, USA
e-mail: petralia@nidcd.nih.gov

Rossier, J., Neurobiologie et Diversité Cellulaire, CNRS,
École Supérieure de Physique et de Chimie Industrielles de la Ville de
Paris 10 rue Vaúquélin F-75231 Paris Cedex 5, France

Roth, A., Max-Planck-Institut für medizinische Forschung, Abt.
Zellphysiologie, Jahnstr. 29, D-69120 Heidelberg, Germany

Rothstein, J.D., Johns Hopkins University,
Dept. of Neurology and Neuroscience, Meyer 6-109, 600 N.Wolfe Str., Baltimore, MD 21287, USA
e-mail: jrothste@welchlink.welch.jhu.edu

Rubio, M.E., NIDCD/NIH, 36/5D08, 36 Convent Drive, MSC 4162, Bethesda, MD 20892-4162, USA
present address: Max-Plank-Institute for Experimental Medicine Department of Molecular Biology of Neuronal Signals, Hermann-Rein-Str. 3, D-37075 Göttingen, Germany

Sakmann, B., Max-Planck-Institut für medizinische Forschung, Abt. Zellphysiologie, Jahnstr. 29, D-69120 Heidelberg, Germany

Seeburg, P.H., Max-Planck-Institut für medizinische Forschung, Abt. Molekulare Neurobiologie, Jahnstr. 29, D-69120 Heidelberg, Germany

Taskin, B., Physiologisches Institut, Universität Freiburg, Hermann-Herder-Str. 7, D-79104 Freiburg, Germany

Trussell, L.O., Oregon Hearing Research Center NRC-04, OHSU, 3181 SW Sam Jackson Park Rd. Portland OR 97201, USA
e-mail: trussell@ohsu.edu

Wenthold, R.J., NIDCD/NIH, 36/5D08, 36 Convent Drive, MSC 4162, Bethesda, MD 20892-4162, USA

Wollmuth, L.P., Max-Planck-Institut für medizinische Forschung, Abt. Zellphysiologie, Jahnstr. 29, D-69120 Heidelberg, Germany – Present address: Department of Neurobiology and Behavior, State University of New York at Stony Brook, Stony Brook, NY 11794-5230, USA

Wyllie, D.J.A., Department of Pharmacology, University College London, Gower Street, London WC1E 6BT, United Kingdom

Contents

Section I: Molecular Structure of Glutamate Receptors

CHAPTER 1

Structure of Ionotropic Glutamate Receptors

CHAPTER 7

Block of AMPA and Kainate Receptors by Polyamines and Arthropod Toxins

CHAPTER 8

Kainate Receptors

CHAPTER 9

Molecular Determinants Controlling Functional Properties of AMPARs and NMDARs in the Mammalian CNS

Section III: Glutamatergic Synaptic Transmission

CHAPTER 12

Physiology of Glutamatergic Transmission at Calyceal and Endbulb Synapses of the Central Auditory Pathway

Section IV: Involvement of Glutamate Receptors and Transporters in Neurological Diseases

CHAPTER 14

Glutamate-Mediated Excitotoxicity

CHAPTER 15

Glutamate Transporter Dysfunction and Neuronal Death

CHAPTER 16

NMDA Recepor Antagonists and Their Potential as Neuroprotective Agents

Section I
Molecular Structure of Glutamate Receptors

CHAPTER 1
Structure of Ionotropic Glutamate Receptors

M. Hollmann

A. Historical Sketch of Glutamate Receptor Cloning

By 1987, all of the major ligand-gated ion-channel families had succumbed to cloning efforts, with one notable exception, the glutamate receptors. Acetylcholine receptors had been cloned first, aided by the fact that these proteins bind α-bungarotoxin with high specificity and affinity, and that they are highly concentrated in the muscle tissue of the electric fish *Torpedo californica and Torpedo marmorata*. This had allowed protein purification followed by partial microsequencing (Devillers-Thiery et al. 1979; Raftery et al. 1980) as well as generation of antibodies (Tzartos and Lindstrom 1980). The resulting information and analytical tools, in turn, had been used to screen cDNA libraries with amino acid sequence-derived oligonucleotide probes and antibodies which, in 1982, had led to the isolation of the first cDNA clones for acetylcholine receptors (Ballivet et al. 1982; Giraudat et al. 1982; Noda et al. 1982; Sumikawa et al. 1982). A glycine receptor cDNA had been isolated in 1987 by taking advantage of the specific high-affinity binding of the competitive antagonist strychnine to the receptor. This allowed isolation and partial microsequencing of a strychnine-binding protein (Pfeiffer et al. 1982), providing sequence information to construct specific oligonucleotide probes, which were then used to isolate the glycine receptor cDNA (Grenningloh et al. 1987). γ-Aminobutyric acid (GABA)A receptors were also cloned in 1987, via low-stringency hybridization screening based on oligonucleotide sequence derived from the partial amino acid sequence of a protein purified by benzodiazepine-affinity chromatography (Schofield et al. 1987).

Glutamate receptors, however, had remained elusive. This family of glutamate-activated cation channels had long been known to exhibit unprecedented diversity, consisting of at least three major pharmacological subfamilies (Table 1), defined by their most selective agonists: α-amino-3-hydroxy-5-methyl-4-isoxazole propionate (AMPA) receptors, kainate (KA) receptors, and *N*-methyl-D-aspartate (NMDA) receptors (Mayer and Westbrook 1987; Collingridge and Lester 1989; Honore 1989; Monaghan et al. 1989). Although all of the aforementioned techniques had been tried on glutamate receptors, due to a lack of high-affinity receptor ligands of sufficiently high specificity, and the non-availability of a rich source of glutamate receptor protein, protein biochemical isolation strategies for functional ion channel-

Table 1. Functional characteristics of glutamate receptor subfamilies. For the vertebrate glutamate receptors, subfamilies have been established based on sequence homology (see Table 14) which, at the same time, represent pharmacological classes; for invertebrate receptors, no clear picture for a sequence-based or pharmacological classification has emerged yet (see Table 16); those six invertebrate receptor subunits for which expression data exist have been listed to illustrate the divergence of properties encountered

Subfamily (based on sequence homology)	Genes	Relative agonist potencies	Homomeric/ heteromeric ion channels	Conductance	Activation kinetics	Desensitization	Ca^{2+} permeability	Distinguishing functional properties
A) Vertebrate Receptors								
AMPA receptors	GluR1–GluR4	QA > AMPA > GLU > KA[a]	+/+	Small	Fast	Fast and pronounced by GLU, not by KA	Yes, in absence of GluR2	High affinity (~10 nM) [^{3}H]AMPA binding[o]
KA receptors (low affinity)	GluR5–GluR7	DOM > KA > QA~GLU >> AMPA[b]	+/+	Very small	Fast	Very fast and pronounced, by GLU and KA	Yes	Low affinity (~75 nM) [^{3}H]KA binding[p]
KA receptors (high affinity)	KA1 – KA2	KA > QA > DOM > GLU >> AMPA[c,d]	–/+	Very small[d]	Fast[c]	Very fast and pronounced, by GLU and KA[c]	Yes[c]	High affinity (5–15 nM) [^{3}H]KA binding[q]; functional only in heteromers with low affinity KA receptors
NMDA receptor (glycine binding subunit)	NMDAR1	GLU>NMDA[e]	+/+	Large	SLOW	Not very pronounced (glycine-dep., Ca^{2+}-dep.)	Yes, high	Glycine-binding subunit of NMDA receptors; voltage-dependent Mg^{2+} block; glycine is co-agonist; MK-801 blocks ion channel
NMDA receptor (glutamate binding subunit)	NR2A–NR2D	GLU> NMDA[f,g]	–/+	Large[f]	Slow[f]	Not very pronounced (glycine-dep., Ca^{2+}-dep.)[f]	Yes, high[f]	Glutamate-binding subunit of heteromeric NMDA receptors; voltage-dependent Mg^{2+} block[f]; glycine is co-agonist[f]; MK-801 blocks ion channel[f]
NMDA receptor (modulatory subunit)	NR3A–NR3B	No data available	–/–	–	–	–	–	Possibly a modulatory subunit of NMDA receptors; ion pore domain is non-functional[h]
Orphan receptors	Delta1–Delta2	Agonist(s) unknown	–/–	–	–	–	–	Possibly a modulatory subunit; ion pore domain is non-functional[h]

Kainate binding proteins (KBPs)	KBP-chick, -frog, -fish-α, -fish-β, -toad	DOM > KA > GLU[i]	–/–	–	–	–	–	Function unknown; ion pore domains are functional[j]
B) Invertebrate receptors								
Drosophila kainate receptor	DGluR-I	KA[k]	+/?	Unknown	Unknown	Unknown	Unknown	KA activates small currents in oocytes which are not inhibited by CNQX[k]
Drosophila glutamate receptor	DGluR-IIA, DGluR-IIB	GLU > QA, AMPA, KA[l]	+/?	Unknown	Unknown	Unknown	Unknown	GLU and ASP activate currents[l]
Drosophila NMDA receptor	DNMDAR-I	GLY[m]	(+)/?	Unknown	Unknown	Unknown	Unknown	Does respond weakly to GLY but not GLU[m]
Lymnaea glutamate receptor 1	Lym-eGluR1	Agonist(s) unknown	–/–	–	–	–	–	Function unknown
Lymnaea glutamate receptor 2	Lym-eGluR2	GLU >> KA > IBO > AMPA[n]	+/?	Unknown	Unknown	Little if any	No[n]	Not activated by ASP, DOM, QA, or NMDA[n]
Other invertebrate receptors	(17 more genes from *Lymnaea*, *C.elegans*, *Schistocerca*; see Table 7)	No data available as expressable full-length clones have not been reported, yet	No data	No data	No data	No data	No data	No data available

GLU glutamate; KA kainate; AMPA α-amino-3-hydroxy-5-methyl-4-isoxazole propionate; QA quisqualate; DOM domoate; ASP aspartate; GLY glycine; IBO ibotenate

[a] Nakanishi et al. (1990); Sakimura et al. (1990)
[b] Egebjerg et al. (1991); Sommer et al. (1992)
[c] In heteromeric assemblies with low affinity kainate receptors
[d] Sakimura et al. (1992); Howe (1996), and ligand binding data (Werner et al. 1991)
[e] Moriyoshi et al. (1991); Nakanishi et al. (1992)
[f] In heteromeric assemblies with NMDAR-1
[g] Ishii et al. (1993)
[h] Villmann et al. (1999)
[i] Order of potencies derived from ligand binding experiments (Wada et al. 1989)
[j] Villmann and Hollmann (1997)
[k] Ultsch et al. (1992)
[l] Schuster et al. (1991)
[m] Ultsch et al. (1993)
[n] Stühmer et al. (1996)
[o] Keinänen et al. (1990)
[p] Bettler et al. (1992); Lomeli et al. (1992); Sommer et al. (1992)
[q] Herb et al. (1992)

forming glutamate receptors had come up empty. At this stage, a newly introduced technique came to the rescue: expression cloning. This technique does not require specific ligands, specific antibodies, or partial sequence information, and makes no assumptions about the structure of the molecule to be cloned. Instead, it relies solely on a biological assay of a distinctive functional property of the target molecule displayed upon expression in a heterologous system, such as *Xenopus* oocytes, for example.

This method had already shown its potential in the cloning of the substance-K receptor (Masu et al. 1987), and it turned out to be the method of choice for glutamate receptors. Not only did expression cloning yield the cDNA clone for the first ionotropic glutamate receptor, GluR1 (Hollmann et al. 1989), it also provided first access to the important NMDA receptor subfamily of glutamate receptors (Moriyoshi et al. 1991), which turned out to be so distantly related to the other glutamate receptor subfamilies (see Table 14) that sequence-based screening techniques were likely to overlook these receptors. Expression cloning thus provided the original sequence information, which other cloning techniques, such as low-stringency hybridization screening and reverse transcriptase polymerase chain reaction (RT-PCR) with degenerate primers, then could exploit. Application of these screening techniques led to the rapid identification of related subunits of the AMPA receptor subfamily (Bettler et al. 1990; Boulter et al. 1990; Keinänen et al. 1990; Nakanishi et al. 1990; Sakimura et al. 1990) as well as the KA receptor subfamily (Bettler et al. 1990; Egebjerg et al. 1991; Werner et al. 1991; Bettler et al. 1992; Herb et al. 1992; Sakimura et al. 1992) and the NMDA receptor subfamily (Ikeda et al. 1992; Kutsuwada et al. 1992; Meguro et al. 1992; Monyer et al. 1992; Yamazaki et al. 1992b).

The power of expression cloning is further illustrated by the fact that the first member of the family of metabotropic glutamate receptors (mGluRs), which are G-protein-coupled 7-transmembrane domain receptors, was also discovered by expression cloning (Masu et al. 1991). The mGluRs are as distinct from other metabotropic receptors, such as muscarinic acetylcholine receptors or serotonin receptors, as ionotropic glutamate receptors are distinct from other ligand-gated ion channels (see Sect. D.I.2.a) and were, therefore, similarly inaccessible to standard sequence-based screening techniques. For comprehensive reviews of the molecular biology of mGluRs, which are not going to be covered in this volume, see Schoepp et al. (1990); Schoepp and Conn (1993); Nakanishi (1994); Schoepp (1994); Pin and Bockaert (1995); Pin and Duvoisin (1995); Pin et al. (1996); Conn and Pin (1997).

The cloning of the ionotropic glutamate receptors and their structural and functional properties have been covered fully or partly in several excellent reviews, in which many aspects discussed in this chapter are treated in far greater detail than is possible here (Darlison 1992; Seeburg 1993; Hollmann and Heinemann 1994; McBain and Mayer 1994; Nakanishi and Masu 1994; Bettler and Mulle 1995; Mori and Mishina 1995; Sprengel and Seeburg 1995; Cleland 1996; Chittajallu et al. 1999).

B. Subunit Diversity of Glutamate Receptors – Molecular Mechanisms

I. Molecular Mechanisms Creating Subunit Diversity

1. Multiple Genes

For neurotransmitters such as acetylcholine, GABA, and glycine, it had been found that their actions were mediated through families of closely related but functionally diverse receptor subunits (Barnard et al. 1987; Betz 1991; Novere and Changeux 1995). Obviously, a single glutamate receptor subunit was not sufficient to account for the well-documented pharmacological and electrophysiological diversity of glutamate binding sites (Watkins and Evans 1981; Cotman et al. 1987; Honore 1989; Wroblewski and Danysz 1989) and glutamate-activated excitatory postsynaptic responses, respectively (Mayer and Westbrook 1987; Collingridge and Lester 1989; McDonald and Johnston 1990), which had been encountered in the central nervous system. Thus, it came as no surprise that, following the expression cloning of GluR1, the first member of the glutamate receptor family (Hollmann et al. 1989), several additional glutamate receptor subunits were discovered in rapid succession. Homology screening methods, such as low-stringency hybridization with probes generated from cloned subunits and PCR-mediated cDNA amplification with degenerate primers based on the sequences of identified genes, were employed independently by several groups in a search for additional, related sequences. Indeed, a large number of candidate cDNA clones were identified as potential members of the family of ionotropic glutamate receptors. Based on their amino acid sequences, those subunits could be grouped into several subfamilies (Tables 1, 14), with members of one subfamily sharing considerably higher sequence identities with each other than with any other subunits. Interestingly, the functional properties of the various subunits, such as ligand binding or the pharmacology of ion channel activation, turned out to be more similar within a subfamily than across subfamilies, underscoring the validity of the use of sequence similarity data for receptor subunit classification. Furthermore, all the major previously identified pharmacological classes of glutamate receptors, such as NMDA, KA and AMPA receptors were found to be represented by separate subfamilies of sequence-related genes (Table 1).

To date, a total of 23 different genes belonging to the family of ionotropic excitatory glutamate receptors have been reported from vertebrates. Eighteen of these subunits occur in mammals (Tables 2–5), while five genes, all of them belonging to the subfamily of kainate binding proteins (KBPs), have been found exclusively in non-mammalian vertebrates (Table 6). In invertebrates, 26 excitatory glutamate receptor genes have been reported (Table 7). In most cases, however, it is unclear whether those genes represent invertebrate counterparts of vertebrate genes (that is, whether they are homologous genes) or

Table 2. Cloned α-amino-3-hydroxy-5-methyl-4-isoxazole propionate (AMPA) receptor subunits from vertebrates. All ionotropic glutamate receptor genes cloned to date (June 1998) are listed, as well as all splice variants reported in the literature. References to the original descriptions and the GenBank accession numbers (if available) are given for each gene and every variant in every species it has been described from. Where redundant sequence reports exist for the same species, all sequences are listed, except for a few partial, incomplete sequences in cases where full-length sequences are available. Receptor subunits are grouped by gene and, within one gene, are initially sorted by animal species, then by splice variant and year of description. Rat sequences are given first, as most subunits were first cloned from rat. When two different GenBank accession numbers are listed, both numbers refer to the same sequence. The nomenclature used by Hollmann and Heinemann (1994) was adopted for the "Gene" and "Gene variant" columns. Alternate original names and subsequently proposed name variations are listed in the forth column. Signal peptide and mature protein sizes are given as supplied by the original authors, except for numbers marked (*), which were derived using the SPSCAN program for signal peptide identification contained in the University of Wisconsin sequence analysis software package (Devereux et al. 1984)

Gene	Gene variant	Species	Alternate name used in original publication	Total protein size (amino acids)	Mature protein size (amino acids)	Signal peptide size (amino acids)	References	GenBank accession number
GluR1	GluR1-flop	Rat (R.n.)	GluR-K1	907	889	18	Hollmann et al. 1989	X17184
	–″–	Rat (R.n.)	GluR-A	907	889	18	Keinänen et al. 1990	M36418
	–″–	Rat (R.n.)	GluR-K1	907	889	18	Nakanishi et al. 1990	–
	GluR1-flip	Rat (R.n.)	GluR-Aflip	907	889	18	Sommer et al. 1990	M38060
	GluR1-flop	Human (H.s.)	KR4	906	888	18	Potier et al. 1992	X58633 [S40299]
	–″–	Human (H.s.)	HBGR1	906	888	18	Sun et al. 1992	M81886
	GluR1-flip	Human (H.s.)	GluHI	907	889	18	Puckett et al. 1991	M64752
	GluR1-flip	Mouse (M.m.)	musGluR1, or GluRα1	907	889	18	Sakimura et al. 1990	X57497
	GluR1-flip	Chick (G.g.)	cGluR1	902	884	18	Paperna et al. 1996	X89510
	GluR1α-flop	Fish (O.n.)	fGluR1α	908	890	18	Wu et al. 1996	L49497[a]
	–″–	Fish (Mo.sp.)	GluR-Morone	(949 ps)		(6 ps)	Ponomareva and McMahon 1996	U44736
	GluR1αc-flop	Fish (O.n.)	fGluR1αc	930	912	18	Wu et al. 1996	L49500[a]
	GluR1β	Fish (O.n.)	fGluR1β	(ps)			Wu et al. 1996	L49498[a]

GluR2	GluR2-flop	Rat (R.n.)	GluR2	883	862	21	Boulter et al. 1990	M85035
	–″–	Rat (R.n.)	GluR-B	883	862	21	Keinänen et al. 1990	M36419
	GluR2-flip	Rat (R.n.)	GluR-Bflip	883	862	21	Sommer et al. 1990	M38061
	–″–	Rat (R.n.)	GluR-K2	883	862	21	Nakanishi et al. 1990	X54655
	GluR2-flip	Human (H.s.)	HBGR2	883	862	21	Sun et al. 1994	L20814
	GluR2-flop	Mouse (M.m.)	musGluR2, or GluRα2	883	862	21	Sakimura et al. 1990	X57498
	GluR2l-flop	Mouse (M.m.)	GluR2 long-flop	901	880	21	Köhler et al. 1994	L32151
	GluR2l-flip	Mouse (M.m.)	GluR2 long-flip	901	880	21	Köhler et al. 1994	L32151
	GluR2-flop	Pigeon (C.l.)	GluP-II	883	862	21	Ottiger et al. 1995	Z29713
	GluR2-flip	Chick (G.g.)	cGluR2	883	862	21	Paperna et al. 1996	X89508
	GluR2α-flop	Fish (O.n.)	fGluR2α-flop	907	878	29	Kung et al. 1996	L34036
	GluR2αc-flop	Fish (O.n.)	fGluR2αc-flop	895	866	29	Kung et al. 1996	L34080
	GluR2αc-flip	Fish (O.n.)	fGluR2αc-flip	895	866	29	Kung et al. 1996	L46366
	GluR2β-flop[b]	Fish (O.n.)	fGluR2β-flop[b]	864	849	15	Kung et al. 1996	L37836
	GluR2β-flip[b]	Fish (O.n.)	fGluR2β-flip[b]	864	849	15	Kung et al. 1996	L34037
GluR3	GluR3-flop	Rat (R.n.)	GluR3	888	866	22	Boulter et al. 1990	M85036
	–″–	Rat (R.n.)	GluR-C	888	866	22	Keinänen et al. 1990	M36420
	GluR3-flip	Rat (R.n.)	GluR-Cflip	888	866	22	Sommer et al. 1990	M38062
	–″–	Rat (R.n.)	GluR-K3	888	866	22	Nakanishi et al. 1990	X54656
	GluR3-flop	Human (H.s.)	hGluR3flop	894	866	28	Rampersad et al. 1994	U10302
	–″–	Human (H.s.)	hGluRC	894	865	29	McLaughlin and Kerwin 1994	X82068
	GluR3-flip	Human (H.s.)	hGluR3flip	894	866	28	Rampersad et al. 1994	U10301
	–″–	Human (H.s.)	hGluR3flip	894	866	28	Varney et al. 1998	–
	GluR3-flop	Pigeon (C.l.)	GluP-III	(270 ps)			Ottiger et al. 1995	Z29714
	GluR3-flop	Chick (G.g.)	cGluR3	888	866	22	Paperna et al. 1996	X89509
	GluR3α-flop[b]	Fish (O.n)	fGluR3α-flop[b]	886	865	21	Chang et al. 1998	AF015253
	GluR3α-flip[b]	Fish (O.n)	fGluR3α-flip[b]	886	865	21	Chang et al. 1998	AF014400
	GluR3β-flop[b]	Fish (O.n)	fGluR3α-flip[b]	(750 ps)			Chang et al. 1998	–
	GluR3-flip	Goldfish (C.a.)	GluRC-flip	(38 ps)			Ueda 1997	D66901
	–″–	Goldfish (C.a.)	GluRC-flip	(82 ps)			Ueda 1997	D66902

Table 2. *Continued*

Gene	Gene variant	Species	Alternate name used in original publication	Total protein size (amino acids)	Mature protein size (amino acids)	Signal peptide size (amino acids)	References	GenBank accession number
GluR4	GluR4-flop	Rat (R.n.)	GluR-D	902	881	21	Keinänen et al. 1990	M36421
	GluR4-flip	Rat (R.n.)	GluR4flip	902	881	21	Bettler et al. 1990	M85037
	–"–	Rat (R.n.)	GluR-Dflip	902	881	21	Sommer et al. 1990	M38063
	GluR4c-flop	Rat (R.n.)	GluR-4cflop	884	863	21	Gallo et al. 1992	S94371
	GluR4c-flip	Rat (R.n.)	GluR-4cflip		863		Gallo et al. 1992	–
	GluR4-flip	Human (H.s.)	hGluR4	902	881	21	Fletcher et al. 1995	U16129
	(unknown)	Pigeon (C.l.)	GluP-IV	(269 ps)			Ottiger et al. 1995	Z29920
	GluR4-flop	Chick (G.g.)	cGluR4	902	881	21	Paperna et al. 1996	X89507
	GluR4c	Chick (G.g.)	GluR4c	(ps)			Ravindranathan et al. 1997	U65992
	GluR4d	Chick (G.g.)	GluR4d	(ps)			Ravindranathan et al. 1997	U65993
	GluR4s	Chick (G.g.)	GluR4s	(ps)			Ravindranathan et al. 1997	U65991
	GluR4-flop	Goldfish (C.a.)	gfGluR4	(900 ps)			Goldman 1994	U12018
	–"–	Goldfish (C.a.)	gfGR52	(900 ps)			Ueda and Hieber 1995	D45381

A.d. Anas domesticus (duck); *A.l. Apteronotus leptorhynchus* (a weakly electric fish); *C.a. Carassius auratus* (goldfish); *C.e. Caenorhabditis elegans* (nematode); *C.l. Columba livia* (pigeon); *D.m. Drosophila melanogaster* (fruit fly); *G.d. Gallus domesticus* (domestic chick); *G.g. Gallus gallus* (chick); *H.s. Homo sapiens* (human); *L.s. Lymnaea stagnalis* (pond snail); *M.m. Mus musculus* (mouse); *M.sp. Mus spec.* (mouse); *Mo.sp. Morone spec.* (fish, a striped bass hybrid); *O.n. Oreochromis nilotica x aureus* (a freshwater fish hybrid); *P.i.* Panulirus interruptus (California spiny lobster); *R.n. Rattus norvegicus* (rat); *R.p. Rana pipiens* (the leopard frog); *S.c. Sus scrofa* (wild boar); *S.g. Schistocerca gregaria* (locust); *X.l. Xenopus laevis* (clawed toad); aa amino acid; ps partial sequence; n.d. not determined

[a] Sequence is not accessible in GenBank yet

[b] fGluR2β is 87.7% identical to fGluR2α, while fGluR3β is ~92% identical to fGluR3α; this means that the two β-subunits presumably represent different genes. Thus, fish appear to possess two different GluR2 and GluR3 genes (Kung et al. 1996; Chang et al. 1998)

Table 3. Cloned kainate receptor subunits from vertebrates. GluR5 to GluR7 represent low-affinity kainate receptors while KA1 and KA2 are high-affinity kainate receptors

Gene	Gene variant	Species	Alternate name used in original publication	Total protein size (amino acids)	Mature protein size (amino acids)	Signal peptide size (amino acids)	References	GenBank accession number
GluR5	GluR5(Q)-1a	Rat (R.n.)	GluR-5-a	871	841	30	Sommer et al. 1992	Z11712
	GluR5(Q)-1b	Rat (R.n.)	GluR5–1	920	890	30	Bettler et al. 1990	M83560
	–″–	Rat (R.n.)	GluR-5-b	920	890	30	Sommer et al. 1992	Z11713
	GluR5(Q)-1c	Rat (R.n.)	GluR-5-c	949	919	30	Sommer et al. 1992	Z11714
	GluR5(Q)-2a	Rat (R.n.)	GluR-5–2a		826		Sommer et al. 1992	–
	GluR5(Q)-2b	Rat (R.n.)	GluR5–2	905	875	30	Bettler et al. 1990	M83561
	–″–	Rat (R.n.)	GluR-5–2b		875		Sommer et al. 1992	–
	GluR5(Q)-2c	Rat (R.n.)	GluR-5–2c		904		Sommer et al. 1992	–
	GluR5(Q)-1d	Human (H.s.)	HGR5–4, = hGluR5–1d	918	888	30	Gregor et al. 1993a	L19058
	GluR5(Q)-2b	Human (H.s.)	humEAA3	905	875	30	Korczak et al. 1995	U16125
	–″–	Human (H.s.)	hEAA3 C/D	905	875*	30*	Kamboj et al. 1996a	I25018
	GluR5(R)-2c	Mouse (M.m.)	musGluR5–2c, or musGluR5–3	934[a]	904[a]	30	Gregor et al. 1993a	X66118[a]
GluR6	GluR6(R)	Rat (R.n.)	GluR6	908[b]	877[b]	31	Egebjerg et al. 1991	Z11548[b]
	GluR6(Q)	Rat (R.n.)	GluR-6	908	877	31	Lomeli et al. 1992	Z11715
	GluR6(Q)	Human (H.s.)	hGRIK2	908	877	31	Paschen et al. 1994	S75105
	–″–	Human (H.s.)	humEAA4	908	877	31	Hoo et al. 1994	U16126
	–″–	Human (H.s.)	hEAA4	908	877*	31*	Kamboj et al. 1996b	I28906
	–″–	Human (H.s.)	hGluR6a	908			Daggett et al. 1996	–
	GluR6(Q)b	Human (H.s.)	hGluR6b	870			Daggett et al. 1996	–
	GluR6(R)	Mouse (M.m.)	$\beta 2$	864[c]	833[c]	31	Morita et al. 1992	D10054[c]
	GluR6(R)c	Mouse (M.m.)	musGluR6–2, or musGluR6c	869	838	31	Gregor et al. 1993a	X66117
	GluR6	Toad (X.l.)	XenGluR6	(285 ps)			Ishimaru et al. 1996	X94116

Table 3. *Continued*

Gene	Gene variant	Species	Alternate name used in original publication	Total protein size (amino acids)	Mature protein size (amino acids)	Signal peptide size (amino acids)	References	GenBank accession number
GluR7	GluR7a	Rat (R.n.)	GluR7		888		Bettler et al. 1992	M83552
	–″–	Rat (R.n.)	GluR-7	919	888	31	Lomeli et al. 1992	Z11716
	GluR7b	Rat (R.n.)	GluR7b	910	879	31	Schiffer et al. 1997	AF027331
	GluR7a	Human (H.s.)	humEAA5	919	888	31	Nutt et al. 1994	U16127
	(Unknown)	Mouse (M.m.)	musGluR7	(133 ps)			Gregor et al. 1993b	–
KA1	KA1	Rat (R.n.)	KA-1	956[d]	936[d]	20	Werner et al. 1991[c]	X59996[d]
	–″–	Rat (R.n.)	KA1	956	936	20	Boulter 1994c	U08257
	KA1	Human (H.s.)	humEAA1	956	936	20	Kamboj et al. 1994	S67803
	–″–	Human (H.s.)	hEAA1	956	936	20	Kamboj et al. 1996c	I28953 [I38875]
KA2	KA2	Rat (R.n.)	KA-2	979	965	14	Herb et al. 1992	Z11581
	–″–	Rat (R.n.)	KA2	979	965	14	Boulter 1994d	U08258
	KA2	Human (H.s.)	humEAA2	980	961	19	Kamboj et al. 1992	S40369
	–″–	Human (H.s.)	hEAA2	980	961	19	Kamboj et al. 1996d	I17703 [I38615]
	KA2	Mouse (M.m.)	γ2	979	961	18	Sakimura et al. 1992	D10011 [D01273, S83752]

[a] This clone has a 295-bp deletion described as a cloning artifact; the start of the coding region given in GenBank is wrong: it should read nt 543 rather than nt 456

[b] In the original publication, the mature protein size was given as 853 amino acids, due to a sequencing error which was later corrected (Bettler et al. 1992)

[c] One missing "A" in a stretch of eight adenosins of the sequence caused a frame shift and led to publication of this truncated protein sequence of only 833 amino acids; the mature protein actually is 877 amino acids

[d] This clone was originally reported with a frameshifted C-terminus which was later corrected (Herb et al. 1992). For abbreviations and further explanations, see legend of Table 2

Table 4. Cloned *N*-methyl-D-aspartate (NMDA) receptor subunits from vertebrates. The NMDAR1 subunit is the glycine-binding subunit, while the four NR2 subunits are the glutmate-binding subunits of the NMDA receptor complex. The NR3 subunits are likely modulatory subunits

Gene	Gene variant	Species	Alternate name used in original publication	Total protein size (amino acids)	Mature protein size (amino acids)	Signal peptide size (amino acids)	References	GenBank accession number
NMDAR1	NR1–1a	Rat (R.n.)	NMDAR1	938	920	18	Moriyoshi et al. 1991	S61907 [X63255]
	–″–	Rat (R.n.)	NMDAR1-SL	938	920	18	Anantharam et al. 1992	–
	–″–	Rat (R.n.)	NMDA-R1A	938	920	18	Nakanishi et al. 1992	S44964
	–″–	Rat (R.n.)	NMDAR1–1a	938	920	18	Hollmann et al. 1993	U08261
	–″–	Rat (R.n.)	NMDAR1	938	920	18	Sullivan et al. 1994	U11418
	NR1–1b	Rat (R.n.)	NMDAR1-LL	959	941	18	Anantharam et al. 1992	X65227
	–″–	Rat (R.n.)	NMDAR1B	959	941	18	Sugihara et al. 1992	S39217
	–″–	Rat (R.n.)	NMDA-R1B	959	941	18	Nakanishi et al. 1992	S45121
	–″–	Rat (R.n.)	NMDAR1–1b	959	941	18	Hollmann et al. 1993	U08263
	NR1–2a	Rat (R.n.)	NMDAR1-SS	901	883	18	Anantharam et al. 1992	–
	–″–	Rat (R.n.)	NMDA-R1C	901	883	18	Sugihara et al. 1992	S39218
	–″–	Rat (R.n.)	NMDA-R1C	901	883	18	Nakanishi et al. 1992	–
	–″–	Rat (R.n.)	NMDAR1–2a	901	883	18	Hollmann et al. 1993	U08262
	NR1–2b	Rat (R.n.)	NMDAR1-LS	922	904	18	Anantharam et al. 1992	–
	–″–	Rat (R.n.)	NMDAR1F	922	904	18	Sugihara et al. 1992	S39218
	–″–	Rat (R.n.)	NMDAR1–2b	922	904	18	Hollmann et al. 1993	U08264
	NR1–3a	Rat (R.n.)	NMDAR1D	922	904	18	Sugihara et al. 1992	S39219
	–″–	Rat (R.n.)	NMDAR1–3a	922	904	18	Hollmann et al. 1993	U08265
	NR1–3b	Rat (R.n.)	NMDAR1–3b	943	925	18	Hollmann et al. 1993	U08266
	NR1–4a	Rat (R.n.)	NMDAR1E	885	867	18	Sugihara et al. 1992	S39217
	–″–	Rat (R.n.)	NMDAR1–4a	885	867	18	Hollmann et al. 1993	U08267
	–″–	Rat (R.n.)	NMDAR1C	885	867	18	Durand et al. 1992	–
	NR1–4b	Rat (R.n.)	NMDAR1G	906	888	18	Sugihara et al. 1992	S39219
	–″–	Rat (R.n.)	NMDAR1B	(26 ps)			Durand et al. 1992	S46393 (S1)
	–″–	Rat (R.n.)	NMDAR1B	(25 ps)			Durand et al. 1992	S46394 (S2)
	–″–	Rat (R.n.)	NMDAR1B	906	884*	22*	Durand et al. 1992	L01632
	–″–	Rat (R.n.)	NMDAR1–4b	906	888	18	Hollmann et al. 1993	U08268
	NR1–4?	Rat (R.n.)	NMDAR1 E/G	(26 ps)			Sugihara et al. 1992	S39220

Table 4. *Continued*

Gene	Gene variant	Species	Alternate name used in original publication	Total protein size (amino acids)	Mature protein size (amino acids)	Signal peptide size (amino acids)	References	GenBank accession number
	NR1–5 (trunc.)	Rat (R.n.)	NMDAR1trunc.	181	163	18	Sugihara et al. 1992	S39221
	–″–	Rat (R.n.)	NMDAR1–5	181	163	18	Hollmann et al. 1993	–
	NR1–1a	Human (H.s.)	hNR1	938	920	18	Karp et al. 1993	D13515
	–″–	Human (H.s.)	hNR1–3	928	919	(9 ps)	Foldes et al. 1993	L13268
	–″–	Human (H.s.)	hNR1a	(n.d.)			Le Bourdelles et al. 1994	–
	–″–	Human (H.s.)	hNMDAR1	938			Hess et al. 1996	–
	NR1–2a	Human (H.s.)	hNR1–2	(602 ps)			Foldes et al. 1993	L13267
	NR1–3a	Human (H.s.)	hNR1–4	(592 ps)			Foldes et al. 1994b	U08106
	–″–	Human (H.s.)	hNR1d	(n.d.)			Le Bourdelles et al. 1994	–
	NR1–3b	Human (H.s.)	hNMDA1–1		925		Nash et al. 1997	–
	NR1–4a	Human (H.s.)	hNR1–1	885	867	18	Foldes et al. 1993	L13266
	–″–	Human (H.s.)	hNR1	885	867	18	Planells-Cases et al. 1993	L05666
	–″–	Human (H.s.)	hNR1e		865		Le Bourdelles et al. 1994	–
	NR1–4b	Human (H.s.)	hNR1N	(195 ps)			Foldes et al. 1994b	U08107
	–″–	Human (H.s.)	hNMDA1–2		888		Nash et al. 1997	–
	NR1-?a	Human (H.s.)	hNR1	(863 ps)			Le et al. 1994a	A38680
	NR1-?a	Pig (S.s.)	NMDAR1-a	(243 ps)			Matteri 1997	AF008560
	NR1–1a	Mouse (M.m.)	ζ1	938	920	18	Yamazaki et al. 1992	D10028 [S88563]
	–″–	Mouse (M.sp.)	ζ1		920		Masami 1994d	E06821
	–″–	Mouse (M.sp.)	ζ1-N598Q		920		Masami 1994c	E06820
	–″–	Mouse (M.m.)	NMDH-18		920		Mishina 1996f	I19109
	–″–	Mouse (M.m.)	NMDH-19		920		Mishina 1996g	I19110
	NR1–2a	Mouse (M.m.)	ζ1–2	901	883	18	Yamazaki et al. 1992	S37525
	NR1–6a	Duck (A.d.)	duck-NMDA-R1	965	947	18	Kurosawa et al. 1994	D83352 [S71540]
	NR1–4?	Goldfish (C.a.)	NMDAR1-gf	(221 ps)			Hieber and Goldman 1995	S78680
	NR1–7b	Fish (A.l.)	aptNR1	966	948	18	Bottai et al. 1998	AF060557[$z]
	NR1–4a	Toad (X.l.)	XNMDAR1E	(ps)			Soloviev et al. 1996	–
	NR1–4b	Toad (X.l.)	XNMDAR1	904	884	20	Soloviev et al. 1996	X94156 [X94081]

NR2A	NR2A	Rat (R.n.)	NR2A	1464	1445	19	Monyer et al. 1992	M91561
	–″–	Rat (R.n.)	NR2A	1464	1442	22	Ishii et al. 1993	D13211
	–″–	Rat (R.n.)	NR2A	1464	1445*	19*	Boulter 1997	AF001423
	–″–	Human (H.s.)	hNR2A	1464	1445*	19*	Le et al. 1994b	A38688
	–″–	Human (H.s.)	hNR2A		1444		Le Bourdelles et al. 1994	–
	–″–	Human (H.s.)	hNR2A	1464	1445*	19*	Foldes et al. 1994a	U09002
	–″–	Human (H.s.)	hNMDAR2A	1464	1444	20	Hess et al. 1996	U90277
	–″–	Mouse (M.m.)	ε1	1464	1445	19	Meguro et al. 1992	D10217
	–″–	Mouse (M.sp.)	ε1	1464	1442*	22*	Masami 1993	E05440
	–″–	Mouse (M.m.)	NMDH-5	1464	1442*	22*	Mishina 1996a	I19101
NR2B	NR2B	Rat (R.n.)	NR2B	1482	1456	26	Monyer et al. 1992	M91562
	–″–	Rat (R.n.)	NR2B		1456		Ishii et al. 1993	–
	–″–	Rat (R.n.)	NR2B	1482	1456	26	Sullivan et al. 1994	U11419
	–″–	Human (H.s.)	hNR3	1484	1458	26	Adams et al. 1995	U11287
	–″–	Human (H.s.)	hNMDAR2B	1484	1458	26	Hess et al. 1996	U90278
	–″–	Mouse (M.m.)	ε2	1482	1456	26	Kutsuwada et al. 1992	D10651
	–″–	Mouse (M.sp.)	ε2		1456		Masami 1994a	E06594
	–″–	Mouse (M.sp.)	ε2-N589Q		1456		Masami 1994b	E06819
	–″–	Mouse (M.m.)	NMDH-6	1482	1456*	26*	Mishina 1996b	I19102
	–″–	Mouse (M.m.)	NMDH-17		1456		Mishina 1996e	I19108
	NR2B-2	Mouse (M.m.)	ε2	(ps)			Klein et al. 1998	AF033356
NR2C	NR2C	Rat (R.n.)	NR2C	1237[a]	1218[a]	19	Monyer et al. 1992	M91563[a]
	–″–	Rat (R.n.)	NR2C	1250	1218	32	Ishii et al. 1993	D13212
	–″–	Rat (R.n.)	NR2C	1250	1218	32	Boulter 1994a	U08259
	NR2C-c[b]	Rat (R.n.)	NR2 C 5′UTR var.	n.d.[b]			Suchanek et al. 1995	–
	NR2C-d[c]	Rat (R.n.)	NR2 C 5′UTR var.	n.d.[c]			Suchanek et al. 1995	–
	NR2C	Human (H.s.)	hNR2C	1233			Lin et al. 1996	L76224
	NR2C-1	Human (H.s.)	hNR2C-1	1236	1215	21	Daggett et al. 1998	U77782
	NR2C-2	Human (H.s.)	hNR2C-2	1231	1210	21	Daggett et al. 1998	–
	NR2C-3	Human (H.s.)	hNR2C-3	1244	1223	21	Daggett et al. 1998	–
	NR2C-4	Human (H.s.)	hNR2C-4	1219	1198	21	Daggett et al. 1998	–
	–″–	Mouse (M.m.)	ε3	1239	1220	19	Kutsuwada et al. 1992	D10694
	–″–	Mouse (M.sp.)	ε3	1239	1220*	19*	Masami 1994e	E06595
	NR2C-a	Mouse (M.m.)	NR2C	n.d.			Suchanek et al. 1995	L350114-28
	NR2C-b[d]	Mouse (M.m.)	NR2 C 5′UTR var.	n.d.[d]			Suchanek et al. 1995	L35017
	NR2C	Mouse (M.m.)	NMDH-7	1239	1220*	19*	Mishina 1996c	I19103

Table 4. *Continued*

Gene	Gene variant	Species	Alternate name used in original publication	Total protein size (amino acids)	Mature protein size (amino acids)	Signal peptide size (amino acids)	References	GenBank accession number
NR2D	NR2D	Rat (R.n.)	NR2D-2	(ps)			Ishii et al. 1993	D13214
	–″–[e]	Rat (R.n.)	NR2D-1[e]	1356	1332	24	Ishii et al. 1993	D13213
	–″–	Rat (R.n.)	NR2D	1323	1263*	60*	Monyer et al. 1994	L31612
	–″–	Rat (R.n.)	NR2D mutated	1323	1263*	60*	Monyer et al. 1994	L31611
	–″–[e]	Rat (R.n.)	NR2D-1[e]	1356			Monyer et al. 1994	–
	–″–[e]	Rat (R.n.)	NR2D-3[e]	(n.d.)			Monyer et al. 1994	–
	–″–	Rat (R.n.)	NR2D	1323	1263*	60*	Boulter 1994b	U08260
	–″–	Human (H.s.)	hNMDAR2D	1336	1313	23	Hess et al. 1998	U77783
	–″–	Mouse (M.m.)	ε4	1323	1296	27	Ikeda et al. 1992	D12822
	–″–	Mouse (M.sp.)	musNR2D	1323	1295	28	Masami 1994e	E07654
	–″–	Mouse (M.m.)	NMDH-16	1323	1263*	60*	Mishina 1996d	I19107
NR3A	NR3A	Rat (R.n.)	χ–1 or chi-1	1115	1089	26	Ciabarra et al. 1995	L34938
	–″–	Rat (R.n.)	NMDAR-L	1115	1082	33	Sucher et al. 1995	U29873
NR3B	NR3B	Rat (R.n.)	χ–2 or chi-2	(ps)			Forcina et al. 1995	–
	NR3B	Human (H.s.)	hNMDAR-L-like	(901 ps)	(879 ps)	22	Lamerdin et al. 1998	AC004528

[a] In the original publication, the mature protein size was given as 943 amino acids, which was later corrected to 1218 amino acids (Burnashev et al. 1992) when a 1-bp deletion was discovered in the clone that had introduced an early stop codon
[b] An additional exon of 58 bp is present in the 5'UTR (Suchanek et al. 1995)
[c] An unspecified additional exon is present in the 5'UTR (Suchanek et al. 1995)
[d] 97 bp of exon 4 is lacking in the 5'UTR
[e] These apparent splice variants were suggested to be PCR artifacts (Monyer et al. 1994); for additional abbreviations and further explanations see legend of Table 2

Table 5. Cloned orphan receptor subunits

Gene	Gene variant	Species	Alternate name used in original publication	Total protein size (amino acids)	Mature protein size (amino acids)	Signal peptide size (amino acids)	References	GenBank accession number
delta1	delta1	Rat (R.n.)	delta-1	1007	992	15	Lomeli et al. 1993	Z17238
	–″–	Rat (R.n.)	delta1	1009	989*	20*	Boulter 1994a	U08255
	–″–	Mouse (M.m.)	δ1	1009	994	15	Yamazaki et al. 1992a	D10171
delta2	delta2	Rat (R.n.)	delta-2	1008	991	17	Lomeli et al. 1993	Z17239
	–″–	Rat (R.n.)	delta2	1007	991	16	Boulter 1994b	U08256
	–″–	Human (H.s.)	h-delta2	1007	990*	17*	Hu et al. 1998	AF009014
	–″–	Mouse (M.m.)	musDelta2	1007	991	16	Araki et al. 1993	D13266

For abbreviations and further explanations see legend of Table 2

Table 6. Cloned kainate binding proteins (KBPs)

Gene	Gene variant	Species	Alternate name used in original publication	Total protein size (amino acids)	Mature protein size (amino acids)	Signal peptide size (amino acids)	References	GenBank accession number
KBP-chick	KBP-chick	Chick (G.d.)	KBP	487	464	23	Gregor et al. 1989	X17700
	–"–	Duck (A.d.)	pDI-11	487	464	23	Kimura et al. 1993	S62230 [D83351]
KBP-frog	KBP-frog	Frog (R.p.)	KBP	487	470	17	Wada et al. 1989	X17314
KBP-fish-α	KBP-fish-α	Goldfish (C.a.)	GFKARα	459	439	20	Wo and Oswald 1994	U08017
KBP-fish-β	KBP-fish-β	Goldfish (C.a.)	GFKARβ	464	439	25	Wo and Oswald 1994	U08016
KBP-toad	KBP-toad	Toad (X.l.)	XenU1	479	462	17	Ishimaru et al. 1996	X93491

For abbreviations and further explanations see legend of Table 2

Table 7. Cloned putative glutamate receptor subunits from invertebrates

Gene	Gene variant	Species	Alternate name used in original publication	Total protein size (aa)	Mature protein size (aa)	Signal peptide size (aa)	References	GenBank accession number
DGluR-I	DGluR-I	Fruit fly (D.m.)	DGluR-I	991	964	27	Ultsch et al. 1992	M97192
DGluR-IIA	DGluR-IIA	Fruit fly (D.m.)	DGluR-II	906	883	23	Schuster et al. 1991	M73271
DGluR-IIB	DGluR-IIB	Fruit fly (D.m.)	DGluRIIB	913	895*	18*	Petersen et al. 1997	AF044202
DNMDAR-I	DNMDAR-I	Fruit fly (D.m.)	DNMDAR-I	997	971	26	Ultsch et al. 1993	X71790
Lym-eGluR1	Lym-eGluR1	Snail (L.s.)	LymGluR		898		Hutton et al. 1991	X60086
Lym-eGluR2	Lym-eGluR2	Snail (L.s.)	InvGluR-K1, or Lym-eGluR2	953	929 (928*)	24 (25*)	Stühmer et al. 1996	X87404
Lym-eGluR3	Lym-eGluR3	Snail (L.s.)	Lym-eGluR3	(135 ps)			Harvey et al. 1997	–
Lym-eGluR4	Lym-eGluR4	Snail (L.s.)	Lym-eGluR4	(151 ps)			Harvey et al. 1997	–
Lym-eGluR5	Lym-eGluR5	Snail (L.s.)	Lym-eGluR5	(144 ps)			Harvey et al. 1997	–
GLR-1	GLR-1A	Worm (C.e.)	GLR-1	962	947	15	Maricq et al. 1995	U34661
	GLR-1B	Worm (C.e.)	C06E1.4	983	968	15	Sulston et al. 1992	L16559 [L18807]
GLR-2	GLR-2	Worm (C.e.)	B0280.12	(735 p?)			Wilson et al. 1994	U10438
GLR-3	GLR-3	Worm (C.e.)	C06A8.9	(479 p?)			Wilson et al. 1994	U39849 (S9)
GLR-4	GLR-4	Worm (C.e.)	C06A8.10	(353 ps)			Wilson et al. 1994	U39849 (S10)
GLR-5	GLR-5	Worm (C.e.)	K10D3.1	956	937	19	Wilson et al. 1994	Z75545 (S1)
GLR-6	GLR-6	Worm (C.e.)	C43H6.9	704	686	18	Wilson et al. 1994	U51999 (S9)
GLR-7	GLR-7	Worm (C.e.)	C08B6.5	(423 ps)	392	31	Wilson et al. 1994	Z72502
GLR-8	GLR-8	Worm (C.e.)	F22A3.3	(398 ps)			Wilson et al. 1994	U41547
GLR-9	GLR-9	Worm (C.e.)	T25E4.2	(396 ps)			Wilson et al. 1994	U23411
GLR-10	GLR-13	Worm (C.e.)	T01C3.10	(563 p?)			Wilson et al. 1994	Z78413
GLR-11	GLR-11	Worm (C.e.)	F14H8.0	(232 ps)			Wilson et al. 1994	Z81061
Loc1	Loc1	Locust (S.g.)	Loc1	(69 ps)			Usherwood et al. 1993	–
Loc2	Loc2	Locust (S.g.)	Loc2	(71 ps)			Usherwood et al. 1993	–
Loc3	Loc3	Locust (S.g.)	Loc3	(71 ps)			Usherwood et al. 1993	–
Pan-eGluR1	Pan-eGluR1a	Lobster (P.i.)	Pan-eGluR1a	914			Krenz et al. 1999	–
	Pan-eGluR1b	Lobster (P.i.)	Pan-eGluR1b	914			Krenz et al. 1999	–
Pan-NMDAR-I	Pan-NMDAR-I	Lobster (P.i.)	Pan-NMDAR-I	(ps)			W. Krenz personal communication	–

GLR-1 is the only excitatory glutamate receptor cDNA clone reported from *C.elegans* to date. GLR-2 through GLR-11 represent 10 sequences which were discovered as potential glutamate receptor homologs in the course of the *C. elegans* genome sequencing project. They have been given arbitrary numbers for the purpose of this compilation

p? This might be a partial clone; the signal peptide search algorithm SPSCAN in the University of Wisconsin sequence analysis software package (Devereux et al. 1984) does not recognize a signal peptide at the translation start site of the clone. For further abbreviations and explanations see legend of Table 2

Table 8. Cloned inhibitory, chloride-conducting glutamate receptor subunits (glutamate-gated chloride channels)

Gene	Gene variant	Species	Alternate name used in original publication	Protein size (aa)[a]	References	GenBank accession number
GluCl-α1	GluCl-α	Nematode (C.e)	GluCl-α	461	Cully et al. 1994	U14524
	–″–	Fruit fly (D.m.)	dGluCl-α	456	Cully et al. 1996	U58776
GluCl-α2	GluCl-α2 A	Nematode (C.e)	GluClα2 A	657	Dent et al. 1997	AJ000538
	GluCl-α2B	Nematode (C.e)	GluClα2B	478	Dent et al. 1997	AJ000537
GluCl-β	GluCl-β	Nematode (C.e)	GluCl-β	434	Cully et al. 1994	U14525
	–″–	Nematode (C.e)	cosmid F25F8	434	Wilson et al. 1994	U88171
	–″–	Nematode (H.c)	GluCl-β	432	Delany et al. 1997	Y09796
GluCl-gbr-2	GluCl-gbr-2 A	Nematode (H.c)	GluCl-gbr-2a	469	Jagannathan et al. 1998a	Y14233
	GluCl-gbr-2B	Nematode (H.c)	GluCl-gbr-2b	438	Jagannathan et al. 1998b	Y14234
	GluCl-gbr-2	Nematode (C.e)	Ce-gbr-2	416	Wolstenholme 1997	–
	GluCl-gbr-3	Nematode (C.e)	Ce-gbr-3	430	Wolstenholme 1997	U41113
GluCl-X	GluCl-X	Nematode (C.e)	cosmid C27H5	1332	Wilson et al. 1994	U14635

[a] Size given in amino acids (aa) includes signal peptide. All glutamate-gated chloride channels identified to date have been cloned either from the nematodes *Caenorhabditis elegans* (C.e.) and *Haemonchus contortus* (H.c.), or the fruit fly *Drosophila melanogaster* (D.m.)

whether they have developed independently and, thus, are merely homoplasous genes. Invertebrates additionally possess at least five genes for inhibitory, chloride-conducting glutamate-gated channels which are only found in invertebrates (CLELAND 1996). These five genes are listed in Table 8 but will not be discussed any further.

2. Alternative Splicing

In addition to the existence of multiple genes, glutamate receptor diversity is further increased through the post-transcriptional mechanism of alternative splicing. All glutamate receptor genes are split genes, consisting of numerous exons interrupted by intronic sequence (see Sect. C.1). This gene structure lends itself to the splicing of alternate exons at certain positions in the genomic sequence (Table 9).

a) Splicing of Extracellular Domains

Splicing of extracellular domains occurs in three different regions: in the far N-terminal domain [or leucine-isoleucine-valine binding protein (LIVBP) homology domain, Fig. 5], as seen in the NMDAR1 subunit and in the KA receptor GluR5; in the S1 domain of the ligand-binding lysine-arginine-ornithine binding protein (LAOBP) homology region (Fig. 5), as seen in the *Caenorhabditis elegans* glutamate receptor gene GLR-1; and in the L3 domain (loop3, between transmembrane domains B and C, Fig. 5), as exemplified by the AMPA receptors GluR1 to GluR4 and the human NMDA receptor subunit NR2C. For an overview of the various splice variants identified to date as well as the pertinent references see Table 9.

b) Splicing of Intracellular Domains

Splicing in intracellular domains is largely restricted to the C-terminal domain and has been observed in the AMPA receptors GluR2 and GluR4, the KA receptors GluR5 to GluR7, the NMDA receptor subunit NMDAR1, and the modulatory NMDA receptor subunit NR3A. C-terminal splice variants reported for the NMDA receptor NR2D subunit (ISHII et al. 1993) were later suggested to be PCR artifacts as no splice sites could be found in the genomic sequence (MONYER et al. 1994). An unusual splicing event has recently been reported for the intracellular L1 domain (loop1, in front of the ion pore domain, Fig. 5) of the human NMDA receptor subunit NR2C; however, it is unclear at present whether this might represent an artifact (DAGGETT et al. 1998). See Table 9 for a listing of splice variants and references.

c) Splicing of the Ion-Pore Domain

In the lobster Pan-eGluR1 subunit, a very interesting splicing event has recently been discovered in which the ion-pore domain, together with the two

Table 9. Variants of excitatory ionotropic glutamate receptor subunits resulting from alternative splicing

Receptor subfamily	Gene	Splice variants	Location of alternative splice site and size difference of variant	References
AMPA receptors	GluR1	GluR1-flop = GluR1	–	Hollmann et al. 1989
		GluR1-flip	"Flip/flop": L3 domain, same size	Sommer et al. 1990
		GluR1α-flop	["α": L1 domain, 25 aa insert][a]	Wu et al. 1996
		GluR1α-flip		Wu et al. 1996
		GluR1αc-flop	"αc": C-terminal domain, 22 aa longer than α	Wu et al. 1996
		GluR1αc-flip		Wu et al. 1996
		GluR1β-flop	["β": L1 domain, 36 aa insert][a]	Wu et al. 1996
		GluR1β-flip		Wu et al. 1996
	GluR2	GluR2-flop = GluR2	–	Keinänen et al. 1990
		GluR2-flip	"Flip/flop": L3 domain, same size	Sommer et al. 1990
		GluR2-flop/long	"Long": C-terminal domain, 18 aa longer	Köhler et al. 1994
		GluR2-flip/long		Köhler et al. 1994
		GluR2α-flop	["α": L1 domain,17 aa shorter than 2β][b]	Kung et al. 1996
		GluR2αc-flop	"αc": C-terminal domain 12 aa shorter than α	Kung et al. 1996
		GluR2αc-flip		Kung et al. 1996
		GluR2β-flop	["β": L1 domain,17 aa longer than 2α][b]	Kung et al. 1996
		GluR2β-flip		Kung et al. 1996
	GluR3	GluR3-flip = GluR3	–	Keinänen et al. 1990
		GluR3-flip	"Flip/flop": L3 domain, same size	Sommer et al. 1990
	GluR4	GluR4-flop = GluR4	–	Keinänen et al. 1990
		GluR4-flip	"Flip/flop": L3 domain, same size	Sommer et al. 1990
		GluR4c-flop	"c": C-terminal domain 18 aa shorter than GluR4	Gallo et al. 1992
		GluR4c-flip		Gallo et al. 1992
		GluR4d-flop	"d": C-terminal domain 38 aa shorter than GluR4	Ravindranathan et al. 1997

		GluR4d-flip		Ravindranathan et al. 1997
		GluR4s	"s": a C-terminal variant truncated in front of the flip/flop domain	Ravindranathan et al. 1997
KA receptors (low affinity)	GluR5	GluR5–1a	"1/2": LIVBP domain (N-terminal); "1"-variants have 15 aa insert	Sommer et al. 1992
		GluR5–1b	"a": C-terminal domain 49 aa shorter than "b"	Bettler et al. 1990
		GluR5–1c	"b": (original cDNA clone)	Sommer et al. 1992
		GluR5–1d	"c": C-terminal domain 29 aa longer than "b"	Gregor et al. 1993
		GluR5–2a	"d": C-terminal domain 2 aa shorter than "b"	Sommer et al. 1992
		GluR5–2b = GluR5		Bettler et al. 1990
		GluR5–2c		Sommer et al. 1992
	GluR6	GluR6a = GluR6	–	Egebjerg et al. 1991
		GluR6b	"b": C-terminal domain 38 aa shorter than "a"	Daggett et al. 1996
		GluR6c	"c": C-terminal domain 39 aa shorter than "a"	Gregor et al. 1993
	GluR7	GluR7a = GluR7	–	Bettler et al. 1992
		GluR7b	"b": C-terminal domain 9 aa shorter than "a"	Schiffer et al. 1997
KA receptors (high affinity)	KA1	None reported	–	–
	KA2	None reported	–	–
NMDA receptors (GLY binding)	NMDAR1	NR1–1a	"a/b": LIVBP domain (N-terminal); "b"-variants have 21 aa insert	Moriyoshi et al. 1991
		NR1–1b	"1": C-terminal domain	Sugihara et al. 1992
		NR1–2a	"2": C-terminal domain 37 aa shorter than "1"	Sugihara et al. 1992
		NR1–2b		Sugihara et al. 1992
		NR1–3a	"3": C-terminal domain 16 aa shorter than "1"	Sugihara et al. 1992

Table 9. *Continued*

Receptor subfamily	Gene	Splice variants	Location of alternative splice site and size difference of variant	References
		NR1–3b		Hollmann et al. 1993
		NR1–4a	"4": C-terminal domain 53 aa shorter than "1"	Sugihara et al. 1992
		NR1–4b		Sugihara et al. 1992
		NR1–5 (trunc.)	"5": a C-terminally truncated variant	Sugihara et al. 1992
		NR1–6a	"6": C-terminal domain [in birds] 27 aa longer than "1"	Kurosawa et al. 1994
		NR1–6b		Kurosawa et al. 1994
		NR1–7a	"7": C-terminal domain [in fish] 7 aa longer than "1"	Bottai et al. 1998
		NR1–7b		Bottai et al. 1998
NMDA receptors (GLU binding)	NR2A	None reported	–	–
	NR2B	NR2B-1 = NR2B	–	–
		NR2B-2	5′-UTR (lacks exon 2, 353 bp)	Klein et al. 1998
	NR2C	NR2C-a = NR2C =	–	Monyer et al. 1992
		NR2C-1	–	–
		NR2C-b	"b": 5′-UTR (97 bp deletion)	Suchanek et al. 1995
		NR2C-c	"c": 5′-UTR (58 bp inserted)	Suchanek et al. 1995
		NR2C-d	"d": 5′-UTR (insert,bp unknown)	Suchanek et al. 1995
		NR2C-2[d]	"2": L1 (5 aa deletion)	Daggett et al. 1998
		NR2C-3[d]	"3": L3 (8 aa insertion)	Daggett et al. 1998
		NR2C-4[d]	"4": L3 (17 aa deletion)	Daggett et al. 1998
	NR2D	(NR2D-1)[c]	["2": C-terminal domain][c] 33 aa longer than NR2D-2	Ishii et al. 1993
		NR2D-2 = NR2D		Ishii et al. 1993
		(NR2D-3)[c]	["3": 5′-UTR and LIVBP domain][c] 85 aa lacking vs. NR2D-2	Monyer et al. 1994

NMDA receptors (modulatory)	NR3A	NR3A-1 = NR3A	–	–
		NR3A-2	["2": C-terminal domain] 20 aa longer than NR3A-1	SUN et al. 1998; PEREZ-OTANO et al. 1998
	NR3B	None reported	–	–
Orphan receptors	delta1	None reported	–	–
	delta2	None reported	–	–
Kainate binding proteins (KBPs)	KBP-chick	None reported	–	–
	KBP-frog	None reported	–	–
	KBP-fish-α	None reported	–	–
	KBP-fish-β	None reported	–	–
	KBP-toad	None reported	–	–
Invertebrate receptors	GLR-1	GLR-1A	–	MARICQ et al. 1995
		GLR-1B	"B": LAOBP domain (N-terminal) 21 aa longer than "A"	SULSTON et al. 1992
	Pan-eGluR1	Pan-eGluR1a	–	KRENZ et al. 1999
		Pan-eGluR1b	"b": L1 + ion pore domain + L2 46 aa, same size as "a"	KRENZ et al. 1999

GLY, glycine; GLU, glutamate; UTR, untranslated region; aa, amino acids. For designation and abbreviations of receptor domains see Fig. 5

[a] GluR1α and GluR1β appear to be two slightly different copies of the GluR1 gene present in fish

[b] GluR2α and GluR2β appear to be two slightly different copies of the GluR2 gene present in fish

[c] These apparent splice variants have been suggested to represent PCR artifacts (MONYER et al. 1994)

[d] These apparent isoforms potentially could have arisen by improper transcript processing (DAGGETT et al. 1998)

intracellular loops flanking this domain (46 amino acids total), is alternatively spliced. This event is tissue specific and selects between two exons of 35% sequence identity containing either a cysteine or a glutamine at the Q/R editing site (see Sect. B.I.3. for a brief discussion of editing). The cysteine-containing variant appears to be restricted to muscle, while the glutamine-containing variant is brain specific. Interestingly, alternative splicing in this case apparently mimics the effect of editing at the Q/R site of Pan-eGluR1 and, thus, might compensate for the absence of an editing mechanism in this subunit (KRENZ et al. 1999).

d) Splicing of Non-Coding Regions

In the NR2C gene, splice variants arising in the 5′ untranslated region have been reported (SUCHANEK et al. 1995). This type of splicing might modulate expression levels of the protein rather than its functional properties.

3. RNA Editing

Another quite unexpected mechanism creating molecular diversity of glutamate receptors was discovered in 1991 in Peter Seeburg's laboratory (SOMMER et al. 1991). This post-transcriptional mechanism, termed RNA editing, had originally been discovered in mitochondrial RNA of trypanosomes (BENNE et al. 1986) and is a predominant feature of plant mitochondrial transcripts (GRAY et al. 1992). RNA editing specifically alters, inserts or deletes a single nucleotide within a specific codon, thereby changing the encoded amino acid residue which usually leads to modified functional properties of the protein. In glutamate receptors, a conversion of specific adenosine (A) residues to inosine (I) by a double-stranded RNA deaminase causes CAG → CIG conversions, which change a glutamine residue to an arginine, or AGA → IGA conversions, which change an arginine to a glycine residue, as inosine will be interpreted as guanosine (G) during translation (reviewed by SEEBURG 1996). For an overview of editing variants identified in the various glutamate receptor subfamilies to date see Table 10.

a) Editing Within the Ion Pore Domain

The most dramatic functional effects of RNA editing have been seen with editing sites located within the ion permeation pathway. At this crucial site, a single amino acid exchange can drastically alter the permeability of the ion pore. This was originally demonstrated for a glutamine (Q) present in all AMPA receptor proteins except GluR2, which instead has the gene-encoded Q altered to an arginine (R) via RNA editing. This change from Q to R at what has been dubbed the Q/R site, which occurs in virtually 100% of the GluR2 subunits, shuts off the calcium permeability of the ion channel (HUME

Table 10. RNA editing events observed in vertebrate ionotropic glutamate receptor subunits. Amino acid positions apply to the mature proteins

Subfamily	Gene	Edited amino acids	Localization in protein	Extent of editing[a]	References
AMPA receptors	GluR1	None found			SOMMER et al. 1991; LOMELI et al. 1994
	GluR2	Q(586) → R (rat)	In ion pore domain	~100%	SOMMER et al. 1991
		Q(586) → R (chick)	In ion pore domain	n.d.	LOWE et al. 1997
		Q(586) → R (human)	In ion pore domain	~100%	PASCHEN et al. 1994
		–″–	In ion pore domain	99.5% (gray matter)	AKBARIAN et al. 1995
		–″–	In ion pore domain	93% (white matter)	AKBARIAN et al. 1995
		–″–	In ion pore domain	72–100% (region-dependent)	NUTT and KAMBOJ 1994a
		No Q/R editing in fish: gene codes for R			KUNG et al. 1996
		R(743) → G (rat)	At start of flip/flop domain		LOMELI et al. 1994
		R(739) → G (fish α)	At start of flip/flop domain	~80% flip; ~85% flop	KUNG et al. 1996
		R(730) → G (fish β)	At start of flip/flop domain		KUNG et al. 1996
	GluR3	R(746) → G (rat)	At start of flip/flop domain	~95% flip; ~100% flop	LOMELI et al. 1994
		R(747) → G (chick)	At start of flip/flop domain	n.d.	PAPERNA et al. 1996
		R(746) → G (fish)	At start of flip/flop domain	n.d.	CHANG et al. 1998
	GluR4	R(744) → G [rat]	At start of flip/flop domain	~55% flip; ~90% flop	LOMELI et al. 1994

Table 10. *Continued*

Subfamily	Gene	Edited amino acids	Localization in protein	Extent of editing[a]	References
KA receptors (low affinity)	GluR5	Q(606) → R (rat GluR5-1)	In ion pore domain	39%	Sommer et al. 1992
			In ion pore domain	36% (hippocampus)	Kamphuis and Da Silva 1995
		Q(606) → R (human)	In ion pore domain	83%	Nutt and Kamboj 1994b
			In ion pore domain	43–91% (gray matter)	Paschen and Djuricic 1994
			In ion pore domain	41% (white matter)	Paschen and Djuricic 1994
	GluR6	Q(588) → R (rat)	In ion pore domain	74.5%	Sommer et al. 1992
			In ion pore domain	75%	Köhler et al. 1993
			In ion pore domain	92% (hippocampus)	Kamphuis and Da Silva 1995
		Q(588) → R (human)	In ion pore domain	89–95% (gray matter)	Paschen et al. 1994
			In ion pore domain	10% (white matter)	Paschen et al. 1994
			In ion pore domain	66–90% (gray matter)	Paschen and Djuricic 1995
			In ion pore domain	55% (white matter)	Paschen and Djuricic 1995
		No editing in chick	–		Lowe et al. 1997
		I(534) → V [rat]	In TMD A	~80%	Köhler et al. 1993
		Y(538) → C (rat)	In TMD A	~80%	Köhler et al. 1993
	GluR7	None found (rat)	–		Bettler et al. 1992; Lomeli et al. 1992
		S(310) → A [human][b]	In LIVBP domain (*N*-terminal)	38%	Nutt et al. 1994
		R(352) → Q [human][b]	In LIVBP domain (*N*-terminal)	4.2%	Nutt et al. 1994

KA receptors (high affinity)	KA1	None reported
	KA2	None reported
NMDA receptors (GLY binding)	NMDAR1	None reported
NMDA receptors (GLU binding)	NR2A	None reported
	NR2B	None reported
	NR2C	None reported
	NR2D	None reported
NMDA receptors (modulatory)	NR3A	None reported
	NR3B	None reported
Orphan receptors	delta1	None reported
	delta2	None reported

AMPA, α-amino-3-hydroxy-5-methyl-4-isoxazole propionate; GLY, glycine; GLU, glutamate; KA, kainate; LIVBP, leucine-isoleucine-valine binding protein; n.d., not determined; NMDA, *N*-methyl-D-aspartate; TMD, transmembrane domain. For designation and abbreviations of receptor domains, see Fig. 5

[a] Numbers apply to adult animals; as editing is developmentally regulated, numbers might be considerably lower or even zero for very young animals

[b] A mechanism other than RNA editing may explain this variant

et al. 1991; Burnashev et al. 1992) and causes a linear instead of the normally inwardly rectifying current–voltage relationship (Hume et al. 1991; Verdoorn et al. 1991). Interestingly, while editing at the Q/R site of GluR2 is found in rat, human and chick, it appears to be absent from fish and invertebrate glutamate receptors. In fish, the "edited" arginine residue is already encoded in the gene (Kung et al. 1996).

In the ion pore domain of certain low affinity KA receptors, a similar editing event is found at the Q/R site. It also alters a gene-encoded glutamine to an arginine and causes similar effects on calcium permeability and rectification properties as in AMPA receptors, but occurs in only 39–75% of KA receptor subunits GluR5 and GluR6 (Sommer et al. 1991; Table 10). Additionally, editing of GluR6 has been shown to reduce single-channel conductivity and convey a distinct chloride permeability to the ion pore (Burnashev et al. 1996).

b) Other Editing Sites

The low affinity KA receptor subunit GluR6, in addition to being edited in the ion pore domain, has two RNA editing sites located in the first transmembrane domain (TMDA) (see Fig. 5 for domain structure of glutamate receptors). These sites are usually only partially edited and display valine to isoleucine (I $\rightarrow$ V) and tyrosine to cysteine (Y $\rightarrow$ C) alterations, respectively. No other glutamate receptor subunit edits sites within a transmembrane domain.

In the AMPA receptors GluR2, GluR3, and GluR4, an additional editing site has been discovered (Lomeli et al. 1994) at the N-terminal end of the flop/flip splice domain in the second extracellular domain (L3, Fig. 5). This site, which has been found in rat, chick and fish, is only partially edited and is characterized by alteration of a genome-encoded arginine to glycine (R $\rightarrow$ G).

It should be noted that two sites in the N-terminal domain of the human KA receptor subunit GluR7 have been reported to express amino acids other than those encoded in the gene (Nutt et al. 1994). However, it is unclear whether these changes represent RNA editing or occur via some other mechanism.

4. Post-Translational Modifications

a) N-Glycosylation

All glutamate receptor subunits are extensively N-glycosylated. Each subunit contains a minimum of 4 extracellular consensus sites for N-glycosylation (N-X-S/T, X $\neq$ P) and can have up to 12 (Everts et al. 1997). In those cases that have been investigated, all potential N-glycosylation sites have been found to be glycosylated (Hollmann et al. 1994; Everts et al. 1999). *O*-glycosylation of glutamate receptors has also been demonstrated (Hullebroeck and Hampson 1992). While the functional role of glutamate receptor glycosylation is still an open question, it has been shown that N-glycosylation is not strictly necessary for ion channel function and does not dramatically affect ligand binding (MUßHOFF et al. 1992; KAWAMOTO et al. 1995; ARVOLA and KEINÄNEN 1996; EVERTS et al. 1997). However, it may be necessary for efficient expression of some subunits, e.g., NMDAR1 receptors (CHAZOT et al. 1995; HALL et al. 1997), and it is indispensable for lectin modulation of glutamate receptors (EVERTS et al. 1997).

b) Phosphorylation

All glutamate receptor subunits cloned so far have numerous putative phosphorylation sites for protein kinase A, protein kinase C, calcium/calmodulin-dependent kinase II, and tyrosine kinase. For a detailed discussion of these sites and the functional consequences of phosphorylation see Chap. 2 by Lee & Huganir.

c) Palmitoylation

In one study, palmitoylation was shown to occur at two C-terminal cysteine residues of the KA receptor GluR6 (PICKERING et al. 1995). While this secondary modification apparently did not directly influence channel function, it was shown to inhibit phosphorylation by protein kinase A. Thus, an indirect modulatory effect is possible.

II. The Cloned Subunits

1. AMPA Receptors

The first glutamate receptor subunit (HOLLMANN et al. 1989), now known as GluR1 (or GluRA in a different nomenclature, KEINÄNEN et al. 1990) originally had been named GluR-K1 to indicate that it was thought to represent the KA receptor subfamily. Indeed, the glutamatergic agonist kainic acid is a more efficacious agonist than glutamate or AMPA in the *Xenopus* oocyte expression system employed for cloning (HOLLMANN et al. 1990). However, it soon became clear that this observation was due to KA causing slower and less complete desensitization than glutamate, and that KA actually has a lower potency than both AMPA and glutamate for activating the GluR1 ion channel. Besides, when ligand binding data became available, they clearly showed that AMPA and glutamate had much higher affinities for this receptor than KA (KEINÄNEN et al. 1990) Therefore, this receptor was later reassigned to the AMPA receptor subfamily (BOULTER et al. 1990; KEINÄNEN et al. 1990). In addition, it rapidly became clear that there were other, sequence-related receptor subunits with properties more fitting to a true KA receptor, such as high affinity for $[^3H]$KA in binding assays and EC_{50} values for channel gating which were considerably lower for KA than for either AMPA or glutamate (BETTLER et al. 1990; EGEBJERG et al. 1991; BETTLER et al. 1992), see Sect. B.II.2).

After the first sequence had been reported, homology screening methods revealed three additional closely related subunits, which were named GluR2, GluR3, and GluR4 (Bettler et al. 1990; Boulter et al. 1990; Nakanishi et al. 1990), or GluRB-D (Keinänen et al. 1990). The four subunits share 68–74% sequence identity at the protein level (Table 14), have a size of approximately 100 kDa (Table 13) and are structurally similar, as suggested by their nearly identical hydropathy profiles (Fig. 4). The high sequence identity among these four sequences, in combination with significantly lower sequence identities with other subunits, sets them apart as a distinct glutamate receptor subfamily. To date, AMPA receptors have been cloned from rat (GluR1–4), human (GluR1–4), mouse (GluR1, GluR2), pigeon (GluR2–4), chick (GluR1–4), and various fish (GluR1–4). Sequence identity across species is better than 98% among mammals and higher than 75% among invertebrates in general (Table 17). For specifics and references regarding the published cDNA clones see Table 2. For the cDNAs from mouse, a different nomenclature has been introduced which uses Greek letters for receptor subfamilies. AMPA receptors GluR1–GluR4 in that nomenclature are called $\alpha1$ to $\alpha4$ (Meguro et al. 1992; Sakimura et al. 1992).

When expressed in *Xenopus* oocytes or human embryonic kidney (HEK293) cells, each subunit responds to AMPA, glutamate, KA, domoate, and quisqualate, but not to NMDA, by opening a cation-conducting pore (Boulter et al. 1990; Keinänen et al. 1990; Nakanishi et al. 1990). Surprisingly, GluR1, GluR3 and GluR4 but not GluR2 were found to be permeable to calcium in addition to sodium and potassium (Hollmann et al. 1991). GluR2 differs from the other AMPA receptor subunits by undergoing RNA editing (see Sect. B.I.3.a) within the ion permeation pathway, thereby changing a key amino acid that is responsible for the calcium permeability of the receptor (Hume et al. 1991; Burnashev et al. 1992) as well as its rectification properties (Verdoorn et al. 1991). Alternative splicing of the flip-flop domain determines the kinetics of receptor desensitization which can be slow (flip) or fast (flop) (Mosbacher et al. 1994). A point mutation in the N-terminal extracellular domain (L485Y in GluR3) has recently been described that abolishes desensitization entirely (Stern-Bach et al. 1998).

2. KA Receptors

In the course of low-stringency hybridization screening with an AMPA receptor GluR1 probe, a receptor subunit was isolated that shared only approximately 40% amino acid sequence identity with AMPA receptors (Bettler et al. 1990; Sommer et al. 1992). Based on its high affinity for $[^3H]$KA ($K_d = 73.3$ nM, Sommer et al. 1992) and a comparatively low affinity for glutamate ($K_I =$ 290 nM, Sommer et al. 1992), this subunit, termed GluR5, was classified as a KA receptor. A related subunit, GluR6, was identified shortly afterwards (Egebjerg et al. 1991), and, finally, a third subunit, GluR7, was established as another member of this subfamily (Bettler et al. 1992). The three proteins share 74–81% amino acid sequence identity (Table 14) and are structurally very similar (Fig. 4), thus defining them as a distinct receptor subfamily. All three have binding sites for KA, with affinities in the range of 60–95 nM (LOMELI et al. 1992; SOMMER et al. 1992), and have been termed "low-affinity KA receptors" (BETTLER et al. 1992).

Two additional subunits with significantly higher affinities for [^{3}H]KA, termed KA1 (WERNER et al. 1991; K_d for KA = 4.7 nM) and KA2 (HERB et al. 1992; SAKIMURA et al. 1992; K_d for KA = 15 nM) were identified and found to form a separate subfamily of high-affinity KA receptors. They share 69% amino acid sequence identity with each other, but only 42–44% with GluR5 through GluR7. Compared with AMPA receptors, sequence identity drops to 35–37% (Table 14).

When expressed in *Xenopus* oocytes or HEK cells, subunits GluR5 and GluR6 respond to glutamate, KA, domoate, and quisqualate, but not AMPA or NMDA, by opening a cation-conducting, calcium-permeable pore (EGEBJERG et al. 1991; MORITA et al. 1992; SAKIMURA et al. 1992; SOMMER et al. 1992). In one report, AMPA very weakly activated GluR5–2a, but with an EC_{50} of only 3 mM (SOMMER et al. 1992). GluR7 does not give measurable currents in *Xenopus* oocytes (BETTLER et al. 1992; LOMELI et al. 1992). Only recently this subunit was shown to exhibit ion channel function in HEK cells when activated with concentrations of KA or glutamate above 10 mM (SCHIFFER et al. 1997). High concentrations such as these cannot be tested in *Xenopus* oocytes as they cause unspecific background currents even in control oocytes (MICHAEL HOLLMANN, unpublished observation). KA1 and KA2 do not form functional ion channels upon homomeric expression in either oocytes or mammalian cells (WERNER et al. 1991; HERB et al. 1992; KAMBOJ et al. 1992; SAKIMURA et al. 1992). They do, however, incorporate into functional heteromeric KA receptor complexes with low-affinity KA receptors GluR5 and GluR6 (see Sect. D.IV.2).

To date, KA receptors reportedly have been cloned from rat (GluR5 to GluR7, KA1, KA2), mouse (GluR5 to GluR7, KA2, sometimes referred to as glutamate receptors $\beta1$–$\beta3$, $\gamma2$; MORITA et al. 1992; SAKIMURA et al. 1992), human (GluR5 to GluR7, KA1, KA2), and toad (GluR6). Sequence identity across species is better than 96% (Table 17). See Table 3 for details on the published cDNA clones and references.

3. NMDA Receptors

The first NMDA receptor cDNA, just like the first AMPA receptor GluR1, was isolated by means of expression cloning (MORIYOSHI et al. 1991). Once the amino acid sequence of this clone, which was named NMDAR1 or, in short, NR1 ($\zeta1$ for the mouse cDNA, YAMAZAKI et al. 1992b), was available and found to share merely 24–30% identity with other glutamate receptor subunits (Table 14), it was obvious why homology screening had failed to identify this clone: sequences with such low homology had usually been disregarded.

Further screening with the NMDAR1 sequence as a probe soon identified four related subunits (Ikeda et al. 1992; Kutsuwada et al. 1992; Meguro et al. 1992; Monyer et al. 1992), which share 46–56% amino acid sequence identity with each other but only 26–29% with NMDAR1. They were named NMDAR2A through NMDAR2D, or, in short, NR2A–NR2D. For the mouse genes, the names $\varepsilon 1$–$\varepsilon 4$ are being used by some authors (Ikeda et al. 1992; Kutsuwada et al. 1992; Meguro et al. 1992).

Another subunit identified by PCR-mediated screening with degenerate primers to NMDA receptor sequences was originally named NMDAR-L for "NMDA receptor-like subunit" by one group (Sucher et al. 1995) and $\chi 1$ (chi-1) by another group (Ciabarra et al. 1995). A second, related subunit ($\chi 2$, or chi-2) was reported, but only in abstract form and as a partial clone, and was never followed up on (Forcina et al. 1995). However, in the process of sequencing human chromosome 19, a sequence with 52.1% identity to NMDAR-L/$\chi 1$ was reported recently (Lamerdin et al. 1998), which likely represents a human $\chi 2$ cDNA. The only initial indication that these subunits might be members of the NMDA receptor family were a weak sequence identity (25.7–29.5%, Table 14) with NMDAR1 and NR2A-D subunits and the presence of an asparagine at the editing site within the ion pore, a feature shared by all NMDA receptor subunits but absent from other subfamilies. Recently, a knock-out mutant of NMDAR-L/$\chi 1$ was reported to result in increased NMDA receptor currents and increased numbers of dendritic spines, suggesting a role for NMDAR-L/$\chi 1$ as a negative modulator of the NMDA receptor complex (Das et al. 1998). NMDAR-L/$\chi 1$ was renamed and is now called NR3A (Das et al. 1998). The subunit $\chi 2$ thus should be called NR3B, a name herewith assigned to the human subunit (GenBank accession number AC004528, Table 4) as the rat sequence has not been published.

To date NMDA receptor subunits reportedly have been cloned from rat (NMDAR1, NR2A–NR2D, and NR3A), mouse (NMDAR1, and NR2A–NR2D), human (NMDAR1 and NR2A–NR2D, NR3B), pig (NMDAR1), duck (NMDAR1), goldfish (NMDAR1), and toad (NMDAR1). Sequence identity between species is better than 90% for NMDAR1 and 89% for NR2A–NR2D (Table 17). For details on the published cDNA clones and references, see Table 4.

NMDAR1 was found to express homomeric functional ion channels, but only in *Xenopus* oocytes (Moriyoshi et al. 1991; Yamazaki et al. 1992b), not in mammalian cells such as HEK293 cells (Monyer et al. 1992; Lynch et al. 1994; Grimwood et al. 1995; Sucher et al. 1996). None of the NR2 subunits, when expressed by themselves, assemble into functional ion channels in either system (Ikeda et al. 1992; Kutsuwada et al. 1992; Meguro et al. 1992; Monyer et al. 1992). However, upon coexpression with NMDAR1, such channels are readily formed by all four NR2 subunits (see Sect. D.IV.2). NMDA receptors in vivo and in vitro are highly regulated and intensely modulated, for example, by pH (Traynelis et al. 1995; Kashiwagi et al. 1996; Kashiwagi et al. 1997; Mott et al. 1998), polyamines (Rock and Macdonald 1995; Gallagher et al.

1997; LEWIN et al. 1998), and zinc (HOLLMANN et al. 1993; PAOLETTI et al. 1997; TRAYNELIS et al. 1998). A key structural determinant for each of these modulatory actions is an alternatively spliced exon (exon 5) in the N-terminal extracellular domain of the NR1 subunit, on which these modulators converge. In addition, modulation depends on the particular NR2 subunit expressed in a given heteromeric complex.

The NR3A subunit, similarly, is incapable of formation of homomeric ion channels. Its ion pore domain has been transplanted into other glutamate receptor subunits, such as GluR6 and NMDAR1, and even in those chimeras the pore failed to support channel function (VILLMANN et al. 1999). NR3A is thought to modulate NMDA receptor function (DAS et al. 1998), but it is unclear how this modulation is effected. No expression studies have been reported for NR3B to date.

4. Orphan Receptors: Delta Subunits

Homology screening with GluR1–GluR4 probes in 1992 identified a subunit that shared ~30% amino acid sequence identity with AMPA receptors and was termed $\delta 1$ (YAMAZAKI et al. 1992a) (or delta1, HOLLMANN and HEINEMANN 1994) to indicate its classification as a member of a separate subfamily. A second subunit, evidently belonging to the same subfamily as suggested by ~56% amino acid sequence identity with delta1, was cloned in 1993 (LOMELI et al. 1993) and named $\delta 2$ (or delta2). Surprisingly, attempts to identify a glutamatergic ligand or, indeed, any ligand that could bind to these two subunits failed. Similarly, attempts to express functional ion channels in *Xenopus* oocytes or HEK cells proved futile. Although receptor protein was clearly expressed, neither the standard agonists glutamate, KA, or AMPA nor endogenous candidate neurotransmitters such as the sulfur-containing amino acids L-homocysteic acid or L-cysteinesulfinic acid were able to activate ionic currents (YAMAZAKI et al. 1992a; LOMELI et al. 1993) Furthermore, coexpression with other glutamate receptor subunits provided no evidence of any functional interaction of delta subunits with AMPA, KA, or NMDA receptor subunits (LOMELI et al. 1993). Recently, transplantation of the ion pore domains of delta1 and delta2 into KA receptor (GluR6) or AMPA receptor subunits (GluR1) failed to produce functional chimeric ion channels (VILLMANN et al. 1999). In a naturally occurring mouse mutant, the *lurcher* mouse, which shows severe ataxia and apoptotic cell death in cerebellar Purkinje cells, it was demonstrated that the defects were caused by a G $\rightarrow$ A point mutation in the delta2 gene, immediately downstream of transmembrane domain B (see Fig. 5 for receptor topology), which changes an alanine residue to a threonine (ZUO et al. 1997). Preliminary evidence from two recently published abstracts suggests that expression in HEK293 cells of mutant delta2 clones with the *lurcher* (A $\rightarrow$ T) mutation causes constituitively open channels with properties reminiscent of AMPA receptors (KOHDA et al. 1998; ZUO et al. 1998). Knocking out the delta2 gene in mice caused deficits in long-term depression in the cerebellum (HIRANO et al. 1995) as well as impaired motoric coordination, motor learning and Purkinje cell synapse formation (FUNABIKI et al. 1995; KASHIWABUCHI et al. 1995; KURIHARA et al. 1997). Taken together, the available evidence suggests a modulatory role of delta subunits rather than active participation in the formation of ion conduction pathways.

To date, delta receptor subunits reportedly have been cloned from rat (delta1 and delta2), mouse (delta1 and delta2, also referred to as receptors δ1 and δ2), and human (delta2). Sequence identity between species is higher than 97% (Table 17). See Table 5 for details on the published cDNA clones and references.

5. KBPs in Non-Mammalian Vertebrates

Simultaneously with the cloning of the AMPA receptor GluR1 in 1989, two cDNA clones for so-called kainate binding proteins (KBPs) had been reported. These clones had been isolated following biochemical purification of the proteins, partial protein sequence analysis, and screening of libraries with oligonucleotide primers deduced from the partial sequence. One of the KBPs was cloned from the chicken *Gallus domesticus* (KBP-chick, GREGOR et al. 1989) and the other from *Rana pipiens berlandieri*, the leopard frog (KBP-frog, WADA et al. 1989). These KBPs turned out to be proteins of only 464 and 470 amino acid residues, respectively, thus being only about half the size of other glutamate receptors (Fig. 4). However, they share between 35% and 41% amino acid sequence identity with AMPA and KA receptors, and, at the same time, are ~55% identical to each other (Table 14), which qualifies them as members of another subfamily of glutamate receptors. Additional KBPs were found in the goldfish *Carassius auratus*, from which two different genes were cloned, GFKARα and GFKARβ (WO and OSWALD 1994), which herein will be called KBP-fish-α and KBP-fish-β. Another gene was discovered in the South African clawed toad *Xenopus laevis*, from which a subunit was cloned and was named XenU1 (ISHIMARU et al. 1996), herein called KBP-toad. Thus, so far, a total of five different KBP genes have been identified from four different vertebrate species (Table 6), and they all show similar amino acid sequence identity with each other (50–68%) and with AMPA or KA receptors (35–42%, Table 14). One additional KBP reported from the duck *Anas domesticus* (KIMURA et al. 1993) represents the duck homolog of KBP-chick (92.8% identity, Table 17). Surprisingly, no KBPs have been found in mammals, despite efforts to isolate such clones (BULL 1998).

As their name suggests, all members of the KBP family bind KA with high affinity. KBP-frog has a K_d for KA of 5.5 nM (WADA et al. 1989), while the K_d for KA of KBP-chick is distinctly lower at only 560 nM (GREGOR et al. 1992). KBP-fish-α, KBP-fish-β and KBP-toad also bind KA with high affinities, 47 nM, 25 nM, and 9.1 nM, respectively (ISHIMARU et al. 1996; WO and OSWALD 1996). Despite the high affinity KA binding, none of the KBPs expresses functional ion channels upon homomeric expression in *Xenopus* oocytes or mammalian cells. However, their ion channel domains are fully functional, as was recently demonstrated by transplanting the ion pore domains of KBPs into KA (GluR6) or AMPA (GluR1) receptors. Functional KA-gated channels were obtained for such chimeric constructs from all five known KBP subunits (VILLMANN et al. 1997). It appears possible that more than one KBP subunit, possibly four or five different ones, need to combine in a receptor complex to form a functional ion channel. However, so far no more than two different KBP subunits have been identified in any given species (Table 6). Alternatively, KBPs might be strictly modulatory subunits with no direct involvement in shaping an ion channel.

Recently, coexpression of KBP-toad with an NMDAR1 subunit cloned from *Xenopus laevis* (SOLOVIEV et al. 1996; SOLOVIEV et al. 1998) was reported to yield functional ion channels with unique pharmacological properties. Those heteromeric receptors, surprisingly, were activated by glutamate, KA, AMPA, and NMDA, and were named "unitary" glutamate receptor to indicate this promiscuous pharmacology. All agonist-activated responses at this peculiar receptor complex, reportedly, were glycine-dependent and blocked by magnesium and the non-competitive NMDA receptor channel blocker MK-801; thus, they showed NMDA receptor-like properties which, in this case, were elicited by AMPA and KA receptor ligands. Further experiments will be necessary to confirm these data.

6. Invertebrate Glutamate Receptors

Invertebrate animals contain abundant glutamate receptors in the central nervous system as well as at their neuromuscular junctions where glutamate receptors substitute for the acetylcholine receptors found in vertebrate muscle (DARLISON 1992; BETZ et al. 1993; USHERWOOD 1994; CLELAND 1996). Some of these invertebrate glutamate receptors are excitatory cation channels (D-type receptors), while others are ligand-gated anion channels (H-type receptors) which are inhibitory.

a) Receptors Forming Cation Channels

Cation channel-forming glutamate receptors were first reported from the fruit fly *Drosophila melanogaster*, as DGluR-IIA (originally called DGluR-II, SCHUSTER et al. 1991) and DGluR-I (ULTSCH et al. 1992), and from the pond snail *Lymnaea stagnalis* as Lym-eGluR1 (formerly named LymGluR, HUTTON et al. 1991). They were found by PCR-mediated DNA amplification of genomic DNA with degenerate primers deduced from sequences conserved between GluR1 and KBPs and are most closely related to AMPA receptors (DGluR-I, 42–45% sequence identity; Lym-eGluR1, 43–46%) and KA receptors (DGluR-II, 30–33%; Table 15). Several additional clones were later identified from *Lymnaea stagnalis* (STÜHMER et al. 1996; HARVEY et al. 1997) and were named Lym-eGluR2 through Lym-eGluR5. Lym-eGluR2 is most closely related to rat low-affinity KA receptors (44–47% sequence identity) and forms functional ion channels activated by glutamate (STÜHMER et al. 1996; Table 15). Lym-eGluR3 through Lym-eGluR5 are partial cDNA clones. Lym-eGluR3 shows remarkably similar sequence identity with at least three subfamilies of mammalian glutamate receptors (AMPA, KA, and NMDA receptors) as well as KBPs from various species (29–38%, Table 15). Lym-eGluR4 and Lym-eGluR5 have high sequence identity with mammalian AMPA receptors (55–65%) and, in addition, have considerable homology with KA receptors (47–52%) and KBPs (36–48%; Table 15). A cDNA most closely related to mammalian NMDA receptor NMDAR1 (46.4% amino acid identity) was discovered in *Drosophila* (ULTSCH et al. 1993). Another Drosophila gene, DGluR-IIB, shows limited sequence identity with KBPs and is related to DGluR-IIA (43.7% identity; Table 16).

From the locust *Schistocerca gregaria*, partial sequence has been obtained of three glutamate receptors, Loc1–Loc3 which, based on sequence identity data, may fall into the KA receptor subfamily (Table 15). In the nematode *C. elegans*, a glutamate receptor named GLR-1 with sequence homology to AMPA receptors (39–40% amino acid sequence identity; Table 15) has been identified by genetic means and shown to be involved in mechanosensory signaling (MARICQ et al. 1995). The *C. elegans* genome sequencing project, in addition, has identified at least ten more candidate genes for glutamate receptors (named GLR-2 through GLR-11 herein) and a splice variant of GLR-1 (GLR-1B). Four of the genes revealed by the *C. elegans* genome sequencing project have been reported in abstract form (*glr-2* and *glr-3*, MELLEM et al. 1997; *nmr-1* and *nmr-2*, BROCKIE et al. 1997); however, it is unclear whether they represent any of the above-mentioned candidate genes GLR-2 to GLR-11, as no reference to the sequences was given in the abstracts. For details on the genes GLR-1 to GLR-11, see Table 7; for sequence comparison with vertebrate receptors (mostly low sequence identity, except for GLR-2, GLR-3, GLR-5, and GLR-10) and other invertebrate receptors (very little homology, except for GLR-2, GLR-3, and GLR-5) see Tables 15 and 16.

Recently, two genes, Pan-eGluR1 and Pan-NMDAR-I, have been cloned from the California spiny lobster, *Panulirus interruptus*. Pan-eGluR1 exists in two tissue-specific splice variants (see Sect. B.I.2.c) and is 38.2% and 43.9% identical to rat GluR1 and GluR6, respectively. Highest sequence identity compared with other invertebrate receptors is found with Lym-eGluR2 from *Lymnaea* (43.4%), DGluR-II and DGluR-I from *Drosophila* (35 and 33.6%, respectively) and GLR-1 from *C. elegans* (31.6%). Pan-NMDAR-I, which is only known from partial sequence across the transmembrane domain regions, is 73% identical to DNMDAR-I from *Drosophila* (KRENZ et al. 1999; W.D. KRENZ, personal communication).

To date no ligand-binding data are available for any of the cloned invertebrate glutamate receptors. DGluR-I has been expressed in *Xenopus* oocytes where ionic currents can be activated by KA with an EC_{50} of ~75 μM (ULTSCH et al. 1992). As no other glutamatergic agonist could mimic the action of the plant metabolite KA, the endogenous agonist for this receptor remains elusive.

DGluR-II could only be activated by unphysiologically high concentrations of glutamate (EC_{50} 35 mM) and L-aspartate (EC_{50} 50 mM; SCHUSTER et al. 1991), casting some doubt on the validity of those measurements. For Lym-eGluR1 and Pan-eGluR1, no ion-channel function could be shown in *Xenopus* oocytes (HUTTON et al. 1991; KRENZ et al. 1999). Lym-eGluR2, however, responds to glutamate, KA, and AMPA, but not aspartate, domoate, or NMDA (STÜHMER et al. 1996; Table 1). The *Drosophila* NMDA receptor DNMDAR-I shows weak responses to glycine, but not to glutamate (ULTSCH et al. 1993); its function remains unclear.

For the *C. elegans* glutamate receptor candidate genes, no expression studies in heterologous systems have been reported to date.

b) Receptors Forming Anion Channels

Invertebrates have long been known to possess glutamate receptors which are ligand-gated anion channels (USHERWOOD and GRUNDFEST 1965). Several such glutamate receptors have been cloned from nematodes, the first two being GluCl-α and GluCl-β (CULLY et al. 1994). A total of five genes have been reported to date (WOLSTENHOLME 1997). For details and references to the published cDNAs, see Table 8. An excellent, comprehensive and recent review summarizes the properties of these receptors and their genes (CLELAND 1996). As these genes are structurally as well as sequence-wise related to the inhibitory $GABA_A$ and glycine receptors, rather than excitatory glutamate receptors, they will not be discussed in this volume.

7. Plant Glutamate Receptors

Recently, in a brief preliminary note (LAM et al. 1998), the cloning of four genes from *Arabidopsis thaliana*, the thalia cresse, which have significant regional sequence identities with animal ionotropic glutamate receptor subunits, was reported. The four receptor subunits, named AtGLR1 through AtGLR4, appear to be structurally similar to animal glutamate receptors and share sequence identities of up to 60% across certain conserved transmembrane domains and ligand binding domains. While these genes may offer interesting clues to the evolutionary history of glutamate receptors, they will not be dealt with in this volume.

8. Subunits of Doubtful Status as Glutamate Receptors

a) A Putative NMDA Receptor-Like Complex

In 1991, a cDNA clone was reported (KUMAR et al. 1991a) which had been isolated by screening a rat cDNA library with an antibody generated against a 71-kDa protein purified on a glutamate affinity column (EATON et al. 1990). The protein encoded by that cDNA was named GBP (glutamate binding protein) and was postulated to represent the glutamate binding subunit of an NMDA receptor complex of four proteins, based on the following rather indirect chain of evidence. The GBP, when expressed as a bacterial fusion protein, showed cross-reactivity with the antibody made against the 71-kDa protein, which is one of four major proteins obtained (among several minor ones) in the eluate of a glutamate-derivatized glass-fiber affinity column (LY and MICHAELIS 1991). This eluate, upon reconstitution into liposomes had generated glutamate- and NMDA-activatable [^{14}C]methylamine influx into liposomes. As this influx had shown pharmacological properties reminiscent of NMDA receptors (LY and MICHAELIS 1991), it was concluded that the four major bands in the affinity column eluate represented the components of an NMDA receptor complex.

The cloned GBP has no homology with any member of the glutamate receptor family (Tables 14, 15). The percentage sequence identity values are not significantly different from those calculated for 100 randomizations of the GBP sequence. Furthermore, the highest sequence similarity reported in the original publication was with the N-terminal domain of collagen, and with two short stretches of sequence in γ-glutamyl-transpeptidase (KUMAR et al. 1991b). The hydropathy plot illustrated in Fig. 4 underscores the absence of any structural similarity with members of the glutamate receptor family.

Apart from generation of a bacterial fusion protein, no expression in any heterologous system has been reported for the GBP to date. No ligand binding or other functional data have been published, apart from a brief statement unaccompanied by a description of methods that a K_d of 263 nM had been estimated for L-glutamate binding (KUMAR et al. 1991b).

Cloning of a hypothetical second protein of the putative four-protein NMDA receptor complex, named GlyBP (glycine binding protein), was reported in 1995 (KUMAR et al. 1995). Similar to the situation with GBP, the protein shows no sequence or structural similarity with members of the glutamate receptor family (Tables 14, 15; Fig. 4). Randomized GlyBP sequence data demonstrate that the 23.9% mathematical sequence identity with NR2C and the 25.4% identity with Loc1 are not statistically significant. Furthermore, GlyBP is not related to GBP (17.3% sequence identity, equal to random similarity). Apart from expression in bacteria, which reportedly produced a protein with low-affinity binding sites for glycine (K_{ds} of 647 nM and 20,000 nM, KUMAR et al. 1995), no expression studies have been published.

Recently, the cloning of a third hypothetical member of the putative four-protein NMDA receptor complex was reported: the so-called CPP-BP [3-((±)-2-carboxypiperazine-4-yl)-propyl-1-phosphonic acid binding protein], a 78.9-kDa protein with three putative transmembrane domains but apparently no signal sequence (KUMAR et al. 1998). Similar to the GBP and GlyBP, this protein shows no sequence or structural similarity with members of the glutamate receptor family, or any other known proteins including GBP and GlyBP. Extracts of bacteria expressing the CPP-BP reportedly bind CPG39653 (D,L-ε-2-amino-4-propyl-5-phosphono-3-pentanoic acid), glutamate, and glycine.

The hypothetical fourth protein of the putative four-protein NMDA receptor complex, a 36-kDa protein of unknown ligand specificity, so far has been reported only in abstract form as a partial clone (MACH et al. 1997).

Table 11 lists the hypothetical members of the putative NMDA receptor complex reported to date. The crucial experiment of coexpressing the four proteins in a heterologous expression system, or reconstituting a mixture of the four separately expressed proteins in artificial bilayers to test directly for NMDA receptor function still has to be carried out. Convincing functional data supporting the existence of an NMDA receptor complex radically different from the well-characterized NMDAR1/NR2 receptor complex has not yet been shown. Specifically, the case for the four proteins GBP, GlyBP, CPP-BP, and 36-kDa protein representing an NMDA receptor complex has yet to be made.

b) Rat cDNA Clone GR33

In 1993, an alleged presynaptic glutamate receptor, named GR33, with the pharmacological and electrophysiological properties of a typical NMDA receptor, was published (SMIRNOVA et al. 1993a, b). The amino acid sequence of this receptor (GenBank accession number A36559), however, is 100% identical to that of the presynaptic plasma membrane protein syntaxin 1b (BENNETT et al. 1992; GenBank accession number M95735). It is hard to imagine how a protein with a single transmembrane domain, no discernible ion channel domain, and without an extracellular domain to bind a ligand could possibly act as a glutamate-gated ion channel (see review by BROSE 1993). Attempts by this author to duplicate the published electrophysiological responses (SMIRNOVA et al. 1993b) of this cDNA clone in the oocyte expression system failed (MICHAEL HOLLMANN, unpublished observation; the GR33 cDNA was provided by T. SMIRNOVA). Likewise, the rat syntaxin1b cDNA clone provided by M.K. Bennett for comparison failed to give any NMDA, glutamate, or KA-activated responses, in the presence as well as in the absence of glycine. It seems possible that, in the original study by SMIRNOVA et al., the GR33 cDNA clone and a rat NMDAR1 cDNA clone used as a control may have been inadvertently switched during preparation of probes for expression studies. This would explain why the published electrophysiological properties of GR33 (SMIRNOVA et al. 1993b) appear to be virtually identical to those of rat NMDAR1.

C. The Gene Structure of Glutamate Receptors

I. Genomic Organization

Glutamate receptor genes are generally large, split genes consisting of multiple exons. Gene size varies from 11 exons for the chick KBP (GREGOR et al. 1992) to 22 for NMDAR1 (HOLLMANN et al. 1993), and from a size of 10.7 kb

Table 11. Cloned protein components of a putative *N*-methyl-D-aspartate receptor-like complex unrelated to any other glutamate receptor

Gene	Gene variant	Species	Alternate name used in original publication	Total protein size (aa)	Mature protein size (aa)	Signal peptide size (aa)	References	GenBank accession number
GBP	GBP	Rat (R.n.)	GBP	516	494	22	KUMAR et al. 1991	S61973
		Human (H.s.)	hNRGW	(208, ps)			WON et al. 1995	U44954
	Dros-GBP[a]	Fruit fly (D.m.)	dNMDARa1[a]	203		p?	PELLICENA-PALLE and SALZ 1995	L37377
GlyBP	GlyBP	Rat (R.n.)	TCP-BP	470		p?	KUMAR et al. 1995	S80259 (U53513)
CPP-BP	CPP-BP	Rat (R.n.)		719	719	None	KUMAR et al. 1998	
36-kDa protein	36-kDa protein	Rat (R.n.)		(ps)			MACH et al. 1997	

CPP-BP, 3-[(±)-2-carboxypiperazine-4-yl]-propyl-1-phosphonate binding protein; GBP, glutamate binding protein; GlyBP, glycine binding protein; p?, this might be a partial clone, as the signal peptide search algorithm SPSCAN in the University of Wisconsin sequence analysis software package (Evereux et al. 1984) does not recognize a signal peptide at the translation start site of the clone. For abbreviations and further explanations see legend of Table 2

[a]This subunit is merely 46.3% identical to rat GBP and thus may represent a different gene

(*C. elegans* GLR-1, MARICQ et al. 1995) to more than 90kb for the AMPA receptor GluR2 (KÖHLER et al. 1994), and possibly even 500kb for human GluR5 (GREGOR et al. 1994). Only a few complete exon–intron structures for glutamate receptor genes have been published to date (Table 12) and will be discussed briefly in the following section. In general, exon–intron borders are not conserved between receptor subfamilies. None of the 16 introns of GluR2, for example, is found at a position homologous to intron positions in NMDAR1. Compared with KBP-chick, only three intron positions of GluR2 are conserved.

In the following discussion, information on the existence of knock-out mouse mutants for the various glutamate receptor genes is also included.

1. AMPA Receptors

The structure of the mouse GluR2 gene has been reported by KÖHLER et al. (1994). The gene consists of 17 exons and covers at least 90kb (exons 7–12, HIGUCHI et al. 1993), but most probably around 200kb. There are no introns in the 5′ untranslated region. The main transcriptional start site is located on exon 1, about 430bp upstream of the translation start site. Additional transcriptional start sites, which are differentially used in cortical and cerebellar neurons, are present in the region –340bp to –481bp, where a silencer and two additional regulatory elements responsible for preferential expression in neuronal cells over glial cells have been detected (MYERS et al. 1998). Exons 14 and 15 are alternatively spliced exons leading to flop and flip variants, respectively. Exon 17 is also alternatively spliced and introduces a C-terminal variant. The three transmembrane domains A–C and the ion pore domain (Fig. 5) are located on three separate exons (nos. 11, 12, and 16), with the ion pore domain and transmembrane domain A sharing exon 11 (KÖHLER et al. 1994). The same arrangement of exons and introns around the transmembrane domains was found for GluR3 and GluR4 (SOMMER et al. 1991). A knock-out mutant for the GluR2 gene has been generated (JIA et al. 1996) as well as a specific knock-out of the editing mechanism of the GluR2 gene (BRUSA et al. 1995).

2. KA Receptors

No complete intron–exon analysis of any of the low affinity KA receptors, GluR5 to GluR7, has been published to date. A partial analysis by SOMMER et al. (1991) found that the three transmembrane domains A–C and the ion pore domain are located on four separate exons. Thus, compared with AMPA receptors, an additional intron separates the ion pore domain and transmembrane domain A. A knock-out mutant of the GluR6 gene has recently been reported, providing evidence for a role of KA receptors in KA-induced seizure activity and synaptic transmission (MULLE et al. 1998).

The gene structure of the high affinity KA receptors KA2 from rat was reported in 1997 (HUANG and GALLO 1997). The gene consists of 20 exons spread out over 54kb of genomic sequence, which means that introns account

Table 12. Data on gene size and structure of excitatory ionotropic glutamate receptor subunits

Receptor subfamily	Gene	Size of gene (kb)	Number of exons	Species	Notes	References
AMPA receptors	GluR1	>50		Human	50 to possibly 250 kb	Puckett et al. 1991
				Mouse	Partial sequence of 4 exons around flip-flop domain	Sommer et al. 1990
	GluR2			Mouse	Partial sequence of 4 exons around flip-flop domain	Sommer et al. 1990
				Mouse	Partial sequence of 3 exons around ion pore domain	Sommer et al. 1991
		>90		Mouse	Partial sequence of 6 exons (exons 7–12)	Higuchi et al. 1993
			17	Mouse	Entire gene reported	Köhler et al. 1994
				Rat	Promoter region analysis	Myers et al. 1998
	GluR3			Mouse	Partial sequence of 4 exons around flip-flop domain	Sommer et al. 1990
				Mouse	Partial sequence of 3 exons around ion pore domain	Sommer et al. 1991
	GluR4			Mouse	Partial sequence of 4 exons around flip-flop domain	Sommer et al. 1990
				Mouse	Partial sequence of 3 exons around ion pore domain	Sommer et al. 1991
KA receptors (low affinity)	GluR5			Mouse	Partial sequence of 3 exons around ion pore domain	Sommer et al. 1991
		400–500		Human		Gregor et al. 1994
	GluR6			Mouse	Partial sequence of 3 exons around ion pore domain	Sommer et al. 1991
	GluR7				No data available	
KA receptors (high affinity)	KA1					
	KA2	54	20	Rat	Entire gene reported	Huang and Gallo 1997

NMDA receptor (GLY binding)	NMDAR1			Rat	Partial sequence of 3 exons in 5' untranslated region	BAI and KUSIAK 1993
		25	22	Rat	Entire gene reported	HOLLMANN et al. 1993
		31	21	Human	Entire gene reported	ZIMMER et al. 1995
				Rat	Promoter region analysis	BAI and KUSIAK 1995
NMDA receptor (GLU binding)	NR2A				No data available	
	NR2B			Mouse	Promoter region analysis; 5' untranslated domain (3 exons) is >20 kb	KLEIN et al. 1998
				Mouse	Promoter analysis; 2 exons in 5' untranslated domain	SASNER and BUONANNO 1996
	NR2C	17.4	14	Mouse	Two exons in 5' untranslated region	NAGASAWA et al. 1996
		~20	15	Mouse	Three exons in 5' untranslated region	SUCHANEK et al. 1995
				Mouse	Promoter region analysis	SUCHANEK et al. 1997
				Human	Three splice variants described	DAGGETT et al. 1998
	NR2D				No data available	
NMDA receptor (modulatory)	NR3A				No data available	
	NR3B				No data available	
Orphan receptors	delta1				No data available	
	delta2				No data available	
KBPs	KBP-chick	11.2	11	Chick	Entire gene reported; promoter region analysis	GREGOR et al. 1992
Invertebrate receptors	DGluR-I			Fly	Partial sequence of 3 exons of transmembrane region	ULTSCH et al. 1992
	DNMDAR-I			Fly	Partial sequence of 2 exons around ion pore domain	ULTSCH et al. 1993
	GLR-1	10.7	14	Worm	Entire gene reported; promoter region identified	MARICQ et al. 1995
	Pan-eGluR1			Lobster	Partial sequence of exons around transmembrane region	KRENZ et al. 1999

AMPA, α-amino-3-hydroxy-5-methyl-4-isoxazole propionate; *GLY*, glycine; *GLU*, glutamate; *KA*, kainate; *KBP*, kainate-binding protein; *NMDA*, *N*-methyl-D-aspartate; *TMD*, transmembrane domain

for 51 kb of the entire genomic sequence. There is a 500-bp negative regulatory element located near the 3′ end of intron 1. The exon–intron arrangement of the transmembrane domains is similar to that of low-affinity KA receptors, in that transmembrane domains A–C and the ion-pore domain are located on four separate exons (nos. 14, 16, 19, and 15, respectively). However, transmembrane domain B is split by an intron such that the first half of this domain shares exon 15 with the ion-pore domain, while the second half is located on exon 16. The translation start codon is located on exon 2 (HUANG and GALLO 1997).

3. NMDA Receptors

The gene for rat NMDAR1 was the first gene of a functional glutamate receptor to be elucidated (HOLLMANN et al. 1993). It consists of 22 exons spanning 25 kb and includes exon 3 which leads to a truncated receptor. Exons 5 and 21 and a portion of exon 22 are alternatively spliced. Exons 11–20 are tightly clustered and encode the transmembrane domain region of the receptor. The three transmembrane domains A–C and the ion pore domain are located on four separate exons. The 5′ untranslated region has been investigated in detail (BAI and KUSIAK 1993; BAI and KUSIAK 1995) and was found to contain two transcription start sites 276 bp and 238 bp upstream of the translation start site on exon 1 and a 356-bp promoter/enhancer region. The human NMDAR1 gene was reported in 1995 (ZIMMER et al. 1995) and consists of 21 exons spanning 31 kb. Its organization is identical to that of the rat gene, except that the human gene lacks exon 3.

Several independent knock-out mouse mutants for the NMDAR1 gene have been generated, and a considerable number of studies has been performed on these mutant animals (FORREST et al. 1994; ITO et al. 1996; MCHUGH et al. 1996; TOKITA et al. 1996; TONEGAWA et al. 1996; TSIEN et al. 1996; FUNK et al. 1997; IWASATO et al. 1997; MESSERSMITH et al. 1997; KIYAMA et al. 1998).

Of the four genes of the NR2 subfamily, only the mouse NR2C gene has been analyzed for its intron–exon structure, but independently by two groups (SUCHANEK et al. 1995; NAGASAWA et al. 1996). The gene is composed of 15 exons spanning ~20 kb (SUCHANEK et al. 1995). NAGASAWA et al. (1996) reported only 14 exons, apparently missing one exon in the 5′ untranslated region which is involved in alternative splicing events. The transcription start site is located 439 bp upstream of the translation start site, and a promoter region of ~400 bp was found 664 bp upstream of the transcription start site. A silencer is present 108 bp upstream of the start codon, which is located on exon 4 (SUCHANEK et al. 1995). The three transmembrane domains A–C and the ion pore domain are located on three separate exons (nos. 10, 11, and 14), with the ion pore domain and transmembrane domain B sharing exon 11.

For mouse NR2B, the large (>20 kb) 5′ untranslated region has been investigated (KLEIN et al. 1998) and was found to closely resemble the genomic organization of NR2C. The transcriptional start site and the promoter region were identified. The start codon was found on exon 4, similar to the NR2C gene.

Knock-out mutants have been reported for NR2A (Sakimura et al. 1995; Kadotani et al. 1996; Takahashi et al. 1996; Kishimoto et al. 1997; Minami et al. 1997), NR2B (Kutsuwada et al. 1996), NR2C (Ebralidze et al. 1996; Kadotani et al. 1996; Kadotani et al. 1998), and NR2D (Ikeda et al. 1995; Minami et al. 1997). In addition, a mouse mutant overexpressing NR2D has been generated (Okabe et al. 1998), as well as mouse mutants expressing C-terminally truncated versions of NR2A, NR2B, and NR2C, which result in defective intracellular signaling (Sprengel et al. 1998).

No genomic sequence information is available for NR3 genes, but a knock-out mutant of NR3A has been reported and utilized to demonstrate a negative modulatory action of NR3A on NMDA receptor currents (Das et al. 1998).

4. Orphan Receptors

The genomic organization of the delta1 and delta2 genes has not yet been reported. However, a knock-out mutation of the mouse delta2 gene has been generated and used in several studies aimed at elucidation of delta subunit function (Funabiki et al. 1995; Hirano et al. 1995; Kashiwabuchi et al. 1995; Kurihara et al. 1997). The data suggest that delta2 may be involved in synapse formation in the cerebellum, motor learning, and postsynaptic downregulation of glutamate sensitivity (see also Sect. B.II.4.).

5. Kainate Binding Proteins

The KBP from chick was the first glutamate receptor gene for which a gene structure was reported (Gregor et al. 1992). The gene consists of 11 exons spread over 11.2 kb of genomic sequence. The promoter region is found within 480 bp of the 5′ end of the coding region. The three transmembrane domains A–C and the ion pore domain are located on four separate exons, nos. 6, 8, 10, and 7, respectively.

6. Invertebrate Glutamate Receptors

Partial genomic structure has been reported for the *Drosophila* glutamate receptor gene DGluR-I. Interestingly, it was found that, unlike any of the vertebrate glutamate receptor genes, the three transmembrane domains A–C and the ion pore domain are located on only two separate exons, with transmembrane domains A and B sharing an exon with the ion pore domain while transmembrane domain C is located on another exon. A fragment of genomic sequence for the DNMDAR-I gene showed that transmembrane domain A is separated by an intron from the ion channel domain and transmembrane domain B, which reside together on another exon.

The *C. elegans* gene GLR-1 consists of 14 exons occupying 10.7 kb of genomic sequence. The translation start codon is located in exon 1. The exon–intron organization of the transmembrane domain region of GLR-1 is unique in that transmembrane domain A is split by an intron, its N-terminal half being located on exon 8 and its C-terminal half sharing exon 9 with the ion pore domain. Transmembrane domains B and C are located on exons 10 and 12, respectively. A knockout mutation has been generated for GLR-1 and used to determine the involvement of this gene in mechanosensory signaling in the nematode (Maricq et al. 1995).

Information regarding the genomic organization of the ten additional putative glutamate receptor genes discovered in the genome of *C. elegans* during the course of the genome sequencing project can be found in GenBank under the respective accession numbers given in Table 7. The lobster glutamate receptor, Pan-eGluR1, has two alternative exons encoding the ion pore domain plus the flanking intracellular loops L1 and L2 (see Fig. 5 for domain nomenclature), which lie adjacent to each other in the gene (Krenz et al. 1999).

II. Chromosomal Localization

Of the 18 mammalian glutamate receptor genes, localization on human chromosomes has been mapped for 16 and inferred for one additional gene. In mouse, 13 of these genes have been mapped, and in rat 7. None of the five known genes for vertebrate KBPs have been mapped so far. Of the 23 invertebrate genes listed in Table 7, only 14 have been mapped. Table 13 lists the reported chromosomal positions with the appropriate references and, in addition, provides the gene locus names, a prediction of the molecular weight of the mature proteins, and information on the availability of gene knock-out mutants.

There had been high hopes that chromosomal localization of glutamate receptor genes would reveal meaningful correlations with known loci involved in neurological diseases. To a large extent these hopes have not been fulfilled. The suspected correlation of GluR5 with familial amyotrophic lateral sclerosis (Eubanks et al. 1993) has been ruled out upon closer examination (Gregor et al. 1994). For the GluR6 gene, a lack of linkage with idiopathic generalized epilepsies has been reported (Sander et al. 1995). However, some other possible correlations with minor neurological disease states, which had been summarized in an earlier review (Hollmann and Heinemann 1994), have not been ruled out. Some additional potential links between glutamate receptor genes and neurological disorders have been uncovered since. Juvenile absence epilepsy, a common subtype of idiopathic generalized epilepsy, has been linked to allelic forms of GluR5 (Sander et al. 1997), and the GluR6 locus has been implicated in the variation in age of onset of Huntington's disease (Rubinsztein et al. 1997). In another linkage study, a positive correlation was demonstrated between NMDAR1, NR2B, and NR2C (but not NR2A) gene loci and schizophrenia in a closely related population (Riley et al. 1997).

Table 13. Chromosomal localization of excitatory ionotropic glutamate receptor genes. In addition to the chromosomal localization, the gene locus names and the predicted molecular weights of the mature proteins are listed. The molecular weights have been calculated for the rat proteins except for NR3B (human), DGluR-I, DGluRII-A (*Drosophila*), and GLR-1 (*Caenorhabditis elegans*). Availability of knock-out mutants is indicated (published knock-outs only). Detailed references to knock-out studies are provided in Sect. C.II

Receptor subfamily	Gene	Molecular weight of mature protein (kDa)	Species	Chromosomal localization	Gene locus	Gene knock-out available
AMPA receptors	GluR1	99,765	Rat			
			Human	5[a]; 5q33[b]; 5q32–33[c]; 5q31.3–33.3[d]	*gria1*	
			Mouse	11 (e)		
	GluR2	96,362	Rat			
			Human	4q32–33[c]; 4q25–34.3[d]	*gria2*	
			Mouse	3[e]		Yes (also editing –/–)
	GluR3	98,022	Rat			
			Human	Xq25–26[c]	*gria3*	
			Mouse	X[e]		
	GluR4	98,375	Rat			
			Human	11q22–23[c]; 11q22[f]	*gria4*	
			Mouse	9[e]		
KA receptors (low affinity)	GluR5	99,198	Rat			
			Human	21q22[g]; 21q21.1–22.1[h]; 21q21.1[i,j]	*grik1*	
			Mouse	16[e]		
	GluR6	98,974	Rat			
			Human	6[k]; 6q16.3–q21[l]	*grik2*	
			Mouse	10[e]		Yes
	GluR7	100,352	Rat			
			Human	1[m]; 1p33–34[n]	*grik3*	
			Mouse	4[e]		

Table 13. *Continued*

Receptor subfamily	Gene	Molecular weight of mature protein (kDa)	Species	Chromosomal localization	Gene locus	Gene knock-out available
KA receptors (high affinity)	KA1	105,073	Rat	8[f,o]		
			Human	11q22.3[f]	*grik4*	
			Mouse	9[f]		
	KA2	107,792	Rat	1[f]		
			Human	19q13.2[f]	*grik5*	
			Mouse	7[f]		
NMDA receptor (GLY binding)	NMDAR1	103,518	Rat	3[p]		
			Human	9q34.3[q,r,s]; 9q35[m]; 9q34[t]; 9q34.3-qter[u]	*grin1*	
			Mouse	2[e]		Yes
NMDA receptors (GLU binding)	NR2A	163,253	Rat	10[p]		
			Human	16[v]; 16p13[t,w]; 16p13.2[x]	*grin2a*	
			Mouse			Yes (also *C*-terminal –/–)
	NR2B	163,385	Rat	4[p]		
			Human	12p12[y]	*grin2b*	
			Mouse	6[z]		Yes (also *C*-terminal –/–)
	NR2C	133,491	Rat	10[p]		
			Human	17q25[t,w]; 17q24–q25[x]	*grin2c*	
			Mouse			Yes (also *C*-terminal –/–)
	NR2D	143,660	Rat	1[p]:		
			Human	19q13.1–qter[x]	*grin2d*	
			Mouse			Yes (also overexpressed)
NMDA receptors (modulatory)	NR3A	121,580[ai]	Rat			
			Human		*grin3a*	
			Mouse			Yes
	NR3B	At least 97,303 (partial clone)	Rat			
			Human	19p13.3[aa]	*grin3b*	
			Mouse			

Orphan receptors	delta1	110,136	Rat			
			Human	10q11–q23[ah, aj]	*grid1*	
			Mouse	14[ah]		
	delta2	111,186	Rat			
			Human	4q22[ab]	*grid2*	
			Mouse	6[ab]		Yes
Invertebrate receptors	DGluR-I	108,495	Fly	3L, 65C[ac]		
	DGluR-IIA	101,801	Fly	2L, 25F[ad]		
	DNMDAR-I	109,510	Fly	3R, 83AB[ae]		
	GLR-1	106,399	Worm	III[af]		Yes
	GLR-2 to -11		Worm	III[af]		
Putative NMDA-receptor-associated protein	GBP	56,964	Rat			
			Human	8[r]; 8q24.3[ag]	*grina*	
			Mouse			

AMPA, α-amino-3-hydroxy-5-methyl-4-isoxazole propionate; GLY, glycine; GLU, glutamate; KA, kainate; NMDA, *N*-methyl-D-aspartate

[a] Potier et al. 1992
[b] Puckett et al. 1991
[c] McNamara et al. 1992
[d] Sun et al. 1992
[e] Gregor et al. 1993
[f] Szpirer et al. 1994
[g] Potier et al. 1993
[h] Eubanks et al. 1993
[i] Gregor et al. 1994
[j] Sander et al. 1997
[k] Paschen et al. 1994
[l] Sander et al. 1995
[m] Hollmann and Heinemann 1994
[n] Puranam et al. 1993
[o] Pravenec et al. 1998
[p] Kuramoto et al. 1994
[q] Karp et al. 1993
[r] Collins et al. 1993
[s] Brett et al. 1994
[t] Takano et al. 1993
[u] Zimmer et al. 1995
[v] Le Bourdelles et al. 1994
[w] Mori and Mishina 1995
[x] Kalsi et al. 1998
[y] Riley et al. 1997
[z] Madarnas et al. 1994
[aa] Lamerdin et al. 1998
[ab] Uz et al. 1998
[ac] Ultsch et al. 1992
[ad] Schuster et al. 1991
[ae] Ultsch et al. 1993
[af] Wilson et al. 1994
[ag] Lewis et al. 1996
[ah] Treadaway and Zuo 1998
[ai] Assuming a mature protein of 1082 amino acids, as reported by Sucher et al. 1995
[aj] Predicted from location of gene in mouse

D. The Protein Structure of Glutamate Receptors

I. Primary Structure: Sequences and Similarities

1. Sequence Similarities Between Subunits

a) Sequence Identity Matrix

All glutamate receptor subunits cloned to date display some degree of amino acid sequence identity with other glutamate receptor subunits. The degree of identity varies widely and, for the AMPA receptor GluR1, for example, ranges between 21.6% (when compared with NR3A) and 69.2% (when compared with GluR2, Table 13). However, even the low sequence identity of 21.6% is statistically significant when compared with sequence identities calculated for 100 randomizations of the GluR1 or NR3A sequences. Comparison of the amino acid sequences of the cloned members of the ionotropic glutamate receptor family revealed overall identities that suggested a natural breakdown into subfamilies of sequence-related subunits. In Table 14, those sequences from vertebrate animals have been grouped together that share more than 45% sequence identity with each other. Interestingly, the groups obtained by this rather arbitrary criterion turn out to match receptor subfamilies as defined by pharmacological and electrophysiological properties (compare with Table 1).

For the majority of invertebrate glutamate receptor subunits, no such "automatic" grouping becomes apparent, as can be gathered from Table 16. However, when compared with vertebrate glutamate receptor subunits (Table 15), some (but not all) of the invertebrate receptor subunits can be classified as potentially "AMPA receptor-like" (DGluR-I, Lym-eGluR1, Lym-eGluR4, Lym-eGluR5, GLR-1, GLR-2), "KA receptor-like" (Lym-eGluR2, GLR-3, Loc1, Loc2, Loc3), or "NMDA receptor-like" (DNMDAR-I, GLR-10). This is also reflected in the dendrogram in Fig. 3.

Sequence identity between receptor subunits is not distributed equally across the length of the proteins. This is illustrated in Fig. 1, in which a comparison of regional sequence identities among eight vertebrate subunits representing the eight receptor subfamilies (as defined in Table 1) is shown. While the average overall sequence identity among these eight subunits is 28.2% (see legend of Fig. 1 for the subunits included in the analysis), it ranges from a minimum of 20.8% in the C-terminal domain to a maximum of 47.0% in transmembrane domain B (see Fig. 5 for domain nomenclature). The two areas of highest sequence identity of a non-transmembrane domain region lie in the S1 domain in front of transmembrane domain A and in the S2 domain between transmembrane domains B and C. These domains (marked "C" and "H" in Fig. 1) comprise the two parts of the agonist binding domain (see Sect. D.III.2.).

b) Phylogenetic Relationships

Using pairwise amino acid sequence comparison of glutamate receptor subunits, it is possible to construct a dendrogram. Such a dendrogram is shown in

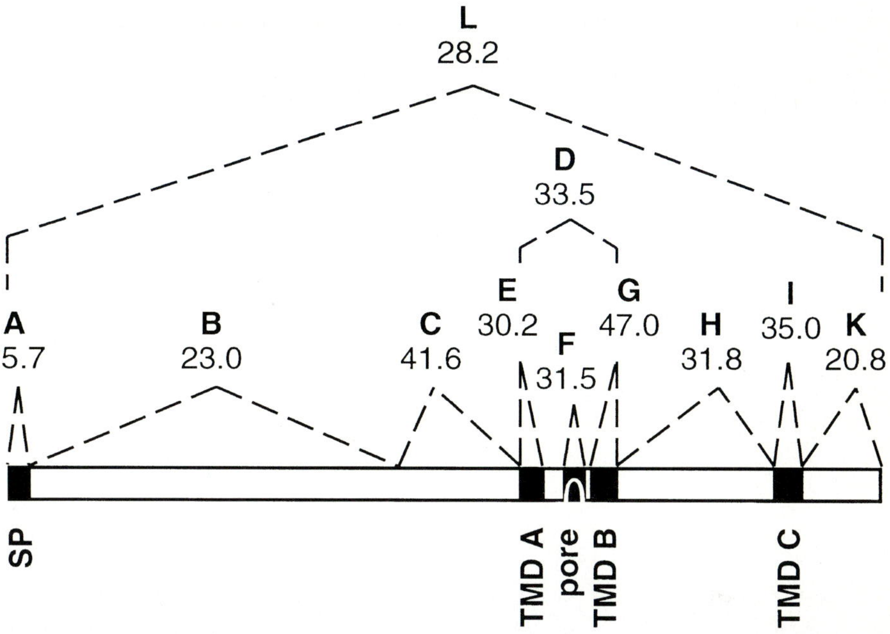

Fig. 1. Regional variation in sequence identity between vertebrate glutamate receptor subunits. For this analysis, pairwise sequence identities were calculated separately for the domains marked A–L of eight glutamate receptor subunits, comprising one representative each of the eight vertebrate receptor subfamilies listed in Table 1. Subunits included in the analysis (GenBank accession numbers given in *brackets*) are rat GluR1 (X17184), GluR6 (Z11548), KA1 (X59996), delta1 (Z17238), NMDAR-1 (S61907), NR2A (M91561), NR3A (U29873), and KBP-frog (X17700). The 12 domains are (for definition see Fig. 5): *A*, signal peptide; *B*, LIVBP domain; *C*, S1 domain; *D*, TMD A–TMD B; *E*, TMD A; *F*, ion pore domain; *G*, TMD B; *H*, S2 domain; *I*, TMD C; *K*, C-terminus; *L*, entire subunit. Calculations were performed using the program "gap" of the University of Wisconsin software package (Devereux et al. 1984). Standard settings were employed except that the amino acid comparison matrix "identpep" was used. All pairwise values obtained for each of the domains (28 total) were averaged. *SP*, signal peptide; *TMD A, B, C,* transmembrane domains A, B, C; *pore* ion pore domain

Fig. 2 for the 23 vertebrate receptor subunits. It is a graphic representation of the relatedness of the various receptor subunits, based on pairwise sequence identity data (Table 14). The longer the horizontal branches are between two subunits and the first node which connects them, the lower is their relatedness (see scale on the X-axis). For example, the tree shows that the delta subunits are more similar to AMPA/KA receptors than to NMDA receptors. It also shows that the NR3 subunits are more similar to NMDAR1 than to NR2 subunits, and that they are quite dissimilar from AMPA and KA receptors (Fig. 2). A dendrogram (sometimes also called a phenogram) such as this should not be confused with a phylogram, or phylogenetic tree, which is constructed to depict inferred evolutionary relationships rather than pairwise sequence identity, as the two are not necessarily identical (Swafford and Olsen 1990).

Table 14. Sequence identities among 23 ionotropic excitatory glutamate receptor subunits cloned from vertebrates (plus two putative receptor proteins of unclear biological significance). Subunits have been grouped into subfamilies (separated by blank lines) based on calculated sequence identities. GBP and GlyBP have no homologies to any glutamate receptor or to each other and are merely shown to illustrate this fact. Percent amino acid sequence identity was computed using the "gap" program in the University of Wisconsin sequence analysis software package (Devereux

	GluR1	GluR2	GluR3	GluR4	GluR5	GluR6	GluR7	KA1	KA2	NMDAR1	NR2A
GenBank accession number	X17184	M85035	M85036	M36421	M83561	Z11548	Z11716	U08257	U08258	X63255	AF001423
GluR1		**69.2**	**68.5**	**68.0**	41.2	39.6	38.8	36.5	36.4	28.8	26.1
GluR2	**69.2**		**73.6**	**72.9**	39.9	40.5	39.9	34.9	37.4	27.2	24.3
GluR3	**68.5**	**73.6**		**73.3**	40.2	40.0	39.8	35.9	37.1	25.9	25.6
GluR4	**68.0**	**72.9**	**73.3**		39.5	39.3	38.1	36.2	35.7	28.1	24.6
GluR5	41.2	39.9	40.2	39.5		**78.2**	**72.0**	42.6	44.3	28.7	24.8
GluR6	39.6	40.5	40.0	39.3	**78.2**		**79.2**	44.0	44.6	29.6	24.7
GluR7	38.8	39.9	39.8	38.1	**72.0**	**79.2**		42.6	43.5	27.0	24.9
KA1	36.5	35.0	35.9	36.2	42.6	44.0	42.6		**69.2**	28.4	23.3
KA2	36.4	37.4	37.1	35.7	44.3	44.6	43.5	**69.2**		27.7	23.9
NMDAR1	28.8	27.2	25.9	28.1	28.7	29.6	27.0	28.4	27.7		28.8
NR2A	26.1	24.3	25.6	24.6	24.8	24.7	24.9	23.3	23.9	28.8	
NR2B	26.1	25.3	22.7	25.0	23.9	24.1	24.6	24.6	21.8	27.7	**56.1**
NR2C	26.4	24.1	23.3	27.0	23.9	24.4	24.8	26.2	26.3	26.8	**46.1**
NR2D	24.4	22.9	24.0	23.9	24.0	23.7	23.4	23.4	25.7	26.4	43.6
NR3A	21.6	24.6	25.0	22.9	24.9	23.1	23.2	24.0	23.3	26.1	28.1
NR3B	23.9	22.3	21.3	21.5	23.7	24.0	24.5	25.6	24.7	25.7	27.1
delta1	30.2	30.5	30.7	30.2	32.1	29.4	31.9	28.9	29.2	24.0	23.5
delta2	28.2	29.7	29.2	29.5	31.6	30.2	29.6	28.0	26.4	26.9	24.6
KBP (chick)	35.4	36.6	36.2	35.4	37.7	40.1	38.4	35.1	35.6	26.9	27.1
KBP (frog)	37.1	36.9	35.8	36.5	40.5	41.2	40.5	35.5	37.0	27.5	22.7
KBP (fish-α)	36.7	37.5	35.5	37.3	39.4	39.2	41.6	37.2	38.4	30.4	24.8
KBP (fish-β)	35.3	35.0	34.9	35.6	37.8	37.6	37.1	36.0	36.7	30.0	27.2
KBP (toad)	36.3	36.9	35.4	35.9	38.2	40.0	37.6	35.6	37.7	28.5	24.7
GBP	19.9	16.8	18.7	19.8	16.7	19.4	19.5	20.5	20.0	18.7	20.2
GlyBP	19.4	20.0	18.9	20.3	18.1	20.1	20.9	18.5	18.4	17.4	21.1

However, in this case a true phylogram turned out to have exactly the same branching pattern when constructed from the same 23 sequences listed in Fig. 2 using the "growtree" program of the University of Wisconsin sequence analysis software package. Therefore, the dendrogram in Fig. 2 indeed represents the phylogenetic relationship between the depicted subunits, and it has the added advantage of allowing estimation of identities between sequences and clusters of sequences directly from the X-axis of the graph. A phylogram, however, would depict the branching pattern normalized to the amount of inferred mutational change between the various subunits which is a far less intuitive measure and therefore was not used in Fig. 2. When the same cluster analysis was performed on only the "core domains" of these 23 receptor subunits instead of the entire sequences, the exact same branching pattern was

et al. 1984). Standard settings were employed except that the amino acid comparison matrix "identpep" was used. Values above 45% (arbitrary cut-off) are shown in *bold*. Values to the lower left of the diagonal are the same as those to the upper right but have been printed anyway to facilitate comparison between subunits. Except for the KBPs and NR3B (human), sequences from rat have been used. For subfamily designations, see Table 1

NR2B	NR2C	NR2D	NR3A	NR3B	delta1	delta2	KBP (chick)	KBP (frog)	KBP (fish-α)	KBP (fish-β)	KBP (toad)	GBP	GlyBP
U11419	U08259	U08260	U29873	AC004528	U08255	U08256	X17700	X17314	U08017	U08016	X93491	S61973	S80259
26.1	26.4	24.4	21.6	23.9	30.2	28.2	35.4	37.1	36.7	35.3	36.3	19.9	19.4
25.3	24.1	22.9	24.6	22.3	30.4	29.7	36.6	36.9	37.5	35.0	36.9	16.8	20.0
22.7	23.3	24.0	25.0	21.3	30.7	29.2	36.2	35.8	35.5	34.9	35.4	18.7	18.9
25.0	27.0	23.9	22.9	21.5	30.2	29.5	35.4	36.5	37.3	35.6	35.8	19.8	20.3
23.9	23.9	24.0	24.9	23.7	32.1	31.6	37.7	40.5	39.4	37.8	38.2	16.7	18.1
24.1	24.4	23.7	23.1	24.0	29.4	30.2	40.1	41.2	39.2	37.6	40.0	19.4	20.1
24.6	24.8	23.4	23.2	24.5	31.9	29.6	38.4	40.5	41.6	37.1	37.6	19.5	20.9
24.6	26.2	23.4	24.0	25.6	28.9	28.0	35.1	35.5	37.2	36.0	35.6	20.5	19.8
21.8	26.3	25.7	23.3	24.7	29.2	26.4	35.6	37.0	38.4	36.7	37.7	20.0	18.4
27.7	26.8	26.4	26.1	25.7	24.0	26.9	26.9	27.5	30.4	30.0	28.5	18.7	17.4
56.1	**46.1**	43.6	28.1	27.1	23.5	24.6	27.1	22.7	24.8	27.2	24.7	20.2	21.1
	45.9	41.9	26.0	28.3	22.3	24.6	24.7	23.6	26.7	26.4	25.4	18.9	18.8
45.9		**52.1**	27.1	29.0	21.8	22.0	27.4	25.8	26.3	25.3	27.2	18.4	23.9
41.9	**52.1**		25.7	29.5	22.7	21.3	28.5	24.4	26.0	24.3	26.6	18.8	19.7
26.0	27.1	25.7		**55.2**	23.0	23.9	28.3	27.8	25.5	25.7	26.5	19.0	19.5
28.3	29.0	29.5	**55.2**		21.7	22.6	29.0	22.6	28.0	25.3	24.7	20.2	21.5
22.3	21.8	22.5	23.0	21.7		**55.9**	29.1	29.4	32.4	31.0	30.2	20.1	19.4
24.6	22.0	21.3	23.9	22.6	**55.9**		29.5	28.6	30.2	27.0	31.3	19.0	21.5
24.7	27.4	28.0	28.3	29.0	29.1	29.5		**55.5**	**51.7**	**53.0**	**56.8**	19.8	21.5
23.6	25.8	24.4	27.8	22.6	29.4	28.6	**55.5**		**50.0**	**49.8**	**67.9**	18.3	20.6
26.7	26.3	26.0	25.5	28.0	32.4	30.2	**51.7**	**50.0**		**60.9**	**51.1**	16.7	17.0
26.4	25.3	24.3	25.7	25.3	31.0	27.0	**53.0**	**49.8**	**60.9**		**51.6**	18.5	19.4
25.4	27.2	26.6	26.5	25.2	30.2	31.3	**56.8**	**67.9**	**51.1**	**51.6**		15.9	19.5
18.9	18.4	18.8	19.0	20.2	20.1	19.0	19.8	18.3	16.7	18.5	15.9		17.3
18.8	23.9	19.7	19.5	21.5	19.4	21.5	21.5	20.6	17.0	19.4	19.5	17.3	

obtained. The "core domain" for this purpose was defined as the sequence between the beginning of the S1 domain and the end of transmembrane domain C (see Fig. 5).

For invertebrate glutamate receptors, it is difficult to generate a meaningful dendrogram as many of the sequences are only partial sequences, and overall sequence identity is much smaller than among vertebrate glutamate receptors. Nevertheless, to graphically present similarities between vertebrate and invertebrate glutamate receptor subunits, a combined tree was constructed from the 23 invertebrate glutamate receptor subunits listed in Table 15 as well as the 23 vertebrate subunits of Fig. 2. The resulting dendrogram (Fig. 3) shows that DGluR-I, Lym-eGluR1, Lym-eGluR4, Lym-eGluR5, GLR-1, and GLR-2 cluster with AMPA receptors, while Lym-eGluR2, GLR-3, Loc1,

Table 15. Sequence identities between 23 vertebrate and 23 invertebrate cloned ionotropic excitatory glutamate receptor subunits (plus two putative receptor proteins of unclear biological significance). Vertebrate subunits have been grouped into subfamilies (separated by blank lines) based on calculated sequence identities (Table 14). GBP and GlyBP have no homologies to any glutamate receptor or to each other and are merely shown to illustrate this fact. Percent amino acid sequence identity was computed using the "gap" program in the University of Wisconsin sequence

	DGluR-I	DGluR-IIA	DGluR-IIB	DNMDAR-I	Lym-eGluR1	Lym-eGluR2	Lym-eGluR3[a]	Lym-eGluR4[a]	Lym-eGluR5[a]
GluR1	**42.4**	29.1	28.0	28.1	**43.8**	38.3	37.8	**63.9**	**60.5**
GluR2	**44.8**	31.1	27.4	29.7	**45.9**	39.1	36.3	**64.6**	**58.3**
GluR3	**44.3**	32.0	28.1	27.4	**43.5**	37.5	34.1	**63.2**	**55.6**
GluR4	**44.7**	31.1	25.4	28.9	**43.0**	37.9	35.6	**63.9**	**59.6**
GluR5	36.2	32.7	28.7	26.9	38.6	**46.8**	36.3	**52.1**	**47.7**
GluR6	36.1	32.3	29.6	28.1	37.9	**45.0**	38.5	**51.4**	**48.3**
GluR7	33.3	30.4	29.9	25.5	35.9	**43.5**	34.8	**50.0**	**50.0**
KA1	33.6	29.7	28.5	27.7	36.5	36.3	34.1	**47.2**	**46.7**
KA2	33.9	30.7	28.6	27.6	34.1	36.5	34.8	**47.9**	**46.7**
NMDAR-1	26.8	19.7	22.1	**46.4**	26.3	22.2	30.4	35.7	36.8
NR2A	23.3	21.9	21.1	28.2	24.7	23.1	31.1	30.8	29.3
NR2B	22.8	21.6	20.0	27.4	25.2	21.5	31.9	32.9	30.6
NR2C	21.5	21.0	22.1	28.8	23.8	24.6	28.9	32.1	34.4
NR2D	23.7	18.8	20.3	27.6	22.1	23.0	30.4	31.5	29.3
NR3A	20.7	21.7	24.2	25.7	23.5	24.0	23.0	25.4	27.3
NR3B	21.4	19.6	23.0	24.7	22.9	25.0	26.7	24.5	28.9
delta1	30.8	27.0	23.7	25.0	27.0	29.7	29.6	37.1	**43.3**
delta2	27.8	25.1	23.8	25.1	28.0	26.6	23.7	38.2	29.5
KBP (chick)	36.3	30.2	30.6	30.3	35.5	36.4	31.1	**44.4**	35.9
KBP (frog)	37.2	32.8	31.6	31.4	34.4	38.8	32.6	**44.4**	**42.3**
KBP (fish-α)	35.8	31.6	28.9	27.3	35.6	39.0	33.6	**45.1**	38.0
KBP (fish-β)	32.4	31.7	31.0	29.5	33.7	38.0	30.4	**41.5**	38.0
KBP (toad)	36.0	29.0	30.0	31.5	37.6	39.6	32.6	**47.9**	**42.3**
GBP	19.3	19.0	19.5	19.0	18.4	20.0	20.9	20.3	16.6
GlyBP	18.8	17.5	19.7	20.6	20.9	19.0	21.6	20.1	19.2

[a] Partial sequences

Loc2, and Loc3 cluster preferentially with KA receptors. DNMDAR-I and GLR-10, however, are more closely related to NMDA receptors. Other invertebrate subunits are uninformative in this dendrogram and their relationships remain unclear.

2. Sequence Similarities to Other Proteins

a) Glutamate Receptors are Unrelated to Other Ligand-Gated Ion Channels

Prior to the cloning of glutamate receptors, it had been assumed that ionotropic acetylcholine, glycine, GABA, serotonin, and glutamate receptors were evolutionarily related and would belong to a superfamily of ligand-gated ion channels (Barnard et al. 1987). However, sequence and structural data of these ion channels revealed that glutamate receptors clearly do not belong to the ion channel superfamily, which originally had been defined for acetyl-

analysis software package (Devereux et al. 1984). Standard settings were employed except that the amino acid comparison matrix "identpep" was used. Values above 40% (arbitrary cut-off) are shown in bold. For subfamily designations see Table 1. GenBank accession numbers are the same as listed in Tables 14 and 16. Except for the KBPs, the vertebrate sequences used are from rat (first full-length sequence reported; Table 14)

GLR-1	GLR-2[a]	GLR-3[a]	GLR-4[a]	GLR-5	GLR-6	GLR-7[a]	GLR-8[a]	GLR-9[a]	GLR-10[a]	GLR-11[a]	Loc1[a]	Loc2[a]	Loc3[a]
39.2	**40.8**	38.6	22.1	32.8	30.0	20.6	23.3	22.1	26.6	18.6	33.3	38.0	**43.7**
38.9	**42.0**	**40.4**	17.6	30.3	30.0	19.7	24.5	21.3	28.9	16.0	33.3	**40.8**	**45.1**
39.6	**41.8**	38.4	19.7	32.7	30.0	18.6	25.5	20.8	26.6	22.9	36.2	**42.3**	**45.1**
39.5	**40.5**	39.1	21.2	30.9	28.6	18.7	26.1	23.3	27.3	22.1	34.8	39.4	**43.7**
33.2	36.3	**43.0**	26.3	35.4	30.9	19.3	21.5	19.6	27.8	22.6	**49.3**	**50.7**	**52.1**
33.4	36.9	**41.8**	22.7	36.2	32.2	20.6	22.7	22.8	28.5	22.2	**52.2**	**50.7**	**52.1**
32.6	35.7	**41.2**	21.2	35.0	33.9	19.1	22.2	19.1	28.3	19.0	**47.8**	**52.1**	**54.9**
32.2	33.3	39.8	29.4	31.9	28.9	18.8	22.3	20.9	27.0	15.9	**44.9**	**42.3**	**52.1**
31.2	34.5	39.4	23.5	31.1	28.8	19.6	22.1	19.9	27.3	19.4	**46.4**	**45.1**	**59.2**
27.8	27.3	26.6	18.1	24.1	28.6	20.3	21.4	22.5	30.9	24.0	25.8	22.5	32.8
22.2	27.3	27.0	20.9	23.6	23.1	18.6	19.8	21.4	36.7	24.9	25.0	15.5	24.3
26.0	24.7	29.8	19.4	23.0	25.5	19.0	20.2	20.5	35.9	26.7	27.5	23.9	28.2
22.8	25.3	25.5	19.3	21.0	20.8	17.7	20.9	20.1	35.7	23.1	26.1	18.3	21.1
22.5	23.6	25.0	15.6	21.0	24.7	20.1	22.4	19.8	36.2	20.8	27.5	23.2	25.4
23.7	22.4	25.6	20.8	24.2	25.8	18.4	21.8	18.6	33.1	21.5	23.2	21.1	28.2
20.4	22.4	25.3	21.7	24.9	25.6	20.6	22.5	19.6	34.4	18.4	26.1	21.1	26.2
28.2	30.6	31.4	19.3	27.7	28.9	20.1	23.7	19.0	23.8	16.8	21.7	18.3	33.3
24.8	29.3	30.5	20.2	26.5	27.4	21.3	23.2	20.5	24.8	21.1	31.9	22.9	33.8
33.4	32.9	33.3	20.8	31.1	30.2	24.7	23.8	18.7	27.8	16.1	**42.9**	39.1	**43.8**
33.4	35.9	34.8	20.2	34.5	29.9	20.5	25.3	20.7	25.6	20.8	39.7	34.4	**51.6**
36.8	36.8	33.0	17.9	35.4	31.6	20.3	25.9	21.6	24.5	17.9	39.7	37.5	**43.8**
34.5	34.6	35.6	20.6	34.4	29.9	17.8	21.7	18.3	26.3	17.6	34.9	34.4	**42.2**
35.6	35.7	34.5	17.3	34.1	29.6	21.5	26.3	16.7	28.4	18.1	34.9	**41.0**	**50.0**
19.2	16.8	18.4	16.9	19.0	17.3	17.2	17.5	17.6	16.7	19.7	21.7	21.1	21.1
18.4	22.2	15.5	18.0	16.3	19.2	17.1	16.1	18.7	21.9	25.1	25.4	20.0	15.5

choline, glycine, and GABA receptors and which also encompasses ionotropic serotonin receptors. Amino acid sequence identity between glutamate receptors and other ligand-gated ion channels was determined to be in the range of only 20–22%, a level which is not significant when recalculated with randomized sequences. Furthermore, structural peculiarities conserved in all other ligand-gated ion channels such as a cysteine–cysteine disulfide bridge in the N-terminal domain (the "ligand-gated ion channel signature sequence" of authors, BARNARD et al. 1987) are absent from all glutamate receptors (COCKCROFT et al. 1993; see also discussion in HOLLMANN 1996). Thus, glutamate receptors define their own independent superfamily of ion channel-forming proteins.

b) Similarities With Bacterial Substrate Binding Proteins

Soon after the cloning of the first glutamate receptor subunit, NAKANISHI et al., in 1990, noticed significant regional sequence identity (30–33% in the S1

Table 16. Sequence identities among 23 ionotropic excitatory glutamate receptor subunits cloned from invertebrates. Subunits have been grouped by species, separated by a blank line. Percent amino acid sequence identity was calculated using the "gap" program in the University of Wisconsin sequence analysis software package (Devereux et al. 1984). Standard settings were employed except

	DGluR-I	DGluR-IIA	DGluR-IIB	DNR-I	Lym-eGluR1	Lym-eGluR2	Lym-eGluR3[a]	Lym-eGluR4[a]	Lym-eGluR5[a]	GLR-1
GenBank accession number	M97192	M73271	AF044202	X71790	X60086	X87404				U34661
DGluR-I		29.4	25.5	25.8	**40.5**	34.3	31.1	**56.3**	**55.6**	34.8
DGluR-IIA	29.4		**43.7**	21.5	30.2	33.0	26.7	38.9	32.2	27.3
DGluR-IIB	25.5	**43.7**		22.7	27.2	30.3	27.4	35.4	32.9	25.3
DNMDAR-I	25.8	21.5	22.7		26.3	27.0	32.6	34.3	37.9	23.9
Lym-eGluR1	**40.5**	30.2	27.2	26.3		36.9	39.3	**68.1**	**56.8**	37.4
Lym-eGluR2	34.3	33.0	30.3	27.0	36.9		35.6	**47.9**	**49.0**	32.0
Lym-eGluR3[a]	31.1	26.7	27.4	32.6	39.3	35.6		37.8	**42.2**	39.3
Lym-eGluR4[a]	**56.3**	38.9	35.4	34.3	**68.1**	**47.9**	37.8		**54.9**	**60.4**
Lym-eGluR5[a]	**55.6**	32.2	32.9	37.9	**56.8**	**49.0**	**42.2**	**54.9**		**58.8**
GLR-1	34.8	27.3	25.3	23.9	37.4	32.0	39.3	**60.4**	**58.8**	
GLR-2[a]	**40.8**	27.7	27.1	25.1	38.6	37.3	37.0	**53.5**	**51.0**	**52.3**
GLR-3[a]	38.1	32.5	30.9	27.6	39.0	39.9	28.9	38.2	38.4	37.4
GLR-4[a]	22.5	25.4	18.5	18.2	20.4	20.8	26.2	21.5	21.5	19.5
GLR-5	31.6	25.9	28.4	24.2	29.1	32.1	32.6	**40.3**	**41.8**	32.6
GLR-6	25.0	24.0	25.9	26.8	26.5	27.9	33.3	36.9	37.4	28.8
GLR-7[a]	20.2	20.0	20.8	19.2	18.2	21.0	26.1	25.2	23.3	19.2
GLR-8[a]	23.1	24.9	21.9	22.9	21.0	23.2	22.0	35.2	29.5	24.5
GLR-9[a]	20.3	21.0	20.8	20.7	21.4	24.4	26.7	21.1	21.2	19.7
GLR-10[a]	27.7	21.2	24.1	31.3	22.4	23.8	24.4	25.4	28.1	24.4
GLR-11[a]	17.2	20.7	18.1	24.5	18.5	19.3	17.0	18.4	19.2	18.4
Loc1[a]	34.8	29.0	27.5	26.9	34.3	**56.5**	34.5	38.5	33.3	35.3
Loc2[a]	36.6	**43.7**	32.4	22.1	39.7	**54.9**	28.8	**43.9**	38.0	37.1
Loc3[a]	38.0	39.4	31.0	29.4	**44.1**	**53.5**	30.8	**45.5**	**43.7**	**42.9**

[a] Partial sequences

domain of Fig. 5) between AMPA receptors and periplasmic bacterial amino acid binding proteins such as the glutamine binding protein and speculated that this region might comprise the agonist binding site (NAKANISHI et al. 1990). They also noticed even higher sequence identity (44–47%) between AMPA receptors and the glutamine binding protein in parts of the S2 domain (Fig. 5), but were unable to offer an explanation, as this domain, at the time, was thought to be intracellular and therefore not a likely candidate for the agonist binding domain (see Sect. D.III.1.). These observations were largely ignored until 1993 when they were re-examined (COCKCROFT et al. 1993) and extended to metabotropic glutamate receptors (O'HARA et al. 1993). The profound impact these sequence similarities turned out to have on the identification of the agonist binding site is discussed in Sect. D.III.2.

3. Sequence Similarities Across Species

Seventeen of the 18 mammalian glutamate receptor subunits cloned to date have been identified in rat cDNA libraries; 16 have also been reported from human cDNA, and 12 from mouse. In addition, certain mammalian subunits have been cloned from pig and various non-mammalian vertebrates such as

that the amino acid comparison matrix "identpep" was used. As in Table 15, values above 40% (arbitrary cut-off) are shown in *bold*. Values to the lower left of the diagonal are the same as those to the upper right but have been printed anyway to facilitate comparison between subunits. For details on the sequences used, see Table 7

GLR-2[a]	GLR-3[a]	GLR-4[a]	GLR-5	GLR-6	GLR-7[a]	GLR-8[a]	GLR-9[a]	GLR-10[a]	GLR-11[a]	Loc1[a]	Loc2[a]	Loc3[a]
U10438	U39849 (S9)	U39849 (S10)	Z75545 (S1)	U51999 (S9)	Z72502	U41547	U23411	Z78413	Z81061			
40.8	38.1	22.5	31.6	25.0	20.2	23.1	20.3	27.7	17.2	34.8	36.6	38.0
27.7	32.5	25.4	25.9	24.0	20.0	24.9	21.0	21.2	20.7	29.0	**43.7**	39.4
27.1	30.9	18.5	28.4	25.9	20.8	21.9	20.8	24.1	18.1	27.5	32.4	31.0
25.1	27.6	18.2	24.2	26.8	19.2	22.9	20.7	31.3	24.5	26.9	22.1	29.4
38.6	39.0	20.4	29.1	26.5	18.2	21.0	21.4	22.4	18.5	34.3	39.7	**44.1**
37.3	39.9	20.8	32.1	27.9	21.0	23.2	24.4	23.8	19.3	**56.5**	**54.9**	**53.5**
37.0	28.9	26.2	32.6	33.3	26.1	22.0	26.7	24.4	17.0	34.5	28.8	30.8
53.5	38.2	21.5	**40.3**	36.9	25.2	35.2	21.1	25.4	18.4	38.5	**43.9**	**45.5**
51.0	38.4	21.5	**41.8**	37.4	23.3	29.5	21.2	28.1	19.2	33.3	38.0	**43.7**
52.3	37.4	19.5	32.6	28.8	19.2	24.5	19.7	24.4	18.4	35.3	37.1	**42.9**
	35.3	20.7	33.2	28.9	22.0	22.9	19.8	27.7	20.7	33.8	34.8	**49.3**
35.3		17.5	**43.4**	35.1	20.0	25.4	18.8	27.5	18.5	**44.9**	**42.3**	**45.1**
20.7	17.5		21.3	16.6	16.8	19.7	15.9	15.0	16.9	17.4	16.9	22.1
33.2	**43.4**	21.3		32.1	21.8	28.0	17.6	27.2	18.4	34.8	38.0	**46.5**
28.9	35.1	16.6	32.1		21.4	22.1	21.2	25.7	21.7	36.2	28.6	37.1
22.0	20.0	16.8	21.8	21.4		19.1	19.2	17.8	19.1	21.7	18.3	16.9
22.9	25.4	19.7	28.0	22.1	19.1		20.2	19.2	18.8	16.7	16.9	22.5
19.8	18.8	15.9	17.6	21.2	19.2	20.2		17.3	20.6	17.4	26.9	25.4
27.7	27.5	15.0	27.2	25.7	17.8	19.2	17.3		14.7	27.7	18.3	15.5
20.7	18.5	16.9	18.4	21.7	19.1	18.8	20.6	14.7		20.6	17.6	21.1
33.8	**44.9**	17.4	34.8	36.2	21.7	16.7	17.4	27.7	20.6		39.1	**47.8**
34.8	**42.3**	16.9	38.0	28.6	18.3	16.9	26.9	18.3	17.6	39.1		**49.3**
49.3	**45.1**	22.1	**46.5**	37.1	16.9	22.5	25.4	15.5	21.1	**47.8**	**49.3**	

pigeon, chick, duck, goldfish, other species of fish, and toad (Table 2). Comparison of amino acid sequences reveals that mammalian glutamate receptor subunits of the AMPA, KA, orphan, and NMDAR1 subfamilies in general are more than 97% identical across species. NR2 subunits fall off to some extent in being merely more than 89% identical across mammalian species. When amino acid sequences from other vertebrate animals are examined, sequence identity drops to more than 91% for birds and more than 75% for fish (Table 17).

For glutamate receptor subunits from invertebrate animals, meaningful cross-species sequence identity estimates cannot be obtained as the potential homologies among the 23 genes identified from a total of four species remain unclear. For example, the *Drosophila* subunit DGluR-I shows considerable sequence identity to the *Lymnaea* subunits Lym-eGluR4 (56.3%) and Lym-eGluR5 (55.6%). However, it is not clear whether DGluR-I is the *Drosophila* homolog of any of the two *Lymnaea* subunits. That ambiguous situation is complicated by the fact that the *C. elegans* subunit GLR-1 also has significant identity with Lym-eGluR4 (60.4%) and Lym-eGluR5 (58.8%), but not with DGluR-I (34.8). As another example, the *Lymnaea* subunit Lym-eGluR2 shares substantial identity (53.5–56.5%) with three subunits from the locust *Schistocerca*., Loc1, Loc2, and Loc3, but the question remains whether these subunits are truly homologous (Table 16).

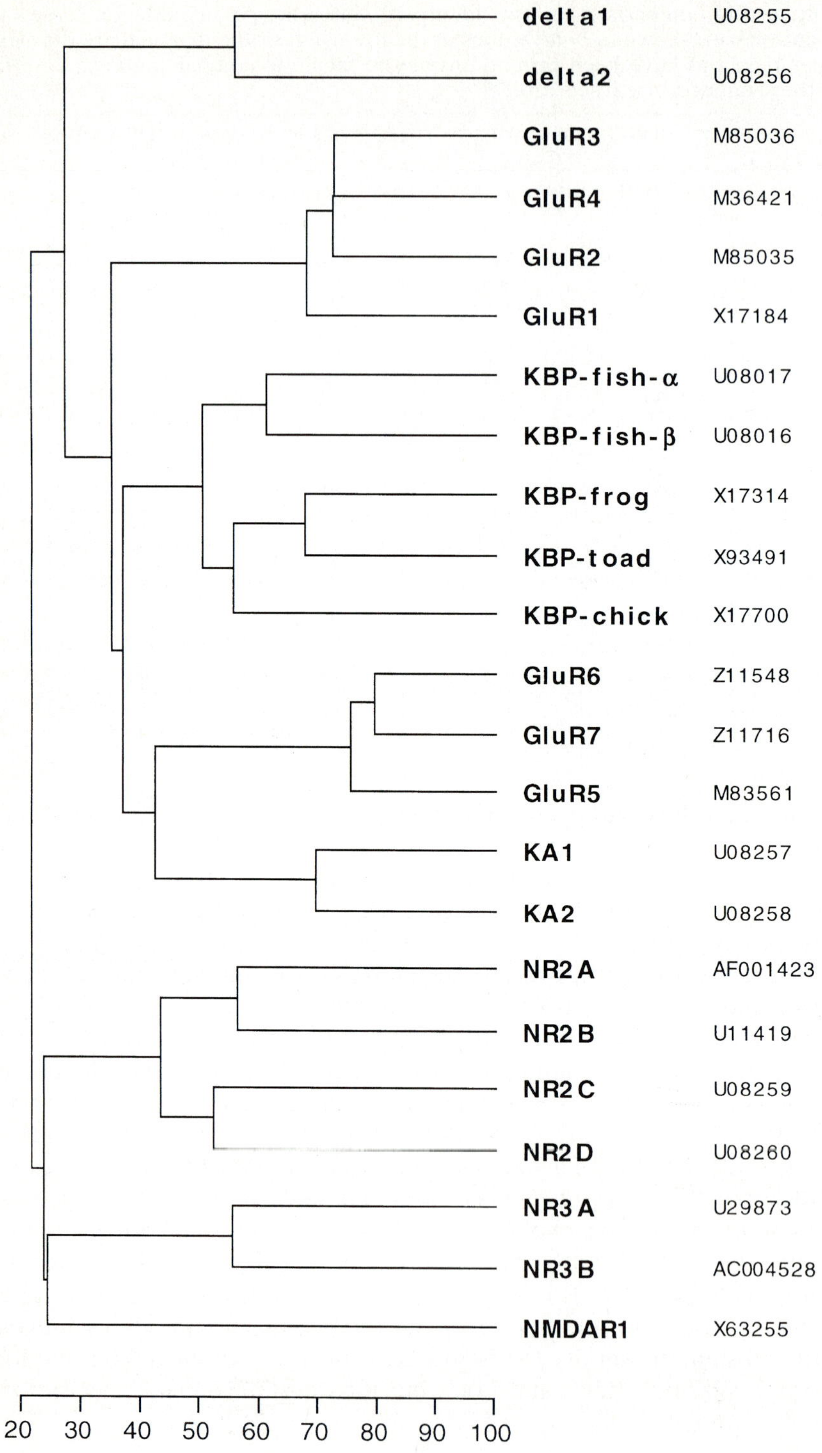
delta1 U08255
delta2 U08256
GluR3 M85036
GluR4 M36421
GluR2 M85035
GluR1 X17184
KBP-fish-α U08017
KBP-fish-β U08016
KBP-frog X17314
KBP-toad X93491
KBP-chick X17700
GluR6 Z11548
GluR7 Z11716
GluR5 M83561
KA1 U08257
KA2 U08258
NR2A AF001423
NR2B U11419
NR2C U08259
NR2D U08260
NR3A U29873
NR3B AC004528
NMDAR1 X63255
20 30 40 50 60 70 80 90 100

II. Secondary Structure: Hydrophobic Domains

Analysis of the hydropathy along the sequence of integral membrane proteins traditionally has been used as a means to predict candidate domains for membrane crossings (KYTE and DOOLITTLE 1982). While putative transmembrane domains identified by this method need to be verified experimentally (see Sect. D.III.1.), hydropathy plots at the very least can serve to rule out certain structures. For example, it would be difficult to misinterpret the hydropathy plot of a one-pass membrane protein as that of a 7-transmembrane receptor. Figure 4 provides an overview of hydropathy plots of all cloned excitatory ionotropic glutamate receptor subunits, plus two potential members of a putative NMDA receptor-like complex (GBP and GlyBP, see Table 11), and one inhibitory ionotropic glutamate receptor (GluCl-α, see Table 8). Plots have been drawn to scale to allow direct comparison of subunits, and get an impression of the overall structure. Inspection of the figure reveals a striking conservation of certain features, such as the three transmembrane domains (see Sect. D.III.1.) indicated by arrows above the scale bar, a hydrophobic signal peptide at the beginning of each subunit, and a characteristic hydrophilic "dip" following transmembrane domain A and preceding the very weakly hydrophobic ion-pore domain (marked "p" in Fig. 4). GBP and GlyBP appear to be structurally totally different from glutamate receptors, and from each other. The hydropathy plots of the partial sequences GLR-8 and GLR-9 from *C. elegans* when compared with GluCl-α suggest that these subunits may be related to inhibitory rather than excitatory glutamate receptors.

III. Tertiary Structure: Receptor Topology

1. The Transmembrane Topology

When the first glutamate receptor sequence was reported in 1989, a four transmembrane domain topology with extracellular N- and C-termini was assumed based on the hydropathy plot of GluR1 (Fig. 4) and a perceived structural homology to the acetylcholine receptor (HOLLMANN et al. 1989). Even at that

◄

Fig. 2. Dendrogram of the 23 known vertebrate glutamate receptor subunits. The tree was generated by the "pileup" program contained in the University of Wisconsin sequence analysis software package (DEVEREUX et al. 1984). Cluster analysis was done on complete amino acid sequences from rat, except for NR3B, where the human sequence was used because a rat sequence is not available. GenBank accession numbers are given in the *right column*. Standard settings were employed except that the amino acid comparison matrix "identpep" was used. The length of the *horizontal branches* connecting any two subunits represents the degree of their relatedness based on pairwise sequence comparison. The *scale bar* at the bottom can be used to estimate the percentage sequence identity at the nodes that connect certain subsets of sequences. While this dendrogram is based on cluster analysis and, therefore, is not a true phylogenetic tree, such a tree when generated from the same set of sequences with the "growtree" program of the University of Wisconsin software package was found to show exactly the same branching pattern

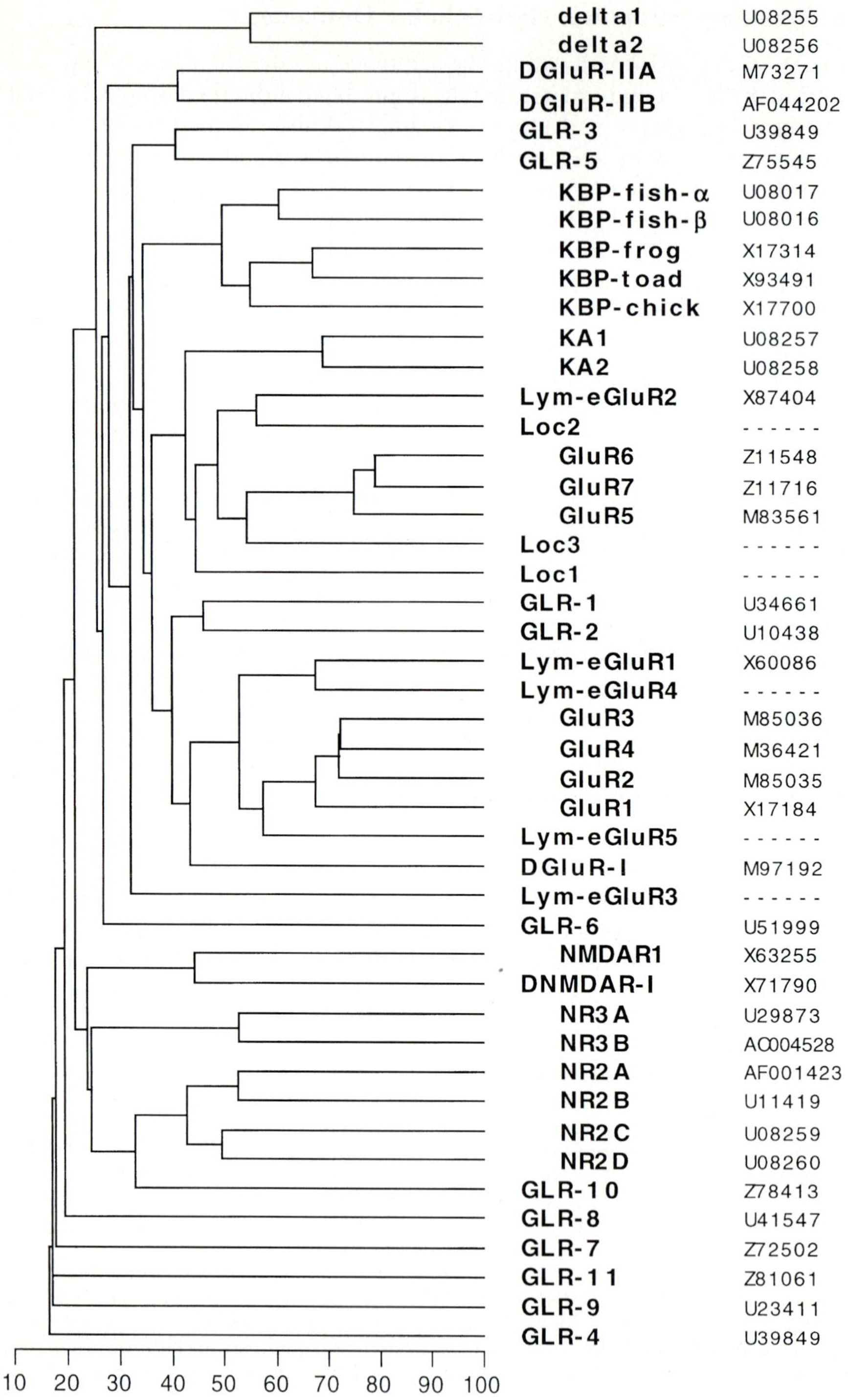
delta1 U08255
delta2 U08256
DGluR-IIA M73271
DGluR-IIB AF044202
GLR-3 U39849
GLR-5 Z75545
KBP-fish-α U08017
KBP-fish-β U08016
KBP-frog X17314
KBP-toad X93491
KBP-chick X17700
KA1 U08257
KA2 U08258
Lym-eGluR2 X87404
Loc2 ------
GluR6 Z11548
GluR7 Z11716
GluR5 M83561
Loc3 ------
Loc1 ------
GLR-1 U34661
GLR-2 U10438
Lym-eGluR1 X60086
Lym-eGluR4 ------
GluR3 M85036
GluR4 M36421
GluR2 M85035
GluR1 X17184
Lym-eGluR5 ------
DGluR-I M97192
Lym-eGluR3 ------
GLR-6 U51999
NMDAR1 X63255
DNMDAR-I X71790
NR3A U29873
NR3B AC004528
NR2A AF001423
NR2B U11419
NR2C U08259
NR2D U08260
GLR-10 Z78413
GLR-8 U41547
GLR-7 Z72502
GLR-11 Z81061
GLR-9 U23411
GLR-4 U39849
10 20 30 40 50 60 70 80 90 100

time, however, the assignment of the four transmembrane domains was controversial (HOLLMANN et al. 1990; KEINÄNEN et al. 1990). Inspection of hydropathy plots of additional glutamate receptors did not help. As illustrated in Fig. 4, some subunits, as exemplified by the NR2 subfamily, seem to support a four transmembrane domain structure, while others, including the low-affinity KA receptors and the delta subunits, appear to show only three transmembrane domains.

Experimental evidence that suggested that the originally proposed four transmembrane domain topology might me wrong first came from immunohistochemical studies with antibodies directed against C-terminal receptor sequences. Those antibodies showed labeling only after permeabilization of the cells, indicating an intracellular localization of the C-terminus (PETRALIA and WENTHOLD 1992; CRAIG et al. 1993; MARTIN et al. 1993; MOLNAR et al. 1993; MOLNAR et al. 1994). As the N-terminus with the same method could be shown to be indeed extracellular, this result was in obvious conflict with a topology comprising four transmembrane domains.

In 1994, studies of receptor N-glycosylation at native and engineered sites of a KBP from goldfish (WO and OSWALD 1994) and of the AMPA receptor GluR1 (HOLLMANN et al. 1994) finally demonstrated convincingly a very different novel topology for glutamate receptors, as depicted in Fig. 5; there are only three true transmembrane domains, A, B, and C, corresponding to the formerly assigned transmembrane domains I, III, and IV (sometimes called M1, M3, and M4; see small insert in Fig. 5). The N-terminus is extracellular and the C-terminus is intracellular. The domain previously regarded as transmembrane domain II (or M2), which lines the ion permeation path (see Chap. 7), does not traverse the membrane. Instead, this domain appears to form a hairpin loop similar to the P-region or H5 domain of potassium channels (YELLEN et al. 1991; YOOL and SCHWARZ 1991). The hairpin loop, which originates at the intracellular side, returns to the intracellular side of the membrane without actually crossing the lipid bilayer. There are three sequence loops which interconnect the three transmembrane domains and the pore domain (L1, L2, and L3, Fig. 5). The large loop L3 is extracellular, in contrast to the old topological model in which this loop was assumed to be intracellular (see insert in Fig. 5, and review by WO and OSWALD 1995a). Further evidence for the novel topology model came from epitope tagging studies on the AMPA receptor GluR3 (BENNETT and DINGLEDINE 1995), N-glycosylation studies on the L3 domain of the NMDAR1 subunit

◄

Fig. 3. Dendrogram of the 23 known vertebrate plus 23 invertebrate glutamate receptor subunits. As this dendrogram is based on simple cluster analysis, it is not a true phylogenetic tree. However, it illustrates to which vertebrate subunits or groups of subunits the various invertebrate glutamate receptor subunits might be most closely related. For further details on the method used to construct the tree, see legend to Fig. 2. GenBank accession numbers are given in the *right column*. The column containing the sequence names has been arranged such that names of invertebrate sequences protrude to the left for easy recognition

Table 17. Amino acid sequence identity of vertebrate ionotropic glutamate receptor subunits across species. Amino acid sequence of glutamate receptor subunits cloned from various species have been compared to the rat sequence. For detailed references to the sequences used, see Tables 2–6. If several sequences were available, the one listed first in Tables 2–6 has been used for this comparison. Percent amino acid sequence identity was calculated using the "gap" program in the University of Wisconsin sequence analysis software package (Devereux et al. 1984). Standard settings were employed except that the amino acid comparison matrix "identpep" was used

	Rat	Human	Mouse	Pig	Pigeon	Chick	Duck	Goldfish	Fish	Toad
GluR1	100	98.2	98.6	n.a.	n.a.	92.0	n.a.	n.a.	75.1	n.a.
GluR2	100	98.5	99.5	n.a.	95.4	94.9	n.a.	n.a.	82.7	n.a.
GluR3	100	99.2	n.a.	n.a.	99.6[b]	95.8	n.a.	91.4[b]	87.7	n.a.
GluR4	100	97.8	n.a.	n.a.	97.4[b]	96.4	n.a.	89.0	n.a.	n.a.
GluR5	100	97.1	99.0	n.a.	n.a.	n.a.	n.a.	n.a.	n.a.	n.a.
GluR6	100	98.8	98.3	n.a.	n.a.	n.a.	n.a.	n.a.	n.a.	96.5[b]
GluR7	100	98.6	n.a.	n.a.	n.a.	n.a.	n.a.	n.a.	n.a.	n.a.
KA1	100	97.7	n.a.	n.a.	n.a.	n.a.	n.a.	n.a.	n.a.	n.a.
KA2	100	98.3	99.6	n.a.	n.a.	n.a.	n.a.	n.a.	n.a.	n.a.
KBP (chick)					n.a.	100	92.8	n.a.	n.a.	n.a.
delta1	100	n.a.	99.5	n.a.	n.a.	n.a.	n.a.	n.a.	n.a.	n.a.
delta2	100	97.3	99.3	n.a.	n.a.	n.a.	n.a.	n.a.	n.a.	n.a.
NMDAR1	100	99.3	99.8	98.4[b]	n.a.	n.a.	91.1	91.0[b]	n.a.[a]	90.2
NR2A	100	94.7	98.6	n.a.	n.a.	n.a.	n.a.	n.a.	n.a.	n.a.
NR2B	100	98.5	96.7	n.a.	n.a.	n.a.	n.a.	n.a.	n.a.	n.a.
NR2C	100	88.7	94.5	n.a.	n.a.	n.a.	n.a.	n.a.	n.a.	n.a.
NR2D	100	96.3	96.8	n.a.	n.a.	n.a.	n.a.	n.a.	n.a.	n.a.

n.a., no sequence available
[a] Sequence has not been entered into GenBank yet
[b] Partial sequence

(Wood et al. 1995), and additional N-glycosylation and domain deletion studies on two goldfish KBPs (Wo et al. 1995; Wo and Oswald 1995b).

A detailed account of the evidence for the three transmembrane domain model can be found in the review by Hollmann (1996), including a discussion of phosphorylation studies, which appeared to favor the original four transmembrane domain topology. Those studies had claimed the presence of protein kinase A phosphorylation sites in the L3 domain of the KA receptor subunit GluR6 (Raymond et al. 1993; Wang et al. 1993), a tyrosine kinase phosphorylation site (Moss et al. 1993), and a calcium/calmodulin-dependent protein kinase-II phosphorylation site in the L3 domain of GluR1 (McGlade-McCulloh et al. 1993; Yakel et al. 1995). However, for none of the sites had phosphorylation been shown directly and unequivocally – for example, by phosphopeptide sequencing. Indeed, more recent studies failed to confirm those phosphorylation sites and, instead, demonstrated phosphorylation sites for protein kinase A (Roche et al. 1996) and calcium/calmodulin-dependent protein kinase II (Barria et al. 1997) in the C-terminal domain of GluR1. Similarly, in the NMDAR1 subunit, phosphorylation sites for protein kinases A and C were found to reside in the C-terminal domain (Tingley et al. 1997), in full agreement with the novel three transmembrane domain topology. For a detailed discussion of glutamate receptor phosphorylation and its functional implications, see Chap. 2 by Lee & Huganir.

The proposed hairpin loop of the ion pore domain (Fig. 5) has recently been confirmed by cysteine-scanning mutagenesis of this domain in the NMDA receptor (Kuner et al. 1996). The microarchitecture of this most important domain is described in detail in Chap. 6 by Kuner, Wollmuth & Sakmann.

2. The Agonist Binding Site

In the original topological model of glutamate receptors, which assumed four transmembrane domains, the N-terminal domain was viewed as the only large extracellular domain. Consequently, the agonist binding site was presumed to be located in this domain, much like it is in the acetylcholine receptor (Devillers-Thiery et al. 1993). However, in 1994, data on the agonist binding properties of chimeric receptors constructed from the AMPA receptor GluR3 and the KA receptor GluR6 revealed that, in glutamate receptors, two non-contiguous domains participate in the formation of the agonist binding site (Stern-Bach et al. 1994). One domain lies in the N-terminal domain in front of transmembrane domain A and was named the S1 domain, the other is located in between transmembrane domains B and C and was named the S2 domain (Fig. 5). Obviously, both domains S1 and S2 had to be extracellular in order to combine to form the agonist binding site, which constituted additional independent proof for the three transmembrane domain topology that had been proposed at the same time (Hollmann et al. 1994; Wo and Oswald 1994). Stern-Bach et al. took up an early observation by Nakanishi et al. (Nakanishi et al. 1990) that there is homology between the S1 and S2 domains of glutamate receptors with the N- and C-terminal halves, respectively, of

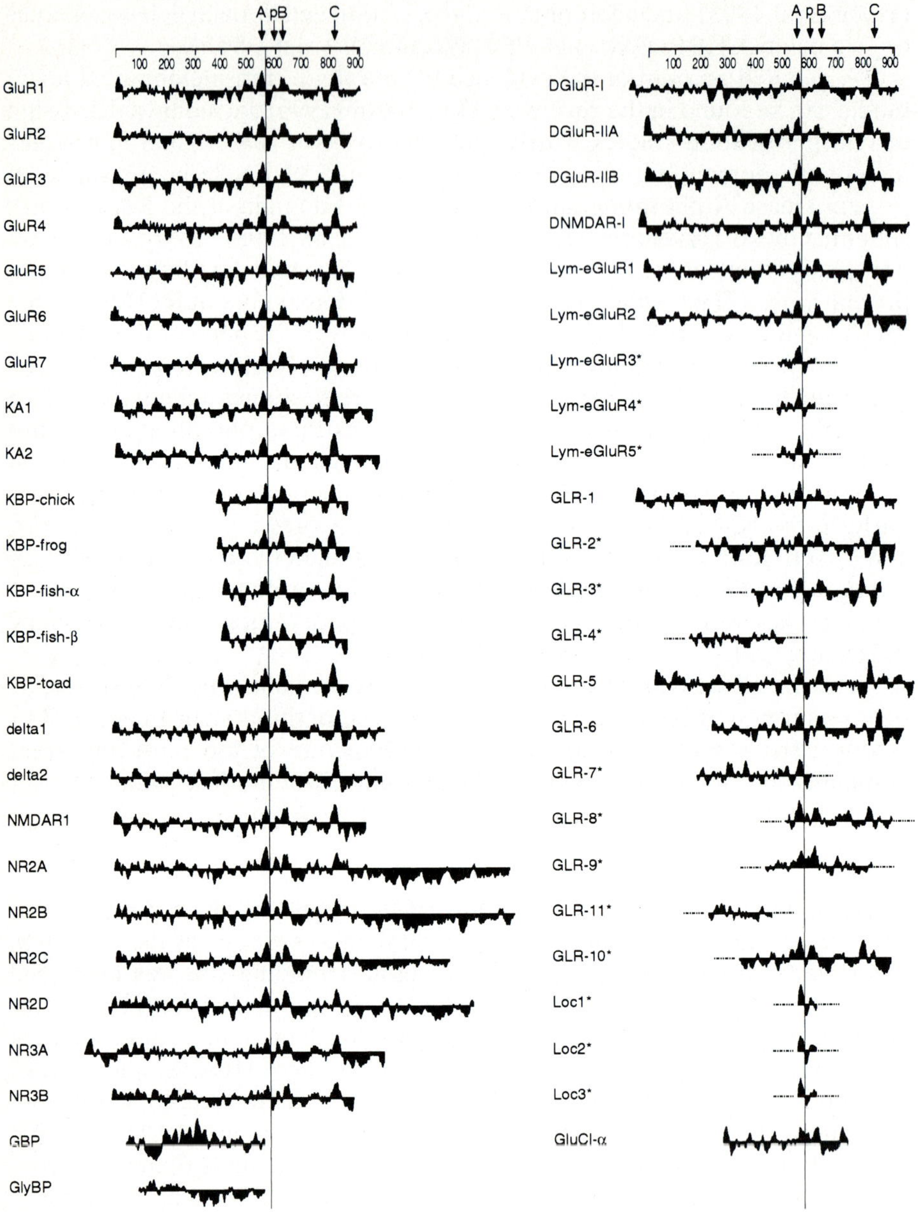
A pB C
100 200 300 400 500 600 700 800 900
GluR1
GluR2
GluR3
GluR4
GluR5
GluR6
GluR7
KA1
KA2
KBP-chick
KBP-frog
KBP-fish-α
KBP-fish-β
KBP-toad
delta1
delta2
NMDAR1
NR2A
NR2B
NR2C
NR2D
NR3A
NR3B
GBP
GlyBP
A p B C
100 200 300 400 500 600 700 800 900
DGluR-I
DGluR-IIA
DGluR-IIB
DNMDAR-I
Lym-eGluR1
Lym-eGluR2
Lym-eGluR3*
Lym-eGluR4*
Lym-eGluR5*
GLR-1
GLR-2*
GLR-3*
GLR-4*
GLR-5
GLR-6
GLR-7*
GLR-8*
GLR-9*
GLR-11*
GLR-10*
Loc1*
Loc2*
Loc3*
GluCl-α

bacterial periplasmic substrate binding proteins. These proteins, in particular the glutamine binding protein (QBP, NOHNO et al. 1986) and the lysine-arginine-ornithine binding protein (LAOBP, KANG et al. 1991) show considerable sequence identity with ionotropic glutamate receptors over limited stretches of sequence. What made this homology so valuable was the fact that the crystal structure of these amino acid binding proteins had already been solved and that their mode of action was fairly well understood (ADAMS and OXENDER 1989; QUIOCHO 1990; OH et al. 1993).

Bacterial amino acid binding proteins are folded such that the first half of the amino acid chain is intertwined with the second half to form two lobes, I and II, with contributions of the first as well as the second half of the protein to both lobes (ADAMS and OXENDER 1989). The agonist binding site is located at the interface of the two lobes. As shown in Fig. 5, the S1 and S2 domains of a glutamate receptor are believed to be likewise intertwined to form two lobes at the interface of which the agonist binding site is thought to be created.

Superimposition of the glutamate receptor sequence on the structural data of the bacterial binding proteins allowed identification of candidate domains and amino acids involved in agonist binding. This information, in turn, suggested sites for targeted mutagenesis to experimentally probe the involvement of certain amino acids in agonist binding. This was done for NMDA receptors (KURYATOV et al. 1994; HIRAI et al. 1996; LAUBE et al. 1997), the KBP from chick (PAAS et al. 1996), AMPA receptors (MANO et al. 1996), and KA receptors (SWANSON et al. 1997). The data from these studies combined with earlier mutagenesis data on GluR1 (UCHINO et al. 1992) and NMDAR1 (LAUBE et al. 1993) as well as computer-assisted modeling studies (SUTCLIFFE et al. 1996; SUTCLIFFE et al. 1998) confirmed the structural and functional homology between bacterial periplasmic binding proteins and the ionotropic glutamate

Fig. 4. Hydropathy plots of cloned vertebrate and invertebrate glutamate receptor subunits to illustrate structural features and similarities. Twenty-three vertebrate subunits are shown in the *left column*, and 23 invertebrate subunits in the *right column*. Additionally, two protein components [glutamate binding protein (*GBP*) and glycine binding protein (GlyBP)] of a putative *N*-methyl-D-aspartate (*NMDA*) receptor-like complex (see Sect. B.II.7.a.) are shown at the *bottom* of the *left column*. Note their lack of structural homology with glutamate receptors. At the *bottom* of the *right column*, a glutamate-gated chloride channel (GluCl-α) has been included to illustrate structural differences between excitatory and inhibitory glutamate receptors. Note the three clustered, strongly hydrophobic transmembrane domains not interrupted by hydrophilic sequences in the middle of the GluCl-α sequence as opposed to only two such domains separated by a hydrophilic stretch at the homologous position in excitatory glutamate receptors. All subunits have been drawn to scale and are aligned along the C-terminal ends of their first transmembrane domains (TMD A) as marked by the *vertical line*. *Dots* preceding or following a plot indicate incomplete, partial sequences, and the names of such sequences are marked with an *asterisk*. Hydrophobic regions are shown as *upward deflections*, hydrophilic stretches as *downward deflections*. The *arrows* above the *scale bars* point to transmembrane domains *A*, *B*, and *C* and the ion-pore domain *p*. Plots were generated using the Kyte-Doolittle algorithm (with a window of 18 residues) as implemented in the Lasergene program package from DNAstar, Madison, Wis.

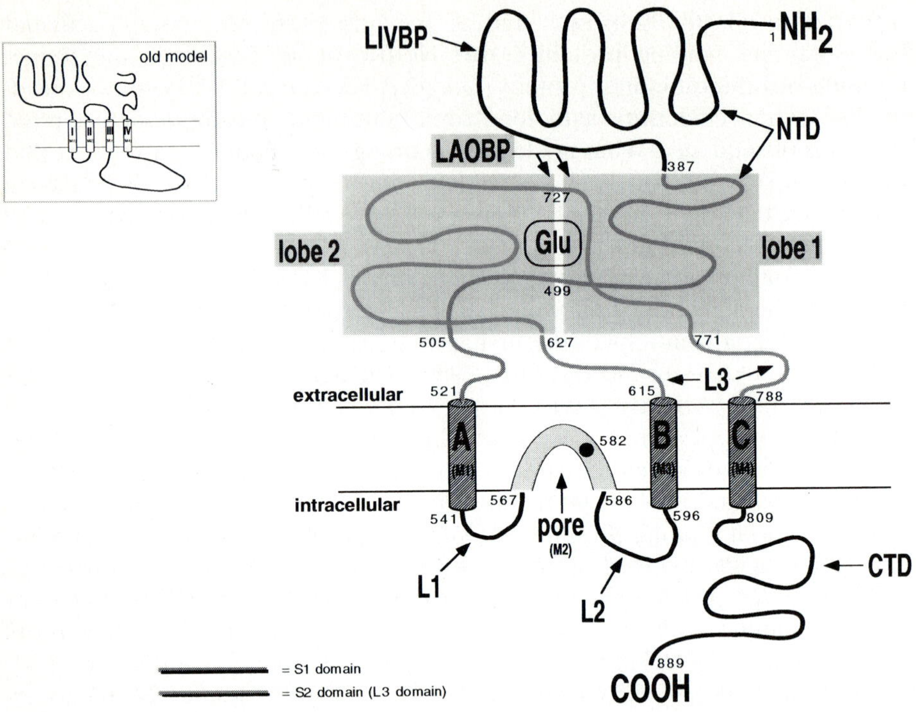

Fig. 5. Topological model of ionotropic glutamate receptors. The α-amino-3-hydroxy-5-methyl-4-isoxazole propionate (*AMPA*) receptor GluR1 is shown as an example for the three transmembrane domain topology and also depicts the schematic arrangement of the extracellular domains, which form the agonist binding site. Note that the two non-contiguous domains, S1 and S2, get intertwined to form the two lobes of the ligand binding domain; thus, lobe 1 consists of a major part of the S1 domain plus a minor contribution by the S2 domain; lobe 2 conversely consists of a major part of the S2 domain with a minor contribution by S1. Starting from the amino terminus, the domains specifically pointed out in the figure are as follows (numbers represent amino acid residues of the mature GluR1 protein, given for the start of each domain or feature; domains are defined as in Paas 1998): *NTD* (1–520, N-terminal domain, extracellular), consisting of the bacterial leucine-isoleucine-valine binding protein (*LIVBP*) homology domain (1–386) and the S1 domain [387–520, homology sequence 1 of the bacterial lysine-arginine-ornithine binding protein (LAOBP), with residues 387–498 located in lobe 1 and residues 499–504 in lobe 2]; *A* (521–540, transmembrane domain A, formerly named transmembrane domain I or M1); L1 (541–566, loop 1, intracellular); pore (567–585, ion pore domain formerly regarded as transmembrane domain II or M2, with Q/R editing site indicated as a *dot* at residue 582); *L2* (586–595, loop 2, intracellular); B (596–614, transmembrane domain B, formerly named transmembrane domain III or M3); *L3* S2 domain (615–787, loop 3, extracellular = homology sequence 2 of LAOBP, with residues 627–726 located in lobe 2 and residues 727–770 in lobe 1); *C* (788–808, transmembrane domain C, formerly named transmembrane domain IV or M4); *CTD* (809–889, C-terminal domain, intracellular). *"Glu"* marks the agonist binding site at the interface of lobes 1 and 2 of the LAOBP homology domain. The small *insert graph* depicts the originally proposed topology for glutamate receptors, which was superseded by the current model

receptor ligand binding domain, and succeeded in identifying a number of key residues involved in agonist binding and agonist-mediated desensitization (for a recent review, see PAAS 1998).

Based on these data, it was suggested that, in glutamate receptors, agonist binding to a low-affinity site on lobe I might initiate channel opening, while additional binding of the same agonist to lobe II brings the two lobes in closer contact and, as a result, shifts the binding to significantly higher affinity, thereby causing desensitization (MANO et al. 1996). This sequence of events has been dubbed the "venus flytrap" mechanism. However, it has been argued that channel opening itself might require binding to both lobes. In that alternative model, the protein movement necessary to bring both lobes into close enough contact to allow the agonist to bind simultaneously to both lobes is believed to constitute the gating signal delivered to the ion pore (LAUBE et al. 1997; PAAS 1998).

Direct demonstration that the S1 and S2 domains form the agonist binding site was achieved in elegant domain deletion experiments. Constructs consisting of only the S1 and S2 domains of AMPA or KA receptors, linked by a hydrophilic spacer while lacking any transmembrane domains or the ion pore domain, were shown to be soluble proteins with fully functional agonist binding sites (Kuusinen et al. 1995; ARVOLA and KEINÄNEN 1996; KEINÄNEN et al. 1997; KEINÄNEN et al. 1998). Mutagenesis experiments implicated an arginine residue in the S1 domain and a glutamic acid residue in the S2 domain of such a construct as crucial for agonist binding (LAMPINEN et al. 1998). Recently, such a soluble construct derived from the AMPA receptor GluR2 was crystalized with the agonist KA in place (CHEN et al. 1998). The structural information obtained from these crystals at a resolution of 1.9 Å confirmed the inferred functional homology with bacterial periplasmic amino acid binding proteins and validated many of the conclusions drawn from earlier mutagenesis data (ARMSTRONG et al. 1998). It revealed the amino acid determinants for agonist binding of GluR2, indicating that eight different amino acids are involved in making up the immediate surroundings of the agonist KA, including the arginine and glutamic acid residues mentioned above. The crystal structure, furthermore, suggests that residues at inter-domain interaction surfaces are involved in long-range allosteric modulation and desensitization.

The far N-terminal domain of ionotropic glutamate receptors (which is present in all glutamate receptors except for the short KBPs, Fig. 4) has been reported to be homologous to another class of bacterial periplasmic solute binding proteins, the leucine-isoleucine-valine binding proteins (LIVBP, O'HARA et al. 1993). Ionotropic glutamate receptors appear to share this LIVBP homology domain with metabotropic glutamate receptors, which are homologous to LIVBP in their entire N-terminal domains and appear to harbor the glutamate binding site in this domain. Indeed, site-directed mutagenesis of residues in mGluR1, which are homologous to residues in LIVBP, known from crystal structure data to be involved in ligand binding altered the agonist binding properties of this metabotropic glutamate receptor (O'HARA et al. 1993). In ionotropic glutamate receptors, however, the key ligand binding residues of the LIVBP domain are not conserved and, so far, no ligand binding properties have been unequivocally associated with this domain. In fact, exchange of the LIVBP domain between GluR6 and NMDAR1 did not affect the basic pharmacological properties (STERN-BACH et al. 1994). Thus, the function of this domain remains elusive. Homology to mammalian fatty acid binding proteins has been noted in the LIVBP domain of the NMDAR1 subunit and was taken as evidence for a possible binding site for the NMDA modulator arachidonic acid (PETROU et al. 1993). However, to date, there is no experimental proof available for this claim.

IV. Quaternary Structure: Oligomerization

1. Subunits Capable of Forming Homooligomeric Ion Channels

Of the 23 vertebrate glutamate receptor subunits, only eight have been proven functional when expressed as homomeric receptors in heterologous systems. Seven additional subunits are functional only in heteromeric assemblies with other subunits, and the remaining eight subunits have not yet been shown to form functional ion channels under any condition. The eight homomerically functional subunits include the four AMPA receptors GluR1 through GluR4, the three KA receptors GluR5 through GluR7, and the NMDA receptor subunit NMDAR1. The expression of homomeric NMDAR1 ion channels, however, is still a matter of debate. While NMDAR1 channels are easily expressed in *Xenopus* oocytes, attempts to express functional ion channels in mammalian cells, such as HEK cells, have not been successful (MONYER et al. 1992; LYNCH et al. 1994; GRIMWOOD et al. 1995; SUCHER et al. 1996). The reasons for this discrepancy between expression systems are still unexplained. Some investigators argue that a NR2 or NR2-like receptor subunit endogenously present in *Xenopus* oocytes may coassemble with NMDAR1 subunits generated from injected cRNA to cause expression of what is in fact a heteromeric NMDA receptor (SOLOVIEV and BARNARD 1997; LAUBE et al. 1998). However, electrophysiological and pharmacological properties of NMDAR1 expressed in oocytes are very different from properties typical of heteromeric NMDA receptors (reviewed by HOLLMANN and HEINEMANN 1994; MCBAIN and MAYER 1994). In addition, evidence for the in vivo expression of homomeric NMDAR1 subunits has been presented (Table 18; GRIMWOOD et al. 1996; WANG and THUKRAL 1996; CHAZOT and STEPHENSON 1997). Additional experiments are necessary to resolve this issue.

2. Subunits Capable of Forming Heterooligomeric Ion Channels

All vertebrate glutamate receptor subunits for which functional expression has been achieved have been shown to be able to coassemble with other glutamate receptor subunits of the same pharmacological class to form heterooligomeric complexes. For some subunits, such as the KA receptors KA1

Table 18. Heteromeric complex formation of ionotropic glutamate receptor subunits. Of the many electrophysiological studies demonstrating heteromeric channel formation of mixtures of recombinant α-amino-3-hydroxy-5-methyl-4-isoxazole propionate, kainate, or *N*-methyl-D-aspartate receptor subunits, only a selection comprising those studies published first is listed

Pharmacological class	Subunit combination	Biochemical evidence	Functional evidence	Reference
AMPA receptors	GluR1/GluR2; GluR2/GluR3; GluR2/GluR4		Electrophysiology in oocytes	Nakanishi et al. 1990
			Electrophysiology in oocytes, calcium permeability, rectification properties	Hollmann et al. 1991
	GluR1/GluR2		Electrophysiology in oocytes; calcium permeability, rectification properties	Hume et al. 1991
	GluR2/GluR4; GluR1/GluR2		Electrophysiology in HEK cells; rectification properties	Verdoorn et al. 1991
	GluR2/GluR4		Electrophysiology in HEK cells; calcium permeability, rectification properties	Burnashev et al. 1992
	Various combinations of AMPA receptors GluR1–4	Co-immunoprecipitation and Western blot analysis		Brose et al. 1994
	Various combinations of AMPA receptors GluR1–4	Co-immunoprecipitation and Western blot analysis		Puchalski et al. 1994
	GluR1/GluR2; GluR2/GluR3	Co-immunoprecipitation and Western blot analysis		Wenthold et al. 1996
	GluR1/GluR4	Co-immunoprecipitation and Western blot analysis		Ripellino et al. 1998
KA receptors	GluR6/KA2		Coexpression induces AMPA response	Herb et al. 1992
	GluR6/KA2		Coexpression induces AMPA response	Sakimura et al. 1992
	Various combinations of KA receptors GluR6-7, KA1-2	Co-immunoprecipitation and Western blot analysis		Brose et al. 1994
	GluR6 or -7/KA2	Co-immunoprecipitation and Western blot analysis		Puchalski et al. 1994
	GluR6/KA2	Co-immunoprecipitation and Western blot analysis		Wenthold et al. 1994

Table 18. *Continued*

Pharmacological class	Subunit combination	Biochemical evidence	Functional evidence	Reference
	GluR5/KA2		RT-PCR, electrophysiology and pharmacology of isolated trigeminal ganglion cells	SAHARA et al. 1997
	GluR6(R)/KA2		Currents and dose-response curves in HEK cells; single channel data	HOWE 1996
	GluR6 or -7/KA2	Co-immunoprecipitation and Western blot analysis		RIPELLINO et al. 1998
NMDA receptors	NR1/NR2A		Electrophysiology in oocytes	MEGURO et al. 1992
	NR1/NR2B; NR1/NR2C		Electrophysiology in oocytes	KUTSUWADA et al. 1992
	NR1/NR2D		Electrophysiology in oocytes	IKEDA et al. 1992
	NR1/NR2A; NR1/NR2B; NR1/NR2C		Electrophysiology in HEK cells	MONYER et al. 1992
	NR1/NR2A; NR1/NR2C		Electrophysiology in oocytes	ISHII et al. 1993
	NR1/NR2A/NR2C		Currents and dose-response curves for GLU and GLY in oocytes	WAFFORD et al. 1993
	NMDA receptors (NR1, NR2A-D	Co-immunoprecipitation and Western blot analysis		BROSE et al. 1994
	NR1a/NR2A (human)		Currents and dose-response curves in oocytes	LE BOURDELLES et al. 1994
	NR1/NR2A/NR2C	Co-immunoprecipitation, Western blot analysis	Pharmacological properties analyzed in HEK cells	CHAZOT et al. 1994
	NMDAR1/NR2A/NR2B	Co-immunoprecipitation, Western blot analysis		SHENG et al. 1994
	NR1 homomeric receptors (rat, presynaptic)		Pharmacological properties	WANG and THUKRAL 1996

	NR1 homomeric receptors		Excess of NR1 ligand binding sites in a cell line stably transfected with NR1a + NR2A	GRIMWOOD et al. 1996
	NR1/NR2A/NR2B	Co-immunoprecipitation, Western blot analysis		CHAZOT and STEPHENSON 1997b
	NR1 homomeric receptors	Differential solubilization		CHAZOT and STEPHENSON 1997a
	NR1/NR2A/NR2B		Pharmacological and single-channel properties in CHO cells	BRIMECOMBE et al. 1997
	NR1/NR2B/NR2D		Pharmacological properties analyzed in oocytes	BULLER and MONAGHAN 1997
	NR1/NR2A/NR2B	Co-immunoprecipitation, Western blot analysis		LUO et al. 1997
	NR1/NR2A/NR2D; NR1/NR2B/NR2D; NR1/NR2D	Co-immunoprecipitation, Western blot analysis		DUNAH et al. 1998
	NR1/KBP-toad (Xenopus)	Co-immunoprecipitation, Western blot analysis	Pharmacological properties analyzed in HEK cells	SOLOVIEV et al. 1996
	NR1/NR2B/NR3A	Co-immunoprecipitation, Western blot analysis	Single-channel recording in oocytes	DAS et al. 1998
Orphan receptors	delta2		Electrophysiology in HEK cells fails to find any signs of functional interaction with other iGluRs	LOMELI et al. 1993
	delta2	Co-immunoprecipitation and Western blot analysis fails to fimd any signs of coassembly with other iGluRs or mGluRs		MAYAT et al. 1995

AMPA, α-amino-3-hydroxy-5-methyl-4-isoxazole propionate; *KA*, kainate; *NMDA*, *N*-methyl-D-aspartate

and KA2 and the NMDA receptor subunits NR2A–NR2D and NR3A, formation of heteromeric complexes is the only way to generate a functional ion channel. Table 18 lists and references the various heterooligomeric subunit combinations that have been observed. It also gives a brief description of the biochemical or functional evidence on which identification of heterooligomers has been based. Biochemical evidence usually consists of demonstration of co-immunoprecipitation of receptor subunits after controled solubilization under mild detergent conditions designed to keep the oligomeric complex intact. Functional evidence usually comprises demonstration of altered electrophysiological and/or pharmacological properties of coexpressed subunits as opposed to the properties of the respective homomerically expressed subunits. It has been shown for AMPA receptors that formation of heteromeric channels is not a requirement for efficient surface expression of the receptor (HALL et al. 1997). The available data suggest that glutamate receptors are forming hetero-oligomeric complexes exclusively within one of the three main pharmacological classes, AMPA, KA, and NMDA receptors. Subunit assemblies across these classes have never been observed at significant levels, with one notable exception.

In a recent study, coexpression of the NMDAR1 subunit from *Xenopus laevis* with KBP-toad, the KBP from *Xenopus* (which is also called XenU1), has been reported to produce a "unitary" receptor with a unique pharmacology (SOLOVIEV et al. 1996; SOLOVIEV et al. 1998; see also Sect. B.II.5.). This receptor complex is apparently activated by all glutamatergic agonists, including NMDA, glutamate, KA, and AMPA. Interestingly, receptor activation for all these agonists, not only for NMDA, was found to be glycine-dependent, and all responses reportedly were blocked by non-competitive NMDA receptor blockers such as Mg^{2+} and MK-801. Thus, KBP-toad appears to behave like a NR2-like subunit. Expression of KBP-toad mRNA has been detected in *Xenopus* oocytes, albeit at extremely low levels (SOLOVIEV and BARNARD 1997). Based on this finding, it has been suggested that endogenous KBP-toad mRNA might be responsible for the observation of functional ion channels after injection of rat, mouse, or human NMDAR1 cRNA into *Xenopus* oocytes (SOLOVIEV and BARNARD 1997). However, this interpretation is in conflict with the "unitary" nature such a heteromeric subunit complex would be expected to display. In fact, neither rat, mouse nor human NMDAR1, when expressed in oocytes, is activated by KA or AMPA (MORIYOSHI et al. 1991; YAMAZAKI et al. 1992b; PLANELLS-CASES et al. 1993). Homomeric expression of the *Xenopus* NMDAR1 cDNA in oocytes to directly test for the predicted unitary, KA- and AMPA-activatable receptor formed upon heteromerization with endogenous KBP-toad has not been reported (SOLOVIEV et al. 1996).

3. Subunit Stoichiometry of the Oligomeric Receptor Complex

Soon after the cloning of the first glutamate receptor subunits, co-expression of multiple subunits had indicated that glutamate receptors were heterooligomeric complexes (see Sect. D.IV.2.). In the absence of any experimental evidence, and based on the (as we now know ill-conceived) notion of structural similarity to acetylcholine receptors, glutamate receptors were assumed to be pentameric complexes. Subunits were believed to assemble in a doughnut-shaped structure with a central pore which, in the case of the NMDA receptor, was estimated to be 5.5 Å in diameter at its narrow constriction (VILLARROEL et al. 1995). Since 1992, a considerable number of studies have tried to experimentally address the question of subunit stoichiometry using recombinant receptors or antibodies generated against recombinant receptors. Table 19 lists those studies with a brief description of the evidence. Data obtained in systems in which the glutamate receptor subunit composition was entirely unknown or remained undefined (mostly prior to 1992) have not been included in this list.

Table 19 has been arranged chronologically to show that previous studies (1992–1995) generally concluded a pentameric structure or even a higher number of subunits, mostly based on cross-linking experiments and size-exclusion chromatography of solubilized receptor complexes. Starting in 1996, evidence for a tetrameric structure began to emerge, and the three most recent studies published in 1998 all arrived at the conclusion that glutamate receptors are tetramers. While this would fit the much-discussed potential evolutionary relationship of glutamate receptors with potassium channels (WOOD et al. 1995; HOLLMANN 1996), which are known to be tetramers (LIMAN et al. 1992), the issue is far from being settled. It is interesting to note that different laboratories arrived at contrary conclusions, while using virtually the same techniques and even investigating the same subunits.

Cross-linking studies on AMPA receptors by WENTHOLD et al. (1992) and BLACKSTONE et al. (1992), for example, suggested a pentameric structure, while a similar study by WU et al. (1996) concluded a tetrameric structure. As another example, dose–response curves of mixtures of mutant and wild-type GluR1 expressed in oocytes were interpreted as indicative of a pentameric complex (FERRER-MONTIEL and MONTAL 1996) or a tetrameric complex (MANO and TEICHBERG 1998). Finally, analysis of subconductance states of NMDA receptor complexes made up of mixtures of mutant and wild-type subunits were interpreted to show either three NMDAR1 subunits in what was concluded to be a pentameric structure (PREMKUMAR et al. 1997) or only two, with no conclusion of the overall stoichiometry offered (BEHE et al. 1995). A very recent study used an ingenious method of stepwise activation of a fully blocked receptor under single-channel recording, basically "counting" the occupation by agonist of the individual subunits in the receptor complex (ROSENMUND et al. 1998). Four such steps were observed, consistent with a tetrameric structure. However, the possibility that two subunits initially may have to bind agonist before a channel-opening step can be recorded could not be ruled out entirely. In summary, while the scales currently appear to tilt towards a tetrameric structure, the question of receptor stoichiometry for the time being will have to remain open.

Table 19. Quaternary structure (stoichiometry) of ionotropic glutamate receptor complexes. Unless noted otherwise, all investigated subunits were from rat

Subunit(s) investigated	Conclusion	Evidence	Reference
AMPA receptor complex	Pentamer	Detergent-solubilized plasma-membrane proteins, which bound [^{3}H]AMPA and were immunoreactive for GluR1, gave complexes of 610 kDa in size-exclusion chromatography and sucrose density centrifugation	Blackstone et al. 1992
	Pentamer	Antibody-purified solubilized receptor as well as dithio-bis(succinimidyl propionate)-crosslinked receptor from plasma membranes gave bands of 108, 290–360, 470, and 590 kDa on Western blots probed with anti-GluR2/3 antibodies	Wenthold et al. 1992
	No conclusion	Chemical cross-linking with dithio-bis(succinimidyl propionate) of solubilized receptor produced complexes of 205, 350, 480, and 550 kDa on Western blots probed with anti-GluR1 antibodies	Brose et al. 1993
	No conclusion	Size-exclusion chromatography of triton-solubilized synaptic plasma membranes produced a 630-kDa complex that was immunoreactive for GluR1	Brose et al. 1993
NMDA receptor complex	No conclusion	Size-exclusion chromatography of triton-solubilized synaptic plasma membranes produced a 630-kDa complex that was immunoreactive for NMDAR1	Brose et al. 1993
	Pentamer, hexamer, or heptamer	Chemical cross-linking with dithio bis(succinimidyl propionate) of solubilized receptor produced a 730-kDa complex on Western blots probed with anti-NMDAR1 antibodies	Brose et al. 1993
NMDAR1/NR2A	No conclusion	Co-immunoprecipitation and analysis on non-denaturing gels gave a complex of 780–850 kDa	Chazot et al. 1994
NMDAR1/NR2A/NR2C	No conclusion	Co-immunoprecipitation and analysis on non-denaturing gels gave a complex of 780–850 kDa	Chazot et al. 1994
NMDAR1/NR2B	No conclusion (two NR1 subunits)	Coexpression of wildtype and ion pore mutant (N598R) NR1 subunits with NR2B yielded three different patterns of subconductance states, indicating the presence of two NR1 subunits in the complex	Behe et al. 1995

GluR4 (human)	Pentamer	Chemical cross-linking with dithio-bis(succinimidyl propionate) yielded a broad band at 550–600 kDa	Fletcher et al. 1995
GluR1	Pentamer	Coexpression of various ratios of two mutant GluR1 subunits with differential sensitivities to a channel blocker gave blocking curves best fit by assuming five subunits	Ferrer-Montiel and Montal 1996
NMDAR1/NR2A/NR2B	Tetramer, pentamer, or hexamer	Chemical cross-linking of the solubilized receptor produced three high molecular weight components of 603, 700, and 750 kDa	Blahos and Wenthold 1996
NMDAR1/NR2A	Pentamer (2+3)	[^{3}H]MK-801 binding to NMDAR-1/NR2A expressed in HEK cells indicated two NR1 subunits. In view of biochemical data suggesting a complex of 700–850 kDa, a pentamer consisiting of two NMDAR1 and three NR2 subunits is proposed	Chazot et al. 1996
AMPA receptors (pig)	Tetramer	Chemical crosslinking of synaptic plasma membranes and solubilized, partially purified AMPA receptors yielded four immunodetectable bands around 100, 200, 300, and 400 kDa; sucrose density gradient centrifugation and gel filtration chromatography yielded complexes of 320–340 and 480–500 kDa	Wu et al. 1996
NMDAR1/NR2B	Pentamer (3+2)	Coexpression of wild-type and ion pore mutant (N598Q) NR1 subunits with NR2B yielded six different patterns of subconductance states, indicating a pentameric structure	Premkumar and Auerbach 1997
GluR1	Tetramer	Coexpression of wild-type and a mutant GluR1 with low sensitivity to quisqualate yield dose-response curves compatible with a tetrameric structure	Mano and Teichberg 1998
NMDAR1/NR2B	Tetramer	Coexpression of wild-type and mutant low agonist affinity subunits (either NR1 or NR2B) yielded dose-response curves with three discrete affinities, indicating two binding sites each for glutamate and glycine, compatible with a tetramer	Laube et al. 1998
GluR3	Tetramer	Fast activation of a non-desensitizing mutant GluR3 from a fully blocked state produced four discrete steps of single-channel conductivity, indicative of a tetramer	Rosenmund et al. 1998

AMPA, α-amino-3-hydroxy-5-methyl-4-isoxazole propionate; NMDA, N-methyl-D-aspartate; NMDAR, NMDA receptor

Acknowledgements. This work was supported by SFB 406. The author would like to apologize to those colleagues whose work may have been overlooked or had to be omitted due to constraints of space. At present, approximately 60 papers on glutamate receptors are published every week, which means that even during the writing of this chapter approximately 500 additional papers appeared, precluding any attempt at exhaustive coverage.

References

Adams MD, Oxender DL (1989) Bacterial periplasmic binding protein tertiary structures. J Biol Chem 264:15739–15742

Adams SL, Foldes RL, Kamboj RK (1995) Human N-methyl-D-aspartate receptor modulatory subunit hNR3: cloning and sequencing of the cDNA and primary structure of the protein. Biochim Biophys Acta 1260:105–108

Akbarian S, Smith MA, Jones EG (1995) Editing for an AMPA receptor subunit RNA in prefrontal cortex and striatum in Alzheimer's disease, Huntington's disease and schizophrenia. Brain Res 699:297–304

Anantharam V, Panchal RG, Wilson A, Kolchine VV, Treistman SN, Bayley H (1992) Combinatorial RNA splicing alters the surface charge on the NMDA receptor. FEBS Lett 305:27–30

Araki K, Meguro H, Kushiya E, Takayama C, Inoue Y, Mishina M (1993) Selective expression of the glutamate receptor channel $\delta 2$ subunit in cerebellar Purkinje cells. Biochem Biophys Res Comm 197:1267–1276

Arvola M, Keinänen K (1996) Characterization of the ligand binding domains of glutamate receptor (GluR)-B and GluR-D subunits expressed in *Escherichia coli* as periplasmic proteins. J Biol Chem 271:15527–15532

Bai G, Kusiak JW (1993) Cloning and analysis of the 5′ flanking sequence of the rat N-methyl-D-aspartate receptor-1 (NMDAR1) gene. Biochim Biophys Acta 1152:197–200

Bai G, Kusiak JW (1995) Functional analysis of the proximal 5′-flanking region of the N-methyl-D-aspartate receptor subunit gene, NMDAR1. J Biol Chem 270: 7737–7744

Ballivet M, Patrick J, Lee J, Heinemann S (1982) Molecular cloning of cDNA coding for the γ subunit of *Torpedo* acetylcholine receptor. Proc Natl Acad Sci (USA) 79:4466–4470

Barnard EA, Darlison MG, Seeburg P (1987) Molecular biology of the $GABA_A$ receptor: The receptor/channel superfamily. Trends Neurosci 10:502–508

Barria A, Derkach V, Soderling T (1997) Identification of the Ca2+/calmodulin-dependent protein kinase II regulatory phosphorylation site in the a-amino-3-hydroxyl-5-methyl-4-isoxazole-propionate-type glutamate receptor. J Biol Chem 272:32727–32730

Behe P, Stern P, Wyllie DJA, Nassar M, Schoepfer R, Colquhoun D (1995) Determination of NMDA NR1 subunit copy number in recombinant NMDA receptors. Proc R Soc Lond Biol 262:205–213

Benne R, Van Den Burg J, Brakenhoff JPJ, Sloof P, Van Boom JH, Tromp MC (1986) Major transcript of the frameshift *coxII* from trypanosome mitochondria contains four nucleotides that are not encoded in the DNA. Cell 46:819–826

Bennett JA, Dingledine R (1995) Topology profile for a glutamate receptor: Three transmembrane domains and a channel-lining reentrant membrane loop. Neuron 14:373–384

Bennett MK, Calakos N, Scheller RH (1992) Syntaxin: A synaptic protein implicated in docking of synaptic vesicles at presynaptic active zones. Science 257:255–259

Bettler B, Boulter J, Hermans-Borgmeyer I, O'Shea-Greenfield A, Deneris ES, Moll C, Borgmeyer U, Hollmann M, Heinemann S (1990) Cloning of a novel glutamate receptor subunit, GluR5: expression in the nervous system during development. Neuron 5:583–595

Bettler B, Egebjerg J, Sharma G, Pecht G, Hermans-Borgmeyer I, Moll C, Stevens CF, Heinemann S (1992) Cloning of a putative glutamate receptor: A low affinity kainate binding subunit. Neuron 8:257–265

Bettler B, Mulle C (1995) Neurotransmitter receptors. 2. AMPA and kainate receptors. Neuropharmacology 34:123–139

Betz H (1991) Glycine receptors: Heterogeneous and widespread in the mammalian brain. Trends Neurosci 14:458–461

Betz H, Schuster C, Ultsch A, Schmitt B (1993) Molecular biology of ionotropic glutamate receptors in Drosophila melanogaster. Trends Pharmacol Sci 14:428–431

Blackstone CD, Moss SJ, Martin LJ, Levey AI, Price DL, Huganir RL (1992) Biochemical characterization and localization of a non-N-methyl-D-aspartate glutamate receptor in rat brain. J Neurochem 58:1118–1126

Blahos JI, Wenthold RJ (1996) Relationship between N-methyl-D-aspartate receptor NR1 splice variants and NR2 subunits. J Biol Chem 271:15669–15674

Bottai D, Maler L, Dunn RJ (1998) Alternative RNA splicing of the NMDA receptor NR1 mRNA in the neurons of the teleost electrosensory system. J Neurosci 18:5191–5202

Boulter J (1994a) Nucleotide sequence of rat glutamate receptor subunit gene Delta1. GenBank U08255. Unpublished sequence

Boulter J (1994b) Nucleotide sequence of rat glutamate receptor subunit gene Delta2. GenBank U08256. Unpublished sequence

Boulter J (1994c) Nucleotide sequence of rat glutamate receptor subunit gene KA1. GenBank U08257. Unpublished sequence

Boulter J (1994d) Nucleotide sequence of rat glutamate receptor subunit gene KA2. GenBank U08258. Unpublished sequence

Boulter J (1994e) Nucleotide sequence of rat NMDA receptor subunit gene NMDAR2 C. GenBank U08259. Unpublished sequence

Boulter J (1994f) Nucleotide sequence of rat NMDA receptor subunit gene NMDAR2D. GenBank U08260. Unpublished sequence

Boulter J (1997) Nucleotide sequence of rat NMDA receptor gene NMDAR2A. GenBank AF001423. Unpublished sequence

Boulter J, Hollmann M, O'Shea-Greenfield A, Hartley M, Deneris ES, Maron C, Heinemann S (1990) Molecular cloning and functional expression of glutamate receptor subunit genes. Science 249:1033–1037

Brett PM, Le Bourdelles B, See CG, Whiting PJ, Attwood J, Woodward K, Robertson MM, Kalsi G, Povey S, Gurling HMD (1994) Genomic cloning and localization by fish and linkage analysis of the human gene encoding the primary subunit NMDAR1 (grin1) of the NMDA receptor channel. Ann Hum Genet 58:95–100

Brimecombe JC, Boeckman FA, Aizenman E (1997) Functional consequences of NR2 subunit composition in single recombinant N-methyl-D-aspartate receptors. Proc Natl Acad Sci (USA) 94:11019–11024

Brockie PJ, Madsen DM, Maricq AV (1997) Genetic analysis of two *C. elegans* putative NMDA receptor subunits, *nmr-1* and *nmr-2*. Soc Neurosci Abstr 23:936

Brose N (1993) Membrane fusion takes excitatory turn: syntaxin, vesicle docking protein, or glutamate receptor. Cell 75:1043–1044

Brose N, Gasic GP, Vetter DE, Sullivan JM, Heinemann SF (1993) Protein chemical characterization and immunocytochemical localization of the NMDA receptor subunit NMDAR1. J Biol Chem 268:22663–22671

Brose N, Huntley GW, Stern-Bach Y, Sharma G, Morrison JH, Heinemann SF (1994) Differential assembly of coexpressed glutamate receptor subunits in neurons of rat cerebral cortex. J Biol Chem 269:16780–16784

Brusa R, Zimmermann F, Koh D-S, Feldmeyer D, Gass P, Seeburg PH, Sprengel R (1995) Early-onset epilepsy and postnatal lethality associated with an editing-deficient GluR-B allele in mice. Science 270:1677–1680

Bull L (1998) Versuche zur Klonierung von Šugetierhomologen der Familie der Kainatbindeproteine sowie Klonierung eines Kainatbindeproteins aus *Xenopus laevis*. Diploma Thesis, University of Göttingen, Göttingen
Buller AL, Monaghan DT (1997) Pharmacological heterogeneity of NMDA receptors: characterization of NR1a/NR2D heteromers expressed in *Xenopus* oocytes. Eur J Pharmacol 320:87–94
Burnashev N, Monyer H, Seeburg PH, Sakmann B (1992) Divalent ion permeability of AMPA receptor channels is dominated by the edited form of a single subunit. Neuron 8:189–198
Burnashev N, Villarroel A, Sakmann B (1996) Dimensions and ion selectivity of recombinant AMPA and kainate receptor channels and their dependence on Q/R site residues. J Physiol (Lond) 496:165–173
Chang HM, Wu YM, Chang YC, Hsu YC, Hsu HY, Chen YC, Chow WY (1998) Molecular and electrophysiological characterizations of fGluR3a, an ionotropic glutamate receptor subunit of a teleost fish. Mol Brain Res 57:211–220
Chazot PL, Cik M, Stephenson FA (1995) An investigation into the role of N-glycosylation in the functional expression of a recombinant heteromeric NMDA receptor. Mol Membr Biol 12:331–337
Chazot PL, Cik M, Stephenson FA (1996) Evidence for at least two NR1 subunits per NMDA receptor as deduced from the radioligand binding properties of wild-type and mutant NR1/NR2a receptors. Br J Pharmacol 117:68P
Chazot PL, Coleman SK, Cik M, Stephenson FA (1994) Molecular characterization of N-methyl-D-aspartate receptors expressed in mammalian cells yields evidence for the coexistence of 3 subunit types within a discrete receptor molecule. J Biol Chem 269:24403–24409
Chazot PL, Stephenson FA (1997a) Biochemical evidence for the existence of a pool of unassembled C2 exon-containing NR1 subunits of the mammalian forebrain NMDA receptor. J Neurochem 68:507–516
Chazot PL, Stephenson FA (1997b) Molecular dissection of native mammalian forebrain NMDA receptors containing the NR1 C2 exon: direct demonstration of NMDA receptors comprising NR1, NR2A, and NR2B subunits within the same complex. J Neurochem 69:2138–2144
Chen GQ, Sun Y, Jin RS, Gouaux E (1998) Probing the ligand binding domain of the GluR2 receptor by proteolysis and deletion mutagenesis defines domain boundaries and yields a crystallizable construct. Protein Science 7:2623–2630
Chittajallu R, Braithwaite SP, Clarke VRJ, Henley JM (1999) Kainate receptors: subunits, synaptic localization and function. Trends Pharmacol Sci 20:19–26
Ciabarra AM, Sullivan JM, Gahn LG, Pecht G, Heinemann S, Sevarino KA (1995) Cloning and characterization of Chi-1: A developmentally regulated member of a novel class of the ionotropic glutamate receptor family. J Neurosci 15:6498–6508
Cleland TA (1996) Inhibitory glutamate receptor channels. Mol Neurobiol 13:97–136
Cockcroft VB, Ortells MO, Thomas P, Lunt GG (1993) Homologies and disparities of glutamate receptors: A critical analysis. Neurochem Int 23:583–594
Collingridge GL, Lester RAJ (1989) Excitatory amino acid receptors in the vertebrate central nervous system. Pharmacol Rev 41:143–210
Collins C, Duff C, Duncan AMV, Planells-Cases R, Sun W, Norremolle A, Michaelis E, Montal M, Worton R, Hayden MR (1993) Mapping of the human NMDA receptor subunit (NMDAR1) and the proposed NMDA receptor glutamate binding subunit (NMDARA1) to chromosome-9q34.3 and chromosome-8, respectively. Genomics 17:237–239
Conn PJ, Pin JP (1997) Pharmacology and functions of metabotropic glutamate receptors. Annu Rev Pharmacol Toxicol 37:205–237
Cotman CW, Monaghan DT, Ottersen OP, Storm-Mathisen J (1987) Anatomical organization of excitatory amino acid receptors and their pathways. Trends Neurosci 10:273–279

Craig AM, Blackstone CD, Huganir RL, Banker G (1993) The distribution of glutamate receptors in cultured rat hippocampal neurons: postsynaptic clustering of AMPA-selective subunits. Neuron 10:1055–1068

Cully DF, Paress PS, Liu KK, Schaeffer JM, Arena JP (1996) Identification of a *Drosophila melanogaster* glutamate-gated chloride channel sensitive to the antiparasitic agent avermectin. J Biol Chem 271:20187–20191

Cully DF, Vassilatis DK, Liu KK, Paress PS, Vanderploeg LHT, Schaeffer JM, Arena JP (1994) Cloning of an avermectin-sensitive glutamate-gated chloride channel from *Caenorhabditis elegans*. Nature 371:707–711

Daggett LP, Jachec C, Lin F-F, C. D, Varney MA, Hess SD, Velicelebi G, Johnson EC (1996) Functional characterization of two isoforms of the human GluR6 receptor and distribution of GluR6 RNA editing sites. Soc Neurosci Abstr 22: 590

Daggett LP, Johnson EC, Varney MA, Lin FF, Hess SD, Deal CR, Jachec C, Lu CC, Kerner JA, Landwehrmeyer GB, Standaert DG, Young AB, Harpold MM, Velicelebi G (1998) The human N-methyl-D-aspartate receptor 2C subunit: genomic analysis, distribution in human brain, and functional expression. J Neurochem 71:1953–1968

Darlison MG (1992) Invertebrate GABA and glutamate receptors: molecular biology reveals predictable structures but some unusual pharmacologies. Trends Neurosci 15:469–474

Das S, Sasaki YF, Rothe T, Premkumar LS, Takasu M, Crandall JE, Dikkes P, Conner DA, Rayudu PV, Cheung W, Chen HSV, Lipton SA, Nakanishi N (1998) Increased NMDA current and spine density in mice lacking the NMDA receptor subunit NR3A. Nature 393:377–381

Delany NS, Laughton DL, Wolstenholme AJ (1997) GenBank Y09796. Unpublished sequence

Dent JA, Davis MW, Avery L (1997) Avr-15 encodes a chloride channel subunit that mediates inhibitory glutamatergic neurotransmission and ivermectin sensitivity in *Caenorhabditis elegans*. EMBO J 16:5867–5879

Devereux J, Haeberli P, Smithies O (1984) A comprehensive set of sequence analysis programs for the VAX. Nucleic Acids Res 12:387–395

Devillers-Thiery A, Changeux JP, Paroutaud P, Strosberg AD (1979) The amino-terminal sequence of the 40000 molecular weight subunit of the acetylcholine receptor protein from *Torpedo marmorata*. FEBS Lett 104:99–105

Devillers-Thiery A, Galzi JL, Eisele JL, Bertrand S, Bertrand D, Changeux JP (1993) Functional architecture of the nicotinic acetylcholine receptor: a prototype of ligand-gated ion channels. J Membr Biol 136:97–112

Dunah AW, Luo JH, Wang YH, Yasuda RP, Wolfe BB (1998) Subunit composition of N-methyl-D-aspartate receptors in the central nervous system that contain the NR2D subunit. Mol Pharmacol 53:429–437

Durand GM, Gregor P, Zheng X, Bennett MVL, Uhl GR, Zukin RS (1992) Cloning of an apparent splice variant of the rat N-methyl-D-aspartate receptor NMDAR1 with altered sensitivity to polyamines and activators of protein kinase C. Proc Natl Acad Sci (USA) 89:9359–9363

Eaton MJ, Chen JW, Kumar KN, Cong Y, Michaelis EK (1990) Immunochemical characterization of brain synaptic membrane glutamate binding proteins. J Biol Chem 265:16195–16204

Ebralidze AK, Rossi DJ, Tonegawa S, Slater NT (1996) Modification of NMDA receptor channels and synaptic transmission by targeted disruption of the NR2C gene. J Neurosci 16:5014–5025

Egebjerg J, Bettler B, Hermans-Borgmeyer I, Heinemann S (1991) Cloning of a cDNA for a glutamate receptor subunit activated by kainate but not AMPA. Nature 351:745–748

Eubanks JH, Puranam RS, Kleckner NW, Bettler B, Heinemann SF, McNamara JO (1993) The gene encoding the glutamate receptor subunit GluR5 is located on human chromosome 21q21.1–22.1 in the vicinity of the gene for familial amyotrophic lateral sclerosis. Proc Natl Acad Sci (USA) 90:178–182

Everts I, Villmann C, Hollmann M (1997) N-glycosylation is not a prerequisite for glutamate receptor function but is essential for lectin modulation. Mol Pharmacol 52:861–873

Everts I, Petroski R, Kizelsztein P, Teichberg VI, Heinemann SF, Hollmann M (1999) Lectin-induced inhibition of desensitization of the kainate receptor GluR6 depends on the activation state and can be mediated by a single native or ectopic N-linked carbohydrate side chain. J Neurosci 19:916–927

Ferrer-Montiel AV, Montal M (1996) Pentameric subunit stoichiometry of a neuronal glutamate receptor. Proc Natl Acad Sci (USA) 93:2741–2744

Fletcher EJ, Nutt SL, Hoo KH, Elliott CE, Korczak B, McWhinnie EA, Kamboj RK (1995) Cloning, expression and pharmacological characterization of a human glutamate receptor: hGluR4. Receptors Channels 3:21–31

Foldes RL, Adams SL, Fantaske RP, Kamboj RK (1994a) Human N-methyl-D-aspartate receptor modulatory subunit hNR2A: cloning and sequencing of the cDNA and primary structure of the protein. Biochim Biophys Acta 1223:155–159

Foldes RL, Rampersad V, Kamboj RK (1993) Cloning and sequence analysis of cDNAs encoding human hippocampus N-methyl-D-aspartate receptor subunits: evidence for alternative RNA splicing. Gene 131:293–298

Foldes RL, Rampersad V, Kamboj RK (1994b) Cloning and sequence analysis of additional splice variants encoding human N-methyl-D-aspartate receptor (hNR1) subunits. Gene 147:303–304

Forcina MS, Ciabarra AM, Sevarino KA (1995) Cloning of chi-2: a putative member of the ionotropic glutamate receptor superfamily. Soc Neurosci Abstr 21:438.433

Forrest D, Yuzaki M, Soares HD, Ng L, Luk DC, Sheng M, Steward CL, Morgan JI, Connor JA, Curran T (1994) Targeted disruption of NMDA receptor 1 gene abolishes NMDA response and results in neonatal death. Neuron 13:325–338

Funabiki K, Mishina M, Hirano T (1995) Retarded vestibular compensation in mutant mice deficient in $\delta 2$ glutamate receptor subunit. Neuroreport 7:189–192

Funk GD, Johnson SM, Smith JC, Dong XW, Lai J, Feldman JL (1997) Functional respiratory rhythm generating networks in neonatal mice lacking NMDAR1 gene. J Neurophysiol 78:1414–1420

Gallagher MJ, Huang H, Grant ER, Lynch DR (1997) The NR2B-specific interactions of polyamines and protons with the N-methyl-D-aspartate receptor. J Biol Chem 272:24971–24979

Gallo V, Upson LM, Hayes WP, Vyklicky L, Jr., Winters CA, Buonanno A (1992) Molecular cloning and developmental analysis of a new glutamate receptor subunit isoform in cerebellum. J Neurosci 12:1010–1023

Giraudat J, Devillers-Thiery A, Auffray C, Rougeon F, Changeux JP (1982) Identification of a cDNA clone coding for the acetylcholine binding subunit of *Torpedo marmorata* acetylcholine receptor. EMBO J 1:713–717

Goldman DJ (1994) GenBank U12018. Unpublished sequence

Gray MW, et al. (1992) Transcription processing and editing in plant mitochondria. Annu Rev Plant Physiol Plant Mol Biol 43:145–175

Gregor P, Gaston SM, Yang XD, Oregan JP, Rosen DR, Tanzi RE, Patterson D, Haines JL, Horvitz HR, Uhl GR, Brown RH (1994) Genetic and physical mapping of the GluR5 glutamate receptor gene on human chromosome-21. Human Genetics 94:565–570

Gregor P, Mano I, Maoz I, McKeown M, Teichberg VI (1989) Molecular structure of the chick cerebellar kainate binding subunit of a putative glutamate receptor. Nature 342:689–692

Gregor P, O'Hara BF, Yang X, Uhl GR (1993a) Expression and novel isoforms of glutamate receptor genes GluR5 and GluR6. Neuroreport 4:1343–1346

Gregor P, Reeves RH, Jabs EW, Yang XD, Dackowski W, Rochelle JM, Brown RH, Haines JL, O'Hara BF, Uhl GR, Seldin MF (1993b) Chromosomal localization of glutamate receptor genes: relationship to familial amyotrophic lateral sclerosis and other neurological disorders of mice and humans. Proc Natl Acad Sci (USA) 90: 3053–3057

Gregor P, Yang XD, Mano I, Takemura M, Teichberg VI, Uhl GR (1992) Organization and expression of the gene encoding chick kainate binding protein, a member of the glutamate receptor family. Mol Brain Res 16:179–186

Grenningloh G, Rienitz A, Schmitt B, Methfessel C, Zensen M, Beyreuther K, Gundelfinger ED, Betz H (1987) The strychnine binding subunit of the glycine receptors shows homology with nicotinic acetylcholine receptors. Nature 328: 215–220

Grimwood S, Le Bourdelles B, Atack JR, Barton C, Cockett W, Cook SM, Gilbert E, Hutson PH, McKernan RM, Myers J, Ragan CI, Wingrove PB, et al. (1996a) Generation and characterization of stable cell lines expressing recombinant human N-methyl-D-aspartate receptor subtypes. J Neurochem 66:2239–2247

Grimwood S, Le Bourdelles B, Cockett W, Atack J, Hutson PH, Whiting PJ (1996b) Homomeric and heteromeric NMDA receptor subunit assemblies can coexist within the same stable cell line. Br J Pharmacol 117:61P

Grimwood S, Le Bourdelles B, Whiting PJ (1995) Recombinant human NMDA homomeric NR1 receptors expressed in mammalian cells form a high-affinity glycine antagonist binding site. J Neurochem 64:525–530

Hall RA, Hansen A, Andersen PH, Soderling TR (1997) Surface expression of the AMPA receptor subunits GluR1, GluR2, and GluR4 in stably transfected baby hamster kidney cells. J Neurochem 68:625–630

Harvey RJ, Stühmer T, Van Minnen J, Darlison MG (1997) Differential patterns of expression of two novel invertebrate (*Lymnaea stagnalis*) ionotropic glutamate receptor genes. Neurosci Res Commun 20:31–40

Herb A, Burnashev N, Werner P, Sakmann B, Wisden W, Seeburg PH (1992) The KA-2 subunit of excitatory amino acid receptors shows widespread expression in brain and forms ion channels with distantly related subunits. Neuron 8:775–785

Hess SD, Daggett LP, Crona J, Deal C, Lu CC, Urrutia A, Chavez-Noriega L, Ellis SB, Johnson EC, Velicelebi G (1996) Cloning and functional characterization of human heteromeric N-methyl-D-aspartate receptors. J Pharmacol Exp Ther 278: 808–816

Hess SD, Daggett LP, Deal C, Lu CC, Johnson EC, Velicelebi G (1998) Functional characterization of human N-methyl-D-aspartate subtype 1a/2D receptors. J Neurochem 70:1269–1279

Hieber VC, Goldman D (1995) Trans-synaptic regulation of NMDA receptor RNAs during optic nerve regeneration. J Neurosci 15:5286–5296

Higuchi M, Single FN, Kohler M, Sommer B, Sprengel R, Seeburg PH (1993) RNA editing of AMPA receptor subunit GluR-B: a base-paired intron-exon structure determines position and efficiency. Cell 75:1361–1370

Hirai H, Kirsch J, Laube B, Betz H, Kuhse J (1996) The glycine binding site of the N-methyl-D-aspartate receptor subunit NR1: Identification of novel determinants of co-agonist potentiation in the extracellular M3-M4 loop region. Proc Natl Acad Sci (USA) 93:6031–6036

Hirano T, Kasono K, Araki K, Mishina M (1995) Suppression of LTD in cultured Purkinje cells deficient in the glutamate receptor $\delta 2$ subunit. Neuroreport 6:524–526

Hollmann M (1996) The topology of glutamate receptors: Sorting through the domains. In: Monaghan DT, Wenthold R (eds) The Ionotropic Glutamate Receptors. Humana Press, Totowa, New Jersey, pp 39–79

Hollmann M, Boulter J, Maron C, Beasley L, Sullivan J, Pecht G, Heinemann S (1993) Zinc potentiates agonist-induced currents at certain splice variants of the NMDA receptor. Neuron 10:943–954

Hollmann M, Hartley M, Heinemann S (1991) Calcium permeability of KA-AMPA gated glutamate receptor channels depends on subunit composition. Science 252:851–853

Hollmann M, Heinemann S (1994) Cloned glutamate receptors. Annu Rev Neurosci 17:31–108

Hollmann M, Maron C, Heinemann S (1994) N-glycosylation site tagging suggests a three transmembrane domain topology for the glutamate receptor GluR1. Neuron 13:1331–1343

Hollmann M, O'Shea-Greenfield A, Rogers SW, Heinemann S (1989) Cloning by functional expression of a member of the glutamate receptor family. Nature 342:643–648

Hollmann M, Rogers SW, O'Shea-Greenfield A, Deneris ES, Hughes TE, Gasic GP, Heinemann S (1990) The glutamate receptor GluR-K1: Structure, function, and expression in the brain. Cold Spring Harbor Symp Quant Biol 55:41–55

Honore T (1989) Excitatory amino acid receptor subtypes and specific antagonists. Med Res Rev 9:1–23

Hoo KH, Nutt SL, Fletcher EJ, Elliott CE, Korczak B, Deverill RM, Rampersad V, Fantaske RP, Kamboj RK (1994) Functional expression and pharmacological characterization of the human EAA4 (GluR6) glutamate receptor: a kainate selective channel subunit. Receptors Channels 2:327–337

Howe JR (1996) Homomeric and heteromeric ion channels formed from the kainate-type subunits GluR6 and KA2 have very small, but different, unitary conductances. J Neurophysiol 76:510–519

Hu W, Zuo J, Dejager PL, Heintz N (1998) The human glutamate receptor delta2 gene (GRID2) maps to chromosome 4q22. Genomics 47:143–145

Huang F, Gallo V (1997) Gene structure of the rat kainate receptor subunit KA2 and characterization of an intronic negative regulatory region. J Biol Chem 272: 8618–8627

Hullebroeck MF, Hampson DR (1992) Characterization of the oligosaccharide side chains on kainate binding proteins and AMPA receptors. Brain Res 590:187–192

Hume RI, Dingledine R, Heinemann SF (1991) Identification of a site in glutamate receptor subunits that controls calcium permeability. Science 253:1028–1031

Hutton ML, Harvey RJ, Barnard EA, Darlison MG (1991) Cloning of a cDNA that encodes an invertebrate glutamate receptor subunit. FEBS Lett 292:111–114

Ikeda K, Araki K, Takayama C, Inoue Y, Yagi T, Aizawa S, Mishina M (1995) Reduced spontaneous activity of mice defective in the ε4 subunit of the NMDA receptor channel. Mol Brain Res 33:61–71

Ikeda K, Nagasawa M, Mori H, Araki K, Sakimura K, Watanabe M, Inoue Y, Mishina M (1992) Cloning and expression of the e4 subunit of the NMDA receptor channel. FEBS Lett 313:34–38

Ishii T, Moriyoshi K, Sugihara H, Sakurada K, Kadotani H, Yokoi M, Akazawa C, Shigemoto R, Mizuno N, Masu M, Nakanishi S (1993) Molecular characterization of the family of the N-methyl-D-aspartate receptor subunits. J Biol Chem 268: 2836–2843

Ishimaru H, Kamboj R, Ambrosini A, Henley JM, Soloviev MM, Sudan H, Rossier J, Abutidze K, Rampersad V, Usherwood PNR, Bateson AN, Barnard EA (1996) A unitary non-NMDA receptor short subunit from *Xenopus*: DNA cloning and expression. Receptors Channels 4:31–49

Ito I, Sakimura K, Mishina M, Sugiyama H (1996) Age-dependent reduction of hippocampal LTP in mice lacking N-methyl-D-aspartate receptor ε1 subunit. Neurosci Lett 203:69–71

Iwasato T, Erzurumlu RS, Huerta PT, Chen DF, Sasaoka T, Ulupinar E, Tonegawa S (1997) NMDA receptor-dependent refinement of somatotopic maps. Neuron 19:1201–1210

Jagannathan S, Laughton DL, Skinner TM, Lunt GG, Wolstenholme AD (1998a) GenBank Y14233. Unpublished sequence

Jagannathan S, Laughton DL, Skinner TM, Lunt GG, Wolstenholme AD (1998b) GenBank Y14234. Unpublished sequence

Jia ZP, Agopyan N, Miu P, Xiong ZG, Henderson J, Gerlai R, Taverna FA, Velumian A, MacDonald J, Carlen P, Abramow-Newerly W, Roder J (1996) Enhanced LTP in mice deficient in the AMPA receptor GluR2. Neuron 17:945–956

Kadotani H, Hirano T, Masugi M, Nakamura K, Nakao K, Katsuki M, Nakanishi S (1996) Motor discoordination results from combined gene disruption of the NMDA receptor NR2A and NR2C subunits, but not from single disruption of the NR2A or NR2C subunit. J Neurosci 16:7859–7867

Kadotani H, Namura S, Katsuura G, Terashima T, Kikuchi H (1998) Attenuation of focal cerebral infarct in mice lacking NMDA receptor subunit NR2C. Neuroreport 9:471–475

Kalsi G, Whiting P, Le Bourdelles B, Callen D, Barnard EA, Gurling H (1998) Localization of the human NMDAR2D receptor subunit gene (GRIN2D) to 19q13.1-qter, the NMDAR2A subunit gene to 16p13.2 (GRIN2A), and the NMDAR2C subunit gene (GRIN2C) to 17q24-q25 using somatic cell hybrid and radiation hybrid mapping panels. Genomics 47:423–425

Kamboj R, Elliott CE, Nutt SL (1996a) Kainate-binding human CNS glutamate receptors EAA3C and EAA3D, DNA encoding them, and expression of the DNA in transformed cells . GenBank I25018. Patent US 5547855-A

Kamboj R, Elliott CE, Nutt SL (1996b) Kainate-binding, human CNS receptors of the EAA4 family. GenBank I28906. Patent US 5574144-A

Kamboj R, Nutt SL, Shekter L, Wosnick MA (1996c) Kainate-binding human CNS receptors of the EAA1 family. GenBank I28953. Patent US 5576205-A

Kamboj R, Nutt SL, Shekter L, Wosnick MA (1996d) Kainate-binding, human CNS receptors of the EAA2 family. GenBank I17703. Patent US 5494792-A

Kamboj RK, Schoepp DD, Nutt S, Shekter L, Korczak B, True RA, Rampersad V, Zimmerman DM, Wosnick MA (1994) Molecular cloning, expression, and pharmacological characterization of humEAA1, a human kainate receptor subunit. J Neurochem 62:1–9

Kamboj RK, Schoepp DD, Nutt S, Shekter L, Korczak B, True RA, Zimmerman DM, Wosnick MA (1992) Molecular structure and pharmacological characterization of humEAA2, a novel human kainate receptor subunit. Mol Pharmacol 42:10–15

Kamphuis W, Da Silva FHL (1995) Editing status at the Q/R-site of glutamate receptor A subunit, B subunit, 5 subunit and 6 subunit messenger RNA in the hippocampal kindling model of epilepsy. Mol Brain Res 29:35–42

Kang C-H, Shin WC, Yamagata Y, Gokcen S, Ames GF-L, Kim SH (1991) Crystal structure of the lysine-, arginine-, ornithine-binding protein (LAO) from *Salmonella typhimurium* at 2.7-Å resolution. J Biol Chem 266:23893–23899

Karp SJ, Masu M, Eki T, Ozawa K, Nakanishi S (1993) Molecular cloning and chromosomal localization of the key subunit of the human N-methyl-D-aspartate receptor. J Biol Chem 268:3728–3733

Kashiwabuchi N, Ikeda K, Araki K, Hirano T, Shibuki K, Takayama C, Inoue Y, Kutsuwada T, Yagi T, Kang Y, Aizawa S, Mishina M (1995) Impairment of motor coordination, Purkinje cell synapse formation, and cerebellar long term depression in GluR $\delta 2$ mutant mice. Cell 81:245–252

Kashiwagi K, Fukuchi J, Chao J, Igarashi K, Williams K (1996) An aspartate residue in the extracellular loop of the N-methyl-D-aspartate receptor controls sensitivity to spermine and protons. Mol Pharmacol 49:1131–1141

Kashiwagi K, Pahk AJ, Masuko T, Igarashi K, Williams K (1997) Block and modulation of N-methyl-D-aspartate receptors by polyamines and protons: role of amino acid residues in the transmembrane and pore-forming regions of NR1 and NR2 subunits. Mol Pharmacol 52:701–713

Kawamoto S, Hattori S, Sakimura K, Mishina M, Okuda K (1995) N-linked glycosylation of the α-amino-3-hydroxy-5-methylisoxazole-4-propionate (AMPA)-selective glutamate receptor channel $\alpha 2$ subunit is essential for the acquisition of ligand binding activity. J Neurochem 64:1258–1266

Keinänen K, Arvola M, Kuusinen A, Johnson M (1997) Ligand recognition in glutamate receptors: insights from mutagenesis of the soluble α-amino-3-hydroxy-5-methyl-4-isoxazole propionic acid (AMPA) binding domain of glutamate receptor type D (GluR-D). Biochem Soc Trans 25:835–838

Keinänen K, Jouppila A, Kuusinen A (1998) Characterization of the kainate-binding domain of the glutamate receptor GluR-6 subunit. Biochem J 330:1461–1467

Keinänen K, Wisden W, Sommer B, Werner P, Herb A, Verdoorn TA, Sakmann B, Seeburg PH (1990) A family of AMPA-selective glutamate receptors. Science 249:556–560

Kimura N, Kurosawa N, Kondo K, Tsukada Y (1993) Molecular cloning of the kainate-binding protein and calmodulin genes which are induced by an imprinting stimulus in ducklings. Mol Brain Res 17:351–355

Kishimoto Y, Kawahara S, Kirino Y, Kadotani H, Nakamura Y, Ikeda M, Yoshioka T (1997) Conditioned eyeblink response is impaired in mutant mice lacking NMDA receptor subunit NR2 A. Neuroreport 8:3717–3721

Kiyama Y, Manabe T, Sakimura K, Kawakami F, Mori H, Mishina M (1998) Increased threshold for long-term potentiation and contextual learning in mice lacking the NMDA-type glutamate receptor e1 subunit. J Neurosci 18:6704–6712

Klein M, Pieri I, Uhlmann F, Pfizenmaier K, Eisel U (1998) Cloning and characterization of promoter and 5′-UTR of the NMDA receptor subunit $\varepsilon 2$: evidence for alternative splicing of 5′ non-coding exon. Gene 208:259–269

Köhler M, Burnashev N, Sakmann B, Seeburg PH (1993) Determinants of Ca2+ permeability in both TM1 and TM2 of high affinity kainate receptor channels: diversity by RNA editing. Neuron 10:491–500

Köhler M, Kornau H-C, Seeburg PH (1994) The organization of the gene for the functionally dominant α-amino-3-hydroxy-5-methyl-isoxazole-4-propionic acid receptor subunit GluR-B. J Biol Chem 269:17367–17370

Kohda K, Kondo M, Tomomura M, Yuzaki M (1998) Characterization of d2 glutamate receptors by a mutation that causes channel activities. Soc Neurosci Abstr 24: 841

Korczak B, Nutt SL, Fletcher EJ, Hoo KH, Elliott CE, Rampersad V, McWhinnie EA, Kamboj RK (1995) cDNA cloning and functional properties of human glutamate receptor EAA3 (GluR5) in homomeric and heteromeric configuration. Receptors Channels 3:41–49

Krenz WD, Boulter J, Selverston AI, Heinemannn SF (1999) Tissue-specific alternative splicing of P loop-encoding exons in an invertebrate glutamate receptor subunit gene. submitted

Kumar KN, Babcock KK, Johnson PS, Chen X, Ahmad M, Michaelis EK (1995a) Cloning of the cDNA for a brain glycine binding, glutamate binding and thienylcyclohexylpiperidine-binding protein. Biochem Biophys Res Comm 216:390–398

Kumar KN, Eggeman KT, Adams JL, Michaelis EK (1991a) Hydrodynamic properties of the purified glutamate-binding protein subunit of the N-methyl-D-aspartate receptor. J Biol Chem 266:14947–14952

Kumar KN, Tilakaratne N, Johnson PS, Allen AE, Michaelis EK (1991b) Cloning of cDNA for the glutamate binding subunit of an NMDA receptor complex. Nature 354:70–73

Kumar KN, Johnson PS, Chen XY, Pal R, Ahmad M, Ragland T, Bigge C, Michaelis EK (1998) Cloning of a brain N-methyl-D-aspartate- and D,L-e-2- amino-4-propyl-5-phosphono-3-pentanoic acid (CGP 39653)-binding protein. Biochem Biophys Res Comm 253:463–469

Kumar VM, John J, Govindaraju V, Khan NA, Raghunathan P (1996) Magnetic resonance imaging of NMDA-induced lesion of the medial preoptic area and changes in sleep, temperature and sex behavior. Neurosci Res 24:207–214

Kuner T, Wollmuth LP, Karlin A, Seeburg PH, Sakmann B (1996) Structure of the NMDA receptor channel M2 segment inferred from the accessibility of substituted cysteines. Neuron 17:343–352

Kung S-S, Wu Y-M, Chow W-Y (1996) Characterization of two fish glutamate receptor cDNA molecules: absence of RNA editing at the Q/R site. Mol Brain Res 35: 119–130

Kuramoto T, Maihara T, Masu M, Nakanishi S, Serikawa T (1994) Gene mapping of NMDA receptors and metabotropic glutamate receptors in the rat (*Rattus norvegicus*). Genomics 19:358–361

Kurihara H, Hashimoto K, Kano M, Takayama C, Sakimura K, Mishina M, Inoue Y, Watanabe M (1997) Impaired parallel fiber → Purkinje cell synapse stabilization during cerebellar development of mutant mice lacking the glutamate receptor δ2 subunit. J Neurosci 17:9613–9623

Kurosawa N, Kondo K, Kimura N, Ikeda T, Tsukada Y (1994) Molecular cloning and characterization of avian N-methyl-D-aspartate receptor type1 (NMDA-R1) gene. Neurochem Res 19:575–580

Kuryatov A, Laube B, Betz H, Kuhse J (1994) Mutational analysis of the glycine binding site of the NMDA receptor: Structural similarity with bacterial amino acid binding proteins. Neuron 12:1291–1300

Kutsuwada T, Kashiwabuchi N, Mori H, Sakimura K, Kushiya E, Araki K, Meguro H, Masaki H, Kumanishi T, Arakawa M, Mishina M (1992) Molecular diversity of the NMDA receptor channel. Nature 358:36–41

Kutsuwada T, Sakimura K, Manabe T, Takayama C, Katakura N, Kushiya E, Natsume R, Watanabe M, Inoue Y, Yagi T, Aizawa S, Arakawa M, Takahashi T, Nakamura Y, Mori H, Mishina M (1996) Impairment of suckling response, trigeminal neuronal pattern formation, and hippocampal LTD in NMDA receptor ε2 subunit mutant mice. Neuron 16:333–344

Kuusinen A, Arvola M, Keinänen K (1995) Molecular dissection of the agonist binding site of an AMPA receptor. EMBO J 14:6327–6332

Kyte J, Doolittle RF (1982) A simple method for displaying the hydrophobic character of a protein. J Mol Biol 157:105–132

Lam TT, Fu J, Li SH, Abler AS, Tso MOM (1995) N-methyl-D-aspartate (NMDA) induced apoptosis in rat retina. Invest Ophthalmol Visual Sci 36:S934–934

Lamerdin JE, McCready PM, Skowronski E, Adamson AW, Burkhart-Schultz K, Gordon L, Kyle A, Ramirez M, Stilwagen S, Phan H, Velasco N, Garnes J, Danganan L, Poundstone P, Christensen M, Georgescu A, Avila J, Liu S, Attix C, Andreise T, Trankheim M, Amico-Keller G, Coefield J, Duarte S, Lucas S, Bruce R, Thomas P, Quan G, Kronmiller B, Arellano A, Montgomery M, Ow D, Nolan M, Trong S, Kobayashi A, Olsen AO, Carrano AV (1998) Sequence analysis of a 3.5 Mb contig in human 19p13.3 containing a serine protease gene cluster. GenBank AC004528. Unpublished sequence

Lampinen M, Pentikainen O, Johnson MS, Keinanen K (1998) AMPA receptors and bacterial periplasmic amino acid-binding proteins share the ionic mechanism of ligand recognition. EMBO J 17:4704–4711

Laube B, Hirai H, Sturgess M, Betz H, Kuhse J (1997) Molecular determinants of agonist discrimination by NMDA receptor subunits: Analysis of the glutamate binding site on the NR2B subunit. Neuron 18:493–503

Laube B, Kuhse J, Betz H (1998) Evidence for a tetrameric structure of recombinant NMDA receptors. J Neurosci 18:2954–2961

Laube B, Kuryatov A, Kuhse J, Betz H (1993) Glycine-glutamate interactions at the NMDA receptor: role of cysteine residues. FEBS Lett 335:331–334

Le BB, Myers JA, Whiting PJ (1994a) cDNAs encoding human NMDA-22A receptor subunit and isoforms of the human NMDA-R1 receptor subunit, transfected cell line expressing them. GenBank A38680. Patent WO 9411501-A

Le BB, Myers JA, Whiting PJ (1994b) cDNAs encoding human NMDA-22A receptor subunit and isoforms of the human NMDA-R1 receptor subunit, transfected cell line expressing them. GenBank A38688. Patent WO 9411501-A

Le Bourdelles B, Wafford KA, Kemp JA, Marshall G, Bain C, Wilcox AS, Sikela JM, Whiting PJ (1994) Cloning, functional coexpression, and pharmacological characterisation of human cDNAs encoding NMDA receptor NR1 and NR2 A subunits. J Neurochem 62:2091–2098

Lewin AH, Sun GB, Fudala L, Navarro H, Zhou LM, Popik P, Faynsteyn A, Skolnick P (1998) Molecular features associated with polyamine modulation of NMDA receptors. J Med Chem 41:988–995

Lewis TB, Wood S, Michaelis EK, Dupont BR, Leach RJ (1996) Localization of a gene for a glutamate binding subunit of a NMDA receptor (GRINA) to 8q24. Genomics 32:131–133

Liman ER, Tytgat J, Hess P (1992) Subunit stoichiometry of a mammalian K+ channel determined by construction of multimeric cDNAs. Neuron 9:861–871

Lin YJ, Bovetto S, Carver JM, Giordano T (1996) Cloning of the cDNA for the human NMDA receptor NR2 C subunit and its expression in the central nervous system and periphery. Mol Brain Res 43:57–64

Lomeli H, Mosbacher J, Melcher T, Hoger T, Geiger JRP, Kuner T, Monyer H, Higuchi M, Bach A, Seeburg PH (1994) Control of kinetic properties of AMPA receptor channels by nuclear RNA editing. Science 266:1709–1713

Lomeli H, Sprengel R, Laurie DJ, Köhr G, Herb A, Seeburg PH, Wisden W (1993) The rat d1 and d2 subunits extend the excitatory amino acid receptor family. FEBS Lett 315:318–322

Lomeli H, Wisden W, Köhler M, Keinänen K, Sommer B, Seeburg PH (1992) High-affinity kainate and domoate receptors in rat brain. FEBS Lett 307:139–143

Lowe DL, Jahn K, Smith DO (1997) Glutamate receptor editing in the mammalian hippocampus and avian neurons. Mol Brain Res 48:37–44

Luo JH, Wang YH, Yasuda RP, Dunah AW, Wolfe BB (1997) The majority of N-methyl-D-aspartate receptor complexes in adult rat cerebral cortex contain at least 3 different subunits (NR1/NR2 A/NR2B). Mol Pharmacol 51:79–86

Ly AM, Michaelis EK (1991) Solubilization, partial purification, and reconstitution of glutamate-activated and N-methyl-D-aspartate-activated cation channels from brain synaptic membranes. Biochemistry 30:4307–4316

Lynch DR, Anegawa NJ, Verdoorn T, Pritchett DB (1994) N-methyl-D-aspartate receptors: different subunit requirements for binding of glutamate antagonists, glycine antagonists, and channel-blocking agents. Mol Pharmacol 45:540–545

Mach J, Kumar KN, Pal R, Huschenbett J, Michaelis E (1997) Cloning of a novel gene for a subunit of an NMDA receptor-like complex. Soc Neurosci Abstr 23:936

Madarnas AR, Henderson JT, Roder JC (1994) The NMDA receptor subunit 2B locus (NMDAR 2B) maps to the distal end of murine chromosome 6. Mamm Genome 5:115–116

Mano I, Lamed Y, Teichberg VI (1996) A venus flytrap mechanism for activation and desensitization of α-amino-3-hydroxy-5-methyl-4-isoxazole propionic acid receptors. J Biol Chem 271:15299–15302

Mano I, Teichberg VI (1998) A tetrameric subunit stoichiometry for a glutamate receptor channel complex. Neuroreport 9:327–331

Maricq AV, Peckol E, Driscoll M, Bargmann CI (1995) Mechanosensory signaling in *C. elegans* mediated by the GLR-1 glutamate receptor. Nature 378:78–81

Martin LJ, Blackstone CD, Levey AI, Huganir RL, Price DL (1993) AMPA glutamate receptor subunits are differentially distributed in rat brain. Neuroscience 53:327–358

Masami M (1993) New protein and gene coding the protein. GenBank E05440. Patent JP 1993239098-A

Masami M (1994a) New protein and gene coding the protein. GenBank E06594. Patent JP 1994014783-A

Masami M (1994b) New protein and gene coding the protein. GenBank E06819. Patent JP 1994062861-A

Masami M (1994c) New protein and gene coding the protein. GenBank E06820. Patent JP 1994062861-A

Masami M (1994d) New protein and gene codin g the protein. GenBank E06821. Patent JP 1994062861-A

Masami M (1994e) New protein and gene coding the protein. GenBank E07654. Patent JP 1994157597-A

Masu M, Tanabe Y, Tsuchida K, Shigemoto R, Nakanishi S (1991) Sequence and expression of a metabotropic glutamate receptor. Nature 349:760–765

Masu Y, Kazuhisa N, Tamaki H, Harada Y, Kuno M, Nakanishi S (1987) cDNA cloning of bovine substance-K receptor through oocyte expression system. Nature 329:836–838

Matteri RL (1997) Partial cDNA sequence of porcine N-methyl-D-aspartate (NMDA) receptor. GenBank AF008560. Unpublished sequence

Mayat E, Petralia RS, Wang YX, Wenthold RJ (1995) Immunoprecipitation, immunoblotting, and immunocytochemistry studies suggest that glutamate receptor δ subunits form novel postsynaptic receptor complexes. J Neurosci 15:2533–2546

Mayer ML, Westbrook GL (1987) The physiology of excitatory amino acids in the vertebrate central nervous system. Prog Neurobiol 28:197–276

McBain CJ, Mayer ML (1994) N-methyl-D-aspartic acid receptor structure and function. Physiol Rev 74:723–760

McDonald JW, Johnston MV (1990) Physiological and pathophysiological roles of excitatory amino acids during central nervous system development. Brain Res Rev 15:41–70

McGlade-McCulloh E, Yamamoto H, Tan SE, Brickey DA, Soderling TR (1993) Phosphorylation and regulation of glutamate receptors by calcium calmodulin-dependent protein kinase II. Nature 362:640–642

McHugh TJ, Blum KI, Tsien JZ, Tonegawa S, Wilson MA (1996) Impaired hippocampal representation of space in CA1-specific NMDAR1 knockout mice. Cell 87:1339–1349

McLaughlin DP, Kerwin RW (1994) GenBank X82068. Unpublished sequence

McNamara JO, Eubanks JH, McPherson JD, Wasmuth JJ, Evans GA, Heinemann SF (1992) Chromosomal localization of human glutamate receptor genes. J Neurosci 12:2555–2562

Meguro H, Mori H, Araki K, Kushiya E, Kutsuwada T, Yamazaki M, Kumanishi T, Arakawa M, Sakimura K, Mishina M (1992) Functional characterization of a heteromeric NMDA receptor channel expressed from cloned cDNAs. Nature 357:70–74

Mellem JE, Zheng Y, Maricq AV (1997) Ionotropic glutamate receptors in *C. elegans*. Soc Neurosci Abstr 23:936

Messersmith EK, Feller MB, Zhang H, Shatz CJ (1997) Migration of neocortical neurons in the absence of functional NMDA receptors. Mol Cell Neurosci 9:347–357

Minami T, Sugatani J, Sakimura K, Abe M, Mishina M, Ito S (1997) Absence of prostaglandin e-2-induced hyperalgesia in NMDA receptor epsilon subunit knockout mice. Br J Pharmacol 120:1522–1526

Mishina M (1996a) NMDH receptor proteins and genes encoding the same. GenBank I19101. Patent US 5502166-A

Mishina M (1996b) NMDH receptor proteins and genes encoding the same. GenBank I19102. Patent US 5502166-A

Mishina M (1996c) NMDH receptor proteins and genes encoding the same. GenBank I19103. Patent US 5502166-A

Mishina M (1996d) NMDH receptor proteins and genes encoding the same. GenBank I19107. Patent US 5502166-A

Mishina M (1996e) NMDH receptor proteins and genes encoding the same. GenBank I19108. Patent US 5502166-A

Mishina M (1996f) NMDH receptor proteins and genes encoding the same. GenBank I19109. Patent US 5502166-A

Mishina M (1996g) NMDH receptor proteins and genes encoding the same. GenBank I19110. Patent US 5502166-A
Molnar E, Baude A, Richmond SA, Patel PB, Somogyi P, McIlhinney RAJ (1993) Biochemical and immunocytochemical characterization of antipeptide antibodies to a cloned GluR1 glutamate receptor subunit: cellular and subcellular distribution in the rat forebrain. Neuroscience 53:307–326
Molnar E, McIlhinney RAJ, Baude A, Nusser Z, Somogyi P (1994) Membrane topology of the GluR1 glutamate receptor subunit: Epitope mapping by site-directed antipeptide antibodies. J Neurochem 63:683–693
Monaghan DT, Bridges RJ, Cotman CW (1989) The excitatory amino acid receptors: Their classes, pharmacology, and distinct properties in the function of the central nervous system. Annu Rev Pharmacol Toxicol 29:365–402
Monyer H, Burnashev N, Laurie DJ, Sakmann B, Seeburg PH (1994) Developmental and regional expression in the rat brain and functional properties of four NMDA receptors. Neuron 12:529–540
Monyer H, Sprengel R, Schoepfer R, Herb A, Higuchi M, Lomeli H, Burnashev N, Sakmann B, Seeburg PH (1992) Heteromeric NMDA receptors: molecular and functional distinction of subtypes. Science 256:1217–1221
Mori H, Mishina M (1995) Structure and function of the NMDA receptor channel. Neuropharmacology 34:1219–1237
Morita T, Sakimura K, Kushiya E, Yamazaki M, Meguro H, Araki K, Abe T, Mori KJ, Mishina M (1992) Cloning and functional expression of a cDNA encoding the mouse $\beta 2$ subunit of the kainate-selective glutamate receptor channel. Mol Brain Res 14:143–146
Moriyoshi K, Masu M, Ishii T, Shigemoto R, Mizuno N, Nakanishi S (1991) Molecular cloning and characterization of the rat NMDA receptor. Nature 354:31–37
Mosbacher J, Schoepfer R, Monyer H, Burnashev N, Seeburg PH, Ruppersberg JP (1994) A molecular determinant for submillisecond desensitization in glutamate receptors. Science 266:1059–1062
Moss SJ, Blackstone CD, Huganir RL (1993) Phosphorylation of recombinant non-NMDA glutamate receptors on serine and tyrosine residues. Neurochem Res 18:105–110
Mott DD, Doherty JJ, Zhang SN, Washburn MS, Fendley MJ, Lyuboslavsky P, Traynelis SF, Dingledine R (1998) Phenylethanolamines inhibit NMDA receptors by enhancing proton inhibition. Nature Neuroscience 1:659–667
Mulle C, Sailer A, Perezotano I, Dickinson-Anson H, Castillo PE, Bureau I, Maron C, Gage FH, Mann JR, Bettler B, Heinemann SF (1998) Altered synaptic physiology and reduced susceptibility to kainate-induced seizures in GluR6-deficient mice. Nature 392:601–605
Mußhoff U, Madeja M, Bloms P, Muschnittel K, Speckmann EJ (1992) Tunicamycin-induced inhibition of functional expression of glutamate receptors in *Xenopus* oocytes. Neurosci Lett 147:163–166
Myers SJ, Peters J, Huang YF, Comer MB, Barthel F, Dingledine R (1998) Transcriptional regulation of the GluR2 gene: neural- specific expression, multiple promoters, and regulatory elements. J Neurosci 18:6723–6739
Nagasawa M, Sakimura K, Mori KJ, Bedell MA, Copeland NG, Jenkins NA, Mishina M (1996) Gene structure and chromosomal localization of the mouse NMDA receptor channel subunits. Mol Brain Res 36:1–11
Nakanishi N, Axel R, Shneider R (1992) Alternative splicing generates functionally distinct N-methyl-D-asparte receptors. Proc Natl Acad Sci (USA) 89:8552–8556
Nakanishi N, Shneider NA, Axel R (1990) A family of glutamate receptor genes: evidence for the formation of heteromultimeric receptors with distinct channel properties. Neuron 5:569–581
Nakanishi S (1994) Metabotropic glutamate receptors: synaptic transmission, modulation, and plasticity. Neuron 13:1031–1037

Nakanishi S, Masu M (1994) Molecular diversity and functions of glutamate receptors. Annu Rev Biophys Biomol Struct 23:319–348

Nash NR, Heilman CJ, Rees HD, Levey AI (1997) Cloning and localization of exon 5-containing isoforms of the NMDAR1 subunit in human and rat brains. J Neurochem 69:485–493

Noda M, Takahashi H, Tanabe T, Toyosato M, Furutani Y, Hirose T, Asai M, Inayama S, Miyata T, Numa S (1982) Primary structure of α-subunit precursor of *Torpedo californica* acetylcholine receptor deduced from cDNA sequence. Nature 299: 793–797

Nohno T, Saito T, Hong J-S (1986) Cloning and complete nucleotide sequence of the Escherichia coli glutamine permease operon (gln HPQ). Molecular Gen Genetics 205:260–269

Novere NL, Changeux J-P (1995) Molecular evolution of the nicotinic acetylcholine receptor: an example of multigene family in excitable cells. J Mol Evol 40:155–172

Nutt SL, Hoo KH, Rampersad V, Deverill RM, Elliott CE, Fletcher EJ, Adams SL, Korczak B, Foldes RL, Kamboj RK (1994) Molecular characterization of the human EAA5 (GluR7) receptor: a high-affinity kainate receptor with novel potential RNA editing sites. Receptors Channels 2:315–326

Nutt SL, Kamboj RK (1994a) Differential RNA editing efficiency of AMPA receptor subunit GluR-2 in human brain. Neuroreport 5:1679–1683

Nutt SL, Kamboj RK (1994b) RNA editing of human kainate receptor subunits. Neuroreport 5:2625–2629

O'Hara PJ, Sheppard PO, Thogersen H, Venezia D, Haldeman BA, McGrane V, Houamed KM, Thomsen C, Gilbert TL, Mulvihill ER (1993) The ligand binding domain in metabotropic glutamate receptors is related to bacterial periplasmic binding proteins. Neuron 11:41–52

Oh B-H, Pandit J, Kang C-H, Nikaido K, Gokcen S, Ames GF-L, Kim S-H (1993) Three-dimensional structures of the periplasmic lysine-, arginine-, ornithine-binding protein with and without a ligand. J Biol Chem 268:11348–11355

Okabe S, Collin C, Auerbach JM, Meiri N, Bengzon J, Kennedy MB, Segal M, McKay RDG (1998) Hippocampal synaptic plasticity in mice overexpressing an embryonic subunit of the NMDA receptor. J Neurosci 18:4177–4188

Ottiger HP, Gerfin-Moser A, Del Principe F, Dutly F, Streit P (1995) Molecular cloning and differential expression patterns of avian glutamate receptor mRNAs. J Neurochem 64:2413–2426

Paas Y (1998) The macroarchitectures and microarchitectures of the ligand binding domain of glutamate receptors. Trends Neurosci 21:117–125

Paas Y, Eisenstein M, Medevielle F, Teichberg VI, Devillers-Thiery A (1996) Identification of the amino-acid subsets accounting for the ligand binding specificity of a glutamate receptor. Neuron 17:979–990

Paoletti P, Ascher P, Neyton J (1997) High-affinity zinc inhibition of NMDA NR1-NR2 A receptors. J Neurosci 17:5711–5725 (+ correction J Neurosci 17:U8)

Paperna T, Lamed Y, Teichberg VI (1996) cDNA cloning of chick brain α-amino-3-hydroxy-5-methyl-4-isoxazolepropionic acid receptors reveals conservation of structure, function and posttranscriptional processes with mammalian receptors. Mol Brain Res 36:101–113

Paschen W, Blackstone CD, Huganir RL, Ross CA (1994a) Human GluR6 kainate receptor (GRIK2): molecular cloning, expression, polymorphism, and chromosomal assignment. Genomics 20:435–440

Paschen W, Djuricic B (1994) Extent of RNA editing of glutamate receptor subunit GluR5 in different brain regions of the rat. Cell Mol Neurobiol 14:259–270

Paschen W, Djuricic B (1995) Regional differences in the extent of RNA editing of the glutamate receptor subunits GluR2 and GluR6 in rat brain. J Neurosci Meth 56:21–29

Paschen W, Hedreen JC, Ross CA (1994b) RNA editing of the glutamate receptor subunits GluR2 and GluR6 in human brain tissue. J Neurochem 63:1596–1602

Pellicena-Palle A, Salz HK (1995) The putative *Drosophila* NMDARa1 gene is located on the 2nd chromosome and is ubiquitously expressed in embryogenesis. Biochim Biophys Acta 1261:301–303

Petersen SA, Fetter RD, Noordermeer JN, Goodman CS, DiAntonio A (1997) Genetic analysis of glutamate receptors in *Drosophila* reveals a retrograde signal regulating presynaptic transmitter release. Neuron 19:1237–1248

Petralia RS, Wenthold RJ (1992) Light and electron immunocytochemical localization of AMPA-selective glutamate receptors in the rat brain. J Comp Neurol 318:329–354

Petrou S, Ordway RW, Singer JJ, Walsh JV (1993) A putative fatty acid binding domain of the NMDA receptor. Trends Biochem Sci 18:41–42

Pfeiffer F, Graham D, Betz H (1982) Purification by affinity chromatography of the glycine receptor of rat spinal cord. J Biol Chem 257:9389–9393

Pickering DS, Taverna FA, Salter MW, Hampson DR (1995) Palmitoylation of the GluR6 kainate receptor. Proc Natl Acad Sci (USA) 92:12090–12094

Pin J-P, Gomeza J, Prezeau L, Joly C, Bockaert J (1996) The metabotropic glutamate receptors: differences and similarities with the other G-protein coupled receptors. In: Schwartz TW, Hjorth SA, Sandholm-Kastrup J (eds) Structure and function of 7TM receptors, vol 39. Munksgaard, Copenhagen, pp 343–354

Pin JP, Bockaert J (1995) Get receptive to metabotropic glutamate receptors. Curr Opin Neurobiol 5:342–349

Pin JP, Duvoisin R (1995) The metabotropic glutamate receptors: Structure and functions. Neuropharmacology 34:1–26

Planells-Cases R, Sun W, Ferrer-Montiel AV, Montal M (1993) Molecular cloning, functional expression, and pharmacological characterization of an N-methyl-D-aspartate receptor subunit from human brain. Proc Natl Acad Sci (USA) 90: 5057–5061

Ponomareva LV, McMahon DG (1996) GenBank U44736. Unpublished sequence

Potier M-C, Dutriaux A, Lambolez B, Bochet P, Rossier J (1993) Assignment of the human glutamate receptor gene GluR5 to 21q22 by screening a chromosome-21 YAC library. Genomics 15:696–697

Potier M-C, Spillantini MG, Carter NP (1992) The human glutamate receptor cDNA GluR1: cloning, sequencing, expression and localization to chromosome 5. DNA Seq 2:211–218

Pravenec M, Kren V, Hope M, Wang JM, Lezin ES (1998) Linkage mapping of the interleukin-1-beta converting-enzyme (il1bc) and the glutamate receptor subunit KA1 (grik4) genes to rat chromosome-8. Folia Biologica 44:107–109

Premkumar L, Erreger K, Auerbach A (1997) NMDA receptor stoichiometry determined by subconductance pattern analysis. Biophys J 72:A334

Premkumar LS, Auerbach A (1997) Stoichiometry of recombinant N-methyl-D-aspartate receptor channels inferred from single channel current patterns. J Gen Physiol 110:485–502

Puchalski RB, Louis J-C, Brose N, Traynelis SF, Egebjerg J, Kukekov V, Wenthold RJ, Rogers SW, Lin F, Moran T, Morrison JH, Heinemann SF (1994) Selective RNA editing and subunit assembly of native glutamate receptors. Neuron 13:131–147

Puckett C, Gomez CM, Korenberg JR, Tung H, Meier TJ, Chen XN, Hood L (1991) Molecular cloning and chromosomal localization of one of the human glutamate receptor genes. Proc Natl Acad Sci (USA) 88:7557–7561

Puranam RS, Eubanks JH, Heinemann SF, McNamara JO (1993) Chromosomal localization of gene for human glutamate receptor subunit 7. Somat Cell Mol Genet 19:581–588

Quiocho FA (1990) Atomic structures of periplasmic binding proteins and the high-affinity active transport systems in bacteria. Philos Trans R Soc Lond (B) 326: 341–351

Raftery MA, Hunkapiller M, Strader CD, Hood LE (1980) Acetylcholine receptor: complex of homologous subunits. Science 208:1454–1457

Rampersad V, Elliott CE, Nutt SL, Foldes RL, Kamboj RK (1994) Human glutamate receptor hGluR3 flip and flop isoforms: cloning and sequencing of the cDNAs and primary structure of the proteins. Biochim Biophys Acta 1219:563–566

Ravindranathan A, Parks TN, Rao MS (1997) New isoforms of the chick glutamate receptor subunit GluR4: molecular cloning, regional expression and developmental analysis. Mol Brain Res 50:143–153

Raymond LA, Blackstone CD, Huganir RL (1993) Phosphorylation and modulation of recombinant GluR6 glutamate receptors by cAMP-dependent protein kinase. Nature 361:637–641

Riley BP, Tahir E, Rajagopalan S, Mogudi-Carter M, Faure S, Weissenbach J, Jenkins T, Williamson R (1997) A linkage study of the N-methyl-D-aspartate receptor subunit gene loci and schizophrenia in Southern African Bantu-speaking families. Psychiatric Genet 7:57–74

Ripellino JA, Neve RL, Howe JR (1998) Expression and heteromeric interactions of non-N-methyl-D-aspartate glutamate receptor subunits in the developing and adult cerebellum. Neuroscience 82:485–497

Roche KW, O'Brien RJ, Mammen AL, Bernhardt J, Huganir RL (1996) Characterization of multiple phosphorylation sites on the AMPA receptor GluR1 subunit. Neuron 16:1179–1188

Rock DM, Macdonald RL (1995) Polyamine regulation of N-methyl-D-aspartate receptor channels. Annu Rev Pharmacol Toxicol 463–482

Rosenmund C, Stern-Bach Y, Stevens CF (1998) The tetrameric structure of a glutamate receptor channel. Science 280:1596–1599

Rubinsztein DC, Leggo J, Chiano M, Dodge A, Norbury G, Rosser E, Craufurd D (1997) Genotypes at the GluR6 kainate receptor locus are associated with variation in the age of onset of Huntington disease. Proc Natl Acad Sci (USA) 94:3872–3876

Sahara Y, Noro N, Lida Y, Soma K, Nakamura Y (1997) Glutamate receptor subunits GluR5 and KA-2 are coexpressed in rat trigeminal ganglion neurons. J Neurosci 17:6611–6620

Sakimura K, Bujo H, Kushiya E, Araki K, Yamazaki M, Yamazaki M, Meguro H, Warashina A, Numa S, Mishina M (1990) Functional expression from cloned cDNAs of glutamate receptor species responsive to kainate and quisqualate. FEBS Lett 272:73–80

Sakimura K, Kutsuwada T, Ito I, Manabe T, Takayama C, Kushiya E, Yagi T, Aizawa S, Inoue Y, Sugiyama H, Mishina M (1995) Reduced hippocampal LTP and spatial learning in mice lacking NMDA receptor $\varepsilon1$ subunit. Nature 373:151–155

Sakimura K, Morita T, Kushiya E, Mishina M (1992) Primary structure and expression of the $\gamma2$ subunit of the glutamate receptor channel selective for kainate. Neuron 8:267–274

Sander T, Hildmann T, Kretz R, Furst R, Sailer U, Bauer G, Schmitz B, Beck-Mannagetta G, Wienker TF, Janz D (1997) Allelic association of juvenile absence epilepsy with a GluR5 kainate receptor gene (*GRIK1*) polymorphism. Am J Medic Genet 74:416–421

Sander T, Janz D, Ramel C, Ross CA, Paschen W, Hildmann T, Wienker TF, Bianchi A, Bauer G, Sailer U, Berek K, Neitzel H, Volz A, Ziegler A, Schmitz B, Beck-Mannagetta G (1995) Refinement of map position of the human GluR6 kainate receptor gene (grik2) and lack of association and linkage with idiopathic generalized epilepsies. Neurology 45:1713–1720

Sasner M, Buonanno A (1996) Distinct N-methyl-D-aspartate receptor 2B subunit gene sequences confer neural and developmental specific expression. J Biol Chem 271:21316–21322

Schiffer HH, Swanson GT, Heinemann SF (1997) R at GluR7 and a carboxy-terminal splice variant, GluR7b, are functional kainate receptor subunits with a low sensitivity to glutamate. Neuron 19:1141–1146

Schoepp D, Bockaert J, Sladeczek F (1990) Pharmacological and functional characteristics of metabotropic excitatory amino acid receptors. Trends Pharmacol Sci 11:508–515

Schoepp DD (1994) Novel functions for subtypes of metabotropic glutamate receptors. Neurochem Int 24:439–449

Schoepp DD, Conn PJ (1993) Metabotropic glutamate receptors in brain function and pathology. Trends Pharmacol Sci 14:13–20

Schofield PR, Darlison MG, Fujita N, Burt DR, Stephenson FA, Rodriguez H, Rhee LM, Ramachandran J, Reale V, Glencorse TA, Seeburg PH, Barnard EA (1987) Sequence and functional expression of the $GABA_A$ receptor shows a ligand-gated receptor super-family. Nature 328:221-

Schuster CM, Ultsch A, Schloss P, Cox JA, Schmitt B, Betz H (1991) Molecular cloning of an invertebrate glutamate receptor subunit expressed in *Drosophila* muscle. Science 254:112–114

Seeburg PH (1993) The TiPS/TINS lecture: The molecular biology of mammalian glutamate receptor channels. Trends Pharmacol Sci 14:297–303

Seeburg PH (1996) The role of RNA editing in controlling glutamate receptor channel properties. J Neurochem 66:1–5

Sheng M, Cummings J, Roldan LA, Jan YN, Jan LY (1994) Changing subunit composition of heteromeric NMDA receptors during development of rat cortex. Nature 368:144–147

Smirnova T, Laroche S, Errington ML, Hicks AA, Bliss TVP, Mallet J (1993a) Transsynaptic expression of a presynaptic glutamate receptor during hippocampal long-term potentiation. Science 262:433–436

Smirnova T, Stinnakre J, Mallet J (1993b) Characterization of a presynaptic glutamate receptor. Science 262:430–433

Soloviev MM, Barnard EA (1997) *Xenopus* oocytes express a unitary glutamate receptor endogenously. J Mol Biol 273:14–18

Soloviev MM, Brierley MJ, Shao ZY, Mellor IR, Volkova TM, Kamboj R, Ishimaru H, Sudan H, Harris J, Foldes RL, Grishin EV, Usherwood PNR, Barnard EA (1996) Functional expression of a recombinant unitary glutamate receptor from *Xenopus*, which contains N-methyl-D-aspartate (NMDA) and non-NMDA receptor subunits. J Biol Chem 271:32572–32579

Soloviev MM, Abutidze K, Mellor I, Streit P, Grishin EV, Usherwood PN, Barnard EA (1998) Plasticity of agonist binding sites in hetero-oligomers of the unitary glutamate receptor subunit XenU1. J Neurochem 71:991–1001

Sommer B, Burnashev N, Verdoorn TA, Keinänen K, Sakmann B, Seeburg PH (1992) A glutamate receptor channel with high affinity for domoate and kainate. EMBO J 11:1651–1656

Sommer B, Keinänen K, Verdoorn TA, Wisden W, Burnashev N, Herb A, Köhler M, Takagi T, Sakmann B, Seeburg PH (1990) Flip and flop: a cell-specific functional switch in glutamate-operated channels of the CNS. Science 249:1580–1585

Sommer B, Köhler M, Sprengel R, Seeburg PH (1991) RNA editing in brain controls a determinant of ion flow in glutamate-gated channels. Cell 67:11–19

Sprengel R, Seeburg PH (1995) Ionotropic glutamate receptors. In: Ligand- and voltage-gated ion channels, vol 2. CRC Press, Boca Raton, pp 213–263

Sprengel R, Suchanek B, Amico C, Brusa R, Burnashev N, Rozov A, Hvalby O, Jensen V, Paulsen O, Andersen P, Kim JJ, Thompson RF, Sun W, Webster LC, Grant SGN, Eilers J, Konnerth A, Li J, McNamara JO, Seeburg PH (1998) Importance of the intracellular domain of NR2 subunits for NMDA receptor function in vivo. Cell 92:279–289

Stern-Bach Y, Bettler B, Hartley M, Sheppard PO, O'Hara PJ, Heinemann SF (1994) Agonist selectivity of glutamate receptors is specified by two domains structurally related to bacterial amino acid-binding proteins. Neuron 13:1345–1357

Stern-Bach Y, Russo S, Neuman M, Rosenmund C (1998) A point mutation in the glutamate binding site blocks desensitization of AMPA receptors. Neuron 21:907–918

Stühmer T, Amar M, Harvey RJ, Bermudez I, van Minnen J, Darlison MG (1996) Structure and pharmacological properties of a molluscan glutamate-gated cation channel and its likely role in feeding behavior. J Neurosci 16:2869–2880

Suchanek B, Seeburg PH, Sprengel R (1995) Gene structure of the murine N-methyl-D-aspartate receptor subunit NR2C. J Biol Chem 270:41–44

Suchanek B, Seeburg PH, Sprengel R (1997) Tissue-specific control regions of the N-methyl-D-aspartate receptor subunit NR2C promoter. Biological Chemistry 378: 929–934

Sucher NJ, Akbarian S, Chi CL, Leclerc CL, Awobuluyi M, Deitcher DL, Wu MK, Yuan JP, Jones EG, Lipton SA (1995) Developmental and regional expression pattern of a novel NMDA receptor-like subunit (NMDAR-L) in the rodent brain. J Neurosci 15:6509–6520

Sucher NJ, Awobuluyi M, Choi YB, Lipton SA (1996) NMDA receptors: from genes to channels. Trends Pharmacol Sci 17:348–355

Sugihara H, Moriyoshi K, Ishii T, Masu M, Nakanishi S (1992) Structures and properties of 7 isoforms of the NMDA receptor generated by alternative splicing. Biochem Biophys Res Comm 185:826–832

Sullivan JM, Traynelis SF, Chen HSV, Escobar W, Heinemann SF, Lipton SA (1994) Identification of 2 cysteine residues that are required for redox modulation of the NMDA subtype of glutamate receptor. Neuron 13:929–936

Sulston J, Du Z, Thomas K, Wilson R, Hillier L, Staden R, Halloran N, Green P, Thierry-Mieg J, Qiu L, Dear S, Coulson A, Craxton M, Durbin R, Berks M, Metzstein M, Hawkins T, Ainscough R, Waterston R (1992) The *C. elegans* genome sequencing project: a beginning. Nature 356:37–41

Sumikawa K, Houghton J, Smith JG, Bell L, Richards BM, Barnard EA (1982) The molecular cloning and characterization of cDNA coding for the α subunit of the acetylcholine receptor. Nucleic Acids Res 10:5809–5822

Sun W, Ferrer-Montiel AV, Montal M (1994) Primary structure and functional expression of the AMPA/kainate receptor subunit 2 from human brain. Neuroreport 5:441–444

Sun W, Ferrer-Montiel AV, Schinder AF, McPherson JP, Evans GA, Montal M (1992) Molecular cloning, chromosomal mapping, and functional expression of human brain glutamate receptors. Proc Natl Acad Sci (USA) 89:1443–1447

Sutcliffe MJ, Wo ZG, Oswald RE (1996) Three-dimensional models of non-NMDA glutamate receptors. Biophys J 70:1575–1589

Sutcliffe MJ, Smeeton AH, Wo ZG, Oswald RE (1998) Three-dimensional models of glutamate receptors. Biochem Soc Trans 26:450–458

Swafford DL, Olsen GJ (1990) Phylogeny reconstruction. In: Hillis DM, Moritz C (eds) Molecular systematics. Sinauer, Sunderland, MA, pp 411–501

Swanson GT, Gereau RW, Green T, Heinemann SF (1997) Identification of amino acid residues that control functional behavior in GluR5 and GluR6 kainate receptors. Neuron 19:913–926

Szpirer C, Molne M, Antonacci R, Jenkins NA, Finelli P, Szpirer J, Riviere M, Rocchi M, Gilbert DJ, Copeland NG, Gallo V (1994) The genes encoding the glutamate receptor subunits KA1 and KA2 (grik4 and grik5) are located on separate chromosomes in human, mouse, and rat. Proc Natl Acad Sci (USA) 91:11849–11853

Takahashi T, Feldmeyer D, Suzuki N, Onodera K, Cull-Candy SG, Sakimura K, Mishina M (1996) Functional correlation of NMDA receptor ε subunits expression with the properties of single-channel and synaptic currents in the developing cerebellum. J Neurosci 16:4376–4382

Takano H, Onodera O, Tanaka H, Mori H, Sakimura K, Hori T, Kobayashi H, Mishina M, Tsuji S (1993) Chromosomal localization of the $\varepsilon 1$, $\varepsilon 3$, and $\zeta 1$ subunit genes of the human NMDA receptor channel. Biochem Biophys Res Comm 197:922–926

Tokita Y, Bessho Y, Masu M, Nakamura K, Nakao K, Katsuki M, Nakanishi S (1996) Characterization of excitatory amino acid neurotoxicity in N-methyl-D-aspartate receptor-deficient mouse cortical neuronal cells. Eur J Neurosci 8:69–78

Tonegawa S, Tsien JZ, McHugh TJ, Huerta P, Blum KI, Wilson MA (1996) Hippocampal CA1-region-restricted knockout of NMDAR1 gene disrupts synaptic plasticity, place fields, and spatial learning. Cold Spring Harbor Symp Quant Biol 61:225–238

Treadaway J, Zuo J (1998) Mapping of the mouse glutamate receptor δ1 subunit (grid1) to chromosome 14. Genomics 54:359–360

Traynelis SF, Burgess MF, Zheng F, Lyuboslavsky P, Powers JL (1998) Control of voltage-independent zinc inhibition of NMDA receptors by the NR1 subunit. J Neurosci 18:6163–6175

Traynelis SF, Hartley M, Heinemann SF (1995) Control of proton sensitivity of the NMDA receptor by RNA splicing and polyamines. Science 268:873–876

Tsien JZ, Huerta PT, Tonegawa S (1996) The essential role of hippocampal CA1NMDA receptor-dependent synaptic plasticity in spatial memory. Cell 87:1327–1338

Tzartos SJ, Lindstrom JM (1980) Monoclonal antibodies used to probe acetylcholine receptor structure: localization of the main immunogenic region and detection of similarities between subunits. Proc Natl Acad Sci (USA) 77:755–759

Uchino S, Sakimura K, Nagahari K, Mishina M (1992) Mutations in a putative agonist binding region of the AMPA-selective glutamate receptor channel. FEBS Lett 308:253–257

Ueda H (1997) Multiple forms of AMPA-type glutamate receptor messenger RNA phenotypes in goldfish retina and tectum. Gen Pharmacol 29:575–581

Ueda H, Hieber V (1995) Down-regulation of AMPA-type glutamate receptor gene expression during goldfish optic nerve regeneration. Mol Brain Res 32:151–155

Ultsch A, Schuster CM, Laube B, Betz H, Schmitt B (1993) Glutamate receptors of *Drosophila melanogaster*: primary structure of a putative NMDA receptor protein expressed in the head of the adult fly. FEBS Lett 324:171–177

Ultsch A, Schuster CM, Laube B, Schloss P, Schmitt B, Betz H (1992) Glutamate receptors of *Drosophila* melanogaster: Cloning of a kainate-selective subunit expressed in the central nervous system. Proc Natl Acad Sci (USA) 89:10484–10488

Usherwood PNR (1994) Insect glutamate receptors. Adv Insect Physiol 24:309–341

Usherwood PNR, Grundfest H (1965) Peripheral inhibition in skeletal muscle of insects. J Neurophysiol 28:497–518

Usherwood PNR, Mellor I, Breedon L, Harvey RJ, Barnard EA, Darlison MG (1993) Channels formed by M2 peptides of a putative glutamate receptor subunit of locust. In: Pichon Y (ed) Comparative Molecular Neurobiology, vol 63. Birkhauser Verlag, Basel, pp 241–249

Varney MA, Rao SP, Jachec C, Deal C, Hess SD, Daggett LP, Lin FF, Johnson EC, Velicelebi G (1998) Pharmacological characterization of the human ionotropic glutamate receptor subtype GluR3 stably expressed in mammalian cells. J Pharmacol Exp Ther 285:358–370

Verdoorn TA, Burnashev N, Monyer H, Seeburg PH, Sakmann B (1991) Structural determinants of ion flow through recombinant glutamate receptor channels. Science 252:1715–1718

Villarroel A, Burnashev N, Sakmann B (1995) Dimensions of the narrow portion of a recombinant NMDA receptor channel. Biophys J 68:866–875

Villmann C, Bull L, Hollmann M (1997) Kainate binding proteins possess functional ion channel domains. J Neurosci 17:7634–7643

Villmann C, Strutz N, Morth T, Hollmann M (1999) Investigation by ion channel domain transplantation of rat glutamate receptor subunits, orphan receptors, and a putative NMDA receptor subunit. Eur J Neurosci 11:1765–1778

Wada K, Dechesne CJ, Shimasaki S, King RG, Kusano K, Buonanno A, Hampson DR, Banner C, Wenthold RJ, Nakatani Y (1989) Sequence and expression of a frog brain complementary DNA encoding a kainate-binding protein. Nature 342:684–689

Wafford KA, Bain CJ, Le Bourdelles B, Whiting PJ, Kemp JA (1993) Preferential co-assembly of recombinant NMDA receptors composed of three different subunits. Neuroreport 4:1347–1349

Wang JKT, Thukral V (1996) Presynaptic NMDA receptors display physiological characteristics of homomeric complexes of NR1 subunits that contain the exon 5 insert in the N-terminal domain. J Neurochem 66:865–868

Wang L-Y, Taverna FA, Huang X-P, MacDonald JF, Hampson DR (1993) Phosphorylation and modulation of a kainate receptor (GluR6) by cAMP-dependent protein kinase. Science 259:1173–1175

Watkins JC, Evans RH (1981) Excitatory amino acid transmitters. Annu Rev Pharmacol Toxicol 21:165–204

Wenthold RJ, Petralia RS, Blahos J, Niedzielski AS (1996) Evidence for multiple AMPA receptor complexes in hippocampal CA1/CA2 neurons. J Neurosci 16:1982–1989

Wenthold RJ, Trumpy VA, Zhu WS, Petralia RS (1994) Biochemical and assembly properties of GluR6 and KA2, 2 members of the kainate receptor family, determined with subunit specific antibodies. J Biol Chem 269:1332–1339

Wenthold RJ, Yokotani N, Doi K, Wada K (1992) Immunochemical characterization of the non-NMDA glutamate receptor using subunit specific antibodies: evidence for a hetero-oligomeric structure in rat brain. J Biol Chem 267:501–507

Werner P, Voigt M, Keinänen K, Wisden W, Seeburg PH (1991) Cloning of a putative high-affinity kainate receptor expressed predominantly in hippocampal CA3 cells. Nature 351:742–744

Wilson R, Ainscough R, Anderson K, Baynes C, Berks M, Bonfield J, Burton J, Connell M, Copsey T, Cooper J, Coulson A, Craxton M, Dear S, Du Z, Durbin R, Favello A, Fulton L, Gardner A, Green P, Hawkins T, Hillier L, Jier M, Johnston L, Jones M, Kershaw J, Kirsten J, Laister N, Latreille P, Lightning J, Lloyd C, McMurray A, Mortimore B, O'Callaghan M, Parsons J, Percy C, Rifken L, Roopra A, Saunders D, Shownkeen R, Smaldon N, Smith A, Sonnhammer E, Staden R, Sulston J, Thierry-Mieg J, Thomas K, Vaudin M, Vaughan K, Waterston R, Watson A, Weinstock L, Wilkinson-Sproat J, Wohldman P (1994) 2.2 Mb of contiguous nucleotide sequence from chromosome III of C. elegans. Nature 368:32–38

Wo ZG, Bian ZC, Oswald RE (1995) Asn-265 of frog kainate binding protein is a functional glycosylation site: implications for the transmembrane topology of glutamate receptors. FEBS Lett 368:230–234

Wo ZG, Oswald RE (1994) Transmembrane topology of two kainate receptor subunits revealed by N-glycosylation. Proc Natl Acad Sci (USA) 91:7154–7158

Wo ZG, Oswald RE (1995a) Unraveling the modular design of glutamate-gated ion channels. Trends Neurosci 18:161–168

Wo ZG, Oswald RE (1996) Ligand-binding characteristics and related structural features of the expressed goldfish kainate receptors: identification of a conserved disulfide bond and 3 residues important for ligand binding. Mol Pharmacol 50:770–780

Wo ZGL, Oswald RE (1995b) A topological analysis of goldfish kainate receptors predicts 3 transmembrane segments. J Biol Chem 270:2000–2009

Wolstenholme AJ (1997) Glutamate-gated Cl^- channels in *Caenorhabditis elegans* and parasitic nematodes. Biochem Soc Trans 25:830–834

Won M, Moon K-M, Lee C-E, Yoo HS (1995) Human NMDA receptor glutamate binding chain (hnrgw). GenBank U44954. Unpublished sequence

Wood MW, VanDongen HMA, VanDongen AMJ (1995) Structural conservation of ion conduction pathways in K channels and glutamate receptors. Proc Natl Acad Sci (USA) 92:4882–4886

Wroblewski JT, Danysz W (1989) Modulation of glutamate receptors: molecular mechanisms and functional implications. Annu Rev Pharmacol Toxicol 29:441–474

Wu TY, Liu CI, Chang YC (1996a) A study of the oligomeric state of the α-amino-3-hydroxy-5-methyl-4-isoxazolepropionic acid-preferring glutamate receptors in the synaptic junctions of porcine brain. Biochem J 319:731–739

Wu YM, Kung SS, Chen JC, Chow WY (1996b) Molecular analysis of cDNA molecules encoding glutamate receptor subunits, fGluR1α and fGluR1β, of *Oreochromis* sp. DNA Cell Biol 15:717–725

Yakel JL, Vissavajjhala P, Derkach VA, Brickey DA, Soderling TR (1995) Identification of a Ca2+ calmodulin-dependent protein kinase II regulatory phosphorylation site in non-N-methyl-D-aspartate glutamate receptors. Proc Natl Acad Sci (USA) 92: 1376–1380

Yamazaki M, Araki K, Shibata A, Mishina M (1992a) Molecular cloning of a cDNA encoding a novel member of the mouse glutamate receptor channel family. Biochem Biophys Res Comm 183:886–892

Yamazaki M, Mori H, Araki K, Mori KJ, Mishina M (1992b) Cloning, expression and modulation of a mouse NMDA receptor subunit. FEBS Lett 300:39–45

Yellen G, Jurman ME, Abramson T, MacKinnon R (1991) Mutations affecting internal TEA blockade identify the probable pore-forming region of a K^+ channel. Science 251:939–942

Yool AJ, Schwarz TL (1991) Alteration of ionic selectivity of a K+ channel by mutation of the H5 region. Nature 349:700–704

Zimmer M, Fink TM, Franke Y, Lichter P, Spiess J (1995) Cloning and structure of the gene encoding the human N-methyl-D-aspartate receptor (NMDAR-1). Gene 159:219–223

Zuo J, Dejager PL, Takahashi KA, Jiang WN, Linden DJ, Heintz N (1997) Neurodegeneration in *lurcher* mice caused by mutation in $\delta 2$ glutamate receptor gene. Nature 388:769–773

Zuo J, Wollmuth L, Beck C, Seeburg P, Heintz N, Kuner T (1998) Glutamate receptor delta 2 subunit (grid2) with lurcher mutation forms an AMPA-like channel. Soc Neurosci Abstr 24:841

CHAPTER 2

Phosphorylation of Glutamate Receptors

H.-K. LEE and R.L. HUGANIR

A. Introduction

Excitatory synaptic transmission between neurons in the central nervous system is mediated mainly by the neurotransmitter glutamate. The glutamate released from the presynaptic neuron diffuses across the synaptic cleft and activates glutamate receptors to complete the process of synaptic transmission. Glutamate receptors can be grouped into two broad categories depending on the signal transduction mechanism. Ionotropic glutamate receptors (iGluR) transduce glutamate binding by opening ion channels permeable to cations, while metabotropic glutamate receptors (mGluRs) activate G proteins, which directly or indirectly regulate ion channels and enzymes. Ionotropic glutamate receptors can be further subdivided into three groups depending on their agonist preferences and biophysical properties. *N*-methyl-D-aspartate (NMDA) receptors preferentially bind NMDA, while α-amino-3-hydroxy-5-methyl-4-isoxazole proprionate (AMPA) receptors and kainate (KA) receptors show high affinity for AMPA and KA, respectively. The function of glutamate receptors can be modulated by various mechanisms, however protein phosphorylation has been shown to be critical in the control of glutamate receptor function.

Protein phosphorylation is a ubiquitous post-translational mechanism for the control of protein function. Phosphorylation occurs when a phosphate group is catalytically transferred from adenosine triphosphate (ATP) to a substrate protein. Attachment of the highly negatively charged phosphate group to a protein can transform protein structure and subsequently alter its function. Proteins undergo phosphorylation at serine, threonine, or tyrosine residues. The phosphorylation reaction is mediated by various protein kinases, which can be grouped into two distinct groups, serine/threonine protein kinases and tyrosine protein kinases, depending on which amino acid residues they phosphorylate. Protein kinases show high substrate specificity, and consensus phosphorylation sites for individual kinases have been characterized (KEMP and PEARSON 1990). In addition to having specific sequences for substrate recognition, the subcellular localization of protein kinases by anchoring proteins (reviewed in MOCHLY-ROSEN 1995) can affect which substrates are phosphorylated upon activation of each protein kinase. Dephosphorylation of substrate proteins is mediated by various protein phosphatases. Substrate specificity for protein phosphatases is not thought to be as stringent as for protein kinases (Shinolikar and Ingebritsen 1984; Cohen 1989). However, recent evidence suggests that protein phosphatases, like protein kinases, may also be localized to specific subcellular compartments (Coghlan et al. 1995; Strack et al. 1997c; Strack et al. 1998). It is now well established that both protein kinases and protein phosphatases are regulated by various signal transduction mechanisms. Because most protein kinases can undergo autophosphorylation and are dephosphorylated by protein phosphatases, the net effect of activating a specific signal transduction pathway will depend on multiple factors.

All three types of glutamate receptor can undergo phosphorylation. However, this review will focus on the phosphorylation of AMPA receptors. This review will cover the changes in AMPA receptor function by activation of protein kinases and protein phosphatases, and the identification of specific phosphorylation sites, using the AMPA receptor GluR1 subunit as an example. In addition, the possible role of AMPA receptor phosphorylation in synaptic plasticity will be discussed.

B. Phosphorylation of AMPA Receptors

AMPA-type glutamate receptors are the major mediators of fast excitatory synaptic transmission in the central nervous system. Hence, changes in AMPA receptor function can alter synaptic strength. There are several ways to alter AMPA receptor function. One is by changing subunit composition of the heteromeric receptors. Currently four different subunits, GluR1–4 (or GluRA-D), have been cloned and characterized (Hollmann et al. 1989; Boulter et al. 1990; Keinanen et al. 1990). The four subunits show differential distribution in the brain (Boulter et al. 1990; Keinanen et al. 1990; Blackstone et al. 1992; Petralia and Wenthold 1992; Martin et al. 1993) and form heteromeric receptors with different functional properties (Nakanishi et al. 1990; Wenthold et al. 1992; Wenthold et al. 1996). The developmental profile and regional expression of each subunit show distinct patterns (Pellegrini-Giampietro et al. 1992; Durand and Zukin 1993; Martin et al. 1993; Martin et al. 1998), suggesting a differential regulation of AMPA receptor properties in different areas of the brain during development. One example of how subunit composition can affect receptor function is the observation that AMPA receptors lacking the GluR2 subunit are calcium permeable (Hollmann et al. 1991; Keller et al. 1992). Given that calcium is an important second messenger involved in a wide variety of cellular functions, AMPA receptors lacking GluR2 can transduce signals in a different manner than those containing the GluR2 subunit. In addition, each subunit can undergo alternative splicing to produce two major isoforms, flip and flop, that show different desensitization characteristics (Sommer et al. 1990; Monyer et al. 1991; Mosbacher et al. 1994; Partin et al. 1994; Partin et al. 1995). Thus, the combinatorial assembly of different subunits and their isoforms can generate extensive diversity in AMPA receptor functions.

Phosphorylation of AMPA receptors is an additional mechanism for the control of receptor function. AMPA receptors are basally phosphorylated in cultured neurons (BLACKSTONE et al. 1994; TAN et al. 1994) and in vivo (MAMMEN et al. 1997; KAMEYAMA et al., 1998; LEE et al. 1998). In addition, changes in protein kinase or protein phosphatase activity can affect AMPA receptor currents (GREENGARD et al. 1991; WANG et al. 1991; KELLER et al. 1992; ROSENMUND et al. 1994; WANG et al. 1994; ROCHE et al. 1996; BARRIA et al. 1997a) and AMPA receptor mediated excitatory postsynaptic potentials (EPSPs) (HU et al. 1987; PETTIT et al. 1994; WYLLIE and NICOLL 1994; LLEDO et al. 1995; WANG and KELLY 1995; KAMEYAMA et al. 1998; LEE et al. 1998). Changes in the phosphorylation state of AMPA receptors can alter receptor function more rapidly than changes in subunit composition, and have an added advantage that they are readily reversible. Therefore, it is likely that phosphorylation reactions may mediate processes that require fast reversible changes in AMPA receptor function, while changes in subunit or isoform composition may mediate processes that require a lasting change in AMPA receptor mediated synaptic transmission. The phosphorylation of AMPA receptors and its possible role in various cellular processes, such as long-term synaptic plasticity, has been discussed extensively (RAYMOND et al. 1993; ROCHE et al. 1994; SODERLING et al. 1994).

I. Modulation of AMPA Receptor Function by Protein Kinases and Protein Phosphatases

It is known from numerous studies that AMPA receptor mediated synaptic current is modulated by regulating the activity of protein kinases or protein phosphatases (KNAPP and DOWLING 1987; GREENGARD et al. 1991; WANG et al. 1991; KELLER et al. 1992; ROSENMUND et al. 1994; WANG et al. 1994; ROCHE et al. 1996; BARRIA et al. 1997a). The importance of phosphorylation in maintaining AMPA receptor function was first noticed in the context of a phenomenon called "run down". When AMPA receptor mediated currents evoked by agonists are measured in a whole-cell recording configuration, there is a gradual "run down" of current amplitude with time (GREENGARD et al. 1991; WANG et al. 1991; ROSENMUND et al. 1994). The "run down" of AMPA receptor current can be prevented by using nystatin perforated patch recordings (WANG et al. 1991), indicating that it is produced by dialysis of soluble intracellular factors. These intracellular factors seem to be important for maintaining phosphorylation reactions in the cells, given that including ATP regenerating solution, the catalytic subunit of protein kinase A (PKA) or an inhibitor of protein phosphatases in the intracellular patch pipette can prevent the "run down" (GREENGARD et al. 1991; WANG et al. 1991, 1994; ROSENMUND et al. 1994). These results indicate that AMPA receptor mediated currents are maintained by a balance between the activity of endogenous protein kinases and protein phosphatases.

1. Cyclic Adenosine Monophosphate-Dependent Protein Kinase (PKA) and AMPA Receptors

Cyclic adenosine monophosphate (cAMP)-dependent protein kinase or protein kinase A (PKA) is composed of two catalytic subunits (C subunits) and two regulatory subunits (R subunits), and is activated by an increase in intracellular cAMP concentration. The importance of PKA activity in maintaining AMPA receptor function was first suggested based on the observation that intracellular perfusion of the catalytic subunit of PKA can prevent the "run down" of AMPA receptor currents (KNAPP and DOWLING 1987; WANG et al. 1991). In addition, intracellular application of PKA inhibitory peptides can depress kainate- or glutamate-induced currents in cultured hippocampal neurons (GREENGARD et al. 1991; ROSENMUND et al. 1994). Taken together, these results indicate that basal PKA activity is necessary for maintaining the basal functional level of AMPA receptor mediated currents. However, some investigators have found that activation of PKA can also potentiate AMPA receptor current. When activators of PKA are applied under conditions that prevent the "run down", they can potentiate kainate- or glutamate-induced AMPA receptor currents (WANG et al. 1991; GREENGARD et al. 1991). A potential drawback in measuring AMPA receptor currents by the application of agonists is that both extrasynaptic and synaptic receptors may be activated. For example, extrasynaptic receptors may not be basally phosphorylated and thus PKA can potentiate their function. In contrast, synaptic receptors may be highly phosphorylated basally by PKA and, thus, activators of PKA do not potentiate receptor function. In support of this idea, the PKA activator, forskolin produces only a small increase in the amplitude of synaptic AMPA receptor mediated responses measured as spontaneous excitatory postsynaptic currents (EPSCs) (ROSENMUND et al. 1994; but see GREENGARD et al. 1991). These results are consistent with data demonstrating that PKA is enriched at excitatory synapses.

It is known that PKAs are anchored to postsynaptic densities by A-kinase-anchoring-proteins (AKAPs) (CARR et al. 1992), which can bind the regulatory subunit of PKA with high affinity (CARR et al. 1991). It has been demonstrated that disrupting AKAP binding to PKA by an inhibitory peptide can cause a "run down" of kainate-induced AMPA receptor currents in cultured hippocampal neurons and decrease spontaneous EPSC amplitude (ROSENMUND et al. 1994). These results indicate that endogenous activity of PKA at postsynaptic sites may be necessary for the basal level of synaptic AMPA receptor function. The observation that PKA activation does not result in a large increase in synaptic AMPA receptor function (ROSENMUND et al. 1994) may indicate that the effect of endogenous PKA activity at PSDs on synaptic AMPA receptors is maximal under basal conditions. Evidence supporting this idea has come from recent studies which show that intracellular application of PKA activator does not increase synaptic responses mediated by AMPA receptors (BLITZER et al. 1995; KAMEYAMA et al. 1998), but application of PKI can depress synaptic responses mediated by AMPA receptors (KAMEYAMA et al. 1998).

The effect of PKA in maintaining AMPA receptor function at synapses could occur through direct phosphorylation of the receptors or through indirect regulation of proteins that interact with AMPA receptors. Biochemical characterization of AMPA receptor phosphorylation has provided seemingly contradictory results. On the one hand, activation of PKA can increase phosphorylation of the GluR1 subunit of AMPA receptors in heterologous expression systems (BLACKSTONE et al. 1994; ROCHE et al. 1996; MAMMEN et al. 1997) and in hippocampal slices (MAMMEN et al. 1997; KAMEYAMA et al. 1998; LEE et al. 1998). On the other hand, in cultured hippocampal neurons, PKA does not seem to affect GluR1 phosphorylation as much (MCGLADE-MCCULLOH et al. 1993; TAN et al. 1994). The discrepancy can be reconciled if there are differences in basal phosphorylation of AMPA receptors by PKA in different preparations. As will be discussed later (Sect. BII), there is strong evidence that PKA directly phosphorylates AMPA receptors.

2. Protein Kinase C and AMPA Receptors

Protein kinase C (PKC) is a family of heterogeneous protein kinases activated by Ca^{2+} and/or phospholipids (reviewed in KIKKAWA et al. 1989; HUANG et al. 1991). An interesting property of PKC is that it can be converted to a persistently active form (PKM) by a limited proteolysis that is mediated by a Ca^{2+}-activated protease, calpain (KISHIMOTO et al. 1989; KLANN et al. 1993). In addition, upon activation, PKC can translocate to membranes (KRAFT and ANDERSON 1983; SACKTOR et al. 1993).

It has been shown, using whole-cell recordings in cultured hippocampal neurons, that activation of PKC can modulate AMPA receptor function (WANG et al. 1994). Intracellular perfusion of PKM resulted in an enhancement of kainate-induced currents, likely mediated by native AMPA receptors. In addition, PKM significantly increased the EC50 for kainate, suggesting a reduction in the apparent affinity for the agonist (WANG et al. 1994). PKM increased the decay time constant and, in some cases, the amplitude of AMPA receptor mediated miniature EPSCs (mEPSCs) (WANG et al. 1994). An increase in the decay time constant by PKM can be a reflection of an increase in mean open time of AMPA receptor channels, which could lead to an increase in synaptic responses mediated by AMPA receptors. Indeed, intracellular perfusion of PKC to neurons in hippocampal slices can increase AMPA receptor mediated EPSPs (HU et al. 1987). Whether these changes in AMPA receptor function by PKC are due to a direct phosphorylation of AMPA receptors by PKC or are an indirect modulation of AMPA receptors by some other phosphoprotein is unknown. In addition, there are some doubts as to whether PKC can potentiate AMPA receptor currents. In trigeminal neurons, intracellular perfusion of PKC did not increase kainate-induced currents (CHEN and HUANG 1992), suggesting that the effect of PKC on AMPA receptors may vary, depending on the specific conditions of the experiment.

3. Ca^{2+}/Calmodulin-Dependent Protein Kinase II and AMPA Receptors

Ca^{2+}/calmodulin-dependent protein kinase II (CaMKII) is a member of a family of Ca^{2+}/calmodulin-dependent protein kinases that are activated by an elevation of intracelluar Ca^{2+} (reviewed in BRAUN and SCHULMAN 1995). CaMKII holoenzyme is composed of two major subunits, α (50kDa) and β (60kDa); the α subunit of CaMKII is highly enriched in the postsynaptic densities (PSDs) (KENNEDY et al. 1983). A key feature of CaMKII is that it can become constitutively active upon autophosphorylation on Threonine-286; this can be reversed by protein phosphatase (PP1 or PP2A) activation (reviewed in DUNKLEY 1991). A recent study suggests that the autophosphorylation of CaMKII can be modulated by the frequency and duration of intracellular Ca^{2+} increases (DE KONINCK and SCHULMAN 1998). In addition, autophosphorylated CaMKII can translocate to the PSD (STRACK et al. 1997b), and hence can phosphorylate various proteins in the PSD. These characteristics of CaMKII make this enzyme an important mediator of intracellular Ca^{2+} signaling in the PSD.

Because synaptic AMPA receptors are also present at PSDs, it is likely that their activity will be modulated by CaMKII. Indeed, intracellular application of active CaMKII increases kainate-induced AMPA receptor currents in cultured hippocampal neurons (MCGLADE-MCCULLOH et al. 1993), and in *Xenopus* oocytes (YAKEL et al. 1995) and HEK293 cells (BARRIA et al. 1997a) expressing GluR1. Given that GluR2 expressed in HEK293 cells is not potentiated by CaMKII (BARRIA et al. 1997a), the effect of CaMKII on AMPA receptor currents may be due to direct phosphorylation on the GluR1 subunit.

In addition to affecting agonist-induced AMPA receptor currents, CaMKII can potentiate AMPA receptor mediated synaptic responses. Intracellular application of constitutively active CaMKII (LLEDO et al. 1995) or transfection of postsynaptic cells with constitutively active CaMKII (PETTIT et al. 1994) can potentiate AMPA receptor mediated EPSCs. These results indicate that CaMKII can potentiate synaptic AMPA receptor function.

4. Protein Phosphatases and AMPA Receptors

The reduction in AMPA receptor mediated currents that is observed when protein kinase activity is inhibited (GREENGARD et al. 1991; ROSENMUND et al. 1994) implies that ongoing protein phosphatase activity may counteract protein kinase mediated phosphorylation of AMPA receptors. Supporting this idea, inhibiting protein phosphatase 1 (PP1) and/or protein phosphatase 2A (PP2A) with okadaic acid or microcystin-LR dramatically increased kainate-induced AMPA receptor currents (WANG et al. 1991; MCGLADE-MCCULLOH et al. 1993; Wang et al. 1994). However, whether these protein phosphatases reduced AMPA receptor currents by directly dephosphorylating AMPA receptors or by regulating protein kinases is unknown. Protein phosphatases, especially PP1 and PP2A, can dephosphorylate autophosphorylated protein kinases (Vereb and Gergely 1989; Strack et al. 1997a, b; Ishida et al. 1998; Ricciarelli and Azzi 1998). Further research should be able to distinguish between these possibilities.

II. Mapping Phosphorylation Sites on AMPA Receptors

Although the observation that AMPA receptor currents can be modulated by various kinases or phosphatases indicates the importance of phosphorylation reactions for maintaining AMPA receptor function, it does not prove that direct phosphorylation of AMPA receptors is responsible. To determine if the activity of kinases and phosphatases affect AMPA receptor function by directly phosphorylating or dephosphorylating AMPA receptors, biochemical analysis of AMPA receptor phosphorylation is necessary.

Studies using heterologous expression systems or neuronal cultures demonstrated that AMPA receptor subunits, GluR1, can be phosphorylated on serine and threonine residues by various protein kinases, such as PKA (Blackstone et al. 1994; Roche et al. 1996), PKC (Blackstone et al. 1994; Tan et al. 1994; Roche et al. 1996), and CaMKII (Mcglade-Mcculloh et al. 1993; Tan et al. 1994; Yakel et al. 1995; Barria et al. 1997a; Hayashi et al. 1997; Mammen et al. 1997). Phospho-amino acid analysis of phosphorylated GluR1 subunits show major phosphorylation on serine, a weaker signal on threonine, and little phosphorylation on tyrosine (Blackstone et al. 1994). However, phospho-tyrosine can be increased when GluR1 is co-expressed with Src tyrosine kinase in human embryonic kidney (HEK) 293 cells (Moss et al. 1993). Whether GluR subunits can be phosphorylated on tyrosine residues in vivo is unclear, given that AMPA receptor subunits (GluR1–4) in cortical synaptic membrane preparations do not show tyrosine phosphorylation (Lau and Huganir 1995). This contrasts with the NMDA receptor NR2 subunits, which are highly phosphorylated on tyrosine residues (Moon et al. 1994; Lau and Huganir 1995). At least for the GluR1 subunit of the AMPA receptors, phosphorylation on serine and threonine residues is more prominent and hence, will be discussed exclusively.

The identification of the phosphorylation sites on AMPA receptors by each kinase was hindered initially by the assumption that GluR subunits had the same membrane topology as the nicotinic acetylcholine receptor (nAChR) (Holmann et al. 1989; Betz 1990). The nAChR has a large N-terminal extracellular domain and four transmembrane domains (TMD1–4), and an extracellular C-terminal domain. According to this topology, several putative phosphorylation sites were deduced to be at the major intracellular loop between TMD3 and TMD4 (Keinanen et al. 1990). However, it is now clear from immunocytochemical studies and from identification of glycosylation sites that AMPA receptors have an extracelluar N-terminal region and a cytoplasmic C-terminal domain with four transmembrane segments in between (Molnár et al. 1993; Roche et al. 1993; Tingley et al.,1993; Hollmann et al. 1994; Wo and Oswald 1994; Bennett and Dingledine 1995; Wo and Oswald 1995). In this new model, only three of the four transmembrane domains (TMD1, 3, and 4) span the membrane, while TMD2 is thought to form a loop that acts as a pore region similar to that of K^+ channels (Wo and Oswald 1995; Wood et al. 1995) (Fig. 1). Therefore, putative phosphorylation sites should be on the intracellular loop between TMD1 and TMD2, the intracellular loop between TMD2 and TMD3, or the intracellular C-terminal tail.

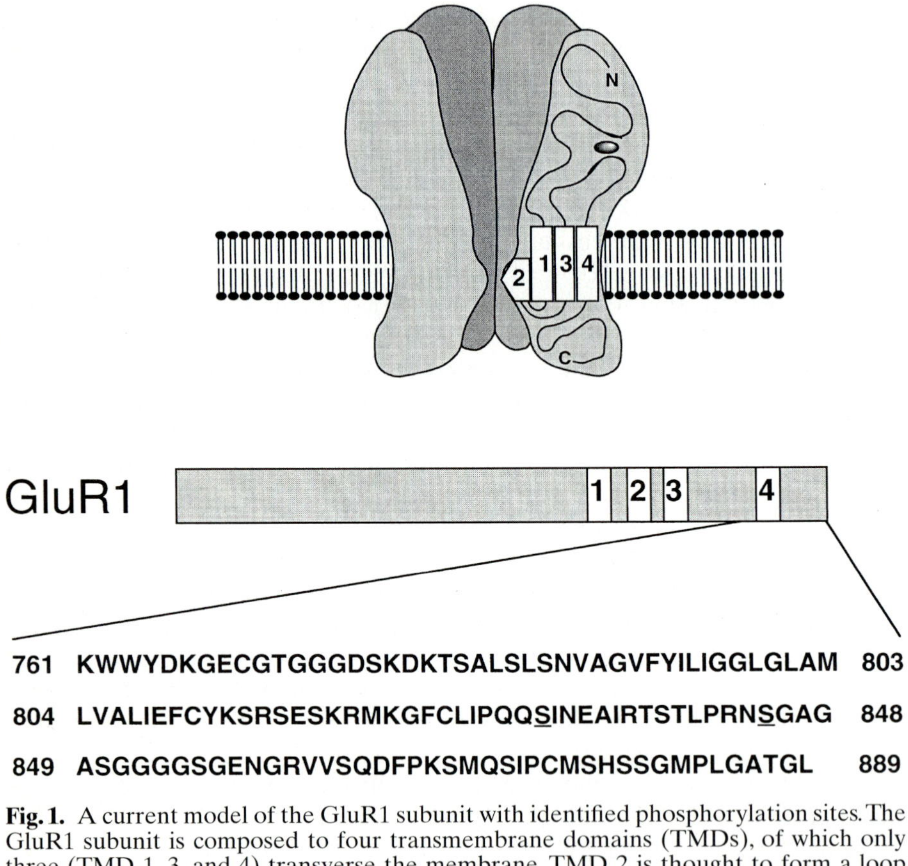

761 KWWYDKGECGTGGGDSKDKTSALSLSNVAGVFYILIGGLGLAM 803

804 LVALIEFCYKSRSESKRMKGFCLIPQQSINEAIRTSTLPRNSGAG 848

849 ASGGGGSGENGRVVSQDFPKSMQSIPCMSHSSGMPLGATGL 889

Fig. 1. A current model of the GluR1 subunit with identified phosphorylation sites. The GluR1 subunit is composed to four transmembrane domains (TMDs), of which only three (TMD 1, 3, and 4) transverse the membrane. TMD 2 is thought to form a loop that lines the pore region, as in K^+ channels. Currently, all the identified phosphorylation sites are mapped on the C-terminal tail, which is located intracellularly. The identified phosphorylation sites, serine-831 and serine-845, are underlined in the sequence. Serine-831 is phosphorylated by protein kinase C (PKC) and Ca^{2+}/calmodulin-dependent protein kinase II (CaMKII), while serine-845 is phosphorylated by protein kinase A (PKA). Phosphorylation at these sites has been shown to increase AMPA receptor mediated currents.

1. Phosphorylation Sites on the GluR1 Subunit

In the GluR1 subunit, the major phosphorylation sites were mapped on the C-terminal intracelluar domain (Roche et al. 1996). This was accomplished by comparing phosphopeptide maps from wild-type GluR1 and GluR2 with a chimeric receptor of GluR2 for which the last 130 amino acids were replaced by those of GluR1. While the phosphopeptide maps of wild-type GluR1 and GluR2 showed distinct patterns, the chimera of GluR2 and GluR1 C-terminal domains produced an identical phosphopeptide map pattern as the wild-type GluR1 (Roche et al. 1996). This suggested that the major phosphorylation sites for GluR1 and GluR2 are on the C-terminal domain (Fig. 1). Phosphorylation sites on the GluR1 C-terminal were further characterized by site-directed mutagenesis (Roche et al. 1996). Using this method, two phosphorylation sites, serine-831 and serine-845, were identified and characterized. On the one hand, phosphorylation at serine-831 can be increased by activation of PKC (Roche et al. 1996; Lee et al. 1998) or CaMKII (Barria et al. 1997a; Mammen et al. 1997) (Fig. 1). On the other hand, activation of cAMP-dependent protein kinase (PKA) was responsible for phosphorylation at serine-845 (Roche et al. 1996; Lee et al. 1998).

Further analysis revealed that the enhancement of AMPA receptor current by PKA (Greengard et al. 1991; Wang et al. 1991) is mediated, at least in part, by phosphorylation of GluR1 at serine-845: PKA potentiated glutamate-induced current was completely abolished by mutating the serine-845 to an alanine (Roche et al. 1996). CaMKII-induced potentiation can be blocked by mutating serine-831 (Barria et al. 1997a), indicating that direct phosphorylation of serine-831 may mediate the potentiation of AMPA receptor currents by CaMKII.

2. Comparison of Phosphorylation of GluR1 In Vitro and In Vivo

In vitro phosphorylation analysis suggests that the sites can be phosphorylated by certain kinases, but does not determine whether the sites are accessible in vivo. Therefore, it is important to confirm the in vitro phosphorylation results in vivo. There are two methods by which to study in vivo phosphorylation. One method uses metabolically labels neurons or brain slices with ^{32}P, and obtains phosphopeptide maps. The other method utilizes phosphorylation site specific antibodies. Each method has its advantages and disadvantages. Metabolic labeling of neurons or brain slices is limited to the comparison of phosphopeptide maps or general levels of phosphorylation change in the whole receptor. In contrast, the use of phosphorylation site specific antibodies allows for the detection of phosphorylation changes at specific sites. Therefore, metabolic labeling is often used to identify sites of phosphorylation, and once phosphorylation sites are identified, phosphorylation site specific antibodies are generated to study these phosphorylation sites in detail.

Phosphopeptide maps of the GluR1 subunit generated from cultured neurons or brain slices reveal patterns that are similar to phosphopeptide maps from in vitro phosphorylation of the GluR1 subunit (BLACKSTONE et al. 1994; ROCHE et al. 1994; BARRIA et al. 1997a; MAMMEN et al. 1997). This confirms, to a first approximation, that GluR1 phosphorylation sites observed in vitro can also be phosphorylated in vivo. Phosphorylation site specific antibodies raised against serine-831 and serine-845 show that, indeed, these two sites are basally phosphorylated in vivo (MAMMEN et al. 1997; KAMEYAMA et al. 1998; LEE et al. 1998). In addition, the phosphorylation at serine-831 in vivo can be altered by manipulations that affect PKC or CaMKII activity (MAMMEN et al. 1997), while serine-845 phosphorylation in vivo is affected only by manipulations that change PKA activity (MAMMEN et al. 1997; KAMEYAMA et al. 1998). However, these in vivo experiments by themselves cannot distinguish between the direct phosphorylation or indirect phosphorylation by each protein kinase. The results from in vivo should be interpreted together with in vitro phosphorylation to determine whether the phosphorylation changes are directly due to the protein kinase of interest. At least for the GluR1 subunit, collective evidence strongly indicates that serine-831 is directly phosphorylated by PKC and CaMKII, while serine-845 is directly phosphorylated by PKA.

C. A Possible Role of AMPA Receptor Phosphorylation in Synaptic Plasticity

Because AMPA receptors mediate the majority of fast excitatory synaptic responses in the brain, their regulation has long been suspected of being important to the control of information flow. In accordance with this idea, any long-term changes in AMPA receptor function could mediate information storage. Long-term potentiation (LTP) and long-term depression (LTD) are well-characterized forms of synaptic plasticity that are thought to be important for information storage in the brain. LTP is a long-lasting increase in synaptic strength that occurs following the application of brief high-frequency stimulation (HFS) to afferents (reviewed in BLISS and COLLINGRIDGE 1993), while LTD is a long-lasting decrease in synaptic strength that occurs following the application of prolonged low-frequency stimulation (LFS) (reviewed in BEAR and ABRAHAM 1996). Synaptic strength is measured as the fast component of EPSPs, which is mediated mainly by AMPA type glutamate receptors.

There are several possible expression mechanisms for LTP (reviewed in MADISON et al. 1991; BLISS and FAZELI 1996) and LTD (reviewed in BEAR and ABRAHAM 1996). On the one hand, a presynaptic expression may be mediated by an increase (BEKKERS and STEVENS 1990) or decrease (BOLSHAKOV and SIEGELBAUM 1994) in glutamate release. On the other hand, a postsynaptic expression could be mediated by changes in glutamate receptors (reviewed in BENKE et al. 1996). There are reports that AMPA receptors may be inserted into synaptic sites after LTP induction (ISAAC et al. 1995; LIAO et al. 1995). In addition, there is evidence suggesting changes in the phosphorylation of

AMPA receptors that may be associated with LTP (BARRIA et al. 1997b) and LTD (LEE et al. 1998; KAMEYAMA et al. 1998).

I. A Possible Model for LTP and LTD Expression by Changes in AMPA Receptor Phosphorylation

Ample evidence corroborates the importance of phosphorylation reactions in LTP and LTD expression (reviewed in BENKE et al. 1996). Recently, a model was proposed to explain LTP and LTD on the basis of changes in AMPA receptor phosphorylation at CaMKII (ser-831) and PKA (ser-845) sites on the GluR1 subunit (KAMEYAMA et al. 1998) (Fig. 2). There is very strong evidence that postsynaptic CaMKII is essential for the induction of LTP (reviewed by LISMAN 1994). As discussed earlier (Sect. B), the AMPA receptor GluR1 subunit is a substrate of CaMKII and phosphorylation of GluR1 by CaMKII can enhance glutamate-evoked currents through AMPA receptors (TAN et al. 1994; YAKEL et al. 1995). This enhancement of AMPA receptor currents by CaMKII is thought to be mediated by phosphorylation at ser-831 on the GluR1 subunit (BARRIA et al. 1997a). In the CA1 region of the hippocampus, LTP-inducing stimulation leads to an increase in phosphorylation of GluR1 by CaMKII (BARRIA et al. 1997b). This indicates that CaMKII phosphorylation of GluR1 represents at least one mechanism for the expression of LTP. In addition, activation of CaMKII can increase synaptic transmission (LLEDO et al. 1995). Therefore, it is suggested that LTP from the baseline state is mediated by CaMKII phosphorylation of the GluR1 subunit of the AMPA receptors (Fig. 2).

Despite this evidence, the role of postsynaptic PKA in the induction and initial expression of LTP is still unclear. The dominant idea about the role that PKA plays in synaptic plasticity suggests that it is involved in the establishment of protein synthesis-dependent, late-phase LTP via regulation of gene expression (FREY et al. 1993; MATTHIES and REYMANN 1993; BOURTCHULADZE et al. 1994; IMPEY et al. 1996). However, PKA inhibitors have been shown to block LTP induction and early expression when certain types of tetanic stimulation are used (BLITZER et al. 1995; THOMAS et al. 1996; BLITZER et al. 1998). This effect was interpreted as being mediated by an indirect activation of protein phosphatases rather than by a direct phosphorylation of substrates by PKA. As discussed in a previous section (Sect. B), PKA phosphorylation of the AMPA receptor GluR1 subunit at ser-845 can increase glutamate-evoked currents in heterologous expression systems (ROCHE et al. 1996). However, when changes in AMPA receptor mediated synaptic responses by PKA activators are monitored in vivo, there are conflicting data regarding the extent to which this effect reflects a postsynaptic modification.

In cultured hippocampal neurons, forskolin has been shown to enhance the amplitude of mEPSCs by increasing the mean open probability of AMPA receptor channels (GREENGARD et al. 1991; WANG et al. 1991); however, in hippocampal slices, forskolin was found to have a negligible postsynaptic effect

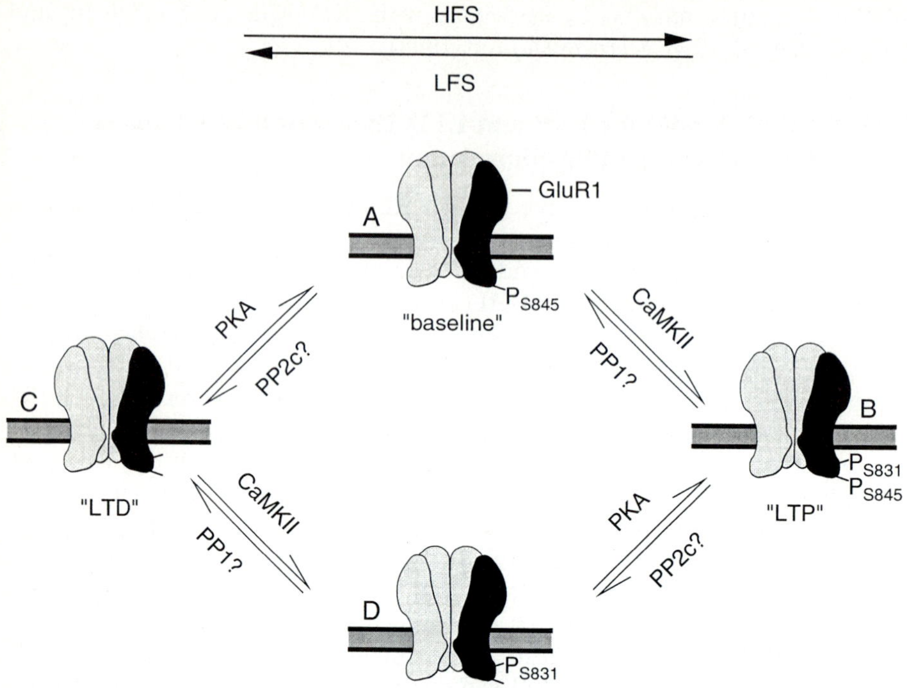

Fig. 2. A possible model for bidirectional synaptic plasticity by changes in GluR1 subunit phosphorylation. The model was designed to take into account the following four observations regarding *N*-methyl-D-aspartate (NMDA) receptor-dependent synaptic plasticity in the CA1 region of the hippocampus. First, from a "baseline" state, synapses can be potentiated in response to high-frequency stimulation (*HFS*) and depressed in response to low-frequency stimulation (*LFS*). Second, long-term potentiation (*LTP*) and long-term depression (*LTD*) can be reversed by LFS and HFS, respectively. Third, the expression of LTP from baseline depends on phosphorylation of a postsynaptic Ca^{2+}/calmodulin-dependent protein kinase II (*CaMKII*) (but not protein kinase A, *PKA*) substrate. Fourth, the expression of LTD from baseline depends on dephosphorylation of a PKA substrate. According to the model, HFS can cause synaptic potentiation in two distinct ways, depending on the initial state of the synapses. For example, from a depressed state, HFS can cause "dedepression" via phosphorylation of a postsynaptic PKA substrate, while from a "baseline" state HFS can cause "potentiation" via phosphorylation of a postsynaptic CaMKII substrate. Similarly, LFS can cause synapses to become weaker in two different ways. For example, from a potentiated state, LFS can cause "depotentiation" via dephosphorylation of a postsynaptic CaMKII substrate, and from a "baseline" state, LFS can cause "depression" via dephosphorylation of a postsynaptic PKA substrate. The mechanism proposed to implement the model is based on observed changes in phosphorylation of the GluR1 subunit of postsynaptic AMPA receptors following induction of LTD and LTP. The phosphatases involved in depression and depotentiation are not known with certainty. (Adapted from Kameyama et al. 1998)

(Chavez-Noriega and Stevens 1994; Rosenmund et al. 1994). Consistent with the latter observations, the injection of cAMP analogs into postsynaptic cells does not increase basal synaptic transmission (Blitzer et al. 1995; Kameyama et al. 1998). These data suggest that PKA sites on postsynaptic AMPA receptors may be saturated under basal conditions. Therefore, LTP of synaptic transmission from the "naive" or "baseline" state is unlikely to be accounted for by PKA phosphorylation of postsynaptic AMPA receptors (Fig. 2). This mechanism for synaptic potentiation may become available only after the PKA sites have been dephosphorylated, which occurs following the induction of LTD (Kameyama et al. 1998; Lee et al. 1998). In accordance with this hypothesis, postsynaptic injection of cAMP does produce potentiation selectively at synapses that had previously undergone LTD (Kameyama et al. 1998).

Collectively, the data support the hypothesis that NMDA receptor-dependent LTD and LTP can result from the bidirectional modification of AMPA receptor phosphorylation. Because different phosphorylation sites are implicated in LTD and LTP, the model suggests that synaptic strength is determined on the basis of four possible phosphorylation states of GluR1 (Fig. 2). If it is assumed that the effects of phosphorylating the PKA site (ser-845) are usually saturated under "baseline" conditions (Fig. 2A), then synaptic transmission may be modified either by the phosphorylation of the CaMKII site, causing potentiation (Fig. 2B), or by dephosphorylation of the PKA site, causing depression (Fig. 2C). It is interesting to note that the model suggests that the reversal of LTP with LFS (depotentiation), and the reversal of LTD with HFS (dedepression) may or may not "reset" the synapse to its original state (Fig. 2D).

While the data clearly support the view that LTP and LTD can result from changes in phosphorylation of the GluR1 subunit of the AMPA receptor, it is equally clear that the model in Fig. 2 is not a complete description of long-term plasticity. For example, the model does not account for the distinct mechanisms of NMDA receptor-independent LTD (Bolshakov and Siegelbaum 1994; Oliet et al. 1997), LTP (Grover and Teyler 1990), and depotentiation (Staubli and Chun 1996). Nonetheless, the model accounts well for the observed biochemical changes in AMPA receptor phosphorylation, as well as for the effects of experimentally manipulated postsynaptic PKA and CaMKII. The model predicts that HFS and LFS can engage different mechanisms, depending on the initial state of the stimulated synapses. These different mechanisms may be differentially regulated, for example, by age or by the history of synaptic activation. Variations in the relative expression of these mechanisms for plasticity could be an important source of variability between laboratories studying LTP and LTD. Perhaps more significantly, such differential regulation would provide a means for gain-control of synaptic plasticity during development or in different behavioral states.

II. Possible Functions of AMPA Receptor Phosphorylation

How changes in phosphorylation of AMPA receptors contribute to synaptic plasticity is still under debate. Several possibilities have been suggested. First, changes in single channel properties of AMPA receptors can mediate changes in synaptic responses. There are reports indicating an increase in mean open probability and/or mean open time of AMPA receptor channels after phosphorylation by PKA (GREENGARD et al. 1991), and PKC (WANG et al. 1994). On the other hand, a recent analysis of LTP suggests an increase in AMPA receptor conductance without much change in mean open probability following LTP induction (BENKE et al. 1998). Whether the change in AMPA receptor conductance is due to phosphorylation is unclear.

Another mechanism by which AMPA receptor phosphorylation can mediate synaptic plasticity is through AMPA receptor clustering at the synapses. In cultured spinal cord neurons, the size of AMPA receptor clusters correlates with the size of AMPA receptor mediated mEPSCs (O'BRIEN et al. 1998). Growing evidence indicates that glutamate receptors are clustered at

Table 1. Changes in α-amino-3-hydroxy-5-methyl-4-isoxazole proprionate (AMPA) receptor function by protein kinases

Protein kinases	Channel property	Agonist-induced current	Synaptic response
cAMP-dependent protein kinase (PKA)	↑ Mean open time and mean open probability[a].	↑ $I_{glutamate}$[a] and I_{KA}[b].	↑ mEPSC amplitude[a]. No change in sEPSC amplitude[c] or evoked EPSPs[d,e].
Protein kinase C (PKC)	↑ Mean open probability[f].	↑ I_{KA}[f]. No change in I_{KA}[g].	↑ mEPSC amplitude[f] and evoked EPSPs[h].
Ca^{2+}/calmodulin-dependent protein kinase II (CaMKII)	–	↑ I_{KA}[i,j,k].	↑ Evoked EPSCs[l,m].

[a] GREENGARD et al. (1991)
[b] WANG et al. (1991)
[c] ROSENMUND et al. (1994)
[d] BLITZER et al. (1995)
[e] KAMEYAMA et al. (1998)
[f] WANG et al. (1994)
[g] CHEN and HUANG (1992)
[h] HU et al. (1987)
[i] MCGLADE-MCCULLOH et al. (1993)
[j] YAKEL et al. (1995)
[k] BARRIA et al. (1997a)
[l] PETTIT et al. (1994)
[m] LLEDO et al. (1995)

Table 2. Changes in GluR1 phosphorylation by protein kinases

Protein kinases	In vitro phosphorylation	In vivo phosphorylation	Phosphorylation sites
cAMP-dependent protein kinase (PKA)	↑ Phosphorylation of GluR1 C-terminal GST fusion protein[a]. No change in GluR1 isolated from PSDs[b].	↑ Phosphorylation in cortical neurons[c] and hippocampal slices[d,e,f]. No change in phosphorylation in hippocampal cultures[g].	Serine-845[a,d,e,f]
Protein kinase C (PKC)	↑ Phosphorylation of GluR1 C-terminal GST fusion protein[a] and GluR1 isolated from PSDs[b].	↑ Phosphorylation in cultured cortical neurons[c] and hippocampal slices[d,f].	Serine-831[a,d,f]
Ca^{2+}/calmodulin-dependent protein kinase II (CaMKII)	↑ Phosphorylation of GluR1 C-terminal GST fusion protein[d] and GluR1 isolated from PSDs[b].	↑ Phosphorylation in primary hippocampal cultures[g] and hippocampal slices[d].	Serine-831[d,h]

[a] Roche et al. (1994)
[b] Mcglade-Mcculloh et al. (1993)
[c] Blackstone et al. (1994)
[d] Mammen et al. (1997)
[e] Kameyama et al. (1998)
[f] Lee et al. (1998)
[g] Tan et al. (1994)
[h] Barria et al. (1997a)

synapses due to interactions with various cellular proteins (Sheng 1997). The C-termini of NMDA receptors and AMPA receptors can bind to synaptic proteins that contain protein – protein interaction motifs called PDZ domains. For instance, NMDA receptors can bind to PDZ domains in the PSD-95/SAP90 family of proteins via their C-terminal T/SXV motif (Kornau et al. 1995; Muller et al. 1996), while AMPA receptor subunit GluR2 binds to GRIP via the C-terminal SVKI sequence (Dong et al. 1997). Interestingly, the C-terminal sequences that interact with PDZ domains contain serine or threonine residues, which could undergo phosphorylation, making these sites candidates for possible modulation of receptor clustering via phosphorylation.

D. Concluding Remarks

Changes in AMPA receptor phosphorylation seem to be responsible for functional changes produced by protein kinases and protein phosphatases. Based on evidence gathered so far, it appears that AMPA receptors can be phosphorylated by PKA, PKC and CaMKII; the phosphorylation potentiates AMPA receptor mediated responses. In the case of the GluR1 subunit, two phosphorylation sites have been identified: serine-831, which is a phosphorylation site for PKC and CaMKII, and serine-845, which is a phosphorylation site for PKA. Differential changes in these two phosphorylation sites are proposed to mediate changes in synaptic strength associated with LTP- and LTD-like processes.

Despite this work, the mapping of phosphorylation sites on AMPA receptors is far from complete. In addition, the function of each phosphorylation site still remains speculative. Furthermore, whether the phosphorylation sites interact with or regulate one another is unknown. Given that AMPA receptor phosphorylation is a likely candidate for the mediation of synaptic changes, further studies clarifying the function of AMPA receptor phosphorylation would lead to a greater insight into mechanisms of synaptic plasticity in the central nervous system.

References

Barria A, Derkach V, Soderling T (1997a) Identification of the Ca^{2+}/calmodulin-dependent protein kinase II regulatory phosphorylation site in the α-amino-3-hydroxyl-5-methyl-4-isoxazole-propionate type glutamate receptor. J Biol Chem 272:32727–32730

Barria A, Muller D, Derkach V, Griffith LC, Soderling TR (1997b) Regulatory phosphorylation of AMPA-type glutamate receptors by CaM-KII during long-term potentiation. Science 276:2042–2045

Baude A, Nusser Z, Molnar E, McIlhinney RA, Somogyi P (1995) High-resolution immunogold localization of AMPA type glutamate receptor subunits at synaptic and non-synaptic sites in rat hippocampus. Neuroscience 69:1031–1055

Bear MF, Abraham WC (1996) Long-term depression in hippocampus. Annu Rev Neurosci 19:437–462

Bekkers JM, Stevens CF (1990) Presynaptic mechanism for long-term potentiation in the hippocampus [see comments]. Nature 346:724–729

Benke TA, Bresink I, Collett VJ, Doherty AJ, Henley JM, Collingridge GL (1996) Post-translational mechanisms which could underlie the postsynaptic expression of LTP and LTD. In: Cortical plasticity (editors: Fazeli MS and Collingridge GL) BIOS Scientific Publishers Ltd., Oxford, p 83–102

Benke TA, Luthi A, Isaac JT, Collingridge GL (1998) Modulation of AMPA receptor unitary conductance by synaptic activity. Nature 393:793–797

Bennett JA, Dingledine R (1995) Topology profile for a glutamate receptor: Three transmembrane domains and a channel-lining reentrant membrane loop. Neuron 14:373–384

Bernard V, Somogyi P, Bolam JP (1997) Cellular, subcellular, and subsynaptic distribution of AMPA-type glutamate receptor subunits in the neostriatum of the rat. J Neurosci 17:819–833

Betz H (1990) Ligand-gated ion channels in the brain: the amino acid receptor superfamily. Neuron 5:383–392

Blackstone C, Murphy TH, Moss SJ, Baraban JM, Huganir RL (1994) Cyclic AMP and synaptic activity-dependent phosphorylation of AMPA-preferring glutamate receptors. J Neurosci 14:7585–7593

Blackstone CD, Moss SJ, Martin LJ, Levey AI, Price DL, Huganir RL (1992) Biochemical characterization and localization of a non-N-methyl-D-aspartate glutamate receptor in rat brain. J Neurochem 58:1118–1126

Bliss TVP, Collingridge GL (1993) A synaptic model of memory: long-term potentiation in the hippocampus. Nature 361:31–39

Bliss TVP, Fazeli MS (1996) The locus of expression of NMDA receptor-dependent LTP in the hippocampus. In: MS Fazeli and GL Collingridge (ed Cortical Plasticity. BIOS Scientific Publishers Ltd., Oxford, p 61–82

Blitzer RD, Connor JH, Brown GP, Wong T, Shenolikar S, Iyengar R, Landau EM (1998) Gating of CaMKII by cAMP-regulated protein phosphatase activity during LTP. Science 280:1940–1943

Blitzer RD, Wong T, Nouranifar R, Iyengar R, Landau EM (1995) Postsynaptic cAMP pathway gates early LTP in hippocampal CA1 region. Neuron 15:1403–1414

Bolshakov VY, Siegelbaum SA (1994) Postsynaptic induction and presynaptic expression of hippocampal long-term depression. Science 264:1148–1152

Boulter J, Hollmann M, O'Shea-Greenfield A, Hartley M, Deneris E, Maron C, Heinemann S (1990) Molecular cloning and functional expression of glutamate receptor subunit genes. Science 249:1033–1037

Bourtchuladze R, Frenguelli B, Blendy J, Cioffi D, Schutz G, Silva AJ (1994) Deficient long-term memory in mice with a targeted mutation of the cAMP-responsive element binding protein. Cell 79:59–68.

Braun AP, Schulman H (1995) The multifunctional calcium/calmodulin-dependent protein kinase: from form to function. Annu Rev Physiol 57:417–445

Carder RK (1997) Immunocytochemical characterization of AMPA-selective glutamate receptor subunits: laminar and compartmental distribution in macaque striate cortex. J Neurosci 17:3352–3363

Carr DW, Hausken ZE, Fraser ID, Stofko-Hahn RE, Scott JD (1992) Association of the type II cAMP-dependent protein kinase with a human thyroid RII-anchoring protein. Cloning and characterization of the RII- binding domain. J Biol Chem 267:13376–13382

Carr DW, Stofko-Hahn RE, Fraser ID, Bishop SM, Acott TS, Brennan RG, Scott JD (1991) Interaction of the regulatory subunit (RII) of cAMP-dependent protein kinase with RII-anchoring proteins occurs through an amphipathic helix binding motif. J Biol Chem 266:14188–14192

Chavez-Noriega LE, Stevens CF (1994) Increased transmitter release at excitatory synapses produced by direct activation of adenylate cyclase in rat hippocampal slices. J Neurosci 14:310–317

Chen L, Huang L-YM (1992) Protein kinase C reduces Mg^{2+} block of NMDA-receptor channels as a mechanism of modulation. Nature 356:521–523

Coghlan VM, Perrino BA, Howard M, Langeberg LK, Hicks JB, Gallatin WM, Scott JD (1995) Association of protein kinase A and protein phosphatase 2B with a common anchoring protein. Science 267:108–111

Cohen P (1989) The structure and regulation of protein phosphatases. Annu Rev Biochem 58:453–508

De Koninck P, Schulman H (1998) Sensitivity of CaM kinase II to the frequency of Ca2+ oscillations [see comments]. Science 279:227–230

Dong H, O'Brien RJ, Fung ET, Lanahan AA, Worley PF, Huganir RL (1997) GRIP: a synaptic PDZ domain-containing protein that interacts with AMPA receptors. Nature 386:279–284

Dunkley PR (1991) Autophosphorylation of neuronal calcium/calmodulin-stimulated protein kinase II. Mol Neurobiol 5:179–202

Durand GM, Zukin RS (1993) Developmental regulation of mRNAs encoding rat brain kainate/AMPA receptors: a Northern analysis study. J Neurochem 61:2239–2246

Frey U, Huang YY, Kandel ER (1993) Effects of cAMP simulate a late stage of LTP in hippocampal CA1 neurons. Science 260:1661–1664

Greengard P, Jen J, Nairn AC, Stevens CF (1991) Enhancement of the glutamate response by cAMP-dependent protein kinase in hippocampal neurons. Science 253:1135–1138

Grover LM, Teyler TJ (1990) Two components of long-term potentiation induced by different patterns of afferent activation. Nature 347:477–479

Hayashi Y, Ishida A, Katagiri H, Mishina M, Fujisawa H, Manabe T, Takahashi T (1997) Calcium- and calmodulin-dependent phosphorylation of AMPA type glutamate receptor subunits by endogenous protein kinases in the post-synaptic density. Mol Brain Res 46:338–342

Hollmann M, Hartley M, Heinemann S (1991) Ca^{2+} permeability of KA-AMPA-gated glutamate receptor channels depends on subunit composition. Science 252:851–853

Hollmann M, Maron C, Heinemann S (1994) N-Glycosylation site tagging suggests a three transmembrane domain topology of the glutamate receptor GluR1. Neuron 13:1331–1343

Hollmann M, O'Shea-Greenfield A, Rogers SW, Heinemann S (1989) Cloning by functional expression of a member of the glutamate receptor family. Nature 342:643–648

Hu G-Y, Hvalby O, Walaas SI, Albert KA, Skjeflo P, Andersen P, Greengard P (1987) Protein kinase C injection into hippocampal pyramidal cells elicits features of long term potentiation. Nature 328:426–429

Huang K-P, Huang FL, Mahoney CW, Chen K-H (1991) Protein kinase C subtypes and their respective roles. Prog Brain Res 89:143–154

Impey S, Mark M, Villacres EC, Poser S, Chavkin C, Storm DR (1996) Induction of CRE-mediated gene expression by stimuli that generate long-lasting LTP in area CA1 of the hippocampus. Neuron 16:973–982

Isaac JTR, Nicoll RA, Malenka RC (1995) Evidence for silent synapses: Implications for the expression of LTP. Neuron 15:427–434

Ishida A, Kameshita I, Fujisawa H (1998) A novel protein phosphatase that dephosphorylates and regulates Ca2+/calmodulin-dependent protein kinase II. J Biol Chem 273:1904–1910

Kameyama K, Lee H-K, Bear MF, Huganir RL (1998) Involvement of a postsynaptic protein kinase A substrate in the expression of homosynaptic long-term depression. Neuron 21:1163–1175

Keinanen K, Wisden W, Sommer B, Werner P, Herb A, Verdoorn TA, Sakmann B, Seeburg PH (1990) A family of AMPA-selective glutamate receptors. Science 249:556–560

Keller BU, Hollmann M, Heinemann S, Konnerth A (1992) Calcium influx through subunits GluR1/GluR3 of kainate/AMPA receptor channels is regulated by cAMP dependent protein kinase. EMBO J 11:891–896

Kemp BE, Pearson RB (1990) Protein kinase recognition sequence motifs. Trends Biochem Sci 15:342–346

Kennedy MB, Bennett MK, Erondu NE (1983) Biochemical and immunochemical evidence that the "major postsynaptic density protein" is a subunit of a calmodulin-dependent protein kinase. Proc Natl Acad Sci USA 80:7357–7361

Kikkawa U, Kishimoto A, Nishizuka Y (1989) The protein kinase C family: Heterogeneity and its implications. Annu Rev Biochem 58:31–44

Kishimoto A, Mikawa K, Hashimoto K, Yasuda I, Tanaka S-I, Tominaga M, Kuroda T, Nishizuka Y (1989) Limited proteolysis of protein kinase C subspecies by calcium-dependent neutral protease (calpain). J Biol Chem 264:4088–4092

Klann E, Chen S-J, Sweatt D (1993) Mechanism of protein kinase C activation during the induction and maintenance of long-term potentiation probed using a selective peptide substrate. Proc Natl Acad Sci USA 90:8337–8341

Knapp AG, Dowling JE (1987) Dopamine enhances excitatory amino acid-gated conductances in cultured retinal horizontal cells. Nature 325:437–439

Kornau HC, Schenker LT, Kennedy MB, Seeburg PH (1995) Domain interaction between NMDA receptor subunits and the postsynaptic density protein PSD-95. Science 269:1737–1740

Kraft AS, Anderson WB (1983) Phorbol esters increase the amount of Ca2+, phospholipid-dependent protein kinase associated with plasma membrane. Nature 301:621–623

Lau L-F, Huganir RL (1995) Differential tyrosine phosphorylation of N-methyl-D-aspartate receptor subunits. J Biol Chem 270:20036–20041

Lee H-K, Kameyama K, Huganir RL, Bear MF (1998) NMDA induces long-term synaptic depression and dephosphorylation of the GluR1 subunit of AMPA receptors in hippocampus. Neuron 21:1151–1162

Liao D, Hessler NA, Malinow R (1995) Activation of postsynaptically silent synapses during pairing-induced LTP in CA1 region of hippocampal slice. Nature 375:400–404

Lisman J (1994) The CaM Kinase II hypothesis for the storage of synaptic memory. Trends Neurosci. 17:406–412

Lledo P-M, Hjelmstad GO, Mukherji S, Soderling TR, Malenka RC, Nicoll RA (1995) Calcium/calmodulin-dependent kinase II and long-term potentiation enhance synaptic transmission by the same mechanism. Proc Natl Acad Sci USA 92:11175–11179

Madison DV, Malenka RC, Nicoll RA (1991) Mechanisms underlying long-term potentiation of synaptic transmission. Annu Rev Neurosci 14:379–397

Mammen AL, Kameyame K, Roche KW, Huganir RL (1997) Phosphorylation of the α-amino-3-hydroxy-5-methylisoxazole-4-propionic acid receptor GluR1 subunit by calcium/calmodulin-dependent kinase II. J Biol Chem 272:32528–32533

Martin LJ, Blackstone CD, Levey AI, Huganir RL, Price DL (1993) AMPA glutamate receptor subunits are differentially distributed in rat brain. Neurosci 53:327–358

Martin LJ, Furuta A, Blackstone CD (1998) AMPA receptor protein in developing rat brain: Glutamate receptor-1 expression and localization change at regional, cellular, and subcellular levels with maturation. Neuroscience 83:917–928

Matthies H, Reymann KG (1993) Protein kinase A inhibitors prevent the maintenance of hippocampal long-term potentiation. Neuroreport 4:712–714

McGlade-McCulloh E, Yamamoto H, Tan S-E, Brickey DA, Soderling TR (1993) Phosphorylation and regulation of glutamate receptors by calcium/calmodulin-dependent protein kinase II. Nature 362:640–642

Mochly-Rosen D (1995) Localization of protein kinases by anchoring proteins: a theme in signal transduction. Science 268:247–251

Molnár E, Baude A, Richmond SA, Patel PB, Somogyi P, McIlhinney RAJ (1993) Biochemical and immunocytochemical characterization of antipeptide antibodies to a cloned GluR1 glutamate receptor subunit: cellular and subcellular distribution in the rat forebrain. Neuroscience 53:307–326

Monyer H, Seeburg PH, Wisden W (1991) Glutamate-operated channels: developmentally early and mature forms arise by alternative splicing. Neuron 6:799–810

Moon IS, Apperson ML, Kennedy MB (1994) The major tyrosine-phosphorylated protein in the postsynaptic density fraction is N-methyl-D-aspartate receptor subunit 2B. Proc Natl Acad Sci USA 91:3954–3958

Mosbacher J, Schoepfer R, Monyer H, Burnashev N, Seeburg PH, Ruppersberg JP (1994) A molecular determinant for submillisecond desensitization in glutamate receptors. Science 266:1059–1062

Moss SJ, Blackstone CD, Huganir RL (1993) Phosphorylation of recombinant non-NMDA glutamate receptors on serine and tyrosine residues. Neurochem Res 18:105–110

Muller BM, Kistner U, Kindler S, Chung WJ, Kuhlendahl S, Fenster SD, Lau LF, Veh RW, Huganir RL, Gundelfinger ED, Garner CC (1996) SAP102, a novel postsynaptic protein that interacts with NMDA receptor complexes in vivo. Neuron 17:255–265

Nakanishi N, Shneider NA, Axel R (1990) A family of glutamate receptor genes: evidence for the formation of heteromultimeric receptors with distinct channel properties. Neuron 5:569–581

O'Brien RJ, Kamboj S, Ehlers MD, Rosen KR, Fischbach GD, Huganir RL (1998) Activity-dependent modulation of synaptic AMPA receptor accumulation. Neuron 21:1067–1078

Oliet SHR, Malenka RC, Nicoll RA (1997) Two distinct forms of long-term depression coexist in hippocampal pyramidal cells. Neuron 18:969–982

Partin KM, Bowie D, Mayer ML (1995) Structural determinates of allosteric regulation in alternatively spliced AMPA receptors. Neuron 14:833–843

Partin KM, Patneau DK, Mayer ML (1994) Cyclothiazide differentially modulates desensitization of alpha-amino-3- hydroxy-5-methyl-4-isoxazolepropionic acid receptor splice variants. Mol Pharmacol 46:129–38

Pellegrini-Giampietro DE, Bennett MV, Zukin RS (1992) Are Ca(2+)-permeable kainate/AMPA receptors more abundant in immature brain? Neurosci Lett 144:65–69

Petralia RS, Wenthold RJ (1992) Light and electron immunocytochemical localization of AMPA-selective glutamate receptors in the rat brain. J Comp Neurol 318:329–354

Pettit DL, Perlman S, Malinow R (1994) Potentiated transmission and prevention of further LTP by increased CaMKII activity in postsynaptic hippocampal slice neurons. Science 266:1881–1885

Raymond LA, Blackstone CD, Huganir RL (1993) Phosphorylation of amino acid neurotransmitter receptors in synaptic plasticity. Trends Neurosci 16:147–153

Ricciarelli R, Azzi A (1998) Regulation of recombinant PKCalpha activity by protein phosphatase 1 and protein phosphatase 2 A [In Process Citation]. Arch Biochem Biophys 355:197–200

Roche KW, O'Brien RJ, Mammen AL, Bernhardt J, Huganir RL (1996) Characterization of multiple phosphorylation sites on the AMPA receptor GluR1 subunit. Neuron 16:1179–1188

Roche KW, Tingley WG, Huganir RL (1994) Glutamate receptor phosphorylation and synaptic plasticity. Curr Opin Neurobiol 4:383–388

Rosenmund C, Carr DW, Bergeson SE, Nilaver G, Scott JD, Westbrook GL (1994) Anchoring of protein kinase A is required for modulation of AMPA/kainate receptors on hippocampal neurons. Nature 368:853–856

Sacktor TC, Osten P, Valsamis H, Jiang X, Naik MU, Sublette E (1993) Persistent activation of the ζ isoform of protein kinase C in the maintenance of long-term potentiation. Proc Natl Acad Sci USA 90:8342–8346

Sheng M (1997) Excitatory synapses. Glutamate receptors put in their place [news; comment]. Nature 386:221, 223

Shinolikar S, Ingebritsen TS (1984) Protein (serine and threonine) phosphate phosphatases. Methods Enzymol 107:102–129

Soderling TR, Tan SE, McGlade-McCulloh E, Yamamoto H, Fukunaga K (1994) Excitatory interactions between glutamate receptors and protein kinases. J Neurobiol 25:304–311

Sommer B, Keinanen K, Verdoorn TA, Wisden W, Burnashev N, Herb A, Kohler M, Takagi T, Sakmann B, Seeburg PH (1990) Flip and flop: a cell-specific functional switch in glutamate-operated channels of the CNS. Science 249:1580–1585

Strack S, Barban MA, Wadzinski BE, Colbran RJ (1997a) Differential inactivation of postsynaptic density-associated and soluble Ca2+/calmodulin-dependent protein kinase II by protein phosphatases 1 and 2 A. J Neurochem 68:2119–2128

Strack S, Choi S, Lovinger DM, Colbran RJ (1997b) Translocation of autophosphorylated calcium/calmodulin-dependent protein kinase II to the postsynaptic density. J Biol Chem 272:13467–13470

Strack S, Westphal RS, Colbran RJ, Ebner FF, Wadzinski BE (1997c) Protein serine/threonine phosphatase 1 and 2 A associate with and dephosphorylate neurofilaments. Brain Res Mol Brain Res 49:15–28

Strack S, Zaucha JA, Ebner FF, Colbran RJ, Wadzinski BE (1998) Brain protein phosphatase 2 A: developmental regulation and distinct cellular and subcellular localization by B subunits. J Comp Neurol 392:515–527

Staubli U, Chun D (1996) Factors regulating the reversibility of long-term potentiation. J Neurosci 15:853–860

Tan S-E, Wenthold RJ, Soderling TR (1994) Phosphorylation of AMPA-type glutamate receptors by calcium/calmodulin-dependent protein kinase II and protein kinase C in cultured hippocampal neurons. J Neurosci 14:1123–1129

Thomas MJ, Moody TD, Makhinson M, O'Dell TJ (1996) Activity-dependent β-adrenergic modulation of low frequency stimulation induced LTP in the hippocampal CA1 region. Neuron 17:475–482

Vereb G, Gergely P (1989) The role of autophosphorylation of cAMP-dependent protein kinase II in the inhibition of protein phosphatase-1. Int J Biochem 21:1137–41

Wang J-H, Kelly PT (1995) Postsynaptic injection of Ca^{2+}/CaM induces synaptic potentiation requiring CaMKII and PKC activity. Neuron 15:443–452

Wang L-Y, Dudek EM, Browning MD, MacDonald JF (1994) Modulation of AMPA/kainate receptors in cultured murine hippocampal neurones by protein kinase C. J Physiol 475.3:431–437

Wang L-Y, Salter MW, MacDonald JF (1991) Regulation of kainate receptors by cAMP-dependent protein kinase and phosphatases. Science 253:1132–1135

Wenthold RJ, Petralia RS, Blahos II J, Niedzielski AS (1996) Evidence for multiple AMPA receptor complexes in hippocampal CA1/CA2 neurons. J Neurosci 16:1982–1989

Wenthold RJ, Tokotani N, Doi K, Wada K (1992) Immunochemical characterization of the non-NMDA glutamate receptor using subunit-specific antibodies. J Biol Chem 267:501–507

Wo ZG, Oswald RE (1994) Transmembrane topology of two kainate receptor subunits revealed by N-glycosylation. Proc Natl Acad Sci USA 91:7154–7158

Wo ZG, Oswald RE (1995) Unraveling the modular design of glutamate-gated ion channels. Trends Neurosci 18:161–168

Wood MW, VanDongen HMA, VanDongen AMJ (1995) Structural conservation of ion conduction pathways in K channels and glutamate receptors. Proc Natl Acad Sci USA 92:4882–4886

Wyllie DJA, Nicoll RA (1994) A role for protein kinases and phosphatases in the Ca^{2+}-induced enhancement of hippocampal AMPA receptor-mediated synaptic responses. Neuron 13:635–643

Yakel JL, Vissavajjhala P, Derkach VA, Brickey DA, Soderling TR (1995) Identification of a Ca^{2+}/calmodulin-dependent protein kinase II regulatory phosphorylation site in non-N-methyl-D-aspartate glutamate receptors. Proc Natl Acad Sci USA 92:1376–1380

CHAPTER 3

The Synaptic Protein Network Associated with Ionotropic Glutamate Receptors

H.-C. Kornau, P.H. Seeburg, and M.B. Kennedy

A. Introduction

Heteromeric glutamate receptor (GluR) channels are integrated into larger protein complexes at specific synaptic locations via protein-interaction modules located at the C-terminal tails of the receptor subunits. Even though the ion-channel characteristics of *N*-methyl-D-aspartate receptors (NMDARs) and α-amino-3-hydroxy-5-methyl-4-isoxazolepropionic acid receptors (AMPARs) are well characterized, much of their cell biology and cell physiology have remained in the dark. The current search for proteins interacting with the major GluRs promises to unravel mechanisms of synapse targeting, clustering in the subsynaptic membrane and linkage to diverse intracellular signaling pathways. This area of research is rapidly evolving. Based on current data, NMDARs and AMPARs appear to be connected to different intracellular proteins. Some of the proteins associated with NMDARs and non-NMDARs have been identified, their corresponding cDNAs have been cloned and their relevance in the function of GluRs has been addressed.

The cloning and characterization of interaction partners of GluRs depended largely on the yeast two-hybrid (Y2H) system, a genetic screen for the identification of interacting proteins (Fields and Song 1989; Fields and Sternglanz 1994; Niethammer and Sheng 1998). Initially used only in searches for binding partners for cytoplasmic or nuclear proteins, it also proved to be useful in finding interacting proteins for cytoplasmic domains of membrane proteins. A major advantage of the system in comparison with conventional biochemical methods is the fact that the identification of a new interaction simultaneously yields the identification of the cDNA for the interaction partner. The cDNA is an essential tool for many additional experiments, such as expression analysis, expression in a heterologous system or the generation of transgenic animals. However, interactions found using the Y2H system must be verified by biochemical and cell-biological experiments.

B. The NMDAR Complex

NMDARs play essential roles in brain development, excitatory neurotransmission, synaptic plasticity and memory formation. Ca^{2+} influx through NMDARs has been found to be essential for the induction of long-term potentiation (LTP) at hippocampal collateral-commissural CA3-CA1 synapses. The steps between the NMDAR-mediated Ca^{2+} current and the strengthening of the synapse have been mysterious, but the presence of a synaptic microdomain in close association with the NMDAR capable of sensing and integrating the change in intracellular Ca^{2+} concentration has been postulated. It has therefore been a common goal of several laboratories to delineate proteins intracellularly associated with NMDARs (Table 1).

Different NMDARs are formed by the assembly of the essential NR1 subunit with any one of four NR2 subunits (NR2A–D). The identity of the NR2 subunit in the receptor determines its electrophysiological properties. The NMDAR is part of the postsynaptic density and is subject to phosphorylation on tyrosine residues in the C-terminus of NR2B (CHO et al. 1992; MOON et al 1994). The biochemical characterization of proteins interacting directly with the NMDAR was hampered by difficulties in solubilizing the receptor under mild conditions. Hence, identification of direct interaction partners for NMDARs has so far depended on the use of the Y2H system. Some of the interacting proteins identified using the Y2H system are prominent components of the PSD fraction and are concentrated at synaptic sites in neurons, consistent with their interaction with the NMDAR in vivo. In addition, some of the interactions could be verified by co-immunoprecipitation from brain extracts after solubilization of the complex in specific detergent combinations. Advances in transgenic mouse technology have made it possible to tackle the significance of protein complexes by genetic means. Mice harboring null mutations or deletions of specific domains prove to be elegant tools for the study of the functional consequences of protein–protein interactions.

I. Direct Interaction with PSD-95 and Related Proteins

The NR2 subunits of the NMDAR contain long C-terminal tails that appear not to influence channel properties in heterologous expression systems, but have been suspected to be important for anchoring the receptor at synaptic microdomains (MONYER et al. 1992). Y2H screens with these long C-terminal domains revealed an interaction with PSD-95, a major component of the postsynaptic density (PSD; CHO et al. 1992; KORNAU et al. 1995; NIETHAMMER et al. 1996). This interaction involves only the very C-terminal four amino acids of the NR2 subunits, which can bind to a PDZ domain, a 90-amino acid motif found in three copies in PSD-95 and, later, also in many other proteins located at membrane specializations (KENNEDY 1995; PONTING and PHILIPS 1995; SHENG 1996; KORNAU et al. 1997). PSD-95 is a member of the MAGUK family of proteins consisting of three PDZ domains, one Src homology 3 (SH3) domain and a region with homology to yeast guanylate kinase (GUK). This protein family is believed to be important for the structural organization of sites of cell-to-cell contact. Within the central nervous system (CNS), both pre- and postsynaptically expressed members have been identified. Among those, SAP102 and Chapsyn110/PSD93 were also found to interact with NMDARs

Table 1. The synaptic protein network associated with ionotropic glutamate receptors (GluRs)

Receptor subtype	Interacting protein	Associated proteins	Putative function	Selected citations
NMDA	PSD-95, SAP102, Chapsyn110/PSD-93	synGAP, nNOS, GKAP/SAPAPs, Neuroligins, Citron, protein kinases, CRIPT	Connection of the NMDA receptor with signal transducers and the cytoskeleton	Kornau et al. 1995; Niethammer et al. 1996; Müller et al. 1996; Kim et al. 1996; Brenman et al. 1996
	Calmodulin		Ca^{2+} dependent channel inactivation	Ehlers et al. 1996
	α-actinin	Actin	Regulated connection to the dendritic cytoskeleton	Wyszinski et al. 1997
	CIPP	Channels, neuroligin		Kurschner et al. 1998
	NF-L			Ehlers et al. 1998
	yotiao			Lin et al. 1998
AMPA	GRIP, ABP	GRASPs?	Connection of the AMPA receptor with signal transducers	Dong et al. 1997; Srivastava et al. 1998
	PICK1	PKCα	AMPA receptor clustering	Xia et al. 1999
	NSF	α/β-SNAPs	Membrane fusion of AMPA receptor containing vesicles	Nishimune et al. 1998; Osten et al. 1998; Song et al. 1998
	SAP97			Leonard et al. 1998
	$G\alpha_{i1}$		Metabotropic signaling	Wang et al. 1997
Kainate	PSD-95, SAP97, SAP102	See above	Clustering	Garcia et al. 1998
Delta2	Chapsyn110/PSD-93	See above	Clustering	Roche et al. 1999

ABP, AMPAR binding protein; AMPA, α-amino-3-hydroxy-5-methyl-4-isoxazolepropionic acid; CIPP, channel-interacting PDZ domain protein; GRASP, GRIP-associated protein; GRIP, glutamate receptor interacting protein; NF-L, neurofilament subunit; NMDA, N-methyl-D-aspartate; NSF, *N*-ethylmaleimide-sensitive factor; PICK1, protein interacting with C kinase 1; PKC, protein kinase C; SAP, synapse-associated protein; SAPAP, SAP-associated protein; SNAP, soluble NSF attachment proteins; Syn GAP, synaptic ras GTPase activating protein

and proved to be postsynaptic proteins (KIM et al. 1996; MUELLER et al. 1996). In addition to these PSD-95-related proteins, the channel-interacting PDZ domain protein (CIPP), a polypeptide with four PDZ domains, can bind to the C-terminal tails of NMDARs, K^+ channels, neurexins and neuroligins in vitro and may also be an element of the NMDAR complex in vivo (KURSCHNER et al. 1998).

What is known about the function of the interaction of PSD-95 and related proteins with NMDARs? They may provide a scaffold for the linkage of several different signal transducers to the NMDAR. This hypothesis is based on the fact that, in addition to the 3 PDZ domains, both the SH3 and GUK domains can subserve protein–protein interactions (KIM et al. 1997; TAKEUCHI et al. 1997). Furthermore, PSD-95 tends to multimerize via disulfide bridges in a head-to-head fashion (HSUEH et al. 1997), allowing for a rather large network of proteins to be localized to the immediate vicinity of NMDARs. Indeed, several partners for PSD-95 have been identified, including ion channels and receptors (SHENG 1996; KORNAU et al. 1997), cell-adhesion molecules (Irie et al. 1997), cytoskeleton-associated factors (see below) and intracellular Ca^{2+}-dependent effectors (BRENMAN et al. 1996; CHEN et al. 1998). A second possible function of this interaction emerges from studies in heterologous cells. Co-expression of NMDARs with PSD-95 in COS cells leads to a clustering of both proteins at discrete spots in the cell membrane (KIM et al. 1996). Therefore PSD family members may be important not only for the connection of NMDARs with other signal-transduction molecules, but also for the localization and clustering of NMDA channels at synaptic sites.

To assess the functional relevance of the interaction between NMDARs and intracellular protein partners in vivo, mice that lack the C-terminal tail of the NR2A, NR2B and NR2C subunits have been generated by gene targeting (SPRENGEL et al 1998). It was known that the truncation of NR2 subunits after transmembrane region 4 did not result in major changes of channel properties after co-expression with NR1 in heterologous expression systems. Hence, phenotypic effects in the mutant mice were expected to result from impaired linkage of the NMDAR to intracellular signal-transduction pathways rather than from changes in NMDAR currents. Surprisingly, however, the phenotypes of the mice carrying truncated NR2 subunits strikingly resemble null mutations of the respective NR2 subunits, suggesting that the interactions mediated by the C-terminal tails are essential for the physiological functions of the NMDAR (SAKIMURA et al. 1995; EBRALIDZE et al. 1996; KUTSUWADA et al. 1996; SPRENGEL et al. 1998). Truncation of the NR2A subunit led to alterations in synaptic connections manifested as changes in LTP, kindling behavior and fear conditioning. Mice carrying the truncated NR2B subunit died prematurely, whereas mice lacking the C-terminus of NR2C showed impaired motor coordination. With respect to all these phenotypes, the mice with the C-terminal deletions are thus far indistinguishable from the respective knockout mice.

One explanation for the similarity of the phenotypes observed between the different mutants would be that the NMDAR lacking its intracellular interaction modules does not reach the postsynaptic membrane specialization where it usually functions. Impaired intracellular receptor targeting might well result in the same deficiencies as complete deletion of the receptor. Unfortunately, the lack of useful antibodies directed against N-terminal regions of the NR2 subunits precluded a direct study of the localization of the truncated subunits. However, synaptic electrophysiological recordings and Ca^{2+} influx measurements indicated that NMDARs with truncated NR2 subunits still reach the postsynaptic membrane specialization (SPRENGEL et al. 1998). These results suggest that the postsynaptic machinery is unable to integrate the information carried by NMDAR currents if the connection between the NMDAR and subsequent signal transducers is missing.

Mice independently generated by gene targeting expressing an NR2B subunit lacking about two-thirds (459 amino acids) of the C-terminal domain (MORI et al. 1998) resulted in neonatal death as described for the NR2B C-terminal deletion (SPRENGEL et al. 1998) as well as the NR2B knock-out mice (KUTSUWADA et al. 1996). However, both electrophysiological and immunocytochemical data suggest an impaired synaptic clustering of NMDARs at postsynaptic sites in the mutants with the comparably smaller deletion in the C-terminus (MORI et al. 1998). Therefore, the C-terminal tails of NR2 subunits may mediate both correct targeting and clustering of NMDARs as well as their connection to signal integration machinery.

Homologous recombination was also employed to generate mice expressing PSD-95 in a truncated form carrying a stop codon behind the second PDZ domain (MIGAUD et al 1998). Homozygous mice did not show obvious neurological abnormalities. However, the frequency sensitivity of LTP induction was altered and the mice showed spatial learning deficits. The truncated PSD-95 protein was not localized to synapses as expected from studies on similar mutants in *Drosophila melanogaster* (TEJEDOR et al. 1997). In contrast, the NMDARs were correctly targeted and their basic biophysical properties remained unchanged in the mutant mice, indicating either that PSD-95 is not essential for the proper targeting of NMDARs or that either SAP102 or Chapsyn110/PSD-93 compensates for the lack of synaptic PSD-95. Interestingly, the changes in LTP and learning suggest that balanced signal integration upon activation of the NMDAR depends on the presence of PSD-95 and its associated proteins, presumably molecules that are able to sense the Ca^{2+} currents entering the NMDAR, such as calmodulin-dependent protein kinase II (CaMKII) and neuronal nitric oxide synthetase (nNOS) (OMKUMAR et al. 1996; BRENMAN et al. 1996), but also their targets (CHEN et al. 1998).

The different phenotypes observed in the NMDAR C-terminal deletion mutants and the PSD-95 mutants imply that the C-terminal domains of the NMDAR subserve more functions than the interaction with PSD-95, including additional protein interactions.

II. Proteins Associated with the NMDAR Via Interaction with PSD-95 and Related Proteins

The PSD-95-related family of proteins contains several potential protein interaction domains in addition to PDZ domains 1 and 2, which interact with the tail of the NR2 subunits of the NMDAR. These include PDZ3, an SH3 domain and the carboxyl terminal GUK domain. As with the interaction between the NMDAR and PSD-95, the Y2H system has been used to identify additional interaction partners for PSD-95 that might be present in a complex with the NMDAR at glutamatergic synapses. A number of candidates has been identified, including the neurexin family of adhesion molecules, GKAP and a family of related proteins termed SAPAPs (synapse-associated protein-associated proteins), nNOS and synaptic ras GTPase activating protein (synGAP).

1. SynGAP

Although synGAP was identified most recently, the evidence that it in fact exists in a complex with the NMDAR at many excitatory synapses in the mammalian forebrain is the strongest. SynGAP was sequenced and cloned as a major protein component of the postsynaptic density fraction (CHEN and KENNEDY 1997; CHEN et al. 1998) and was found in a Y2H screen for proteins interacting with SAP-102 (KIM et al. 1998). Subsequent Y2H screens established that it interacts with PDZ3 of PSD-95 (CHEN et al. 1998; KIM et al. 1998) and of SAP-102 (KIM et al. 1998). SynGAP, like PSD-95 and the subunits of the NMDAR, persists as a major component of the PSD fraction after extraction with relatively harsh detergents, indicating the three proteins are located in a detergent-resistant complex in intact neurons (CHEN et al.1998). Furthermore, SynGAP can be co-immunoprecipitated from brain homogenates with PSD-95 after solubilization of the PSD complex by detergent (CHEN et al. 1998; KIM et al. 1998) and is highly concentrated at synaptic sites in cultured hippocampal neurons as revealed by immunocytochemistry (CHEN et al. 1998; KIM et al. 1998). SynGAP co-localizes with PSD-95 and the NMDAR at essentially all glutamatergic synapses made onto excitatory pyramidal neurons in the cultures, but it is expressed in only a small portion of inhibitory neurons where it also co-localizes with the NMDAR (ZHANG et al. 1999).

SynGAP interacts with the activated GTP-bound form of Ras and greatly stimulates its GTPase activity (CHEN et al. 1998; KIM et al. 1998). Its action is similar to the more familiar Ras GAPs, including p120 RasGAP and neurofibromin (NF1) (WIGLER 1990). Within the PSD fraction, SynGAP is phosphorylated by Ca^{2+}/CaMKII, which rapidly and reversibly inhibits its GAP activity (CHEN et al. 1998). Since CaMKII is associated with the PSD and is a likely target for Ca^{2+} influx through the NMDAR, SynGAP may confer an additional coincidence detection ability on the NMDAR by linking its activation to phosphorylation and inhibition of SynGAP and, thus, to enhanced activation of the mitogen-activated protein (MAP) kinase. Future studies will focus on the role this phosphorylation may play in synaptic and neuronal regulation by the NMDAR.

2. Neuronal Nitric Oxide Synthetase

PSD-95 and a related protein termed PSD-93 (BRENMAN et al. 1996) or Chapsyn-110 (KIM et al. 1996) were identified in a Y2H screen for proteins interacting with the N-terminal PDZ domain of nNOS (BRENMAN et al. 1996). The PDZ domain of nNOS interacts directly with the second PDZ domain of PSD-95 and PSD-93 in a novel form of interaction that is not analogous to the binding of C-terminal motifs to the PDZ domains.

The evidence that nNOS is linked to NMDARs via PSD-95 or PSD-93 is still largely indirect. nNOS is regulated by Ca^{2+}/calmodulin and, thus, is a potential target for NMDAR-induced increases in neuronal Ca^{2+} concentration. In some neuronal populations, activation of NMDARs indeed stimulates production of nitric oxide (GARTHWAITE 1991). Both PSD-95 and PSD-93 are co-expressed in several neuronal populations with nNOS (BRENMAN et al. 1996). nNOS has been detected by immunocytochemistry at the electron microscope level at type-II (glutamatergic) postsynaptic densities in certain cortical structures; however, it does not appear to be specifically concentrated at postsynaptic sites (AOKI et al. 1993). Complexes of PSD-95 and nNOS can be co-immunoprecipitated from brain (BRENMAN et al. 1996), but the abundance and location of these complexes is unknown. It is possible that nNOS is assembled by PSD-95 into a complex with the NMDAR at certain postsynaptic sites in neurons that express large amounts of nNOS. However, it is not yet clear how well nNOS competes with other potential PDZ-binding proteins at various glutamatergic postsynaptic sites.

3. GKAP and the SAPAP Family

A Y2H screen for proteins interacting with the GUK domain of PSD-95 resulted in the isolation of a novel interacting protein termed GKAP (KIM et al. 1997). Subsequent screens revealed a family of four related proteins, including GKAP. The family was termed SAPAPs (TAKEUCHI et al. 1997). In cultured hippocampal neurons, GKAP is concentrated in excitatory synapses, as is PSD-95, supporting the notion that GKAP interacts preferentially with PSD-95 (KIM et al. 1997) and is concentrated along with it at synapses. Furthermore, immunocytochemistry at the electron microscopic (EM) level reveals that GKAP is associated with the postsynaptic density in cerebral cortical synapses (KIM et al. 1997).

Despite the strong evidence that GKAP associates with PSD-95 and, thus, with the NMDAR complex at glutamatergic synapses, its function is still unknown. A recent experiment suggests that the SAPAP family may link PSD-95 to the cytoskeleton (DEGUCHI et al. 1998). When PSD-95 is overexpressed in COS cells, it fractionates with membranes, but can be solubilized by extraction with 1% triton. However, co-expression with GKAP renders about half of PSD-95 resistant to solubilization by triton.

4. Neuroligins

Neuroligins are a family of neuronal membrane proteins, the extracellular domains of which interact with the neurexin proteins (ICHTCHENKO et al. 1995). Neurexins are neuronal surface-adhesion molecules, first identified as the target of α-latrotoxin, or Black-Widow-spider venom (USHKARYOV et al. 1992). The current hypothesis is that interactions between neurexins and neuroligins help to form an intercellular junction, but the relationship of that junction to synapses is uncertain (NGUYEN and SUDHOF 1997). A search using the Y2H screen for proteins interacting with the cytosolic C-terminal tail of neuroligins resulted in the selection of clones encoding the PDZ domains of PSD-95 (IRIE et al. 1997). The interaction between PSD-95 and the C-terminal tail of neuroligins occurs via the third PDZ-domain of PSD-95 and not via PDZ domains 1 or 2.

A complex of neuroligin bound to PSD-95 can be isolated from rat brain after homogenization in 1% NP-40 detergent, substantiating the hypothesis that neuroligins interact with PSD-95 in vivo. However, since neurexins and neuroligins are somewhat enriched at synaptic sites, but are not exclusively located there (USHKARYOV et al. 1992), the cellular location of the complexes with PSD-95 is still unclear. Another protein that interacts with the third PDZ domain of PSD-95, SynGAP (see Sect. B.II.1), is a major constituent of the PSD fraction, whereas neuroligin is not a prominent protein in this fraction (CHEN et al. 1998). This observation implies that the neuroligin-PSD-95 complexes are either a minor constituent of most glutamatergic PSDs or are a major constituent only in a small subset of PSDs. Alternatively, they may be located primarily at non-synaptic sites.

5. Citron

A third protein that interacts with PDZ-3 of PSD-95 was originally isolated in a Y2H screen designed to identify proteins interacting with the small GTP-binding protein Rac and was named citron (MADAULE et al. 1995). Citron was subsequently isolated and sequenced from the PSD fraction and shown to interact primarily with the third PDZ domain of PSD-95 (ZHANG et al. 1999). Citron is concentrated at synaptic sites and co-localizes tightly with the NMDAR at synapses, but it is found in a distinct set of neuronal subtypes in brain (FURUYASHIKI et al. 1999; ZHANG et al. 1999). It is highly expressed in thalamic neurons, most of which are excitatory (FURUYASHIKI et al. 1999; ZHANG et al. 1999); however, in the hippocampus, its expression is limited to inhibitory γ-amino butyric acid-type (GABAergic) neurons (ZHANG et al. 1999). Citron is believed to be a target for regulation by Rac and Rho GTPases; however, like many PSD proteins, its precise function remains unknown (however, see MADAULE et al. 1998). Its limited distribution may reflect the presence of a distinct regulatory mechanism that is associated only with a subset of glutamatergic terminals. The possibility that postsynaptic regulatory mechanisms at glutamatergic synapses are different in otherwise similar neuronal types throughout the brain is an intriguing one that would add further subtlety and complexity to information processing in the brain.

III. Protein Kinases Associated with the NMDAR Complex

A wide variety of protein kinases have been implicated in intracellular signaling through the NMDAR (Bliss and Collingridge 1993), including protein kinase C (PKC) (Malinow et al. 1989), Ca^{2+}/CaMKII (Malinow et al. 1989; Fukunaga et al. 1993; Barria et al. 1997; Ouyang et al. 1997), MAP kinases (English and Sweatt 1996; Xia et al. 1996), the src family of protein tyrosine kinases (Grant et al. 1992; Yu et al. 1997; Lu et al. 1998), and the cyclic adenosine monophosphate (cAMP)-dependent protein kinases (Blitzer et al. 1998). Available evidence suggests that two of these are closely associated with an NMDAR signaling complex in vivo – CaMKII and the src family of protein tyrosine kinases.

1. Ca^{2+}/Calmodulin-Dependent Protein Kinase II

CaMKII was the first signaling molecule to be identified as a major constituent of the PSD (Kennedy et al. 1983; Goldenring et al. 1984; Kelly et al. 1984; Kennedy et al., 1990; Kennedy, 1998). More recently, the NMDAR itself was identified as a target of CaMKII in the PSD (Omkumar et al. 1996) and at least two PSD proteins have been recognized that bind CaMKII with high affinity, the NR2B subunit of the NMDAR (Omkumar et al. 1996; Strack and Colbran 1998) and an unidentified protein of 190kDa (Sahyoun et al. 1986). In addition to NR2B, SynGAP (Chen et al. 1998) and densin-180 (Apperson et al. 1996), both abundant proteins in the PSD fraction, are good targets for phosphorylation by the endogenous CaMKII in the PSD fraction and have moderately high affinities for the kinase in vitro. Therefore, they may also act as binding sites for CaMKII in the PSD complex. In the forebrain, the CaMKII holoenzyme is composed of approximately nine catalytic α-subunits and three closely related catalytic β-subunits (Bennett et al. 1983). Shen et al. (1998) showed recently that the β-subunit, but not the α-subunit, targets the holoenzyme to actin filaments within the neuronal soma and in spines. The authors speculate that additional binding sites further target the kinase holoenzyme to the PSD structure itself.

2. The src Family of Small Protein Tyrosine Kinases

The NMDAR can be regulated by phosphorylation on tyrosine residues (Moon et al. 1994; Wang and Salter 1994; Lau and Huganir 1995; Kohr and Seeburg 1996; Yu et al. 1997). In patches excised from CNS neurons, activation of src family kinases by application of a specific activating peptide increased the open probability of the NMDAR channel more than twofold (Yu et al. 1997). Src itself was identified as the activated kinase by application of a specific inhibiting antibody. Src kinase can be co-immunoprecipitated with the NMDAR under non-denaturing conditions, suggesting that it is present in a complex with the NMDAR (YU et al. 1997). The site of interaction of src with the complex has not been investigated.

IV. Interaction of the NMDAR Complex with the Cytoskeleton

Actin filaments are highly concentrated at postsynaptic densities of central synapses and their state of polymerization influences NMDAR channel activity (ROSENMUND et al. 1993). Using the Y2H system with the C-terminus of NR1 as a bait, α-actinine 2 has been revealed as a candidate interaction partner (WYSZYNSKI et al. 1997). This actin-binding protein is a member of the spectrin/dystrophin family, co-localizes with NMDARs in dendritic spines and can be co-immunoprecipitated with NMDARs from brain. The membrane proximal region in NR1 that mediates the binding to α-actinine 2 is also able to bind to Ca^{2+}/calmodulin (EHLERS et al. 1996; ZHANG et al. 1998). The latter interaction leads to an inactivation of the NMDAR. Interestingly, α-actinine 2 and calmodulin compete for the interaction site at the NMDAR in a Ca^{2+}-sensitive manner, implying that the local concentration of Ca^{2+} regulates the state of coupling of the NMDAR to the cytoskeleton.

Another way that the NMDAR can interact with the cytoskeleton depends on alternative splicing of the NR1 pre-mRNA. The C1 cassette in the NR1 cytoplasmic tail has been shown to mediate an association with the neurofilament subunit NF-L in the Y2H system as well as in co-fractionation and co-localization studies (EHLERS et al. 1998). In addition, yotiao, a protein of unknown function capable of binding to the NR1 subunit and fractionating with cytoskeleton-associated proteins, may increase the complexity of the NMDAR/cytoskeleton interaction (LIN et al. 1998). Furthermore, NMDARs may be indirectly connected with microtubules via a novel postsynaptic protein called CRIPT, specifically interacting with the third PDZ domain in PSD-95 (NIETHAMMER et al. 1998). Immunoprecipitates with anti-CRIPT antibodies from rat brain membrane solubilizates contain β-tubulin, PSD-95, PSD-93/Chapsyn110 and NR2B. Interestingly, upon heterologous expression, CRIPT recruits PSD-95 into the microtubulin-based cytoskeleton, suggesting that CRIPT may represent a link between NMDARs and microtubules in dendritic shafts of excitatory synapses (NIETHAMMER et al. 1998). Another link between NMDARs and microtubules can be mediated by the GK domain of Chapsyn110/PSD-93, which is able to bind microtubule-associated protein 1A, MAP1A (BRENMAN et al. 1998).

C. AMPARs and Their Interacting Proteins

AMPARs are assembled from subsets of four subunits, GluR-A to -D or GluR-1 to -4 (Chap. 1), which are expressed at different levels in different neuronal populations (KEINÄNEN et al. 1990). The subunit stoichiometry of AMPARs is most likely four, with most native AMPARs containing two different subunits per functional receptor channel (e.g., GluR-A/B; GluR-A/D; GluR-B/C) (WENTHOLD et al. 1996). AMPAR channels formed without GluR-B participation are Ca^{2+} permeable, but those containing the GluR-B subunit are impermeable to Ca^{2+} (reviewed in Seeburg et al. 1998). Indeed, heteromeric, GluR-B-containing AMPARs appear to constitute the majority of AMPARs in all principal excitatory neurons (GEIGER et al. 1995; VISSAVAJJHALA et al. 1996).

Many synapses contain both AMPARs and NMDARs. In these synapses, sufficient, controlled Ca^{2+} influx through NMDARs can augment the AMPAR-mediated component of excitatory postsynaptic currents, resulting in a long-lasting increase in synaptic strength, referred to as LTP. There is dispute as to whether the increase in the AMPAR-mediated excitatory synaptic current upon an identical test stimulus results from the insertion of additional AMPARs into the postsynaptic membrane (ISAAC et al. 1995; LIAO et al. 1995; LLEDO et al. 1998) or from a structural modification of AMPARs already postsynaptically located. Modification might occur by phosphorylation (BARRIA et al. 1997), which could increase the probability of opening upon glutamate binding, mean channel open time or single-channel conductance. It is possible that both insertion and modification are used, either side by side, in different synapses or at different developmental times (DURAND et al. 1996).

The intracellular domains of AMPAR subunits are considerably smaller than those of the NMDAR subunits (70 vs 500 residues; compiled by Sprengel and Seeburg 1995) and, hence, provide less surface for interacting with different partners. PDZ-domain-containing protein partners binding to the C-terminal domain [consensus sequence (S/TXV/I/L)] have been identified for GluR-A, GluR-B and possibly GluR-C, and a minor C-terminal splice form of GluR-D (GALLO et al. 1992). These are the MAGUK SAP97, a protein termed GRIP (short for Glutamate Receptor Interacting Protein) and PICK1 (short for Protein Interacting with C Kinase). There is also interaction of GluR-B with NSF (*N*-ethylmaleimide-sensitive factor), mediated by an internal domain within the intracellular C-terminal sequence of these AMPAR subunits. In addition, an indirect interaction between AMPARs and a G_i-protein has been revealed. These interactions, summarized in Table 1, are described and commented on below.

I. SAP97

GluR-A was shown to bind by its very C-terminus to one of the PDZ domains of the synapse-associated protein SAP97 (LEONARD et al. 1998), which is a member of the MAGUK family of closely related proteins, along with PSD-95, PSD-93/chapsyn-110 and SAP102. These latter three members can bind the C-termini of the NMDAR NR2 subunits (see above), but NMDARs do not co-immunoprecipitate with SAP97. Among AMPAR subunits, SAP97 interaction is specific for GluR-A. One problem with this interaction is that SAP97 appears to be predominantly presynaptically located (MÜLLER et al. 1995) and, hence, this interaction may involve only a minor AMPAR population.

II. Glutamate Receptor Interacting Protein

GluR-B and GluR-C as well as AMPARs containing these subunits bind to GRIP, a 1100-residue protein with seven PDZ domains (DONG et al. 1997). Interaction requires both PDZ4 and PDZ5 plus extra residues outside of these domains. This is inconsistent with structural data demonstrating that the C-terminus binds a single PDZ domain (DOYLE et al. 1996), as exemplified by NMDAR-PSD-95 interaction (KORNAU et al. 1995). A serine residue at position –4 from the C-terminal end of the GluR-B and -C subunits is vital for GRIP interaction, which may render this interaction subject to modulation by serine/threonine kinases. A second family member has been characterized (SRIVASTAVA et al. 1998) that appears to be expressed in splice forms with and without PDZ7. Interaction of GluR-B with GRIP2, also termed ABP (AMPAR binding protein) required only PDZ5, in contrast to the more extended sequence requirement for GRIP1. However, these differences might be artifactual, reflecting the particular folding of different GRIP domains expressed in heterologous systems.

As of today, the in vivo relevance for AMPAR–GRIP interaction remains unknown. GRIPs may not be essential for AMPAR clustering at, or targeting to, excitatory synapses, because these proteins are not present in all neuronal populations endowed with GluR-B and -C containing AMPARs. Immunocytochemistry shows both proteins co-localized in many neurons and many synapses. GRIP1 appears to be expressed in both principal excitatory neurons and GABAergic interneurons and is located in excitatory as well as inhibitory synapses. Co-transfection into cultured neurons of a construct to express the C-terminal domain of GluR-B, but not of GluR-A, disrupted AMPAR clustering. When co-expressed in vitro, GRIP does not generate AMPAR clusters (XIA et al. 1999). The current model is that GRIP functions as a molecular scaffold, organizing spatial proximity of AMPARs with other synaptic proteins that can interact with the PDZ domains in GRIP. The identity of these other proteins, termed GRASPs (short for GRIP-ASSOCIATED PROTEINS), should be revealed shortly.

III. Protein Interacting with C Kinase 1

PICK1 (STAUDINGER et al. 1995), a 417-residue protein (mouse) containing a PDZ domain towards the N-terminus and a coiled-coil domain in the middle, is a substrate for PKC. The carboxylate-binding loop of its PDZ domain binds the C-terminus of PKCα (STAUDINGER et al. 1997) and, as revealed by Y2H (XIA et al. 1999), also of AMPARs containing subunits GluR-B, GluR-C or

GluR-D in its short C-terminal splice form (GALLO et al. 1992). The coiled-coil domain in PICK1 leads to homo-oligomerization, which should allow PICK1 aggregates to bind both AMPARs and PKCα and, hence, permit the necessary spatial proximity for PKC-mediated phosphorylation of AMPARs (see also Chap. 2). In agreement with this view, PKCα, AMPARs and PICK1 are found co-localized in excitatory synapses (XIA et al. 1999). Furthermore, when co-expressed in vitro, PICK1 leads to the clustering of AMPARs (XIA et al. 1999). The latter finding may indicate that PICK1 participates in AMPAR clustering in central excitatory synapses.

IV. *N*-Ethylmaleimide-Sensitive Factor

The interaction of GluR-B with NSF, first observed by J. Henley and S. Nakanishi, was independently discovered and studied by three groups (NISHIMUNE et al. 1998; OSTEN et al. 1998; SONG et al. 1998). This interaction and its implications for synaptic physiology are as exciting as they are unexpected. NSF, a homohexameric ATPase, is a constituent of all cells and plays an essential role in intracellular membrane fusion, including intercisternal Golgi protein transport and exocytosis, for example of synaptic vesicles (ROTHMAN 1994). Use of the Y2H system demonstrated that NSF binds a ten-residue sequence within the C-terminal domain of GluR-B, upstream of the very C-terminus that interacts with GRIP. NSF also interacts with the short version of GluR-D (GluR4c, eight of the ten residues identical with the NSF interaction domain of GluR-B), but not with GluR-C (seven residues identical with GluR-B). AMPARs solubilized from brain are complexed with NSF, as shown by immunoblotting, and NSF co-localizes with AMPARs in synaptic boutons and dendritic membranes of low-density cultured neurons. Biochemical evidence further indicates that NSF and GluR-B exist together with SNAPs (soluble NSF Attachment proteins) in a complex in brain extracts, which is stable in the absence of hydrolyzable forms of ATP. Quantitative analysis of GluR-B/NSF immunoprecipitates demonstrated an approximately one-to-one stoichiometry for NSF and GluR-B containing AMPARs (OSTEN et al. 1998).

The interaction of NSF with AMPARs has novel, but not fully understood, functional aspects (JAHN 1998; LIN and SHENG 1998). Infusion of a peptide having the sequence of the NSF interaction domain in GluR-B, via a patch pipette into hippocampal CA1 neurons in an acute brain slice, decreased the amplitude of stimulus-evoked AMPAR-mediated excitatory postsynaptic currents (EPSCs). EPSC reduction was also achieved by infusing an antibody to NSF. The time-course of reduction was relatively rapid, in particular for the small peptide, which decreased EPSC amplitudes by approximately one half within 10 min, with no further subsequent reduction. Similar observations were made when recording miniature EPSCs, the amplitudes of which were reduced by 15 min after peptide infusion. These results were interpreted as resulting from a disruption of NSF–AMPAR complexes, although strictly, other NSF interactions might also have been affected by the infused peptides and antibody.

The rapid, plateauing decline of AMPAR-mediated currents upon disrupting NSF–AMPAR interaction may reveal a rapid turnover of a particular pool of synaptic AMPARs. This could be relevant to the hypothesis that vesicle fusion and AMPAR turnover at the postsynaptic membrane might play a role in synaptic plasticity (LLEDO et al. 1998). Additional suggestive evidence for the necessity of vesicle fusion (possibly vesicles containing AMPARs) in the modifiability of synaptic strength comes from the postsynaptic injection of α-SNAP (LTP increase), of an α-SNAP-derived peptide (LTP decrease) and botulinum toxin, which cleaves synaptobrevin (LTP inhibition) (LLEDO et al. 1998). Alternatively, NSF might exert chaperone activity (FENG and GIERASCH 1998) at synaptic AMPARs and alter the conformation of C-terminal domains, allowing phosphorylation or dephosphorylation to take place and disassembling functional protein complexes, such as those formed by AMPARs with GRIP. Clearly, more work is required to reveal the full physiological relevance of the interaction of AMPARs with NSF.

V. G-Protein Interaction

In addition to these direct interactions of AMPAR subunits with GRIP and NSF, there is evidence that AMPARs interact indirectly with the G-protein subunit, $G\alpha_{i1}$ (WANG and DURKIN 1995; WANG et al. 1997). It appears that in intact neurons, AMPAR activation inhibited pertussis-toxin-mediated ADP-ribosylation of this G_i-protein, as well as forskolin-stimulated adenylyl cyclase activity. The biochemical studies further show that modulation of the G_i-protein arose from its association with GluR-A, most likely via an unknown intermediate partner between GluR-A and the G_i-protein. The experiments, however, do not exclude that other AMPAR subunits also interact with the G protein. Moreover, this 'metabotropic' activity by AMPARs does not appear to require channel activation/ion flux. It is tempting to speculate that activity-dependent phosphorylation of GluR-A C-terminal sequences by PKA, PKC or CaMKII might be involved in linking the AMPAR to this cellular pathway (ANDERSEN and F.-SOLENG 1999).

It is too early to attach the proper physiological significance to this perhaps provocative finding. We are aware that G-protein interaction is also observed for the high-affinity kainate receptor (KAR) (RODRIGUEZ-MORENO and LERMA 1998) although, there too, the observation is in its early stages and the molecular players need delineating.

D. Kainate Receptors, the Orphan GluR Delta2 and Their Interacting Proteins

In contrast to NMDARs and AMPARs, the physiological role of KARs in the brain is little understood. The five subunits from which KARs can be assembled are GluR5–7 and KA1 and 2. The latter two do not form functional channels either as homomers or as KA1/KA2 heteromers. Recent evidence suggests that heteromers can however be formed within the GluR5–7 group of subunits, and previous, recombinant studies showed that subunits of the GluR5–7 group can form heteromers with KA1 and with KA2. In each instance the evidence is derived from emergent properties of heteromeric versus homomeric channels. There are also a few orphan GluRs, of which Delta2 has recently been linked to neurodegeneration in the *lurcher* mouse (Zuo et al. 1997). The currently known interactions with KARs and Delta2 are listed in Table 1 and discussed below.

I. Kainate Receptors and MAGUKs

A recent elegant and in-depth study has indicated that KARs, like NMDARs, also interact with members of the MAGUK family of PDZ-domain-containing proteins (Garcia et al. 1998). GluR6 was found to co-immunoprecipitate from brain with PSD-95, SAP97 and SAP102, and KA2 co-immunoprecipitated with PSD-95 and SAP102, but not with SAP97. Recombinant expression of PSD-95 demonstrated that GluR6 and KA2 interact directly with PSD-95. However, GluR6 and KA2 were found to bind, via their intracellular C-terminal tails, to distinct domains of PSD-95. GluR6 interacted by its C-terminal ETMA sequence with PDZ domain 1, but not to PDZ2, and KA2 with both the SH3 and GK domains of PSD-95. Two proline-rich domains in the C-terminal sequence of KA2 mediate binding to the SH3 domain. Co-expression in cultured cells of PSD-95 and either KA2 or GluR6 leads to receptor clustering. In the case of KA2, specific PSD-95 deletions show that PDZ domains and the presence of either the SH3 domain or the GK domain suffices for clustering. For GluR6 clustering, neither the SH3 nor the GK domain need be contained on a truncated PSD-95 molecule, consistent with distinct interaction domains for KA2 and GluR6 within PSD-95. Very importantly, the same study shows that functional properties of KARs change upon association with SAPs, yielding whole-cell currents that desensitize incompletely to L-glutamate. Incompletely desensitizing KAR-mediated currents have indeed been observed in hippocampal cultures prepared from postnatal rats (Wilding and Huettner 1997). It is possible that PSD-95-mediated clustering of KARs accelerates recovery from desensitization.

II. Delta2 and PSD93

The Delta2 subunit is at present an orphan receptor, primarily expressed in cerebellar Purkinje cells where, in the adult, it is relegated to parallel fiber synapses (Lomeli et al. 1992; Landsend et al. 1997). Mice deficient in Delta2 exhibit cerebellar deficits (Kashiwabuchi et al. 1995). Its partners for channel assembly are unknown. A partner might be suggested by the failure of recombinant Delta2 expression to generate functional, L-Glu-gateable ion channels.

Nevertheless, a spontaneous mutation in its M3 segment underlies the phenotype of the *lurcher* mutant mouse (ZUO et al. 1997) whose Purkinje cells degenerate from a constitutively open cation channel. A closely related subunit is Delta1, also an orphan GluR subunit, expressed more widespread in the brain than Delta2.

Purkinje cells express the MAGUK PSD-93/chapsyn-110, which was shown recently to interact with Delta2 (ROCHE et al. 1999). Delta2 binding is mediated by its very C-terminal -TSI sequence and requires PDZ domains 1 and 2 in PSD-93. Curiously, no binding to single PDZ domains was seen in vitro. When co-expressed recombinantly with PSD-93, Delta2 clusters, suggesting that this might happen also in parallel fiber synapses. Notably, PSD-93 is also found in climbing fiber synapses and, hence, in Purkinje cells, subserves the clustering of Delta2 and other GluRs.

E. Outlook

Signal transduction at the postsynaptic glutamatergic membrane is far more complex than simply the generation of EPSPs by binding of glutamate to its receptors. The strength of individual synapses and, indeed, the state of the postsynaptic neuron can be biochemically modified by regulatory pathways that are triggered by activation of NMDARs. The synaptic protein network associated with these receptors and other GluRs, and visible in the electron microscope as the postsynaptic density, contains a diverse and often highly interacting set of signaling molecules. Of the molecules discussed above, only PSD-95 and its family members appear to be present at essentially all synapses that contain NMDARs. It is not entirely clear as yet which of the other molecules are also present at all of the NMDAR- and/or AMPAR-containing synapses and which are present only in certain subsets of these synapses. However, early indications are that we will soon discern specialized subsets of glutamatergic synapses containing unique complements of signaling molecules (ZIFF 1997; RAO et al. 1998; FURUYASHIKI et al. 1999; ZHANG et al. 1999). This should perhaps not be surprising given that the vast majority of excitatory synapses in the brain are glutamatergic and the functions controlled by these synapses, including perception, thinking, remembering and coordinated movement, are diverse and subtle. In the near future, we will learn much more about the role that these signaling complexes play in brain function as a whole. It will be an exciting and challenging task.

References

Andersen P, Figenschou Soleng A (1999) A thorny question: how does activity maintain dendritic spines? Nature Neurosci 2:5–6

Aoki C, Fenstemaker S, Lubin M, Go CG (1993) Nitric oxide synthase in the visual cortex of monocular monkeys as revealed by light and electron microscopic immunocytochemistry. Brain Res 620:97–113

Apperson ML, Moon I-S, Kennedy MB (1996) Characterization of densin-180, a new brain-specific synaptic protein of the O-sialoglycoprotein family. J Neurosci 16:6839–6852

Barria A, Muller D, Derkach V, Griffith LC, Soderling TR (1997) Regulatory phosphorylation of AMPA type receptors by CaMKII during long-term potentiation. Science 276:2042–2044

Bennett MK, Erondu NE, Kennedy MB (1983) Purification and characterization of a calmodulin-dependent protein kinase that is highly concentrated in brain. J Biol Chem 258:12735–12744

Bliss TVP, Collingridge GL (1993) A synaptic model of memory: long-term potentiation in the hippocampus. Nature 361:31–39

Blitzer RD, Connor JH, Brown GP, Wong T, Shenolikar S, Iyengar R, Landau EM (1998) Gating of CaMKII by cAMP-regulated protein phosphatase activity during LTP. Science 280:1940–1942

Brenman JE, Chao DS, Gee SH, McGee AW, Craven SE, Santillano DR, Wu Z, Huang F, Xia H, Peters MF, Froehner SC, Bredt DS (1996) Interaction of nitric-oxide synthase with the postsynaptic density protein PSD-95 and α-1-syntrophin mediated by PDZ domains. Cell 84:757–767

Brenman JE, Topinka JR, Cooper EC, McGee AW, Rosen J, Milroy T, Ralston HJ, Bredt DS (1998) Localization of postsynaptic density-93 to dendritic microtubules and interaction with microtubule-associated protein 1 A. J Neurosci 18: 8805–8813

Chen H-J, Kennedy MB (1997) Identification and cloning of a novel 130 kd protein containing a ras GTPase-activating domain from the rat forebrain postsynaptic density fraction. Abstr Soc Neurosci 23:1466

Chen H-J, Rojas-Soto M, Oguni A, Kennedy MB (1998) A synaptic Ras-GTPase activating protein (p135 SynGAP) inhibited by CaM Kinase II. Neuron 20:895–904

Cho KO, Hunt CA, Kennedy MB (1992) The rat brain postsynaptic density fraction contains a homolog of the Drosophila discs-large tumor suppressor protein. Neuron 9:929–942

Deguchi M, Hata Y, Takeuchi M, Ide N, Hirao K, Yao I, Irie M, Toyoda A, Takai Y (1998) BEGAIN (brain-enriched guanylate kinase-associated protein), a novel neuronal PSD-95/SAP90-binding protein. J Biol Chem 273:26269–26272

Dong H, O'Brien RJ, Fung ET, Lanahan AA, Worley PF, Huganir RL (1997) GRIP: a synaptic PDZ domain-containing protein that interacts with AMPA receptors. Nature 386:279–284

Doyle DA, Lee A, Lewis J, Kim E, Sheng M, MacKinnon R (1996) Crystal structure of a complexed and peptide-free membrane protein-binding domain: molecular basis of peptide recognition by PDZ. Cell 85:1067–1076

Durand G, Kovalchuk Y, Konnerth A (1996) Long term potentiation and functional synapse induction in developing hippocampus. Nature 381:71–75.

Ebralidze AK, Rossi DJ, Tonegawa S, Slater NT (1996) Modification of NMDAR channels and synaptic transmission by targeted disruption of the NR2C gene. J Neurosci 16:5014–5025

Ehlers MD, Fung ET, O'Brien RJ, Huganir RL (1998) Splice variant-specific interaction of the NMDAR subunit NR1 with neuronal intermediate filaments. J Neurosci 18:720–730

Ehlers MD, Zhang S, Bernhadt JP, Huganir RL (1996b) Inactivation of NMDARs by direct interaction of calmodulin with the NR1 subunit. Cell 84:745–755

English JD, Sweatt JD (1996) Activation of p42 mitogen-activated protein kinase in hippocampal long term potentiation. J Biol Chem 271:24329–24332

Feng H-P, Gierasch LM (1998) Molecular chaperones: clamps for the Clps? Curr Biol 8:R464-R467

Fields S, Song O (1989) A novel genetic system to detect protein-protein interactions. Nature 340:245–246

Fields S, Sternglanz R (1994) The two-hybrid system: an assay for protein-protein interactions. Trends Genet 10:286–292

Fukunaga K, Stoppini L, Miyamoto E, Muller D (1993) Long-term potentiation is associated with an increased activity of Ca^{2+}/calmodulin-dependent protein kinase II. J Biol Chem 268:7863–7867

Furuyashiki T, Fujisawa K, Fujita A, Madaule P, Uchino S, Mishina M, Bito H, Narumiya S (1999) Citron, a rho-target, interacts with PSD-95/SAP-90 at glutamatergic synapses in the thalamus. J Neurosci 19:109–118

Gallo V, Upson LM, Hayes WP, Vyklicky L, Winters CA, Buonanno A (1992) Molecular cloning and developmental analysis of a new glutamate receptor subunit isoform in cerebellum. J. Neurosci 12:1010–1023

Garcia EP, Mehta S, Blair LAC, Wells DG, Shang J, Fukushima T, Fallon JR, Garner CC, Marshall J (1998) SAP90 binds and clusters kainate receptors causing incomplete desensitization. Neuron 21:727–739

Garthwaite H (1991) Glutamate, nitric oxide and cell–cell signalling in the nervous system. Trends Neurosci 14:60–67

Geiger JRP, Melcher T, Koh DS, Sakmann B, Seeburg PH, Jonas P, Monyer H (1995) Relative abundance of subunit mRNAs determines gating and Ca^{2+} permeability of AMPA receptors in principal neurons and interneurons in rat CNS. Neuron 15:193–204

Goldenring JR, McGuire JS, DeLorenzo RJ (1984) Identification of the major postsynaptic density protein as homologous with the major calmodulin-binding subunit of a calmodulin-dependent protein kinase. J Neurochem 42:1077–1084

Grant SG, O'Dell TJ, Karl KA, Stein PL, Soriano P, Kandel ER (1992) Impaired long-term potentiation, spatial learning, and hippocampal development in fyn mutant mice. Science 258:1903–1910

Hsueh YP, Kim E, Sheng M (1997) Disulfide-linked head-to-head multimerization in the mechanism of ion channel clustering by PSD-95. Neuron 18:803–814

Ichtchenko K, Hata Y, Nguyen T, Ullrich B, Missler M, Moomaw C, Südhof TC (1995) Neuroligin 1: a splice site-specific ligand for β-Neurexins. Cell 81:435–443

Irie M, Hata Y, Takeuchi M, Ichtchenko A, Toyoda A, Hirao K, Takai Y, Rosahl TW, Sudhof TC (1997) Binding of neuroligins to PSD-95. Science 277:1511–1515

Isaac JT, Nicoll RA, Malenka RC (1995) Evidence for silent synapses: implications for the expression of LTP. Neuron 15:427–434

Jahn R (1998) Synaptic transmission: Two players team up for a new tune. Curr Biol 8:R856–R858

Kashiwabuchi N, Ikeda K, Araki K, Hirano T, Shibuki K, Takayama C, Inoue Y, Kutsuwada T, Yagi T, Kang Y, Aizawa S, Mishina M (1995) Impairment of motor coordination, Purkinje cell synapse formation, and cerebellar long term depression in GluR2 $\delta 2$ mutant mice. Cell 81:245–252

Keinänen K, Wisden W, Sommer B, Werner P, Herb A, Verdoorn TA, Sakmann B, Seeburg PH (1990) A family of AMPA-selective glutamate receptors. Science 249:556–560

Kelly PT, McGuinness TL, Greengard P (1984) Evidence that the major postsynaptic density protein is a component of a Ca2+/calmodulin-dependent protein kinase. Proc Natl Acad Sci USA 81:945–949

Kennedy MB (1995) Origin of PDZ (DHR, GLGF) domains. Trends Biochem Sci 20:350–350

Kennedy MB (1998) Signal transduction molecules at the glutamatergic postsynaptic membrane. Brain Research Reviews 26:243–257

Kennedy MB, Bennett MK, Bulleit RF, Erondu NE, Jennings VR, Miller SM, Molloy SS, Patton BL, Schenker LJ (1990) Structure and regulation of type II calcium/calmodulin-dependent protein kinase in central nervous system neurons. Cold Spring Harbour Sympos Quant Biol 55:101–110

Kennedy MB, Bennett MK, Erondu NE (1983) Biochemical and immunochemical evidence that the "major postsynaptic density protein" is a subunit of a calmodulin-dependent protein kinase. Proc Natl Acad Sci USA 80:7357–7361

Kim E, Cho KO, Rothschild A, Sheng M (1996) Heteromultimerization and NMDAR-clustering activity of chapsyn-110, a member of the PSD-95 family of proteins. Neuron 17:103–113

Kim E, Naisbitt S, Hsueh YP, Rao. A, Rothschild A, Craig AM, Sheng M (1997) GKAP, a novel synaptic protein that interacts with the guanylate kinase-like domain of the PSD-95/SAP90 family of channel clustering molecules. J Cell Biol 136: 669–678

Kim JH, Liao D, Lau L-F, Huganir RL (1998) SynGAP: a synaptic RasGAP that associates with the PSD-95/SAP90 protein family. Neuron 20:683–691

Kohr G, Seeburg PH (1996) Subtype-specific regulation of recombinant NMDAR-channels by protein tyrosine kinases of the src family. J Physiol (Lond) 492: 445–452

Kornau HC, Schenker LT, Kennedy MB, Seeburg PH (1995) Domain interaction between NMDAR subunits and the postsynaptic density protein PSD-95. Science 269:1737–1740

Kornau HC, Seeburg PH, Kennedy MB (1997) Interaction of ion channels and receptors with PDZ domain proteins. Curr Opin Neurobiol 7:368–373

Kurschner C, Mermelstein PG, Holden WT, Surmeier DJ (1998) CIPP, a novel multivalent PDZ domain protein, selectively interacts with Kir4.0 family members, NMDAR subunits, neurexins, and neuroligins. Mol Cell Neurosci 11: 161–172

Kutsuwada T, Sakimura K, Manabe T, Takayama C, Katakura N, Kushiya E, Natsume R, Watanabe M, Inoue Y, Yagi T, Aizawa S, Arakawa M, Takahashi T, Nakamura Y, Mori H, Mishina M (1996) Impairment of suckling response, trigeminal neuronal pattern formation, and hippocampal LTD in NMDAR epsilon 2 subunit mutant mice. Neuron 16:333–344

Landsend AS, Amiry-Moghaddam M, Matsubara A, Bergersen L, Usami S, Wenthold RJ, Ottersen OP (1997) Differential localization of delta GluRs in the rat cerebellum: coexpression with AMPA receptors in parallel fiber-spine synapses and absence from climbing fiber-spine synapses. J Neurosci 17:834–842

Lau LF, Huganir RL (1995) Differential tyrosine phosphorylation of N-methyl-D-aspartate receptor subunits. J Biol Chem 270:20036–20041

Leonard AS, Davare MA, Horne MC, Garner CC, Hell JW (1998) SAP97 is associated with the (-amino-3-hydroxy-5-methylisoxazole-4-propionic acid receptor GluR1 subunit. J Biol Chem 273:19518–19524

Liao D, Hessler NA, Malinow R (1995) Activation of postsynaptically silent synapses during pairing-induces LTP in CA1 region of hippocampal slice. Nature 375: 400–404

Lin JW, Sheng M (1998) NSF and AMPA receptors get physical. Neuron 21:267–270

Lin JW, Wyszynski M, Madhavan R, Sealock R, Kim JU, Sheng M (1998) Yotiao, a novel protein of neuromuscular junction and brain that interacts with specific splice variants of NMDAR subunit NR1. J Neurosci 18:2017–2027

Lledo PM, Zhang X, Sudhof TC, Malenka RC, Nicoll RA (1998) Postsynaptic membrane fusion and long-term potentiation. Science 279:399–403

Lomeli H, Wisden W, Köhler M, Keinänen K, Sommer B, Seeburg PH (1992) High-affinity kainate and domoate receptors in rat brain. FEBS Lett 307:139–145

Lu YM, Roder JC, Davidow J, Salter MW (1998) Src activation in the induction of long-term potentiation in CA1 hippocampal neurons. Science 279:1363–1367

Madaule P, Eda M, Watanabe N, Fujisawa K, Matsuoka T, Bito H, Ishizaki T, Narumiya S (1998) Role of citron kinase as a target of the small GTPase Rho in cytokinesis. Nature 394:491–494

Madaule P, Furuyashiki T, Reid T, Ishizaki T, Watanabe G, Morii N, Narumiya S (1995) A novel partner for the GTP-bound forms of *rho* and *rac*. FEBS Lett 377:243–248

Malinow R, Schulman H, Tsien RW (1989) Inhibition of post-synaptic PKC or CaMKII blocks induction but not expression of LTP. Science 245:862–866

Migaud M, Charlesworth P, Dempster M, Webster LC, Watabe AM, Makhinson M, He Y, Ramsay MF, Morris RG, Morrison JH, O'Dell TJ, Grant SG (1998) Enhanced long-term potentiation and impaired learning in mice with mutant postsynaptic density-95 protein [see comments]. Nature 396:433–439

Monyer H, Sprengel R, Schoepfer R, Herb A, Higuchi M, Lomeli H, Burnashev N, Sakmann B, Seeburg PH (1992) Heteromeric NMDARs: molecular and functional distinction of subtypes. Science 256:1217–1221

Moon IS, Apperson ML, Kennedy MB (1994) The major tyrosine-phosphorylated protein in the postsynaptic density fraction is *N*-methyl-D-aspartate receptor subunit 2B. Proc Natl Acad Sci USA 91:3954–3958

Mori H, Manabe T, Watanabe M, Satoh Y, Suzuki N, Toki S, Nakamura K, Yagi T, Kushiya E, Takahashi T, Inoue Y, Sakimura K, Mishina M (1998) Role of the carboxy-terminal region of the GluRe2 subunit in synaptic localization of the NMDAR channel. Neuron 21:571–580

Müller BM, Kistner U, Kindler S, Chung WJ, Kuhlendahl S, Fenster SD, Lau LF, Veh RW, Huganir RL, Gundelfinger ED, Garner CC (1996) SAP102, a novel postsynaptic protein that interacts with NMDAR complexes in vivo. Neuron 17:255–265

Müller BM, Kistner U, Veh RW, Cases-Langhoff C, Becker B, Gundelfinger ED, Garner CG (1995) Molecular characterization and spatial distribution of SAP97, a novel presynaptic protein homologous to SAP90 and the *Drosophila* discs-large tumor suppressor protein. J Neurosci 15:2354–2366

Nguyen T, Sudhof TC (1997) Binding properties of neuroligin 1 and neurexin 1-beta reveal function as heterophilic cell adhesion molecules. J Biol Chem 272: 26032–26039

Niethammer M, Sheng M (1998) Identification of ion channel-associated proteins using the yeast two-hybrid system. Methods Enzymol 293:104–122

Niethammer M, Kim E, Sheng M (1996) Interaction between the C terminus of NMDAR subunits and multiple members of the PSD-95 family of membrane-associated guanylate kinases. J Neurosci 16:2157–2163

Niethammer M, Valtschanoff JG, Kapoor TM, Allison DW, Weinberg TM, Craig AM, Sheng M (1998) CRIPT, a novel postsynaptic protein that binds to the third PDZ domain of PSD-95/SAP90. Neuron 20:693–707

Nishimune A, Isaac JTR, Molnar E, Joel J, Nash SR, Tagaya M, Collingridge GL, Nakanishi S, Henley JM (1998) NSF binding to GluR2 regulates synaptic transmission. Neuron 21:7–97

Omkumar RV, Kiely MJ, Rosenstein AJ, Min K-T, Kennedy MB (1996) Identification of a phosphorylation site for calcium/calmodulin-dependent protein kinase II in the NR2B subunit of the N-methyl-D-aspartate receptor. J Biol Chem 271: 31670–31678

Osten P, Srivastava S, Imman GJ, Vilim FS, Khatri L, Lee LM, States BA, Einheber S, Milner TA, Hanson PI, et al (1998) The AMPA receptor GluR2 C terminus can mediate a reversible, ATP-dependent interaction with NSF and alpha and beta SNAPs. Neuron 21:99–110

Ouyang Y, Kantor D, Harris KM, Schuman EM, Kennedy MB (1997) Visualization of the distribution of autophosphorylated calcium/calmodulin-dependent protein kinase II after tetanic stimulation in the CA1 area of the hippocampus. J Neurosci 17:5416–5427

Ponting CP, Phillips C (1995) DHR (sic) domains in syntrophins, neuronal NO synthases and other intercellular proteins. Trends Biochem Sci 20:102–103

Rao, A, Kim E, Sheng M, Craig AM (1998) Heterogeneity in the molecular composition of excitatory postsynaptic sites during development of hippocampal neurons in culture. J Neurosci 18:1217–1229

Roche KW, Ly CD, Petralia RS, Wang YX, McGee AW, Bredt DS, Wenthold RJ (1999) PSD-93 Interacts with the delta2 Glutamate Receptor Subunit at Parallel Fiber Synapses. J Neurosci (in press)

Rodriguez-Moreno A, Lerma J (1998) Kainate receptor modulation of GABA release involves a metabotropic function. Neuron 20:1211–1218

Rosenmund C, Westbrook GL (1993) Calcium-induced actin depolymerization reduces NMDA channel activity. Neuron 10:805–814

Rothman JE (1994) Mechanisms of intracellular protein transport. Nature 372:55–63
Sahyoun N, LeVine III H, McDonald OB, Cuatrecasas P (1986) Specific postsynaptic density proteins bind tubulin and calmodulin-dependent protein kinase type II. J Biol Chem 261:12339–12344
Sakimura K, Kutsuwada T, Ito I, Manabe T, Takayama C, Kushiya E, Yagi T, Aizawa S, Inoue Y, Sugiyama H, et al. (1995) Reduced hippocampal LTP and spatial learning in mice lacking NMDAR epsilon 1 subunit. Nature 373:151–155
Seeburg PH, Higuchi M, Sprengel R (1998) RNA editing of brain glutamate receptor channels: mechanism and physiology. Brain Res Rev 267:217–229
Shen K, Teruel MN, Subramanian K, Meyer T (1998) CaMKIIbeta functions as an F-actin targeting module that localizes CaMKIIalpha/beta heterooligomers to dendritic spines. Neuron 21:593–606
Sheng M (1996) PDZs and receptor/channel clustering: rounding up the latest suspects [comment]. Neuron 17:575–578
Song I, Kamboj S, Xia J, Dong H., Liao D, Huganir RL (1998) Interaction of the N-Ethylmaleimide-Sensitive Factor with AMPA Receptors. Neuron 21:393–400
Sprengel R, Seeburg PH (1995) Ionotropic GluRs. In: Ligand- and voltage-gated ion channels, vol 2. CRC Press, Boca Raton, pp 213–263
Sprengel R, Suchanek B, Amico C, Brusa R, Burnashev N, Rozov A, Hvalby O, Jensen V, Paulsen O, Andersen P, Kim JJ, Thompson RF, Sun W, Webster LC, Grant SG, Eilers J, Konnerth A, Li J, McNamara JO, Seeburg PH (1998) Importance of the intracellular domain of NR2 subunits for NMDAR function in vivo. Cell 92: 279–289
Srivastava S, Osten P, Vilim FS, Kathri L, Inman G, States B, Daly C, DeSouza S, Abagyan R, Valtschanoff JG, Weinberg RJ, Ziff EB (1998) Novel anchorage of GluR2/3 to the postsynaptic density by the AMPA receptor-binding protein ABP. Neuron 21:581–591
Staudinger J, Lu J, Olsen EN (1997) Specific interaction of the PDZ domain of the PDZ domain protein PICK1 with the COOH terminus of protein kinase C-alpha. J Biol Chem 172:32019–32024
Staudinger J, Zhuo J, Burgess R, Elledge SJ, Olsen EN (1995) PICK1: a perinuclear binding protein and substrate for protein kinase C isolated by the yeast two-hybrid system. J Cell Biol 128:263–271
Strack S, Colbran RJ (1998) Autophosphorylation-dependent targeting of calcium/calmodulin- dependent protein kinase II by the NR2B subunit of the N-methyl-D-aspartate receptor. J Biol Chem 273:20689–20692
Takeuchi M, Hata Y, Hirao K, Toyoda A, Irie M, Takai Y (1997) SAPAPs. A family of PSD-95/SAP90-associated proteins localized at postsynaptic density. J Biol Chem 272:11943–11951
Tejedor FJ, Bokhari A, Rogero O, Gorczyca M, Zhang J, Kim E, Sheng M, Budnik V (1997) Essential role for dlg in synaptic clustering of Shaker K+ channels in vivo. J Neurosci 17:152–159
Ushkaryov YA, Petrenko AG, Geppert M, Sudhof TC (1992) Neurexins: synaptic cell surface proteins related to the alpha-latrotoxin receptor and laminin. Science 257:50–56
Vissavajjhala P, Janssen WG, Hu Y, Gazzaley AH, Moran T, Hof PR, Morrison JH (1996) Synaptic distribution of the AMPA-GluR2 subunit and its colocalization with calcium-binding proteins in rat cerebral cortex: an immunohistochemical study using a GluR2-specific monoclonal antibody. Exp Neurol 142:296–312
Wang Y, Durkin JP (1995) Alpha-amino-3-hydroxy-5-methyl-4-isoxazolepropionic acid, but not N-methyl-D-aspartate, activates mitogen-activated protein kinase through G-protein beta gamma subunits in rat cortical neurons. J Biol Chem 270:22783–22787
Wang Y, Small DL, Stanimirovic DB, Morley P, Durkin JP (1997) AMPA receptor-mediated regulation of a Gi-protein in cortical neurons. Nature 389:502–504
Wang YT, Salter MW (1994) Regulation of NMDARs by tyrosine kinases and phosphatases. Nature 369:233–235

Wenthold RJ, Petralia RS, Blahos J, Niedzielski AS (1996) Evidence for multiple AMPA receptor complexes in hippocampal CA1/CA2 neurons. J Neurosci 16:1982–1989
Wigler MH (1990) GAPs in understanding Ras. Nature 346:696–697
Wilding TJ, Huettner JE (1997) Activation and desensitization of hippocampal kainate receptors. J Neurosci 17:2713–2721
Wyszynski M, Lin J, Rao A, Nigh E, Beggs AH, Craig AM, Sheng M (1997) Competitive binding of alpha-actinin and calmodulin to the NMDAR. Nature 385: 439–442
Xia J, Zhang X, Staudinger J, Huganir RL (1999) Clustering of AMPA receptors by the synaptic PDZ domain-containing protein PICK1. Neuron 22:179–187
Xia X, Dudek H, Miranti CK, Greenberg ME (1996) Calcium influx via the NMDAR induces immediate early gene transcription by a MAP kinase /ERK-dependent mechanism. J Neurosci 16:5425–5436
Yu XM, Askalan R, Keil GJ, 2nd, Salter MW (1997) NMDA channel regulation by channel-associated protein tyrosine kinase Src. Science 275:674–678
Zhang S, Ehlers MD, Bernhardt JP, Su CT, Huganir RL (1998) Calmodulin mediates calcium-dependent inactivation of N-methyl-D-aspartate receptors. Neuron 21: 443–453
Zhang W, Vazquez L, Apperson M, Kennedy MB (1999) Citron binds to PSD-95 at glutamatergic synapses on inhibitory neurons in the hippocampus [in process citation]. J Neurosci 19:96–108
Ziff EB (1997) Enlightening the postsynaptic density. Neuron 19:1163–1174
Zuo J, Dejager PL, Takahashi KA, Jiang WN, Linden, DJ, Heintz N (1997) Neurodegeneration in *lurcher* mice caused by mutation in $\delta 2$ gluatmate receptor gene. Nature 388:769–773

CHAPTER 4

Cellular and Subcellular Distribution of Glutamate Receptors

R.S. Petralia, M.E. Rubio, and R.J. Wenthold

A. General Distribution

The general distribution of ionotropic glutamate receptors has been described in numerous reviews (Hollmann and Heinemann 1994; Petralia and Wenthold 1996; Bahn and Wisden 1997; Petralia 1997; Watanabe 1997) and will be mentioned only briefly here. Glutamate receptors are found in nearly all neurons and in many types of glia in the central nervous system (CNS), as well as in many cells in the peripheral nervous system and in other structures. Each ionotropic glutamate receptor subunit shows a distinct pattern of distribution in the CNS. Some, such as the α-amino-3-hydroxy-5-methyl-4-isoxazole proprionate (AMPA) receptor subunits, GluR2 and GluR3, and the *N*-methyl-D-aspartate (NMDA) receptor subunit NR1 are both abundant and widespread. Others, such as the AMPA receptor subunits, GluR1 and GluR4, the NMDA receptor subunit NR2B, and the kainate receptor subunits, GluR5 and GluR6, have a restricted distribution and are abundant in some areas of the brain and populations of neurons. For example, GluR1 receptor subunits are abundant in most neurons of the hippocampus while they are rare in neurons of the cerebellum (although abundant in Bergmann glia; Fig. 1). Some glutamate receptor subunits, such as NR2C and $\delta 2$ which are expressed in cerebellar granule cells and Purkinje cells, respectively, are abundant only in one or a few structures. Finally, some ionotropic glutamate receptor subunits, such as the kainate receptor subunit, KA1, and $\delta 1$, are expressed only at low levels. While the relative expression levels vary, most neurons express multiple subtypes and subunits of glutamate receptors. The variability in combinations and expression levels of receptors suggests that the properties of the physiological responses to glutamate, which are dependent on the composition of receptors, differ from neuron–to–neuron and synapse–to–synapse. For example, the virtual absence of GluR2 in AMPA receptors of most interneurons of the hippocampus and cerebral cortex indicates that these neurons express calcium-permeable AMPA receptors.

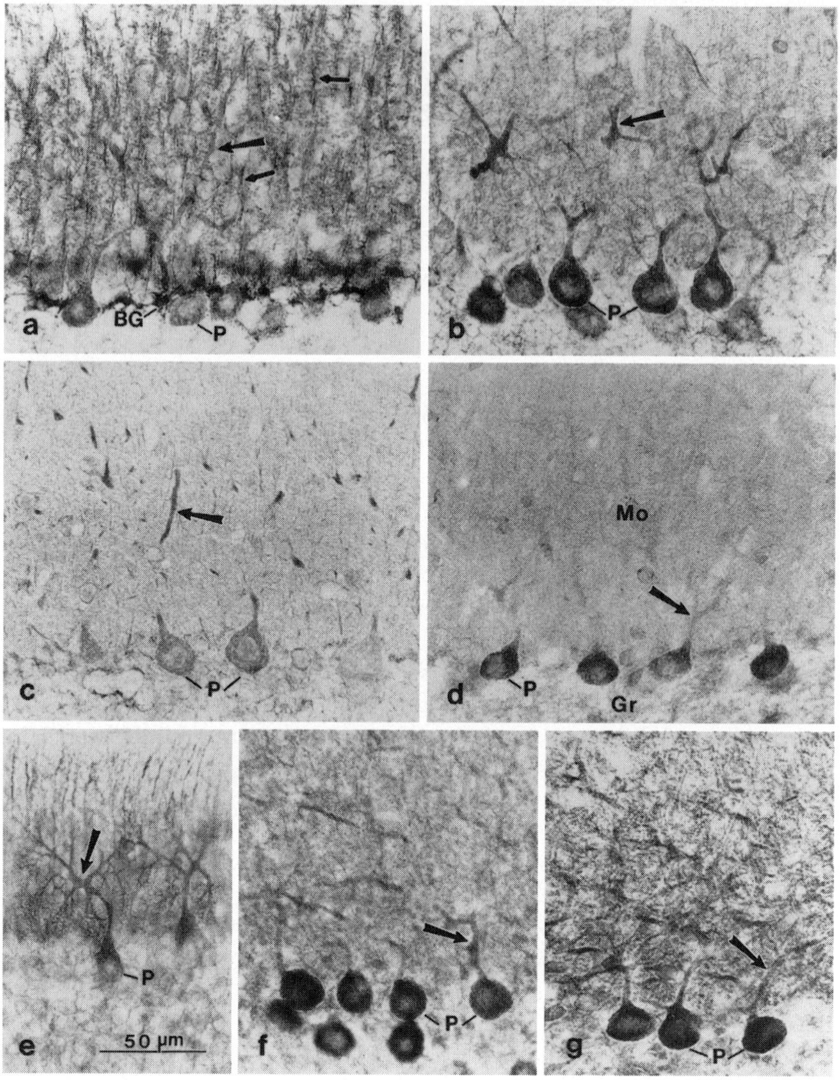

Fig. 1. Sagittal sections of cerebellar cortex of the rat labeled with antibodies to (a) GluR1, (b) GluR2/3, (c) GluR5–7, (d) NR1, (e) δ1/2 at P10, (f) δ1/2 in adult, and (g) mGluR1α. The *large arrow* indicates Purkinje cell dendrite; the *small arrow* indicates Bergmann glial process. *Mo,* molecular layer; *Gr,* granular layer; *P,* Purkinje cell; *BG,* Bergmann glial cell. (Published in Zhao et al. 1997)

B. Subcellular Distribution

I. Postsynaptic Receptors

Within a neuron, glutamate receptors are distributed, to varying degrees, on presynaptic or postsynaptic membranes, on nonsynaptic portions of the cell membrane, and in the cytoplasm. Ionotropic glutamate receptors most often are found on the postsynaptic membrane, adjacent to the postsynaptic density (Fig. 2). Tangential distribution (along the postsynaptic membrane) varies with the type of receptor and with the synapse population, as revealed by immunogold labeling methods (Figs. 2–4). In some cases, the distribution is uniform, as is seen for δ1/2 immunolabeling in parallel fiber synapses of the cerebellum (Fig. 3; LANDSEND et al. 1997; PETRALIA et al. 1998; ZHAO et al. 1998) and AMPA receptor immunolabeling in endbulb synapses of the anteroventral cochlear nucleus (Fig. 4; WANG et al. 1998). In contrast, there are several cases for which the AMPA receptor tangential distribution is uneven, with higher labeling in the outer portions of the synapse, as noted in the cerebral cortex (KHARAZIA et al. 1996b; KHARAZIA and WEINBERG 1997), cochlea (MATSUBARA et al. 1996), neostriatum (BERNARD et al. 1997), and cerebellum (PETRALIA et al. 1998). In the cerebral cortex, NMDA receptor immunolabeling has been reported to be more prominent in the outer (KHARAZIA et al. 1996a) or central portions (KHARAZIA and WEINBERG 1997) of the synapse. The significance of

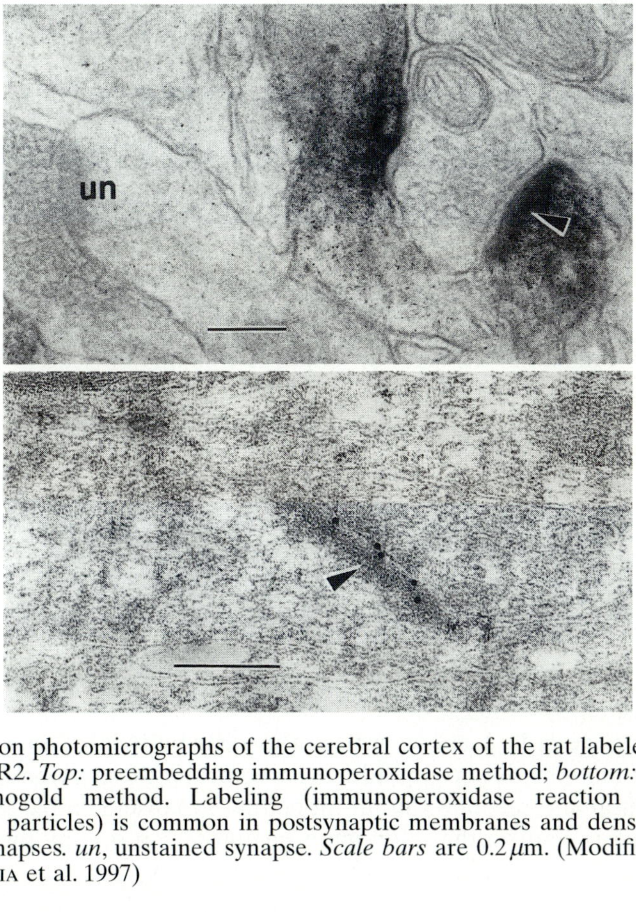

Fig. 2. Electron photomicrographs of the cerebral cortex of the rat labeled with antibody to GluR2. *Top:* preembedding immunoperoxidase method; *bottom:* postembedding immunogold method. Labeling (immunoperoxidase reaction product or immunogold particles) is common in postsynaptic membranes and densities (*arrowheads*) of synapses. *un*, unstained synapse. *Scale bars* are 0.2 μm. (Modified from Fig. 13 in PETRALIA et al. 1997)

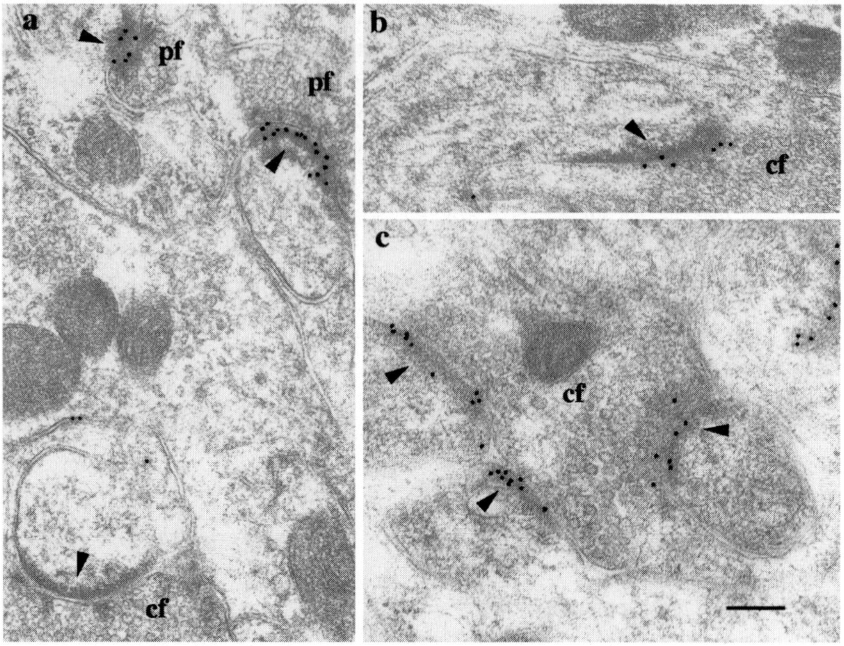

Fig. 3a–c. Immunogold labeling (10 nm gold) for δ1/2 (**a, c**) and GluR2/3 (**b**) in the adult (**a, b**) and postnatal day 14 (**c**) rat cerebellum. *Arrowheads* indicate the postsynaptic densities/membranes (found on the heads of spines). Abundant labeling is seen with δ1/2 antibody in parallel fiber synapses (*pf*) in adults (**a**), in contrast to little or no labeling in climbing fiber synapses (*cf*) in adults. Compare this to the abundant labeling for δ1/2 seen in climbing fiber synapses in the second postnatal week (**c**) and for GluR2/3 in climbing fiber synapses in adults (**b**). *Scale bar* is 0.2 μm. (Modified from Figures in ZHAO et al. 1998)

the variable tangential distribution patterns remains unclear, given that the patterns appear to be both synapse and receptor dependent; the distributions may be functionally relevant, with an arrangement of receptor molecules that relates to their interaction with glutamate release from the presynaptic terminal. Most synapses in the CNS probably have multiple release sites (HARRIS and SULTAN 1995), and the distribution of postsynaptic glutamate receptors may be related to the distribution of these release sites, as proposed by XIE et al. (1997). Alternatively, the distribution pattern may be related to the insertion and removal of receptors from the postsynaptic membrane, which may occur predominantly at the outer portions of the synapse. Finally, some of the variability may reflect epitope availability, rather than actual receptor distributions.

The data described above reflect the distributions of immunogold in synapse populations but do not address specifically the relationship between different glutamate receptors in an individual synapse. Double immunogold labeling studies confirm that glutamate receptors can occupy the same indi-

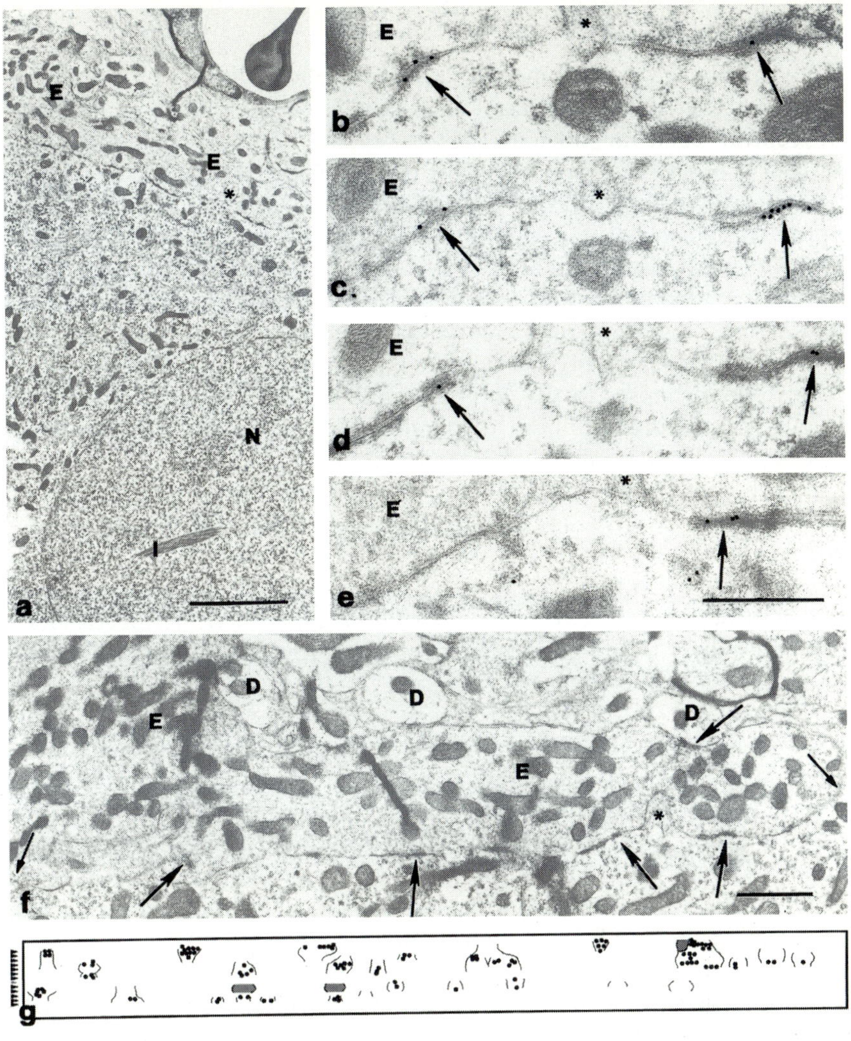

Fig. 4. a–e Electron micrographs of the anteroventral cochlear nucleus of the rat immunolabeled (immunogold method; 10 nm gold) with antibody to GluR2/3. Sections were taken through the endbulb on a spherical bushy cell (**a**), and the positions of active zones and gold particles (**b–f**) were mapped in two dimensions (**g**). **a** Seventh section in series shown at low magnification; this micrograph is included to illustrate the overall structure of the endbulb in relation to the spherical cell soma. **b–e** Four sections from part of the endbulb synapse (sections 1–4 from the series). *Asterisk* demarcates the same position in all micrographs. **f** Third section in the series, printed about the same size as the diagram in **g**. *Small arrows* denote the beginning and end of the endbulb region mapped in **g**. **g** Diagram of the map of 13 serial sections of the endbulb. The beginning of each section is marked by an *arrowhead* on the *far left*; the eighth section was lost and is marked with a *lighter arrowhead*. Sections varied in thickness from silver to yellow, with an estimated average thickness of 70 nm. This is represented in the diagram, with a total of 910 nm for the 13 sections. The synaptic active zones (all are in the same endbulb) are *outlined*. Three attachment plaques are *shaded* more densely. Gold particles are indicated by *dots*. The junctional complex seen in the *far left* of **b–d** (including a single gold particle in **c**) is unclear and is not included in **g**. *D*, dendrite forming synapse with endbulb (i.e., on the side opposite that of the somal/endbulb synapse), *E*, endbulb; *I*, intranuclear rod; *N*, nucleus; *large arrows*, postsynaptic densities with gold. *Scale bars*: **a** 3 μm; **b–e** 0.5 μm; **f** 1 μm. (Modified from Fig. 7 of WANG et al. 1998)

vidual synapse. Examples include GluR2/3 and GluR4 colocalization in cochlear hair cell synapses (MATSUBARA et al. 1996), endbulb synapses in the anteroventral cochlear nucleus (WANG et al. 1998), and auditory nerve synapses on basal dendrites of fusiform cells in the dorsal cochlear nucleus (RUBIO and WENTHOLD 1997). While this may reflect the presence of AMPA receptor complexes containing multiple subunits, in the latter study, only GluR2/3 was found at parallel fiber synapses on apical dendrites of fusiform cells; this is consistent with a differential distribution of GluR4 in fusiform cells, as discussed below. GluR2/3 immunolabeling colocalizes with mGluR1α (NUSSER et al. 1994) and δ1/2 immunolabeling (LANDSEND et al. 1997) in Purkinje cell spine synapses in the cerebellum (Fig. 3). In the latter case, double immunogold-labeling is limited to parallel fiber synapses, whereas climbing fiber synapses contain only GluR2/3 immunolabeling, consistent with findings, described below, showing that δ receptors are rare or absent from climbing fiber synapses in adults (Figs. 3, 5, 6; LANDSEND et al 1997; ZHAO et al. 1997, 1998). Finally, GluR2/3 and NR1 immunogold labeling colocalize in cerebral cortex synapses (KHARAZIA et al. 1996a). A separation can be seen between AMPA and metabotropic receptor distribution, such that mGluR1α labeling is confined mainly to the perisynaptic region. Ionotropic glutamate receptors are uncommon in the perisynaptic region of the membrane (NUSSER et al. 1994; PETRALIA et al. 1998). The presence of glutamate receptors (GluR1, MOLNÁR et al. 1993; NR1, HUNTLEY et al. 1994) in other areas of the neuron cell membrane has been described (reviewed in PETRALIA 1997) and is consistent with the results of physiological studies (e.g. CLARK et al. 1997; HÄUSSER and ROTH 1997). Such nonsynaptic receptors could represent receptors in transit to or from the synapses (see below). An alternative is that their nonsynaptic location is functionally relevant, as has been proposed for certain γ-aminobutyric acid $(GABA)_A$ receptors in cerebellar granule cells (NUSSER et al. 1998). In this case, δ $GABA_A$-containing $GABA_A$ receptor complexes are exclusively extrasynaptic and may mediate tonic inhibition, while δ $GABA_A$-lacking $GABA_A$ receptor complexes are found at synapses and may mediate phasic inhibition.

II. Presynaptic Receptors

There is some evidence for presynaptic ionotropic glutamate receptors, especially for NMDA receptors (AOKI et al. 1994; LIU et al. 1994; SIEGEL et al. 1994; WANG and THUKRAL 1996) and kainate receptors (LERMA 1997; MALVA et al. 1998). Among AMPA receptors, presynaptic immunogold labeling is described most commonly for GluR4 (e.g. MATSUBARA et al. 1996; WANG et al. 1998). Functional data supporting a presynaptic location exists only for kainate receptors (CHITTAJALU et al. 1996; RODRIGUEZ-MORENO et al. 1997). Presynaptic labeling is best characterized for those metabotropic glutamate receptors for which a presynaptic localization is established both immunocytochemically and functionally (OHISHI et al. 1994; PETRALIA et al. 1996a, b; SHIGEMOTO et al. 1996).

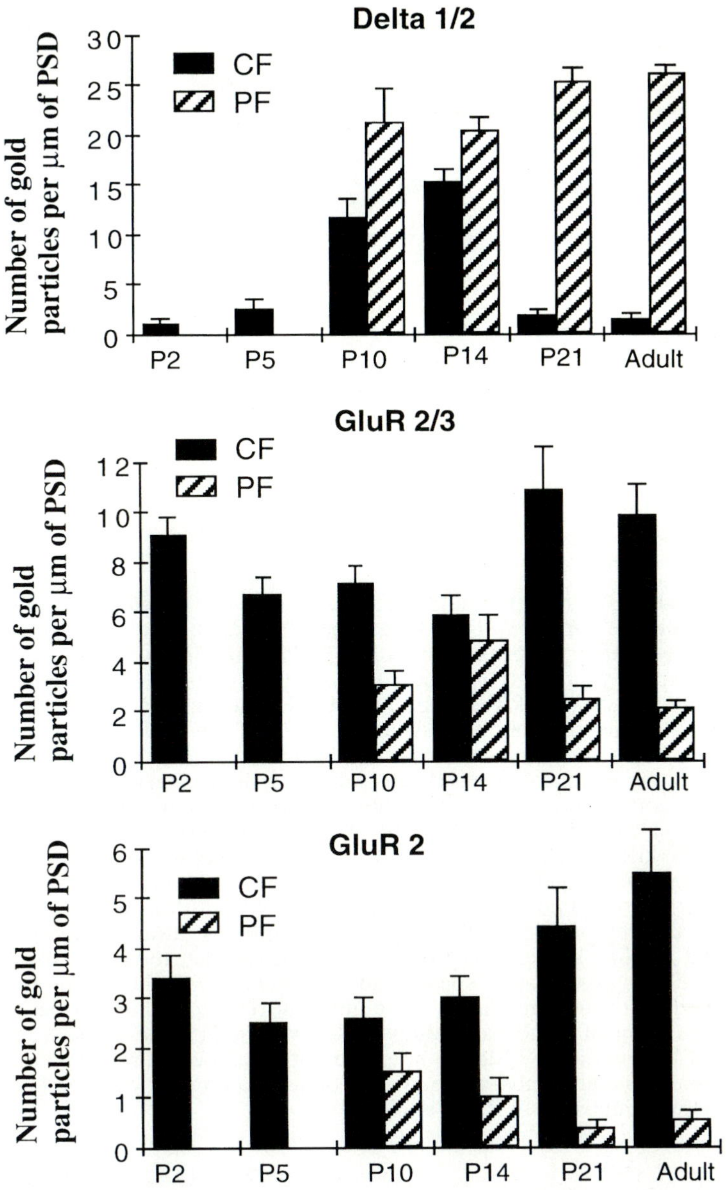

Fig. 5. Histograms illustrating the changes in immunogold labeling for δ1/2 (*top*), GluR2/3 (*middle*), and GluR2 (*bottom*) in the postsynaptic density/membrane (PSD) of parallel (*PF*) and climbing (*CF*) fiber synapses on Purkinje cells during development of the cerebellum. Note especially the large differences in the pattern of immunolabeling between δ1/2 and GluR2/3 and GluR2 at P2–5 and P21-adult. Lower levels of immunogold labeling for CF versus PF (δ1/2) or for PF versus CF (GluR2/3; GluR2) were statistically significant ($P < 0.01$) at P10 (GluR2/3 only), P14 (GluR2 only), and at P21 and in adults for all three antibodies. For δ1/2, statistical significance also was found for the following: P5 CF versus P10 CF, P14 CF versus either P21 or adult CF, and P14 PF versus P21 PF (Published in Zhao et al. 1998)

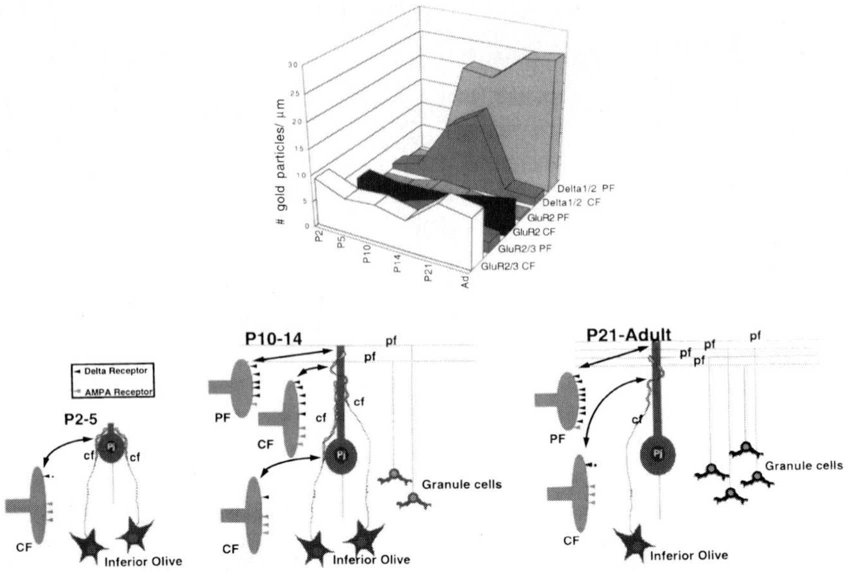

Fig. 6. Summary histogram (*top*) and diagrams (*bottom*) of development of glutamate receptors at parallel (*P10–adult*) and climbing (*P2–adult*) fiber synapses. In the histogram, note especially the peak in immunogold labeling of the δ receptors at P10–14 in climbing fiber synapses (*CF*), the peaks of the α-amino-3-hydroxy-5-methyl-4-isoxazole proprionate (AMPA) receptors at P2–5, and the inverse patterns of peaks for parallel (*PF*) and climbing fiber synapses in adults for AMPA versus δ receptors. Diagrams: climbing fibers (*cf*) innervate the Purkinje cell (*Pj*) body up to about the end of the second postnatal week. By P21, climbing fiber innervation is reduced to a single fiber per Purkinje cell. Climbing fiber synapses on Purkinje cell bodies (early postnatal ages) have many postsynaptic AMPA receptors (based on labeling for GluR2/3 and GluR2) and few δ receptors. Climbing fiber synapses on Purkinje cell dendrites (later postnatal ages to adult) have many AMPA receptors; they have many δ receptors in the second postnatal week, but very few from P21 to adult. Immunogold labeling for δ receptors at parallel fiber (*pf*) synapses is always abundant, but less immunogold labeling is seen for AMPA receptors. Labeled terminals are illustrated diagramatically as postsynaptic spine heads and necks, with the receptors arranged along the surface of the spine head. The number of receptors shown is based roughly on the values of mean number of gold particles/synapse in Tables 1 and 2 and Results published in ZHAO et al. (1998); it is intended only to show the relative amounts and not to represent actual numbers. The *asterisk* denotes a level of less than half a receptor per synapse (Published in ZHAO et al. 1998)

III. Intracellular Glutamate Receptors

In addition to their distribution in various synaptic and nonsynaptic plasma membranes, ionotropic glutamate receptors are often distributed widely in the neuron cytoplasm. Examples include the abundant immunoperoxidase and immunofluorescence cytoplasmic labeling of GluR1, GluR2, GluR2/3, NR1, and NR2 A/B antibodies in hippocampal pyramidal cells, and for GluR2, GluR2/3, GluR5–7, and NR1 antibodies in cerebellar Purkinje cells (Fig. 1;

PETRALIA and WENTHOLD 1992; MARTIN et al. 1993; PETRALIA et al. 1994a,b, 1997; PETRALIA 1997; ZHAO et al. 1997). In spinal cord cultures, a large intradendritic pool of GluR1 receptors is seen with immunofluorescence both before and after synapse formation (MAMMEN et al. 1997). Immunoperoxidase labeling in the cytoplasm usually shows distinctive patterns, with patches of staining concentrated on various organelles such as endoplasmic reticulum, Golgi, and mitochondria (e.g. ESHHAR et al. 1993; KHARAZIA et al. 1996a). Labeling of mitochondria typically is limited to a patch on one pole of the mitochondrion, often associated with a reticular membrane element (see reviews of PETRALIA 1997; RIZZUTO et al. 1998). These distributions represent pathways of glutamate receptor transport through the neuron and, in some cases, suggest that a substantial cytoplasmic pool exists. Similar distribution patterns in the cytoplasm have been shown using immunogold techniques, although labeling density usually is low due to the lower sensitivity of the technique (see below). Recently, we analyzed the distribution of gold labeling in dendrites of fusiform cells of the dorsal cochlear nucleus, pyramidal cells of the hippocampus (Fig. 7), and Purkinje cells of the cerebellum, using several AMPA and metabotropic glutamate receptor antibodies. In these three cell types, the intracellular pool of glutamate receptors appears to be organized into clusters of gold particles. These clusters are distributed in the dendrite cyotoplasm, and are associated with a smooth endoplasmic reticulum membrane and cytoskeleton.

In addition to the immunocytochemical data based on analyses using several different antibodies and multiple detection techniques, recent biochemical studies also show a pool of receptors that is not present on the surface of the cell. In cultured hippocampal neurons, about half of the total AMPA receptors are found on the surface of neurons, as shown using proteolysis, crosslinking and biotinylation (HALL and SODERLING 1997b). For NMDA receptors, about half of the NR1 subunit is expressed on the surface, while much more of the NR2 is on the surface (HALL and SODERLING 1997a). Similar results have been obtained in studies of cultured spinal cord neurons (MAMMEN et al. 1997) and cultured cerebellar granule cells (HUH and WENTHOLD 1999). Finally, green fluorescent protein-tagged AMPA receptors expressed in cultured hippocampal neurons are seen throughout the dendrites of living neurons (DOHERTY et al. 1997).

As discussed below, intracellular glutamate receptors may play a critical role as a reserve pool, facilitating a rapid up (or down) regulation of surface receptors. Although similar mechanisms may be envisioned for all synaptic receptors, it is interesting to note that the glycine receptor has a very different distribution with little detectable intracellular pool. Studies of many different neurons using several different antibodies show labeling predominantly at postsynaptic sites (TRILLER et al. 1985; WENTHOLD et al. 1988). Even immunofluorescence analyses of brain tissue show a punctate arrangement of glycine receptors, which is a pattern never reported for glutamate receptors. In contrast, low levels of cytoplasmic glutamate receptors usually indicate cor-

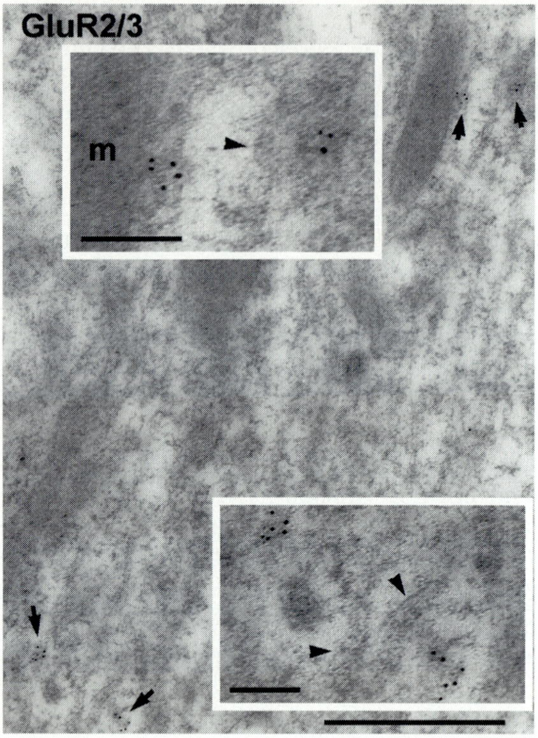

Fig. 7. Post-embedding immunogold labeling (5 nm gold) with an antibody selective for the α-amino-3-hydroxy-5-methyl-4-isoxazole proprionate (AMPA) receptor subunits GluR2/3 in a dendrite in the stratum radiatum of the CA1 region of the hippocampus. *Arrows* indicate clusters of gold particles in the cytoplasm. These clusters are shown at high magnification in the insets. They appear to label reticular structures associated with mitochondria (*m*) or microtubules (*arrowheads*). *Scale bars* are 0.5 μm for the low magnification and 0.1 μm for the *insets*. (RUBIO and WENTHOLD, unpublished data)

respondingly low levels of postsynaptic receptors, although there are exceptions. For example, in some cases, immunoperoxidase labeling for δ1/2 glutamate receptors can be abundant in synapses yet low in the cytoplasm (see Fig. 11 in PETRALIA 1997).

C. Differential Distribution of Glutamate Receptors Within Neurons

Because most neurons receive several different excitatory and inhibitory inputs, it might be assumed that the neurons have developed a mechanism for selectively targeting their multiple glutamate receptors to various populations of synapses, such that they differ in their receptor composition and, consequently, in their physiological responses to neurotransmitters. Indeed, results from several studies support a differential distribution of both glutamate (e.g. LANDSEND et al. 1997; RUBIO and WENTHOLD 1997; ZHAO et al. 1997, 1998) and GABA (NUSSER et al. 1996a,b) receptors, and demonstrate that the expression of receptors can be controlled at the level of the individual synapse.

I. Fusiform Neurons of the Cochlear Nucleus

One type of glutamate receptor organization involves synapse populations on two separate dendrite arbors, typically apical and basal dendrites arranged on opposite poles of the cell; this type of dendrite morphology is typical of large pyramidal cells of the cerebral cortex and hippocampus, although definitive evidence for a differential distribution of this kind has not been published for these neurons. However, this type of differential distribution has been described in detail for the highly-polarized fusiform cells of the dorsal cochlear nucleus (Fig. 8, Table 1; RUBIO and WENTHOLD 1997). These neurons receive two major glutamatergic inputs – granule cell-parallel fiber synapses on their apical dendrites and auditory nerve fiber synapses on their basal dendrites. Glutamate receptor composition for these two synapse populations is distinctly different. Immunogold labeling for GluR2/3, GluR2, NR2A/B and δ1/2 antibodies is present at both apical and basal dendrite synapses. In contrast, immunogold labeling for GluR4 and the metabotropic glutamate receptor, mGluR1α, is present only at basal dendrite synapses. Physiological studies confirm that metabotropic glutamate receptors are present in fusiform cells

Table 1. Summary of the post-embedding immunoreactivity for glutamate receptor subunits at the auditory nerve and parallel fiber synapses

Receptors	Auditory nerve synapses (basal dendrites)		Parallel fiber synapses (apical dendrites)	
	Number of PSDs	Number of gold particles per μm of PSD ± SE	Number of PSDs	Number of gold particles per μm of PSD ± SE
GluR2/3 [a,b]	18	17.7 ± 4.0	17	16.5 ± 3.2
GluR2[b]	25	9.1 ± 1.1	17	7.2 ± 1.2
GluR4[b]	17	19.1 ± 2.2	17	0
GluR4 (10 nm gold)	9	9.2 ± 1.9	8	0
NR2 A/B[b]	19	6.4 ± 1.4	14	9.8 ± 1.3
mGluR1α[a,b]	35	8.0 ± 1.3	25	0
δ1/2[b]	31	8.3 ± 1.2	25	33.9 ± 3.1

[a] Monoclonal antibodies
[b] 5-nm gold was used for immunogold labeling quantification with all the antibodies selective for the glutamate receptor subunits, except for GluR4 which was analysed using 5 nm and 10 nm. Table modified from Table 2 in RUBIO and WENTHOLD (1997)

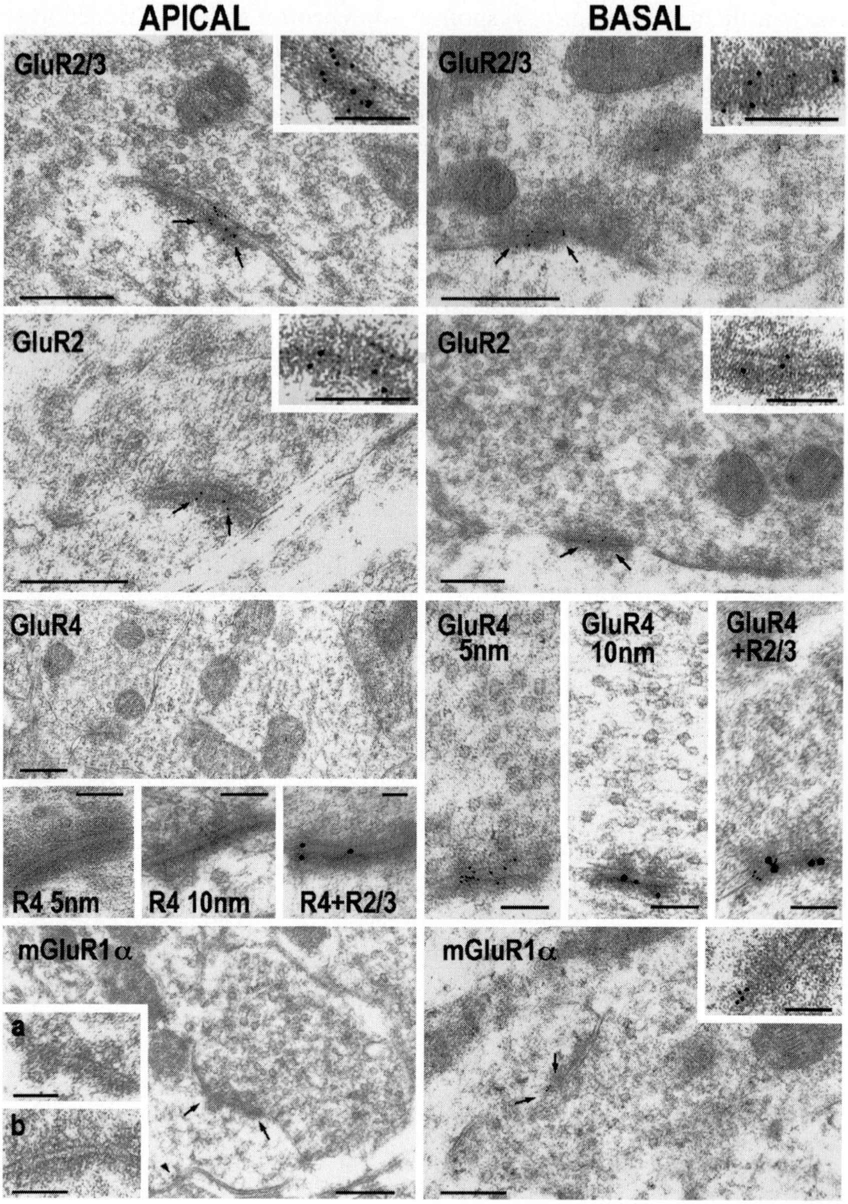
APICAL
BASAL
GluR2/3
GluR2/3
GluR2
GluR2
GluR4
GluR4 5nm
GluR4 10nm
GluR4 +R2/3
R4 5nm
R4 10nm
R4+R2/3
mGluR1α
mGluR1α
a
b

but do not modulate responses evoked by parallel fiber stimulation (Molitor and Manis 1997); this indicates that functional metabotropic glutamate receptors are absent from synapses on apical dendrites. The preferential presence of GluR4 at auditory synapses on basal dendrites may reflect the faster neurotransmission known to occur at auditory synapses compared to most other synapses in the CNS. AMPA receptors with a high concentration of GluR4 have fast responses. The rapid firing of the auditory inputs to the basal dendrites of the fusiform cells presumably is necessary for accurate sound localization by the auditory nuclei.

II. Climbing Fiber and Parallel Fiber Synapses of Purkinje Neurons

As discussed above, studies of fusiform neurons of the dorsal cochlear nucleus demonstrate that glutamate receptors can selectively target synapses on different dendrites. However, because synapse populations are most often found on the same dendrite, specific targeting also must occur for glutamate receptors that are not spatially separate. Purkinje neurons of the cerebellum receive two glutamatergic inputs – climbing fibers and parallel fibers – to the same dendrite arborization. In comparing the distribution of glutamate receptors at these two synapse populations in the adult, parallel fiber/Purkinje cell spine synapses were found to have a high density of postsynaptic $\delta 2$ glutamate receptors, while climbing fiber/Purkinje cell spine synapses were found to have few or none (Figs. 3, 5, 6; Landsend et al. 1997; Zhao et al. 1997). In contrast, AMPA receptors (GluR2 and GluR2/3 antibodies) were expressed at both populations, but had a four-fold higher concentration in climbing fiber

◄

Fig. 8. Post-embedding immunogold labeling with antibodies selective for the α-amino-3-hydroxy-5-methyl-4-isoxazole proprionate (AMPA) receptor subunits GluR2/3 (5 nm), GluR2 (5 nm gold), and GluR4 (5 nm and 10 nm gold) and the metabotropic mGluR1α (*insets*: a 5 nm gold; b 10 nm gold) in apical and basal dendrites of fusiform cells. On postsynaptic membranes of parallel fiber synapses with apical dendrites (panels on the *left*), immunogold labeling is seen only with antibodies to GluR2/3 (monoclonal antibody) and GluR2, but not with antibodies to GluR4 and mGluR1α. In contrast, postsynaptic membranes of auditory nerve synapses on basal dendrites (panels on the *right*) show immunogold labeling for all four antibodies. *Insets* of GluR2/3, GluR2 and mGluR1α show higher magnification of the postsynaptic densities (*arrows*). Insets at the bottom of the left GluR4 micrograph (apical dendrite synapses) show postsynaptic membranes of parallel fiber synapses at apical dendrites of fusiform cells after immunogold labeling for GluR4 using 5 nm and 10 nm gold, and double labeling for GluR4 (5 nm gold) plus GluR2/3 (monoclonal antibody, 15 nm gold). Note that only the 15 nm gold is seen, showing that only GluR2/3 immunolabeling is present at postsynaptic membranes of parallel fiber synapses on apical dendrites. In the three GluR4 micrographs on the *right* (basal dendrite synapses), postsynaptic membranes of the auditory nerve are immunolabeled for GluR4 (5 nm, 10 nm gold), and double labeled for GluR4 (5 nm gold) and GluR2/3 (monoclonal, 15 nm gold). *Scale bars*: GluR2/3, GluR2, mGluR1α, 0.25 μm; *insets*, 0.1 μm; GluR4 low magnification, 0.5 μm; *insets*, 0.1 μm (modified from figures in Rubio and Wenthold, 1997)

synapses (ZHAO et al. 1998). Based on findings of impaired long-term depression in $\delta 2$ knockout mice (KASHIWABUCHI et al. 1995), the differential distribution of $\delta 2$ receptor subunits in Purkinje cell synapses may reflect a specific role of these receptors in the control of synaptic plasticity at the parallel fiber synapse.

III. Pyramidal Neurons of the Hippocampus

A second example of a neuron showing a differential distribution of glutamate receptors among synapse populations in the same dendrite arborization is the CA3 pyramidal neuron of the hippocampus (reviewed in PETRALIA 1997). Physiological/pharmacological studies indicate that long-term potentiation (LTP) is NMDA receptor-dependent in most synapses on the apical dendrites (as well as on the basal dendrites) of these neurons, while LTP is NMDA receptor-independent in the specialized mossy terminal synapses on the proximal portion of the apical dendrite (ZALUTSKY and NICOLL 1990; DERRICK et al. 1991); this suggests that NMDA receptors may be absent from mossy terminal synapses on CA3 pyramidal neurons. Immunoperoxidase labeling for the NR1 receptor supports this idea, showing substantial postsynaptic staining in small spine synapses of the CA3 region, but little definitive staining in the mossy terminal synapses (PETRALIA et al. 1994a; SIEGEL et al. 1994). The presence of NMDA receptors was confirmed later by showing that the mossy terminal synapses do have some functional NMDA receptors (SPRUSTON et al. 1995). Indeed, immunolabeling with NR2A/B antibody shows substantial postsynaptic staining in some mossy terminal synapses (PETRALIA et al. 1994b). A more recent study using antibodies specific to either NR2A or NR2B shows that NR2A may be present throughout the dendrite arborization of CA3 pyramidal neurons (including the mossy fiber synapses), while NR2B appears to be present in most CA3 pyramidal neuron synapses but absent from mossy fiber synapses (FRITSCHY et al. 1998). Thus, the presence of functionally different NMDA receptors within populations of CA3 pyramidal neuron synapses appears to be due to variations in receptor subunit composition; most synapses in the CNS have high NR1 content and both NR2A and NR2B, while mossy fiber synapses have low NR1 content and only NR2A.

Other examples of differential distribution of ionotropic glutamate receptors exist in the hippocampus. CA3 interneurons show a differential distribution of functional AMPA receptors, with calcium-permeable (GluR2 lacking?) AMPA receptors apposed to mossy fiber terminals, and calcium-impermeable (GluR2 containing?) AMPA receptors apposed to commissural/associational axon terminals (TOTH and MCBAIN 1998). In addition, at least three populations of AMPA receptors (GluR1+GluR2; GluR2+GluR3; GluR1 only) have been detected in the CA1/CA2 region, although whether these different receptor complexes exist at different or at the same synapses is not known (WENTHOLD et al. 1996). The differential distribution of AMPA receptors versus

NMDA receptors may be modified during development and induced by functional changes such as LTP, and has most often been studied in the apical dendrites of the pyramidal neurons of the CA1 region of the hippocampus. In this region, all synapses on these dendrites may have NMDA receptors, but not AMPA receptors, at an early age. These "silent synapses" can acquire AMPA receptors following adequate activation, so that during development the ratio of pure NMDA receptor synapses to NMDA/AMPA receptor synapses decreases (see above; ISAAC et al. 1995; WU et al. 1996; DURAND et al. 1996; PETRALIA et al. 1999; review by MALENKA and NICOLL 1997).

D. Receptor Distribution During Development

The general distribution of ionotropic glutamate receptors during development is reviewed by others (e.g. BAHN and WISDEN 1997; WATANABE 1997) and will not be discussed in detail. The early postnatal period (0–25 days in the rat) is characterized by extensive changes in the expression of several glutamate receptor subunits, implying developmental changes in the functional properties of the receptors. A primary example is the ontogeny of NR2 subunits (reviewed by WATANABE 1997). NR2B is abundant in the forebrain during both immature and adult stages, whereas NR2A is abundant only in adults; in the cerebellum there is a developmental switch in NR2 subunits, with the NR2B subunit being replaced by NR2A and NR2C. NR2D is common throughout the brain stem and spinal cord during development but is uncommon in the adult. Presumably, both NR2B and NR2D play important roles in neural development. In fact, mutant mice that lack NR2B (KUTSUWADA et al. 1996; review by WATANABE 1997) or that express NR2B without the intracellular C-terminal domain (SPRENGEL et al. 1998) die perinatally. The changeover from immature to mature NMDA receptor composition is mediated by presynaptic activity (VALLANO et al. 1996; GOTTMANN et al. 1997; OZAKI et al. 1997).

I. Development of Glutamate Receptors in Cultured Neurons

The development of ionotropic glutamate receptors in individual synapse populations has been studied in vitro and in vivo, and provides clues to understanding synapse formation and function in the immature CNS. In cultured rat ventral spinal cord (MAMMEN et al. 1997; O'BRIEN et al. 1997), AMPA receptors cluster at very immature synapses. This phenomenon appears to be independent of NMDA receptor clustering, although the presence of some NMDA receptor molecules could not be ruled out. In hippocampal cultures (CRAIG et al. 1994; RAO et al. 1998), AMPA receptors and GABA receptors cluster independently at glutamatergic and GABAergic terminals, respectively. Formation of postsynaptic structures at glutamatergic synapses involves an initial clustering of scaffolding proteins which may be necessary for anchoring glutamate receptors in the postsynaptic membrane (see below). Receptor activation also plays a role in receptor clustering. Modifications in activity may affect the initial clustering of anchoring/scaffolding proteins, thus leading to changes in receptor clustering, as shown for glycine receptors (KIRSCH and BETZ 1998). However, the role that activity plays in determining the clustering of glutamate receptors is not clear; glutamate receptor activation does not induce receptor clustering in spinal neuron cultures (O'BRIEN et al. 1997) and a chronic blockade of NMDA receptor activity promotes NMDA receptor clustering in hippocampal neuron cultures (RAO and CRAIG 1997). AMPA and NMDA receptors cluster at synapses in different sequences. AMPA receptors first cluster at dendrite spines (but see MARTIN et al. 1998); colocalization of AMPA and NMDA receptors at these spines occurs later. NMDA receptors form early synaptic and nonsynaptic clusters on dendrite shafts. These latter synapses may, by definition, be "silent synapses," although the presence of some AMPA receptors in these synapses could not be ruled out. Together, these in vitro studies suggest the following: early localization of ionotropic glutamate receptors at glutamatergic synapses is highly specific and may be regulated by anchoring proteins; AMPA receptors may cluster at early synapses independent of NMDA receptors, and this phenomenon may be common throughout the CNS; NMDA receptors may cluster at early synapses lacking AMPA receptors in some cases, and these could be functionally "silent synapses." These in vitro studies suggest that at least two different sequences of glutamate receptor accumulations are involved in the formation of glutamatergic synapses and that these may vary with the brain structure. However, these in vitro studies relied on light microscope immunocytochemistry, for which the exact localization of labeling in relation to the postsynaptic membrane cannot be determined.

II. In Vivo Development of Synaptic Glutamate Receptors in the Cerebellum

Recently, we have performed in vivo electron microscope immunocytochemical studies that examine the development of glutamate receptors in postsynaptic membranes of tissue sections from the cerebellum (ZHAO et al. 1997, 1998; PETRALIA et al. 1998) and hippocampus (PETRALIA et al. 1999). As noted above, parallel fiber/Purkinje cell spine synapses in the adult cerebellar cortex have a high density of postsynaptic δ glutamate receptors, whereas climbing fiber/Purkinje cell spine synapses have few or none (Figs. 3, 5, 6). In contrast, AMPA receptors (labeled with GluR2 and GluR2/3 antibodies) occur in higher concentrations in climbing fiber synapses than in parallel fiber synapses. Immunogold studies show that early synapses (first postnatal week), presumably all from climbing fibers, have high levels of AMPA receptors (as seen in adults) but low levels of δ receptors ($\delta 2$ immunoperoxidase labeling is detected in synapses by P0; TAKAYAMA et al. 1996). Thus, as indicated by in vitro studies of spinal cord and hippocampus (discussed above), AMPA receptors may arrive early at newly forming synapses (NMDA receptors were not studied in detail). By the second postnatal week, δ receptors occur in high concentrations in both parallel and climbing fiber synapses on dendrites (AMPA receptor densities show only relatively small changes from this age to adult). The second postnatal week is a crucial time, during which the adult dendrites form, parallel fiber synapses form, and climbing fiber innervation is reduced to a single climbing fiber (with multiple synaptic contacts) for each Purkinje cell. The localization of δ receptors to both parallel and climbing fiber synapses at this time is highly specific; δ receptors remain infrequent at somal climbing fiber synapses, which are destined to be lost, and δ receptors are absent from GABAergic synapses. Thus, these data indicate that a major switch occurs in glutamate receptor composition at climbing fiber synapses in the second postnatal week. These findings support suggestions (Kashiwabuchi et al. 1995; Kurihara et al. 1997) that δ receptors have two functions in Purkinje cells: they are involved in the initial formation of adult parallel and climbing fiber synapses and, in adults, they are involved in parallel fiber synapse function and probably play a direct role in long-term depression of parallel fiber synapse responses. In addition, as suggested by the in vitro studies discussed above, these findings also indicate that specific targeting mechanisms develop early in synapse formation and may be responsible for precise control of developmental changes in glutamate receptor composition of synapses (see below).

III. In Vivo Development of Synaptic Glutamate Receptors in the Hippocampus

Distinctive differences in the development of glutamate receptor distribution patterns are seen between the cerebellum and the hippocampus. In the hippocampus (CA1 stratum radiatum), AMPA receptors are low in density or absent from most synapses at postnatal days 2 and 10, while they are abundant at most synapses in adults, suggesting that there is a dramatic rise in AMPA receptors during later stages in the development of these synapses (Petralia et al. 1999). In contrast, NMDA receptors are found at moderate levels at all three ages studied and show only a modest increase in adults, suggesting that many of the synapses seen at postnatal days 2 and 10 contain NMDA receptors but lack AMPA receptors. The latter synapses probably would be similar to NMDA+/AMPA-synapses seen in hippocampal neurons in vitro (Rao et al. 1998). These findings correspond remarkably well with physiological studies that suggest the presence of NMDA receptor-containing "silent synapses" that lack functional AMPA receptors (Durand et al. 1996; Wu et al. 1996; Malenka and Nicoll 1997). Thus, the developing hippocampus may have "silent synapses," which probably are not present in developing cerebellar Purkinje cells. Early Purkinje cell synapses may be similar to the "AMPA-first" early synapses seen in cultured neurons from the hippocampus and spinal cord (described above). The function of NMDA receptors in Purkinje cell development is less clear, although experimental studies suggest that they play some role in synapse maturation (Rabacchi et al. 1992; Vallano et al. 1996).

E. Mechanisms of Receptor Targeting

The complex, changing pattern of glutamate receptor distribution during development and the precise pattern of differential distribution of these receptors in populations of synapses in the adult imply that targeting of glutamate receptors in neurons is a precisely regulated process (reviews by Ehlers et al. 1996; Kirsch et al. 1996). While the nature of these processes is not well understood, two mechanisms can be postulated: 1) glutamate receptors are synthesized and assembled in the cell body and are transported via separate routes to different synaptic populations; 2) synthesis and assembly occurs in the cell body as in the first category, but the receptor molecules are transported indiscriminately throughout the somatodendritic compartment and are sorted only at the individual synapse. It appears that glutamate receptors are synthesized and assembled predominantly in the cell body. Neurons retain most of their organelles for membrane protein synthesis and assembly in the cell body, including the rough endoplasmic reticulum, Golgi apparatus and trans-Golgi network (Peters et al. 1991). Also, glutamate receptor messenger ribonucleic acid (mRNA) is confined largely to the cell body, according to numerous in situ hybridization studies (e.g. Craig et al. 1993; Eshhar et al. 1993; see discussion in Gazzaley et al. 1997). Therefore, receptors will need to be actively transported from the cell body to the synapse with various molecular motors, such as different kinds of kinesins (review by Hirokawa 1998) or unconventional myosins (review by Mermall et al. 1998). Kinesins, as well as cytoplasmic dyneins, are associated with the transport of organelles along microtubules. For example, KIFC2 may transport certain kinds of multivesicular bodies in dendrites. Myosins are associated with a number of organelles and may transport organelles and other cell components along actin filaments; they may also be involved in exocytosis and endocytosis.

The first model, differential transport of glutamate receptor molecules from the cell body to the synapse along separate pathways for each synapse population, is supported by numerous studies of polarized epithelial cells, in which some proteins are targeted directly to the apical or basolateral membrane (e.g. Dotti and Simons 1990; Perez-Velazquez and Angelides 1993; Drubin and Nelson 1996; Wozniak and Limbird 1996). Recent studies of glutamate receptor distributions in the fusiform cells of the dorsal cochlear nucleus support this model. As discussed above, these neurons have GluR4 and mGluR1α receptor molecules associated with auditory synapses on the basal dendrites but not with parallel fiber synapses on the apical dendrites.

Examination of receptor immunolabeling in the cytoplasm shows that these receptor molecules are found in high concentrations throughout the basal dendrites but are uncommon in the finer apical dendrite branches (RUBIO and WENTHOLD 1999). In contrast, considerable immunolabeling of receptor molecules (such as with GluR2/3 antibody) that go to both apical and basal dendrite synapse populations can be found throughout the cytoplasm in apical and basal dendrites. Thus, in this case, the simplest explanation is that molecules of GluR4 and mGluR1α are transported mainly within the basal dendrites (or excluded from the apical dendrites), thereby explaining their selective association with auditory synapses on basal dendrites and corresponding absence from parallel fiber synapses on apical dendrites. Synapse specificity could be obtained if the intracellular organelles which contain the receptors have a molecular signal specifying their destination. This could be the receptor molecule itself with the carboxy terminus being the likely domain carrying this information, given that it is exposed to the cytoplasm.

In the second model, glutamate receptor molecules are transported throughout the somatodendritic compartment indiscriminately and are selected at the synapse. In polarized epithelial cells, some proteins are transported initially throughout the cell but retained selectively at one pole (apical or basolateral) (e.g. WOZNIAK and LIMBIRD 1996). If targeting mechanisms function at the synapse, they may occur at one or more levels. Receptors destined for a particular synapse could be selected and inserted into the membrane of the dendrite near the synapse spine (or near a collection of spines), within the extrasynaptic membrane of the spine head, or directly into the postsynaptic density/membrane area (see below). Selective targeting may be determined at the postsynaptic density/membrane by selective attachment of glutamate receptors to specific scaffolding/anchoring proteins (Fig. 9). A large number of these proteins is known to be associated with various glutamate receptor types. The best known include PSD-95 and associated proteins (SAP97, SAP-102, PSD-93) for NMDA receptors, GRIP and ABP for AMPA receptors, and Homer for type I metabotropic glutamate receptors (reviews by EHLERS et al. 1996; KENNEDY 1997; ZIFF 1997). These proteins attach to part of the glutamate receptor C-terminal and, in most or all cases, link to elements of the cytoskeleton of the postsynaptic density. Various factors control these attachments. For example, calcium entry through NMDA receptors may result ultimately in cleavage of links to the cytoskeleton and even cleavage of part of the AMPA receptor C-terminus (BI et al. 1997; review by ZIFF 1997). These effects can result in changes in receptor function and distribution and ultimately lead to major changes in synapse structure. In addition to anchoring receptors to the PSD, associations between glutamate receptors and anchoring proteins probably play key roles in linking receptors to transduction mechanisms that mediate the neuron's responses to activation of the glutamate receptors (SPRENGEL et al. 1998; reviews by PAWSON and SCOTT 1997; CRAVEN

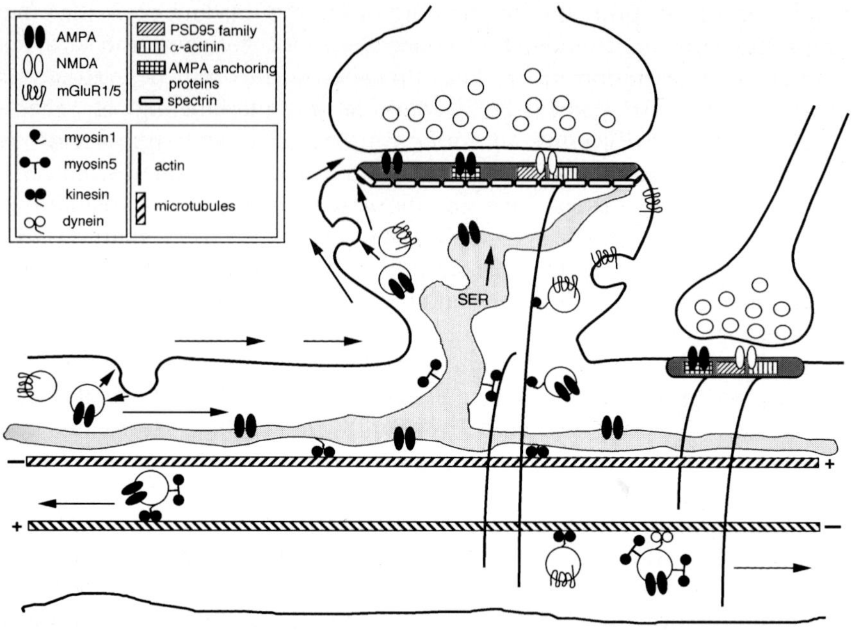

Fig. 9. Diagram summarizing the possible mechanisms which regulate distribution of glutamate receptors at synapses. For transport along dendrites, glutamate receptors may be incorporated in the membrane of transport vesicles or within reticular membranes (*SER*). Transport along dendrites presumably involves kinesin or dynein motors that can move the membrane-associated receptor molecules along microtubules. Subsequently, receptors could be moved directly to the synapses by traveling along actin pathways for short distances, using various myosin motors. Alternatively, the receptors could be incorporated into the nonsynaptic cell membrane (presumably via a similar mechanism) of the dendrite shaft or spine by exocytosis. Glutamate receptors would be bound in place at the synapse using various anchoring/scaffolding proteins which also may be linked to subsynaptic elements such as actin filaments. In some cases, glutamate receptors may be retained loosely within the postsynaptic membrane using a corral made up of cytoskeletal elements such as spectrin, which appears to be juxtaposed to the postsynaptic density in spine synapses (see discussion in ZIFF, 1997). It is likely that several mechanisms of targeting and retention of glutamate receptors may operate at a synapse so that different types of glutamate receptors can be regulated independently. See text for details

and BREDT 1998). An abnormal loss of this association between glutamate receptors and their anchoring proteins can lead to a disruption in the cellular signaling pathways and reduce stability of the signaling components, as suggested for mutant mice expressing NMDA receptors without the intracellular C-terminal domain (SPRENGEL et al. 1998).

It also is possible that anchoring mechanisms vary according to the type of synapse or functional state of the synapse. ALLISON et al. (1998) find that detergent extraction of cultured hippocampal neurons leads to a complete loss of GluR1 immunoreactive "clusters" from pyramidal neuron spines but has no effect on GluR1 clusters on dendrite shafts of GABAergic neurons. They suggest that the AMPA receptors in dendrite shaft synapses of GABAergic neurons may be anchored by a strong mechanism such as a GRIP protein, while the AMPA receptors in the pyramidal neuron spine synapses may be held in place by a weak mechanism such as a spectrin "corral" (Fig. 9). At present, anchoring mechanisms are poorly understood. It remains to be demonstrated that any of the "anchoring" proteins are actually involved in the anchoring of receptors at the synapse. As noted above, studies with mutant mice indicate that NMDA receptors require their C-termini for the maintenance of proper signaling pathways. However, these mutant mice have functional synaptic NMDA receptors, even though they lack binding sites of the anchoring proteins. Thus, the apparent synaptic expression of these mutant NMDA receptors suggests that other factors are involved in the anchoring of glutamate receptors at synapses.

Based on present data, there is evidence that both "selective targeting" and "selective anchoring" are involved in obtaining synapse-specific expression of receptors. For example, PSD95 can interact with both NMDA receptors and the shaker class of potassium channels; specificity could be obtained if potassium channels are always presynaptic and NMDA receptors are always postsynaptic. As noted, "selective targeting" may be the most efficient mechanism for differential distribution in apical versus basal dendrites. However, a greater degree of complexity is necessary to explain a differential distribution of glutamate receptors to two synapse populations on the same dendrite; in this case, "selective anchoring" or some combination of "selective targeting" and "selective anchoring" may be necessary.

Finally, some types of glutamate receptor may be synthesized and assembled within the dendrite near the synapse (review by Kirsch et al. 1996). Dendrites possess functional elements equivalent to the endoplasmic reticulum, Golgi, and trans-Golgi network (e.g. Torre and Steward 1996). However, as noted above, most evidence supports somal localization of glutamate receptor mRNAs. Miyashiro et al. (1994) used a very sensitive technique to detect glutamate receptor mRNA in dendrites, but it is not clear whether this mRNA is functionally significant or accidental (see discussions in Steward 1994, 1997). Gazzaley et al. (1997) report NR1 mRNA in dendrites of cultured hippocampal neurons, and possibly in dendrites in sections from the dentate gyrus molecular layer (following perforant path transection). They suggest that NR1 mRNA normally exists at very low levels in dendrites and is upregulated during denervation-induced synaptic plasticity. It is possible that some glutamate receptors are synthesized in the cell body and that others are synthesized in the dendrites, so that receptor density and composition at synapses can be regulated at two levels. In any case, if dendrite synthesis of glutamate receptors is a significant process, then it will be necessary to consider differential targeting mechanisms for glutamate receptor mRNAs.

F. Mechanisms of Synaptic Receptor Insertion and Removal

Although there is little direct evidence to explain how glutamate receptors are inserted and removed from the synaptic membrane, this likely occurs through the general mechanisms of endocytosis and exocytosis. The two major routes for receptor incorporation into synapses are: 1) insertion into nonsynaptic membrane followed by lateral diffusion to the synapse (Baude et al. 1995), and 2) direct incorporation from cytoplasmic tubulovesicular compartments into the synapse (Fig. 9). In the first case, receptor molecules would be transported to the plasma membrane of the synaptic spine or to the plasma membrane along the dendrite shaft near the spine. Here they would enter the plasma membrane via an exocytotic process. Spacek and Harris (1997), using serial reconstruction ultrastructural analysis, show examples of smooth vesicle fusion to the sides of synaptic spines of the CA1 region of the hippocampus, and suggest that these represent an exocytotic event that brings receptor molecules to the synapse. This could be a very efficient and rapid method of receptor incorporation into synapses. For example, Benke et al. (1993), using fluorescence imaging of live hippocampal neurons labeled with the NMDA receptor-specific ligand conantokin-G, estimate that NMDA receptors can diffuse across the surface of a spine head in two seconds. This model of glutamate receptor incorporation at membranes also is consistent with studies of acetylcholine receptor clustering at developing neuromuscular junctions (e.g. Ruegg and Bixby 1998).

The second major mechanism for the incorporation of glutamate receptors would be a route from the tubulovesicular elements in the spine head directly into the postsynaptic membrane. Such a mechanism might transport receptors through the postsynaptic density at any point in the plane of the synaptic active zone. Alternatively, receptor transport might occur near the edge of the postsynaptic density. In support of this, Spacek and Harris (1997) describe bridges of reticular membrane that connect the spine apparatus, which is a complex arrangement of reticular elements in some spine heads, to the edges of the postsynaptic density and membrane (Fig. 10). Because the spine apparatus also is connected to the endoplasmic reticulum of the dendrite shaft, a continuous pathway is available for transport of receptors from the dendrite directly into the postsynaptic density or membrane (see below; also, for general discussions of trafficking along reticular tubules, see Mironov et al. 1997 and Nakata et al. 1998).

The tangential distribution of glutamate receptors may offer clues to the mechanism employed for incorporation of these receptors into the postsynaptic membrane. The common presence of higher densities of ionotropic glutamate receptors along the outer portions of the postsynaptic membrane suggests that this site of receptor accumulation marks a point of entry (or removal) from the synapse, either via membrane diffusion or direct insertion. The very different distribution of Group I metabotropic glutamate receptors, which are concentrated in the perisynaptic regions of some synapses, suggests

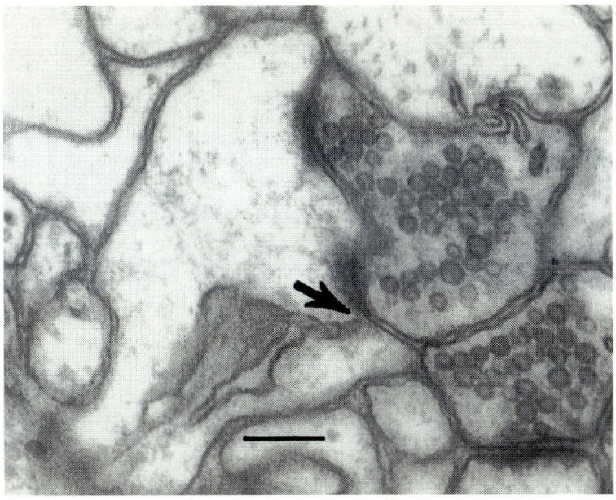

Fig. 10. Association of the spine apparatus with the postsynaptic density, in large, mushroom dendritic spines of the adult hippocampus (CA1) of the rat. A tubule of SER from the spine apparatus is directed to the outer edge of the perforated postsynaptic density (*arrow*) (modified from Fig. 4a, SPACEK and HARRIS 1997). *Scale bar* is 0.2 μm

that a different mechanism may exist for incorporation of these receptors into the synapse.

It is not clear whether the insertion and removal of receptors from synapses occur via different mechanisms. Coated pits can be found in the spine head membrane near the synapse (SPACEK and HARRIS 1997) and may be more common during development of the synapses (e.g. ESHHAR et al. 1993; PETRALIA et al. 1998). Presumably, coated pits are involved in endocytosis of at least some membrane elements in synapses (for a general review of endocytosis, see RIEZMAN et al. 1997). Clathrin-coated pits internalize human muscarinic (metabotropic) cholinergic receptors in transfected culture cells (TOLBERT and LAMEH 1996), although other internalization mechanisms are known for muscarinic receptors (FERON et al. 1997). Internalized glutamate receptor molecules would be transported through endosomal compartments (PARTON et al. 1992) where they could be differentially sorted, as seen with other membrane proteins (FISCHER et al. 1997), and transported to the cell body for degradation or perhaps, in some cases, returned to the membrane. Future studies that concentrate on immunogold labeling in coated pits (endocytotic events?), smooth vesicle-membrane associations (exocytotic events?), and reticular connections below the synapse may elucidate the mechanisms of glutamate receptor incorporation and removal.

References

Allison DW, Gelfand VI, Spector I, Craig AM (1998) Role of actin in anchoring postsynaptic receptors in cultured hippocampal neurons: Differential attachment of NMDA versus AMPA receptors. J Neurosci 18:2423–2436

Aoki C, Venkatesan C, Go C-G, Mong JA, Dawson TM (1994) Cellular and subcellular localization of NMDA-R1 subunit immunoreactivity in the visual cortex of adult and neonatal rats. J Neurosci 14:5202–5222
Bahn S, Wisden W (1997) A map of non-NMDA receptor subunit expression in the vertebrate brain derived from *in situ* hybridization histochemistry. In: Monaghan DT, Wenthold RJ (eds) The ionotropic glutamate receptors. Humana Press Inc., Totowa, NJ, pp 149–187
Baude A, Nusser Z, Molnár E, McIlhinney RAJ, Somogyi P (1995) High-resolution immunogold localization of AMPA type glutamate receptor subunits at synaptic and non-synaptic sites in rat hippocampus. Neuroscience 69:1031–1055
Benke TA, Jones OT, Collingridge GL, Angelides KJ (1993) *N*-methyl-D-aspartate receptors are clustered and immobilized on dendrites of living cortical neurons. Proc Natl Acad Sci USA 90:7819–7823
Bernard V, Somogyi P, Bolam JP (1997) Cellular, subcellular, and subsynaptic distribution of AMPA-type glutamate receptor subunits in the neostriatum of the rat. J Neurosci [Correction (1997) 17:7180] 17:819–833
Bi X, Chen J, Dang S, Wenthold RJ, Tocco G, Baudry M (1997) Characterization of calpain-mediated proteolysis of GluR1 subunits of α-amino-3-hydroxy-5-methylisoxazole-4-propionate receptors in rat brain. J Neurochem 68:1484–1494
Chittajallu R, Vignes M, Dev KK, Barnes JM, Collingridge GL, Henley JM (1996) Regulation of glutamate release by presynaptic kainate receptors in the hippocampus. Nature 379:78–81
Clark BA, Farrant M, Cull-Candy SG (1997) A direct comparison of the single-channel properties of synaptic and extrasynaptic NMDA receptors. J Neurosci 17:107–116
Craig AM, Blackstone CD, Huganir RL, Banker G (1993) The distribution of glutamate receptors in cultured rat hippocampal neurons: Postsynaptic clustering of AMPA-selective subunits. Neuron 10:1055–1068
Craig AM, Blackstone CD, Huganir RL, Banker G (1994) Selective clustering of glutamate and γ-aminobutyric acid receptors opposite terminals releasing the corresponding neurotransmitters. Proc Natl Acad Sci USA 91:12373–12377
Craven SE, Bredt DS (1998) PDZ proteins organize synaptic signaling pathways. Cell 93:495–498
Derrick BE, Weinberger SB, Martinez JL (1991) Opioid receptors are involved in an NMDA receptor-independent mechanism of LTP induction at hippocampal mossy fiber-CA3 synapses. Brain Res Bull 27:219–223
Doherty AJ, Collingridge GL, Henley JM (1997) GFP fusion proteins and AMPA receptor trafficking. Biochem Soc Trans 25:540S
Dotti CG, Simons K (1990) Polarized sorting of viral glycoproteins to the axon and dendrites of hippocampal neurons in culture. Cell 62:63–72
Drubin DG, Nelson WJ (1996) Origins of cell polarity. Cell 84:335–344
Durand GM, Kovalchuk Y, Konnerth A (1996) Long-term potentiation and functional synapse induction in developing hippocampus. Nature 381:71–75
Ehlers MD, Mammen AL, Lau L-F, Huganir RL (1996) Synaptic targeting of glutamate receptors. Curr Opin Cell Biol 8:484–489
Eshhar N, Petralia RS, Winters CA, Niedzielski AS, Wenthold RJ (1993) The segregation and expression of glutamate receptor subunits in cultured hippocampal neurons. Neuroscience 57:943–964
Feron O, Smith TW, Michel T, Kelly RA (1997) Dynamic targeting of the agonist-stimulated m2 muscarinic acetylcholine receptor to caveolae in cardiac myocytes. J Biol Chem 272:17744–17748
Fischer Y, Thomas J, Sevilla L, Muñoz P, Becker C, Holman G, Kozka IJ, Palacín M, Testar X, Kammermeier H, Zorzano A (1997) Insulin-induced recruitment of glucose transporter 4 (GLUT4) and GLUT1 in isolated rat cardiac myocytes. J Biol Chem 272:7085–7092

Fritschy JM, Weinmann O, Wenzel A, Benke D (1998) Synapse-specific localization of NMDA and GABA(A) receptor subunits revealed by antigen-retrieval immunohistochemistry. J Comp Neurol 390:194–210
Gazzaley AH, Benson DL, Huntley GW, Morrison JH (1997) Differential subcellular regulation of NMDAR1 protein and mRNA in dendrites of dentate gyrus granule cells after perforant path transection. J Neurosci 17:2006–2017
Gottmann K, Mehrle A, Gisselmann G, Hatt H (1997) Presynaptic control of subunit composition of NMDA receptors mediating synaptic plasticity. J Neurosci 17:2766–2774
Hall RA, Soderling TR (1997a) Differential surface expression and phosphorylation of the *N*-methyl-D-aspartate receptor subunits NR1 and NR2 in cultured hippocampal neurons. J Biol Chem 272:4135–4140
Hall RA, Soderling TR (1997b) Quantitation of AMPA receptor surface expression in cultured hippocampal neurons. Neuroscience 78:361–371
Harris KM, Sultan P (1995) Variation in the number, location and size of synaptic vesicles provides an anatomical basis for the nonuniform probability of release at hippocampal CA1 synapses. Neuropharmacology 34:1387–1395
Häusser M, Roth A (1997) Dendritic and somatic glutamate receptor channels in rat cerebellar Purkinje cells. J Physiol 501:77–95
Hirokawa N (1998) Kinesin and dynein superfamily proteins and the mechanism of organelle transport. Science 279:519–526
Hollmann M, Heinemann S (1994) Cloned glutamate receptors. Annu Rev Neurosci 17:31–108
Huh KH, Wenthold RJ (1999) Turnover analysis of glutamate receptors identifies a rapidly degraded pool of the N-methyl-D-aspartate receptor subunit, NRI, in cultured cerebellar granule cells. J Biol Chem 274:151–157
Huntley GW, Vickers JC, Janssen W, Brose N, Heinemann SF, Morrison JH (1994) Distribution and synaptic localization of immunocytochemically identified NMDA receptor subunit proteins in sensory-motor and visual cortices of monkey and human. J Neurosci 14:3603–3619
Isaac JTR, Nicoll RA, Malenka RC (1995) Evidence for silent synapses: implications for the expression of LTP. Neuron 15:427–434
Kashiwabuchi N, Ikeda K, Araki K, Hirano T, Shibuki K, Takayama C, Inoue Y, Kutsuwada T, Yagi T, Kang Y, Aizawa S, Mishina M (1995) Impairment of motor coordination, Purkinje cell synapse formation, and cerebellar long-term depression in GluRδ2 mutant mice. Cell 8:245–252
Kennedy MB (1997) The postsynaptic density at glutamatergic synapses. Trends Neurosci 20:264–268
Kharazia VN, Phend KD, Rustioni A, Weinberg RJ (1996a) EM colocalization of AMPA and NMDA receptor subunits at synapses in rat cerebral cortex. Neurosci Lett 210:37–40
Kharazia VN, Wenthold RJ, Weinberg RJ (1996b) GluR1-immunopositive interneurons in rat neocortex. J Comp Neurol 368:399–412
Kharazia VN, Weinberg RJ (1997) Tangential synaptic distribution of NMDA and AMPA receptors in rat neocortex. Neurosci Lett 238:41–44
Kirsch J, Betz H (1998) Glycine-receptor activation is required for receptor clustering in spinal neurons. Nature 392:717–720
Kirsch J, Meyer G, Betz H (1996) Synaptic targeting of ionotropic neurotransmitter. Mol Cell Neurosci 8:93–98
Kurihara H, Hashimoto K, Kano M, Takayama C, Sakimura K, Mishina M, Inoue Y, Watanabe M (1997) Impaired parallel fiber-Purkinje cell synapse stabilization during cerebellar development of mutant mice lacking the glutamate receptor δ2 subunit. J Neurosci 17:9613–9623
Kutsuwada T, Sakimura K, Manabe T, Takayama C, Katakura N, Kushiya E, Natsume R, Watanabe M, Inoue Y, Yagi T, Aizawa S, Arakawa M, Takahashi T, Nakamura Y, Mori H, Mishina M (1996) Impairment of suckling response, trigeminal neu-

ronal pattern formation, and hippocampal LTD in NMDA receptor ε2 subunit mutant mice. Neuron 16:333–344

Landsend AS, Amiry-Moghaddam M, Matsubara A, Bergersen L, Usami S, Wenthold RJ, Ottersen OP (1997) Differential localization of δ glutamate receptors in the rat cerebellum: Coexpression with AMPA receptors in parallel fiber-spine synapses and absence from climbing fiber-spine synapses. J Neurosci 17:834–842

Lerma J (1997) Kainate reveals its targets. Neuron 19:1155–1158

Liu H, Wang H, Sheng M, Jan LY, Jan YN, Basbaum AI (1994) Evidence for presynaptic *N*-methyl-D-aspartate autoreceptors in the spinal cord dorsal horn. Proc Natl Acad Sci USA 91:8383–8387

Malenka RC, Nicoll RA (1997) Silent synapses speak up. Neuron 19:473–476

Malva JO, Carvalho AP, Carvalho CM (1998) Kainate receptors in hippocampal CA3 subregion: evidence for a role in regulating neurotransmitter release. Neurochem Int 32:1–6

Mammen AL, Huganir RL, O'Brien RJ (1997) Redistribution and stabilization of cell surface glutamate receptors during synapse formation. J Neurosci 17:7351–7358

Martin LJ, Blackstone CD, Levey AI, Huganir RL, Price DL (1993) AMPA glutamate receptor subunits are differentially distributed in rat brain. Neuroscience 53:327–358

Martin LJ, Furuta A, Blackstone CD (1998) AMPA receptor protein in developing rat brain: Glutamate receptor-1 expression and localization change at regional, cellular, and subcellular levels with maturation. Neuroscience 83:917–928

Matsubara A, Laake JH, Davanger S, Usami S, Ottersen OP (1996) Organization of AMPA receptor subunits at a glutamate synapse: A quantitative immunogold analysis of hair cell synapses in the rat organ of Corti. J Neurosci 16:4457–4467

Mermall V, Post PL, Mooseker MS (1998) Unconventional myosins in cell movement, membrane traffic, and signal transduction. Science 279:527–533

Mironov AA, Weidman P, Luini A (1997) Variations on the intracellular transport theme: Maturing cisternae and trafficking tubules. J Cell Biol 138:481–484

Miyashiro K, Dichter M, Eberwine J (1994) On the nature and differential distribution of mRNAs in hippocampal neurites: Implications for neuronal functioning. Proc Natl Acad Sci USA 91:10800–10804

Molitor SC, Manis PB (1997) Evidence for functional metabotropic glutamate receptors in the dorsal cochlear nucleus. J Neurophysiol 77:1889–1905

Molnár E, Baude A, Richmond SA, Patel PB, Somogyi P, McIlhinney RAJ (1993) Biochemical and immunocytochemical characterization of antipeptide antibodies to a cloned GluR1 glutamate receptor subunit: Cellular and subcellular distribution in the rat forebrain. Neuroscience 53:307–326

Nakata T, Terada S, Hirokawa N (1998) Visualization of the dynamics of synaptic vesicle and plasma membrane proteins in living axons. J Cell Biol 140:659–674.

Nusser Z, Mulvihill E, Streit P, Somogyi P (1994) Subsynaptic segregation of metabotropic and ionotropic glutamate receptors as revealed by immunogold localization. Neuroscience 61:421–427

Nusser Z, Sieghart W, Benke D, Fritschy J-M, Somogyi P (1996a) Differential synaptic localization of two major γ-aminobutyric acid type A receptor α subunits on hippocampal pyramidal cells. Proc Natl Acad Sci USA 93:11939–11944

Nusser Z, Sieghart W, Stephenson FA, Somogyi P (1996b) The α6 subunit of the $GABA_A$ receptor is concentrated in both inhibitory and excitatory synapses on cerebellar granule cells. J Neurosci 16:103–114

Nusser Z, Sieghart W, Somogyi P (1998) Segregation of different $GABA_A$ receptors to synaptic and extrasynaptic membranes of cerebellar granule cells. J Neurosci 18:1693–1703

O'Brien RJ, Mammen AL, Blackshaw S, Ehlers MD, Rothstein JD, Huganir RL (1997) The development of excitatory synapses in cultured spinal neurons. J Neurosci 17:7339–7350

Ohishi H, Ogawa-Meguro R, Shigemoto R, Kaneko T, Nakanishi S, Mizuno N (1994) Immunohistochemical localization of metabotropic glutamate receptors, mGluR2 and mGluR3, in rat cerebellar cortex. Neuron 13:55–66

Ozaki M, Sasner M, Yano R, Lu HS, Buonanno A (1997) Neuregulin-β induces expression of an NMDA-receptor subunit. Nature 390:691–694

Parton RG, Simons K, Dotti CG (1992) Axonal and dendritic endocytic pathways in cultured neurons. J Cell Biol 119:123–137

Pawson T, Scott JD (1997) Signaling through scaffold, anchoring, and adaptor proteins. Science 278:2075–2080

Perez-Velazquez JL, Angelides KJ (1993) Assembly of $GABA_A$ receptor subunits determines sorting and localization in polarized cells. Nature 361:457–460

Peters A, Palay SL, Webster HD (1991) The fine structure of the nervous system, 3rd edn. Oxford University Press, New York

Petralia RS (1997) Immunocytochemical localization of ionotropic glutamate receptors (GluRs) in neural circuits. In: Monaghan DT, Wenthold RJ (eds) The ionotropic glutamate receptors. Humana Press, Totowa, NJ, pp 219–263

Petralia RS, Wenthold RJ (1992) Light and electron immunocytochemical localization of AMPA-selective glutamate receptors in the rat brain. J Comp Neurol 318:329–354

Petralia RS, Yokotani N, Wenthold RJ (1994a) Light and electron microscope distribution of the NMDA receptor subunit NMDAR1 in the rat nervous system using a selective anti-peptide antibody. J Neurosci 14:667–696

Petralia RS, Wang Y.-X, Wenthold RJ (1994b) The NMDA receptor subunits NR2A and NR2B show histological and ultrastructural localization patterns similar to those of NR1. J Neurosci 14:6102–6120

Petralia RS, Wenthold RJ (1996) Types of excitatory amino acid receptors and their localization in the nervous system and hypothalamus. In: Brann DW, Mahesh VB (eds) Excitatory amino acids: Their role in neuroendocrine function. CRC, Boca Raton, FL, pp 55–101

Petralia RS, Wang Y-X, Niedzielski AS, Wenthold RJ (1996a) The metabotropic glutamate receptors, mGluR2 and mGluR3, show unique postsynaptic, presynaptic and glial localizations. Neuroscience 71:949–976

Petralia RS, Wang Y-X, Zhao H-M, Wenthold RJ (1996b) Ionotropic and metabotropic glutamate receptors show unique postsynaptic, presynaptic and glial localizations in the dorsal cochlear nucleus. J Comp Neurol 372:356–383

Petralia RS, Wang Y-X, Mayat E, Wenthold RJ (1997) Glutamate receptor subunit 2-selective antibody shows a differential distribution of calcium-impermeable AMPA receptors among populations of neurons. J Comp Neurol 385:456–476

Petralia RS, Zhao H-M, Wang Y-X, Wenthold RJ (1998) Variations in the tangential distribution of postsynaptic glutamate receptors in Purkinje cell parallel and climbing fiber synapses during development. Neuropharmacology 37:1321–1334

Petralia RS, Esteban J, Wang Y-X, Partridge JG, Zhao H-M, Wenthold RJ, Malinow R (1999) Selective acquisition of AMPA receptors over postnatal development suggests a molecular basis for silent synapses. Nature Neurosci 2:31–36

Rabacchi S, Bailly Y, Delhaye-Bouchaud N, Mariani J (1992) Involvement of the N-methyl-D-aspartate (NMDA) receptor in synapse elimination during cerebellar development. Science 256:1823–1825.

Rao A, Craig AM (1997) Activity regulates the synaptic localization of the NMDA receptor in hippocampal neurons. Neuron 19:801–812

Rao A, Kim E, Sheng M, Craig AM (1998) Heterogeneity in the molecular composition of excitatory postsynaptic sites during development of hippocampal neurons in culture. J Neurosci 18:1217–1229

Riezman H, Woodman PG, van Meer G (1997) Molecular mechanisms of endocytosis. Cell 91:731–738

Rizzuto R, Pinton P, Carrington W, Fay FS, Fogarty KE, Lifshitz LM, Tuft RA, Pozzan

T (1998) Close contacts with the endoplasmic reticulum as determinants of mitochondrial Ca^{2+} responses. Science 280:1763–1766
Rodríguez-Moreno A, Herreras O, Lerma J (1997) Kainate receptors presynaptically downregulate GABAergic inhibition in the rat hippocampus. Neuron 19:893–901
Rubio ME, Wenthold RJ (1997) Glutamate receptors are selectively targeted to postsynaptic sites in neurons. Neuron 18:939–950
Rubio ME, Wenthold RJ (1999) Differential distribution of intracellular glutamate receptors in dendrites. J Neurosci 19:5549–5562
Ruegg MA, Bixby JL (1998) Agrin ochestrates synaptic differentiation at the vertebrate neuromuscular junction. Trends Neurosci 21:22–27
Shigemoto R, Kulik A, Roberts JDB, Ohishi H, Nusser Z, Kaneko T, Somogyi P (1996) Target-cell-specific concentration of a metabotropic glutamate receptor in the presynaptic active zone. Nature 381:523–525
Siegel SJ, Brose N, Janssen WG, Gasic GP, Jahn R, Heinemann SF, Morrison JH (1994) Regional, cellular, and ultrastructural distribution of *N*-methyl-D-aspartate receptor subunit 1 in monkey hippocampus. Proc Natl Acad Sci USA 91:564–568
Spacek J, Harris KM (1997) Three-dimensional organization of smooth endoplasmic reticulum in hippocampal CA1 dendrites and dendritic spines of the immature and mature rat. J Neurosci 17:190–203
Sprengel R, Suchanek B, Amico C, Brusa R, Burnashev N, Rozov A, Hvalby Ø, Jensen V, Paulsen O, Andersen P, Kim JJ, Thompson RF, Sun W, Webster LC, Grant SGN, Eilers J, Konnerth A, Li J, McNamara JO, Seeburg PH (1998) Importance of the intracellular domain of NR2 subunits for NMDA receptor function in vivo. Cell 92:279–289
Spruston N, Jonas P, Sakmann B (1995) Dendritic glutamate receptor channels in rat hippocampal CA3 and CA1 pyramidal neurons. J Physiol 482:325–352
Steward O (1994) Dendrites as compartments for macromolecular synthesis. Proc Natl Acad Sci. USA 91:10766–10768
Steward O (1997) mRNA localization in neurons: A multipurpose mechanism? Neuron 18:9–12
Takayama C, Nakagawa S, Watanabe M, Mishina M, Inoue Y (1996) Developmental changes in expression and distribution of the glutamate receptor channel $\delta 2$ subunit according to the Purkinje cell maturation. Dev Brain Res 92:147–155
Tolbert LM, Lameh J (1996) Human muscarinic cholinergic receptor Hm1 internalizes via clathrin-coated vesicles. J Biol Chem 271:17335–17342
Torre ER, Steward O (1996) Protein synthesis within dendrites: Glycosylation of newly synthesized proteins in dendrites of hippocampal neurons in culture. J Neurosci 16:5967–5978
Toth K, McBain CJ (1998) Afferent specific innervation of two distinct AMPA receptor subtypes on single hippocampal interneurons. Nature Neurosci 1:572–578
Triller A, Cluzeaud F, Pfeiffer F, Betz H (1985) Distribution of glycine receptors at central synapses: An immunoelectron microscopy study. J Cell Biol 101:683–688
Vallano ML, Lambolez B, Audinat E, Rossier J (1996) Neuronal activity differentially regulates NMDA receptor subunit expression in cerebellar granule cells. J Neurosci 16:631–639
Wang JKT, Thukral V (1996) Presynaptic NMDA receptors display physiological characteristics of homomeric complexes of NR1 subunits that contain the exon 5 insert in the N-terminal domain. J Neurochem 66:865–868
Wang Y-X, Wenthold RJ, Ottersen OP, Petralia RS (1998) Endbulb synapses in the anteroventral cochlear nucleus express a specific subset of AMPA-type glutamate receptor subunits. J Neurosci 18:1148–1160
Watanabe M (1997) Developmental dynamics of gene expression for NMDA receptor channel. In: Monaghan DT, Wenthold RJ (eds) The ionotropic glutamate receptors. Humana Press Inc., Totowa, NJ, pp 189–218
Wenthold RJ, Parakkal MH, Oberdorfer MD, Altschuler RA (1988) Glycine receptor

immunoreactivity in the ventral cochlear nucleus of the guinea pig. J Comp Neurol 276:423–435
Wenthold RJ, Petralia RS, Blahos J II, Niedzielski AS (1996) Evidence for multiple AMPA receptor complexes in hippocampal CA1/CA2 neurons. J Neurosci 16:1982–1989
Wozniak M, Limbird LE (1996) The three α_2-adrenergic receptor subtypes achieve basolateral localization in Madin-Darby canine kidney II cells via different targeting mechanisms. J Biol Chem 271:5017–5024
Wu G-Y, Malinow R, Cline HT (1996) Maturation of a central glutamatergic synapse. Science 274:972–976
Xie X, Liaw J-S, Baudry M, Berger TW (1997) Novel expression mechanism for synaptic potentiation: Alignment of presynaptic release site and postsynaptic receptor. Proc Natl Acad Sci USA 94:6983–6988
Zalutsky RA, Nicoll RA (1990) Comparison of two forms of long-term potentiation in single hippocampal neurons. Science 248:1619–1624
Zhao H-M, Wenthold RJ, Wang Y-X, Petralia RS (1997) Delta glutamate receptors are differentially distributed at parallel and climbing fiber synapses on Purkinje cells. J Neurochem 68:1041–1052
Zhao H-M, Wenthold RJ, Petralia RS (1998) Glutamate receptor targeting to synaptic populations on Purkinje cells is developmentally regulated. J Neurosci 18: 5517–5528.
Ziff EB (1997) Enlightening the postsynaptic density. Neuron 19:1163–1174

Section II
Functional Properties
of Glutamate Receptors

CHAPTER 5

Activation of Single AMPA- and NMDA-Type Glutamate-Receptor Channels

P. Béhé, D. Colquhoun, and D.J.A. Wyllie

A. Introduction

I. Rates of Physiological Events

In real life, the thing that matters about glutamate receptors is their response to the brief pulse of agonist (glutamate) that is released by a presynaptic nerve during synaptic transmission. The brief nature of this pulse makes it likely that postsynaptic channels will experience only a single activation (see below). The receptor is not at equilibrium, and it is the rate at which it works that controls its physiological properties. The study of rates (kinetics) is, therefore, crucial to the understanding of physiology. For some purposes, an empirical description of how things are observed to change is quite sufficient, and such empirical descriptions may be described appropriately as models; but, in order to understand how the receptor actually works, this is not good enough. For this purpose, we must aim to describe the receptor in terms of physical realities. The states in a kinetic mechanism must not be convenient abstractions but actual physical structures.

There is some hope that we may be near to this ideal for the (muscle-type) nicotinic receptor (Colquhoun and Sakmann 1985; Sine et al. 1995), but it has not yet been achieved for *N*-methyl-D-aspartate (NMDA) or α-amino-3-hydroxy-5-methyl-4-isoxazole propionic acid (AMPA) receptors. To the extent that we can succeed in this aim, the description is better defined as a mechanism than as a model because the latter term (purposely) evades the crucial question of physical reality. The attempt to do so is important, not just for intellectual curiosity and not just for physiology. It is also essential for understanding how proteins work, the effects of mutations on receptors and the relationship (insofar as one is discernible) between the structure of drugs and their effects. Many attempts have, of course, been made to understand these questions in the absence of much knowledge about mechanisms. But, it is now clear that these attempts are, at best, inspired guesswork (which does not, of course, mean that they are always wrong). This question has recently been reviewed in detail (Colquhoun 1998).

It has been a long-standing goal of physiologists to explain the time course of synaptic currents. In most cases, and certainly in the case of synaptic currents mediated by NMDA receptors, the time course is determined by the rate at which channels close after the removal of free agonist (often referred to as their "deactivation" rate) (ANDERSON and STEVENS 1973). The next natural question concerns the nature of the single-channel events that underlie the synaptic current.

In the simplest case (latency followed by single opening) the macroscopic time course of the synaptic current would be the convolution of the distributions of first latency and of the length of the channel opening (COLQUHOUN and HAWKES 1995a), as used, for example, by ALDRICH et al. (1983) for sodium channels. This description works, however, only if there is a single sort of open state and the latency is followed by a single opening. This is not the case for the NMDA or AMPA receptors (or for most other agonist-activated channels).

The general relationship that is to be expected between single-channel activations and the macroscopic time course has only recently been given (COLQUHOUN et al. 1997), and experimental results on NMDA receptors have been found to be consistent with this theory (WYLLIE et al. 1998). The latter paper also serves to emphasise the high resolving power of single-channel measurements compared with those of macroscopic currents. For example, the distribution of the length of individual channel activations for the recombinant NR1a/NR2D receptor contains (at least) six exponential components, so it would be expected that the macroscopic response to a fast concentration jump would also contain six exponential components with very similar time constants (COLQUHOUN et al. 1997; WYLLIE et al. 1998). The observed responses to concentration jumps were consistent with this being correct, but a free fit to the macroscopic response could not resolve more than two or, at best, three of the components. This example makes it clear that it will generally be hard to discern physically realistic mechanisms from macroscopic responses alone.

II. The Definition of a Channel Activation

The idea of an individual receptor activation provides the crucial link between single-channel events and macroscopic (synaptic) currents. The term 'activation' defines the sequence of events that take place between the first opening that follows binding of agonist and the last opening before complete dissociation of agonist. In general, this definition implies that partially liganded shut states may form part of an activation. The property that defines activations is that they are separated by one or more sojourns in the resting state (no agonist bound), and so they must become progressively further and further apart as the agonist concentration is reduced. Experimentally, they are defined as bursts of openings (for NMDA receptors, these bursts are so long, and have so much sub-structure that they have been referred to as super-clusters).

The most direct way to see an individual receptor activation is to do a very brief concentration jump on a one-channel patch (WYLLIE et al. 1997); insofar as the agonist concentration is zero after the jump, no re-association can take place and the whole of the observed signal is one activation (Fig. 5). It is also possible to define individual activations in steady-state recordings if they are made at a sufficiently low concentration such that activations are well separated. For NMDA receptors, the activations are so long that this means using very low concentrations, usually 5–100 nM glutamate (Gibb and Colquhoun 1991, 1992; Wyllie et al. 1998). Insofar as the concentration is near zero in these experiments, the distributions of activation lengths should have time constants similar to those seen in the jump experiments, but the initial condition at the start of an activation may be very different for the two sorts of experiments so the areas attached to these components may be quite different, as discussed below.

III. The NMDA Receptors

The NMDA receptor has been the most intensely studied of all agonist-activated ion channels over the past decade and has been demonstrated to play a pivotal role in a variety of physiological and pathophysiological functions, not to mention an ever-increasing number of 'phenomena'. Furthermore, it is clear that NMDA receptors can be modulated by factors ranging from protein kinases to protons and from polyamines to simple membrane stretch. Two properties of the NMDA receptor are, however, likely to give rise to this macromolecule being so widely involved in biological responses. First, the NMDA receptor is highly permeable to Ca^{2+} and, second, glutamate remains bound to and activates NMDA receptors for several hundreds of milliseconds. Molecular biology and site-directed mutagenesis have allowed us to gain insights into what particular amino acids control the Ca^{2+} permeability of these channels (see Chaps. 1 and 7). Our knowledge of the factors that give rise to the very complex activation structure of these channels is far from being complete, but it is fundamental to a full understanding of the NMDA receptor.

IV. The AMPA Receptors

AMPA receptors mediate the fast component of glutamate-evoked synaptic currents in the central nervous system (CNS). Functionally, therefore, this receptor class can be considered to be similar to nicotinic acetylcholine receptors that are found at the endplate of skeletal muscle; they provide a fast depolarisation of the postsynaptic membrane. One should, however, be careful not to extend this analogy too far because, unlike nicotinic receptors at the endplate, AMPA receptors in the CNS are a heterogeneous class of ligand-gated ion channels. Such heterogeneity does not arise simply from the fact that there are four different AMPA-receptor subunits (GluR1, 2, 3 and 4, also termed GluR-A, B, C and D) each of which can form homomeric channels, but also from the fact that these four can form many heteromeric combinations. The existence of cellular mechanisms that generate different splice variants for each subunit, and of ribonucleic acid (RNA) editing, complicate still further the number of different AMPA subunits that can be synthesised, and means that potentially thousands of possible receptor combinations could be present in the CNS (see Chaps. 1 and 9). Given this, it is perhaps not surprising that, as discussed below, there have been a wide variety of single-channel conductances and kinetic parameters reported for both native and recombinant AMPA receptors.

B. Comparison of Native and Recombinant Receptors

The question of the subunit composition of native receptors continues to be of paramount interest, and, although progress has been made, the job is very far from being completed. An important tool in this quest is to make comparisons between native receptors and recombinant receptors in some heterologous expression system.

It is the bane of all work on recombinant receptors that one can never be sure exactly what proteins one's expression system has produced, and to what extent their behaviour will mimic quantitatively the behaviour of native receptors. This has been a serious problem for neuronal nicotinic receptors, for which even single-channel conductances are unreliable (Lewis et al. 1997; Sivilotti et al. 1997). To a lesser extent, it has also been a problem for muscle-type nicotinic receptors, for which conductances seem to be reliable, but kinetic behaviour may not be (Gibb et al. 1990; and unpublished observations, discussed in Edmonds et al. 1995).

A recent review (Sucher et al. 1996) emphasised the differences between recombinant and native NMDA receptors. However, many of the studies that were cited used shortcut methods that do not allow firm conclusions to be drawn on this question. It has been a basic principle of pharmacology for many years that, if one wishes to compare two receptors, it is necessary to measure some quantity (classically, the equilibrium constant for binding of a competitive antagonist) that should be characteristic of the receptor, but independent of differences between one tissue and another, and of irrelevant experimental variables, e.g. nature and concentration of the agonist. Differences in EC_{50} values often cannot be interpreted unambiguously because of differences in conditions such as calcium and glycine concentration and different application methods (with consequent differences in desensitisation). Even more worrying is the widespread use of the IC_{50} as a measure of the activity of antagonists like 2-amino-5-phosphonovaleric acid (APV).

It has been well known, at least since 1949, that it is impossible to estimate real equilibrium constants from IC_{50} curves, and, consequently, such measurements cannot be used for quantitative comparisons of receptors in different tissues or expression systems. Leff and Dougall (1993) tried to resuscitate the IC_{50}, but their attempt to modify the Hill equation to allow for competitive antagonism is, like the Hill equation itself, entirely empirical; it cannot be derived from any physical mechanism. It merely describes parallel-shifted Hill curves and, as such, is useful only as a means of implementing the Schild method. The "general Cheng-Prusoff equation" which they derive from it has no sound physical basis and so cannot be relied on to estimate real equilibrium constants, though it has been seized upon by some authors as a convenient shortcut. There is, however, one condition under which the IC_{50} should give a good estimate of the antagonist equilibrium constant, and that is when the application of agonist is so brief that there is insufficient time for the occupancy by antagonist to change perceptibly (Colquhoun et al. 1992). In general, though, the correct method will be the Schild analysis (Anson et al. 1998); this takes a little longer to do, and there are few reliable measurements in the literature.

The use of IC_{50} measurements to compare magnesium sensitivities is equally unsound. It has been well known since the work of Adams (1976) that the action of a channel blocker is expected to depend on the fraction of channels that are open, e.g. on the agonist concentration. The fact that this varies from one set of experiments to the other makes it impossible to compare the IC_{50} values. If magnesium were a simple open-channel blocker, it would be possible to measure the association and dissociation rate constants for its combination with the open channel, values that would be characteristic of the channel. However, magnesium does not act like a simple open-channel blocker (Nowak et al. 1994), and its action is not fully understood, so there is no known method of extracting from experimental results a quantity, such as an equilibrium constant or rate constant, that could reasonably be expected to be independent of things like the nature of the agonist and its concentration.

When single-channel properties of NMDA receptors are measured, a rather different picture emerges from that presented by Sucher et al. (1996). The conductances and subconductances of NMDA receptors expressed in oocytes and in mammalian cell lines are quantitatively very similar (Stern et al. 1994), and there is good reason to think that they are similar to native receptors too, as discussed below.

C. Single-Channel Conductances of NMDA Receptors

Measurements of the single-channel conductance of NMDA-receptor channels have proved useful for identification of native receptors. For a comprehensive review of these single-channel conductance values, the reader is referred to a recent review by Cull-Candy et al. (1995). NMDA-receptor channel conductances do not appear to depend on the glutamate concentration, though no systematic study has appeared, and judged by results obtained before the requirement for glycine (Johnson and Ascher 1987; Kleckner and Dingledine 1988) was known, they appear not to depend on the glycine concentration either. However, the conductances do depend on the extracellular calcium concentration. The main single-channel conductance level for the most common sort of NMDA-receptor channel is about 50 pS in 1 mM calcium, but 42 pS in 2.5 mM calcium and 66 pS in ethylenediamine tetraacetate (EDTA)-buffered solution (Gibb and Colquhoun 1992), and a similar dependence is also found in a "low-conductance" channel (recombinant NR1a/NR2D; Wyllie et al. 1996). The values given below all refer to 1 mM calcium in bicarbonate buffer (about 0.85 mM free Ca^{2+}), or 0.85 mM Ca^{2+} in hydroxyethylpiperazine ethanesulfonic acid (HEPES) buffer.

In outside-out patches from *Xenopus* oocytes, the combinations NR1/NR2A and NR1/NR2B both give single-channels with a main conductance level of about 50 pS and a subconductance level of about 40 pS (Fig. 1) (Stern et al. 1992; also see Tsuzuki et al. 1994). These channels resemble closely the native receptors that have been described in cerebellar granule cells as well as the very similar channels in hippocampal neurones (Howe et al. 1991; Gibb and Colquhoun 1991, 1992). This similarity was not restricted only to the two conductance levels, but it was also found that the frequency of transitions between different conductance levels was in close quantitative agreement with the measurements of Howe et al. (1991) in cultured cerebellar granule cells. Furthermore, the quantitative similarity extended to the distributions of open times, shut times and burst lengths, etc. The conclusion of this work was that low-concentration equilibrium single-channel records are indistinguishable for (a) NR1/NR2A channels expressed in oocytes, (b) NR1/NR2B channels expressed in oocytes, and (c) the '50-pS' channels observed in various cultured, dissociated or brain-slice neurones. However, recombinant NR1/NR2A and NR1/NR2B channels are clearly distinguishable on the basis of their deactivation rates (Monyer et al. 1994; Vicini et al. 1998), their glycine sensitivity (which is ten times lower for the NR1/NR2A combination than for NR1/NR2B (Kutsuwada et al. 1992; Stern et al. 1992), as well as by antagonists such as ifenprodil (Williams 1993).

The NR1/NR2C channels, however, have lower conductance. They have a main conductance level of 36 pS with a sublevel about half the amplitude of the main level and with a briefer open time (Stern et al. 1992). These channels resemble closely the low conductance channels observed in brain slices by Farrant et al. (1994). They described the development of NMDA receptors in rat cerebellum, and found that, in 14-day-old rat cerebellar slices, the channels all appeared to be of the '50-pS' type, described above. However, in older animals they found that mature post-migratory granule cells contain a subtype of receptor that shows conductance levels of 33 pS and 20 pS in 1 mM calcium; these values are similar to those found with the NR1/NR2C combination. This result is in beautiful agreement with in situ hybridisation studies (Watanabe et al. 1992; Monyer et al. 1994), which show that there is little messenger RNA for NR2C present in the cerebellum at 7 days after birth, but after 14 days the amount of NR2C message has increased to something near to the adult level.

The NR1/NR2D channels have very similar conductance levels (35 pS and 17 pS) to NR1/NR2C but differ from them in two ways: firstly, the mean duration of the subconductance level is longer than that of the main level, and,

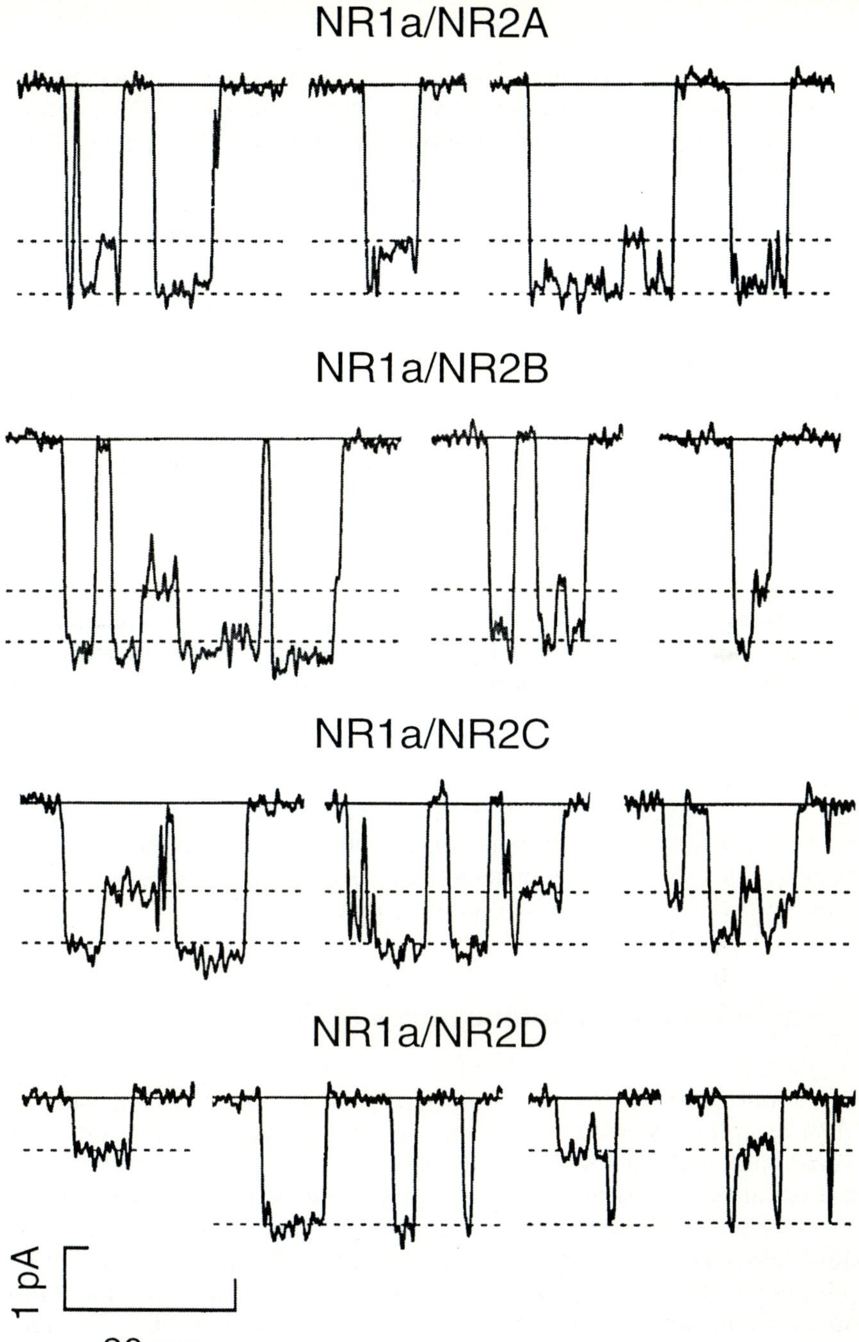

Fig. 1. Recombinant *N*-methyl-D-aspartate (NMDA) receptor channel currents. Examples of recombinant NMDA single-channel currents recorded in outside-out patches held at –60 mV. NR1a/NR2A and NR1a/NR2B channels give 'high conductance' channels with a main level at about 50 pS and a sublevel at 40 pS. In contrast NR1a/NR2C and NR1a/NR2D channels are lower in conductance, possessing a main level at around 35 pS and a sublevel at 18 pS

secondly, the records show temporal asymmetry. This asymmetry is manifested by the fact that direct transitions from the 35-pS level to the 17-pS level are more common than transitions in the opposite direction (Wyllie et al. 1996). Such asymmetry is not observed with any of the other subunit combinations or with native 'high conductance' channels. A similar asymmetry was, however, reported by Cull-Candy and Usowicz (1987). They reported single channels in cultured "large cerebellar neurones". Their records apparently contained two sorts of channels: one of which was the usual 50/40-pS type, but the other had conductance levels of 18 pS and 38 pS, which are similar values to those found in the NR1/NR2C and NR1/NR2D combinations, but the lower conductance type showed asymmetry which resembles only the NR1/NR2D combination. Momiyama et al. (1996) investigated Purkinje cells in cerebellar slices and found, up to P12, very similar low conductance channels (only); after this time, NMDA-receptor channels are absent from Purkinje cells. The suggestion that these channels are the NR1/NR2D type agrees very well with the results of in situ hybridisation studies, which show that only NR1 and NR2D mRNAs are detectable in Purkinje cells at this age (Akazawa et al. 1994). Very similar low-conductance channels, along with typical high conductance channels, were detected in deep cerebellar nuclei and in spinal cord, dorsal-horn neurones by Momiyama et al. (1996), and these too may be NR1/NR2D channels (Ishii et al. 1993; Tölle et al. 1993; Monyer et al. 1994).

An additional NMDA-receptor subunit has been cloned, which was termed initially χ-1 (Ciabarra et al. 1995) or NMDA-L (Sucher et al. 1995), but is now referred to as NR3A (Das et al. 1998). This subunit, which has a low homology with the others, when expressed together with NR1 and NR2A subunits, seems to give recombinant NMDA-receptor channels with conductances lower than those of NR1/NR2A channels alone.

D. Single-Channel Conductances of AMPA Receptors

It would appear that there are very few values of single-channel conductance, up to about 35 pS, that have *not* been reported to be the conductance of AMPA receptors. As with all electrophysiological measurements, differences in recording conditions undoubtedly play a major role in generating such a range of values. This is clearly not the only factor because, even under nominally constant conditions, AMPA receptors still exhibit a variety of unitary conductances. There has been no systematic study of the dependence of their conductance on calcium concentration. Furthermore, the determination of channel conductances is hindered by the fact that openings of AMPA receptors are very brief [as, of course, is to be expected because these channels mediate a rapidly decaying excitatory post-synaptic conductance (EPSC)], as well as having predominantly small amplitudes. Recently, however, a clearer picture has begun to emerge as the single-channel conductances of some recombinant AMPA receptors have been reported.

I. AMPA-Receptor Channels in Cerebellar Granule Cells

One of the best-characterised cell types is the cerebellar granule cell, so we shall concentrate on AMPA-receptor channels in these cells. Single-channel conductances have been measured (CULL-CANDY et al. 1988; HOWE et al. 1991; WYLLIE et al. 1993), and the AMPA subunit mRNA present in these cells has been analysed (MONYER et al. 1991; MOSBACHER et al. 1994). Recordings of single-channel currents in outside-out membrane patches from HEK 293 cells expressing recombinant AMPA receptors (SWANSON et al. 1997) has permitted the elucidation of some of the factors that contribute to determining the conductance of AMPA-receptor channels.

In recordings from membrane patches excised from granule cells, activation of non-NMDA-receptor channels produces three types of response. The 'high-conductance' response is composed of channel openings to three conductance levels – 12, 21 and 31 pS for glutamate, and is similar for AMPA and kainate (WYLLIE et al. 1993). The conductances are thought to represent multiple conductances of a single receptor type as direct transitions are seen occasionally between the three levels and the relative proportion of each type of opening is constant between different patches (WYLLIE et al. 1993). The 'low-conductance' response again gives discrete channel openings, but this time to levels of around 6 pS and 10 pS, again with transition between these levels being observed. Channels with extremely low conductances mediate the third type of response. No discrete channel openings are observed in this 'femtosiemens' response, which is characterised by a small noise increase in the glutamate-evoked membrane current. In situ hybridisation of mRNAs coding for AMPA subunits in granule cells indicate that both the 'flip' and 'flop' isoforms of GluR2 and GluR4 exist in these cells. Homomeric $GluR4(Q)_{flip}$ receptors ("Q" here refers to the fact that a glutamine residue is to be found at the Q/R site) give channels with multiple conductance levels of 8, 15 and 24 pS when activated by glutamate (similar conductances are observed if AMPA is used to activate the channel). Heteromeric channels comprised of $GluR2(Q)_{flop}/GluR4(Q)_{flip}$ also give channels with the same multiple conductance levels as is seen with homomeric $GluR4_{flip}$. However, as GluR2 subunits are most likely to be fully edited, that is contain an arginine (R) residue at the Q/R site, such heteromeric channels are unlikely to exist in vivo.

The question arises then, are the 'high-conductance' channels seen in granule cell patches composed of homomeric $GluR4(Q)_{flip}$ subunits? Similarities in conductance levels alone should not, of course, be the sole basis for such a proposition, but other pieces of evidence are also consistent with such a conclusion. The relative proportions of each of the three conductance levels are similar for both native and $GluR4(Q)_{flip}$ channels. $GluR4(Q)_{flip}$ channels and native high-conductance channels are sensitive to intracellular polyamines and, thus, are likely to be Ca^{2+} permeable (SWANSON et al. 1997). Although it is true that 'high-conductance' channels in granule cells have linear single-channel current-voltage relationships (see WYLLIE et al. 1993), which would not be expected of homomeric GluR4(Q)$_{flip}$ channels, it is likely that this resulted from the 'wash-out' in outside-out patches of the intracellular factor(s) responsible for generating the inward rectification seen with such channels. One clear discrepancy does remain, however – native 'high-conductance' channels give 10, 20 and 30 pS openings when kainate, instead of glutamate or AMPA, is used to activate the channel. With homomeric GluR4(Q)$_{flip}$ channels, kainate does not give discrete channel openings, but rather spectral analysis of the kainate-evoked current suggests the channels underlying the response have a weighted mean conductance of around 2.5 pS (Swanson et al. 1997). This is a clear difference between homomeric GluR4(Q)$_{flip}$ channels and 'high-conductance' channels in granule cells. It also appears to be a rare example of the channel conductance being determined by the nature of the agonist used to open the channel (though if kainate openings were extremely brief, spectral analysis could underestimate the single-channel conductance). To date, there have been no reports as to the single-channel conductance of GluR4(Q)$_{flop}$ channels other than for kainate, which, once again, produces a noise increase with no discernible channel openings (weighted conductance ~4 pS). Given that the flip and flop isoforms of GluR subunits differ in their amino acid sequences in the extracellular region between the third and fourth transmembrane segment, it would be surprising if alterations in this area of the protein had a major effect on the conductance of the channel.

What then of the 'low-conductance' channels? It appears that the co-expression of GluR4(Q)$_{flip}$ with either GluR2(R)$_{flip}$ or GluR2(R)$_{flop}$ results in a reduction of single-channel conductance. The 4-pS and 9-pS events seen when such heteromeric channels are activated by AMPA are very reminiscent of 'low-conductance' events seen in native channels.

Finally, it would appear that 'femtosiemens' channels are formed by homomeric GluR2(R)$_{flip}$ and GluR2(R)$_{flop}$ channels. Thus, in addition to affecting Ca^{2+} permeability and its influence on the current–voltage, the presence of a GluR2(R) subunit reduces the single-channel conductance of AMPA receptors (Swanson et al. 1997) (in a manner that is analogous to the introduction of N $\rightarrow$ R mutated subunits in the NMDA receptor; Béhé et al. 1995).

E. Structure and Stoichiometry of Glutamate Receptors

I. Location of the Agonist-Binding Sites

Since the AMPA receptors, GluR1 to GluR4, will all form homomeric channels, there is, presumably, one glutamate-binding site for each subunit (unless the subunits are arranged asymmetrically with the binding sites at interfaces). When the NR1 subunit of the NMDA receptor was first cloned (Moriyoshi et al. 1991), the fact that it was found by expression cloning in oocytes led to the presumption that the glutamate-binding site must be on the NR1 subunit. This has, subsequently, turned out not to be the case (the response seen with

NR1 alone in oocytes may depend on an endogenous subunit that has been found in oocytes; SOLOVIEV and BARNARD 1997). In fact it is the glycine-binding site that appears to be on the NR1 (KURYATOV et al. 1994; WAFFORD et al. 1995; HIRAI et al. 1996; WILLIAMS et al. 1996), as judged by the fact that single amino acid mutations reduced glycine potency by up to four orders of magnitude, with only small effects on glutamate potency. These studies suggested that the glycine-binding site has a bi-lobar structure, similar to that found by crystallography for a family of bacterial periplasmic amino acid-binding proteins (OH et al. 1993, 1994), in which the ligand is bound in a "Venus-flytrap" mechanism. One lobe (the 'S1 domain') is part of the pre-M1 N-terminal region; the other is C terminal of M3 (the S2 domain). More recently, similar experiments have located the glutamate-binding site on the homologous regions of the NR2 subunit (LAUBE et al. 1997; ANSON et al. 1998; LUMMIS et al. 1998).

Single amino acid mutations in NR2A or NR2B produced increases in the EC_{50} for glutamate of up to 1000-fold, with little or no change in Hill slope and little effect on glycine potency. The most effective mutant found by ANSON et al. (1998), NR2A(T671A), was in the S2 domain, which had a 1000-fold increase in EC_{50} for glutamate, and they adduced several lines of evidence to support the view that such a large change could not result primarily from effects on gating (it would require something of the order of a million-fold reduction in the gating constant; COLQUHOUN 1998). They also found that the T671A mutant had a 250-fold increase in the equilibrium constant for binding of the competitive antagonist APV, which again suggests a change in the binding site itself. The T671A mutant produced single-channel openings that were similar to the wild-type NR1/NR2A, but had a much faster deactivation rate after 100 ms concentration jump, which is consistent with the same interpretation. LUMMIS et al. (1998) also found a 1000-fold increase in EC_{50} with the nearby NR2A(G669S) mutation in the S2 domain. LAUBE et al. (1997) found a 120-fold increase in EC_{50} with NR2B(S664G) in the S2 domain (homologous with S670G in NR2A), but, oddly, the IC_{50} for APV was decreased, not increased, with this mutation. Similar effects were found by mutations in the S1 domain, the most effective being NR2B (E387A), which is homologous with E394A in NR2A (240-fold increase in glutamate EC_{50}; LAUBE et al. 1997), and NR2A(H466 A) (220-fold increase in glutamate EC_{50}; ANSON et al. 1998).

Although, in principle, the use of competitive antagonists is a good way of avoiding the binding-gating problem (COLQUHOUN 1998), the results found by LAUBE et al. (1997) and ANSON et al. (1998) suggest that the correlation between effects of mutations on agonists and antagonists may be imperfectly correlated and further investigation of this would be interesting.

II. Which Subunits Co-Assemble in NMDA Receptors?

Almost all of the native NMDA-receptor channels that have been described in sufficient detail so far (including the two cases discussed above) can be closely matched by expression of NR1 with only one of the NR2 subunit types. The only obvious exception, so far, is a preliminary report by Paleček et al. (1998) that motor neurones conta[illegible] NMDA-receptor channel with a conductance that is larger than that [illegible] any recombinant receptor. Nevertheless, there are now quite good [illegible] to believe that more than two sorts of subunit can assemble to form [illegible]t" receptors. As with other aspects of stoichiometry, there is little una[illegible] about the details, perhaps because of the indirectness of the methods [illegible]ve to be used. The majority of studies use immunoprecipitation. Seve[illegible]dies have found coprecipitation of NR1/NR2A/NR2B, with less ag[illegible]t about NR1/NR2A/NR2C (Chazot et al. 1994; Sheng et al. 1994; Di[illegible]l. 1995), but Chazot and Stephenson (1997) found that in mouse forebrain 46% of the NR1 immunoreactivity was associated with the NR2B subunit, and, of this, only 13% (6% of total) was NR1/NR2A/NR2B. Dunah et al. (1998) found in the thalamus that anti-NR2D precipitates 93% of NR2D and 48% of NR1, but only 25% of NR2A and 36% of NR2B, so binary NR1/NR2A was also present, but in the midbrain NR2A or NR2B were always co-precipitated. The immunoprecipitation method has the benefit of identifying proteins, but has the disadvantage that there is no way to be sure that the protein that is detected has been correctly inserted into the membrane and is functional. Blahos and Wenthold (1996) used isolated synaptic membranes (from rat forebrain), and they found that most receptors appeared to be either NR1/NR2A or NR1/NR2B, with little evidence for much triplet formation. They, and Chazot and Stephenson (1997), also found that more than one splice variant of NR1 can co-assemble in one receptor, with little preference for any particular NR2 subunit.

Electrophysiological studies have mostly been carried out on recombinant receptors. They have the advantage that they demonstrate that the receptors are inserted in the membrane and are functional, but the disadvantage that identification of which subunits are involved is indirect. No good single-channel studies have yet been published; inferences so far have been based on whole-cell responses. The first report by Wafford et al. (1993) suggested, on the basis of glycine sensitivity curves, that NR1, NR2A and NR2C could co-assemble, because co-expression produced an intermediate EC_{50} with little change in Hill slope for glycine. Köhr and Seeburg (1996) suggested co-assembly of NR1/NR2A/NR2B on the basis of intermediate deactivation rates when all three were transfected in HEK293 cells. However, Vicini et al. (1998), also using transfected HEK293 cells, found that most cells showed responses to short concentration jumps that were like those of either NR1/NR2A or of NR1/NR2B, with only a low proportion of putative triplets (NR1/NR2A/NR2B). Buller and Monaghan (1997) suggested co-assembly, in *Xenopus* oocytes, of NR1/NR2A/NR2D on the basis of IC_{50} values for antagonists, but they did not determine equilibrium constants for the antagonists (instead, a correction of IC_{50} values proposed by Durand et al. 1992, was used, but this correction, though soundly-based for binding experiments, is not valid for responses).

In summary, there is good evidence that some triplets can form, but considerable uncertainty about how many native receptors are triplets.

III. Stoichiometry of Recombinant NMDA Receptors

Despite the intense interest in glutamate receptors over the last decade or so, there is still serious doubt about whether they are tetramers or pentamers. The proposed membrane topology, which is now generally accepted, bears a superficial resemblance to that of a potassium channel, and that has made the idea of a tetrameric structure popular, but hard evidence is still lacking. There have been three electrophysiological approaches to the problem, each using somewhat different approaches, but with little agreement in the results.

BÉHÉ et al. (1995) used an N → R mutation in the NR1a and NR2A subunits, at the "QRN site". This mutation causes a great reduction in single-channel conductance. When wild-type and mutant NR1 subunits were expressed with wild-type NR2A, only one intermediate conductance level was observed, so it was inferred that there are probably two NR1 subunits in the receptor. However, when the same mutation was made at the homologous position in the NR2A subunit, the co-expression of wild-type and mutant NR2A together with wild-type NR1 gave a baffling array of conductance levels, which had no easy interpretation.

PREMKUMAR and AUERBACH (1997) used a similar approach but with quite different results. They used the NR2B subunit and an N → Q mutation rather than N → R. Furthermore, the wild-type and mutant NR2 subunits were coexpressed, not with wild-type NR1 but with mutant NR1. This approach gave relatively simple results for the NR2 subunit; one intermediate was found, so it was suggested that there were two NR2 subunits in the receptor. However, coexpression of wild-type and mutant NR1 (with mutant NR2) produced a complex mixture of types, which the authors interpreted as suggesting the presence of three NR1 subunits in direct contradiction with the results of BÉHÉ et al. (1995). They thus concluded that the receptor is a pentamer.

This sort of approach is clearly susceptible to errors in either direction. It is quite possible that all possible subunit combinations will not be expressed in sufficient amounts to be detected, or that some combinations are indistinguishable, so causing an underestimate of the number of subunits present. It is equally possible that the properties of the receptor will depend on the order in which the subunits are arranged round the channel, so causing an overestimate of the number of subunits. It was an advantage of the N → R mutation used by BÉHÉ et al. (1995) that the large differences in conductance made it possible to show the entire results, and there was no need for subtle kinetic analysis. However, the complex results obtained by PREMKUMAR and AUERBACH (1997) for the NR1 subunit were separated into different classes by hidden Markov methods, the reliability of which (through no fault of the authors) is impossible for the reader to assess. The matter remains undecided.

A study by LAUBE et al. (1998) used a different method. They too used coexpression of wild-type and mutant subunits, but they used NR1(Q387K), which has a low sensitivity to glycine, and NR2B(E387A), which has a low sensitivity to glutamate. They then attempted to assess the number of components in macroscopic concentration–response curves, by fitting multiple Hill-equation components to them. The results were compatible with the view that the NR1/NR2B receptor was a tetramer that contained two subunits of each type. The resolving power of this method is much lower than that of single-channel analysis, and the Hill equation is empirical. Although the Hill equation often provides a tolerable fit to concentration–response curves, it is obviously not the correct equation, and it is not clear that it is good enough for distinguishing subtle differences in a multicomponent fit. Unfortunately, in the absence of a good kinetic mechanism, the equation that should be fitted is not known. The effect, if any, of these problems on the conclusions cannot realistically be assessed at the moment.

IV. Stoichiometry of Recombinant AMPA Receptors

There is also disagreement about stoichiometry in the case of AMPA receptors. FERRER-MONTIEL and MONTAL (1996) and MANO and TEICHBERG (1998) both used the approach of MACKINNON (1991), which is now known to have given the correct tetrameric stoichiometry of a potassium channel (DOYLE et al. 1998). This method uses coexpression of wild-type subunits with mutated subunits that have a reduced affinity for an inhibitor. In both studies, GluR1 receptors were used. FERRER-MONTIEL et al. (1996) used a mutation that reduces the sensitivity of the GluR1 receptor to the inhibitors PCP and MK801, whereas MANO and TEICHBERG (1998) used, more dubiously on theoretical grounds, a mutation that reduced the sensitivity of the receptor to desensitisation by quisqualate. The former study concluded that the GluR1 receptor contains five subunits, and the latter concluded that it contained four.

The Mackinnon method involves plotting, against concentration of inhibitor, of the measured values of $\ln(R_{mix}/R_{mut})/\ln(p_{mut})$, where R_{mix} and R_{mut} are the responses to (an arbitrary concentration of) glutamate in the presence of the specified inhibitor concentration for mixed and all wild-type receptors respectively, and p_{mut} is the overall fraction of mutant subunits that were transfected into the cell. The asymptote of this plot, for high inhibitor concentration, is taken as an estimate of n, the number of subunits. It does not seem to have been widely noticed that this asymptote is, in general, less than n. The expression for the asymptote is actually

$$\left(\frac{1}{\ln(p_{mut})}\right)\ln\left[\sum_{i=0}^{n-1}\frac{K_i}{K_n}\binom{n}{i}p_{mut}^{i}(1-p_{mut})^{n-i}+p_{mut}^{n}\right],$$

where K_i is the equilibrium constant for binding of the antagonist to a receptor that contains i mutant subunits (and hence n–i wild-type subunits). This asymptote is always less than n. It will be close to n if the all mutant receptor has a much lower affinity (large K_n) for antagonist than any other receptor (say 1000-fold higher to be on the safe side), so the first term in square brackets becomes negligible relative to the second. In general, though, the value for the asymptote is quite sensitive to errors in the estimation of p_{mut}, the fraction of subunits that contain the mutation in the assembled receptors. Because it has to be assumed that p_{mut} is the same as the ratio of the amounts of RNA that are injected, accuracy is not easy to guarantee.

As an example, consider the case where $n = 4$ and $p_{mut} = 0.9$, as in MacKinnon (1991); with his values for K_i this gives an asymptote of 3.91, close enough to 4 to enable MacKinnon to get the correct answer. If, however, n had actually been 5, but p_{mut} had been overestimated and was actually 0.45 rather than 0.9, then the asymptote of the plot would have been exactly 4 despite the fact that $n = 5$, and a quite incorrect conclusion might have been drawn. Apart from this hazard, there are several other potential problems. For example, the whole argument depends on random assembly of subunits, i.e. assembly and insertion must be unaffected by the mutation. It is also assumed that the response that is measured is directly proportional to the number of unblocked channels. This is reasonable with a high-affinity blocker in the potassium channel case, but not so for an agonist-activated channel; indeed, it will not be true unless the response can be tested with a pulse of agonist so short that no blocker dissociates (not always easy in a whole oocyte).

A third study, by Rosenmund et al. (1998) used a different approach. They used mainly a chimaeric AMPA/kainate channel (GluR3/GluR6). They stepped from a high antagonist concentration into a high agonist concentration in order to slow the activation of the channel. They found, unusually, that the single-channel conductance depended on the number of agonist molecules that are bound (a phenomenon that has been shown even more clearly for the ion channels that are activated by intracellular cyclic guanosine monophosphate (cGMP); Ruiz and Karpen 1997). The results were interpreted in terms of a simple mechanism, which suggested that the largest conductances were produced when four agonist molecules were bound. If the mechanism they used is sufficiently close to the truth, this implies that four agonist molecules are needed to open the channel optimally (see below). They interpreted this result as meaning that the channels were tetrameric, though it could also be that a fifth molecule produced no further increase in conductance. As the authors are careful to point out, the inference of a tetrameric structure depends on an assumption of the simplest binding mechanism. Whether or not this has foiled this ingenious approach to the stoichiometry problem remains to be seen.

Electrophysiological methods, such as those just described, face many problems of interpretation (though they have given correct answers in other cases). In the face of the many uncertainties, it could well be that only biochemical, and ultimately structural, studies will provide unambiguous answers. One recent biochemical study suggested a tetrameric structure, but several others have been ambiguous (Wu et al. 1996, and references therein).

V. How Many Agonist Molecules Are Needed to Activate Glutamate Receptors?

One of the first questions that one would like to answer concerns the number of agonist molecules that must be bound in order to open the ion channel (which is not necessarily the same as the number of agonist-binding subunits). There may, of course, be no unique answer to this question. In principle, the channel might be expected to open more efficiently as more ligands become bound (RUIZ and KARPEN 1997; COLQUHOUN 1998).

AMPA receptors must have an agonist-binding site on each subunit because all the subunits (GluR1–4) can form homomeric receptors, but it is important to know how many of these sites must be occupied in order to open the channel efficiently. In the case of NMDA receptors, the number of glutamate-binding sites is presumably the same as the number of NR2 subunits, i.e. two or three (see above). The Hill slope of equilibrium concentration–response curves provides a clue about the minimum number of agonist molecules that are needed. Many such curves have been published for AMPA and NMDA receptors and there is general agreement that Hill slopes range from 1 to 2, which implies that more than one agonist must be bound before the channel can open. Further interpretation is complicated because of desensitisation and the fact that Hill slopes are dependent not only on binding of agonist but also on their ability to open the channel once bound (COLQUHOUN 1998; LEWIS et al. 1998). Alternatives to the concentration–response curve approach have, however, provided additional insight.

By studying the kinetics of AMPA receptor responses (in cultured hippocampal neurones) to rapid applications of low concentrations of quisqualate, CLEMENTS et al. (1998) have demonstrated clearly that the current elicited by this agonist has a sigmoid rising phase, thus suggesting at least two agonists need to be bound for the channel to open. Fit of simple sequential mechanisms suggests a (modest) improvement of the fit when it is assumed that two rather than three bindings are needed to open the channel. Even if the simple models used are adequate, this does not tell us much about what happens when three, four or possibly five agonist molecules are bound (as is presumably possible in a homomeric receptor). However, the time course of recovery after removal of agonist, shows little sigmoidicity. Some sigmoidicity would be expected if it was necessary to wait for two or three agonist molecules to dissociate before the channel could shut. The authors suggest that this may mean that such dissociations are too fast to be seen, and that this means that agonist binding shows negative co-operativity, such that the affinity for bindings beyond the second is low.

The recent report by ROSENMUND et al. (1998) suggested that up to four agonist molecules must be bound to open an AMPA-receptor channel (actually a GluR3/GluR6 chimaera) optimally (see above). This appears to disagree with the results of CLEMENTS et al. (1998), though the two studies were on different sorts of channel. However, the results could be compatible if the last two bindings in the study by CLEMENTS et al. (1998) showed sufficient negative co-operativity. The two studies were interpreted on the basis of quite different postulates about the channel mechanisms. CLEMENTS et al. (1998) supposed that all subunits underwent a concerted conformation change so that there was only one sort of open conformation, whereas the results of ROSENMUND et al. (1998) were clearly inconsistent with a concerted conformation change. Once again, the interpretation of observations is hindered by lack of knowledge about the physical reaction mechanism.

BENVENISTE and MAYER (1991) measured the rate of antagonist binding to NMDA receptors and compared the results with reaction schemes that involved the simple sequential binding of either one or two molecules; they concluded that two binding sites fitted better than one. CLEMENTS and WESTBROOK (1991) looked at NMDA receptors in outside-out patches from cultured hippocampal neurones. They measured the time course of the averaged response that followed a step into various glutamate concentrations (with fixed glycine concentration). To interpret the results, they postulated a simple linear reaction scheme in which only fully-liganded channels were able to open. On this assumption, the results were fitted better by the assumption that there were two rather than three binding sites for glutamate (though the distinction depended on quite small differences in shape between the predicted curves). A similar conclusion was reached for glycine sites by stepping into low glycine concentrations (with a fixed glutamate concentration). The problem is, once again, that it is really possible to distinguish subtle changes of shape (in this case, of the sigmoidicity of the onset of the response) if one has an adequate physical description of the channel. It is obvious from single-channel results that a simple linear binding scheme is not an adequate description of the NMDA receptor, but it is not clear at all whether this is a sufficiently serious problem to invalidate the conclusions. It is, for example, very unlikely that it would be possible to distinguish in this way between the case where there are two binding sites that must both be occupied to open the channel, and the case were there are three binding sites, occupation of two (or more) of which is sufficient to open the channel with reasonable efficiency.

The mechanisms that are used do, of course, provide an adequate fit of macroscopic currents, despite the fact that single-channel results show they cannot really be 'right'. This is, perhaps, an illustration of the point mentioned in the introduction. An empirical description, one that is not based on physical reality, may well provide curves that go through the data, but one cannot expect such descriptions to have much predictive ability.

Although it is not easy to tell whether binding of a third agonist molecule occurs or not (ROSENMUND et al. 1998), it does seem very likely that two glutamate molecules are sufficient to open both NMDA- and AMPA-receptor channels quite effectively. There is, however, no suggestion thus far that singly-liganded channels can open (GIBB and COLQUHOUN 1992), though this is quite well documented now for muscle-type nicotinic receptors (COLQUHOUN and SAKMANN 1981; JACKSON 1986).

F. Activation of NMDA Receptors

I. Steady-State Recordings: Can We Reach a Steady State?

The theory on which the interpretation of experiments with constant agonist concentration is based (Colquhoun and Hawkes 1982, 1995a) assumes that the channel has reached a steady state. It does not assume equilibrium, so it is still applicable to channels such as NR1/NR2D, which show non-equilibrium behaviour in the steady state (see below). In fact, judging by stability plots (Colquhoun and Sigworth 1995), it does seem to be possible to achieve something close to a steady state for quite long periods in the best patches (though certainly not in all). Despite the numerous reports describing rundown of NMDA receptor-channel activity in dialysed cells (MacDonald et al. 1989; Medina et al. 1995, 1996) and the evidence for regulation of channel activity by phosphorylation (Lieberman and Mody 1994; Wang and Salter 1994; Wang et al. 1994), the single-channel activity assessed on excised patches appears remarkably stable from a few minutes to hours after excision, even when the cytoplasmic side of the channel is exposed to nothing else than salts and ethyleneglycol tetraacetic acid (EGTA). Perhaps this favourable circumstance is because of the constancy of the milieu, the fact that patch excision isolates the channel from cytoskeletal influences (Rosenmund and Westbrook 1993), and because the use of very low agonist concentration which ensures that the internal calcium level will remain well buffered. Cell-attached single-channel recordings resolve receptor activations that are similar to those in outside-out patches (Gibb and Colquhoun 1992; Kleckner and Pallota 1995), so we may hope that results on the latter are relevant to real cells.

Several authors have reported the rare occurrence of 'high P_{open}' periods in both native and recombinant NR1/NR2A receptors (Jahr and Stevens 1987; Howe et al. 1991; Gibb and Colquhoun 1992; Stern et al. 1992). During these episodes, in which shut times suddenly become shorter, the channel is open for most of the time for periods that can be hundreds of milliseconds. Their cause is still not known, but they occur in both cell-attached and excised patches, in the absence of adenosine triphosphate (ATP) etc., and with or without added glycine. In our steady-state recordings, such periods are excised (by visual inspection of stability plots) to avoid distortion of shut time distributions. However, their occurrence during synaptic currents and in concentration-jump experiments would not be obvious and could provide an additional complicating factor.

II. Activations in Low-Concentration Steady-State Records

Steady-state recordings of single-channel activity from central neurone patches exposed to a low concentration of NMDA in the presence of glycine revealed the complex bursting behaviour of this glutamate-receptor subtype

(ASCHER et al. 1988; HOWE et al. 1988; CULL-CANDY and USOWICZ 1989; HOWE et al. 1991). These studies were all carried out before the requirement for glycine was known. As a consequence, the effective agonist concentration was uncontrolled, and, although bursting behaviour was obvious on a short time scale, it was not possible to identify the channel activations, which are long.

LESTER et al. (1990) used brief concentration jumps to show elegantly that the time course of the synaptic current was determined by the rate at which the channel shuts in the absence of agonist. This was followed by more detailed studies using both steady-state single-channel recording (GIBB and COLQUHOUN 1992; KLECKNER and PALLOTA 1995) and concentration jumps (EDMONDS and COLQUHOUN 1992; LESTER and JAHR 1992; DZUBAY and JAHR 1996; WYLLIE et al. 1998). We shall review the contribution made by each method.

One complicating factor in investigation of NMDA receptors is the fact that glycine is also required for the channel to open (JOHNSON and ASCHER 1987; KLECKNER and DINGLEDINE 1988). The EC_{50} for glycine (at an arbitrary but high glutamate concentration) has been estimated as 2 μM for NR1/NR2A receptors, and about tenfold less for the other three combinations (IKEDA et al. 1992; KUTSUWADA et al. 1992). Therefore, it is now usual to include 10–20 μM of glycine in the recording solution. At the synapse, the resting glycine concentration is not really known. ATTWELL et al. (1993) showed that, in theory, glycine transporters might keep the glycine concentration as low as 0.2 μM. The glycine concentration is hard to measure experimentally, but microdialysis has suggested a value around 5 μM (rising during ischaemia) (BAKER et al. 1991), so this is a reasonable upper limit for the concentration in the synaptic cleft. Notice that even this concentration is barely enough to saturate NR1/NR2A receptors. It is, therefore, quite possible that changes in cleft glycine concentration could affect synaptic transmission. Such changes could occur as a result of reverse transport consequent on a rise in intracellular sodium concentration (ATTWELL et al. 1993), but their actual importance is not known, and experimental evidence (THOMSON et al. 1989) is difficult to obtain and inconsistent.

In order to resolve individual channel activations, it proved necessary to use very low concentrations of glutamate (GIBB and COLQUHOUN 1991, 1992). This is because individual activations may contain quite long shut times (mean length 20–250 ms, depending on subunit combination). Therefore, to distinguish these from the shut times that separate one activation from the next, the activations must be several seconds apart (on average). It is difficult to show a typical recording because a single-channel activation can consist of as few as one-channel opening, or as many as several tens of openings. Figure 2 shows some activations of the recombinant NR1a/NR2A receptor channel. In this experiment, in which the glutamate concentration was 100 nM (+20 μM glycine), the longest and second longest components of the shut-time distribution had "time constants" of 3211 ms and 23.9 ms, so bursts ("super-clusters") were defined by a critical shut time of 88 ms, and these were taken as

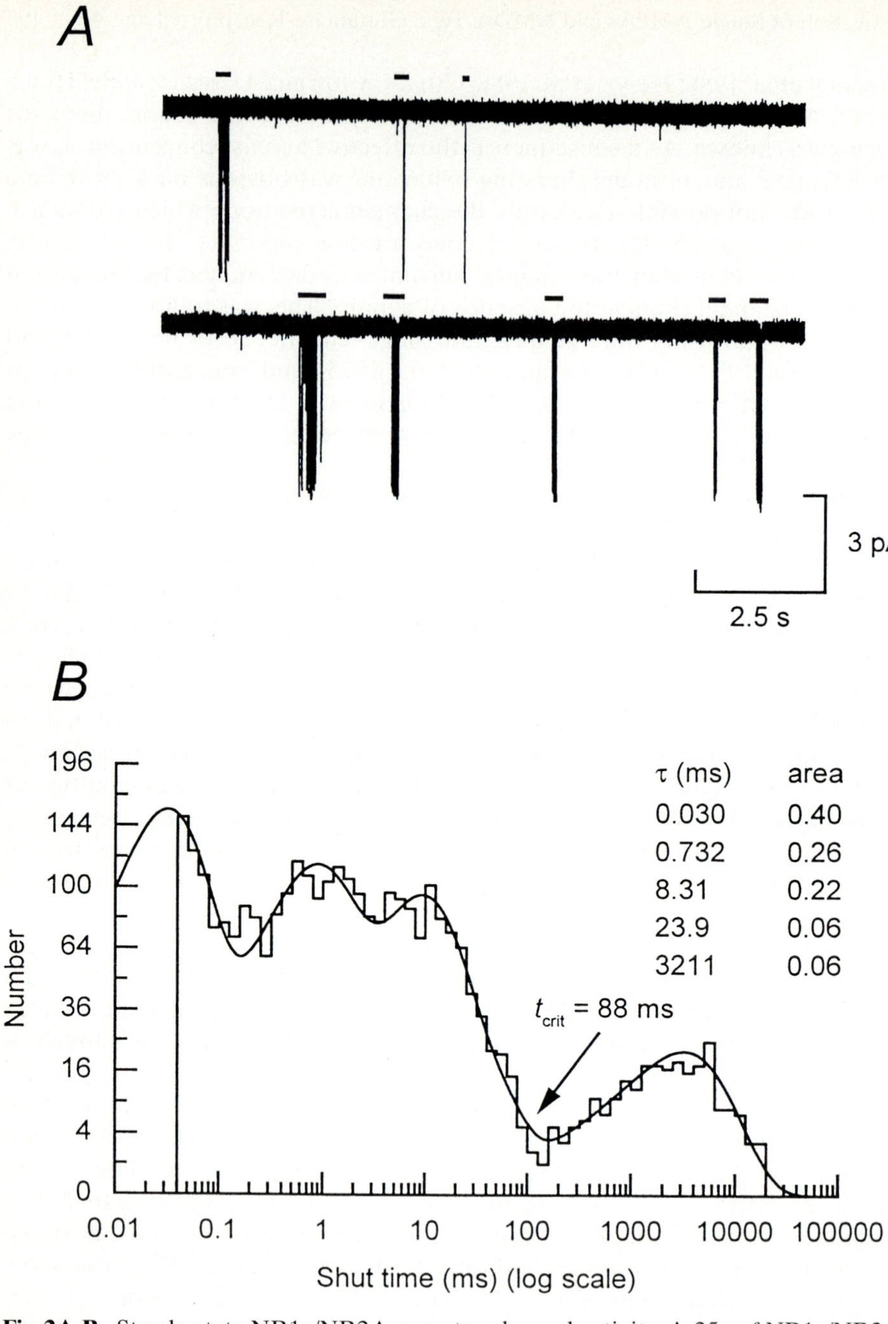

Fig. 2A,B. Steady-state NR1a/NR2A receptor channel activity. **A** 25 s of NR1a/NR2A receptor channel activity (evoked by 100 nM glutamate plus 20 μM glycine) taken from a recording lasting 20 min in total. The lines above each series of openings indicate separate channel activations. **B** Shut-time distribution for the whole recording fitted with a mixture of five exponential components with means and areas as indicated. A critical gap length (t_{crit}) of 88 ms was calculated that achieved optimal separation of gaps contained in the fourth and fifth components of the distribution. This t_{crit} was then used to identify the channel activations in **A**. Adapted from WYLLIE et al. 1998

estimates of the individual activations of the channel, each being marked with a line above the trace.

Figure 3 shows examples of distribution of the durations of channel activations. They are shown for recombinant NR1a/NR2A and NR1a/NR2D. These two receptor subtypes have vastly different kinetic parameters (WYLLIE et al. 1998). The NR1a/NR2A activations are relatively short (overall mean length 36 ms), with an average of seven (6.78 ± 1.01) openings per activation and are open for a substantial fraction of the time ($P_{open} = 0.36$). In contrast, NR1a/NR2D receptor activations are extremely long (overall mean 1602 ms) with an average of 40 (40.19 ± 8.61) openings per activation and have a very low $P_{open} = 0.04$. These means are, however, not very informative because the range is enormous – for NR1a/NR2D the number of openings per activation may be anything from one to several hundred.

One question that has yet to be addressed is regarding the extent to which the long shut times that often occur with a single activation can be considered as short-lived desensitised states, in the way that has been suggested for gamma aminobutyric acid (GABA) receptors (JONES and WESTBROOK 1995). Desensitisation of NMDA receptors has proved to be a complex and controversial matter. It is not even settled whether desensitisation occurs mainly from open or shut states (LIN and STEVENS 1994; COLQUHOUN and HAWKES 1995b). From a single channel perspective, it would be interesting to look at the activation structure in which macroscopic glycine-independent desensitisation has been altered by mutation of amino acids in the *N*-terminal region of the NR2A subunit (KRUPP et al. 1998).

III. Are the Rate Constants Constant?

The conventional way of analysing single-channel records, just like the conventional way of analysing any chemical kinetic problem, assumes that the system exists in a smallish number of discrete states, and that the rate constants that describe the frequency of transitions between states really are constant (do not vary with time). There are two sorts of ways in which this may not be true. The first, and trivial, way is that rate constants that depend on membrane potential and rate constants for association reactions, which depend on ligand concentration, will not be constants if potential or concentration, respectively, vary with time. That is why we always try to keep them constant by using voltage clamps and by doing fast concentration jumps. There is, however, a much more fundamental way in which the rate constants could vary. It could, for example, be the case that the rate constant for dissociation of a bound molecule was not constant, but depended on the length of time for which the molecule had been bound. In this case, the system would be described as non-Markovian. The normal assumption is that the tendency of the molecule to dissociate in the next microsecond (say) is the same, regardless of how long the molecule has been bound already – this is the Markov property. It implies that ligand–receptor complex "has no memory", so its

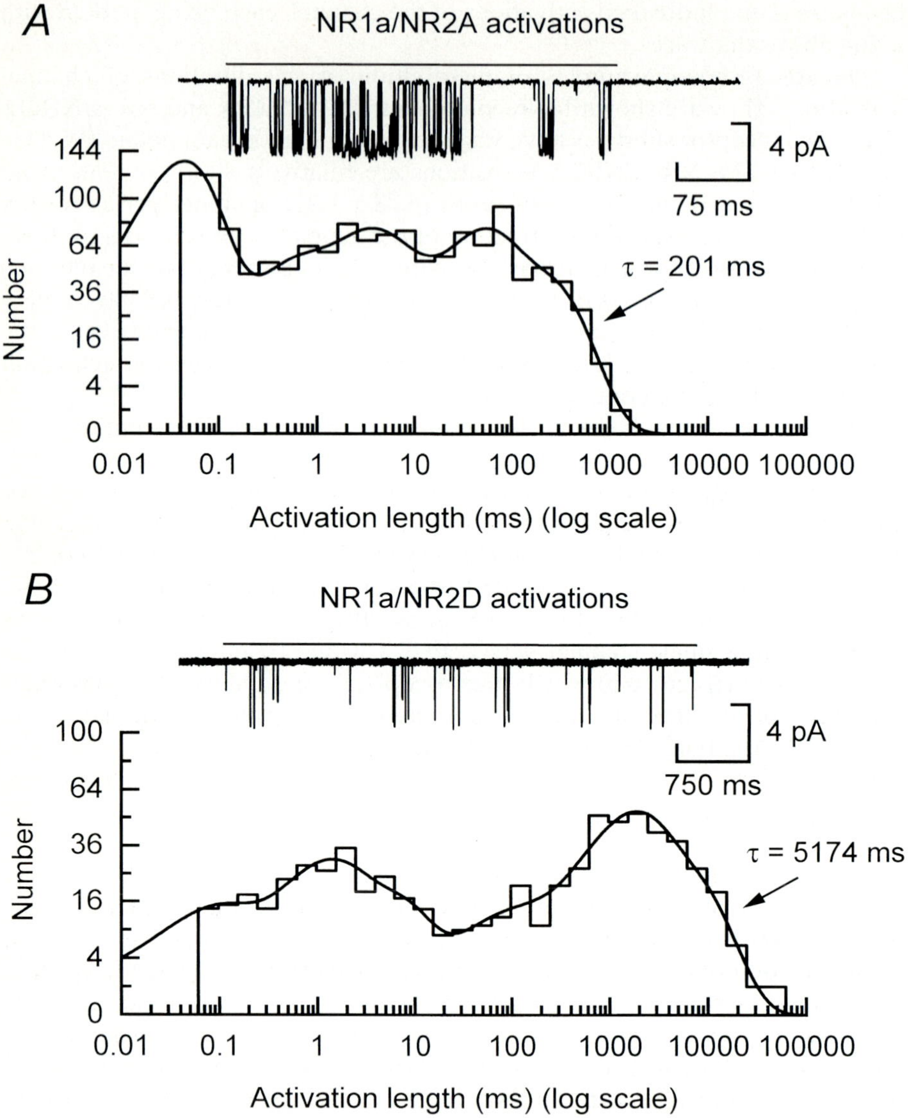

Fig. 3A,B. Activation distributions for NR1a/NR2A and NR1a/NR2D receptor channels. **A** Distribution of NR1a/NR2A receptor channel activations fitted with a mixture of six exponentials with means and areas of 42 μs (39%), 0.380 ms (8%), 1.88 ms (8%), 4.08 ms (14%), 40.6 ms (17%) and 201 ms (14%). The overall mean of the distribution is 35.8 ms. **B** Distribution of NR1a/NR2D receptor channel activations fitted with a mixture of six exponentials with means and areas of 71 μs (8%), 1.03 ms (18%), 4.71 ms (14%), 65.6 ms (6%), 1405 ms (31%) and 5174 ms (23%). In each case the slowest component of each distribution carries the majority of the charge. Examples of channel activations found in the slowest of each distribution are shown. Adapted from WYLLIE et al. 1998

future behaviour depends only on its present state, not on its past history (Colquhoun and Hawkes 1995a). There is little concrete reason to believe that there is anything seriously wrong with the Markov assumptions.

Obviously, a protein molecule can, in principle, adopt an essentially infinite number of conformations, but all the evidence suggests that a small number of conformations are predominant. The single-channel record is probably one of the most direct and striking demonstrations of this fact, together with crystal structures that have been determined for many proteins. Nevertheless, it has been suggested (Liebovitch 1989) that fractal analysis might be more appropriate (despite the fact that the results it provides are quite uninformative about physical mechanisms). There is much evidence against this view (McManus et al. 1988; McManus and Magleby 1989; Gibb and Colquhoun 1992; Wyllie et al. 1998).

IV. In How Many States Can the Receptor Exist?

The number of shut states of a channel is at least the number of components in its shut-time distribution, and the number of open states is at least the number of components in the open-time distribution or the distribution of the total open time per burst (Colquhoun and Hawkes 1982). The NMDA receptor is a great deal more complicated than the nicotinic acetylcholine receptor. The latter appears to exist, predominantly, in exactly the number of states that would be expected on physical grounds: three shut states (with zero, one and two agonist molecules bound) and two open states (with one and two agonist molecules bound). Its activation consists of two or three openings in quick succession, separated by openings that are mostly very short.

In contrast, the NMDA receptor shows five (NR1a/NR2A) or six (NR1a/NR2D) components in the shut-time distribution so there are probably at least six shut states. The components are not all easy to separate and show no very obvious concentration dependence, but the distribution of log(shut time) is certainly not unimodal as expected for a fractal process. Both native and recombinant NMDA receptors seem to have four detectable open states, two at each conductance level. These all seem to be interconnected, and at least two of them are directly accessible from the shut states. Thus, we are probably dealing with at least ten states altogether. All of the shut-time components, except the slowest, appear to be 'within an activation'. Thus there are four or five shut-time components 'within an activation', and these, together with the four open states suggest that there should be eight or nine components in the distribution of the length of the activation (Colquhoun and Hawkes 1982; Wyllie et al. 1998). In fact, only six components could be resolved in the distribution of activation lengths (Wyllie et al. 1998). It is, of course, only too easy to miss components that have either small areas, or time constants that are not well separated from others. In any case, it is clear that at least ten states must be postulated to account for the single-channel properties of the NMDA receptors. At present it is not known what physical structures these states correspond with, and it is a future challenge to attempt to clarify this question. Clearly it will be a much harder job than for the muscle-nicotinic receptor.

A major step forward in this task would be to discover how the states are connected. Single-channel analysis offers a tool for doing this that is not available in any other approach, the investigation of correlations between successive events. This will be discussed next.

V. Correlations in Single-Channel Records – How Are the States Connected?

The fact that a Markov process is (in the sense mentioned above) without memory does not mean that there cannot be correlations between, for example, the length of one opening and the next (Fredkin et al. 1985; Blatz and Magleby 1989; Colquhoun and Hawkes 1987; Colquhoun and Hawkes 1995a; Colquhoun et al. 1996). Steady-state, low-concentration recordings (cell-attached or outside-out) from NMDA receptors show a negative correlation between the length of an opening and the length of the adjacent shut time, and a positive correlation between the length of one opening and the next (Gibb and Colquhoun 1992). The distributions of the open-time durations for openings that are adjacent to short shuttings had the same time constants as for those that were adjacent to long shuttings, as predicted for a Markov process. The fact that the mean of the former was greater than that of the latter was entirely a result of the different areas that were associated with each time constant.

These sorts of correlation are qualitatively similar to those seen with the muscle-nicotinic receptor (Colquhoun and Sakmann 1985; Labarca et al. 1985), and with large-conductance, calcium-dependent, potassium channels (McManus and Magleby 1989). Qualitatively, they can occur when a long-lived open state is connected with a short-lived shut state, which in turn connects to a longer shut state from which brief openings can occur. We have also found that the first opening in an activation tends to be shorter than average, and this too places a constraint on how states are connected. It is likely that further investigation of the correlation structure will provide further information about connections between states, along the lines described by Blatz and Magleby (1989), Magleby and Weiss (1990), Magleby and Song (1992) and Colquhoun et al. (1996). At the moment, we are still a long way from having a complete kinetic mechanism for any NMDA receptor. Perhaps the best attempt so far is that of Kleckner and Pallotta (1995), but this was designed to describe only the short bursts that occur within an activation, not the entire channel activation, and it also fails to account for the observed correlations.

At present, the prospects for being able to identify structural counterparts of the states that are inferred from kinetic analysis, do not seem as good as for the nicotinic receptor, but that, nevertheless must be the aim.

VI. Temporal Asymmetry

Any reaction scheme that is capable of reaching true equilibrium must obey the principle of microscopic reversibility (see, for example COLQUHOUN and HAWKES 1982). In single molecule terms, this implies that the behaviour is symmetrical in time. If a single-channel record were recorded on tape, this means that it would be impossible to tell whether the tape was being played forwards or backwards. Several cases have been reported in which channels do not behave in this way (HAMILL and SAKMANN 1981; TRAUTMANN 1982; CULL-CANDY and USOWICZ 1987). In all of these cases, the phenomenon has been manifested as unequal transition frequencies for transitions from a main conductance level to a subconductance level. It is also possible (but thus far unreported) that this sort of asymmetry could be manifested as, for example, a difference between the mean duration of the first and last opening in a burst that contained three openings (COLQUHOUN and HAWKES 1982).

In the case of the NMDA receptor, no sign of asymmetry is seen with recombinant NR1a/NR2A, NR1a/NR2B or NR1a/NR2C receptors. However, recombinant NR1a/NR2D receptors do show the phenomenon (WYLLIE et al. 1996), as do some native receptors that, by other criteria, are very likely to be NR1/NR2D receptor channels (MOMIYAMA et al. 1996). At the more fundamental level, it implies that the channel does not obey the principle of microscopic reversibility and cannot come to a true equilibrium. It is perhaps surprising that this phenomenon seems, so far, to be relatively rare. As pointed out by LÄUGER (1983, 1985), the ion flux through a channel is not at equilibrium, so if there is any interaction between the flow of ions and the opening and shutting of the channel, the process as a whole cannot be at equilibrium. This appears to be what is happening with the NR1a/NR2D channel for which the extent of asymmetry depends strongly on the ionic composition of the recording solutions on each side of the membrane (unpublished data of PB). An elegant analysis of asymmetry in a mutant NMDA receptor has been provided by SCHNEGGENBURGER and ASCHER (1997) based on the principles enunciated by LÄUGER.

VII. Concentration Jumps: Differences Between Subunit Combinations

Studies by MONYER et al. (1994) and KÖHR and SEEBURG (1996) established that the deactivation of NR1/NR2A receptors is faster than that of NR1/NR2B receptors, whereas that of NR1/NR2D receptors is very slow. More recently, brief (1 ms) pulses of agonist have been used in order to mimic more closely synaptic conditions (VICINI et al. 1998 in HEK cells, and WYLLIE et al. 1998 in oocytes). These studies confirmed that NR1/NR2A receptors have the fastest deactivation rate. VICINI et al. found the rate to be much the same for four different splice variants of the NR1 subunit, and that NR1/NR2C showed a deactivation rate comparable with that of NR1/NR2B, whereas NR1/NR2D was much slower. The decay (deactivation) at zero agonist concentration, after the end of the 1 ms pulse of agonist, usually shows more than one exponential component. Vicini et al. (1998) state their results as weighted mean constants, the individual time constants being weighted with the relative amplitudes of the components. This makes the results a bit hard to interpret since it is not clear at which time point the relative amplitudes were measured (the results will be strongly dependent on this).

In both studies, a substantial amount of variability in decay rates was found from one patch to another. Wyllie et al. (1998) observed, with NR1a/NR2A, a tendency for patches that contained more channels to show a slower deactivation rate. The reason for this is not known.

The study by Wyllie et al. (1998) concentrated on a comparison of the fastest and slowest types. They found that the current through NR1a/NR2A channels following a concentration jump from zero to 1 mM glutamate for 1 ms was well fitted by three exponential components with time constants of 13 ms (rising phase), 70 ms and 350 ms (decaying phase). Similar concentration jumps on NR1a/NR2D channels were well fitted by two exponentials with means of 45 ms (rising phase) and 4408 ms (decaying phase) components. During prolonged exposure to glutamate, NR1a/NR2A channels desensitised with a time-constant of 649 ms while NR1a/NR2D channels exhibited no decline during the prolonged exposure (though this is not sufficient to conclude that there is no desensitisation; see Feltz and Trautmann 1982).

VIII. Latency to the First Opening

The distribution of the length of time that elapses (the first latency) between applying a pulse of agonist and the first opening of the channel is potentially very informative about the reaction mechanism. Jahr (1992) and Dzubay and Jahr (1996) reported an indirect estimate of the mean first latency for NMDA receptors in primary cultures of hippocampal neurones. Their method was to use the channel-blocking antagonist, MK-801; if this blocks only open channels, and is essentially irreversible, then it is possible to estimate the first latency and the probability of a channel being open. To do this, it is also necessary to assume that every channel that opens will become blocked during its first opening. Under these conditions, Dzubay and Jahr (1996) assume that the time course of the current will be approximately proportional to the distribution of first latencies. They suppose that any channel that opens for more than about 2 ms will become irreversibly blocked, and so they estimate that the mean first latency is of the order of 10 ms. This value is in fair agreement with the directly-observed value for recombinant NR1a/NR2 A channels (see below). The time course of the current cannot, however, be expected to give a very precise estimate of the shape of the first latency distribution, even if its mean is roughly correct. This is illustrated in Fig. 4, which shows two examples of the calculated time course of the probability of being open (solid lines)

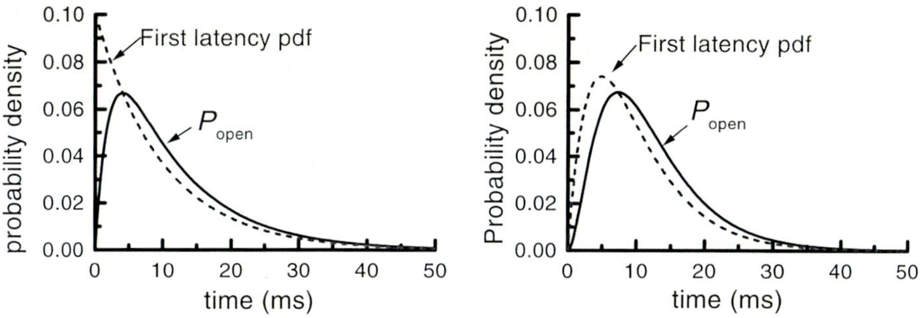

Fig. 4. Estimation of first latency by use of a channel blocker. The *dashed lines* show two assumptions about the shape of the first latency distribution, the mean first latency being 10 ms in both cases. The *solid lines* show the corresponding probability of being open, which is calculated on the assumption that the latency is followed by a single opening of mean duration 2 ms. The mean first latency inferred from the *solid curve* would be 12 ms in both cases (not very different from the correct value of 10 ms), but the *solid curves* give little idea of the shape of the first latency distribution

when the mean first latency is 10 ms and the mean open time is 2 ms. These are calculated by convolving the first latency distribution with the open time distribution on the assumption that the channel is blocked the first time it opens so there is only ever one opening (as in Colquhoun and Hawkes 1995a). The left-hand graph assumes a single exponential first latency (mean 10 ms), and the right-hand graph, more realistically, assumes a double exponential first latency (found by convolving shut times with means of 4 ms and 6 ms, so the mean latency is still 10 ms). Clearly the P_{open} curves (solid lines) in Fig. 4 do not describe the shape of the latency distribution well, but they do give a reasonable idea of the average latency (the mean inferred from the solid lines is actually the mean latency plus the mean open time, i.e. 12 ms in both cases). Their results are also consistent with those of Benveniste and Mayer (1995) who used a related method.

Direct observation of the first latency distribution requires a patch that contains only one channel, and which lasts for a long time (there is only one latency per concentration jump, and data acquisition following jumps has to be long for NMDA receptors). This has been achieved only rarely, but Wyllie et al. (1997) found, in this way, a mean first latency of 20–30 ms for NR1a/NR2A receptors expressed in oocytes. Figure 5 shows three channel activations elicited by a 1 ms pulse of glutamate to an outside–out patch containing a single NR1a/NR2A NMDA-receptor channel. It is clear that the channel does not open immediately following glutamate application but rather does so after a delay (first latency) of up to a few tens of milliseconds. The average of 248 such activations gives a slowly decaying current that has a time-course resembling that of a NMDA synaptic current.

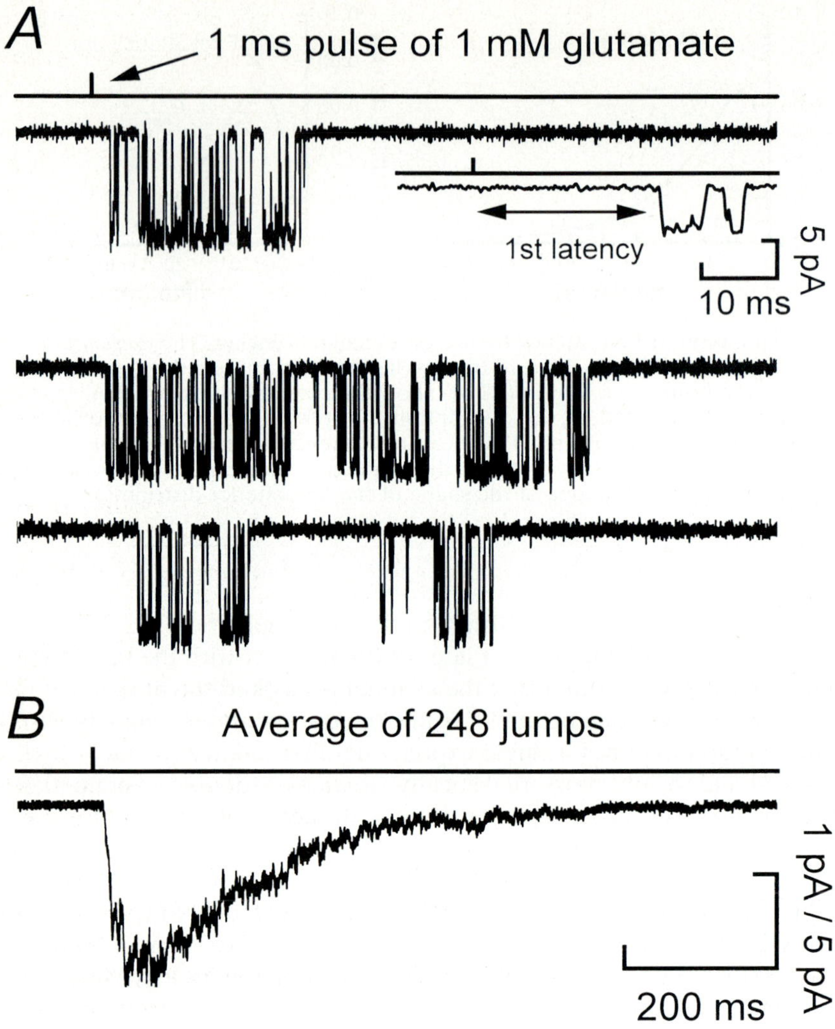

Fig. 5A,B. NR1a/NR2A receptor channel activations evoked by 1 ms concentration jumps. **A** Examples of three channel activations recorded in an outside–out patch containing only one channel. Channel openings were evoked by exposing the patch to 1 mM glutamate for 1 ms. The *inset* in the first trace shows the activation on an expanded time base to illustrate more clearly the first latency. **B** Average of 248 jumps gives a mean current with a peak amplitude of 1.7 pA and a slowly decaying time course. Single-channel currents in this patch had a mean amplitude of around 5.5 pA giving an estimate of the P_{open} at the peak of 0.3. Adapted from Wyllie et al. 1997

IX. The Relationship Between Steady-State Activity and Concentration Jumps

It is only recently that a fairly complete theoretical background has been given for the behaviour of non-stationary single-channel records, such as the channels that are elicited by a jump in agonist concentration (COLQUHOUN et al. 1997). It is shown there (see also WYLLIE et al. 1998) that the following measurements should all have the same number of exponential components and the time constants for these components should be essentially the same (for the precise assumptions see Appendix to WYLLIE et al. 1998).

1. The macroscopic response that follows a concentration jump to zero concentration (NMDA receptor channels hardly ever open within a 1-ms pulse of agonist, so the entire response to a 1-ms pulse is measured at zero agonist concentration).
2. The distribution of the length of activations measured in steady-state records at very low agonist concentrations (estimated as bursts or super-clusters).
3. The macroscopic current synthesised by adding such low-concentration activations after aligning them on their first openings.

WYLLIE et al. (1998) showed that observations on NR1a/NR2A and NR1a/NR2D receptors are consistent with these predictions (and this constitutes additional evidence that the Markov framework is valid). When the macroscopic currents are fitted with the number of time constants (five or six) needed to fit the distribution of activation length, with the time constants fixed at the values found from that distribution, a good fit can be obtained in most cases. A free fit to the macroscopic currents can, of course, produce good estimates of only two or three of the time constants, but this is merely a reflection of the poor resolving power of macroscopic measurements. Figure 6 compares the time-course of aligned activations and macroscopic currents evoked by 1 ms pulses of agonist for NR1a/NR2A and NR1a/NR2D channels.

The fact that the time constants agree means that it is indeed possible to predict from steady-state measurements that a jump response will have a slow component (as in GIBB and COLQUHOUN 1991, 1992). What we cannot predict is how big that slow component will be. The relative amplitudes or areas that are attached to each time constant differ vastly between jumps and steady-state results. This is because the initial vector (the fraction of receptors in each state) at the start of an activation in the steady state is quite different from that for a concentration jump. The amplitudes of the components in a macroscopic jump cannot be calculated from the steady-state results without detailed knowledge of the kinetic mechanism. It would not be correct simply to convolve the observed first latency distribution (WYLLIE et al. 1997) with the distribution of activation lengths. If there is only one open state, it should be convolved with the macroscopic probability, $P_{11}(t)$, that the channel is open at time t, given that it is open at $t = 0$, but if, as for the NMDA receptor, there

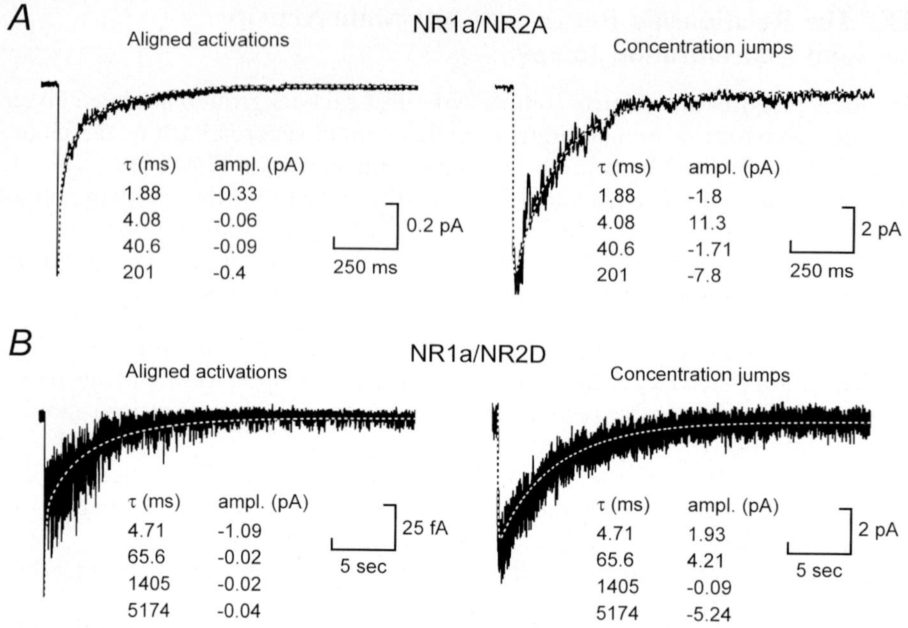

Fig. 6A,B. Comparison of aligned activations and concentration jumps on NR1a/NR2A and NR1a/NR2D receptor channels. *Left panels*, currents obtained from the alignment and averaging of NR1a/NR2A receptor channel activations (**A**) or NR1a/NR2D receptor channel activations (**B**). *Right panels*, examples of macroscopic currents mediated by NR1a/NR2A (**A**) or NR1a/NR2D (**B**) receptor channels and recorded following a 1 ms pulse of 1 mM glutamate. In each case the *white dashed line* shows the fit of these currents with a sum of exponentials with time constants fixed at the values obtained from the distributions of activation length shown in Fig. 3. Adapted from Wyllie et al. 1998

is more than one open state, the first latency distribution, although it contains valuable kinetic information, cannot be used directly in this way at all (see Colquhoun and Hawkes 1995a).

It is, presumably, this difference in the initial vector that accounts for the fact that one-opening activations are more frequent in steady-state activations than in activations elicited by a concentration jump (unpublished data). Despite this fact, the probability of being open is similar (around 0.3) at the peak of the jump response and for steady-state activations.

In an earlier study, Edmonds and Colquhoun (1992) found longer first latencies in dentate granule cells. The subunit composition of these receptors is not known, so it may be that some receptors have longer first latencies. It must also be noted that it is an ever-present hazard of latency measurements that a small background contamination by agonist (or spontaneous openings) could give rise to the appearance of long latencies. They also found that the decay of aligned super-clusters and super-cluster length distributions did not appear to share the same set of time-constants. Their suggestion that this discrepancy could be explained in part by the existence of a long first latency was incorrect (see WYLLIE et al. 1998). It now seems likely that small amplitude of the slowest component in the aligned super-cluster decay prevented its detection.

G. Activation of AMPA Receptors

In contrast to NMDA single-channel activations, AMPA receptor activations would appear to be remarkably simple, consisting of bursts of one or two individual openings (for example see WYLLIE et al. 1993). As AMPA receptors mediate the fast component of glutamatergic EPSCs that last at most a few milliseconds, such brief activations are perhaps to be expected. However, despite the fact that the channel activation are brief, many factors contribute to determine the precise nature and duration of these events.

I. Macroscopic Currents from Recombinant AMPA Receptor Channels

Our understanding of the factors that control AMPA receptor channel kinetics has been advanced greatly from studies of recombinant receptor whole-cell currents and macroscopic currents in outside-out membrane patches. As mentioned above, mRNA coding for AMPA subunits exist in two splice variants which have been termed "flip" and "flop" (SOMMER et al. 1990). In this paper, it was suggested that some aspects of the gating kinetics of AMPA receptors were dependent on the particular splice variants of the AMPA subunits forming the receptor-channel complex. However, in this study, whole-cell recordings from HEK 293 cells expressing recombinant AMPA receptors would be unlikely to resolve fast components of the activation kinetics of these channels due to the fact that perfusion of these cells by agonist was relatively slow. This problem was overcome in a later study (MOSBACHER et al. 1994), which made outside-out patch recordings from *Xenopus* oocytes expressing AMPA receptors and used a piezo-electric device to achieve rapid solution exchanges. In response to brief (1 ms) pulses of glutamate homomeric GluR-D_{flip} and GluR-D_{flop} channels, each exhibit rapid decay kinetics ($\tau_{decay} = 0.6$ ms). Similarly, homomeric GluR-A_{flip} and GluR-A_{flop} channels each show similar deactivation kinetics albeit with slower decay time constants ($\tau_{decay} = 1.1$ ms). However, during sustained agonist exposure, differences in the response of flip and flop channels become apparent. Homomeric GluR-D_{flop} channels desensitise almost as rapidly as they deactivate following a brief concentration jump ($\tau_{desens} = 0.9$ ms) whereas GluR-D_{flip} channels desensitise more slowly ($\tau_{desens} = 3.6$ ms). Similarly, homomeric GluR-C_{flop} channels desensitise more rapidly than their homomeric flip counterparts. This difference between flip and flop desensitisation rates is not seen with homomeric GluR-A channels. As many native AMPA receptors are likely to be heteromers containing GluR-B subunits, it is interesting to note that GluR-D_{flop} channels still desensitise faster than GluR-D_{flip} channels when expressed with GluR-B subunits. The developmental regulation of expression levels of various splice variants of AMPA subunits (Monyer et al. 1991) provides a mechanism by which the kinetics of AMPA receptors may be controlled. The fact that certain AMPA receptors desensitise in less than a millisecond has important implications for synaptic transmission. The decay of the AMPA component of glutamatergic EPSCs has been suggested to reflect the closing rates of the AMPA receptor channels following exposure to glutamate (Colquhoun et al. 1992). However, under certain circumstances the desensitisation rate of AMPA receptor channels may contribute significantly to the time course of the synaptic current (Trussell et al. 1993). Clearly in the case of GluR-D_{flop} channels, a change in the concentration profile of glutamate is unlikely to have a substantial effect on the decay of the synaptic current.

In addition to the well-documented RNA editing of the Q/R site of AMPA receptor subunits (Sommer et al. 1991), a second site of RNA editing also exists (Lomeli et al. 1994). The last codon of exon 13 of the genes encoding the GluR-B, C and D subunits can undergo editing to switch the codon at this position from one that encodes an arginine residue (AGA) to one that gives a glycine residue (GGA). This residue precedes immediately the alternatively spliced flip and flop sequences of amino acids. GluR-A subunits appear to exist only in the non-edited arginine version. The presence of arginine or glycine residues does not appear to affect the rise time or decay time of AMPA currents following brief agonist exposure. The main kinetic difference between the two forms of subunit appears to be in the rate of recovery from desensitisation. In experiments where pairs of brief (1 ms) pulses of glutamate were applied to outside-out patches, AMPA receptor channels containing edited (glycine) residues showed less depression of the second response and faster rates of recovery than AMPA receptor channels containing non-edited residues. This is true for both homomeric and heteromeric channel combinations. Such findings have led to the proposal that in the CNS postsynaptic AMPA receptor channels that are edited at this site may be better able to convey high frequency presynaptic activity than non-edited AMPA receptor channels (Lomeli et al. 1994).

II. Identifying Native AMPA Receptors on the Basis of Macroscopic Deactivation and Desensitisation Kinetics

The characterisation of the kinetic properties of recombinant AMPA receptors together with in situ mRNA hybridisation studies has provided valuable information concerning the subunit combination of native AMPA receptors. Studies that combine electrophysiological recordings with reverse-transcription polymerase chain reaction of mRNAs present in individual cells may give further insights into how edited and differentially spliced subunits affect the kinetics of native AMPA receptor channels (LAMBOLEZ et al. 1992, 1996; BOCHET et al. 1994; JONAS et al. 1994; GEIGER et al. 1995; ANGULO et al. 1997; GÖTZ et al. 1997).

Several studies have shown that AMPA receptor channels in principal (excitatory) neurones in the hippocampus and cortex display slower deactivation and desensitisation rates than those found in inhibitory interneurones (for example, see COLQUHOUN et al. 1992; JONAS and SAKMANN 1992; HESTRIN 1993; JONAS et al. 1994; GEIGER et al. 1995; LAMBOLEZ et al. 1996; GÖTZ et al. 1997; see MONYER et al. Chap. 9, and GEIGER et al. Chap. 11; this volume). How do these kinetic differences correlate with levels of expression of different AMPA subunits? In the hippocampus and neocortex, principal neurones contain high levels of mRNA for the GluR-B subunit mainly in its flip form (GEIGER et al. 1995; LAMBOLEZ et al. 1996). In addition to conferring linearity to AMPA receptor channel current-voltage relationships in these cells, as well as their low Ca^{2+} permeability, it was suggested that the presence of this subunit gave rise to channels with relatively slow gating kinetics. The faster gating kinetics of AMPA receptors found in interneurones has been suggested to result from (a) the presence of higher levels of GluR-D (mainly in the flop version) in these cells (GEIGER et al. 1995), or (b) a more general increase in the levels of flop variants of each of the four AMPA receptor subunits (LAMBOLEZ et al. 1996). In cultures of hippocampal neurones, two types of neurone have been classified on the basis of current-voltage relationship and Ca^{2+} permeability of their AMPA receptors (OZAWA et al. 1991). AMPA receptor channels in type II neurones (putative GABAergic interneurones) have inwardly rectifying current-voltage relationships, high Ca^{2+} permeability, and contain only GluR1(A) and GluR4(D) subunits each in the flop variant (BOCHET et al. 1994). This result, therefore, supports both the proposal that GluR-D_{flop} specifically is an important regulator of fast-gating kinetics in interneurones as well as the proposal that it is the increased levels of flop variants of any AMPA subunits that determines rapid kinetics. More recently, however, a study of fast spiking and regular spiking non-pyramidal cells of the neocortex (ANGULO et al. 1997) has provided evidence that no single molecular parameter (e.g. ratio of flip to flop, presence/absence of particular subunit, R/G editing of subunits) is the sole determinant of the AMPA receptor channel kinetics. Furthermore, this study showed it was possible to obtain inwardly rectifying current-voltage relationships from cells with slow desensitisation kinetics and conversely fast desensitisation from patches with weak rectification. This latter observation has also been documented for GABAergic neurones in the substantia nigra (see also GÖTZ et al. 1997). These findings suggest, therefore, that the absence of GluR-B subunits in a receptor complex does not necessarily mean that receptors will have fast desensitisation kinetics. Thus, it is unlikely that one will be able to correlate the presence of specific subunits with particular kinetic properties in the same way that it has been possible to relate the presence of GluR-B subunits with Ca^{2+}-impermeable receptors with linear current-voltage relationships.

III. Single-Channel Kinetics

The fact that the non-NMDA receptor mediated component of glutamatergic EPSCs lasts at most a few milliseconds tells us that activations of AMPA receptors are brief. Unlike NMDA receptors, single-channel analysis of AMPA receptors has proved remarkably unpopular. This is no doubt due to the fact that the channel openings are so brief and open to poorly defined conductance levels (see above). Even among the reports that have appeared, few have used glutamate as an agonist on these channels. In cerebellar granule cells 'high conductance' AMPA receptor channels activated by glutamate have a mean apparent open time of 660 μs and a mean burst length of just 920 μs (Wyllie et al. 1993). Given the similarity of the mean apparent open time and burst length, it is unsurprising that such bursts contain few resolvable brief gaps (on average 0.17 per burst). Homomeric $GluR4_{flip}$ channels also give briefing openings with glutamate (170 μs) and short bursts (overall mean 1.0 ms) (see Swanson et al. 1997). As mentioned above, fast concentration jumps on $GluR4_{flip}$ channels (Mosbacher et al. 1994) give rapidly deactivating macroscopic currents whose time constant is very similar to the mean burst length of 'high conductance' channels. However, it should be noted that, although the burst length of 'high conductance' channels is very similar to decay-time constant of the AMPA component of the EPSC at the mossy fibre-granule cell synapse (Silver et al. 1992), it is unclear whether these 'high conductance' channels mediate the synaptic current (see above). Other reports of glutamate activation of AMPA receptor single-channel currents also indicate that openings are brief (for example see Greengard et al. 1991; Tang et al. 1991).

IV. Models of AMPA Receptor Activation

Several kinetic schemes have been proposed for AMPA-receptor activation (for example see Raman and Trussell 1992; Jonas et al. 1993; Häusser and Roth 1997). Each of the schemes proposed has been based on observations made from the study of macroscopic currents and, to a large extent, are able to predict accurately the properties of currents evoked by short and long pulses of agonist application, recovery of receptors from desensitisation and the EC_{50} and Hill slopes of dose-response curves. How well these schemes can account for single-channel data is less clear as none of the models put forward has been based on data obtained from single-channel experiments. For the purposes of this review, therefore, we have simulated some single-channel data based on one of the published models for AMPA receptor channel activation. The model chosen is that from a study of AMPA receptors in cerebellar Purkinje cells (Häusser and Roth 1997). This model contains one open state and eight shut states, and is shown below as Scheme 1. Calculations were done with the rate constants specified by Häusser and Roth (1997). With a low glutamate concentration (1 μM), it is predicted that openings (mean duration 0.26 ms) occur in well-separated bursts (30.7 s apart for one channel), each

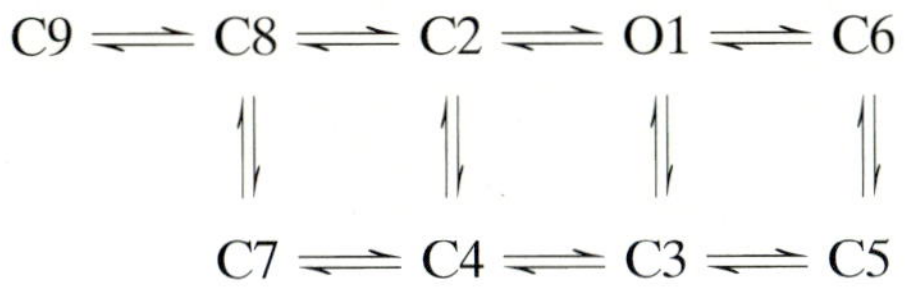

Scheme 1

burst containing on average 4.1 openings. The shut times within a burst (spent in states 2 to 8) are mostly short (43 μs, 96.5% of area), with some longer gaps (9.1 ms, 3.2% of area; the other five components have negligible area). The short gaps result almost entirely from oscillations between states 1 and 2 (fully-liganded open and shut, just as for the nicotinic receptor). The longer gaps within a burst result mostly from single sojourns in state 6, with some visits to states 3 and 5 too. The burst length distribution has a predominant component with a mean of 1.1 ms (88% of area) and a component of 10.2 ms (11.5% of area), the other six components being negligible; the overall mean is 2.3 ms. The macroscopic response to a very short (0.1 ms) pulse of 1 mM glutamate is predicted to have seven exponential components but the decay is dominated by a single exponential (1.1 ms). There is, as expected from the burst length distribution, a component with a time constant of 10.2 ms, but although this accounts for about 7% of the area, it has an amplitude only 1.4% of the 1 ms component and would be undetectable in practice.

The best resolution that can be achieved for small-amplitude events such as AMPA receptor channels, even with time-course fitting, is about 100 μs. When this resolution is imposed on the simulated record (for both openings and closings) most of the short gaps within bursts are undetected, resulting in a mean of only 1.4 openings per burst instead of the predicted 4.1 (see above).

The distribution of shut times and apparent burst lengths after such a resolution has been imposed is shown in Fig. 7A. A free maximum likelihood fit to this shut time distribution resolves only three of the eight components – the fitted time constants of 66 μs (53.1%), 9.3 ms (4.2%) and 34.4 s (42.7%) may be compared with three largest components of the true shut-time distribution, namely 44 μs (73.2%), 9.1 ms (2.45%) and 30.1 s (24.2%). This sort of distortion caused by limited time resolution can be dealt with by, for example, the methods described by Hawkes et al. (1992) and exemplified in Colquhoun et al. (1996).

The distribution of the apparent lengths of activations with 100 μs resolution is shown in Fig. 7B. Two components could be fitted, as might be expected from the very small areas (less than 1%) that are predicted for the other six components. The fitted components, 1.09 ms (89.8%) and 10.2 ms (10.2%), may be compared with the two main components of the true distribution, 1.09 ms (87.8%) and 10.2 ms (11.5%). This agreement is a good example of the fact that burst length distributions are far less susceptible to errors resulting from missed events than are open-time or shut-time distributions.

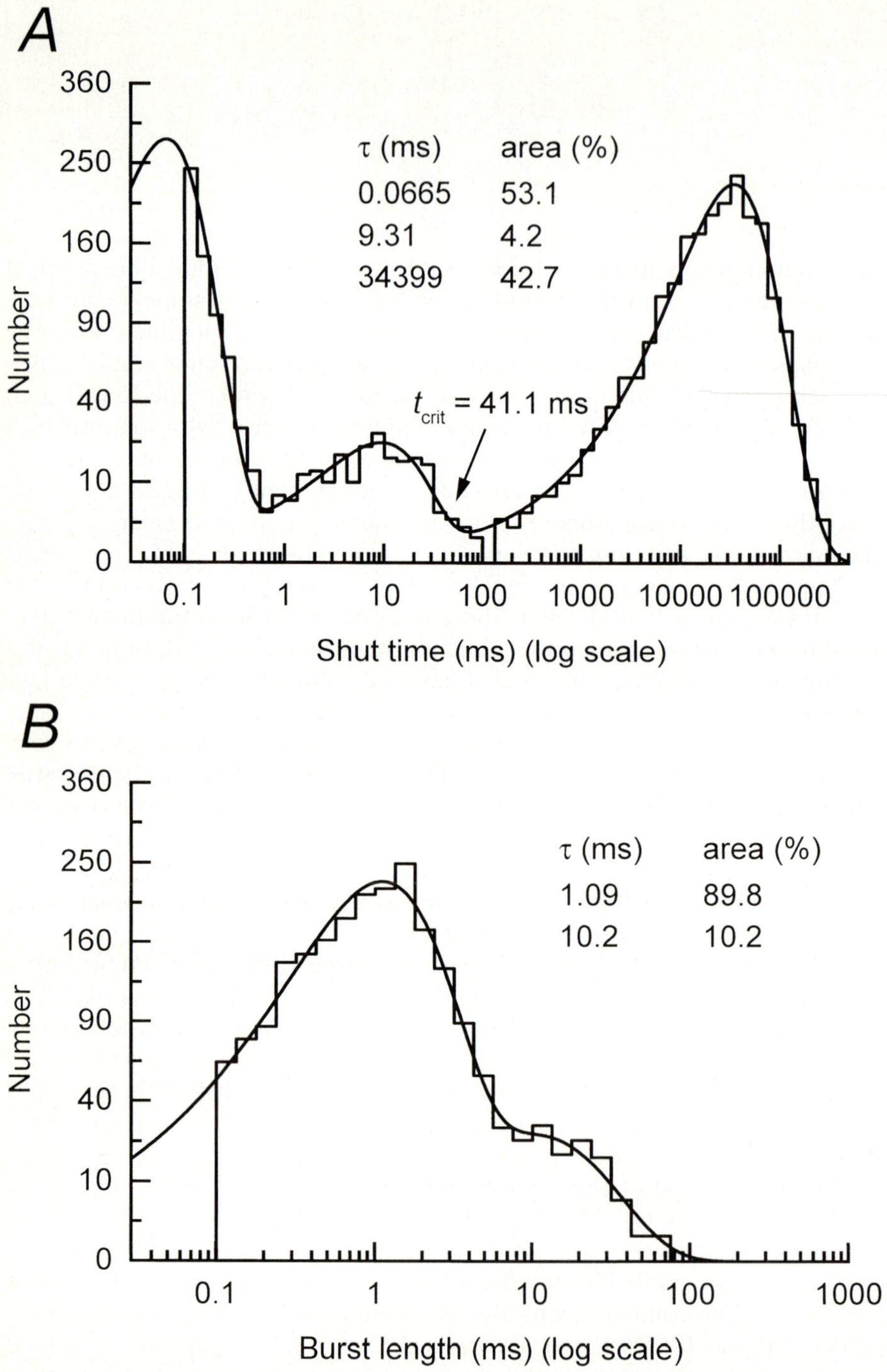
A
360
250
160
90
40
10
0
Number
τ (ms) area (%)
0.0665 53.1
9.31 4.2
34399 42.7
t_{crit} = 41.1 ms
0.1 1 10 100 1000 10000 100000
Shut time (ms) (log scale)
B
τ (ms) area (%)
1.09 89.8
10.2 10.2
0.1 1 10 100 1000
Burst length (ms) (log scale)

Fig. 7. Shut time and burst length distributions for AMPA receptor channel activity simulated using the model shown in Scheme 1. **A**, Shut time distribution obtained from the simulation of AMPA receptor channel activity by 1 μM glutamate and based on the kinetic scheme proposed by Häusser and Roth (1997). Despite the fact that there are eight shut states in the scheme, only three components are visible in the steady-state shut-time distribution. This distribution gave a t_{crit} of 41.1 ms for the separation of the second and third components, and this value was used to calculate the burst length distribution which is shown in **B**

H. Conclusions

As stated in the introduction, one the goals for those of us working in this field is to generate kinetic mechanisms of channel activations which reproduce faithfully the nature of single-channel activity seen for both NMDA and AMPA receptor-channels. Understanding the single-channel activation structure of both these types of ligand-gated ion channel and how these activations can be controlled and/or modified will provide additional insights into synaptic physiology that can not be obtained from the study of macroscopic currents alone.

Acknowledgements. Work in the authors" laboratory is supported by the Medical Research Council, The Wellcome Trust and The Royal Society. This review was submitted on 8-Jan-99.

References

Adams PR (1976) Drug blockade of open end-plate channels. J Physiol 260:531–552
Akazawa C, Shigemoto R, Bessho Y, Nakanishi S, Mizuno N (1994) Differential expression of five N-methyl-D-aspartate receptor subunit mRNAs in the cerebellum of developing and adult rats. J Comp Neurol 347:150–160
Aldrich RW, Corey DP, Stevens CF (1983) A reinterpretation of mammalian sodium channel gating based on single channel recording. Nature 306:436–441
Anderson CR, Stevens CF (1973) Voltage clamp analysis of acetylcholine produced end-plate current fluctuations at frog neuromuscular junction. J Physiol 235: 655–691
Angulo MC, Lambolez B, Audinat E, Hestrin S, Rossier J (1997) Subunit composition, kinetic and permeation properties of AMPA receptors in single neocortical nonpyramidal cells. J Neurosci 17:6685–6696
Anson LC, Chen PE, Wyllie DJA, Colquhoun D, Schoepfer R (1998) Identification of amino acid residues of the NR2A subunit which control glutamate potency in recombinant NR1/NR2A NMDA receptors. J Neurosci 18:581–598
Ascher P, Bregestovski P, Nowak L (1988) N-methyl-D-aspartate-activated channels of mouse central neurones in magnesium-free solutions. J Physiol 399:207–226
Attwell D, Barbour B, Szatkowski M (1993) Nonvesicular release of neurotransmitter. Neuron 11:401–407
Baker AJ, Zornow MH, Scheller MS, Yaksh TL, Skilling SR, Smullin DH, Larson AA, Kuczenski R (1991) Changes in extracellular concentrations of glutamate, aspartate, glycine, dopamine, serotonin, and dopamine metabolites after transient global ischemia in the rabbit brain. J Neurochem 57:1370–1379

Béhé P, Stern P, Wyllie DJA, Nassar M, Schoepfer R, Colquhoun D (1995) Determination of the NMDA NR1 subunit copy number in recombinant NMDA receptors. Proc R Soc Lond B 262:205–213

Benveniste M, Mayer ML (1991) A kinetic analysis of antagonist action at NMDA receptors: two binding sites each for glutamate and glycine. Biophysical J 59: 560–573

Benveniste M, Mayer ML (1995) Trapping of glutamate and glycine during open channel block of rat hippocampal neuron NMDA receptors by 9-aminoacridine. J Physiol 483:367–384

Blahos J, Wenthold RJ (1996) Relationship between N-Methyl-D-aspartate receptor NR1 splice variants and NR2 subunits. J Biol Chem 271:15669–15674

Blatz AL, Magleby KL (1989) Adjacent interval analysis distinguishes among gating mechanisms for the fast chloride channel from rat skeletal muscle. J Physiol 410:561–585

Bochet P, Audinat E, Lambolez B, Crépel F, Rossier J, Iino M, Tsuzuki K, Ozawa S (1994) Subunit composition at the single-cell level explains the functional properties of a glutamate-gated channel. Neuron 12:383–388

Buller AL, Monaghan DT (1997) Pharmacological heterogeneity of NMDA receptors: characterization of NR1a/NR2D heteromers expressed in *Xenopus* oocytes. Eur J Pharmacol 320:87–94

Chazot PL, Coleman SK, Cik M, Stephenson FA (1994) Molecular characterization of N-methyl-D-aspartate receptors expressed in mammalian cells yields evidence for the coexistence of three subunit types within a discrete receptor molecule. J Biol Chem 269:24403–24409

Chazot PL, Stephenson FA (1997) Molecular dissection of native mammalian forebrain NMDA receptors containing the NR1 C2 exon: direct demonstration of NMDA receptors comprising NR1, NR2A, and NR2B subunits within the same complex. J Neurochem 69:2138–2144

Ciabarra AM, Sullivan JM, Gahn LG, Pecht G, Heinemann S, Sevarino KA (1995) Cloning and characterization of χ-1: A developmentally regulated member of a novel class of the ionotropic glutamate receptor family. J Neurosci 15:6498–6508

Clements JD, Feltz A, Sahara Y, Westbrook GL (1998) Activation kinetics of AMPA receptor channels reveal the number of functional agonist binding sites. J Neurosci 18:119–127

Clements JD, Westbrook GL (1991) Activation kinetics reveal the number of glutamate and glycine binding sites on the N-methyl-d-aspartate receptor. Neuron 7:605–613

Colquhoun D (1998) Binding, gating, affinity and efficacy. The interpretation of structure-activity relationships for agonists and of the effects of mutating receptors. Br J Pharmacol 125:923–948

Colquhoun D, Hawkes AG (1982) On the stochastic properties of bursts of single ion channel openings and of clusters of bursts. Phil Trans R Soc Lond B 300:1–59

Colquhoun D, Hawkes AG (1987) A note on correlations in single ion channel records. Proc R Soc Lond B 230:15–52

Colquhoun D, Hawkes AG (1995a) The principles of the stochastic interpretation of ion channel mechanisms. In Single channel recording, Sakmann B and Neher pp 397–482 Plenum Press, New York

Colquhoun D, Hawkes AG (1995b) Desensitization of N-methyl-D-aspartate receptors: a problem of interpretation. Proc Natl Acad Sci USA 92:10327–10329

Colquhoun D, Hawkes AG, Srodzinski K (1996) Joint distributions of apparent open times and shut times of single ion channels and the maximum likelihood fitting of mechanisms. Phil Trans R Soc Lond A 354:2555–2590

Colquhoun D, Hawkes AG, Merlushkin A, Edmonds B (1997) Properties of single ion channel currents elicited by a pulse of agonist concentration or voltage. Phil Trans R Soc Lond A 355:1743–1786

Colquhoun D, Jonas P, Sakmann B (1992) Action of brief pulses of glutamate on AMPA/kainate receptors in patches from different neurones of rat hippocampal slices. J Physiol 458:261–287

Colquhoun D, Sakmann B (1981) Fluctuations in the microsecond time range of the current through single acetylcholine receptor ion channels. Nature 294:464–466

Colquhoun D, Sakmann B (1985) Fast events in single-channel currents activated by acetylcholine and its analogues at the frog muscle end-plate. J Physiol 369:501–557

Colquhoun D, Sigworth FJ (1995) Analysis of single ion channel data. In Single channel recording, Sakmann B and Neher E pp 483–587 Plenum Press, New York

Cull-Candy SG, Farrant M, Feldmeyer D (1995) NMDA channel conductance: a user's guide. In Excitatory Amino Acids and Synaptic Transmission, eds Wheal HV and Thomson AM pp 121–132 Academic Press Limited

Cull-Candy SG, Howe JR, Ogden DC (1988) Noise and single-channels activated by excitatory amino acids in rat cultured cerebellar granule cells. J Physiol 400: 189–222

Cull-Candy SG, Usowicz MM (1987) Multiple-conductance channels activated by excitatory amino acids in cerebellar neurones. Nature 325:525–528

Cull-Candy SG, Usowicz MM (1989) On the multiple conductance channels activated by excitatory amino acids in large cerebellar neurones of the rat. J Physiol 415: 555–582

Das S, Sasaki YF, Rothe T, Premkumar LS, Takasu M, Crandall JE, Dikkes P, Conner DA, Rayudu PV, Cheung W, Chen HS, Lipton SA, Nakanishi N (1998) Increased NMDA current and spine density in mice lacking the NMDA receptor subunit NR3 A. Nature 393:377–381

Didier M, Xu M, Berman SA, Bursztajn S (1995) Differential expression and co-assembly of NMDA zeta 1 and epsilon subunits in the mouse cerebellum during postnatal development. Neuroreport 6:2255–2259

Doyle DA, Cabral JM, Pfuetzner RA, Kuo A, Gulbis JM, Cohen SL, Chait BT, MacKinnon R (1998) The Structure of the Potassium channel: Molecular Basis of K^+ Conduction and Selectivity. Science 280:69–77

Dunah AW, Luo J, Wang YH, Yasuda RP, Wolfe BB (1998) Subunit composition of N-Methyl-D-aspartate receptors in the central nervous system that contain the NR2D subunit. Mol Pharmacol 53:429–437

Durand GM, Gregor P, Zheng X, Bennett MVL, Uhl GR, Zukin RS (1992) Cloning of an apparent splice variant of the rat N-methyl-D-asparate receptor NMDAR1 with altered sensitivity to polyamines and activators of protein kinase C. Proc Natl Acad Sci USA 89:9359–9363

Dzubay JA, Jahr CE (1996) Kinetics of NMDA channel opening. J Neurosci 16: 4129–4134

Edmonds B, Colquhoun D (1992) Rapid decay of averaged single-channel NMDA receptor activations recorded at low agonist concentration. Proc R Soc Lond B 250:279–286

Edmonds B, Gibb AJ, Colquhoun D (1995) Mechanisms of activation of muscle nicotinic acetylcholine receptors, and the time course of endplate currents. Ann Rev Physiol 57:469–493

Farrant M, Feldmeyer D, Takahashi T, Cull-Candy SG (1994) NMDA-receptor channel diversity in the developing cerebellum. Nature 368:335–339

Feltz A, Trautmann A (1982) Desensitization at the frog neuromuscular junction: a biphasic process. J Physiol 322:257–272

Ferrer-Montiel AV, Montal M (1996) Pentameric subunit stoichiometry of a neuronal glutamate receptor. Proc Natl Acad Sci USA 93:2741–2744

Fredkin DR, Montal M, Rice JA (1985) Identification of aggregated Markovian models: application to the nicotinic acetylcholine receptor. In Proceedings of the Berkeley Conference in Honor of Jerzy Neyman and Jack Kiefer, Le Cam, L. M. and Olshen, R. A., pp 269–289, Wadsworth, Monterey

Geiger JRP, Melcher T, Koh DS, Sakmann B, Seeburg PH, Jonas P, Monyer H (1995) Relative abundance of subunit mRNAs determines gating and Ca^{2+} permeability of AMPA receptors in principal neurons and inter neurons in rat CNS. Neuron 15:193–204

Gibb AJ, Colquhoun D (1991) Glutamate activation of a single NMDA receptor-channel produces a cluster of openings. Proc R Soc Lond B 243:39–45

Gibb AJ, Colquhoun D (1992) Activation of N-methyl-D-aspartate receptors by L-glutamate in cells dissociated from adult rat hippocampus. J Physiol 456:143–179

Gibb AJ, Kojima H, Carr JA, Colquhoun D (1990) Expression of cloned receptor subunits produces multiple receptors. Proc R Soc Lond B 242:108–112

Götz T, Kraushaar U, Geiger J, Lübke J, Berger T, Jonas P (1997) Functional properties of AMPA and NMDA receptors expressed in identified types of basal ganglia neurons. J Neurosci 17:204–215

Greengard P, Jen J, Nairn AC, Stevens CF (1991) Enhancement of the glutamate response by cAMP dependent protein kinase in hippocampal neurons. Science 253:1135–1138

Hawkes AG, Jalali A, Colquhoun D (1992) Asymptotic distributions of apparent open times and shut times in a single channel record allowing for the omission of brief events. Phil Trans R Soc Lond B 337:383–404

Hamill OP, Sakmann B (1981) Multiple conductance states of single acetylcholine receptor channels in embryonic muscle cells. Nature 294:462–464

Häusser M, Roth A (1997) Dendritic and somatic glutamate receptors channels in rat cerebellar Purkinje cells. J Physiol 501:77–95

Hestrin S (1993) Different glutamate receptors channels mediate fast excitatory synaptic currents in inhibitory and excitatory cortical neurons. Neuron 11:1083–1091

Howe JR, Colquhoun D, Cull-Candy SG (1988) On the kinetics of large-conductance glutamate-receptor ion channels in rat cerebellar granule neurons. Proc R Soc Lond B 233:407–422

Howe JR, Cull-Candy SG, Colquhoun D (1991) Currents through single glutamate-receptor channels in outside-out patches from rat cerebellar granule cells. J Physiol 432:143–202

Hirai H, Kirsch J, Laube B, Betz H, Kuhse J (1996) The glycine binding site of the N-methyl-D-aspartate receptor subunit NR1: Identification of novel determinants of co-agonist potentiation in the extracellular M3-M4 loop region. Proc Natl Acad Sci USA 93:6031–6036

Ikeda K, Nagasawa M, Mori H, Araki K, Sakimura K, Watanabe M, Inoue Y, Mishina M (1992) Cloning and expression of the ε (epsilon) 4 subunit of the NMDA receptor channel. FEBS Letters 313:34–38

Ishii T, Moriyoshi K, Sugihara H, Sakurada K, Kadotani H, Yokoi M, Akazawa C, Shigemoto R, Mizuno N, Masu M, Nakanishi S (1993) Molecular characterization of the family of the N-methyl-D-aspartate receptor subunits. J Biol Chem 268: 2836–2843

Jackson MB (1986) Kinetics of unliganded acetylcholine receptor channel gating. Biophys J 49:663–672

Jahr CE, Stevens CF (1987) Glutamate activates multiple single channel conductances in hippocampal neurons. Nature 325:522–525

Jahr CE (1992) High probability opening of NMDA receptor channels by L-glutamate. Science 255:4702–4255

Johnson JW, Ascher P (1987) Glycine potentiates the NMDA response in cultured mouse brain neurons. Nature 325:529–531

Jonas P, Major G, Sakmann B (1993) Quantal components of unitary EPSCs at the mossy fibre synapse on CA3 pyramidal cells of rat hippocampus. J Physiol 472: 615–663

Jonas P, Racca C, Sakmann B, Seeburg PH, Monyer H (1994) Differences in Ca^{2+} permeability of AMPA-type glutamate receptor channels in neocortical neurons caused by differential GluR-B subunit expression. Neuron 12:1281–1289

Jonas P, Sakmann B (1992) Glutamate receptor channels in isolated patches from CA1 and CA3 pyramidal cells of rat hippocampal slices. J Physiol 455:143–171
Jones MV, Westbrook GL (1995) Desensitized states prolong $GABA_A$ channel responses to brief agonist pulses. Neuron 15:181–191
Kleckner NW, Dingledine R (1988) Requirements for glycine in activation of NMDA-receptors expressed in *Xenopus* oocytes. Science 241:835–837
Kleckner NW, Pallotta BS (1995) Burst kinetics of single NMDA receptor currents in cell-attached patches from rat brain cortical neurons in culture. J Physiol 486: 411–426
Köhr G, Eckardt S, Luddens H, Monyer H, Seeburg PH (1994) NMDA receptor channels: subunit-specific potentiation by reducing agents. Neuron 12:1031–1040
Köhr G, Seeburg PH (1996) Subtype-specific regulation of recombinant NMDA receptor-channels by protein tyrosine kinases of the src family. J Physiol 492:445–452
Krupp JJ, Vissel B, Heinemann SF, Westbrook GL (1998) N-terminal domains in the NR2 subunit control desensitization of NMDA receptors. Neuron 20:317–327
Kuryatov A, Laube B, Betz H, Kuhse J (1994) Mutational analysis of the glycine-binding site of the NMDA receptor: Structural similarity with bacterial amino acid-binding proteins. Neuron 12:1291–1300
Kutsuwada T, Kashiwabuchi N, Mori H, Sakimura K, Kushiya E, Araki K, Meguro H, Masaki H, Kumanishi T, Arakawa M, Mishina M (1992) Molecular diversity of the NMDA receptor channel. Nature 358:36–41
Labarca P, Rice JA, Fredkin DR, Montal M (1985) Kinetic analysis of channel gating: application to the cholinergic receptor channel and the chloride channel from Torpedo californica. Biophys J 47:469–478
Lambolez B, Audinat E, Bochet P, Crépel F, Rossier J (1992) AMPA receptor subunits expressed by single Purkinje cells. Neuron 9:247–258
Lambolez B, Ropert N, Perrais D, Rossier J, Hestrin S (1996) Correlation between kinetics and RNA splicing of α-amino-3-hydroxy-5-methylisoxazole-4-propionic acid receptors in neocortical neurons. Proc Natl Acad Sci USA 93:1797–1802
Laube B, Hirai H, Sturgess M, Betz H, Kuhse J (1997) Molecular determinants of agonist discrimination by NMDA receptor subunits: Analysis of the glutamate binding site on the NR2B subunit. Neuron 18:493–503
Laube B, Kuhse J, Betz H (1998) Evidence for tetrameric structure of recombinant NMDA receptors. J Neurosci 18:2954–2961
Läuger P (1983) Conformational transitions of ionic channels. In Single Channel Recording, eds. Sakmann, B and Neher, E, pp 177–189, Plenum Press, New York.
Läuger P (1985) Ionic channels with conformational substates. Biophys J 47:581–591
Leff P, Dougall IG (1993) Further concerns over Cheng-Prusoff analysis. Trends Pharmacol Sci 14:110–112
Lester RAJ, Clements JD, Westbrook GL, Jahr CE (1990) Channel kinetics determine the time course of NMDA receptor-mediated synaptic currents. Nature 346: 565–567
Lester RAJ, Jahr CE (1992) NMDA channel behavior depends on agonist affinity. J Neurosci 12:635–643
Lewis TM, Harkness PC, Sivilotti LG, Colquhoun D, Millar NS (1997) The ion channel properties of a rat recombinant neuronal nicotinic receptor are dependent on the host cell type. J Physiol 505:299–306
Lewis TM, Sivilotti LG, Colquhoun D, Gardiner RM, Schoepfer R, Rees M (1998) Properties of human glycine receptors containing the hyperekplexia mutation α1(K276E), expressed in *Xenopus* oocytes. J Physiol 507:25–40
Lieberman DN, Mody I (1994) Regulation of NMDA channel function by endogenous Ca^{2+}-dependent phosphatase. Nature 369:235–239
Liebovitch LS (1989) Testing fractal and Markov models of ion channel kinetics. Biophys J 55:373–377
Lin F, Stevens CF (1994) Both open and closed NMDA receptor channels desensitize. J Neurosci 14:2153–2160

Lomeli H, Mosbacher J, Melcher T, Höger T, Geiger JRP, Kuner T, Monyer H, Higuchi M, Bach A, Seeburg PH (1994) Control of kinetic properties of AMPA receptor channels by nuclear RNA editing. Science 266:1709–1713

Lummis SCR, Fletcher EJ, Green T (1998) NMDA receptor NR2 subunits contain amino acids involved in glutamate binding. J Physiol 506P:76P

MacDonald JF, Mody I, Salter MW (1989) Regulation of N-methyl-D-aspartate receptors revealed by intracellular dialysis of murine neurones in culture. J Physiol 414:17–34

MacKinnon R (1991) Determination of the subunit stoichiometry of a voltage-activated potassium channel. Nature 350:232–235

Magleby KL, Weiss DS (1990) Identifying kinetic gating mechanisms for ion channels by using two-dimensional distributions of simulated dwell times. Proc R Soc Lond B 241:220–228

Magleby KL, Song L (1992) Dependency plots suggest the kinetic structure of ion channels. Proc R Soc Lond B 249:133–142

Mano I, Teichberg VI (1998) A tetrameric subunit stoichiometry for a glutamate receptor-channel complex. Neuroreport 9:327–331

McManus OB, Weiss DS, Spivak CE, Blatz AL, Magleby KL (1988) Fractal models are inadequate for the kinetics of four different ion channels. Biophys J 54:859–870

McManus OB, Magleby KL (1989) Kinetic time constants independent of previous single-channel activity suggest Markov gating for a large conductance Ca-activated K channel. J Gen Physiol 94:1037–1070

Medina I, Filippova N, Bakhramov A, Bregestovski P (1996) Calcium-induced inactivation of NMDA receptor-channels evolves independently of run-down in cultured rat brain neurones. J Physiol Lond. 495: 411–427

Medina I, Filippova N, Charton G, Rougeole S, Ben-Ari Y, Khrestchatisky M, Bregestovski P (1995) Calcium-dependent inactivation of heteromeric NMDA receptor-channels expressed in human embryonic kidney cells. J Physiol 482: 567–573

Momiyama A, Feldmeyer D, Cull-Candy SG (1996) Identification of a native low-conductance NMDA-channel with reduced sensitivity to Mg^{2+} in rat central neurones. J Physiol 494:479–492

Monyer H, Seeburg PH, Wisden W (1991) Glutamate-operated channels: developmentally early and mature forms arise by alternative splicing. Neuron 6:799–810

Monyer H, Burnashev N, Laurie DJ, Sakmann B, Seeburg PH (1994) Developmental and regional expression in the rat brain and functional properties for four NMDA receptors. Neuron 12:529–540

Moriyoshi K, Masu M, Ishii T, Shigemoto R, Mizuno N, Nakanishi S (1991) Molecular cloning and characterization of the rat NMDA receptor. Nature 354:31–37

Mosbacher J, Schoepfer R, Monyer H, Burnashev N, Seeburg PH, Sakmann B (1994) A molecular determinant for submillisecond desensitization in glutamate receptors. Science 266:1059–1062

Nowak L, Bregestovski P, Ascher P, Herbet A, Prochiantz A (1984) Magnesium gates glutamate-activated channels in mouse central neurones. Nature 307:462–463

Oh BH, Pandit J, Kang CH, Nikaido K, Gokcen S, Ames GF, Kim SH (1993) Three-dimensional structures of the periplasmic lysine/arginine/ornithine-binding protein with and without a ligand. J Biol Chem 268:11348–11355

Oh BH, Kang CH, De-Bondt H, Kim SH, Nikaido K, Joshi AK, Ames GF (1994) The bacterial periplasmic histidine-binding protein. structure/function analysis of the ligand-binding site and comparison with related proteins. J Biol Chem 269: 4135–4143

Ozawa S, Iino M, Tsuzuki K (1991) Two types of kainate response in cultured rat hippocampal neurons. J Neurophysiol 66:2–11

Paleček J, Abdrachmanova G, Vyklicky L (1998) Glutamate receptor development and single channel properties in rat spinal cord motoneurones. J Physiol 511P:35S

Premkumar LS, Auerbach A (1997) Stoichiometry of recombinant N-methyl-D-aspartate receptor channels inferred from single-channel current patterns. J Gen Physiol 110:485–502
Raman IM, Trussell LO (1992) The kinetics of the response to glutamate and kainate in neurons of the avian cochlear nucleus. Neuron 9:173–186
Ruiz ML, Karpen JW (1997) Single cyclic nucleotide-gated channels locked in different ligand-bound states. Nature 389:389–392
Rosenmund C, Westbrook GL (1993) Calcium-induced actin depolymerization reduces NMDA channel activity. Neuron 10:805–814
Rosenmund C, Stern-Bach Y, Stevens CF (1998) The tetrameric structure of a glutamate receptor channel. Science 280:1596–1599
Schneggenburger R, Ascher P (1997) Coupling of permeation and gating in an NMDA-channel pore mutant. Neuron 18:167–177
Sheng M, Cummings J, Roldan LA, Jan YN, Jan LY (1994) Changing subunit composition of heteromeric NMDA receptors during development of rat cortex. Nature 368:144–147
Silver RA, Traynelis SF, Cull-Candy SG (1992) Rapid-time-course miniature and evoked excitatory currents at cerebellar synapses in situ. Nature 355:163–166
Sine SM, Ohno K, Bouzat C, Auerbach A, Milone M, Pruitt JN, Engel AG (1995) Mutation of the acetylcholine receptor α subunit causes a slow-channel myasthenic syndrome by enhancing agonist binding affinity. Neuron 15:229–239
Sivilotti LG, McNeil DK, Lewis, TM, Nassar M, Schoepfer R, Colquhoun D (1997) Recombinant neuronal nicotinic receptors, expressed in *Xenopus* oocytes, do not resemble native receptors of the rat superior cervical ganglion J Physiol 500: 123–138
Soloviev MM, Barnard EA (1997) *Xenopus* oocytes express a unitary glutamate receptor endogenously. J Molec Biol 273:14–18
Sommer B, Keinänen K, Verdoorn TA, Wisden W, Burnashev N, Herb A, Köhler M, Takagi T, Sakmann B, Seeburg PH (1990) Flip and flop; A cell-specific functional switch in glutamate-operated channels of the CNS. Science 249:1580–1585
Stern P, Béhé P, Schoepfer R, Colquhoun D (1992) Single channel conductances of NMDA receptors expressed from cloned cDNAs: comparison with native receptors. Proc R Soc Lond B 250:271–277
Stern P, Cik M, Colquhoun D, Stephenson FA (1994) Single channel properties of cloned NMDA receptors in HEK 293 cells: comparison with results from *Xenopus* oocytes. J Physiol 476:391–397
Sucher NJ, Akbarian S, Chi CL, Leclerc CL, Awobuluyi M, Deitcher DL, Wu MK, Yuan JP, Jones EG, Lipton SA (1995) Developmental and regional expression pattern of a novel NMDA receptor-like subunit (NMDAR-L) in the rodent brain. J Neurosci 15:6509–6520
Sucher NJ, Awobuluyi MN, Choi Y-B, Lipton SA (1996) NMDA receptors: from genes to channels. Trends Pharmacol Sci 17:348–355
Swanson GT, Kamboj SK, Cull-Candy SG (1997) Single-channel properties of recombinant AMPA receptors depends on RNA editing, splice variation, and subunit composition. J Neurosci 17:58–69
Tang CM, Shi QY, Katchman A, Lynch G (1991) Modulation of the time course of fast EPSCs and glutamate channel kinetics by aniracetam. Science 254:288–290
Thomson AM, Walker VE, Flynn DM (1989) Glycine enhances NMDA-receptor mediated synaptic potentials in neocortical slices. Nature 338:422–424
Tölle TR, Berthele A, Zieglgansberger W, Seeburg PH, Wisden W (1993) The differential expression of 16 NMDA and Non-NMDA receptor subunits in the rat spinal cord and in periaqueductal gray. J Neurosci 13:5009–5028
Trautmann A (1982) Curare can open and block ionic channels associated with cholinergic receptors. Nature 298:272–275
Trussell LO, Zhang S, Raman IM (1993) Desensitization of AMPA receptors upon multiquantal neurotransmitter release. Neuron 10:1185–1196

Tsuzuki K, Mochizuki S, Iino H, Mishina M, Ozawa S (1994) Ion permeation properties of the cloned mouse $\varepsilon 2/\zeta 1$ NMDA receptor channel. Molec Brain Res 26:37–46
Vicini S, Wang J-F, Li JH, Zhu WJ, Wang YH, Luo JH, Wolfe BB, Grayson DR (1998) Functional and pharmacological differences between recombinant N-methyl-D-aspartate receptors. J Neurophysiol 79:555–566
Wafford KA, Bain CJ, Le Bourdelles B, Whiting PJ, Kemp JA (1993) Preferential co-assembly of recombinant NMDA receptors composed of three different subunits. Neuroreport 4:1347–1349
Wafford KA, Kathoria M, Bain CJ, Marshall G, Le-Bourdelles B, Kemp JA, Whiting PJ (1995) Identification of amino acids in the N-methyl-D-aspartate receptor NR1 subunit that contribute to the glycine binding site. Mol Pharmacol 47:374–380
Wang L-Y, Orser BA, Brautigan DL, MacDonald JF (1994) Regulation of NMDA receptors in cultured hippocampal neurons by protein phosphatases 1 and 2A. Nature 369:230–232
Wang YT, Salter MW (1994) Regulation of NMDA receptors by tyrosine kinases and phosphatases. Nature 369:233–235
Watanabe M, Inoue Y, Sakimura K, Mishina M (1992) Developmental changes in distribution of NMDA receptor channel subunit mRNAs. Developmental Neurosci 3:1138–1140
Williams K (1993) Ifenprodil discriminates subtypes of the N-methyl-D-aspartate receptor: selectivity and mechanisms at recombinant heteromeric receptors. Mol Pharmacol 44:851–859
Williams K, Chao J, Kashiwagi K, Masuko T, Igarashi K (1996) Activation of N-Methyl-D-Aspartate receptors by glycine: Role of an aspartate residue in the M3-M4 loop of the NR1 subunit. Mol Pharmacol 30:701–708
Wu T-Y, Liu C, Chang Y-C (1996) A study of the oligomeric state of the α-amino-3-hydroxy-5-methyl-4-isoxazolepropionic acid-preferring glutamate receptors in the synaptic junctions of porcine brain. Biochem J 319:731–739
Wyllie DJA, Béhé P, Nassar M, Schoepfer R, Colquhoun D (1996) Single-channel currents from recombinant NMDA NR1a/NR2D receptors expressed in *Xenopus* oocytes. Proc R Soc Lond B 263:1079–1086
Wyllie DJA, Edmonds B, Colquhoun D (1997) Single activations of recombinant NMDA NR1a/NR2A receptors recorded in one-channel patches. J Physiol 501P:13P
Wyllie DJA, Béhé P, Colquhoun D (1998) Single-channel activations and concentration jumps: comparison of recombinant NR1a/NR2A and NR1a/NR2D NMDA receptors. J Physiol 510:1–18 (Erratum J Physiol 512:939)
Wyllie DJA, Traynelis SF, Cull-Candy SG (1993) Evidence for more than one type of non-NMDA in outside-out patches from cerebellar granule cells of the rat. J Physiol 463:193–226

CHAPTER 6

The Ion-Conducting Pore of Glutamate Receptor Channels

T. KUNER, L.P. WOLLMUTH, B. SAKMANN

Abbreviations

3D	Three dimensional
3TM	Three transmembrane
5-HT_3R	5-hydroxytryptamine (serotonin) type-3 receptor
AChR	Acetylcholine receptor
AMPA	α-Amino-3-hydroxy-5-methylisoxazole-4-propionic acid
AMPAR	AMPA receptor
BU	Bead units
Ca channel	Voltage-gated calcium channel
CNG	Cyclic-nucleotide-gated
EDTA	Ethylene diamine tetraacetic acid
$GABA_AR$	γ-Aminobutyric acid type A receptor
GluR	Glutamate receptor
GlyR	Glycine receptor
nAChR	Nicotinic acetylcholine receptor
MTSES	2-Sulfonatoethyl-methanethiosufonate
MTSET	2-Trimethylammonioethyl-methanethiosufonate
NMDAR	*N*-methyl-D-aspartate receptor
SCAM	Substituted-cysteine-accessibility method
SEM	Standard error of the mean

A. Introduction

Ionotropic glutamate receptors (GluRs) mediate the postsynaptic response at most excitatory synapses in the brain. A variety of GluR channel subtypes provides these synapses with a repertoire of distinct computational properties, many of which arise from ionic interactions with the channel pore. In both *N*-methyl-D-aspartate receptor (NMDAR) and α-amino-3-hydroxy-5-methylisoxazole-4-propionic acid receptor (AMPAR) channels, a voltage-dependent blocking process renders the response of the channel dependent on the concurrent activity of the synapse. The voltage dependence of extracellular Mg^{2+} block in NMDAR (MAYER et al. 1984; NOWAK et al. 1984) and cytoplasmic polyamine block in AMPAR channels (BOWIE and MAYER 1995; DONEVAN and ROGAWSKI 1995; ISA et al. 1995; KAMBOJ et al. 1995; KOH et al.

1995) allows these transmembrane proteins to integrate pre- and postsynaptic signals. Furthermore, Ca^{2+} influx through GluR channels can trigger long-lasting changes in synaptic strength (BLISS and COLLINGRIDGE 1993) but can also lead to cell death under pathological conditions (CHOI 1988; MELDRUM and GARTHWAITE 1990). The amount of Ca^{2+} entering a neuron during receptor activation may determine the nature and persistence of the postsynaptic response (BERRIDGE 1998). Thus, the control of Ca^{2+} influx by voltage-dependent blocking mechanisms is essential for the role of GluRs in synaptic plasticity as a basis of higher brain functions, such as learning and memory. Given the physiological and pathophysiological relevance of ionic mechanisms in GluR function, understanding the structural basis of ion permeation and blocking in these receptor channels is of the utmost interest.

Molecular identification of the GluR-subunit genes (HOLLMANN and HEINEMANN 1994; NAKANISHI and MASU 1994; SEEBURG 1993) set the stage for studies aimed at correlating the molecular structure of GluR channels with their biophysical properties. Considerable progress has been made in defining determinants of ion permeation and blocking, laying the foundation for the engineering of gene-targeted mice expressing GluRs with a known change in function (BRUSA et al. 1995; SINGLE et al. 1997). Such studies promise to enhance our understanding of how distinct biophysical properties of GluR channels contribute to the physiology and behavior of the animal. Eventually, structural information will aid in the development of new drugs designed to specifically interfere with the role of GluR channels under disease conditions.

B. Concepts and Methods

GluR channels, like all ion channels, carry out a basic function: in the open state, they form a water-filled pore allowing ions to selectively and rapidly cross the membrane. A common structural motif of ion channels resembles an hourglass, with two funnels connected by a narrow constriction. At the constriction, the diameter of the pore approaches its smallest dimension, typically that of a single amino acid side chain or a partially hydrated ion. Thus, residues positioned at the channel's narrow constriction can directly interact with ions, thereby defining the selectivity of the pore. From these considerations, the narrow constriction of a channel can be defined as a structural and functional module, the selectivity filter (HILLE 1992). The funnels or vestibules connecting the selectivity filter to the extracellular and cytoplasmic bulk solution can be defined as another "structure–function module", since they may enrich the concentration of certain ions via electrostatic mechanisms and provide local binding sites for ions and modulatory drugs (DANI 1986).

Correlating ion channel function with three-dimensional (3D) molecular structures has proven difficult, because ion channels, like other integral membrane proteins, are hard to crystallize and, hence, are not amenable to X-ray crystallography. Recently, the first high-resolution structure of an ion channel was resolved (Doyle et al. 1998). This protein, however, is small in size (about 10 kD), and the fundamental problem of crystallizing large transmembrane proteins, including GluR subunits, remains; this has prompted the development of alternative strategies to elucidate the structure of ion channels. Cryoelectron microscopy has revealed direct insights into acetylcholine-receptor (AChR) structure (Unwin et al. 1988, 1995) and, recently, has provided the first glimpses of a GluR (Madden, personal communication). These methods provide direct data about the structure of channel proteins, but they have not yet reached the resolution required to detect single amino acid residues.

Electrophysiological techniques, in combination with molecular, genetic approaches to manipulation of channel proteins, have allowed the functional consequences of a mutation to be assessed (Miller 1989). Structural inferences based on the replacement of a single amino acid residue are often imprecise and speculative. An extension of this approach has been to substitute multiple residues at a given position and to correlate the functional consequences with the chemical nature of the side chain (Wollmuth et al. 1998b). A powerful tool, the substituted-cysteine-accessibility method (SCAM; Karlin and Akabas 1998 for a review), allows the identification of channel-lining residues and, based on the pattern of exposed residues, inferences on the structures of channel-lining segments. This method has produced valuable new insights into structural and functional aspects of a variety of ion channels, including nicotinic acetylcholine receptors (nAChR), γ-aminobutyric acid type-A receptor ($GABA_AR$), GluRs, K channels, Na channels, cyclic-nucleotide-gated (CNG) channels, and other proteins (Akabas et al. 1992; Chiamvimonvat et al. 1996; Kuner et al. 1996; Kürz et al. 1995; Pascual et al. 1995; Sun et al. 1996; Xu and Akabas 1993).

The tremendous amount of data generated by these different approaches is difficult to combine into a simple, structural picture of an ion channel. Molecular modeling of channel structures (Guy and Seetharamulu 1986; Sankararamakrishnan et al. 1996; Sutcliffe et al. 1996), as a complementary approach, is useful for integration of structure–function data and for constraint of structural conclusions within the rules of protein structure. The success of this approach can be appreciated from a comparison of the K-channel structures modeled by Guy and coworkers (Durell and Guy 1996) with the structure determined by the laboratory of MacKinnon (Doyle et al. 1998). However, it should be kept in mind that models derived from functional data are much less precise than models generated on the basis of X-ray diffraction patterns. In this chapter, we summarize results obtained from systematic mutagenesis and functional evaluation, SCAM and molecular modeling of GluR subunits.

C. Design of the GluR Channel Domain

GluR subunits are proposed to have a modular design (Wo and Oswald 1995b) consisting of a large *N*-terminal domain, a bipartite ligand-binding domain, a channel domain and a *C*-terminal domain of variable size. The channel domain of GluRs, encompassing M1 through M3 and the M4 segment, consists of only about 150 amino acid residues out of a total of about 900–1500, depending on the type of subunit (Fig. 1A). Segments within this domain are highly conserved across GluR channels (SPRENGEL and SEEBURG 1995) and may contain most of the structural determinants of ion permeation and blocking (KUNER and SCHOEPFER 1996). GluR subunits possess a distinct membrane topology, having three membrane-spanning segments (M1, M3, M4), with M2 forming a re-entrant loop, as depicted in Fig. 1B (BENNETT and DINGLEDINE 1995; HIRAI et al. 1996; HOLLMANN et al. 1994; WO and OSWALD 1994, 1995a; WOOD et al. 1995). Although the precise folding pattern of the channel domain remains unknown, identification of channel-lining residues using SCAM (BECK et al. 1999; KUNER et al. 1996; KUNER et al. 1997) has provided constraints for the development of molecular models of the hydrophobic segments (GUY and KUNER, unpublished results). These models will be used here to illustrate structural aspects of the GluR channel domain.

I. Three Transmembrane Segments Plus Pore-Inserted Loop

A molecular model of a single GluR-subunit channel domain is shown in Fig. 1C. The side view shows the longitudinal profile of a subunit, and the ion's view shows the pore-lining segments from the perspective of the central axis of the channel. The transmembrane segments M1, M3 and M4 are positioned at different angles and in close proximity, with the M4 segment bracketed by the M1 and M3 segments. The segments are positioned in a staggered fashion, with M1 oriented most intracellularly and M3 most extracellularly. In an oligomeric assembly of tetrameric (LAUBE et al. 1998; MANO and TEICHBERG 1998; ROSENMUND et al. 1998) or pentameric (BÉHÉ et al. 1995; FERRER-MONTIEL and MONTAL 1996; PREMKUMAR and AUERBACH 1997) stoichiometry, the three transmembrane (3TM) segments would form a barrel-like structure separating the ion-conducting pore from the surrounding lipids. The M2 loop is inserted from the cytoplasmic side into the 3TM barrel, with its helical part pointed towards the barrel's inner wall and the random-coil part oriented towards the central axis of the channel. The tip of the M2 loop defines the narrow constriction of the channel, which is flanked by two vestibules. The lining of the extracellular vestibule is formed by the *C*-terminal part of the M3 segment, the *N*-terminal part of the M4 segment and by the region preceding the M1 segment (BECK et al. 1999). The intracellular vestibule is formed by the M2 segment (KUNER et al. 1996).

II. Structure of the M2 Loop

The structure of the M2 segment has been inferred from the accessibility of substituted cysteines to sulfhydryl-specific reagents (KUNER et al. 1996, 1997). Channel-lining residues of the NR1, NR2 and GluR-D subunits are shown in

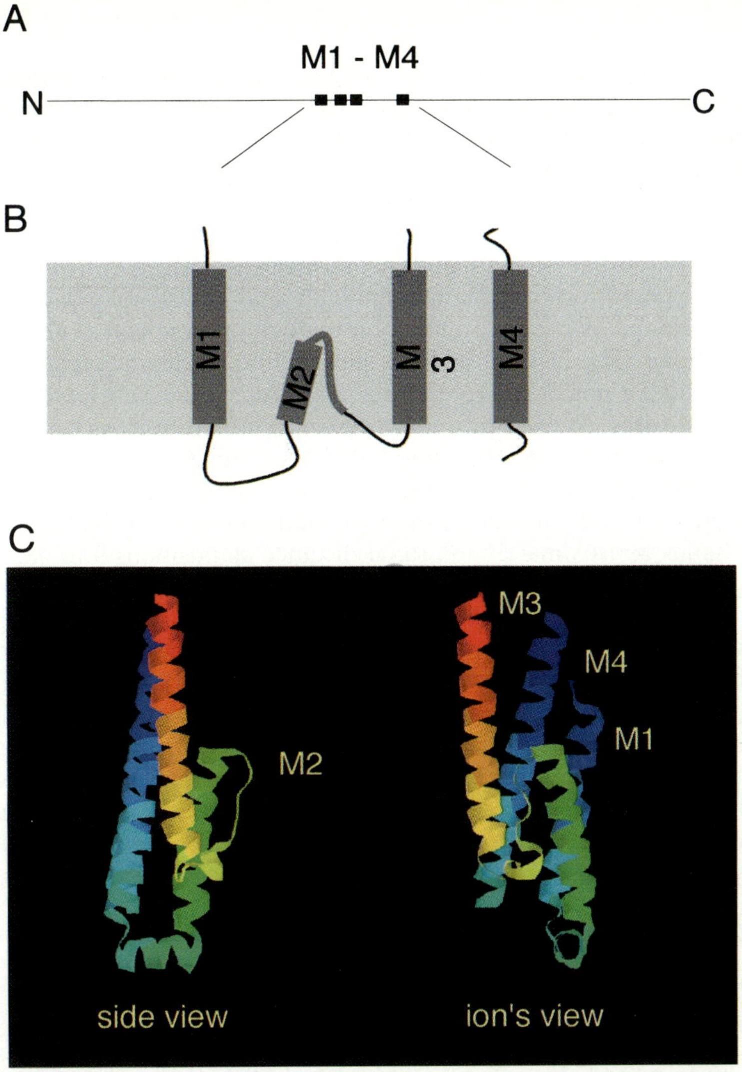

Fig. 1A–C. Structural design of the glutamate receptor (GluR) channel domain. **A** Schematic representation of a GluR subunit. *Black boxes* represent the hydrophobic segments M1 through M4. In most subunits, the amino-terminal domain is about 600 amino acid residues long. The length of the carboxy-terminal domain ranges from 49 to 645 residues. The core region of about 280 residues (M1 to M4) contains part of the glutamate-binding site (about 90 residues) located within the extracellular loop connecting M3 and M4. **B** Membrane topology of a GluR subunit. The segments included in this graph represent the channel domain as defined here. Hydrophobic segments are shown in *grey*, *lines* denote flanking sequences and connecting loops. **C** Molecular model illustrating a possible folding pattern of the NR1 subunit channel domain (Guy and Kuner, unpublished results). The continuous M1-to-M3 region is coloured, starting with *blue* in M1 and ending with *red* at the end of M3. M2 is colored *green*. M4 is modeled as a separate segment inserted between M1 and M3 and is colored *blue*. For simplicity, the transmembrane-spanning regions are modeled as α-helices. The image was rendered with RasMol (Roger Sayle)

Fig. 2A. The differential accessibility of residues to extra- or intracellularly applied, impermeant reagents demonstrated that the M2 segment forms a loop with both of its limbs originating on the intracellular side of the channel (KUNER et al. 1996). Further, the residue at position 0, designated as the *Q/R-site* or *N-site*, was accessible only from the extracellular side, whereas flanking residues were accessible only from the intracellular side, indicating that, in all subunits, this residue is positioned at the tip of the M2 loop (Fig. 2A). The pattern of exposed residues is consistent with an α-helical secondary structure for the *N*-terminal part of M2 and a non-regular, random-coil structure for its *C*-terminal part (Fig. 2B). In the NR1 subunit, the M2 helix extends from the most intracellular position (–18) to the most extracellular position (0), thereby making four turns. At the tip of the loop, represented by position 0, a three-residue hairpin kink (Fig. 2B, blue) forms the transition into a random-coil structure, descending towards the intracellular channel entrance. In this model, the random-coil configuration of the descending limb, comprised of positions 0 to +4, spans approximately the same distance as positions 0 to –8, thereby aligning positions –8 and +4 at about the same level in the vertical axis of the channel. The overall configuration of the M2 segments of the three different GluR subunits is essentially identical, although there are subtle differences consistent with their different functional properties. A similar structure presumably holds for the kainate receptors, since homologous residues to those that are exposed in AMPAR channels influence blocking of GluR-6 receptors by intracellular polyamines (MAYER, personal communication). Hence, evidence from three main members of the GluR family – AMPA, kainate, and NMDAR channels – converges on the concept of M2 forming a pore loop structure, as depicted in Fig. 2B.

Fig. 2A,B. Structure of the M2 loop. **A** Channel-lining residues of the M2 segment. The M2 loop is shown in the same orientation as in Fig. 1C. Only residues tested for accessibility to sulfhydryl-specific reagents (NR1 and NR2: 2-trimethylammonioethyl-methanethiosufonate; glutamate receptor (GluR)-D_i: 2-sulfonatoethyl-methanethiosufonate) are shown. Exposed residues point towards the central axis of the channel (*right border* of the figure). *Upward-pointing triangles* indicate accessibility to impermeant reagents from the extracellular side, whereas *downward-pointing triangles* denote accessibility from the intracellular side. The residue at the tip of the M2 loop is designated as position "0" and corresponds to the Q/R/N site. Positions on the amino-terminal side have negative signs, whereas those on the carboxy-terminal side have positive signs. Position 0 corresponds to residue 598 in the NR1 subunit, 595 in the NR2A subunit, 593 in the NR2C subunit, and 587 in the GluR-D subunit (position in the mature protein). Homologous residues of the ascending and descending limbs are aligned horizontally to simplify comparison. **B** Ribbon diagram of the M2 loop in a side view and from the ion's perspective. The ascending helical segment is coloured in *pink*, the descending random coil in *white*, and the kink formed by residues 0, +1 and +2 is shown in *blue*. Note that position +2 is located approximately at the same level as position –4 and position +4 at the level of –8. The length of the M2 helix is approximately 2.2 nm

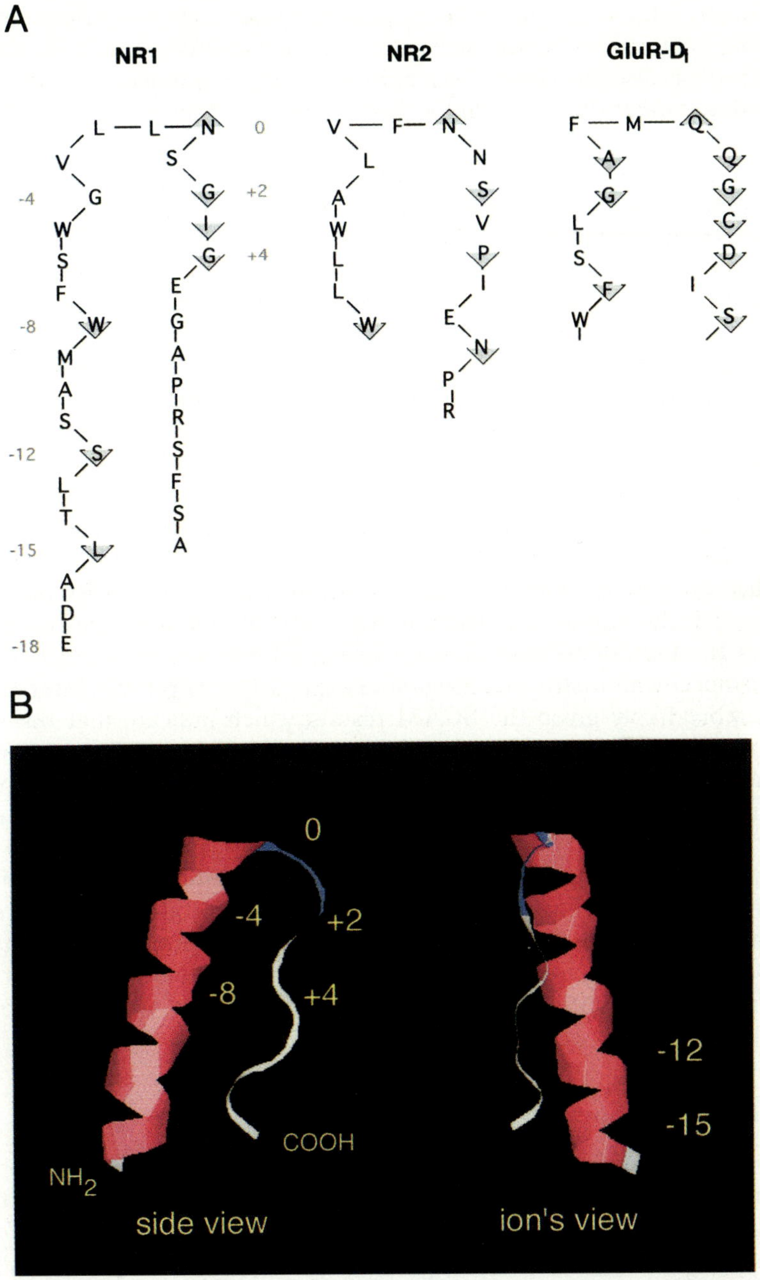
A
NR1
NR2
GluR-D$_{\text{I}}$
0
+2
+4
-4
-8
-12
-15
-18
B
0
-4
+2
-8
+4
COOH
NH$_2$
side view
-12
-15
ion's view

Consistent with the model presented above, secondary-structure-prediction algorithms (CHOU and FASMAN 1978; GARNIER et al. 1978; ROST and SANDER 1994) also classify M2 as a helix followed by a short stretch of random coil with a kink at the identical location. Analyses of the secondary structures of peptides derived from transmembrane segments yield results consistent with helical structures of parts of the M2 segment (MONTAL 1995, 1996). Furthermore, the general conclusion that M2 folds into a loop structure is also supported using blocking by extracellular and intracellular Mg^{2+} as an indicator of sidedness (KUPPER et al. 1996). However, inferring localization relative to a narrow constriction from a functional parameter like blocking by Mg^{2+} is difficult, since residues on both sides of the loop affect blocking by extracellular Mg^{2+} (KUNER et al. 1996). Recently, the highly conserved tryptophan residue at position –8 has been suggested to be positioned close to the tip of the loop (WILLIAMS et al. 1998). Again, these conclusions were based on the effects of mutant channels on blocking by extracellular Mg^{2+}; blocking by Mg^{2+} was affected by substituting smaller amino acid residues for the tryptophan, whereas substitutions of large residues with aromatic functional groups for this tryptophan left blocking by Mg^{2+} unaffected. These results were interpreted as evidence for a π-electron interaction of the aromatic groups with permeating Mg^{2+} ions at the narrow constriction. Alternatively, and as suggested by these authors, the mutations could cause a structural rearrangement of residues at the tip, thereby indirectly affecting blocking by Mg^{2+}. This latter interpretation seems more likely given the SCAM results, which indicate that residues *C*-terminal to this tryptophan are accessible to intracellular application. Thus, the residue at position –8 may stabilize the random-coil part of the M2 loop, with substitutions of it disrupting, in a remote fashion, the structure of the narrow constriction.

In summary, the structure of M2 depicted in Fig. 2B is consistent with the available data. The exact positioning of the M2 loop, however, remains unclear. It could be positioned in the channel in an upright position, as depicted in Figs. 1C and 2B, but could also be positioned at a flatter angle, similar to its orientation in a model proposed for the CNG channel (SUN et al. 1996).

III. Intrinsic Properties of the M2 Loop

The pore loop structure of the M2 segment, as presented in the previous section, bears several functional implications. First, short, three-residue hairpin kinks are often found at the ends of helices, providing binding sites for substrates in enzymes, receptors, transporters, immunoglobulins and other proteins (BRANDEN and TOOZE 1991; STRYER 1995). Amino acid residues positioned at such loops (*blue stretch* in Fig. 2B, including positions 0 to +2) have a high degree of conformational freedom and can form unique structures. This correlates well with their critical functional roles, as described below. Second, a unique structural property of random-coil segments is that the carbonyl and amide groups of the main chain are not hydrogen bonded, in contrast to segments with a regular secondary structure. This renders them available for interactions with permeating ions. Third, the macrodipole of the M2-pore helix is directed towards the extracellular vestibule of the channel and may, therefore, attract cations, adding to the effects of charged residues in the extracellular vestibule. Finally, loop structures can be flexible, which might be an important property for specific functional effects that they mediate. In the case of NMDAR channels, this may contribute to the mechanism for the strong voltage dependence of blocking by Mg^{2+}.

D. The Selectivity Filter

I. Pore Size of GluR Channels

The dimensions of the smallest diameters of various GluR channels have been estimated from the permeability of differently sized organic cations (VILLARROEL et al. 1995; ZAREI and DANI 1995; BURNASHEV et al. 1996; WOLLMUTH et al. 1996). This approach is based on the assumption that the permeation of organic cations through the channel is predominantly governed by their cross-sectional area rather than any specific interactions they have with the pore wall (DWYER et al. 1980).

1. NMDAR Channels

The narrow constriction of recombinant NMDAR channels assembled from NR1 and NR2A subunits has a mean diameter of approximately 0.55 nm (VILLARROEL et al. 1995). Assuming a circular shape of the narrow constriction, this value corresponds to a cross-sectional area of 0.24 nm^2 (Fig. 3, *top left*). Native NMDAR channels in cultured hippocampal neurons have a comparable cross-sectional diameter of 0.26 nm^2. However, the shape of the narrow constriction was suggested to be rectangular, measuring 0.45×0.57 nm (ZAREI and DANI 1995). Both studies conclude that the narrow constriction is flanked by vestibules or funnels of a much larger size (>0.4 nm^2), with the extracellular vestibule larger than the intracellular vestibule.

2. Non-NMDAR Channels

In comparison with the NMDAR channel, AMPAR and kainate-receptor channels have larger pore dimensions (BURNASHEV et al. 1996). In homomeric channels containing the unedited kainate-receptor or AMPAR subunits, the mean diameters were 0.75 and 0.78 nm, respectively (Fig. 3, *top right*). Sur-

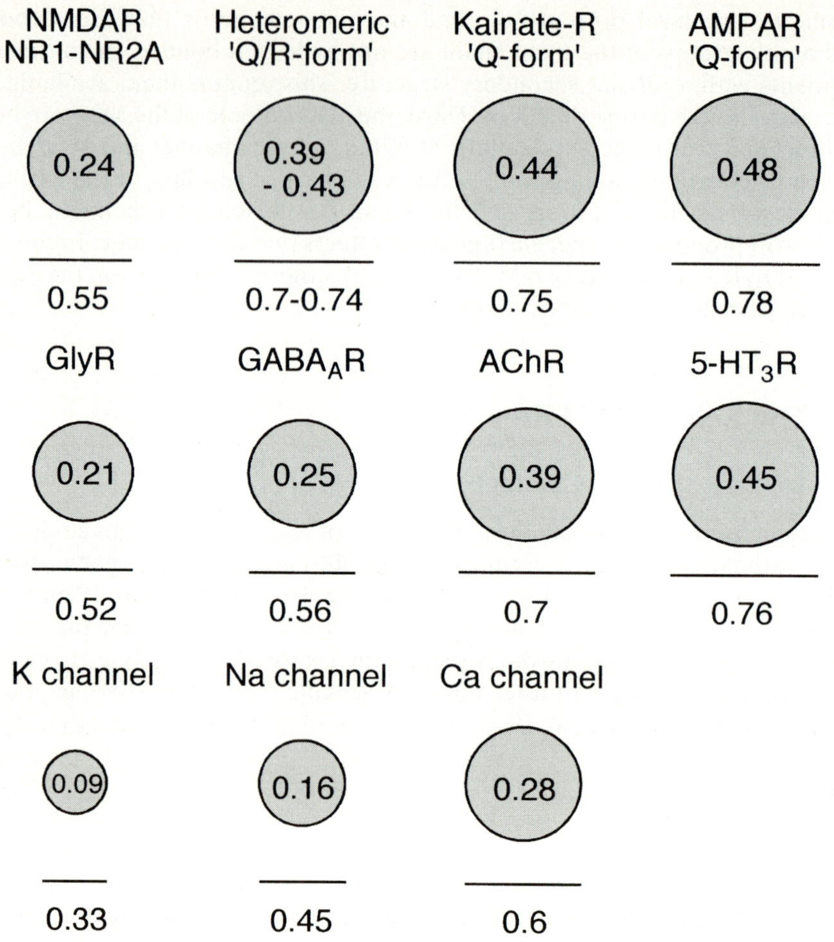

Fig. 3. Dimensions of the narrowest part of the glutamate receptor (GluR) pore in comparison with that in other ion channels. All pores are shown as cylindrical structures to simplify comparison. *Numbers* given within the pore denote the cross-sectional area (nm^2), whereas *lines* indicate the mean diameter (in nm). Values for the non-GluR ligand-gated channels are from: GlyR and GABA$_A$ (BORMANN et al. 1987); nAChR (DWYER et al. 1980; WANG and IMOTO 1992) and 5-HT$_3$ (YAKEL et al. 1990; YANG 1990). Those for voltage-gated channels are from (HILLE 1992)

prisingly, a comparable diameter was found in heteromeric channels coassembled from unedited "Q-form" and edited "R-form" subunits (0.7 and 0.74 nm) and in homomeric kainate-receptor channels containing the edited GluR-6(R) subunit (0.76 nm). The finding that channels containing one or more arginine residues at the *Q/R site* have roughly the same dimensions as channels containing only the much smaller glutamine residue is unexpected and difficult to explain, given that the residue at position 0 of the loop (the *Q/R/N site*) is exposed to the lumen of the channel (KUNER et al. 1996, 1997). A possible structural explanation for this observation is discussed below.

3. Comparison with Other Ion Channels

Figure 3 contrasts the dimensions of the narrow constriction in GluR channels composed of different subunits to that in other ion channels. Based on pore size, three groups can be distinguished: (a) non-NMDAR, nAChR and 5-hydroxy-tryptamine receptor ($5\text{-}HT_3R$) channels forming large pores; (b) NMDAR, $GABA_AR$, glycine receptor (GlyR) and Ca channels forming pores with an intermediate size; and (c) K channels and Na channels forming small pores. At first glance, the size of the narrow constriction seems to correlate with the selectivity of a channel. Large-sized channels of the first group are selective for cations but display only moderate selectivity between different monovalent and divalent cations. In contrast, small sized channels of group c are highly selective for a single ionic species, K^+ or Na^+. Consistent with the small pore size, the selectivity mechanism in K channels depends critically on the precisely defined geometry of residues lining the narrow portion of the pore (DOYLE et al. 1998). Channels with pores of intermediate size are either anion selective (GlyR, $GABA_AR$) or Ca^{2+} selective (Ca channels, NMDAR). Despite the similar pore size of Ca channels and NMDAR channels, they use different mechanisms of Ca^{2+} permeation (YANG et al. 1993; ZAREI and DANI 1994). The functional significance of the smaller pore size in NMDAR channels compared with that in other cation-selective, ligand-gated ion channels remains unknown but might be related to the mechanism of voltage-dependent Mg^{2+} block (see below).

In terms of properties of the ion-conducting pore, NMDAR channels can be classified as either "high"- or "low"-affinity channels (MONYER et al. 1994; KUNER and SCHOEPFER 1996). "High"-affinity channels, encompassing channels composed of NR2A or NR2B, display identical permeation and blocking properties and may have a similar pore size. Consistent with this idea, the pore size of native NMDAR channels from rat hippocampal CA1 neurons, which predominantly contain NR2A and NR2B subunits (ZAREI and DANI 1995), are indistinguishable from that of recombinant NR1–NR2A channels (VILLARROEL et al. 1995). However, "low"-affinity channels containing NR2C or NR2D subunits may differ in their pore sizes from those containing NR2A or NR2B, possibly providing a basis for the differences in Ca^{2+} permeation and Mg^{2+} block that they mediate. However, this idea remains untested.

II. Determinants of the Selectivity Filter

Ionic selectivity arises from the specific interactions of permeating ions with the pore wall within the narrow region of a channel, where ions and side chains or main-chain atoms of residues come close enough to create a permeation barrier. Such a barrier could be created by structural (sieving), electrostatic and/or chemical mechanisms.

1. NMDAR Channels

Two complementary approaches led to the identification of residues forming the narrow constriction in NMDAR channels. With SCAM (KUNER et al. 1996), the impermeant probe 2-trimethylammonioethyl-methanethiosufonate (MTSET), applied from the extra- or intracellular side of the membrane, revealed an inaccessible residue bracketed by residues which were accessible from either the external or internal sides of the channel; position 0 was accessible from the extracellular side only, position +1 was inaccessible from either side and position +2 was accessible from the intracellular side only (Fig. 2A, *grey triangles*). This pattern would be consistent with a narrow constriction formed by residues situated around position +1. The second approach was based on the idea that a systematic variation of the volume of pore-lining residues at the narrow constriction should correlate with the permeability of organic cations. Based on this approach, position 0 in the NR1 subunit, position +1 and, to a lesser extent +2 in the NR2 subunit, were found to form the narrow constriction (WOLLMUTH et al. 1996). Figure 4A illustrates such results for one of the substitutions. Substituting glycine for the asparagine residues at positions NR1(0) or NR2(+1) results in an increase of the mean diameter by 0.2 or 0.12 nm, respectively. If both mutant subunits were coexpressed, the pore size increased by 0.32 nm to a mean diameter of 0.87 nm. Hence, the contribution of the individual mutations to pore size was exactly additive in the double mutant, strongly supporting the view that the nonhomologous asparagine residues at positions NR1(0) and NR2(+1) are the major determinants of the narrow constriction, and that they are positioned at similar levels along the vertical axis of the channel. It is interesting to note that, in the NR2 subunit, substitutions at position +2 also altered pore size. This observation is relevant when considering the determinants of pore size in AMPAR channels

Fig. 4A,B. Determinants of the narrow constriction. **A** A close-up view of the narrow constriction of the *N*-methyl-D-aspartate receptor (NMDAR) channel. The tips of the M2 loops are shown (also in Figs. 1C, 8), including positions –1 to +1 (L–1, N0, S+1) of the NR1 subunit and positions –1 to +3 (F–1, N0, N+1, S+2, V+3) of the NR2 subunit. The predicted structural changes resulting from glycine (G) substitutions at the NR1(N0) and/or NR2(N+1) positions are lined up below the wild-type constriction. Pore diameter and the relative increase caused by a mutation is shown in the *right panel*. *Dots* represent surfaces, as defined by van der Waals radii. Virtual mutations and images were created with SwissPDBViewer (Nicolas Guex). **B** Comparison between NMDAR and α-amino-3-hydroxy-5-methylisoxazole-4-propionic-acid receptor (AMPAR) channels. Primary determinants of the narrow constriction are highlighted. Note the asymmetric positioning of position 0 in the NMDAR channel [NR2A(N0), *italics*] and the more symmetric positioning in AMPAR channels. Except for the NR1 subunit, the residue at position 0 is external to the narrow constriction

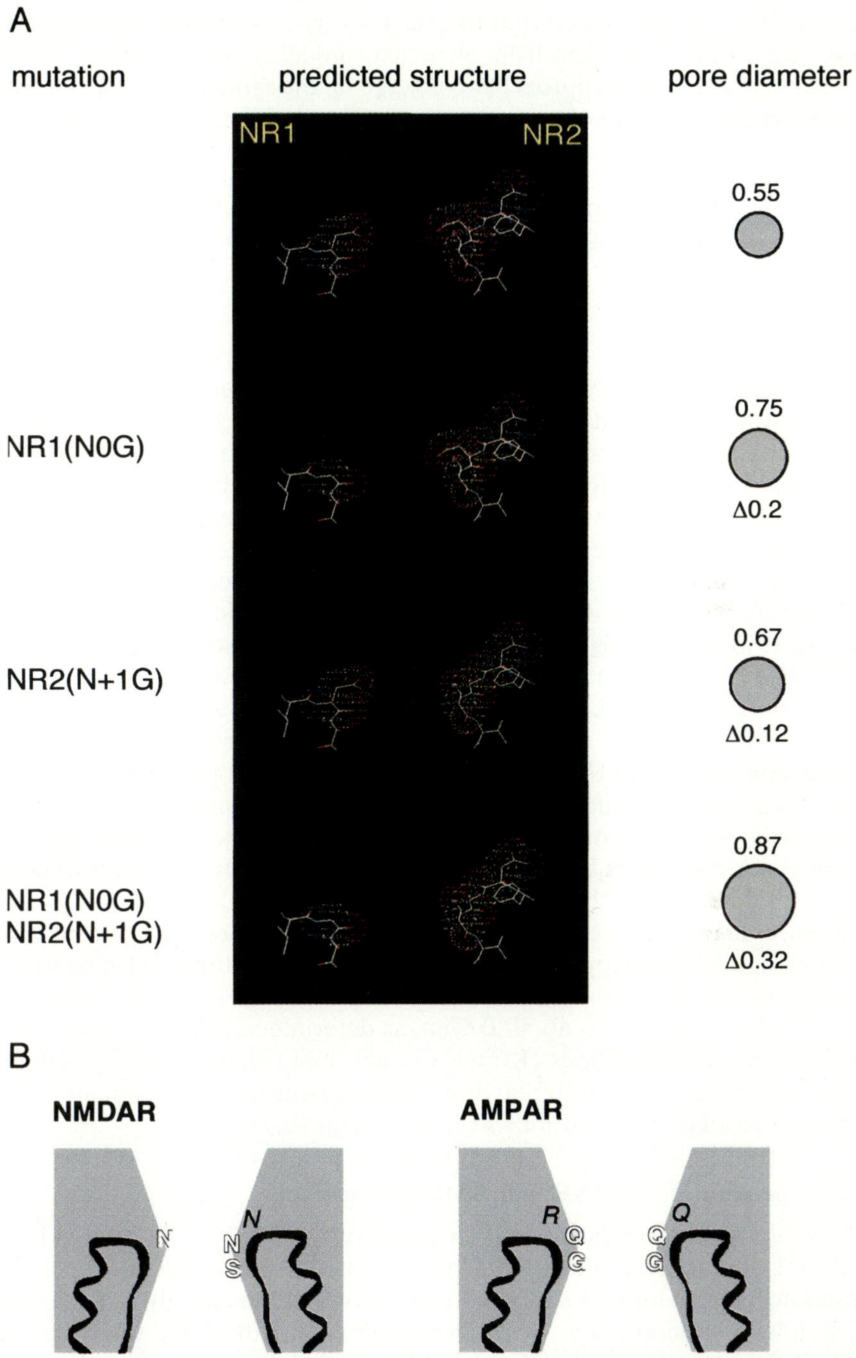
A
mutation
predicted structure
pore diameter
NR1
NR2
0.55
NR1(N0G)
0.75
Δ0.2
NR2(N+1G)
0.67
Δ0.12
NR1(N0G)
NR2(N+1G)
0.87
Δ0.32
B
NMDAR
AMPAR
N
N
S
R
Q
G
Q
Q
G

(see below) and may suggest that the narrow region of the channel extends at least 0.3 nm (measure taken from molecular model shown in Fig. 8).

In conclusion, both approaches converge on the same result, one approach relying on the presence of a barrier preventing reaction with a reporter cysteine and the other defining the disruption of a barrier. Combining the results of both approaches, a picture of the narrow constriction emerges, as schematized in Fig. 4B (left), with the NR1(N0), NR2(N + 1) and NR2(S + 2) residues defining the narrow constriction.

2. AMPAR Channels

After the early discovery of the functionally critical *Q/R site* in AMPAR channels (Hume et al. 1991; Verdoorn et al. 1991), the structural environment of this site was not studied in detail apart from the study by Dingledine et al. (1992). The *Q/R site* was often assumed to contribute to the narrow constriction of the channel. However, channels containing either arginine or the smaller glutamine at position 0 do not differ in the dimensions of their pores, suggesting that other residues define the constriction. Using 2-aminoethyl-methanethiosufonate or MTSET as a probe, SCAM revealed that cysteines at positions 0, +2 and +4 were accessible from either side of the membrane (Kuner et al. 1997). This result is not unexpected given the much larger pore of AMPAR channels in comparison with NMDAR channels (Fig. 3). However, 2-sulfonatoethyl-methanethiosufonate (MTSES), a negatively charged reagent of intermediate size, could be used as a probe, since it does not cross the selectivity filter. MTSES reacted with a cysteine at position 0 only from the extracellular side and with cysteines at positions +1 and +2 only from the intracellular side. The latter two positions were not accessible when expressed as homomeric channels but were accessible when coexpressed with wild-type subunits (Kuner, unpublished results). This result could be explained by steric hindrance due to a decrease in pore size created by the cysteine substitutions. Additional evidence supporting this idea is that substitution of glycine with an alanine residue at positon +2 resulted in a strong decrease of the mean pore diameter from 0.75 nm to about 0.6 nm, as determined from the relative permeability of organic cations (Kuner, unpublished results). Again, combining the results from the two different approaches confines the determinants of the narrow constriction to positions +1 and +2 (Fig. 4B, *right*).

3. Comparison of NMDAR and AMPAR Channels

In addition to the similar overall structure of the M2 loop in NMDAR and AMPAR channels, the selectivity filter is formed at almost identical positions within the hairpin kink of the pore loop (Fig. 4B). Interestingly, in this regard, the GluR-D subunit seems to be more similar to the NR2 than the NR1 subunit. In both the GluR-D and NR2 subunits, the narrow constriction is formed by positions +1 and +2, but not by position 0. This bears important consequences for the positioning of the functionally critical residue at the 0 position, which, relative to the constriction, seems to be oriented towards the extracellular vestibule. As discussed in detail below, such a conformation could explain how an arginine residue, "mutated" into place by RNA editing, would abolish Ca^{2+} permeability without affecting pore size.

In NMDAR channels, the NR1 and NR2 subunits have a slightly different arrangement of the constriction-forming residues. In the NR1 subunit, position 0 appears to be the primary determinant, in contrast to the NR2 subunit, where residues at positions +1 and +2 define the constriction. Nevertheless, although side chains of residues at position +1 in the NR1 subunit are not exposed to the pore, there remains the possibility that its main-chain carbonyl group, as well as that of position +2, might form part of the constriction. In any case, the narrow constriction of NMDAR channels is formed by the asymmetric contribution of the NR1 and NR2 subunits, a fact that may contribute to the different functional roles medinted by the subunits.

III. Determinants of Ion Permeation and Blocking

1. Ca^{2+} Permeation

Under physiological conditions, GluR channels conduct a mixture of monovalent cations (K^+ and Na^+) and Ca^{2+} (MacDermott et al. 1986; Mayer and Westbrook 1987). Indeed, Ca^{2+} permeation through these channels is an important, synaptically controlled mechanism of post-synaptic Ca^{2+} influx and mediates many of the biological functions triggered by GluR activation. Although the Ca^{2+} flux through NMDAR channels has received considerable attention, recent work has indicated that, at some synapses in the brain, AMPAR channels are also Ca^{2+} permeable and contribute to synaptic signaling (Otis et al. 1995; Geiger et al. 1995; Gu et al. 1996). A measure of the Ca^{2+} influx mediated by a GluR-channel subtype is provided by the fractional Ca^{2+} current. The fraction of the total current carried by Ca^{2+} can be calculated from a simultaneous measurement of the total charge carried by monovalent ions and Ca^{2+} and the charge carried by Ca^{2+} alone (Fig. 5A, *left*). GluR subtypes differ in terms of the fractional Ca^{2+} current (Fig. 5A, *right*); non-NMDAR channels cover a range of 0.5 to 4% whereas, in NMDAR channels, it is considerably higher, between 8 and 14% (Schneggenburger et al. 1993; Burnashev et al. 1995; Wollmuth and Sakmann 1998).

a) NMDAR Channels

An asparagine residue positioned at the tip of the pore loop of the NR1 subunit, N0, is the primary determinant of Ca^{2+} permeation at the narrow constriction. In contrast, the adjacent asparagine residues in the NR2 subunit only weakly affect Ca^{2+} permeation, a situation inverse to that found for blocking by Mg^{2+} (Burnashev 1993; Wollmuth et al. 1998b). In addition, the mechanism of Ca^{2+} permeation in NMDAR channels appears to depend on an

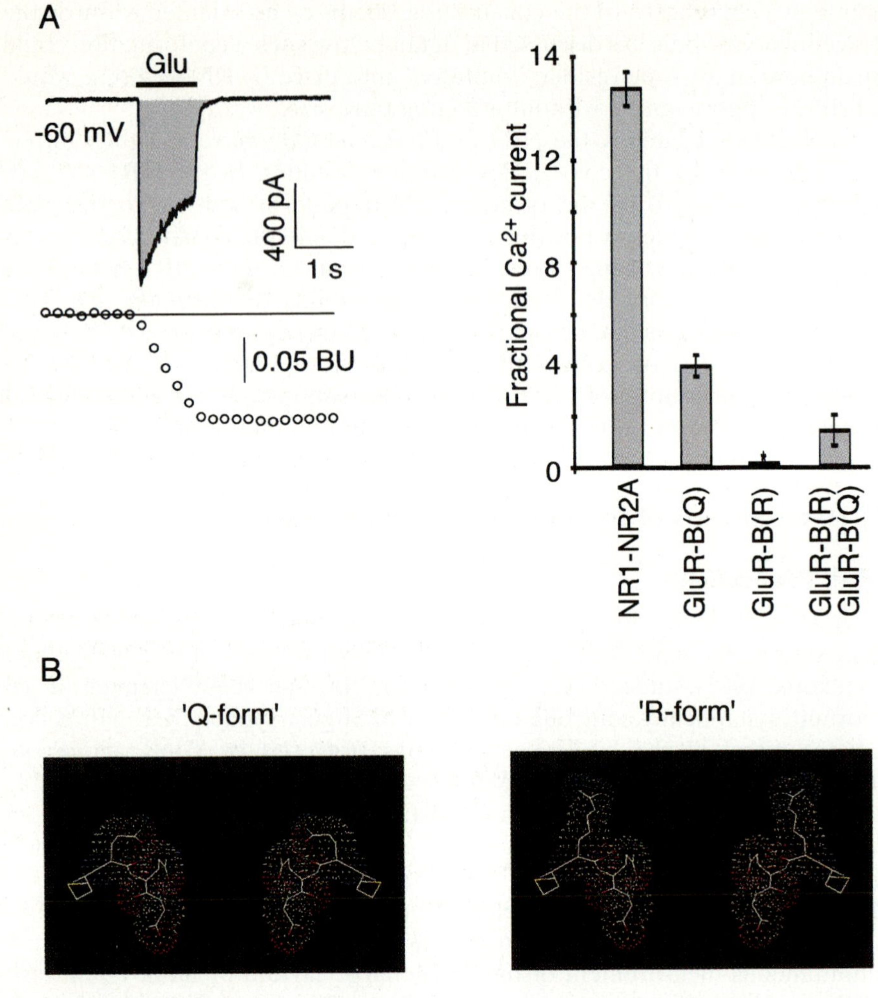

Fig. 5A,B. Determinants of Ca^{2+} permeation. **A** Fractional Ca^{2+} currents in glutamate receptor channels. The *left panel* shows a simultaneous measurement of whole-cell current and Ca^{2+} influx in *N*-methyl-D-aspartate receptor channels composed of NR1 and NR2A subunits. The fluorescence intensity (*circles*) as a function of Ca^{2+} influx is given in bead units (BU; WOLLMUTH and SAKMANN 1998). The fraction of the total current carried by Ca^{2+} was calculated from total charge determined from the whole-cell current response (*shaded*) and the charge carried by Ca^{2+} derived from the fluorescence signal. The r*ight panel* summarizes fractional Ca^{2+} currents determined for different GluR subunits. Values shown are the mean ±2 times the standard error of the mean and were recorded at –60 mV with 1.8 mM $CaCl_2$, 140 mM NaCl extracellularly and 140 mM CsCl, 2 mM fura-2 intracellularly. Homomeric GluR-B(R) channels yield only minute currents; the *bar* indicates the result of a single experiment (*asterisk*). **B** Structural model of α-amino-3-hydroxy-5-methylisoxazole-4-propionic-acid receptor "Q-form" and "R-form" channels. The arginine residue is shown in a rotamer position, with the guanidinium group oriented towards the extracellular vestibule. The side chain of the arginine residue is likely to fluctuate between different rotamer positions. However, none of them result in a decrease in pore size

additional site in the extracellular vestibule of the channel at an electrical distance of approximately 0.1 (PREMKUMAR and AUERBACH 1996; SHARMA and STEVENS 1996). Such an external binding site for Ca^{2+} has been suggested to be the main determinant of the high fractional Ca^{2+} currents in NMDAR channels (8–14%; Fig. 5A, *right*) and may mediate the differences in Ca^{2+} permeation properties between GluR subtypes (WOLLMUTH and SAKMANN 1998). In this view, residues positioned at the channel's narrow constriction, primarily the asparagine at NR1(N0), would act as a simple permeation barrier, providing an energetic environment favorable for passage of Ca^{2+}. Candidate residues of an extracellular Ca^{2+} binding site have been traced to the sequence motif "DRPEER", located *C*-terminal to the M3 segment of the NR1 subunit (WOLLMUTH et al. 1998a). The carboxyl groups of D640 and E643 may form a coordinative binding site for Ca^{2+}. Thus, the presence of a unique sequence motif in the NR1 subunit may generate a different mechanism of Ca^{2+} permeation in NMDAR channels compared with AMPAR channels, thereby increasing the fraction of the total current carried by Ca^{2+}.

b) AMPAR Channels

Ca^{2+} permeability of AMPAR channels is regulated by RNA editing of the GluR-B subunit (SOMMER et al. 1991), redefining the identity of a single residue located at position 0′ of the pore loop, known as *Q/R site* (BURNASHEV et al. 1992a). The presence of an arginine residue at this position (GluR-B) renders AMPAR channels impermeable to Ca^{2+}, whereas a glutamine (GluR-A, -C, -D) produces Ca^{2+}-permeable channels (HUME et al. 1991; BURNASHEV et al. 1992a). Ca^{2+}-permeable AMPAR channels exhibit only a weak selectivity for Ca^{2+} over monovalent cations, in comparison with other Ca^{2+}-permeable channels, such as NMDAR channels or voltage-gated Ca channels (BURNASHEV et al. 1995; WOLLMUTH and SAKMANN 1998). Consistent with this observation, only a small fraction of the total current, about 3 to 4% in nonedited AMPAR channels, is carried by Ca^{2+} under physiological conditions (WOLLMUTH and SAKMANN 1998). The structural basis for the differences between "Q-form" and "R-form" channels, in terms of Ca^{2+} permeability, cannot be due to differences in pore size since, in both channel types, the size of the pore is identical. Rather, positive charges at the narrow constriction of "R-form" channels may, solely through an electrostatic mechanism, prevent permeation of Ca^{2+} (BURNASHEV et al. 1996). Figure 5B illustrates a simple approximation of the structural differences between nonedited and edited AMPAR channels. In both channels, the size of the pore is identical. In homomeric "R-form" channels, the guanidinium groups may project into the extracellular vestibule and form a ring of positive charges. Although the presence of a single arginine residue in the channel complex attenuates single-channel conductance of monovalent ions, it abolishes Ca^{2+} permeability consistent with the stronger interaction of the positively charged arginine with divalent ions compared with monovalent ions. Homomeric "R-form" channels, but not channels containing a smaller number of arginines, lose their cation selectivity and are slightly permeable to Cl^-, compatible with a high density of positive charges created by the arginine residues (BURNASHEV et al. 1996).

2. Mg^{2+} Block of NMDAR Channels

The strongly voltage-dependent blocking by extracellular Mg^{2+} (MAYER et al. 1984; NOWAK et al. 1984) endows the NMDAR channel with computational properties (SEEBURG et al. 1995). Activation of the receptor by presynaptic release of glutamate elicits current flow only with concurrent or near-concurrent postsynaptic activity. Hence, the NMDAR functions as a coincidence detector. The physiological relevance of voltage-dependent blocking by intracellular Mg^{2+} remains uncertain.

A key feature of blocking by extracellular Mg^{2+} is its strong voltage dependence (Fig. 6, *upper left panel*). The increase of blocking with more negative potentials, manifested at the single-channel level as increasingly long dwell times, arises from an interaction of Mg^{2+} with its blocking site in a region of the channel influenced by the transmembrane electric field. The relative location of a blocking site in the pore over which the voltage drop occurs can be described by the Woodhull model (WOODHULL 1973). A number of studies have indicated that the voltage dependence of blocking by extracellular Mg^{2+} gives an apparent site of interaction between 80 and 100% across the transmembrane electric field (ASCHER and NOWAK 1988; JAHR and STEVENS 1990; KUNER and SCHOEPFER 1996; WOLLMUTH et al. 1998b). However, intracellular Mg^{2+} blocks the channel at an electrical distance of 30% (JOHNSON and ASCHER 1990), positioning its apparent blocking site external to that of external Mg^{2+}, a discrepancy difficult to reconcile with an impermeant blocker. The identification of the molecular determinants of these blocking sites should help to clarify this issue and to explain the basis of the strong voltage dependence of blocking by external Mg^{2+}.

a) Extracellular Mg^{2+} Block

Based on sequence conservation among different members of the GluR family and the observation that the *Q/R site* in AMPAR channels affected Ca^{2+} permeation, mutations at the homologous position in NMDAR channels identified the N site of the NR2 subunit as an important factor in Mg^{2+} block (BURNASHEV et al. 1992b; MORI et al. 1992; SAKURADA et al. 1993). Subsequently, an approach involving the systematic swapping of residues at homologous positions of the M2 segments in the NR1 and NR2 subunits (KUNER, unpublished results) revealed that mutations of the asparagine residue adjacent to the N site in the NR2 subunit exhibited much stronger effects on Mg^{2+} block than mutations of the N site itself (RUPPERSBERG et al. 1994). A more detailed comparison using substitutions of any one of the three asparagines at the tip of the loop, NR1(N0), NR2A(N0) and NR2A(N+1), with a set of amino acid residues, including glycine, serine, glutamine, asparagine and arginine, revealed

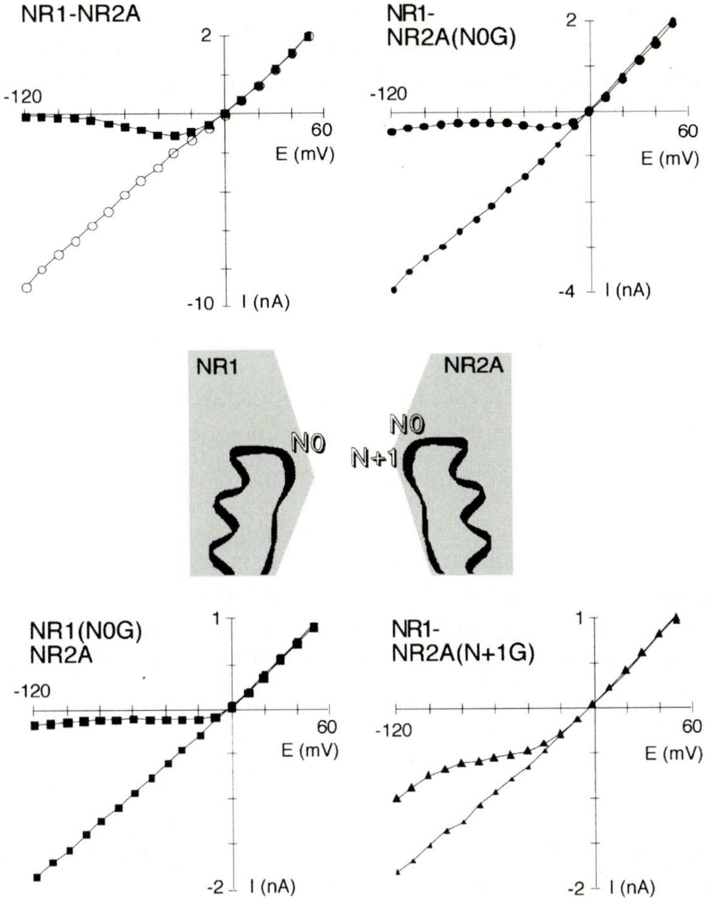

Fig. 6. Determinants of blocking of *N*-methyl-D-aspartate receptor (NMDAR) channels by extracellular Mg^{2+} reside in the NR2 subunit. Current–voltage relationships illustrating the blocking, by extracellular Mg^{2+}, of wild-type NR1–NR2A and mutant NMDAR channels. The mutant channels have glycine substituted in positions at [NR1(N0G) and NR2A(N+1G)] or near [NR2A(N0G)] the narrow constriction (*inset*). The voltage dependence of blocking, analyzed over physiological potentials (–60 mV to –10 mV), was indistinguishable in NR1(N0G)–NR2A channels compared to wild-type, slightly reduced in NR1–NR2A(N0G) channels and strongly attenuated in NR1–NR2A(N+1G) channels. The current–voltage relationships are the peak glutamate-activated currents plotted against membrane voltage in the absence (2 mM ethylene diamine tetraacetic acid) or presence of extracellular Mg^{2+} (0.7 mM). Currents were generated by voltage steps of 10 mV increments and recorded in HEK 293 cells bathed in symmetrical KCl with no extracellular Ca^{2+}

that Mg^{2+} interacts with both asparagines of the NR2A subunit; however, the *N*+1 *site* represents the primary determinant (WOLLMUTH et al. 1998b). Figure 6 illustrates the dominant effect on Mg^{2+} block of glycine substitution at position NR2(N+1) in comparison with glycine substitutions at NR1(N0) and NR2(N0).

The above experiments demonstrated that the adjacent asparagine residues in the NR2 subunit, but not the NR1 N-site asparagine, determine the extent of blocking by extracellular Mg^{2+}. Still, at least two fundamental questions remain. First, how does Mg^{2+} interact with the adjacent asparagines? Although Mg^{2+} is a small ion, its effective diameter, which includes a tight shell of waters of hydration, is large, at least 0.7 nm (WOLLMUTH et al. 1998b). Given a pore diameter of 0.55 nm, Mg^{2+} could simply block the channel by virtue of its size without specifically binding to residues at the narrow constriction. Indeed, the main determinants of the narrow constriction and Mg^{2+} block converge onto a single amino acid residue: the N+1 site of the NR2 subunit. Nevertheless, comparing Mg^{2+} block of mutant channels with known changes in pore size revealed that the diameter of the pore and the degree of Mg^{2+} block did not correlate, suggesting that blocking is not determined by an occlusion mechanism (Fig. 7A) (WOLLMUTH et al. 1998b). The most likely alternative, therefore, is that Mg^{2+} directly binds to the adjacent asparagines.

The second question pertains to the underlying mechanism of the strong voltage dependence of blocking. In Ca^{2+}-free conditions and within potentials ranging from –60 mV to –10 mV, Mg^{2+} block follows the Woodhull model for a "simple" pore blocker. At more negative potentials, the block deviates from this simple scheme, probably reflecting that Mg^{2+} weakly permeates NMDAR channels (WOLLMUTH et al. 1998b). Studies examining blocking by impermeant organic cations suggest that the narrow constriction is positioned about 50–60% of the way across the transmembrane electric field (VILLARROEL et al. 1995; ZAREI and DANI 1995). Consistent with this idea is the fact that

Fig. 7A,B. Possible mechanism of blocking of *N*-methyl-D-aspartate receptor (NMDAR) channels by extracellular Mg^{2+}. **A** The narrow constriction is not a steric barrier for Mg^{2+} block. Fraction blocked by 0.07 mM Mg^{2+} at –120 mV plotted against pore diameter for wild-type (*asterisk*) and mutant NR2A channels [except for NR1(N0G)]. The *dashed lines* are shown to indicate the values that would be observed if only one property, pore size (*horizontal line*) or extent of blocking by Mg^{2+} (*vertical line*), was changed. **B** Movement of the NR2 M2 segment or the hairpin kink may possibly mediate the strong voltage dependence of blocking by extracellular Mg^{2+}. In wild-type channels, the voltage dependence of blocking suggests that Mg^{2+} nearly completely blocks the transmembrane electric field. One possibility is that, when Mg^{2+} is bound, the NR2-pore loop moves deeper into the pore. **C** Comparison of extracellular and intracellular Mg^{2+} block. Analysis of blocking by extracellular Mg^{2+} in mutant NMDAR channels suggests that the deepest site of interaction for Mg^{2+} in the channel is about 50–60% across the transmembrane electric field. Intracellular Mg^{2+}, which primarily interacts with the NR1(N0) position, blocks about 40% of the channel (from the internal side) across the membrane

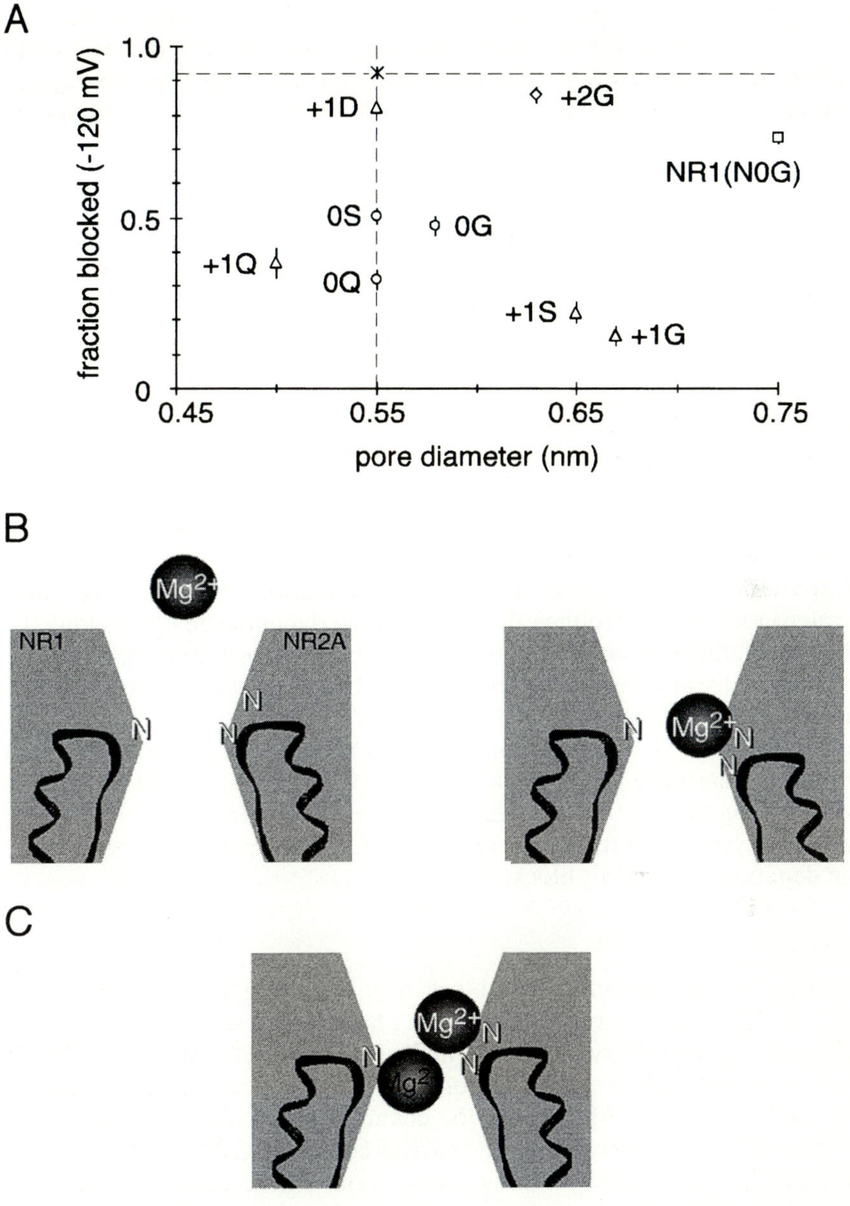
A
1.0
0.5
0
fraction blocked (-120 mV)
0.45
0.55
0.65
0.75
pore diameter (nm)
+1D
+2G
NR1(N0G)
0S
0G
+1Q
0Q
+1S
+1G
B
Mg2+
NR1
NR2A
N
C

intracellular Mg^{2+}, which also blocks the channel at the narrow constriction (albeit at the NR1 N -site, see below), shows a voltage dependence of around 40% (WOLLMUTH et al. 1998c). Since the adjacent NR2-subunit asparagines seem to form a barrier to Mg^{2+} influx, it is unlikely that, in terms of simple blocking, Mg^{2+} penetrates any deeper into the channel than to the narrow constriction. Hence, following a simple Woodhull model, Mg^{2+} occupying its blocking site at the narrow constriction underlies a major part of the strong voltage dependence, perhaps around 50–60% of the 80% electrical distance observed. Still, what mechanism creates the additional ~30% electrical distance? One possibility is that Mg^{2+} binds at the N0 and N+1 positions, thereby producing a conformational change of parts of the M2 loop, moving the structure a short distance towards the intracellular vestibule (Fig. 7B). In this model, the additional voltage dependence would be an intrinsic property of the N+1 site, consistent with the observation that mutations at this site produced channels with a blocking distance for Mg^{2+} of 50% (WOLLMUTH et al. 1998b).

Alternative hypotheses have been proposed. The strong voltage dependence of the block may arise from an interaction of Mg^{2+} blocking the pore with monovalent ions simultaneously present at a "permeant-ion site" (ZAREI and DANI 1995). In this view, electrostatic interactions between Mg^{2+} and monovalent ions would produce the high voltage dependence of blocking. In the context of the structural framework presented above, mutations at the N+1 site might disrupt the "permeant-ion site". In yet another model, the additional voltage dependence was proposed to arise from the excessive ionic pressure of monovalent ions in the pore of the NMDAR channel exerted onto Mg^{2+} blocking the pore (RUPPERSBERG et al. 1994). Because of the inferred physiological significance of blocking by extracellular Mg^{2+}, a major goal of channel biophysics is to further define the mechanism producing the strong voltage dependence of Mg^{2+} block.

b) *Intracellular Mg^{2+} Block*

Determinants of intracellular Mg^{2+} block have been mapped to the narrow constriction as well. For intracellular Mg^{2+}, however, residue NR1(N0) has been defined as the primary determinant (KUPPER et al. 1998; WOLLMUTH et al. 1998c). Consistent with the proposal of JOHNSON and ASCHER (1990), extracellular and intracellular Mg^{2+} block the channel at different sites. However, they do so in close proximity within the narrow constriction. Hence, the narrow constriction forms a pivotal blocking site for both extracellular and intracellular ions (Fig. 7C).

3. Blocking of AMPAR Channels by Intracellular Polyamines

Ca^{2+}-permeable AMPAR channels are blocked in a strongly voltage-dependent manner by intracellular polyamines, such as spermine (BOWIE and MAYER 1995; DONEVAN and ROGAWSKI 1995; ISA et al. 1995; KAMBOJ et al. 1995;

KOH et al. 1995). Although the physiological significance of this blocking is unknown, it may act in an analogous manner to extracellular Mg^{2+} block of NMDAR channels, by controlling Ca^{2+} influx in an activity-dependent manner (Rozov et al. 1998).

Channels containing the GluR-B subunit are not blocked by polyamines, defining the *Q/R site* as a first structural element involved in polyamine block. AMPAR channels with an engineered asparagine residue at the Q/R site show a strong reduction of polyamine block as well (DINGLEDINE et al. 1992). In addition to position 0, almost all exposed residues in the vicinity of the narrow constriction affect polyamine block (KUNER, unpublished results; MAYER, personal communication). For example, substituting the glycine residue at position +2 with cysteine abolishes blocking by intracellular spermine. Substitution of an alanine residue at the same position results in channels with a markedly attenuated spermine block. These mutations also strongly decrease the size of the pore. Therefore, at this position, polyamine block seems to be a function of pore size (KUNER, unpublished results). A different mechanism may hold for each individual site affecting polyamine block. At the Q/R site, the presence of arginine residues may prevent polyamines from blocking the channel via an electrostatic mechanism although, for many other residues hydrophobic interactions may dominate (CUI et al. 1998).

4. Comparison of NMDAR and AMPAR Channels

In both AMPAR and NMDAR channels, the narrow constriction serves as a pivotal blocking site, providing interaction sites for physiologically relevant ions from either side of the conduction pathway. However, differences exist in the structure of the M2 loops, reflecting the functional differences between GluR subunits and the modest degree of sequence conservation of residues forming the selectivity filter. NMDAR channels differ from the other GluR channels because they are assembled from two different types of subunits, whereas AMPAR or kainate-receptor channels arise from subunits of a single subfamily sharing a high degree of sequence identity. The M2 loops of the NR1 and NR2 subunits differ in their geometry, especially around position 0, thereby contributing differently to the formation of the narrow constriction. This may result in an asymmetric cross section of the pore – for example, a rectangular shape (ZAREI and DANI 1995). Such an asymmetric shape may create the structural basis for the contributions of distinct parts of the constriction to Ca^{2+} permeation and blocking by extracellular and intracellular Mg^{2+} within a highly confined region of the pore, as described above. In contrast to the NMDAR channel, AMPAR channels are more likely to have a symmetrical pore, as they arise from a single subunit subtype. In conclusion, structural differences of the selectivity filter within the GluR family are likely to reflect their specialized functional properties as variations of a more global theme.

IV. Structural Model of the Selectivity Filter

Figure 8 shows a molecular model of the NMDAR channel domain integrating structural and functional aspects of the ion-conducting pore, as presented above. A longitudinal section of the channel reveals two vestibules connected by a narrow constriction (*upper panel*). Consistent with experimental observations (VILLARROEL et al. 1995; ZAREI and DANI 1995), the extracellular vestibule appears wider than the intracellular vestibule. The contribution to the selectivity filter of three residues forming the hairpin kink is shown in the lower panel. The size of the narrow constriction is determined mainly by asparagine residues at position 0 of the NR1 subunit and position +1 of the NR2 subunit. The asparagine residue at position 0 of the NR2 subunit projects into the external vestibule and does not contribute to the constriction. The primary determinants of Mg^{2+} block are the N0 and N+1 positions of the NR2 subunit. The precise geometry of these side chains might be important for the formation of a Mg^{2+}-binding site. The asparagine at position 0 of the NR1 subunit controls Ca^{2+} permeation and cytoplasmic Mg^{2+} block. The two subunits differ in the arrangement of the homologous N0 positions, possibly providing the structural basis for their different contributions to Ca^{2+} permeation and Mg^{2+} block.

E. Summary: Correlating Structure and Function of GluR Channels

Although many of the details of GluR-channel structure remain to be discovered, a molecular picture of the channel domain is emerging. The channel domain forms the interface between the lipid environment of the membrane and the ion-conducting pore. The lumenal surface of the channel domain defines the conduction pore, which consists of an external vestibule lined by the preM1, M3 and M4 segments and an internal vestibule lined by the M2 segment. The two vestibules are connected by a narrow constriction formed

Fig. 8. Structural model of the *N*-methyl-D-aspartate receptor (NMDAR) channel: selectivity filter and vestibules. *Upper panel*: longitudinal section of a NMDAR-channel domain showing the NR1 subunit on the *left* and the NR2 subunit on the *right*. A narrow constriction connects the large extracellular vestibule with the smaller cytoplasmic vestibule. The channel lumen spans a vertical distance of approximately 5 nm. The NR1 subunit is shown in the same orientation as in Fig. 1C (*side view*). *Lower panel*: residues forming the selectivity filter, NR1(N0) and NR2(N+1). Side chains and main-chain groups of adjacent residues form a narrow region below the selectivity filter (NR1: S+1, G+2; NR2: S+2, P+4). Residue N0 of the NR2 subunit projects into the external vestibule of the channel. Extracellular Mg^{2+} block is controlled by asparagine residues at positions 0 and +1 of the NR2 subunit. Ca^{2+} permeation and intracellular Mg^{2+} block are controlled by an asparagine residue at position 0 of the NR1 subunit. *Dots* represent van der Waals surfaces. Images were created with SwissPDBViewer (Nicolas Guex)

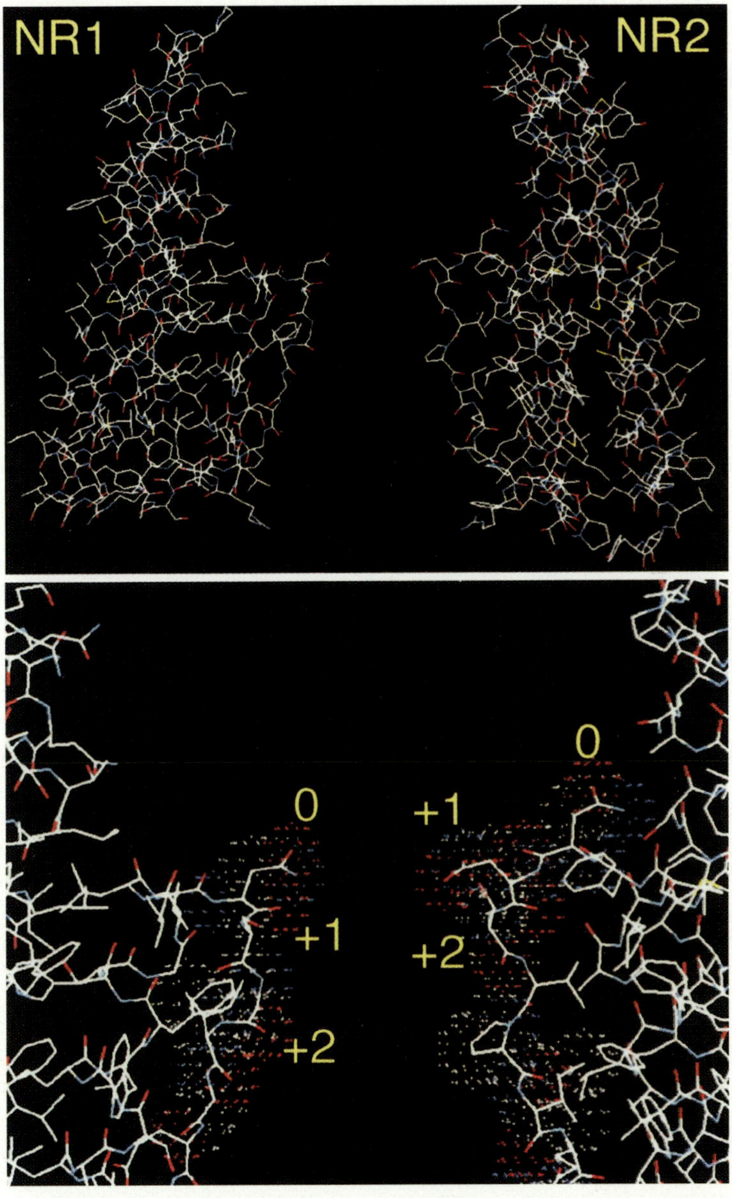
NR1
NR2
0
+1
+2
0
+1
+2

by hydrophilic residues positioned within a three-residue hairpin kink located at the tip of the M2 loop (Figs. 1C, 8).

Many biophysical properties of GluR channels can be correlated with such a structural design. Determinants of pore size, Ca^{2+} permeation, voltage-dependent block by external and internal Mg^{2+} (NMDAR) and voltage-dependent block by intracellular polyamines (AMPAR) are mainly provided by a subunit-specific set of amino acid residues located within the three-residue kink at the tip of the M2 loop (positions 0 to +2, Figs. 2B, 8). Interestingly, such three-residue hairpin kinks often define the active sites of proteins, suggesting that, in GluR channels as well, the unique structural properties of the hairpin kink may be intimately related with the functional properties of the resulting pore. In addition to the selectivity filter as the primary structural determinant of ion permeation in GluRs, the external vestibule plays an important role in modulating GluR-channel function.

Comparison of the structural design of the GluR channel domain to that of other ion channels reveals intriguing similarities but also fundamental differences. A sequence similarity between the P loop of K channels and the M2 segment of GluR channels was taken as evidence that both segments fold into similar 3D structures (WO and OSWALD 1995b; WOOD et al. 1995). This notion is supported by the fact that homologous residues within both segments are exposed to the pore. Comparison of the structural model of M2 shown in Fig. 2B with the structure of the P segment determined by DOYLE et al. (1998) reveals striking similarities. A large and heterogeneous group of voltage-gated ion channels, channels gated by intracellular ligands and the GluR family possess pore loops. In contrast, members of the Cys-loop family (KARLIN and AKABAS 1995) of ion channels (nAChR, 5-HT_3R, $GABA_A$R, GlyR) exhibit a different design of their pores: a transmembrane-spanning α-helix forms the selectivity filter and lining of the vestibules in these channels (KARLIN and AKABAS 1995; UNWIN 1995). Thus, based on the pore-forming motif, ion channels may be generally classified into pore-loop-containing ion channels and transmembrane-helix-containing ion channels.

References

Akabas MH, Stauffer DA, Xu M, Karlin A (1992) Acetylcholine receptor channel structure probed in cysteine-substitution mutants. Science 258:307–310

Ascher P, Nowak L (1988) The role of divalent cations in the *N*-methyl-D-aspartate responses of mouse central neurones in culture. J Physiol (Lond) 399:247–266

Beck C, Wollmuth LP, Seeburg PH, Sakmann B, Kuner T (1999) NMDAR channel segments forming the extracellular vestibule inferred from the accessibility of substituted cysteines. Neuron 22:559–570

Behe P, Stern P, Wyllie D, Nassar M, Schoepfer R, Colquhoun D (1995) Determination of NMDA NR1 subunit copy number in recombinant NMDA receptors. Proceedings Royal Society London B. Biol Sci 262:205–213

Bennett JA, Dingledine R (1995) Topology profile for a glutamate receptor: three transmembrane domains and a channel-lining reentrant membrane loop. Neuron 14:373–384

Berridge MJ (1998) Neuronal Calcium signaling. Neuron 21:13–26
Bliss TVP, Collingridge GL (1993) A synaptic model of memory: long-term potentiation in the hippocampus. Nature 361:31–39
Bormann J, Hamill OP, Sakmann B (1987) Mechanism of anion permeation through channels gated by glycine and γ-aminobutyric acid in mouse cultured spinal neurones. J Physiol (Lond) 385:243–286
Bowie D, Mayer ML (1995) Inward rectification of both AMPA and kainate subtype glutamate receptors generated by polyamine-mediated ion channel block. Neuron 15:453–462
Branden C, Tooze J (1991) Introduction to Protein Structure. Garland Publishing, Inc., New York
Brusa R, Zimmermann F, Koh DS, Feldmeyer D, Gass P, Seeburg PH, Sprengel R (1995) Early-onset epilepsy and postnatal lethality associated with an editing- deficient GluR-B allele in mice. Science 270:1677–1680
Burnashev N (1993) Recombinant ionotropic glutamate receptors: functional distinction imparted by different subunits. Cellular Physiology and Biochemistry 3: 318–331
Burnashev N, Monyer H, Seeburg PH, Sakmann B (1992a) Divalent ion permeability of AMPA receptor channels is dominated by the edited form of a single subunit. Neuron 8:189–198
Burnashev N, Schoepfer R, Monyer H, Ruppersberg JP, Günther W, Seeburg PH, Sakmann B (1992b) Control by asparagine residues of calcium permeability and magnesium blockade in the NMDA receptor. Science 257:1415–1419
Burnashev N, Villarroel A, Sakmann B (1996) Dimensions and ion selectivity of recombinant AMPA and kainate receptor channels and their dependence on Q/R site residues. J Physiol (Lond) 496:165–173
Burnashev N, Zhou Z, Neher E, Sakmann B (1995) Fractional calcium currents through recombinant GluR channels of the NMDA, AMPA and kainate receptor subtypes. J Physiol (Lond) 485:403–418
Chiamvimonvat N, Perez-Garcia MT, Ranjan R, Marban E, Tomaselli GF (1996) Depth asymmetries of the pore-lining segments of the Na^+ channel revealed by cysteine mutagenesis. Neuron 16:1037–1047
Choi DW (1988) Glutamate neurotoxicity and diseases of the nervous system. Neuron 1:623–634
Chou PY, Fasman GD (1978) Prediction of the secondary structure of proteins from their amino acid sequence. Adv Enzymol Relat Areas Mol Biol 47:45–148
Cui C, Bahring R, Mayer ML (1998) The role of hydrophobic interactions of polyamines to non NMDA receptor ion channels. Neuropharmacology 37: 1381–1389
Dani JA (1986) Ion-channel entrances influence permeation. Net charge, size, shape, binding considerations. Biophys J 49:607–618
Dingledine R, Hume RI, Heinemann SF (1992) Structural determinants of barium permeation and rectification in non-NMDA glutamate receptor channels. J Neurosci 12:4080–4087
Donevan SD, Rogawski MA (1995) Intracellular polyamines mediate inward rectification of Ca^{2+}-permeable alpha-amino-3-hydroxy-5-methyl-4-isoxazolepropionic acid receptors. Proc Natl Acad Sci USA 92:9298–9302
Doyle DA, Cabral JM, Pfuetzner RA, Kuo A, Gulbis JM, Cohen SL, Chait BT, MacKinnon R (1998) The structure of the potassium channel: molecular basis of K^+ conduction and selectivity. Science 280:69–77
Durell SR, Guy HR (1996) Structural model of the outer vestibule and selectivity filter of the Shaker voltage-gated K^+ channel. Neuropharmacology 35:761–773
Dwyer TM, Adams DJ, Hille B (1980) The permeability of the endplate channel to organic cations in frog muscle. J Gen Physiol 75:469–492
Ferrer-Montiel AV, Montal M (1996) Pentameric subunit stoichiometry of a neuronal glutamate receptor. Proc Natl Acad Sci USA 93:2741–2744

Garnier J, Osguthorpe DJ, Robson B (1978) Analysis of the accuracy and implications of simple methods for predicting the secondary structure of globular proteins. J Mol Biol 120:97–120

Geiger JR, Melcher T, Koh DS, Sakmann B, Seeburg PH, Jonas P, Monyer H (1995) Relative abundance of subunit mRNAs determines gating and Ca^{2+} permeability of AMPA receptors in principal neurons and interneurons in rat CNS. Neuron 15: 193–204

Gu JG, Albuquerque C, Lee CJ, MacDermott AB (1996) Synaptic strengthening through activation of Ca^{2+}-permeable AMPA receptors. Nature 381:793–796

Guy HR, Seetharamulu P (1986) Molecular model of the action potential sodium channel. Proc Natl Acad Sci USA 83:508–512

Hille B (1992) Ionic Channels of Excitable Membranes, Vol. 2. Sinauer Assiociates, Inc.)

Hirai H, Kirsch J, Laube B, Betz H, Kuhse J (1996) The glycine binding site of the *N*-methyl-D-aspartate receptor subunit NR1: identification of novel determinants of co-agonist potentiation in the extracellular M3-M4 loop region. Proc Natl Acad Sci USA 93:6031–6036

Hollmann M, Heinemann S (1994) Cloned Glutamate Receptors. Annual Review of Neuroscience 17:31–108

Hollmann M, Maron C, Heinemann S (1994) *N*-glycosylation site tagging suggests a three transmembrane domain topology for the glutamate receptor GluR1. Neuron 13:1331–1343

Hume RI, Dingledine R, Heinemann SF (1991) Identification of a site in glutamate receptor subunits that controls calcium permeability. Science 253:1028–1031

Isa T, Iino M, Itazawa S, Ozawa S (1995) Spermine mediates inward rectification of Ca^{2+}-permeable AMPA receptor channels. Neuroreport 6:2045–2048

Jahr CE, Stevens CF (1990) A quantitative description of NMDA receptor-channel kinetic behavior. J Neurosci 10:1830–1837

Johnson JW, Ascher P (1990) Voltage-dependent block by intracellular Mg^{2+} of *N*-methyl-D-aspartate-activated channels. Biophys J 57:1085–1090

Kamboj SK, Swanson GT, Cull-Candy SG (1995) Intracellular spermine confers rectification on rat calcium-permeable AMPA and kainate receptors. J Physiol (Lond) 297–303

Karlin A, Akabas MH (1995) Toward a structural basis for the function of nicotinic acetylcholine receptors and their cousins. Neuron 15:1231–1244

Karlin A, Akabas MH (1998) Substituted-cysteine-accessibility method. In: Methods in Enzymology, Vol. 293. P. M. Conn, eds. (San Diego, CA: Academic Press Inc.), pp. 123–136

Koh DS, Burnashev N, Jonas P (1995) Block of native Ca^{2+} permeable AMPA receptors in rat brain by intracellular polyamines generates double rectification. J Physiol (Lond) 486:305–312

Kuner T, Beck C, Seeburg PH, Sakmann B (1997) Pore-lining residues of the AMPA receptor channel M2 segment. Neuroscience Abstract 23:925

Kuner T, Schoepfer R (1996) Multiple structural elements determine subunit-specificity of Mg^{2+} block in NMDA receptor channels. J Neurosci 16:3549–3558

Kuner T, Wollmuth LP, Karlin A, Seeburg PH, Sakmann B (1996) Structure of the NMDA receptor channel M2 segment inferred from the accessibility of substituted cysteines. Neuron 17:343–352

Kupper J, Ascher P, Neyton J (1996) Probing the pore region of recombinant *N*-methyl-D-aspartate channels using external and internal magnesium block. Proc Natl Acad Sci USA 93:8648–8653

Kupper J, Ascher P, Neyton J (1998) Internal Mg^{2+} block of recombinant NMDA channels mutated within the selectivity filter and expressed in Xenopus oocytes. J Physiol (Lond) 1–12

Kürz LL, Zühlke RD, Zhang H-J, Joho RH (1995) Side-Chain Accessiblilties in the Pore of a K^+ Channel Probed by Sulfhydryl-Specific Reagents after Cysteine-Scanning Mutagenesis. Biophysical Journal 68:900–905

Laube B, Kuhse J, Betz H (1998) Evidence for a tetrameric structure of recombinant NMDA receptors. J Neurosci 18:2954–2961
MacDermott AB, Mayer ML, Westbrook GL, Smith SJ, Barker JL (1986) NMDA receptor activation increases cytoplasmic calcium concentration in cultured spinal cord neurons. Nature 321:261–263
Mano I, Teichberg VI (1998) A tetrameric subunit stoichiometry for a glutamate receptor-channel complex. Neuroreport 9:327–331
Mayer ML, Westbrook GL (1987) Permeation and Block of *N*-methyl-D-aspartic acid receptor channels by divalent cations in mouse cultured central neurons. J Physiol (Lond) 394:501–527
Mayer ML, Westbrook GL, Guthrie PB (1984) Voltage-dependent block by Mg^{2+} of NMDA responses in spinal cord neurons. Nature 309:261–263
Meldrum B, Garthwaite J (1990) Excitatory amino acid neurotoxicity and neurodegenerative disease. Trends Pharmacol. Sci. 11:379–387
Miller C (1989) Genetic manipulation of ion channels: a new approach to structure and mechanism. Neuron 2:1195–1205
Montal M (1995) Molecular mimicry in channel-protein structure. Curr Opin Struct Biol 5:501–506
Montal M (1996) Protein folds in channel structure. Curr Opin Struct Biol 6:499–510
Monyer H, Burnashev N, Laurie DJ, Sakmann B, Seeburg PH (1994) Developmental and regional expression in the rat brain and functional properties of four NMDA receptors. Neuron 12:529–540
Mori H, Masaki H, Yamakura T, Mishina M (1992) Identification by mutagenesis of a Mg^{2+}-block site of the NMDA receptor channel. Nature 358:673–675
Nakanishi S, Masu M (1994) Molecular diversity and functions of glutamate receptors. Annual Review of Biophysics and Biomolecular Structure 23:319–348
Nowak L, Bregestovsky P, Ascher P, Herbet A, Prochiantz A (1984) Magnesium gates glutamate-activated channels in mouse central neurons. Nature 307:462–465
Otis TS, Raman IM, Trussell LO (1995) AMPA receptors with high Ca^{2+} permeability mediate synaptic transmission in the avian auditory pathway. J Physiol (Lond) 482:309–315
Pascual JM, Shieh C-C, Kirsch GE, Brown AM (1995) K^{+} pore structure revealed by reporter cysteines at inner and outer surfaces. Neuron 14:1055–1063
Premkumar LS, Auerbach A (1996) Identification of a high affinity divalent cation binding site near the entrance of the NMDA receptor channel. Neuron 16:869–880
Premkumar LS, Auerbach A (1997) Stoichiometry of recombinant *N*-methyl-D-aspartate receptor channels inferred from single-channel current patterns. J Gen Physiol 110:485–502
Rosenmund C, Stern BY, Stevens CF (1998) The tetrameric structure of a glutamate receptor channel. Science 280:1596–1599
Rost B, Sander C (1994) Combining evolutionary information and neural networks to predict protein secondary structure. Proteins 19:55–72
Rozov A, Zilberter Y, Wollmuth LP, Burnashev N (1998) Facilitation of currents through rat Ca^{2+}-permeable AMPA receptor channels by activity-dependent relief from polyamine block. J Physiol (Lond) 361–377
Ruppersberg JP, v. Kitzing E, Schoepfer R (1994) The mechanism of magnesium block of NMDA receptors. Seminars in the Neurosciences 6:87–96
Sakurada K, Masu M, Nakanishi S (1993) Alteration of Ca^{2+} permeability and sensitivity to Mg^{2+} and channel blockers by a single amino acid substitution in the *N*-methyl-D-aspartate receptor. J Biol Chem 268:410–415
Sankararamakrishnan R, Adcock C, Sansom MS (1996) The pore domain of the nicotinic acetylcholine receptor: molecular modeling, pore dimensions, electrostatics. Biophys J 71:1659–1671
Schneggenburger R, Zhou Z, Konnerth A, Neher E (1993) Fractional contribution of calcium to the cation current through glutamate receptor channels. Neuron 11: 133–143

Seeburg PH (1993) The TINS/TiPS Lecture. The molecular biology of mammalian glutamate receptor channels. Trends Neurosci 16:359–365
Seeburg PH, Burnashev N, Kohr G, Kuner T, Sprengel R, Monyer H (1995) The NMDA receptor channel: molecular design of a coincidence detector. Recent Prog Horm Res 50:19–34
Sharma G, Stevens CF (1996) Interactions between two divalent ion binding sites in *N*-methyl-D-aspartate receptor channels. Proc Natl Acad Sci USA 93:14170–14175
Single FN, Rozov A, Burnashev N, Zimmermann M, Hanley DF, Sakmann B, Seeburg PH, Sprengel R (1997) Effects of targeted point mutation at the N-site of the NMDA receptor channel M2 domain in mice. Society for Neuroscience Abstracts 23:1134
Sommer B, Köhler M, Sprengel R, Seeburg PH (1991) RNA editing in brain controls a determinant of ion flow in glutamate-gated channels. Cell 67:11–19
Sprengel R, Seeburg PH (1995) Ionotropic Glutamate Receptors. In: Ligand and voltage-gated ion channels. R. A. North, eds. (Boca Raton, Florida: CRC Press), pp. 213–263
Stryer L (1995) Biochemistry. (New York: W. H. Freeman)
Sun Z-P, Akabas MH, Goulding EH, Karlin A, Siegelbaum SA (1996) Exposure of Residues in the Cyclic Nucleotide-Gated Channel Pore: P- Region Structure and Function in Gating. Neuron 16:141–149
Sutcliffe MJ, Wo ZG, Oswald RE (1996) Three-dimensional models of non-NMDA glutamate receptors. Biophys J 70:1575–1589
Unwin N, Toyoshima C, Kubalek E (1988) Arrangement of the acetylcholine receptor subunits in the resting and desensitized states, determined by cryoelectron microscopy of crystallized Torpedo postsynaptic membranes. J Cell Biol 107: 1123–1138
Unwin N (1995) Acetylcholine receptor channel imaged in the open state. Nature 373:37–43
Verdoorn TA, Burnashev N, Monyer H, Seeburg PH, Sakmann B (1991) Structural determinants of ion flow through recombinant glutamate receptor channels. Science 252:1715–1718
Villarroel A, Burnashev N, Sakmann B (1995) Dimensions of the narrow portion of a recombinant NMDA receptor channel. Biophys J 68:866–875
Wang F, Imoto K (1992) Pore size and negative charge as structural determinants of permeability in the Torpedo nicotinic acetylcholine receptor channel. Proceedings Royal Society London Ser. B. Biol. Sci. 250:11–17
Williams K, Pahk AJ, Kashiwagi K, Masuko T, Nguyen ND, Igarashi K (1998) The selectivity filter of the *N*-methyl-D-aspartate receptor: a tryptophan residue controls block and permeation of Mg^{2+}. Mol Pharmacol 53:933–941
Wo ZG, Oswald RE (1994) Transmembrane topology of two kainate receptor subunits revealed by *N*-glycosylation. Proc Natl Acad Sci USA 91:7154–7158
Wo ZG, Oswald RE (1995a) A topological analysis of goldfish kainate receptors predicts three transmembrane segments. J Biol Chem 270:2000–2009
Wo ZG, Oswald RE (1995b) Unraveling the modular design of glutamate-gated ion channels. Trends Neurosci 18:161–168
Wollmuth LP, Beck C, Seeburg PH, Sakmann B (1998a) Determinants of Ca^{2+} transport in NMDA receptor channels reside in the M3 segment of the NR1-subunit. Society for Neuroscience Abstracts 24:840
Wollmuth LP, Kuner T, Sakmann B (1998b) Adjacent asparagines in the NR2-subunit of the NMDA receptor channel control the voltage-dependent block by extracellular Mg^{2+}. J Physiol (Lond) 506:13–32
Wollmuth LP, Kuner T, Sakmann B (1998c) Intracellular Mg^{2+} interacts with structural determinants of the narrow constriction contributed by the NR1-subunit in the NMDA receptor channel. J Physiol (Lond) 506:33–52
Wollmuth LP, Kuner T, Seeburg PH, Sakmann B (1996) Differential contribution of the

NR1- and NR2A-subunits to the selectivity filter of recombinant NMDA receptor channels. J Physiol (Lond) 491:779–797
Wollmuth LP, Sakmann B (1998) Different Mechanisms of Ca^{2+} Transport in NMDA and Ca^{2+}-Permeable AMPA Glutamate Receptor Channels. J Gen Physiol 112: 623–636
Wood MW, VanDongen HM, VanDongen AM (1995) Structural conservation of ion conduction pathways in K channels and glutamate receptors. Proc Natl Acad Sci USA 92:4882–4886
Woodhull AM (1973) Ionic blockage of sodium channels in nerve. J Gen Physiol 61:687–708
Xu M, Akabas MH (1993) Amino acids lining the channel of the γ-aminobutyric acid type A receptor identified by cysteine substitution. J Biol Chem 268:21505–21508
Yakel JL, Shao XM, Jackson MB (1990) The selectivity of the channel coupled to the 5-HT3 receptor. Brain Res 533:46–52
Yang J (1990) Ion permeation through 5-hydroxytryptamine-gated channels in neuroblastoma N18 cells. J Gen Physiol 96:1177–1198
Yang J, Ellinor PT, Sather WA, Zhang JF, Tsien RW (1993) Molecular determinants of Ca^{2+} selectivity and ion permeation in L-type Ca^{2+} channels. Nature 366:158–161
Zarei MM, Dani JA (1994) Ionic permeability characteristics of the *N*-methyl-D-aspartate receptor channel. J Gen Physiol 103:231–248
Zarei MM, Dani JA (1995) Structural basis for explaining open-channel blockade of the NMDA receptor. J Neurosci 15:1446–1454

CHAPTER 7

Block of AMPA and Kainate Receptors by Polyamines and Arthropod Toxins

D. BOWIE, R. BÄHRING, and M.L. MAYER

Abbreviations

AMPA	α-amino-3-hydroxy-5-methyl-4-isoxazolepropionic acid
ATX	Argiotoxin
Ca^{2+}	Calcium
CNS	Central nervous system
Glu	L-glutamate
GluR	Glutamate receptor
G–V	Conductance–voltage
JSTX	Joro spider toxin
$[K^+]_{ext}$	External potassium concentration
K_{ir} channels	Inwardly rectifying potassium channels
k_{off}	Rate constant of unbinding
k_{on}	Rate constant of binding
k_{perm}	Rate constant of permeation
Mg^{2+}	Magnesium
$[Na^+]_{ext}$	External sodium concentration
nAChR	Neuronal nicotinic acetylcholine receptor
NMDA	*N*-methyl-D-aspartate
PhTX	Philanthotoxin-343
PPS	*N*-(4-hydroxy-phenylpropanoyl)-spermine
Put	Putrescine
Spd	Spermidine
Spm	Spermine
$z\delta$	Charge times electrical distance from the cytoplasmic face of membrane for blocking

A. Introduction

In a letter describing spermatozoa to the Royal Society in 1678, Anthonii Lewenhoeck noted crystal formation in drying semen and, in doing so, rather unwittingly made the earliest formal identification of polyamines (LEWENHOECK 1678). Although others subsequently confirmed his findings, it was not until 1888 that the familiar term "spermine" (Spm) was used to describe these crystals (LADENBURG and ABEL 1888). The other naturally occurring polyamines, spermidine (Spd) and putrescine (Put), were later discovered in ox pancreas and decomposing tissues, respectively (Dudley et al. 1927), and, like Spm, were shown to be ubiquitous cell components that had changed little throughout evolution, each being essential for normal growth of eukaryotes as well as prokaryotes (Tabor and Tabor 1965).

Ornithine decarboxylase, the rate-limiting enzyme for biosynthesis of polyamines, catalyses the formation of Put, which, together with *S*-adenosylmethionine, is the substrate for synthesis of Spd and Spm (Marton and Pegg 1995). Spd and Spm are found at millimolar levels in cells, the majority complexed with nucleic acids, proteins and phospholipids. This association has underscored the traditionally accepted role of polyamines in cell growth and differentiation (Marton and Pegg 1995). More recently, however, rectification of inwardly rectifying potassium (K_{ir}) channels as well as kainate, α-amino-3-hydroxy-5-methyl-4-isoxazolepropionic acid (AMPA) and neuronal nicotinic acetylcholine receptors (nAChR; Haghighi and Cooper 1998) has been shown to be due to voltage-dependent blocking by micromolar levels of polyamines in the cytoplasm. These apparently disparate roles for polyamines in cells can be partly accounted for by their positively charged nature at physiological pH, which endows them with a high affinity for a variety of cellular polyanions that exert pronounced effects even at low concentrations. The specificity for interaction with target macromolecules is probably conveyed by the long, rod-shaped polyamine structure that is interspersed with multiple amine and methylene groups that are likely to favor multiple contact points through electrostatic and hydrophobic interactions, respectively (Sugiyama et al. 1996; Cui et al. 1998). In view of their chemical properties, it is not surprising that the electrostatic environment of cation-permeable pores, such as that found in AMPA and kainate-receptor channels, is particularly attractive to polyamine ions, which can enter and block the narrow pathway taken by permeating ions. However, the mechanisms that render AMPA and kainate-receptor channels more sensitive to polyamine block than *N*-methyl-D-aspartate (NMDA) receptors are not yet well understood.

In this review, we will focus on the nature and mechanism of voltage-dependent block of AMPA and kainate receptors by cytoplasmic polyamines. During their evolution, the venoms of spiders and wasps have evolved toxins that capitalize on specific chemical properties of polyamines. Although designed to disarm their prey by binding to peripheral sites, such as glutamate receptors (GluRs) at the neuromuscular junction, polyamine-conjugated toxins have selective effects on L-glutamate (Glu)-activated receptors in the vertebrate central nervous system (CNS). We will discuss the use of arthropod toxins as selective pharmacological tools to study AMPA and kainate receptors and their mechanism of block. Finally, recent studies suggest that cytoplasmic polyamines block both closed and open non-NMDA receptor channels (Bowie et al. 1998; Rozov et al. 1998). We will discuss the possibility that activity-dependent modulation of polyamine block may impart short-term plasticity on neurones expressing calcium-permeable non-NMDA receptors.

B. Emergence of Polyamines as Ubiquitous Channel Blockers

I. Historical Perspective

Strong inward rectification was first described for K^+ conductance in skeletal muscle (KATZ 1949). Sir Bernard Katz described the conductance as an "anomalous" rectifier, the anomaly being that, unlike previously studied K^+ conductances, the membrane conductance decreased with depolarization. Progress in understanding the molecular nature of inward rectification was stimulated by the observation that block by internal tetraethylammonium of delayed rectifier K^+ channels resembled the apparent intrinsic gating behavior of K_{ir}-channels (ARMSTRONG 1969). Although the identity of the blocking molecule(s) was still unknown, HILLE and SCHWARZ (1978) proposed that an impermeant blocker in a permeation model based on Eyring rate theory could adequately account for strong inward rectification, including its strict coupling with external K^+ ions. Taken together, these observations provided a framework for ongoing studies aimed at determining the identity of the blocking molecule.

Rectification of K_{ir} channels and nAChRs was first proposed to be a composite of voltage-dependent block by Mg^{2+} with an as yet unidentified but strongly voltage-dependent gating mechanism (MATSUDA et al. 1987; VANDENBERG 1987; SILVER and DECOURSEY 1990; IFUNE and STEINBACH 1990, 1991, 1992; MATHIE et al. 1990; SANDS and BARISH 1992). Rectification of non-NMDA receptors was also described in early studies (PARKER et al. 1986; RANDLE et al. 1988; BOWIE and SMART 1989; IINO et al. 1990; GILBERTSON et al. 1991); however, a structural basis for the mechanism emerged only with the identification of an amino acid residue, the Q/R site (HUME et al. 1991; VERDOORN et al. 1991), which, through RNA editing, was shown to control rectification as well as Ca^{2+} permeability (SOMMER et al. 1991; HIGUCHI et al. 1993). Although, a report that rectification was lost in outside-out patches suggested that a diffusible, cytoplasmic factor may account for non-NMDA receptor rectification (BURRILL et al. 1993), it was not until cytoplasmic polyamines were shown to block K_{ir} channels in a voltage-dependent manner that a likely candidate was identified (FAKLER et al. 1995; FICKER et al. 1994; LOPATIN et al. 1994).

II. Linking Inward Rectification to Polyamine Block

Consistent with the idea that rectification of non-NMDA receptors is also due to a freely diffusible modulator in the cytoplasm, the complex rectification observed under whole-cell recording conditions is lost in excised outside-out patches but restored by the inclusion of Spm in the patch pipette (Fig. 1) (Bowie and Mayer 1995; Koh et al. 1995; Kamboj et al. 1995; Donevan and Rogawski 1995; Isa et al. 1995). Spm block is concentration- and voltage-dependent and occurs selectively in M2 (Q) unedited recombinant AMPA and kainate receptors (Bowie and Mayer 1995; Kamboj et al. 1995) and in neurones possessing calcium-permeable AMPA receptors which lack GluR-B

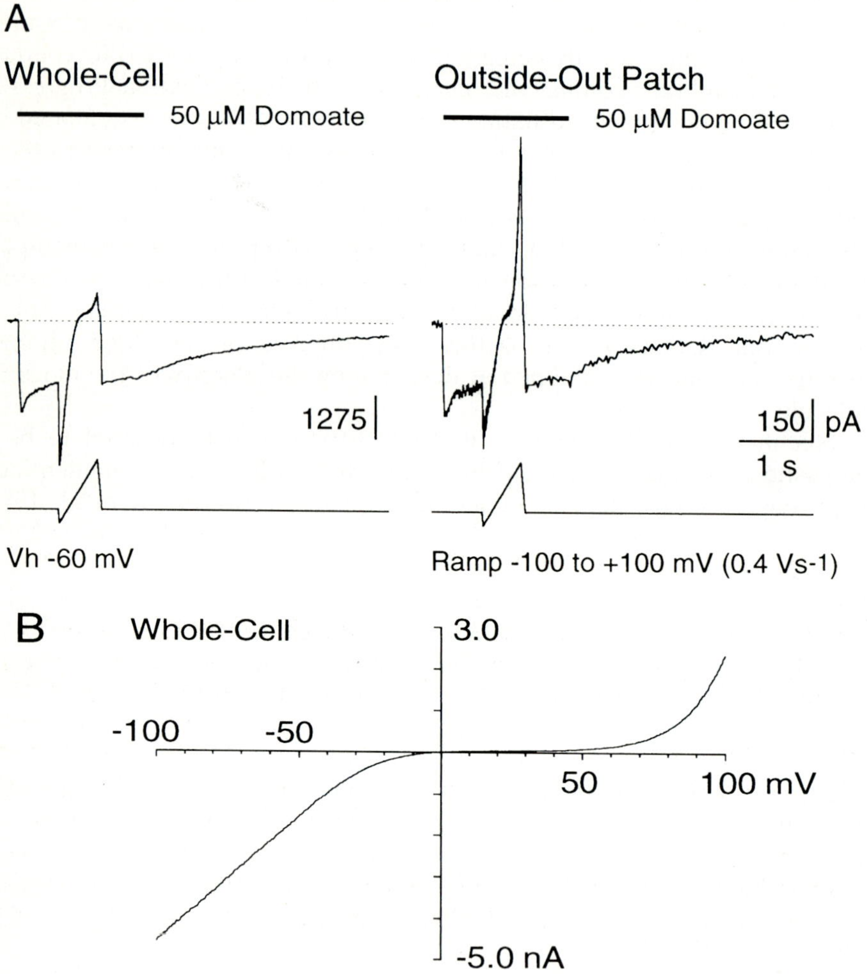

Fig. 1A–D. Inward rectification of kainate receptors is due to block by cytoplasmic polyamines. Strong inward rectification observed under whole-cell (**A**, *left panel*, and **B**) recording conditions is lost in a patch excised from the same HEK 293 cell expressing GluR6(Q) channels (**A**, *right panel*, and **C**). **D** The inclusion of 100 μM spermine to the patch pipette solution restored inward rectification similar to that of whole-cell recordings. Taken from Bowie and Mayer (1995) with permission

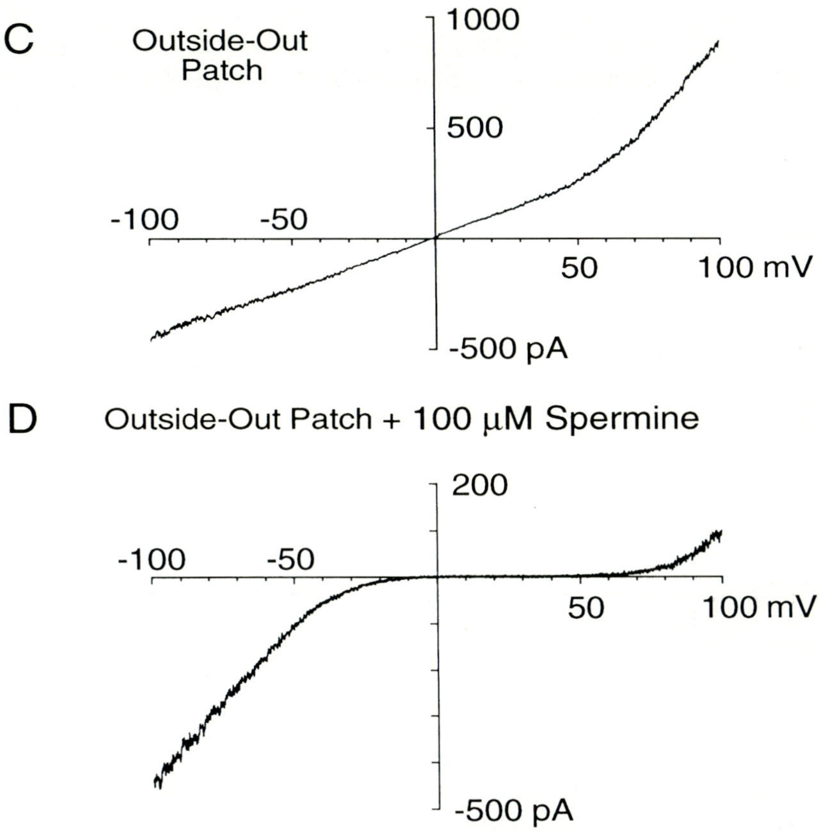

Fig. 1A–D. *Continued*

subunits (Jonas et al. 1994; Koh et al. 1995; Kamboj et al. 1995; Donevan and Rogawski 1995; Isa et al. 1995). Spm is ineffective on M2 (R) edited receptors (Bowie and Mayer 1995) and in neurones expressing calcium-impermeable AMPA receptors that contain the highly edited GluR-B subunit (Jonas et al. 1994; Koh et al. 1995; Kamboj et al. 1995; Donevan and Rogawski 1995). In this latter case, replacement of glutamine residues with positively-charged arginines at the Q/R site is believed to render the electrostatic environment of the pore unfavorable for polyamine block (Dingledine et al. 1992). Other naturally occurring polyamines, such as Spd (K_d at 0 mV = 25 μM) and Put (K_d at 0 mV = 1.2 mM), also cause voltage-dependent block but with a lower affinity than Spm (K_d at 0 mV = 5 μM) (Fig. 2A) (Bowie and Mayer 1995). Knowledge of the affinities of Spm and Spd for AMPA and kainate-receptor channels, combined with biochemical estimates of free cytoplasmic polyamine concentrations (Watanabe et al. 1991), suggests that both Spm and Spd contribute to block observed under whole-cell recording conditions. The

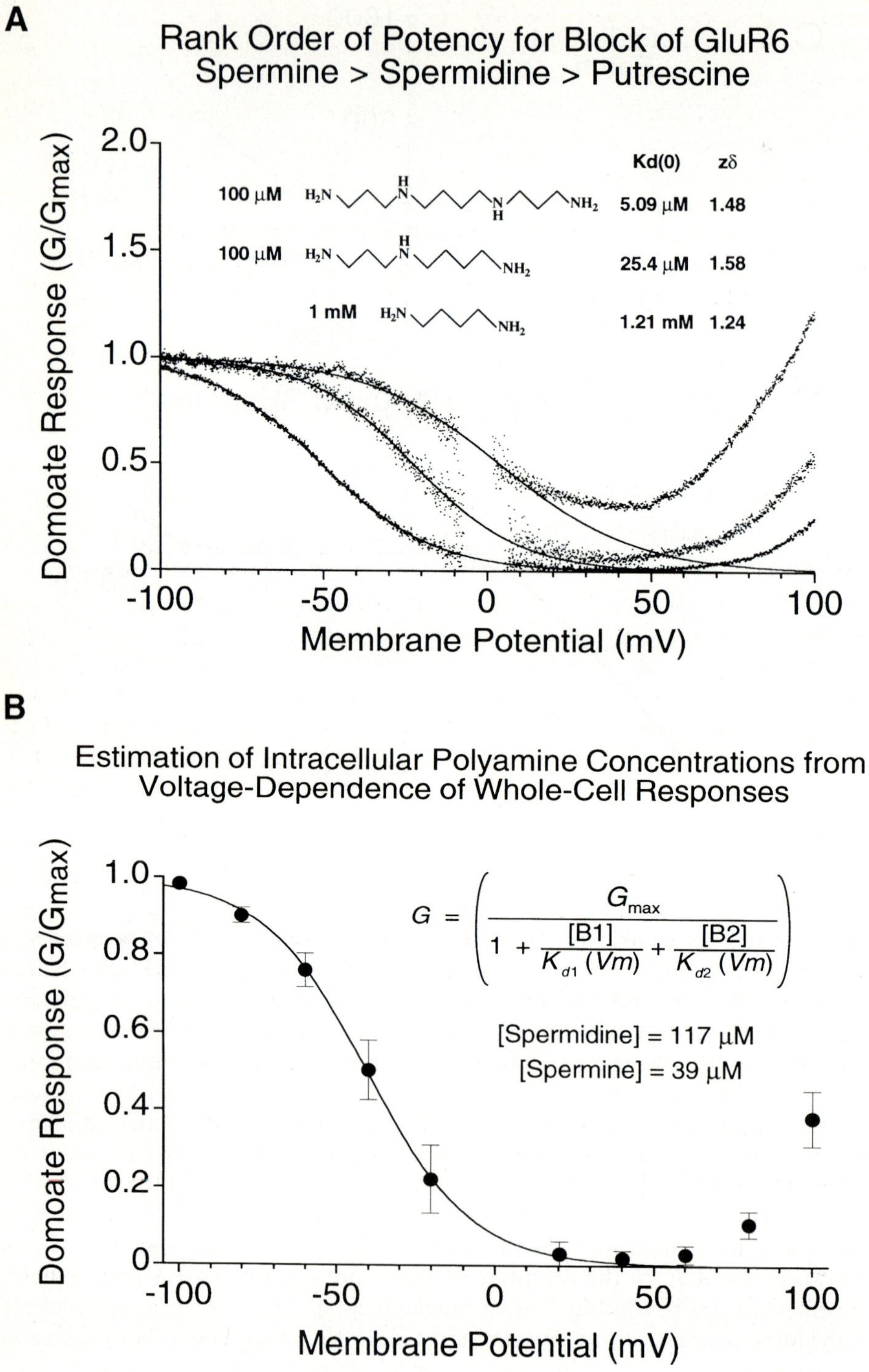
A
Rank Order of Potency for Block of GluR6
Spermine > Spermidine > Putrescine
Kd(0)
zδ
100 μM
5.09 μM
1.48
100 μM
25.4 μM
1.58
1 mM
1.21 mM
1.24
Domoate Response (G/Gmax)
2.0
1.5
1.0
0.5
0
-100
-50
0
50
100
Membrane Potential (mV)
B
Estimation of Intracellular Polyamine Concentrations from Voltage-Dependence of Whole-Cell Responses
Domoate Response (G/Gmax)
1.0
0.8
0.6
0.4
0.2
0
G = (Gmax / (1 + [B1]/Kd1 (Vm) + [B2]/Kd2 (Vm)))
[Spermidine] = 117 μM
[Spermine] = 39 μM
-100
-50
0
50
100
Membrane Potential (mV)

voltage dependences of block observed with Spm ($z\delta = 1.48$) and Spd ($z\delta = 1.58$) are similar to that observed in whole-cell recording conditions, while block by Put is weaker ($z\delta = 1.24$), suggesting that Put does not play a major role (Fig. 2A, inset). The ratio of the free concentrations of Spm to Spd in biochemical estimates from bovine lymphocytes and rat liver cells is 1 :3 (WATANABE et al. 1991). When constrained to this ratio, the free concentrations of Spm and Spd predicted from analysis of the voltage-dependence of whole-cell responses in HEK 293 cells are 40 μM and 120 μM, respectively (Fig. 2B), which is in good agreement with biochemical estimates (WATANABE et al. 1991). As yet, it is not known if the regulation of free-polyamine concentration is a dynamic process; extrusion and uptake mechanisms may be important to consider in this context (KHAN et al. 1994). As discussed in Sect. D, however, the nature of polyamine block is clearly dynamic, but this process relies instead on the activation frequency of the receptor.

III. Relief of Block by Polyamine Permeation

A major difference between polyamine block of K_{ir}-channels and non-NMDA receptors is that relief of block is observed at extreme positive membrane potentials in non-NMDA receptors (Fig. 3). An early explanation for relief of block was that, at high membrane electric-field strengths, cytoplasmic polyamines permeate non-NMDA receptors. The relief of channel block is observed experimentally as the appearance of outward current flow (BOWIE and MAYER 1995; KOH et al. 1995). The different behavior of GluR and K^+ channels is indirectly supported by estimates of their pore dimensions, generally accepted to be smaller for K^+ channels (3-Å diameter) (HILLE 1992) than non-NMDA receptors (7.0–7.8 Å in diameter) (BURNASHEV et al. 1996).

Direct evidence in support of permeation of polyamines through GluR channels has been obtained by comparing internal block by polyamines of differing cross-sectional diameter and in experiments with polyamines as single-charge carriers (BÄHRING et al. 1997). In the first case, the possibility that relief of Spm block is due to its permeation was realized by comparison of Spm block with block produced by the synthetic toxin analogues *N*-(4-hydroxyphenylpropanoyl)-spermine (PPS) and philanthotoxin-343 (PhTX), which have larger cross-sectional diameters (7 Å and 7.5 Å, respectively) than Spm

Fig. 2A,B. Spermine (*Spm*) and spermidine (*Spd*) are responsible for rectification observed in intact cells. **A** Conductance–voltage plots for 100 μM Spm (*left*), 100 μM Spd (*middle*) and 1 mM putrescine (Put; *right*) with fits of the Woodhull model of channel block for a non-permeant blocker. Each polyamine produces voltage-dependent block of GluR6(Q) channels, but with different affinities. Extended structures of Spm (*upper*), Spd (*middle*) and Put (*lower*) are shown in the *inset*. **B** Analysis of whole-cell responses, assuming that Spm and Spd are present in the cytoplasm at a ratio of 1:3, show that the rectification observed in intact HEK 293 cells is due to 40 μM Spm and 120 μM Spd. Adapted from BOWIE and MAYER (1995) with permission

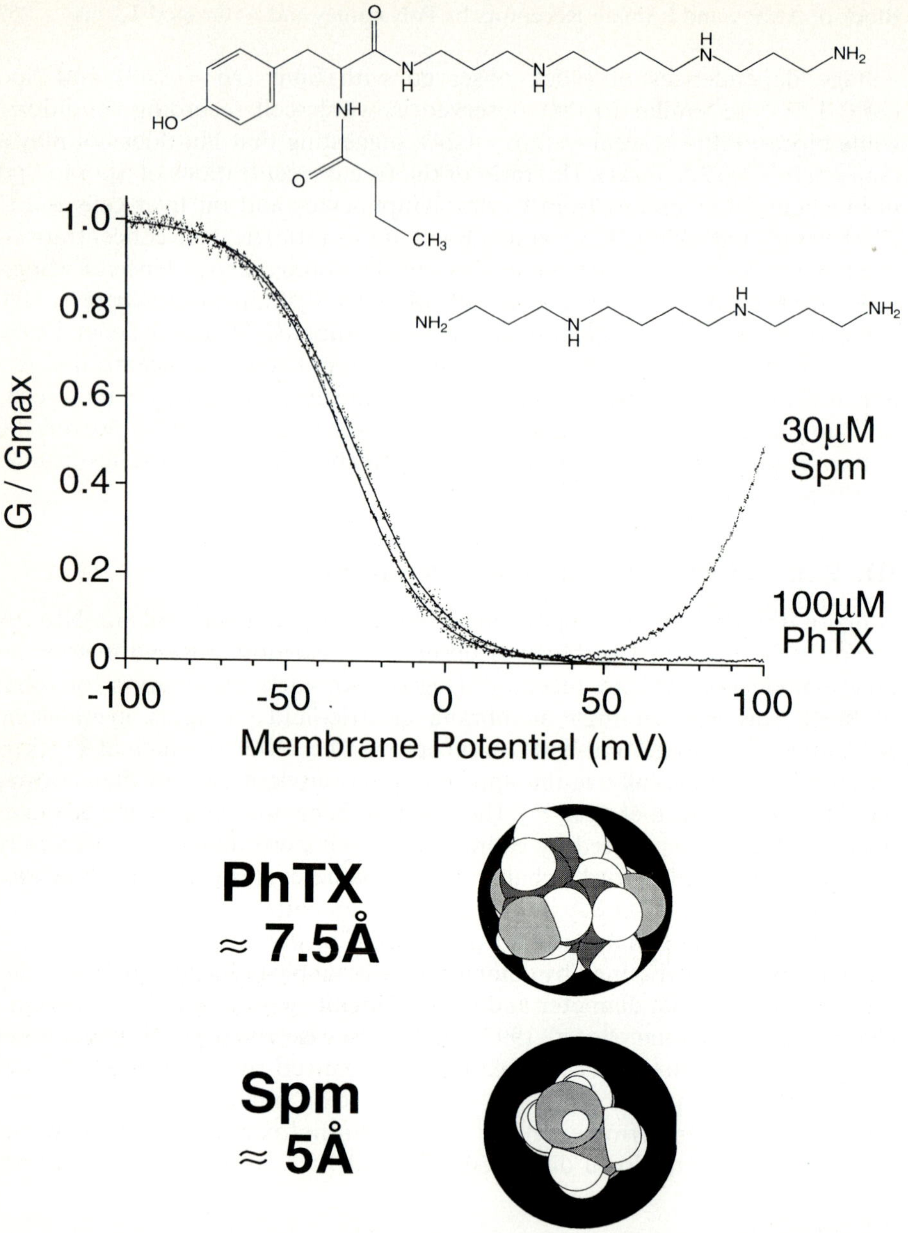

Fig. 3. Relief from block by internal polyamines varies with molecular size. Shown are conductance–voltage plots for internal spermine (Spm; 30 μM) and philanthotoxin-343 (PhTX; 100 μM) with each blocker concentration applied at concentrations giving similar block at negative membrane potentials. Relief from block at positive membrane potentials for Spm was not detectable for PhTX. *Inset* shows the extended structure of PhTX (*upper*) compared with Spm (*lower*). Consistent with these observations, superimposition of Corey–Pauling–Koltum models of PhTX and Spm on *circles* that represent non-NMDA channel dimensions suggest that relief from block is determined by the molecular size of the polyamines. Adapted from Bähring et al. (1997) with permission

(5 Å) and, thus, are less likely to permeate the channel. When Spm, PPS and PhTX concentrations were matched to achieve similar block at negative membrane potentials, the relief of block was much greater for Spm than PPS, and immeasurable for PhTX, consistent with the idea that relief of block for Spm is due to permeation of the blocker molecule (Fig. 3). In the second case, polyamine-ion permeability was ascertained under bi-ionic conditions where all external permeant ions were replaced by either Spm, Spd or Put (BÄHRING et al. 1997). The measurement of inward currents confirmed that Put, Spd and Spm permeate non-NMDA receptor channels with permeability ratios (relative to Na^+) of 0.42, 0.07 and 0.02, respectively (BÄHRING et al. 1997).

C. Internal and External Block by Polyamines and Arthropod Toxins

I. Molecular Determinants of Internal Block

1. Polyamine Structure

Unlike conventional channel blockers, the distributed charges and hydrophobic moieties along the linear structure of polyamines suggests that the blocker molecule may, in fact, interact with multiple contact points on the channel wall as it advances through the permeation pathway. Progress towards identifying which, if not all, of the hydrophilic or hydrophobic groups on the blocker molecule interact with amino acid residues or with other ions in the pore first requires a knowledge of the block mechanism.

Recent voltage-jump-relaxation experiments and dose-response analyses suggest that open-channel block by polyamines exhibits first-order kinetics at a single site. This suggests that, unlike some K_{ir} channels (YANG et al. 1995), the open-channel pore accommodates, at least to a first approximation, only one blocker molecule (BOWIE and MAYER 1995; BOWIE et al. 1998) (Fig. 4A). Transitions into and out of a single blocked state are governed by three molecular events according to the scheme below (BÄHRING et al. 1997; BOWIE et al. 1998):

$$X_{\text{Inside}} \underset{k_{off}}{\overset{[B]k_{on}}{\rightleftharpoons}} X_{\text{Site}} \overset{k_{perm}}{\rightleftharpoons} X_{\text{Outside}}$$

where the rate of binding from the cytoplasm is determined by [B] times the rate constant of binding (k_{on}) and the exit rates are dependent on the rate constant for the return of the blocker to the cytoplasm (k_{off}) or, as described above, the rate constant for the permeation of blocker (k_{perm}). Fitting data with this model for Spm, Spd and PhTX has yielded estimates for their rates of binding and dissociation from the open channel. These estimates indicate that the voltage dependence of block is governed largely by the exit or permeation of the blocker rather than its entry into the channel (BOWIE et al. 1998). Similar to other open-channel blockers, such as local anesthetics at muscle nAChRs

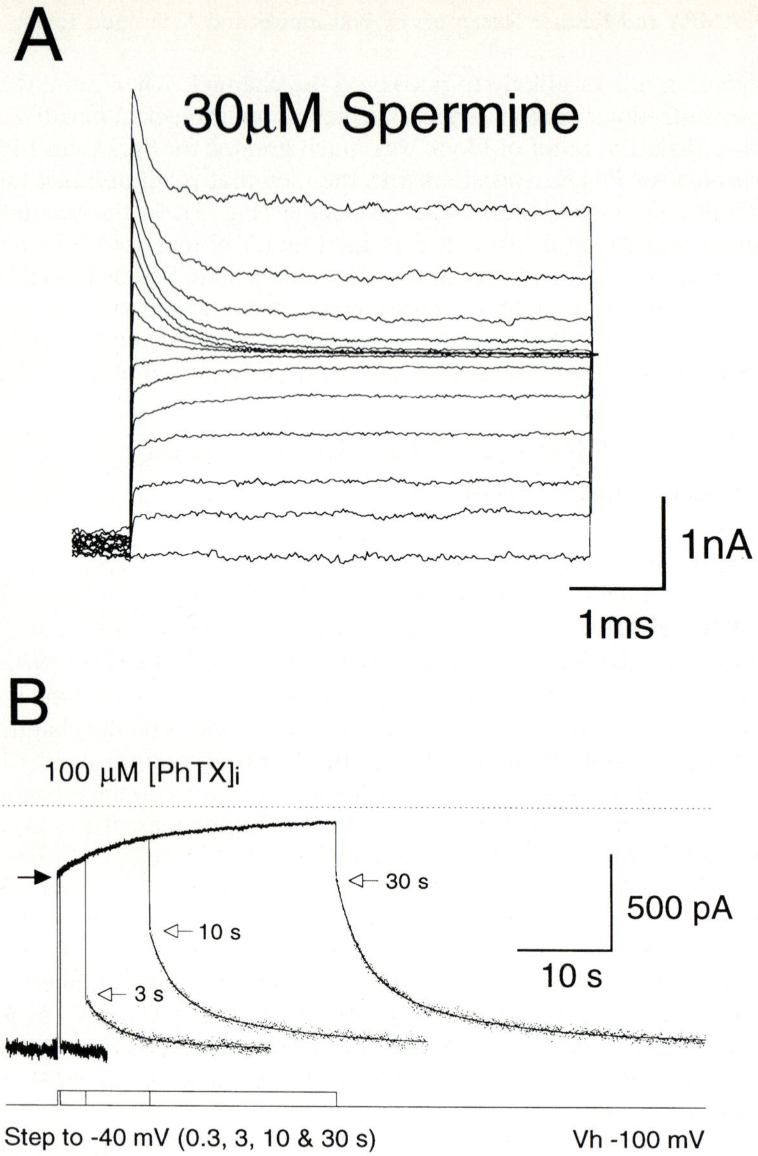

Fig. 4A,B. Open-channel block and trapping mechanisms occur on different time scales. **A** Current relaxations observed following voltage steps from −100 mV to +110 mV (in 15-mV increments), with 30 μM internal spermine, exhibit rapid first order kinetics and reflect the time course of open-channel block for glutamate receptor (GluR) 6(Q). **B** A series of superimposed GluR6(Q) responses following voltage steps from −100 mV to −40 mV in the presence of 100 μM internal philanthotoxin-343 (PhTX); the *black arrow* indicates the level of block predicted by fast ramps like that shown in Fig. 3. Successive voltage steps of increasing duration permit PhTX to remain bound to the channel long enough to destabilize the open state. PhTX is subsequently trapped in the channel, as evidenced by the increasing amplitude of the slow component of block (*open arrows*) and the prolonged period required for complete recovery following repolarization to −100 mV. Taken from Bowie et al. (1998) and Bähring and Mayer (1998) with permission

(NEHER and STEINBACH 1978), the binding rate for Spm ($k_{on} = 48.5\,\mu M^{-1}s^{-1}$) suggests that this process is probably diffusion limited (BOWIE et al. 1998). The slower binding rate for PhTX ($k_{on} = 14.1\,\mu M^{-1}s^{-1}$) suggests that the toxin must enter the channel in a particular conformation for block to occur. Once bound, however, PhTX resides longer than Spd ($k_{on} = 32.6\,\mu M^{-1}s^{-1}$), which is also a trivalent polyamine, suggesting that, in addition, to the polyamine chain, the aromatic moiety of the toxin molecule participates in the stabilizing of the blocked state (BÄHRING and MAYER 1998; BOWIE et al. 1998). As discussed later, in addition to producing low-affinity open-channel block, PhTX is able to trigger entry into a more stable, blocked state, from which recovery requires many seconds.

In addition to blocking the ion-permeation pathway, polyamines also affect gating behavior by destabilizing the open state of the channel (BÄHRING and MAYER 1998; BOWIE et al. 1998). When PhTX, for example, is permitted to remain bound to the open state, the channel enters a conformational state where the toxin molecule becomes trapped, requiring extended recovery periods for complete dissociation (BÄHRING and MAYER 1998) (Fig. 4B). Spm and Spd may also trigger entry into closed blocked states, since channel closure is accelerated by internal polyamines following the removal of Glu (BOWIE et al. 1998). As yet, it is not known if this effect occurs through an allosteric mechanism similar to that reported for PhTX and other sequential channel blockers or if the open state is destabilized by the long, rod-like polyamine structure emptying the pore of permeant ions, similar to the effect of use-dependent blockers on K^+ channels (BAUKROWITZ and YELLEN 1996).

Surprisingly, the role of hydrophobic binding is more important than might be at first anticipated for the strongly voltage-dependent block produced by highly charged, naturally occurring polyamines. A recent comparison of block produced by a series of tetravalent polyamines with different numbers and spacings of methylene groups revealed that hydrophobic interactions strongly influence the blockers' affinities, whereas different spacing, between amine groups had little effect (Fig. 5) (CUI et al. 1998). These results suggest that, in addition to charged residues in the permeation pathway, the hydrophobic environment of the pore may be a major factor in determining the extent of polyamine block.

2. Critical Pore-Lining Residues

Although recent evidence suggests that hydrophobic amino acids in the pore may be important in polyamine binding, most mutagenesis studies have concentrated on two residues: the Q/R site and a conserved ring of negative charges four amino acid residues downstream. Insight into structural elements important for polyamine block followed the identification of RNA editing in M2 at the Q/R site, which contains either a neutral glutamine or a positively charged arginine (HUME et al. 1991; VERDOORN et al. 1991). The Q/R site is believed to contribute to the channel's selectivity filter, since it determines a

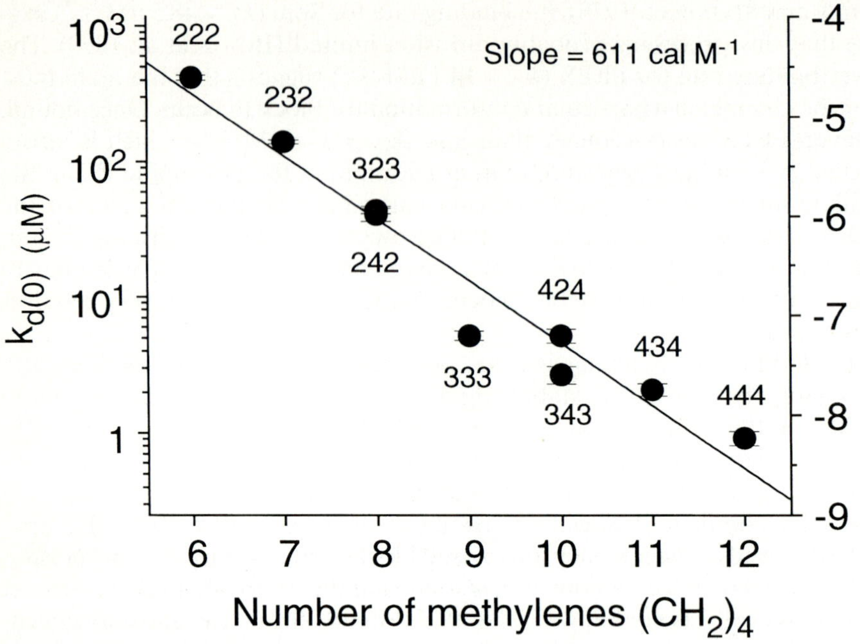

Fig. 5. Plot of the affinity of a series of tetravalent polyamines for glutamate receptor (GluR) 6(Q) block versus the number of methylene groups. The three digits above each data point denote the number of methylenes separating each amine group. A linear fit through the data suggests that polyamine block varies with the number of methylene groups, corresponding to a 2.9-fold increase in affinity per methylene group. Taken from Cui et al. (1998)

number of permeation properties, including block by cytoplasmic polyamines (Bowie and Mayer 1995), calcium (Ca^{2+}) permeability (Jonas and Burnashev 1995), anion versus cation permeability (Burnashev et al. 1996) and single-channel conductance (Howe 1996; Swanson et al. 1996, 1997). Nevertheless, some of these properties can be uncoupled from one another, since mutagenesis of the Q/R site has been shown to remove inward rectification whilst preserving divalent permeability (Burnashev et al. 1992; Curutchet et al. 1992; Dingledine et al. 1992). Consistent with this idea, divalent permeability in heteromeric receptors is more sensitive to replacement of glutamine residues by arginines than internal polyamine block (Washburn et al. 1997).

In addition to the Q/R site, a negatively charged aspartate, four amino acid residues downstream, is also important for internal polyamine block and divalent permeability (Dingledine et al. 1992). Replacement of the negative charge of aspartate with a neutral asparagine of similar size reduces internal polyamine block, with no obvious influence on divalent permeability (Dingledine et al. 1992). A model involving the Q/R site and the downstream aspartate residue has been proposed to account for the differential dependence of divalent permeability and internal polyamine block of open non-

NMDA receptor channels (WASHBURN et al. 1997). In unedited channels, glutamine residues at the Q/R site are believed to form a ring of carbonyl oxygens that contributes to or constitutes the binding site for external divalent ions. Cytoplasmic polyamines entering the channel also interact with this ring and, through electrostatic interactions with amine groups on the blocker molecule, with the carboxyl groups of the downstream aspartate. In heteromeric receptors, the sequential replacement of glutamine residues by arginines at the Q/R site has been proposed to neutralize negative charges in the pore by the formation of salt bridges between the guanidinium group of arginine and the carboxyl group of aspartate. This, apparently, disrupts the divalent ion-binding site to a greater extent than for polyamines but, as more arginines replace each glutamine, the electrostatic environment of the pore becomes increasingly unfavorable for polyamine block (WASHBURN et al. 1997).

3. Interaction with Permeant Ions

Internal polyamines entering the pore interact strongly with permeant ions entering the channel from either the cytoplasm or from the external side of the membrane (BÄHRING et al. 1997; BOWIE et al. 1998) (Fig. 6). A similar interaction between the external potassium concentration ($[K^+]_{ext}$) and K_{ir} channels has been known for some time; in this interaction, the degree of inward rectification, or the blocker's affinity, was found to depend on $[K^+]_{ext}$ rather than the membrane potential *per se* (HILLE 1992). The reduction in polyamine affinity with increasing $[K^+]_{ext}$ can be explained by Eyring rate theory, assuming a strong electrostatic interaction between an external binding site for K^+-ions and an inner site for an impermeable blocker (HILLE 1992). A similar explanation for the reduction in internal polyamine block observed when $[Na^+]_{ext}$ is increased is less obvious for non-NMDA receptors since, unlike K_{ir} channels, they do not appear to act as multi-ion pores (WOLLMUTH and SAKMANN 1998). Recently, BOWIE et al. (1998) described the effect of membrane potential on polyamine block of kainate receptors linked to ion flux and suggested that the development of strong voltage-dependent block following depolarization results from the reduction in driving force of inward ion flux through open channels (Fig. 6C). As yet, these observations have not been modeled quantitatively; however, unlike the case for conventional permeation models, which treat the blocker as a point charge, it may be necessary to consider that the energy transmitted between permeant ions and blocker molecules is positively correlated with the volume of the ion absorbing the energy. Furthermore, a large part of the voltage dependence of polyamine block might be indirect and may reflect coupling to the movement of permeant ions (OLIVER et al. 1998).

II. External Block by Spermine and Arthropod-Toxin Molecules

One prediction from the observation that internal polyamines permeate non-NMDA channels is that their binding site(s) can be accessed by polyamines

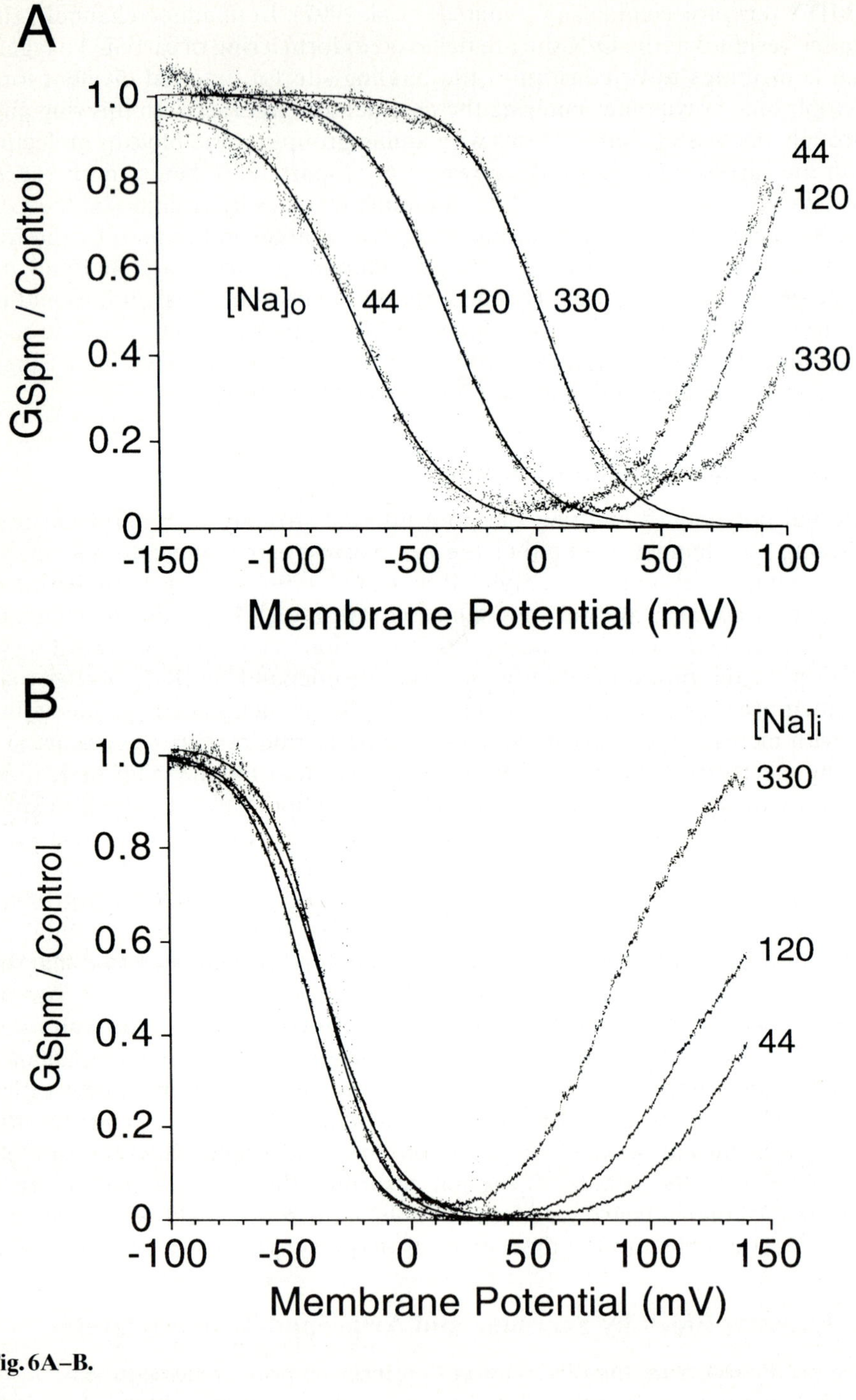
A
1.0
0.8
0.6
0.4
0.2
0
GSpm / Control
[Na]o
44
120
330
44
120
330
-150
-100
-50
0
50
100
Membrane Potential (mV)
B
[Na]i
330
120
44
1.0
0.8
0.6
0.4
0.2
0
GSpm / Control
-100
-50
0
50
100
150
Membrane Potential (mV)

Fig. 6A–B.

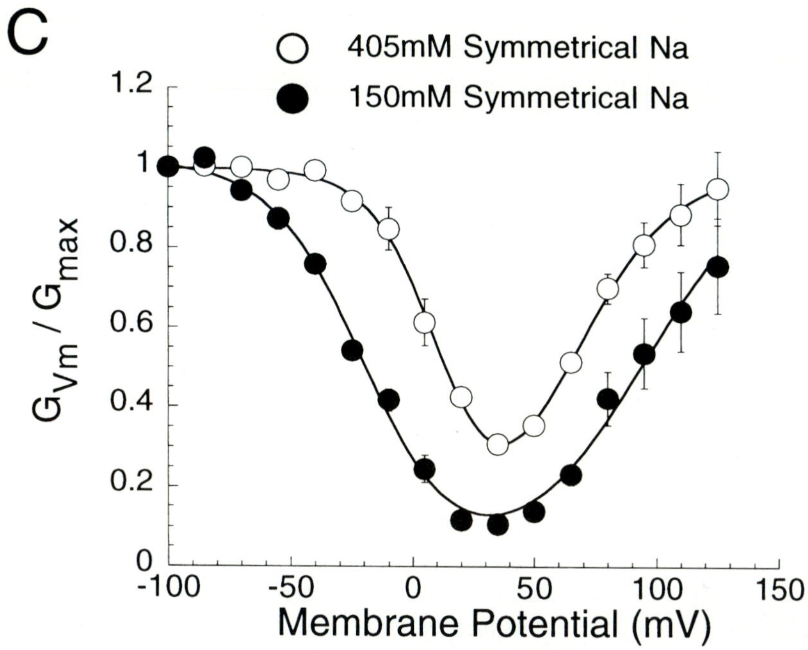

Fig. 6A–C. External and internal permeant ions strongly influence block by internal spermine (Spm). **A** Conductance–voltage relationship for block by 30 μM internal Spm is shifted to the right, with increasing external sodium concentration reflecting a decrease in Spm affinity at negative potentials. **B** In contrast, increases in internal sodium concentration have almost no effect on block at negative potentials, but increase relief from block at positive potentials. **C** Symmetrical increases in external and internal Na from 150 mM to 405 mM cause a concomitant decrease in block for 20 μM Spm at negative and positive potentials. Taken from Bähring et al. (1997) and Bowie et al. (1998) with permission

entering the channel from either side of the membrane. In agreement with this prediction, externally applied Spm produces voltage-dependent block of both AMPA and kainate receptors (Washburn and Dingledine 1996; Washburn et al. 1997; Bähring et al. 1997). Block is controlled by editing of the Q/R site (Washburn and Dingledine 1996; Washburn et al. 1997), as expected if some of the molecular determinants of internal and external block are shared. External block, however, is weaker and much less voltage dependent, which may reflect greater permeation of external polyamines (Bähring et al. 1997). At this time, the molecular basis of asymmetry between internal and external block is not fully understood; however, it is likely to be governed by the location of the barriers and wells that control the binding of permeant ions and polyamines (Bähring et al. 1997).

Although influx and efflux pathways for polyamines from mammalian cells have been described (Khan et al. 1994), a clear physiological role for block by external polyamines remains to be elucidated. In contrast, externally applied polyamine-amide toxins contained in the venoms of spiders and wasps are employed as offensive agents to kill or paralyze other insect prey via their actions on invertebrate ion channels (Usherwood and Blagbrough 1991). Although these toxin molecules are aimed at peripheral targets, such as the invertebrate glutamatergic neuromuscular junction and cholinergic synapses, polyamine amide toxins have subsequently been shown to block GluRs in the mammalian CNS (Usherwood and Blagbrough 1991; Jackson and Usherwood 1988; Jackson and Parks 1989). To date, three main toxin molecules and their synthetic analogues have been used by a variety of investigators to block vertebrate GluRs at nanomolar concentrations: argiotoxin (ATX) and joro spider toxin (JSTX) from orb-web and joro spiders, respectively, and PhTX from the digger wasp, *Philanthus triangulum*. Early studies of native GluRs suggested that ATX was selective for NMDA receptors (Priestley et al. 1989; Draguhn et al. 1991), whereas PhTX was selective for non-NMDA receptors (Jones et al. 1990). Subsequent analysis of recombinant AMPA receptors revealed, however, that both toxins have similar affinities and that, similar to external Spm block, toxin affinity is dependent on editing at the Q/R site (Herlitze et al. 1993; Washburn and Dingledine 1996; Blaschke et al. 1993; Brackley et al. 1993). In retrospect, the selectivity of ATX for native NMDA versus non-NMDA receptors probably reflects widespread expression of the highly edited GluR-B subunit. In view of the cell-specific expression of edited and unedited AMPA receptors in the CNS, polyamine-amide toxins have since proved useful pharmacological tools in determining the subunit composition of native receptors (Iino et al. 1996; Tóth and McBain 1998; Haverkampf et al. 1997).

Polyamine-amide toxins block AMPA- and kainate receptors primarily by a use- and voltage-dependent mechanism (Herlitze et al. 1993; Brackley et al. 1993; Bähring and Mayer 1998); however, some studies have also reported a weak potentiating action (Brackley et al. 1990; Ragsdale et al. 1989). Bähring and Mayer (1998) have shown that the use- and voltage-dependent action of external PhTX is due to an open-channel block mechanism, where access into and out of the blocking site in the pore is possible only when the channel is in the open state. The open state, however, becomes less stable if PhTX remains bound to the channel, and this allosteric mechanism has been proposed to account for closing and subsequent trapping of PhTX in the channel (Bähring and Mayer 1998).

D. Physiological Implications of Cytoplasmic Polyamine Block

Recent studies suggest that polyamines also access the pore in the closed conformation (Bowie et al. 1998; Rozov et al. 1998). Such a blocking mechanism may provide the molecular basis for novel activity regulation of calcium-permeable AMPA and kainate receptors in the CNS. Evidence suggesting that cytoplasmic polyamines may access closed AMPA and kainate receptors was proposed from the observation that the rise times of responses to rapid application of Glu were slowed in a voltage-dependent manner in the presence of Spm (Bowie et al. 1998; Rozov et al. 1998), and that paired responses to brief applications of Glu showed facilitation, reflecting relief from closed-channel block (Rozov et al. 1998). Bowie et al. (1998) suggested that the slowing of rise times reflected voltage-dependent re-equilibration of Spm binding with the open state of the channel, since rise times were faster in high-permeant-ion solutions, which lower Spm's affinity and, thus, curtail its residency time on the binding site. Furthermore, these observations were faithfully repro-

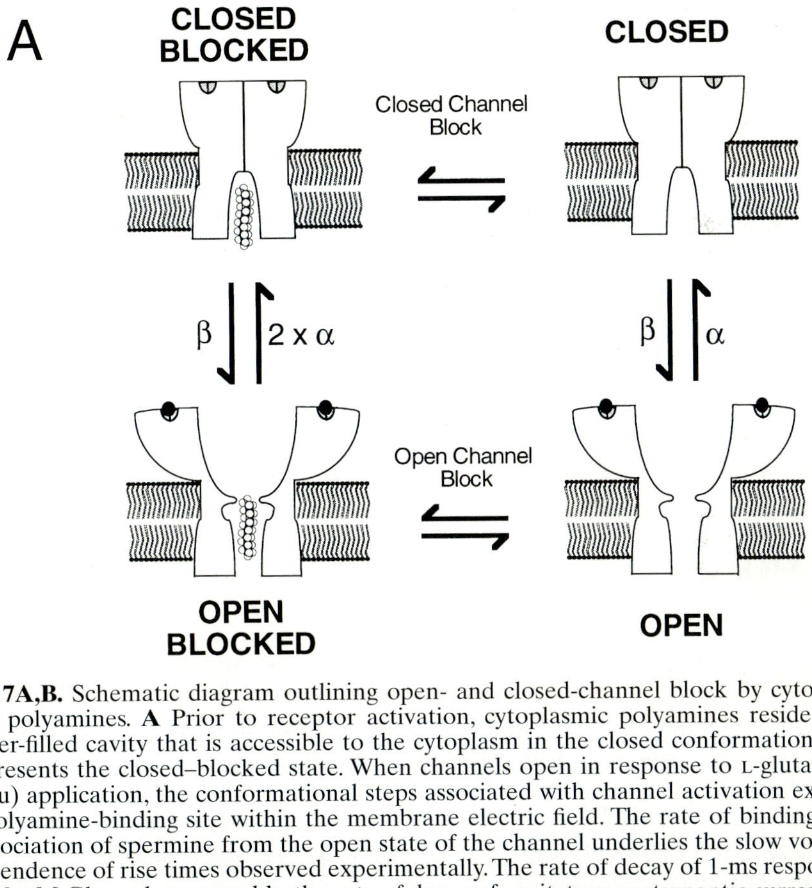

Fig. 7A,B. Schematic diagram outlining open- and closed-channel block by cytoplasmic polyamines. **A** Prior to receptor activation, cytoplasmic polyamines reside in a water-filled cavity that is accessible to the cytoplasm in the closed conformation; this represents the closed–blocked state. When channels open in response to L-glutamate (Glu) application, the conformational steps associated with channel activation expose a polyamine-binding site within the membrane electric field. The rate of binding and dissociation of spermine from the open state of the channel underlies the slow voltage dependence of rise times observed experimentally. The rate of decay of 1-ms responses to 10 mM Glu and, presumably, the rate of decay of excitatory postsynaptic currents at synapses with calcium-permeable non-*N*-methyl-D-aspartate receptors reflects primarily the closing rate of open and open–blocked channels as they relax into closed and closed–blocked states, respectively. **B** Simulations from a kinetic model show that repetitive stimulation leads to potentiation of the Glu-response amplitude as more channels are recruited from the closed–blocked state (*RB*) into the closed state (*R*). Taken from Bowie et al. (1998) with permission

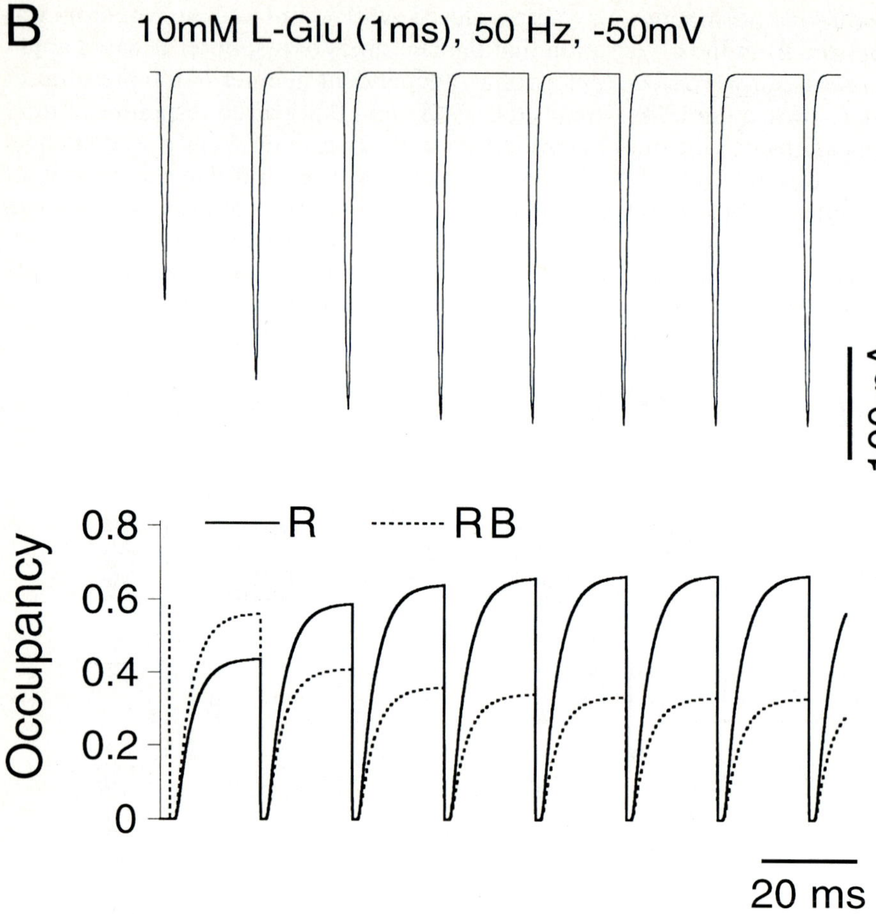

Fig. 7A,B. *Continued*

duced in a kinetic model that included closed-channel block (BOWIE et al. 1998). Similar experimental observations were made by Rozov et al. (1998) when desensitization of AMPA receptors was removed by cyclothiazide. Occupancy of closed states by Spm appeared to be insensitive to membrane potential, which may suggest that polyamines bind to a water-filled cavity in the channel and are, therefore, shielded from the potential drop across the membrane (Fig. 7A) (BOWIE et al. 1998). In contrast, re-block of closed channels is voltage-sensitive, following first-order kinetics with a time constant (at –60 mV) of 375 ms (Rozov et al. 1998). The relatively slow re-block of closed AMPA receptors by polyamines accounts for the facilitation of response amplitude observed following a train of pulses to 10 mM Glu (Rozov et al. 1998); this effect can be reproduced in a kinetic model of open- and closed-channel block of kainate receptors (Fig. 7A, B) (BOWIE et al. 1998). It is not yet known whether activity-dependent modulation of polyamine block may represent a novel form of short-term plasticity in the CNS. Basket cells of the dentate gyrus, for example, which express Ca^{2+}-permeable AMPA receptors, exhibit rapid gating characteristics (GEIGER et al. 1995; GEIGER et al. 1997). These characteristics have been proposed to be pivotal in defining the functional roles fulfilled by basket cells in hippocampal-network activity, such as oscillations (JEFFERYS et al. 1996) and feed-forward and feedback inhibition (BUZSÁKI and CHROBAK 1995). It is interesting that interneuron-doublet firing, which permits long-range synchronization of gamma oscillations in the cortex, is strongly dependent on the unitary AMPA-receptor conductance of basket cells (TRAUB et al 1996). In view of this, one attractive goal of future studies would be to determine whether the short-term increase in AMPA-receptor amplitude, due to polyamine unblock, modulates the generation of long-range, synchronous oscillations in the cortex. Although Ca^{2+} permeability and gating properties are undoubtedly important factors in sculpting neuronal behavior, these recent findings may suggest that cytoplasmic polyamines also fulfill a novel role in synaptic plasticity.

References

Armstrong CM (1969) Inactivation of the potassium conductance and related phenomena caused by quarternary ammonium ion injected in squid axons. J Gen Physiol 54:553–575

Bähring R, Bowie D, Benveniste M, Mayer ML (1997) Permeation and block of rat GluR6 glutamate receptor channels by internal and external polyamines. J Physiol (Lond) 502:575–589

Bähring R, Mayer ML (1998) An analysis of philanthotoxin block for recombinant rat GluR6(Q) glutamate receptor channels. J Physiol (Lond) 509: 635–650.

Baukrowitz T, Yellen G (1996) Use-dependent blockers and exit rate of the last ion from the multi-ion pore of a K+ channel. Science 271:653–656

Blaschke M, Keller BU, Rivosecchi R, Hollmann M, Heinemann S, Konnerth A (1993) A single amino acid determines the subunit-specific spider toxin block of alpha-amino-3-hydroxy-5-methylisoxazole-4-propionate/kainate receptor channels. Proc Natl Acad Sci USA 90:6528–6532

Bowie D, Smart TG (1989) Properties of a kainate receptor expressed in Xenopus laevis oocytes with bovine brain mRNA. J Physiol (Lond) 418:196P(Abstract)

Bowie D, Mayer ML (1995) Inward rectification of both AMPA and kainate subtype glutamate receptors generated by polyamine-mediated ion channel block. Neuron 15:453–462

Bowie D, Lange GD, Mayer ML (1998) Activity-dependent modulation of glutamate receptors by polyamines. J Neurosci 18:8175–8185

Brackley P, Goodnow R, Jr, Nakanishi K, Sudan HL, Usherwood PN (1990) Spermine and philanthotoxin potentiate excitatory amino acid responses of Xenopus oocytes injected with rat and chick brain RNA. Neurosci Lett 114:51–56

Brackley PT, Bell DR, Choi SK, Nakanishi K, Usherwood PN (1993) Selective antagonism of native and cloned kainate and NMDA receptors by polyamine-containing toxins. J Pharmacol Exp Ther 266:1573–1580

Burnashev N, Monyer H, Seeburg PH, Sakmann B (1992) Divalent ion permeability of AMPA receptor channels is dominated by the edited form of a single subunit. Neuron 8:189–198

Burnashev N, Villarroel A, Sakmann B (1996) Dimensions and ion selectivity of recombinant AMPA and kainate receptor channels and their dependence on Q/R site residues. J Physiol (Lond) 496:165–173
Burrill JD, McCullum CB, Hume RI (1993) Mechanism of rectification in the current-voltage relationship of non-NMDA receptor channels. Soc Neurosci Abstr 19:279(Abstract)
Buzsáki G, Chrobak JJ (1995) Temporal structure in spatially organized neuronal ensembles: a role for interneuronal networks. Curr Opin Neurobiol 5:504–510
Cui C, Bähring R, Mayer ML (1998) The role of hydrophobic interactions in binding of polyamines to non NMDA receptor ion channels. Neuropharm 37:1381–1391
Curutchet P, Bochet P, Prado de Carvalho L, Lambolez B, Stinnakre J, Rossier J (1992) In the GluR1 glutamate receptor subunit a glutamine to histidine point mutation suppresses inward rectification but not calcium permeability. J Neurosci 182: 1089–1093
Dingledine R, Hume RI, Heinemann SF (1992) Structural determinants of barium permeation and rectification in non-NMDA glutamate receptor channels. J Neurosci 12:4080–4087
Donevan SD, Rogawski MA (1995) Intracellular polyamines mediate inward rectification of Ca^{2+}-permeable α-amino-3-hydroxy-5-methyl-4-isoxazoleproprionic acid receptors. Proc Natl Acad Sci USA 92:9298–9302
Draguhn A, Jahn W, Witzemann V (1991) Argiotoxin636 inhibits NMDA-activated ion channels expressed in Xenopus oocytes. Neurosci Lett 132:187–190
Dudley HW, Rosenheim O, Starling WW (1927) The constitution and synthesis of spermidine, a newly discovered base isolated from animal tissues. Biochem J 21:97–103
Fakler B, Brändle U, Glowatzki E, Weidemann S, Zenner HP, Ruppersberg JP (1995) Strong voltage-dependent inward rectification of inward rectifier K+ channels is caused by intracellular spermine. Cell 80:149–154
Ficker E, Taglialatela M, Wible BA, Henley CM, Brown AM (1994) Spermine and spermidine as gating molecules for inward rectifier K+ channels. Science 266:1068–1072
Geiger JRP, Melcher T, Koh DS, Sakmann B, Seeburg PH, Jonas P, Monyer H (1995) Relative abundance of subunit mRNAs determines gating and Ca^{2+} permeability of AMPA receptors in principle neurons and interneurons in rat CNS. Neuron 15:193–204
Geiger JRP, Lübke J, Roth A, Frotscher M, Jonas P (1997) Submillisecond AMPA receptor-mediated signaling at a principal neuron-interneuron synapse. Neuron 18:1009–1023
Gilbertson TA, Scobey R, Wilson M (1991) Permeation of calcium ions through non-NMDA glutamate channels in retinal bipolar cells. Science 251:1613–1615
Haghighi AP, Cooper E (1998) Neuronal nicotinic acetylcholine receptors are blocked by intracellular spermine in a voltage-dependent manner. J Neurosci 18:4050–4062
Haverkampf K, Lübke J, Jonas P (1997) Single-channel properties of native AMPA receptors depend on the putative subunit composition. Soc Neurosci Abstr 478.23(Abstract)
Herlitze S, Raditsch M, Ruppersberg JP, Jahn W, Monyer H, Schoepfer R, Witzemann V (1993) Argiotoxin detects molecular differences in AMPA receptor channels. Neuron 10:1131–1140
Higuchi M, Single FN, Köhler M, Sommer B, Sprengel R, Seeburg PH (1993) RNA editing of AMPA receptor subunit GluR-B: a base-paired intron-exon structure determines position and efficiency. Cell 75:1361–1370
Hille B, Schwarz W (1978) Potassium channels as multi-ion single pores. J Gen Physiol 72:409–442
Hille, B (1992) Ionic channels of excitable membranes, Sunderland: Sinauer Associates, Ed. 2
Howe JR (1996) Homomeric and heteromeric ion channels formed from kainate-type subunits GluR6 and KA2 have very small, but different, unitary conductances. J Neurophysiol 76:510–519

Hume RI, Dingledine R, Heinemann SF (1991) Identification of a site in glutamate receptor subunits that controls calcium permeability. Science 253:1028–1031

Ifune CK, Steinbach JH (1990) Rectification of acetylcholine-elicited currents in PC12 phaeochromocytoma cells. Proc. Natl. Acad. Sci. USA 87:4794–4798

Ifune CK, Steinbach JH (1991) Voltage-dependent block by magnesium of neuronal nicotinic acetylcholine receptor channels in rat phaeochromocytoma cells. Proc. Natl. Acad. Sci. USA 87:4794–4798

Ifune CK, Steinbach JH (1992) Inward rectification of acetylcholine-elicited currents in PC12 phaeochromocytoma cells. J Physiol (Lond) 457:143–165

Iino M, Ozawa S, Tsuzuki K (1990) Permeation of calcium through excitatory amino acid receptor channels in cultured rat hippocampal neurons. J Physiol (Lond) 424:151–165

Iino M, Koike M, Isa T, Ozawa S (1996) Voltage-dependent blockage of $Ca(^{2+})$-permeable AMPA receptors by joro spider toxin in cultured rat hippocampal neurons. J Physiol (Lond) 496:431–437

Isa T, Iino M, Itazawa S, Ozawa S (1995) Spermine mediates inward rectification of $Ca(^{2+})$-permeable AMPA receptor channels. Neuroreport 6:2045–2048

Jackson H, Usherwood PNR (1988) Spider toxins as tools for dissecting elements of excitatory amino acid transmission. Trends Neurosci 11:278–283

Jackson H, Parks TN (1989) Spider toxins: recent applications in neurobiology. Ann Rev Neurosci 12:405–414

Jefferys JGR, Traub RD, Whittington MA (1996) Neuronal networks for induced '40kHz' rhythms. Trends Neurosci 19:202–208

Jonas P, Racca C, Sakmann B, Seeburg PH, Monyer H (1994) Differences in Ca^{2+} permeability of AMPA-type glutamate receptor channels in neocortical neurons caused by differential GluR-B subunit expression. Neuron 12:1281–1289

Jonas P, Burnashev N (1995) Molecular mechanisms controlling calcium entry through AMPA-type glutamate receptor channels. Neuron 15:987–990

Jones MG, Anis NA, Lodge D (1990) Philanthotoxin blocks quisqualate-, AMPA- and kainate-, but not NMDA-, induced excitation of rat brainstem neurons in vivo. Br J Pharmacol 101:968–970

Kamboj SK, Swanson GT, Cull-Candy SG (1995) Intracellular spermine confers rectification on rat calcium-permeable AMPA and kainate receptors. J Physiol (Lond) 486:297–303

Katz B (1949) Les constantes électriques de la membrane du muscle. Arch Sci Physiol 2:285–299

Khan NA, Quemener V, Moulinoux JP (1994) In: Carter C (ed) The Neuropharmacology of Polyamines. Academic Press, London, pp 37–60

Koh DS, Burnashev N, Jonas P (1995) Block of native Ca^{2+}-permeable AMPA receptors in rat brain by intracellular polyamines generates double rectification. J Physiol (Lond) 486:305–312

Ladenburg A, Abel J (1888) Ueber das Aethylenimin (Spermin?). Ber dtsch chem Ges 21:758–766

Lewenhoeck A (1678) Observationes D. Anthonii Lewenhoeck, de Natis è semine genitali Animalculis. Phil Trans R Soc Lond 12:1040–1042

Lopatin AN, Makhina EN, Nichols CG (1994) Potassium channel block by cytoplasmic polyamines as the mechanism of intrinsic rectification. Nature Lond 372:366–369

Marton LJ, Pegg AE (1995) Polyamines as targets for therapeutic intervention. Ann Rev Pharmacol Toxicol 35:55–91

Mathie A, Colquhoun D, Cull-Candy SG (1990) Rectification of currents activated by nicotinic acetylcholine receptors in rat sympathetic ganglion neurones. J. Physiol (Lond.) 427:625–655

Matsuda H, Saigusa A, Irisawa H (1987) Ohmic conductance through the inwardly rectifying K channel and blocking by internal Mg^{2+}. Nature Lond 325:156–159

Neher E, Steinbach JH (1978) Local anaesthetics transiently block currents through single acetylcholine-receptor channels. J Physiol (Lond) 277:153–176

Oliver D, Hahn H, Antz C, Ruppersberg JP, Fakler B (1998) Interaction of permeant and blocking ions in cloned inward-rectifier K+ channels. Biophys J 74:2318–2326

Parker I, Sumikawa K, Miledi R (1986) Messenger RNA from retina induces kainate and glycine receptors in Xenopus oocytes. Proc. R. Soc. Lond. B 225:99–106

Priestley T, Woodruff GN, Kemp JA (1989) Antagonism of responses to excitatory amino acids on rat cortical neurons by spider toxin, argiotoxin636. Br J Pharmacol 97:1315–1323

Ragsdale D, Gant DB, Anis NA, Eldefrawi AT, Eldefrawi ME, Konno K, Miledi R (1989) Inhibition of rat brain glutamate receptors by philanthotoxin. J Pharmacol Exp Ther 251:156–163

Randle JCR, Vernier P, Garrigues AM, Brault E (1988) Properties of the kainate channel in rat brain mRNA injected Xenopus oocytes: ionic selectivity and blockage. Mol Cell Biochem 80:121–132

Rozov A, Zilberter Y, Wollmuth LP, Burnashev N (1998) Facilitation of currents through Ca2+-permeable AMPAR channels by activity-dependent relief from polyamine block. J Physiol (Lond) 511.2:361–377

Sands SB, Barish ME (1992) Neuronal nicotinic acetylcholine receptor currents in phaeochromocytoma (PC12) cells: dual mechanisms of rectification. J Physiol (Lond) 447:467–487

Silver MR, DeCoursey TE (1990) Intrinsic gating of inward rectifier in bovine pulmonary artery endothelial cells in the presence or absence of Internal Mg^{2+}. J Gen Physiol 96:109–133

Sommer B, Köhler M, Sprengel R, Seeburg PH (1991) RNA editing in brain controls a determinant of ion flow in glutamate-gated channels. Cell 67:11–19

Sugiyama S, Matsuo Y, Maenaka K, Vassylyev DG, Matsushima M, Kashiwagi K, Igarashi K, Morikawa K (1996) The 1.8-Å X-ray structure of the *Escherichia coli* PotD protein complexed with spermidine and the mechanism of polyamine binding. Protein Science 5:1984–1990

Swanson GT, Feldmeyer D, Kaneda M, Cull-Candy SG (1996) Effect of RNA editing and subunit co-assembly on single-channel properties of recombinant kainate receptors. J Physiol (Lond) 492:129–142

Swanson GT, Kamboj SK, Cull-Candy SG (1997) Single-channel properties of recombinant AMPA receptors depend on RNA editing, splice variation, and subunit composition. J Neurosci 17:58–69

Tabor H, Tabor CW (1965) Spermidine, spermine, and related amines. Pharmacol Rev 245–300

Tóth K, McBain CJ (1998) Afferent specific innervation of two distinct AMPA receptor subtypes on single hippocampal interneurons. Nat. Neurosci 1:572–577

Traub RD, Whittington MA, Stanford IM, Jeffreys JGR (1996) A mechanism for generation of long-range synchronous fast oscillations in the cortex. Nature 383: 621–624

Usherwood PNR, Blagbrough IS (1991) Spider toxins affecting glutamate receptors: polyamines in therapeutic neurochemistry. Pharmac Ther 52:245–268

Vandenberg CA (1987) Inward rectification of a potassium channel in cardiac ventricular cells depends on internal magnesium ions. Proc Natl Acad Sci USA 84: 2560–2564

Verdoorn TA, Burnashev N, Monyer H, Seeburg PH, Sakmann B (1991) Structural determinants of ion flow through recombinant glutamate receptor channels. Science 252:1715–1718

Washburn MS, Dingledine R (1996) Block of alpha-amino-3-hydroxy-5-methyl-4-isoxazolepropionic acid (AMPA) receptors by polyamines and polyamine toxins. J Pharmacol Exp Ther 278:669–678

Washburn MS, Numberger M, Zhang S, Dingledine R (1997) Differential dependence on GluR2 expression of three characteristic features of AMPA receptors. J Neurosci 17:9393–9406

Watanabe S, Kusama-Eguchi K, Kobayashi H, Igarashi K (1991) Estimation of polyamine binding to macromolecules and ATP in bovine lymphocytes and rat liver. J Biol Chem 266:20803–20809

Wollmuth LP, Sakmann B (1998) Different mechanisms of Ca^{2+}-transport in NMDA and Ca^{2+}-permeable AMPA glutamate receptor channels. J. Gen. Physiol. 112: 623–636

Yang J, Jan YN, Jan LY (1995) Control of rectification and permeation by residues in two distinct domains in an inward rectifier K+ channel. Neuron 14:1047–1054

CHAPTER 8

Kainate Receptors

J. Lerma

A. Introduction

The kainate receptor is a component of the glutamate signaling system that has remained elusive to investigators over the years. The lack of specific pharmacological tools has hampered the detection of these receptors in neurons of the central nervous system (CNS) and the determination of their physiological role. Until the cloning of the subunits that make up the kainate receptors, the evidence of their existence as independent receptors in neurons was weak, and it is only recently that we have become able to define the processes in which these receptors are involved (Lerma 1997). Indeed, in spite of considerable evidence from binding and autoradiographic experiments indicating the existence of high-affinity binding sites for kainate in the brain, with a distribution different from that of α-amino-3-hydroxy-5-methylisoxazole-4-propionic-acid (AMPA)-binding sites (Young and Fagg 1990), they were not convincingly detected in brain neurons until more recently (Lerma et al. 1993).

In situ hybridization studies have revealed the anatomical distribution of kainate-receptor subunits (Bahn et al. 1994; Bischoff et al. 1997). The majority of brain cells express some of the subunits of kainate receptors, whilst there are specific cell types (Purkinje cells; cerebellar granule cells) preferentially expressing particular subunits. For instance, GluR5 is abundantly expressed in dorsal root ganglion (DRG) cells. Several years ago, Huettner observed that, in DRG neurons, kainate induced electrophysiological responses that differed from those recorded in CNS neurons. The receptors mediating these responses presented a higher apparent affinity for either kainate or domoate, and underwent marked desensitization (Fig. 1A; Huettner 1990). Functional receptors selective for kainate have been demonstrated in cultured hippocampal neurons (Fig. 1B; Lerma et al. 1993; Wilding and Huettner 1997) and glial cells (Patneau et al. 1994). In these two cell types, kainate induced rapidly activating and inactivating currents. These responses can also be elicited by glutamate, quisqualate and the potent mussel toxin, domoate. By using a multiplex reverse-transcriptase–polymerase-chain-reaction (RT-mPCR) approach at the single cell level, Ruano et al. (1995) showed that these specific responses were present in those hippocampal cells expressing the GluR6 subunit (Fig. 2). The kainate receptor of cultured embryonic hippocampal cells is insensitive to AMPA although, in cultures of postnatal neurons, a high concentration

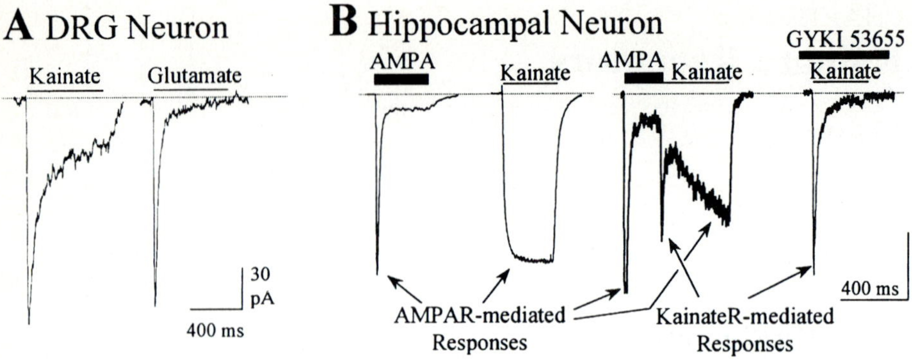

Fig. 1A,B. Responses induced by glutamate-receptor agonists. **A** Activation of kainate receptors in dorsal root ganglion (DRG) cells following rapid application of kainate (300 μM) and glutamate (1 mM; *bars* above the *recordings*). These kainate receptors mostly correspond to GluR5 formations. **B** In mature hippocampal cells, α-amino-3-hydroxy-5-methylisoxazole-4-propionic acid (AMPA; 200 μM) activates a response that largely desensitizes. The same neuron develops a slowly rising, nondesensitizing current when the agonist is kainate (300 μM; *second recording*). Both responses are due to the activation of AMPA receptors. The existence of kainate receptors in this cell may be revealed after desensitizing AMPA receptors with a high concentration of AMPA (*third recording*; the initial response is truncated), and rapidly jumping into a kainate-containing solution. The peak response at the beginning of the kainate perfusion reveals the presence of kainate receptors. The subsequent, slowly developing current represents the activation by kainate of AMPA receptors as they recover from the desensitized state. The total antagonism of AMPA receptors by GYKI 53655 (100 μM) unmasks the presence of the kainate-induced response, which consists of a rapid activating and inactivating current (*fourth recording*). *Vertical calibrations* in this panel are 300, 400, 50 and 30 pA, respectively. Modified from LERMA (1998)

of AMPA alone elicited a response compatible with the inclusion of additional subunits in the receptor (WILDING and HUETTNER 1997). A variability in the expression of distinct subunits during development may account for this observation. More recently, functional kainate receptors with properties compatible with GluR5/KA2 heteromers have been found in rat trigeminal neurons (SAHARA et al. 1997). The purpose of this review is to summarize recent research aimed at determining the functional properties of these elusive receptors and to present those results that are helping us both to clarify the participation of kainate receptors in synaptic transmission at the pre- and postsynaptic levels and to identify behavioral processes in which kainate receptors may have a relevant role.

B. Molecular Biology of Kainate Receptors

I. The Kainate-Receptor-Subunit Family: GluR5–7, KA1 and KA2

The family of subunits that can contribute to the native receptors can be divided into two subfamilies. The first subfamily includes the subunits GluR5,

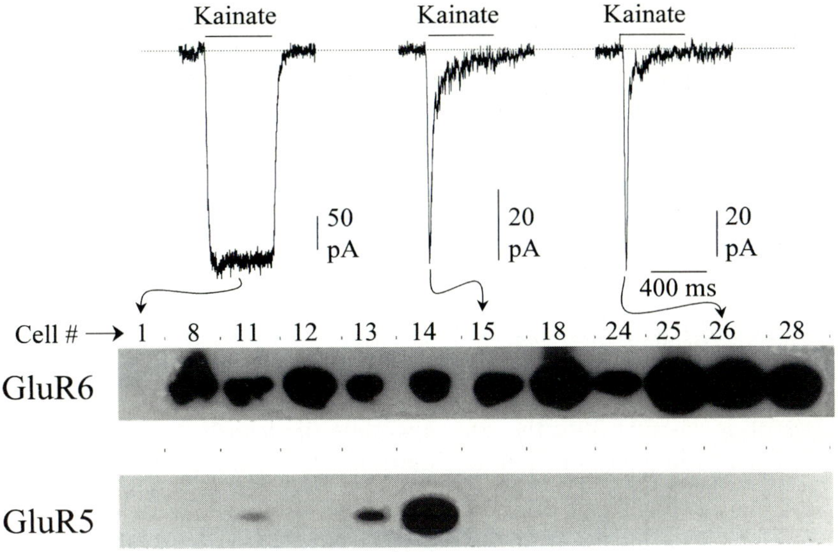

Fig. 2. Rapidly desensitizing, kainate-induced responses are found in cultured hippocampal neurons expressing kainate-receptor subunits. Whole-cell currents evoked by rapid application of kainate (300 μM) are illustrated for three cultured hippocampal cells voltage clamped at –60 mV. These cells were subjected to reverse transcriptase–multiplex polymerase chain reaction (RT-mPCR) analysis. The cyclic DNA fragments obtained after RT-mPCR corresponding to the GluR5–7 and KA-1/KA-2 subunits were hybridized with the GluR6-, GluR5-, GluR7- and KA-1/KA-2-specific probes. All cells showing transient responses expressed the GluR6 subunit (*first row*), and some of them expressed GluR5 (*second row*). No signal was obtained with GluR7 or KA-1/KA-2 probes in any cell analyzed (not shown). Other neurons, like cell 1, did not respond to kainate with transient currents but did with responses of the α-amino-3-hydroxy-5-methylisoxazole-4-propionic acid-receptor-mediated type. No signal corresponding to the GluR5–6 (*column 1*), nor GluR7 and KA-1/KA-2 subunit mRNAs (not shown) were detected. From Ruano et al. 1995, with modifications

GluR6 and GluR7, which have 75–80% homology with each other. All these subunits generate functional receptor channels (Fig. 3), which contain low-affinity kainate-binding sites. The other subfamily is made up of KA1 and KA2, which generate high-affinity binding sites for kainate when recombinantly expressed in mammalian cells although, by themselves, they do not form functional homomeric channels. Whilst the amino acid sequence of these two subunits is approximately 70% homologous, they only display approximately 45% homology to members of the other subfamily. Similar to other glutamate-receptor subunits, kainate-receptor subunits are comprised of approximately 900 amino acids with a total molecular mass of ~100 kDa and have a membrane topology similar to that described for AMPA- and *N*-methyl-D-aspartate (NMDA)-receptor subunits.

Some kainate-receptor subunits may be present as isoforms generated by alternative splicing (Sommer and Seeburg 1992). The GluR5 subunit was initially reported to exist in two molecular forms differing in the presence (GluR5–1) or absence (GluR5–2) of a 15-amino-acid insert in the *N*-terminal extracellular domain (BETTLER et al. 1990). Subsequent analysis revealed the existence of two additional molecular forms (SOMMER et al. 1992) differing in regions downstream of the fourth membrane segment such that the GluR5 subunit may have three alternative *C*-terminal domains. These isoforms are called GluR5–2a (the shortest), GluR5–2b (originally named GluR5–2) and GluR5–2c (the longest). Although they have the same pharmacological behavior, GluR5–2a seems to be more efficiently expressed when transfected in mammalian cells, since it gives larger maximum binding values (SOMMER et al. 1992). No alternative splicing has been reported for GluR6, KA1 and KA2 subunits. In contrast, it has recently been described that the GluR7 subunit has a *C*-terminal splice variant, GluR7b, which results from the insertion of a 40 nucleotide cassette 3′ to the zone encoding the fourth membrane domain. Such an insertion generates a new *C*-terminal 55 residues longer, which does not present sequence homology with other glutamate-receptor subunits (SCHIFFER et al. 1997).

Although splice variants of other glutamate-receptor subunits have been demonstrated to be fundamental for receptor function, the functional significance of kainate-receptor-subunit variants generated by alternative splicing remains to be determined. It could be hypothesized that these splice variants with different *C*-terminal sequences may interact with different cytoskeletal proteins, providing a mechanism for either specific targeting or receptor signaling (NISHIMUNE et al. 1998; OSTEN et al. 1998; EHLERS et al. 1996 for a review).

Similarly, like many other transmitter receptors (SWOPE et al. 1992 for a review), kainate-receptor subunits contain phosphorylation consensus sites for diverse protein kinases. Indeed, recombinant GluR6-mediated responses have been shown to increase by including protein kinase A (PKA) in the recording pipette (RAYMOND et al. 1993; WANG et al. 1993; TRAYNELIS and WAHL 1997). A serine residue at position 684 seems to be the principal PKA phosphorylation site, although the simultaneous elimination of Ser666 was required for the complete elimination of the PKA-induced potentiation (WANG et al. 1993). Despite the striking effect of phosphorylation on channel activity (TRAYNELIS and WAHL 1997), its physiological function in this type of receptor remains a mystery.

Kainate-receptor subunits have been cloned from different mammalian species, including humans. They are nearly identical but have been named differently. For instance, the mouse $\gamma 2$, the human humEAA2 and the rat KA2 kainate-receptor subunits share ~98% of their amino acid sequence. Table 1 shows the names used for different kainate-receptor subunits in different species. Several other subunits have been cloned from non-mammalian vertebrates, including chick, frog and goldfish, which present homology of the mammalian kainate-receptor subunits (GREGOR et al. 1989; WADA et al. 1989;

Table 1. Mammalian kainate-receptor subunits

Rat	Mouse	Human
GluR5[a]	β1[f]	GRIK1[h]/EAA3[i]
GluR6[b]	β2[f]	GRIK2[j]/EAA4[k]
GluR7[c]		EAA5[l]
KA1[d]		HumEAA1[m]
KA2[e]	γ2[g]	HumEAA2[n]

[a] BETTLER et al. 1990
[b] EGEBJERG et al. 1992
[c] BETTLER et al. 1992; LOMELI et al. 1992
[d] WERNER et al. 1991
[e] HERB et al. 1991
[f] MORITA e al. 1992
[g] SAKIMURA et al. 1992
[h] EUBANKS et al. 1993; POTIER et al. 1993
[i] KORCZAK et al. 1995
[j] PASCHEN et al. 1994b
[k] HOO et al. 1994
[l] NUTT et al. 1994
[m] KAMBOJ et al. 1994
[n] KAMBOJ et al. 1992

WO and OSWALD 1994; ISHIMARU et al. 1996; WENTHOLD et al. 1990). These have been called kainate-binding proteins (KBPs), since they show high-affinity kainate-binding activity in cells transfected with cyclic DNAs encoding KBPs. However, they do not form functional channels, nor do they form heteromeric receptors with other subunits (HENLEY 1994). Recently, by constructing chimeric receptors between KBPs and GluR6, it has been demonstrated that all five known KBPs have functional ion-permeation properties in addition to their functional ligand-binding sites. These data have been used to suggest that the lack of ion-channel function in KBPs may result from a failure to translate ligand binding into channel opening (VILLMANN et al. 1997).

II. Editing at the Q/R Site and Current–Voltage Relationship

RNA editing is a source of molecular heterogeneity in a variety of mRNAs encoding for several proteins (CATTANEO 1991; SOMMER and SEEBURG 1992). GluR5 and GluR6, but not the other kainate-receptor subunits, can undergo post-transcriptional mRNA editing at a site located at the position 590 (GluR6), known as the glutamamine/arginine (Q/R) site. The rectification properties of kainate-receptor channels seem to be exclusively controlled by the Q/R site, in the same way as has been observed for the AMPA receptors (HERB et al. 1992; EGEBJERG and HEINEMANN 1993; KÖHLER et al. 1993). The unedited version encodes a glutamine (Q) and exhibits strong inward rectification, while the edited version encodes an arginine (R) at this position and does not rectify (EGEBJERG and HEINEMANN 1993). GluR6 may also be edited at two additional sites located in the first putative membrane domain, where isoleucine can be substituted by valine (the I/V site) and a cysteine may substitute for a tyrosine (the Y/C site). RNA editing at these sites has functional consequences in homomeric GluR6 channels. It seems that the calcium permeability of the GluR6 channels is modulated by the Q/R site whenever the I/V and Y/C sites of the first transmembrane domain are edited (KÖHLER et al. 1993).

The extent of editing is developmentally regulated. Unlike the GluR-2 subunit of AMPA receptors, a significant proportion of unedited kainate subunits are present in both the embryonic and adult brain (PASCHEN et al. 1997; BERNARD and KHRESTCHATISKY 1994). However, an important question is whether different editing variants of the kainate-receptor subunits are coexpressed within a single cell. RUANO et al. (1995) examined the editing of GluR6 subunits in single cultured neurons (from the hippocampus) that expressed native kainate receptors. This analysis demonstrated that different edited variants can be coexpressed by a single cell and that the extent of RNA editing was site-dependent, showing clear differences between the three sites. In most cells, the edited variant at the I/V site was exclusively expressed, whereas the Q/R site was mostly unedited. Interestingly, the cells studied showed edited and unedited mRNAs at the Y/C site to a similar extent. These results indicate that molecular variants generated by RNA editing of the GluR6 subunit are not expressed randomly in single neurons, suggesting the existence of different site- and cell-specific regulators of editing (BELCHER and HOWE 1997).

The functional properties of native kainate-receptor channels (i.e. rectification properties) also seem to be controlled by editing at the Q/R site. A clear relationship between the rectification properties of native kainate receptors and RNA editing at the Q/R site in GuR6-subunit RNA from single neurons could be demonstrated (Fig. 3). Indeed, a linear current–voltage relationship was observed when edited forms (R) predominated, whereas inward rectification was associated with the expression of unedited (Q) RNAs. Thus, the molecular heterogeneity resulting from RNA editing does, indeed, appear to have physiological consequences in native glutamate receptors. Furthermore, the rectification properties seem to be exclusively controlled by the editing of the Q/R site independently of editing sites within the region encoding the first transmembrane domain.

Incomplete editing of GluR5 subunits at the Q/R site has also been found in single hippocampal neurons in adult rats (MACKLER and EBERWINE 1993). Approximately 60% of the GluR5 mRNA is found edited in adult tissue, whilst for GluR6 it is estimated that about 80% of the subunits are edited.

It is worth noting that the inward rectification of kainate-(Q) receptors is not an intrinsic property of the ion channel but rather is due to an open-channel blockade by internal polyamines at inside positive potentials. This

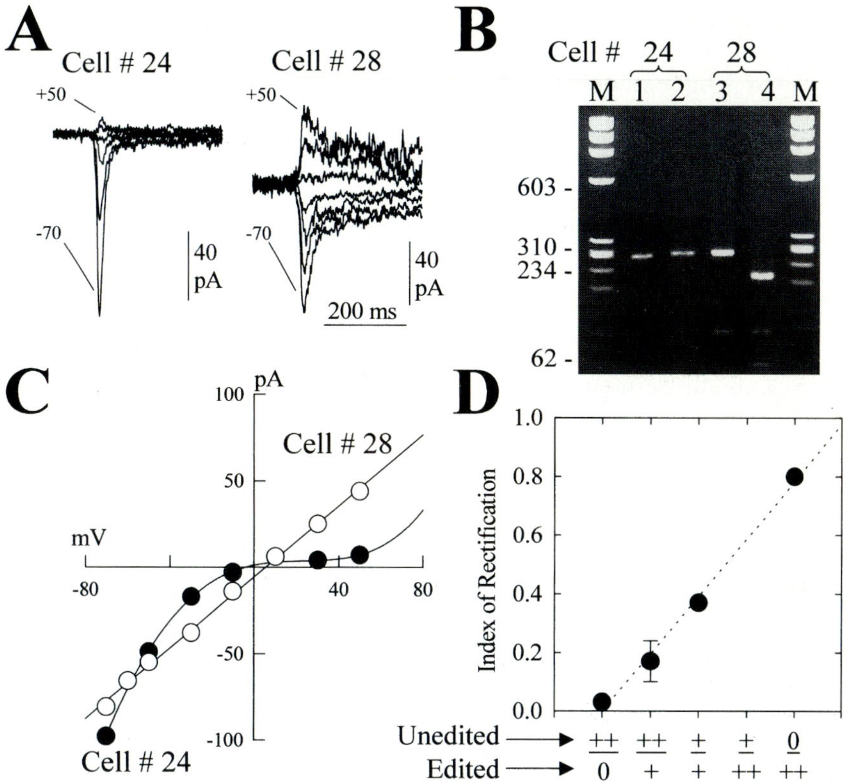

Fig. 3A–D. The Q/R site of the GluR6 subunit accounts for rectification properties of native kainate-receptor channels in cultured hippocampal neurons. **A**, **C** Current–voltage (I–V) relationships of the kainate-induced responses in two different cultured hippocampal neurons. Cell 24 shows strong inward rectification, while cell 28 has an approximately linear I–V relationship. **B** RNA-editing analysis of the Q/R site of the GluR6 fragment generated by polymerase-chain-reaction amplification of material isolated from cells 24 and 28 (*lanes 1 and 3*, respectively). *Lanes 2 and 4* correspond to these fragments after incubation with the Aci-I restriction enzyme. Note the absence of digestion in cell 24 and the full digestion in cell 28, which indicates the exclusive presence of Q and R variants, respectively. **D** The relative amount of edited mRNA was estimated as a function of unedited mRNA and plotted vs the index of rectification of kainate-induced currents, calculated as the ratio of slope conductance at +30 mV and –60 mV. *(++/–, –/++)*, one variant not detected; *(+/+)*, both variants detected at similar leves; *(+/++, ++/+)*, both variants detected, one of them at higher level. From RUANO et al. (1995)

intracellular polycation also confers inward rectification on AMPA-(Q) receptors by a mechanism that is reminiscent to the voltage-dependent channel blocking of inward-rectifying K^+ channels (BOWIE and MAYER and references therein) and may be common in several ligand-gated channels showing inward rectification.

C. Pharmacological Properties of Kainate Receptors

I. Agonist Pharmacology

The development of specific agonists and antagonists for NMDA receptors has allowed us to determine many of the processes in which these receptors are involved. However, the separation of kainate and AMPA receptors in native cells has been difficult, since both receptors are activated by the same collection of exogenous agonists, including kainate, domoate and quisqualate. Indeed, kainate activates AMPA receptors, inducing a large, nondesensitizing response in CNS neurons (Fig. 1B). Experiments carried out using recombinant receptors have demonstrated that GluR5 forms channels that can be activated by kainate, domoate and AMPA (Fig. 4; BETTLER et al. 1990; SOMMER et al. 1992). In contrast, GluR6 generates homomeric channels that can be activated by kainate and domoate but not by AMPA (Fig. 4; Table 2; EGEBJERG et al. 1991; HERB et al. 1992). Recently, it has been demonstrated that GluR7 subunits can also form functional homomeric receptors (SCHIFFER et al. 1997). Interestingly, these receptors have very low affinity for glutamate and are insensitive to AMPA and domoate (Fig. 4). Combinations of GluR5 with GluR6 or GluR7 that form heteromeric channels in heterologous systems and nerve cells have so far not been reported. However, recent experiments in my

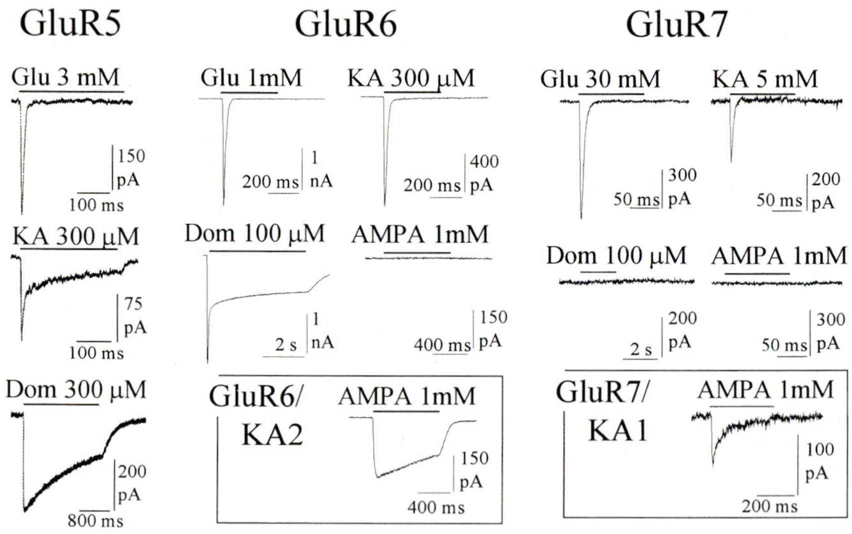

Fig. 4. Responses of recombinant kainate receptors with different molecular compositions expressed in human embryonic kidney 293 cells for different agonists. In all cases, cells were voltage-clamped, and agonists were applied by a rapid-perfusion system. GluR5 responses have been taken with permission from SOMMER et al. 1992; GluR7 traces are from SCHIFFER et al. 1997; GluR6 recordings are unpublished data (PATERNAIN and LERMA). See also Table 2

Table 2. Agonist sensitivity of recombinant kainate receptors

	GluR5	GluR5/KA	GluR6	GluR6/KA	GluR7	GluR7/KA
Glutamate	+	+	+	+	+ (low)	+ (low)
Kainate	+	+	+	+	+ (low)	+ (low)
Domoate	+	+	+	+	−	+ (?)
AMPA	+ (low)	+ (low)	−	+ (low)	−	+ (low)
5-IW	+	+	−	+ (low)	−	+ (low)
Me-Glu	+		+			
ATPA	+		−	+ (low)		
ConA	+	+	+	+	Modest	

AMPA, α-amino-3-hydroxy-5-methylisoxazole-4-propionic acid; ATPA, 2-amino-3-(3-hydroxy-5-tert-butylisoxazol-4-yl) propanoate; ConA, concanavalin A; IW, S-5-iodowillardiine; Me-Glu, methyl glutamate
+, sensitive
−, insensitive
+ (low), sensitive but with low efficiency

laboratory have indicated that GluR5 and GluR6 form functional heteromeric receptors in human embryonic kidney (HEK) 293 cells (Paternain and Lerma, in preparation). Each of these subunits also forms heteromeric receptors when coexpressed with KA1 or KA2, giving rise to ion channels with distinct properties. The inclusion of KA subunits in GluR6 or GluR7 receptors makes them sensitive to AMPA (Fig. 4; Herb et al. 1991; Schiffer et al. 1997), while heteromeric GluR5–KA2 receptors present faster desensitization kinetics than homomeric GluR5 channels.

Binding experiments on isolated brain membranes have documented the existence of low-affinity kainate-binding sites, with a K_d of ~50 nM, as well as a population of sites with a high affinity for kainate, having a K_d of ~5 nM (Young and Fagg 1990). Binding assays performed on recombinantly expressed kainate receptors indicate that the receptors made up of GluR5–7 have a K_d for kainate of 50–100 nM (Bettler et al. 1992; Sommer et al. 1992), whilst the homomeric KA1 or KA2 present a K_d of 4–15 nM (Werner et al. 1991; Herb et al. 1992). Consequently, it has been assumed that GluR5–7 may correspond to low-affinity sites in the brain, whereas KA subunits contribute to high-affinity sites for kainate. However, there are some inconsistencies regarding receptor affinities calculated from radioligand binding and functional experiments. For instance, if KA subunits have a high affinity for kainate, one would expect a higher kainate affinity in heteromers that include KA subunits. In contrast, it seems that the inclusion of KA subunits into functional receptors makes them less sensitive to the agonist (Howe 1996).

Comparing binding affinity constants with the EC_{50} values obtained from functional assays (Table 3), it can be seen that there is a 10–100-fold difference in magnitude. Although this is often the case for binding constants and functional measurements in ligand-gated channels, the differences appear extremely exaggerated in the case of GluR7. To date, researchers have been

Table 3. Agonist pharmacology of native and recombinant kainate receptors (EC_{50}, in μM)

	Hippocampal neurons	DRG	GluR5	GluR6	GluR7
Glutamate	310–330[a,b]	58[d]	631[h]	270–762[a,k,l,m]	5,900[n]
Kainate	22–23[b,c]	12–16[d,f]	33.6[h]	299[a], 1.8[j]	mM range[n]
Domoate		0.7[d]	1.2[h]		Not active[n]
(*S*)-5-IW		0.14[e]	83[o]	Not active[o]	Not active[o]
(2*S*-4*R*)-4-Me-Glu				1[i,j]	
ATPA		0.6[g]	2.1[g]	Not active[g]	
AMPA	Not active[c,d]	260–520[d,e]	3,000[h]	Not active	Not active[n]

Determinations carried out after concanavalin A (ConA) treatment are not included, except for 2-amino-3-(3-hydroxy-5-tert-butylisoxazol-4-yl) propanoate (ATPA) and *S*-5-iodowillardiine (IW) in DRG cells

AMPA, α-amino-3-hydroxy-5-methylisoxazole-4-propionic acid; *DRG*, dorsal root ganglion; *Me-Glu*, methyl glutamate

[a] PATERNAIN et al. 1998
[b] WILDING and HUETTNER 1997
[c] LERMA et al. 1993
[d] HUETTNER 1990
[e] WONG et al. 1994
[f] SAHARA et al. 1997
[g] CLARKE et al. 1997
[h] SOMMER et al. 1992
[i] RAYMOND et al. 1993
[j] JONES et al. 1997
[k] ZHOU et al. 1997
[l] TRAYNELIS and WAHL 1997
[m] HECKMANN et al. 1996
[n] SCHIFFER et al. 1997
[o] SWANSON et al. 1998

unable to account for such an extreme difference. Whatever the explanation, the conclusion is that, from a functional point of view, native kainate receptors do not have a high affinity for glutamate or kainate. Similarly, recombinant kainate receptors do not show high affinity for either glutamate or kainate, although an EC_{50} value as low as 1 μM has been reported for kainate on recombinant GluR6 receptors (JONES et al. 1997). This value of EC_{50} coincides with the one found by others (EGEBGERG et al. 1991; PATERNAIN et al. 1998) in GluR6 recombinant receptors after concanavalin-A (ConA) treatment, which drastically increases receptor affinity (PATERNAIN et al. 1998). The reason for such a disparity remains to be elucidated (see below).

Several molecules behaving as selective agonists for particular kainate receptor subunits have become available. For instance, ATPA [*R,S*-2-amino-3-(5-tert-butyl-3-hydroxy-4-isoxazolyl)propionic acid; LAURIDSEN et al. 1985] has been described as a selective agonist for recombinant GluR5 receptors, being much less active at GluR1–4 receptors (it presents an EC_{50} two orders of magnitude higher). The same conclusion has been obtained by comparing the potency of ATPA for activating DRG receptors (ConA-treated) with its potency for activating receptors expressed by Purkinje cells (mostly of the AMPA type) (CLARKE et al. 1997). Similar data have been reported for the willardine derivative *S*-5-iodo-willardine (IW in Table s 2, 3) (WONG et al. 1994; SWANSON et al. 1998). These results identify ATPA and IW as selective ligands for GluR5 receptors.

II. Antagonist Pharmacology

The prototypical non-NMDA receptor antagonist 6-cyano-7-nitroquinoxaline-2,3-dione (CNQX; HONORÉ et al. 1988) is unable to discriminate between AMPA receptors and kainate receptors (PATERNAIN et al. 1996; WILDING and HUETTNER 1996). Indeed, the dose–inhibition curve for the action of CNQX on steady responses induced by 300 μM kainate (i.e. AMPA-receptor mediated) revealed an IC_{50} of ~1 μM. However, CNQX also inhibits kainate-induced responses in cells with no AMPA receptors, although it does so less potently (IC_{50} of ~6 μM; Table 4). Another non-NMDA receptor antagonist, NS102, was described as a compound that, in cortical membranes, completely inhibits ^{3}H-kainate binding to its low-affinity site whilst essentially showing no effect on high-affinity ^{3}H-kainate binding (JOHANSEN et al. 1993). In addition, NS102 was a potent antagonist of responses evoked at recombinant GluR6

Table 4. Antagonist pharmacology of kainate receptors. Numbers in parenthesis indicate the corresponding pK_b values when available

	Native kainate receptors		GluR5 IC_{50} (μM)	GluR6 IC_{50} (μM)	AMPA receptors Activity
	IC_{50} (μM)	pK_b (μM)			
CNQX	6.1[d(Hip)]	6.5[b(DRG)]			++
NBQX	2.9[a(DRG)]	6–7[a,b(DRG)]	11 (5.4)[a]	2.8[b]	++
NS-102	2.2[d(Hip)]	5[b(DRG)]		~5	+
LY 294486	0.6[e(DRG)]		3.9[e]	Inactive	+
LY 293558	1[a(DRG)]	6.2[a(DRG)]	2.5 (6.2)[a]	>>300[a]	GluR2
LY 382884			K_i = 6.8[e]	Inactive	Inactive

Letters in superscripted parentheses indicate the type of cells from which the data were derived

AMPA, α-amino-3-hydroxy-5-methylisoxazole-4-propionic acid; *CNQX*, 6-cyano-7-nitroquinoxaline-2,3-dione; *DRG*, dorsal root ganglion; *NBQX*, 2,3-dihydroxy-6-nitro-7-sulfamoyl-benzo(f)quinoxaline

[a] BLEAKMAN et al. 1996
[b] WILDING and HUETTNER 1996
[c] PATERNAIN et al. 1998
[d] PATERNAIN et al. 1996
[e] SIMMONS et al. 1998

receptors in HEK cells (Verdoorn et al. 1994). Accordingly, NS102 blocked transient, kainate-induced responses in hippocampal cells in a dose-dependent and reversible manner (IC_{50} = 2.2 μM). However, NS102 is also active on AMPA receptors, as evaluated by steady kainate-induced currents, although it is less potent (IC_{50} for steady responses was estimated to be ~4 μM). This is in agreement with the fact that NS102 was observed to weakly inhibit ^{3}H-AMPA binding (Johansen et al. 1993). At 10 μM, close to the limit of solubility, NS102 was unable to completely inhibit either response (~24% and ~32% of the control peak and steady currents remained, respectively, with 100 μM kainate). Consequently, it proved to be inefficient as a drug to separate the activities mediated by the two receptors (Paternain et al. 1996; Wilding and Huettner 1996).

Fortunately, selective antagonists of the kainate receptors that minimally act on AMPA receptors are being developed. However, the early development of the 2,3-benzodiazepines as antagonists of AMPA receptors (Vizi et al. 1996) and the further demonstration that these compounds presented selectivity for AMPA over kainate receptors (Paternain et al 1995; Wilding and Huettner 1995) has been crucial to the progress in our understanding of kainate-receptor function. In particular, GYKI 53655 (Lilly code number LY300168) is a noncompetitive antagonist of AMPA receptors able to completely block these receptors at 100 μM. At this concentration, it presents no antagonism of native kainate receptors and, therefore, could be used to unmask kainate-receptor currents in neurons (Fig. 1B; Paternain et al. 1995; Wilding and Huettner 1997). GYKI 53655 is a racemic mixture with an IC_{50} value of ~1 μM for antagonizing neuronal AMPA receptors (Paternain et al. 1995; Wilding and Huettner 1995). The active isomer, LY303070, is also available and is, obviously, more potent.

Recently, a new series of compounds that displays different sensitivities for kainate and AMPA receptors has been developed. The decahydroisoquinoline derivatives, such as LY293558 [(3S,4aR,6R,8aR)-6-(2-([^{1}H]tetrazol-5-yl)ethyl)decahydroiso-quinoline-3-carboxylic acid], inhibit GluR5 but not GluR6. However, this compound is also active in antagonizing AMPA-receptor-mediated responses (Bleakman et al. 1996b). Another derivative, LY294486 [(3*SR*,4a*RS*,6*SR*,8a*RS*)-6-((((^{1}H-tetrazol-5-yl)methyl)oxy)methyl)-1,2,3,4,4a,5,6,7,8,8a-decahydroisoquinoline-3-carboxylic acid], has greater selectivity for GluR5 over AMPA receptors, not showing antagonist activity on GluR6. More recently, it has been demonstrated that the compound LY382884 [(3*S*,4a*R*,6*S*,8a*R*)-6-(4-carboxyphenyl)methyl-1,2,3,4,4a,5,6,7,8,8a-deca-hydroisoquinoline-3-carboxylic acid] (Simmons et al. 1998) is specific for antagonizing binding to GluR5 receptors, showing no affinity for the AMPA-receptor subunits GluR1, 2 or 3 nor the kainate-receptor subunits GluR6 and 7 and KA2. Although these are the preliminary steps, these compounds will be of enormous help in determining the role of kainate receptors in the pathophysiology of the CNS.

III. A Bit of Caution: ConA May Alter Considerably Kainate-Receptor Affinity

Many studies, mainly those involving the use of *Xenopus* oocytes as the expression system, use treatment with ConA to avoid desensitization (WONG and MAYER 1993) and to facilitate current detection, as rapid response desensitization limits the magnitude of the current recorded during slow application of agonist. The use of ConA, however, may be a source of confusion regarding affinity values calculated for kainate receptors. PATERNAIN et al. (1998) calculated EC_{50} values for both glutamate and kainate in HEK 293 cells expressing recombinant GluR6 before and after ConA treatment (300 μg/ml, 10 min). The affinity of GluR6 receptors dramatically increased following exposure to ConA, more markedly for kainate than for glutamate. The EC_{50} of glutamate for nondesensitizing receptors (i.e. those treated with ConA) was almost 25-fold lower (762 μM vs 32 μM), whereas the EC_{50} for kainate decreased almost 250-fold (from 299 μM to 1.2 μM). The slope coefficient increased from 0.7 to 1 and from 0.9 to 1.3 for kainate and glutamate, respectively. Similarly, JONES et al. (1997) found that the GluR6-receptor affinity for 2*S*,4*R*-4-methylglutamate increased drastically after ConA treatment (EC_{50} was reduced approximately by 11-fold). The molecular mechanism by which ConA reduces the desensitization of glutamate receptors remains to be elucidated, but it is possible that the higher affinity values previously estimated for kainate and other agonists were due to the uncovering of a high-affinity desensitized state that becomes conductive after ConA treatment. This assumption would account, at least in part, for the higher affinity values observed in radioligand binding assays since, in these experiments, agonist binding to the (higher affinity?) desensitized state is preferentially measured. However, ConA treatment does not seem to change the affinity in DRG kainate responses (HUETTNER 1990). Since GluR5 is the main component of kainate receptors in DRG cells (SOMMER et al. 1992), it is possible that ConA action depends on the receptor type. Thus, ConA treatment to reduce desensitization should be taken into account when comparing affinity values for kainate receptors of the kainate type obtained in different studies. More importantly, the treatment with ConA may alter the affinity and/or pharmacological profile of molecules designed to specifically affect a particular type of kainate receptor.

D. Properties of Kainate Receptors at the Single-Channel Level

The single-channel properties of homomeric channels composed of GluR5 and GluR6 subunits and heteromeric receptors made up of these subunits plus the KA2 subunit have been determined (Table 5). These studies indicated that both homomeric GluR6 (R) and heteromeric GluR5 (R) channels have a

Table 5. Single channel conductance of recombinant kainate receptors (in pS)

GluR6(R)	GluR6(Q)	GluR6(R)/ KA2	GluR6(Q)/ KA2	GluR5(R)	GluR5(Q)	GluR5(R)/ KA2	GluR5(Q)/ KA2
0.23–0.26[a,b]; 0.4[d]	5.4 (8/15/25)[c]; 17[d]	0.57–0.7[a,b]	7.1 (7/13/20)[c]	<0.2[b]	2.9 (5/9/14)[c]	0.95[b]	4.5 (5/9/17)[c]

[a] Single-channel values calculated from noise analysis in human embryonic kidney (HEK) cells transiently or stably transfected with kainate receptor subunits and treated with concanavalin A (Howe 1996)
[b] Single-channel values calculated from noise analysis in HEK cells transiently transfected (Swanson et al. 1996)
[c] Single-channel conductance values estimated directly from resolved openings (Swanson et al. 1996)
[d] Estimated from non-stationary variance analysis in excised outside-out patches from HEK cells (Traynelis and Wahl 1997). The larger single-channel conductance here may reflect a greater proportion of high subconductance levels

unitary conductance in the femtosiemens range. Editing of the Q/R site drastically alters single-channel conductance, such that unedited forms present a larger single-channel conductance. The combination of either GluR5 and GluR6 with subunits such as KA2 produces heteromeric receptors that have 2–3-fold larger single-channel conductances than the respective homomeric channels in their R form; however, these receptors show a lower affinity for their agonists (HOWE 1996; SWANSON et al. 1996). However, GluR6(Q)/KA2 presented single-channel events that were indistinguishable from those of homomeric receptors, although coexpression of KA2 with GluR5(Q) shortened the channel burst length when compared with the homomeric channel (SWANSON et al. 1996). Such an effect is in keeping with the fact that KA and KA2 subunit mRNAs are not susceptible to editing, and both carry a Q residue at the Q/R site. The effect of the Q/R editing site on single-channel behavior is not surprising, since this residue is thought to be part of the channel lumen in both kainate and AMPA receptors. Actually, it has also been observed that the Q forms of AMPA receptors have a larger single-channel conductance than the R forms (SWANSON et al. 1997).

Three subconductance levels could be directly resolved from outside-out patches excised from HEK cells expressing the Q form of either GluR6 or GluR5 in its homomeric configuration or with KA2 in its heteromeric configuration (SWANSON et al. 1996). In all cases, the values were twice or three times the value of the smallest level (Table 5). Recently, it has been observed that the mean single-channel current depends on how many of a receptor's binding sites have agonists bound to them (ROSEMUND et al. 1998) whenever the receptor does not enter into a desensitized state. This notion has been used to conclude that AMPA receptors are tetrameric rather that pentameric structures. Therefore, the three conductance levels observed in kainate receptors at low concentrations of a partially desensitizing agonist (i.e., at low receptor occupancy) is consistent with a tetrameric conformation of the kainate receptors. This may be the origin of the discrepancy between the mean single-channel conductances calculated from stationary analysis (0.22 pS and 5.4 pS for GluR6(R) and GluR6(Q), respectively; SWANSON et al. 1996) and those estimated by non-stationary analysis at a saturating concentration of glutamate (0.4 pS and 17 pS, respectively; TRAYNELIS and WAHL 1997). Regarding native receptors, HUETTNER (1990) estimated single-channel conductances in DRG cells. The estimated unitary conductance by whole-cell current fluctuations was 2–4 pS. However, openings of 8 pS and 15–18 pS were occasionally observed. These results (i.e., 4/8/15 pS of multiple conductance values) agree well with those reported for recombinant receptors made of either GluR5(Q) and/or GluR5(Q)/KA2. More recently, PEMBERTON et al. (1998) have estimated the single-channel conductance of kainate-receptor channels expressed by cerebellar granule cells. Spectra-density analysis of kainate-induced currents in ConA-treated cells gave a mean value of 1 pS.

E. Activation–Desensitization Properties of Kainate Receptors

I. Desensitization

A hallmark of kainate receptors is that, in the continuous presence of agonist, the current flowing through activated channels decays rapidly owing to the desensitization of the receptor (Fig. 3). Depending on the type of receptor, the degree of desensitization may vary. In hippocampal neurons, the onset of desensitization follows a single-exponential time course (LERMA et al. 1993; PATERNAIN et al. 1998), but double-exponential kinetics have also been observed (LERMA et al. 1993; WILDING and HUETTNER 1997). Table 6 presents a compendium of the time constants measured by different authors in different preparations at room temperature. The main conclusion is that the process of desensitization of kainate receptors is very rapid. How rapid is it? There are several problems in resolving this question. As binding occurs prior to desensitization, the binding rate should affect the onset kinetics when estimated from the current decay upon agonist perfusion. In addition, since the relaxation time constants are in the range of resolution of rapid perfusion systems, at least at the whole-cell level, the observed desensitization rate should also be affected by solution changes. Indeed, the time constant of desensitization in both native and recombinant receptors appears to be concentration dependent, decreasing as agonist concentration increases for a range of agonist concentrations (HECKMANN et al. 1996; PATERNAIN et al. 1998). The actual rate of the onset of kainate-receptor desensitization could be estimated by plotting the time constant of current decay versus the concentration of agonist inducing such a process. This type of calculation has been used to obtain an estimate of the desensitization rate constants of kainate receptors in both hippocampal cells (Fig. 5A) and HEK cells expressing recombinant GluR6 channels. Theoretically, the apparent rate of desensitization will increase with the agonist concentration, such that the asymptote of the function should coincide with the rate of desensitization when the binding rate of the agonist is no longer the limiting step. It is worth noting that estimating the

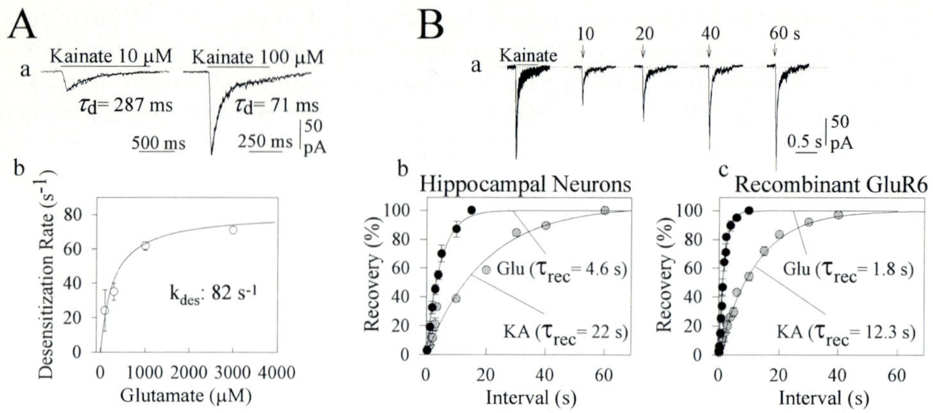

Fig. 5A,B. Desensitization properties of kainate receptors in cultured hippocampal cells. **A** Onset of desensitization is speeded by increasing agonist concentration. The *superimposed lines* on the records represent single exponential fits to the desensitization process; the fits have the indicated time constants (τ_d). Single values collected from different cells were normalized and plotted against the concentration of glutamate. The onset rate, measured as $1/\tau$ at any concentration of glutamate, can be fitted by a rectangular hyperbola in which the asymptotic value corresponds to the onset rate constant for desensitization (k_{des}). **B** Kinetics of desensitization recovery of kainate receptors. Responses were completely desensitized by applying a 1-s pulse of either kainate (300 μM; *top recordings*) or glutamate (300 μM). Conditioning pulses were followed by test pulses of the same duration delivered at different intervals. In the *bottom panels*, the fractional recovery is plotted against the interval between the end of the application of the first pulse of agonist and the beginning of the test response. The data are fitted with single exponential functions (*solid lines*) with the illustrated time constants of recovery (τ_{rec}). From PATERNAIN et al. (1998)

desensitization rate constant in this way makes the value largely independent on the exchange rate and, therefore, is more accurate than if calculated by the direct measurement of current decay. Interestingly, the value calculated in this way is very similar in native and recombinant channels, irrespective of the agonist used, either kainate or glutamate (PATERNAIN et al. 1998). This implies that desensitization follows a single-exponential decay course with a time constant of 11–13 ms in both hippocampal kainate receptors and recombinant GluR6 channels. The desensitization rate estimated from recombinant GluR6 in outside-out patches is slightly faster (5–8 ms; HECKMANN et al. 1996; TRAYNELIS and WAHL 1997), but not very different from the estimation in whole-cell conditions. In some cases, the decaying phase is better fitted by the sum of two exponentials, although the faster component is predominant. This is particularly relevant for DRG cells and recombinant GluR5 receptors (Table 6; HUETTNER 1990; SAHARA et al. 1997; WONG et al. 1994).

When compared to other glutamate receptors, recovery from desensitization of kainate receptors is slow. Full recovery of the glutamate-induced response in hippocampal neurons is obtained 15 s after the initial pulse, although the recovery rate depends on the agonist used to desensitize the

Table 6. Time constants for desensitization onset of native and recombinant kainate receptors

	Native receptors		Recombinant receptors		
	Hippocampus	DRG	GluR5	GluR6	GluR7
Glutamate	8–10 ms[a] 38/400 ms[b]	33/239 ms[k] 16/150 ms[j]	9/68 ms[e] 4.1 ms[l]	4–12 ms[a,d,f,g,l]	9 ms[h]
Kainate	8–12 ms[a] 17/146 ms[c] 27/400 ms[b]	35/400 ms[j] 61/695 ms[i]	15/281 ms[e]	4 ms[g]	3.2 ms[h]
Domoate	160 ms[c]		2.2 s[e]		

DRG, dorsal root ganglion
[a] Paternain et al. 1998
[b] Wilding and Huettner 1997
[c] Lerma et al. 1993
[d] Price and Raymond 1996
[e] Sommer et al. 1992
[f] Traynelis and Wahl 1997
[g] Heakmann et al. 1996
[h] Schiffer et al. 1997
[i] Wong et al. 1994
[j] Huettner 1990
[k] Sahara et al. 1997
[l] Swanson et al. 1997

receptor. Indeed, it is much slower for kainate in both native and recombinant receptors. It is necessary to separate both pulses for about 1 min in hippocampal cells to regain the initial amplitude of the response after desensitization with kainate (Fig. 5B). In general, the recovery from desensitization follows a single exponential with a time constant that also depends on the type of receptor. Indeed, it seems that the insertion of KA subunits into the functional receptor influences multiple aspects of desensitization kinetics. For instance, GluR5 homomeric receptors recover from IW-induced desensitization on a time scale of minutes (2.5 min; Swanson et al. 1998), whereas KA2/GluR5 heteromeric receptors recover faster (τ_{rec} = 12 s). Whatever the type of subunit studied, these data imply that kainate receptors tend to spend long periods of time in the desensitized state, and that desensitization may be considered as an absorbing state, since the equilibrium is almost completely displaced towards the desensitized state.

As mentioned previously, recovery from desensitization is agonist-dependent (Table 7). It may be possible that recovery from desensitization would only depend on the rate of agonist unbinding. However, this is unlikely, since domoate, an agonist that has a higher affinity than kainate (i.e. a slower unbinding rate) and that partially desensitizes kainate receptors (Lerma et al. 1993 for native and Swanson et al. 1997 for recombinant receptors), dissociates on a faster time scale than the recovery from desensitization. Therefore,

Table 7. Kinetics of recovery from desensitization of kainate receptors

	Native receptors		Recombinant	
	Hippocampus	DRG	GluR5	GluR6
Glutamate	4.6 s[a] 5.8/34 s[b]	18 s[e] 36 ms/4.7 s[h]	50 ms/5 s[h]	1.6–1.8 s[a,c,d,f] 4 s[g]
Kainate	22 s[a] 7/44 s[b]			12.3 s[a] 5 s[g]

DRG, dorsal root ganglion
[a] PATERNAIN et al. 1998
[b] WILDING and HUETTNER 1997
[c] HECKMANN et al. 1996
[d] RAYMOND et al. 1996
[e] HUETTNER 1990
[f] TRAYNELIS and WAHL 1997
[g] JONES et al. 1997
[h] SWANSON and HEINEMANN 1998

the recovery from desensitization probably involves a slow equilibration between receptors in the desensitized and activatable states (HECKMANN et al. 1996 for a reaction scheme of GluR6-receptor activation and desensitization).

II. Rapid Desensitization Shapes Dose–Response Curves of Kainate Receptors and Sets Steady-State Receptor Activity

After observing the dose–response curves of either native or recombinant receptors for the agonist glutamate (Figs. 6B, C), it could be concluded that kainate receptors do not have a high affinity for glutamate. Rather, they present an EC_{50} value similar to those found for AMPA receptors in several types of neurons. However, in terms of steady-state desensitization, kainate receptors are very sensitive to ambient concentration of agonist. The sensitivity of native and recombinant receptors to steady-state desensitization has been recently calculated. Kainate receptors are about two orders of magnitude more sensitive to the agonist for desensitization than for activation. The calculation of steady-state desensitization for kainate receptors expressed by cultured hippocampal neurons yielded a value of IC_{50} of about $3\,\mu M$ for glutamate (Figs. 6A, C; PATERNAIN et al. 1998). This means that, at a glutamate concentration that is unable to produce activation, the receptors are readily inactivated. The same result was found when kainate was used as an agonist. Again, a difference of two orders of magnitude in the sensitivity to activation and desensitization of kainate receptors was demonstrated ($0.31\,\mu M$ vs $22\,\mu M$ of kainate). Different sensitivities to activation and desensitization have also been observed in GluR6 recombinant receptors (HECKMANN et al. 1996; PATERNAIN et al. 1998). Interestingly, GluR6 recombinant receptors are much more sensitive to desensitization (IC_{50} of $0.3\,\mu M$) than hippocampal receptors

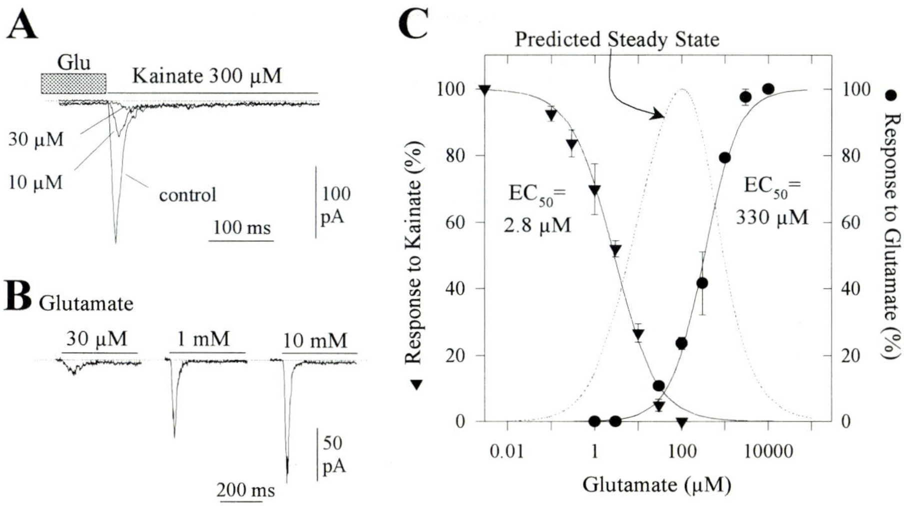

Fig. 6. A–C. Activation–desensitization properties of native kainate receptors in hippocampal cells. **A** The effect of low steady-state concentrations of glutamate on a test response to a saturating concentration of kainate. The cell was exposed to different concentrations of glutamate for 10s before jumping to a solution containing kainate. **B** Dose-dependent activation of kainate receptors by glutamate. **C** Activation (*circles*) and desensitization (*triangles*) curves of kainate receptors. The *solid lines* through the points represent the best least-square fit of the logistic equation to pooled data and presented the indicated half-saturated concentrations. Some steady-state activity is predicted from the fact that both curves cross over. The theoretical steady-state current (*dotted line*) has been calculated as the fraction of non-inactivated channels that would be activated at each agonist concentration, taking into account the activation and desensitization curves, and presents a peak activity at ~100 μM glutamate. From Paternain et al. (1998)

but, in contrast, are less sensitive to glutamate activation (EC_{50} of 500–700 μM).

The activation and desensitization curves for kainate receptors overlap over a range of agonist concentrations (Fig. 6C). According to models of voltage-dependent channels with similar overlapping curves (Hodgkin and Huxley's model; Hille 1992), the existence of a steady-state channel activity would be predicted. For voltage-dependent channels, this steady response is known as the "window current". Theoretically, any ion channel whose activation and desensitization curves overlap should present such a window of steady-state activation. By analogy, steady-state activity could be implied for any ligand-gated channel presenting activation and desensitization properties. As in the Hodgkin and Huxley model, we must assume that opening and desensitization "gates" are independent. The theoretical steady current may be calculated as the fraction of noninactivated channels that would be activated at each agonist concentration, taking into account the activation and desensitization curves. Thus, some current should persist as long as the concentration of agonist remains within certain values. The theoretical curve for kainate receptors is illustrated in Fig. 6C considering constants estimated from hippocampal neurons in culture. For glutamate, this window presents a maximum at around 100 μM. The functional consequence of a concentration window for the actions of kainate or glutamate might be important, since it implies a similar activity at low and high concentrations of agonist, with intermediate concentrations being the most efficient. That this property is physiologically relevant has been illustrated by the fact that the kainate-receptor-mediated modulation of γ-aminobutyric acid (GABA) release follows a bell-shaped dose–response curve, which could be predicted by taking into account previously calculated activation and desensitization curves in cultured hippocampal cells (RODRÍGUEZ-MORENO et al. 1997; PATERNAIN et al. 1998).

Additionally, this aspect of kainate-receptor modulation may also be relevant in terms of the pathological activity of these receptore, since variations in the extracellular concentration of glutamate within this range might occur during intense neuronal activity and/or ischemic episodes (ATTWELL and MOBBS 1994). Indeed, while the resting extracellular concentration of glutamate may be ~3 μM (LERMA et al. 1986), a concentration of about 100 μM has been calculated as the value reached by glutamate in the extracellular fluid under transient ischemic conditions (BENVENISTE et al. 1984). Whether or not this might be relevant for lesions induced by ischemia remains to be investigated. Although no definitive proof is available, there are some indications that suggest a role for kainate receptors in triggering cell death. For instance, kainate is able to kill HEK 293 cells when they have been transfected with GluR6/KA2 (CARVER et al. 1996). In addition, acute and chronic exposure to kainate caused extensive optic-nerve-oligodendrocyte death in culture and in vivo (MATUTE et al. 1997), an effect that is only partially prevented by the AMPA-receptor antagonist GYKI 52466 but is completely abolished by CNQX. Altogether, these data suggest that both AMPA and kainate receptors mediate at least some of the toxicity observed during exposure to kainate.

Another consideration that can be made regarding steady-state activation is that any modification of receptor sensitivity to activation or desensitization will have a striking impact on the steady-state receptor activity. For instance, a shift to the right of the desensitization curve (by the expression of accessory subunits, drug exposure, etc.) would significantly increase the steady-state receptor activity. The larger glutamate IC_{50} measured in postnatal hippocampal cells predicts larger steady-state currents for kainate receptors in these cells, as has been observed (WILDING and HUETTNER 1997). The existence of a clear, bell-shaped curve is unusual but not surprising from a pharmacological point of view and may be important for the estimation of receptor affinity. If only the rising phase of the window curve is considered, it indicates an artifactually high potency for the agonist. Indeed, a bell-shaped dose–response curve rather than a sigmoidal curve would be expected for the steady action of kainate on physiological processes whenever mediated by kainate receptors.

Although this bell-shaped curve has been demonstrated to be physiologically relevant for both voltage- and ligand-gated channels (WILLIAMS et al. 1997), measuring window currents experimentally has proved to be a difficult task. In the case of voltage-dependent channels in smooth muscle cells, average sustained Ca^{2+} currents of the L-type are <2 pA (FLEISCHMANN et al. 1994; SMIRNOV and AARONSON 1992) and can often not be distinguished from the zero-current level. However, such currents have been associated with a substantial Ca^{2+} rise in the cytoplasm, the magnitude of which is consistent with the theoretical window calculated from measurements of desensitization–activation curves.

F. Pre- vs Postsynaptic Localization of Kainate Receptors. Roles in Synaptic Transmission

The role of AMPA and NMDA receptors in fast synaptic transmission is well characterized. However, the demonstration of synaptic responses due to the activation of kainate-receptor channels has proven difficult (LERMA et al. 1997). Recently, it has been shown that responses with a pharmacological profile consistent with that of kainate receptors can be synaptically triggered, generating a slow excitatory postsynaptic current in CA3 pyramidal neurons of the hippocampus upon repetitive stimulation of the mossy fibers (Fig. 7B; CASTILLO et al. 1997; VIGNES and COLLINGRIDGE 1997). This means that synaptically released glutamate has access to postsynaptically localized kainate receptors in addition to AMPA and NMDA receptors. However, this seems to occur only under circumstances involving repetitive presynaptic activity and might reflect the glutamate spill-over from adjacent synapses. The finding that only repetitive stimulation of mossy fibers evokes a kainate-receptor-mediated response on CA3 pyramidal cells is consistent with the high levels of kainate-receptor-subunit expression (GluR6 and KA2) in these cells. Indeed, kainate selectively depolarizes CA3 neurons when it is applied to the stratum lucidum. Interestingly, the synaptic current induced by a train of stimuli was not apparent in mice in which the GluR6 gene had been disrupted (Fig. 7C; MULLE et al. 1998). This result indicates that kainate receptors containing the GluR6 subunit participate in the synaptic transmission at the mossy fiber–CA3 pyramidal neuron synapse. However, it has been recently reported that such synaptic responses are reduced in amplitude by the nearly selective GluR5 antagonists LY293558 and LY294486 (VIGNES et al. 1997), indicating that GluR5 subunits may also contribute to these receptors. The question remains as to whether these synaptic receptors are formed by GluR5 and GluR6 arranged in an unknown configuration. The existence in vivo of heteromeric GluR5/GluR6 receptors, as suggested by our results in recombinant systems (PATERNAIN and LERMA, in preparation) may serve to reconcile these two observations. Efforts to reveal a similar function for kainate receptors in other subfields of the hippocampus have so far failed (LERMA et al. 1997; CASTILLO

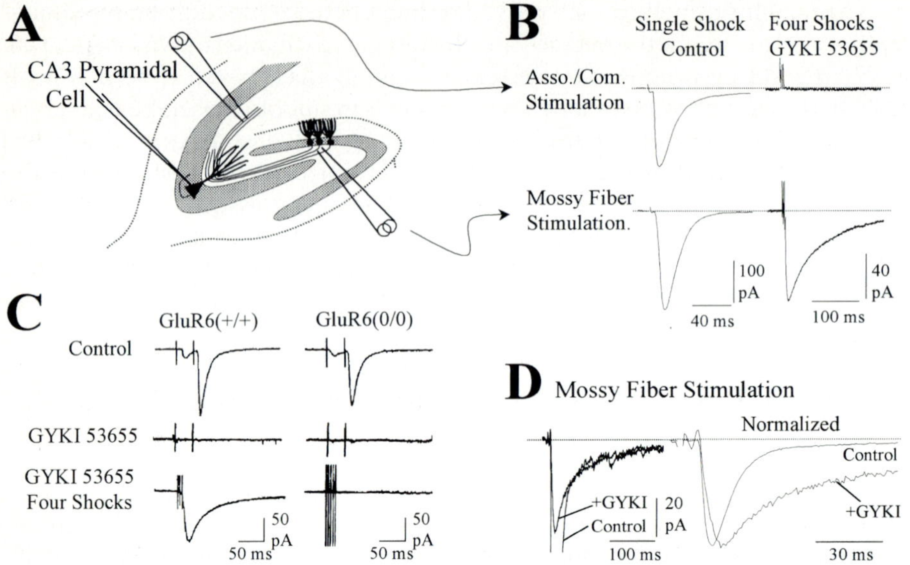

Fig 7. A–D. Kainate receptors mediate a slow postsynaptic current in hippocampal CA3 neurons. **A** Schematic arrangement of the stimulating and recording electrodes. **B** Responses of a CA3 pyramidal neuron to two frequencies of stimulation in the presence and in the absence of GYKI are shown. The inclusion of the α-amino-3-hydroxy-5-methylisoxazole-4-propionic-acid (AMPA)-receptor blocker, GYKI 53655, abolishes responses to single mossy fiber and associational/commissural (Asso/Com) stimulation, unmasking a resistant component specifically activated by repetitive activation by mossy fibers. **C** The synaptic response induced by repetitive stimulation of mossy fibers is absent in the mutant mice lacking the kainate-receptor subunit GluR6 [GluR6(0/0)], whilst fast synaptic transmission is preserved. **D** The kainate-receptor-mediated response (GYKI) is shown normalized to the AMPA-receptor-mediated response (control) (*right* records) to show its slower time course. Data in **B** and **D** are from CASTILLO et al. 1997. *Recordings* in **C** are from MULLE et al. (1998)

et al. 1997), suggesting that it might be a specific property of this type of synapse. These data provide a physiological foundation for the selective distribution of high-affinity kainate-binding sites in this area of the hippocampus. Very recently, however, two studies have provided compelling evidence for the activation of kainate receptors (likely of the GluR5 type) at glutamatergic synapses of CA1 interneurons (COSSART et al. 1998; FRERKING et al. 1998). At the peak synaptic response, the kainate-receptor-mediated current is less than a tenth the size of the AMPA-receptor-mediated response but lasts significantly longer, allowing the interneurons long-lasting integration of their excitatory inputs.

It has been shown that kainate produced a decrease in ^{3}H-glutamate release from hippocampal synaptosomes (CHITTAJALLU et al. 1996). A reduction in the NMDA-receptor-mediated component of CA1 synaptic responses when hippocampal slices were perfused with a low concentration of kainate was also noted. More recently, it has been suggested that such presynaptic inhibitory action of kainate in hippocampal CA1 synapses is primarily due to the inhibition of Ca^{2+} influx into the presynaptic terminals (KAMIYA and OZAWA 1998). Taking into account the fact that kainate acts as a potent convulsing excitotoxin in the CNS, these results, indicating a participation of kainate receptors in inhibiting the synaptic release of glutamate, are somewhat surprising. Indeed, several authors had reported that kainate, even at low concentrations, induced a general increase of excitability, sometimes conveying epilepsy (CHERUBINI et al. 1990; FISHER and ALGER 1994; RODRIGUEZ-MORENO et al. 1997; SLOVITER and DAMIANO 1981; WESTBROOK and LOTHMAN 1983).

Taking advantage of the availability of GYKI 53655 and other specific compounds, two reports have shown that the activation of kainate receptors downregulates GABAergic inhibition in hippocampal CA1 pyramidal neurons (CLARK et al. 1997; RODRIGUEZ-MORENO et al. 1997). A detailed analysis revealed a presynaptic site of action for kainate (Fig. 8). The same result was obtained by using a new compound, ATPA, which seems to selectively activate GluR5-containing receptors (CLARKE et al. 1997; RODRIGUEZ-MORENO and LERMA, unpublished). The combined use of ATPA and the GluR5-specific antagonist LY294486 (CLARKE et al. 1997) indicates that GluR5 subunits form presynaptic kainate receptors that inhibit GABA release in the hippocampus. Whether or not other subunits are involved in this action remains to be determined.

Recent experiments have indicated that the kainate receptors regulating GABA release act as metabotropic receptors rather that ion channels (RODRIGUEZ-MORENO and LERMA 1998). The activation of kainate receptors in GABAergic neurons of the hippocampus results in the initiation of a second messenger cascade following the activation of a Pertussis-toxin-sensitive G protein. The pathway involves the activation of phospholipase C and protein kinase C, establishing a metabotropic action of kainate receptors, well-known ion-channel-forming glutamate receptors that are independent of their ion-channel activity (Fig. 9). There is evidence of AMPA-receptor-mediated regulation of G_i-proteins in cortical neurons (WANG et al. 1997), although no evidence exists regarding the physiological relevance of such interactions. These results indicate that, in addition to forming ion channels, ionotropic glutamate receptors can exhibit metabotropic activity. In the case of kainate receptors, such an activity profoundly affects neuronal function as it results in the modulation of GABAergic inhibition. The G-protein-coupled receptors share the so-called 7-transmembrane (TM) structure. How a 3-TM-domain ion-channel structure couples to G proteins remains to be established.

G. What Should be Called a Kainate Receptor?

After the demonstration that kainate activates all types of AMPA receptors, these receptors started to be referred to as AMPA/kainate receptors. Similarly,

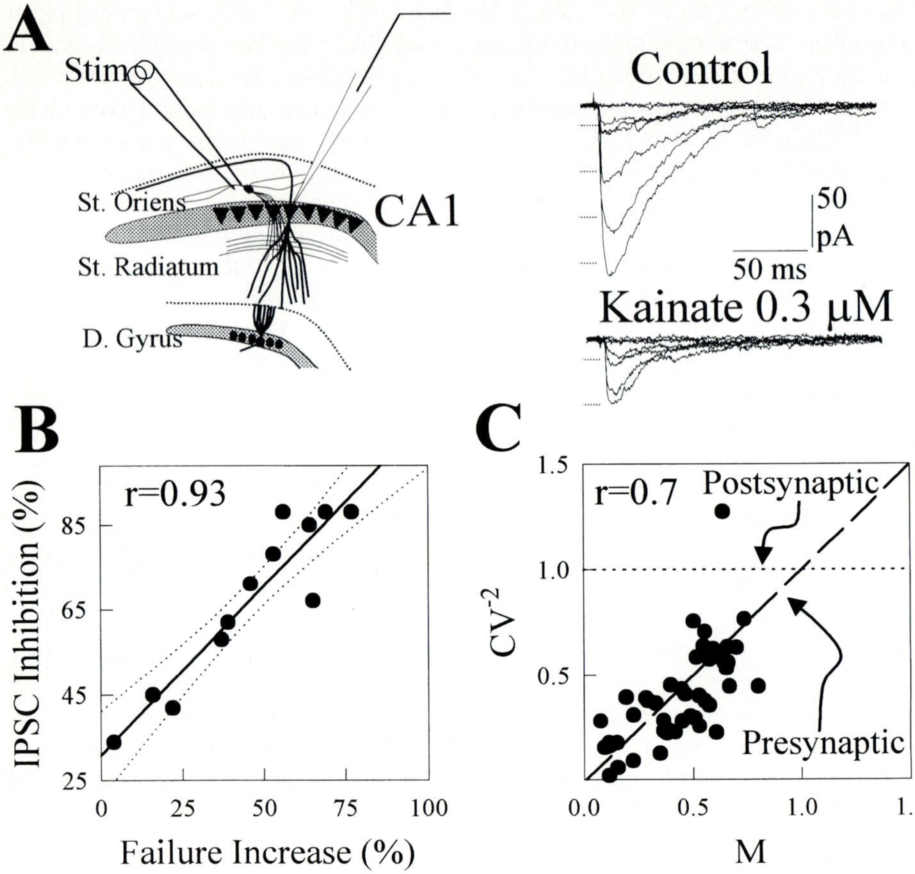

Fig. 8A–C. Kainate modulates the reliability of γ-aminobutyric acid (GABA)ergic transmission in the hippocampus. **A** Schematic representation of the experimental setup: electrical stimuli were delivered via a bipolar electrode on the stratum oriens, while evoked inhibitory synaptic currents were recorded from identified pyramidal neurons, which were kept at −60 mV holding membrane potential. Note how, in the presence of kainate, the mean amplitude of the inhibitory postsynaptic current (IPSC) decreases and the number of transmission failures increases. The increase in failures was correlated with the magnitude of IPSC reduction (**B**; $P < 0.01$). *Dashed lines* correspond to the 95% confidence interval. **C** The effect of kainate on the trial to trial fluctuation of the evoked IPSCs. Mean IPSC amplitude and its coefficient of variation (CV) were measured during kainate application and normalized by the respective control value in each cell. The fractional variation in amplitude (*M*) vs the fractional variation in CV^{-2} are plotted irrespective of the concentration of kainate used. Note that experimental data follow the predicted relation for a purely presynaptic (*dashed line*) rather than postsynaptic (*dotted line*) action. Adapted from RODRÍGUEZ-MORENO et al. (1997)

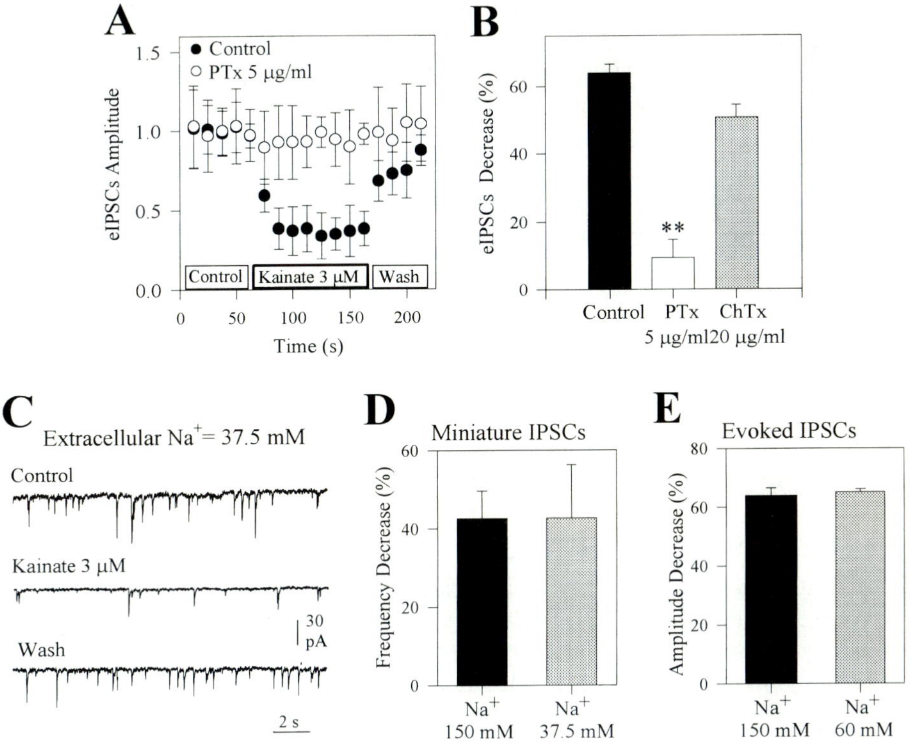

Fig. 9A–E. Kainate receptor modulation of γ-aminobutyric acid (GABA) release involves a metabotropic action and is independent of channel activity. **A** A typical experiment, showing the time course of evoked inhibitory postsynaptic current (eIPSC) amplitude before, during and after bath application of kainate in an untreated (*black circles*) and in a Pertussis-toxin (PTx)-treated (*open circles*) slice. **B** Effect of 3 μM kainate on eIPSC amplitude in normal slices (control) and slices treated with PTx or Cholera (ChTx) toxin. (**C–E**). The downregulation of GABA release by kainate is independent of ion-channel activity, since the reduction in extracellular Na^+ does not affect the degree of reduction in either miniature IPSC frequency induced by kainate (**C, D**) or in the evoked IPSC (E). Adapted from RODRIGUEZ-MORENO et al. (1998)

kainate and AMPA receptors were denominated as high-affinity and low-affinity kainate receptors, respectively, since subunits involved in each type presented high and low affinity for kainate when expressed in heterologous systems. However, as can be appreciated from the pharmacological data (Table 3), kainate receptors do not show an extremely high affinity for kainate, and even heteromeric formations including the high-affinity kainate-binding subunits KA1 and KA2 show a lower affinity than if these subunits were not included. Consequently, most likely low-affinity kainate-binding sites in the brain do not correspond to AMPA receptors but to kainate receptors. Taking into consideration that either type of receptor is able to be activated by kainate (all the kainate and AMPA receptors) and AMPA (GluR5 and any other kainate receptor, including KA subunits), it would be more appropriate to denominate them as AMPA-preferring and kainate-preferring, reflecting the lack of specificity for the ligands after which the receptors were named. However, taking into account that new and selective ligands are being developed and that the subunits forming each receptor have been identified, it is simpler just to use the terms AMPA and kainate receptors. Functionally, a rapid desensitizing response will always be induced by kainate at kainate receptors, whilst a weakly desensitizing response is expected for the activation of AMPA receptors. Similarly, kainate-receptor-mediated activation will predominate at low concentrations of kainate ($<3\,\mu M$), whilst AMPA-receptor activation will predominate at low concentrations of AMPA ($<10\,\mu M$).

H. Possible Pathological Implications for Kainate Receptors

Glutamate has been traditionally associated with neurodegenerative diseases and nerve injury (CHOI 1992). The ability of glutamic acid to kill neurons (excitotoxicity) seems to be mediated by interaction with NMDA receptors, leading to an uncontrollable rise in intracellular calcium concentrations and, from there, cell lysis and death. Consequently, most of the neurotoxic actions of glutamate have been ascribed to the overstimulation of NMDA receptors. Although weaker, there is also evidence that other glutamate receptors different from the NMDA type may also be involved in brain pathologies. However, in which processes are kainate receptors implicated? Molecular mapping of the human kainate receptors has revealed GluR5 located on chromosome 21 (EUBANKS et al. 1993; POTIER et al. 1993), suggesting a hypothetical role of GluR5 in Down's syndrome (GREGOR et al. 1994). Indeed, the region of chromosome 21 containing the gene encoding for GluR5 (GRIK1) has been associated with several phenotypic features found in partial monosomy 21, including several facial features, hypertonia and mental retardation (CHETTOUH et al. 1995). The same region that contains the GluR5 gene coincides with the localization of a mutant gene causing familial amyotrophic lateral sclerosis, leading to the hypothesis that a mutated GluR5 may be responsible for this disease (EUBANKS et al. 1993). However, direct proof of the involvement of kainate receptors in diseases like these is still lacking.

Perhaps the strongest indication available so far is that kainate receptors may play a role in epilepsy. Since interictal discharges can be systematically obtained following blockade of $GABA_A$ receptors, an impairment of GABAergic inhibition is thought to be one of the causes of some epilepsies. Thus, one rationale of epilepsy research assumes that, between inhibition and excitation, there is an imbalance favoring excitation. There is evidence supporting this assumption, the most important perhaps being the fact that potentiators of GABA function, such as benzodiazepines or barbiturates, are effective antiepileptic drugs. Consequently, results showing that kainate receptors control the efficiency of GABAergic synapses account for the potent action of kainate as an epileptogenic substance and point toward kainate receptors as potential targets for antiepileptic therapy. In keeping with this, by using advanced DNA technology and analytical methods, it has been demonstrated that allelic variants of GluR5 (GRIK1) but not GluR6 (GRIK2) confer susceptibility to juvenile absence epilepsy (SANDER et al. 1995; 1997), a common subtype of idiopathic, generalized epilepsy. Additional indications implicating kainate receptors in susceptibility towards epilepsy come from the fact that GluR6-deficient mice are much less susceptible to development of seizures induced by systemic injections of kainate than are normal animals (MULLE et al. 1998).

The possible involvement of kainate receptors in pain processing is derived from the observation that the primary afferent C-fiber neurons abundantly express GluR5 subunits and that the selective GluR5 antagonist, LY382884, which has no affinity for the AMPA receptors attenuates the paw-licking behavior induced by an injection of formalin under the skin, an animal model of persistent nociceptive stimulation (SIMMONS et al. 1998). Thus, these data implicate the kainate receptor GluR5 in mediating persistent spinal algesia.

I. Final Remarks

The story surrounding kainate receptors is not yet finished; on the contrary, it is just starting. There is a great diversity of subunit isoforms, and several of them could be expressed by one neuron at a time. In addition, it is very likely that all the subunits of kainate receptors have not yet been cloned, and associated proteins may also exist. The combination of pharmacological tools (drug design) and molecular-biology tools (transgenic animals) will produce interesting results that will certainly clarify the mechanisms triggered upon kainate-receptor activation in the near future. This knowledge will be crucial for our understanding of nervous-system function.

Acknowledgements. Work in my laboratory has been supported by grants from the Spanish Ministry of Education and Culture (DGICYT # PB93/0150, UE96/0007 and PM-0008/96), the Ministry of Health (FISSS 95/0869) and the European Union (BIO2-CT930243). I thank my collaborators, D. Guinea, J. C. López-García, M. Morales, A. V. Paternain, M. P. Regalado, A. Rodriguez-Moreno and A. Villarroel for their help with the work related to the subject of this chapter, and M. Sefton for editorial assistance. I also appreciate the generous gift of GYKI 53655 from Elli Lilly and Co. (Indianapolis, IN, USA). I am indebted to Drs. P. Castillo and G. Swanson for kindly providing me with their original records illustrated in Figs. 4 and 7.

References

Attwell D, Mobbs P (1994) Neurotransmitter transporters. Curr Op Neurobiol 4: 353–359

Bahn S, Volk B, Wisden W (1994) Kainate receptor gene expression in the developing rat brain. J Neurosci 14:5525–5547

Belcher SM, Howe JR (1997) Characterization of RNA editing of the glutamate-receptor subunits GluR5 and GluR6 in granule cells during cerebellar development. Brain Res Mol Brain Res 52:130–138

Benveniste H, Drejer J, Schousboe A, Diemer NH (1984) Elevation of extracellular concentration of glutamate and aspartate in rat hippocampus during transient cerebral ischemia monitored by intracerebral microdialysis. J Neurochem 43: 1369–1374

Bernard A, Khrestchatisky M (1994) Assessing the extent of RNA editing in the TMII regions of GluR5 and GluR6 kainate receptors during the rat brain development. J Neurochem 62:2057–2060

Bettler B, Boulter J, Hermans-Borgmeyer I, O'Shea-Greenfield A, Deneris ES, Moll C, Borgmeyer U, Hollmann M, Heinemann S (1990) Cloning of a novel glutamate receptor subunit, GluR5: expression in the nervous system. Neuron 5:583–595

Bettler B, Egebjerg J, Sharma G, Pecht G, Hermans-Borgmeyer I, Moll C, Stevens CF, Heinemann S (1992) Cloning of a putative glutamate receptor: a low affinity kainate-binding subunit. Neuron 8:257–265

Bischoff S, Barhanin J, Bettler B, Mulle C, Heinemann S (1997) Spatial distribution of kainate receptor subunit mRNA in the mouse basal ganglia and ventral mesencephalon. J Comp Neurol 379:541–562

Bleakman D, Ballyk B, Schoepp DD, Palmer AJ, Bath KP, Sharpe E, Wooley ML, Bufton H, Kamboj K, Tarnawa I, Lodge D (1996a) Activity of 2,3-benzodiazepines at native rat and recombinant human glutamate receptors in vitro: stereospecificity and selectivity profiles. Neuropharmacology 35:1689–1702

Bleakman D, Schoepp DD, Ballyk B, Bufton H, Sharpe E, Thomas K, Ornstein PL, Kamboj K (1996b) Pharmacological discrimination of GluR5 and GluR6 kainate receptor subtypes by (3S,4aR,6R,8aR)-6-[2-(1(2)H-tetrazole-5-yl)ethyl]decahydroisdoquinoline-3 carboxylic-acid. Mol Pharmacol 49:581–585

Carver JM, Mansson PE, Cortes-Burgos L, Zhou LM, Howe JR, Giordano T (1996) Cytotoxic effects of kainate ligands on HEK cell lines expressing recombinant kainate receptors. Brain Res 720:69–74

Castillo PE, Malenka RC, Nicoll RA (1997) Kainate receptors mediate a slow postsynaptic current in hippocampal CA3 neurons. Nature 388:182–186

Cattaneo R (1991) Different types of messenger RNA editing. Annu Rev Genet 25:71–88

Cherubini E, Rovira C, Ben-Ari Y, Nistri A (1990) Effects of kainate on excitability of rat hippocampal neurons. Epilepsy Res 5:18–27

Chettouh Z, Croquette MF, Delobel B, Gilgenkrants S, Leonard C, Maunoury C, Prieur M, Rethore MO, Sinet PM, Chery M, et al (1995) Molecular mapping of 21 features associated with partial monosomy 21: involvement of the APP-SOD1 region. Am J Hum Genet 57:62–71

Chittajallu R, Vignes M, Dev KK, Barnes JM, Collingridge GL, Henley JM (1996) Regulation of glutamate release by presynaptic kainate receptors in the hippocampus. Nature 379:78–81

Choi DW (1992) Bench to bedside: the glutamate connection. Science 258:241–243

Clarke VRJ, Ballyk BA, Hoo KH, Mandelzys A, Pellizari A, Bath CP, Thomas J, Sharpe EF, Davies, CH, Ornstein PL, Schoepp DD, Kamboj RK, Collingridge GL, Lodge D, Bleakman D (1997) A hippocampal GluR5 kainate receptor regulating inhibitory synaptic transmission. Nature 389:599–602

Cossart R, Esclapez M, Hirsch JC, Bernard C, Ben-Ari Y (1998) GluR5 kainate receptor activation in interneurons increases tonic inhibition of pyramidal cells. Nature Neurosci 1:470–478

Coyle JT (1983) Neurotoxic actions of kainic acid. J Neurochem 41:1–11

Egebjerg J, Heinemann SF (1993) Ca^{2+} permeability of unedited and edited versions of the kainate selective glutamate receptor GluR6. Proc Natl Acad Sci USA 90:755–759

Egebjerg J, Bettler B, Hermans-Borgmeyer I, Heinemann S (1991) Cloning of a cDNA for a glutamate receptor subunit activated by kainate but not by AMPA. Nature 351:745–748

Ehlers MD, Mammen AL, Lau LF, Huganir RL (1996) Synaptic targeting of glutamate receptors. Curr Opin Cell Biol 8:484–489

Eubanks JH, Puranm RS, Kleckner NW, Bettler B, Heinemann SF, McNamara JO (1993) The gene encoding the glutamate receptor subunit GluR5 is located on human chromosome 21q21.1–22.1 in the vicinity of the gene for familial amyotrophic lateral sclerosis. Proc Natl Acad Sci USA 90:178–182

Fisher RS, Alger BE (1984) Electrophysiological mechanisms of kainic-acid-induced epileptiform activity in the rat hippocampal slice. J Neurosci 4:1312–1323

Fleischmann BK, Murray RK, Kotlikoff MI (1994) Voltage window for sustained elevation of cytosolic calcium in smooth muscle cells. Proc Natl Acad Sci USA 91:11914–11918

Frerking M, Malenka RC, Nicoll RA (1998) Synaptic activation of kainate receptors on hippocampal interneurons. Nature Neurosci 1:479–486

Gregor P, Gaston SM, Yang X, O'Regan JP, Rosen DR, Tanzi RE, Patterson D, Haines JL, Horvitz HR, Uhl GR, et al (1994) Genetic and physical mapping of the GLUR5 glutamate receptor gene on human chromosome 21. Hum Genet 94:565–570

Gregor P, Mano I, Maoz I, McKeown M, Teichberg VI (1989) Molecular structure of the chick cerebellar kainate-binding subunit of a putative glutamate receptor. Nature 342:689–692

Heckmann M, Bufler J, Franke C, Dudel J (1996) Kinetics of homomeric GluR6 glutamate receptor channels. Biophys J 71:1743–1750

Henley JM (1994) Kainate-binding proteins: phylogeny, structures and possible functions. Trends Pharmacol Sci 15:182–90

Herb A, Burnashev N, Werner P, Sakmann B, Wisden W, Seeburg PH (1992). The KA-2 subunit of excitatory amino acid receptors shows widespread expression in brain and forms ion channels with distantly related subunits. Neuron 8:775–785

Hille B (1992) G-protein-coupled mechanisms and nervous signaling. Neuron 9: 187–195

Honore T, Davies SN, Drejer J, Fletcher EJ, Jacobsen P, Lodge D, Nielsen FE (1988) Quinoxalinediones: potent competitive non-NMDA glutamate receptor antagonists. Science 241:701–703

Hoo KH, Nutt SL, Fletcher EJ, Elliott CE, Korczak B, Deverill RM, Rampersad V, Fantaske RP, Kamboj RK (1994) Functional expression and pharmacological characterization of the human EAA4 (GluR6) glutamate receptor: a kainate selective channel subunit. Receptors Channels 2:327–337

Howe JR (1996) Homomeric and heteromeric ion channels formed from the kainate-type subunits GluR6 and KA2 have very small, but different, unitary conductances. J Neurophysiol 76:510–519

Huettner JE (1990) Glutamate receptor channels in DRG neurons-activation by kainate and quisqualate and blockade of desensitization by Con-A. Neuron 5: 255–266

Ishimaru H, Kamboj R, Ambrosini A, Henley JM, Soloviev MM, Sudan H, Rossier J, Abutidze K, Rampersad V, Usherwood PN, Bateson AN, Barnard EA (1996) A unitary non-NMDA receptor short subunit from Xenopus: DNA cloning and expression. Receptors Channels 4:31–49

Johansen TH, Drejer T, Wätjen F, Nielsen EO (1993) A novel non-NMDA receptor antagonist shows selective displacement of low-affinity ^{3}H-kainate binding, Eur J Pharmacol-Mol. Pharmacol Sect 246:195–204
Jones KA, Wilding TJ, Huettner JE, Costa AM (1997) Desensitization of kainate receptors by kainate, glutamate and diastereisomers of 4-methylglutamate. Neuropharmacology 36:853–863
Kamboj RK, Schoepp DD, Nutt S, Shekter L, Korczak B, True RA, Zimmerman, Wosnick MA (1992) Molecular structure and pharmacological characterization of humEAA2, a novel human kainate receptor subunit. Mol Pharmacol 42:10–15
Kamboj RK, Schoepp DD, Nutt S, Shekter L, Korczak B, True RA, Rampersad V, Zimmerman DM, Wosnick MA (1994) Molecular cloning, expression, and pharmacological characterization of humEAA1, a human kainate receptor subunit. J Neurochem 62:1–9
Kamiya H, Ozawa S (1998) Kainate receptor-mediated inhibition of presynaptic Ca^{2+} influx and EPSP in area CA1 of the rat hippocampus. J Physiol (Lond) 509:833–845
Köhler M, Burnashev N, Sakmann B, Seeburg PH (1993) Determinants of Ca^{2+} permeability in both TM1 and TM2 of high affinity kainate receptor channels: diversity by RNA editing. Neuron 10:491–500
Korczak B, Nutt SL, Fletcher EJ, Hoo KH, Elliott CE, Rampersad V, McWhinnie EA, Kamboj RK (1995) cDNA cloning and functional properties of human glutamate receptor EAA3 (GluR5) in homomeric and heteromeric configuration. Receptors Channels 3:41–49
Lauridsen J, Honore T, Krogsgaard-Larsen P (1985) Ibotenic acid analogues. Synthesis, molecular flexibility, and in vitro activity of agonists and antagonists at central glutamic acid receptors. J Med Chem 28:668–672
Lerma J (1997) Kainate reveals its targets. Neuron 19:1155–1158
Lerma J (1998) Kainate receptors: an interplay between excitatory and inhibitory synapses. FEBS Letters 430:100–104
Lerma J, Herranz AS, Herreras O, Abraira V, Martin del Rio R (1986). In vivo determination of extracellular concentration of amino acids in the rat hippocampus. A method based on brain dialysis and computerized analysis. Brain Res 384:145–155
Lerma J, Morales M, Vicente MA, Herreras O (1997) Glutamate receptors of the kainate type and synaptic transmission. Trends in Neurosci 20:9–12
Lerma J, Paternain AV, Naranjo JR, Mellström B (1993) Functional kainate-selective glutamate receptors in cultured hippocampal neurons. Proc Natl Acad Sci USA 90:11688–11692
Lomeli H, Wisden W, Köhler M, Keinänen K, Sommer B, Seeburg PH (1992) High-affinity kainate and domoate receptors in rat brain. FEBS Lett 307:139–143
Mackler SA, Eberwine JH (1993) Diversity of glutamate receptor subunits mRNA expression within live hippocampal CA1 neurons. Mol. Pharmacol 44:308–315
Matute C, Sanchez-Gomez MV, Martinez-Millan L, Miledi R (1997) Glutamate receptor-mediated toxicity in optic nerve oligodendrocytes. Proc Natl Acad Sci USA 94:8830–8835
Morita T, Sakimura K, Kushiya E, Yamazaki M, Meguro H, Araki K, Abe T, Mori KJ, Mishina M (1992) Cloning and functional expression of a cDNA encoding the mouse beta 2 subunit of the kainate-selective glutamate receptor channel. Brain Res Mol-Brain Res 14:143–146
Mulle C, Sailer A, Perez-Otaño I, Dickinson-Anson H, Castillo PE, Bureau I, Maron C, Gage FH, Mann JR, Bettler B, Heinemann SF (1998) Altered synaptic physiology and reduced susceptibility to kainate-induced seizures in GluR6-deficient mice. Nature 392:601–605
Nishimune A, Isaac JT, Molnar E, Noel J, Nash SR, Tagaya M, Collingridge GL, Nakanishi S, Henley JM (1998) NSF binding to GluR2 regulates synaptic transmission. Neuron 21:87–97
Nutt SL, Hoo KH, Rampersad V, Deverill RM, Elliott CE, Fletcher EJ, Adams SL, Korczak B, Foldes RL, Kamboj RK (1994) Molecular characterization of the

human EAA5 (GluR7) receptor: a high-affinity kainate receptor with novel potential RNA editing sites. Receptors Channels 2:315–326

Osten P, Srivastava S, Inman GJ, Vilim FS, Khatri L, Lee LM, States BA, Einheber S, Milner TA, Hanson PI, Ziff EB (1998) The AMPA receptor GluR2 C terminus can mediate a reversible, ATP-dependent interaction with NSF and α- and β-SNAPs. Neuron 21:99–110

Paschen W, Blackstone CD, Huganir RL, Ross CA (1994) Human GluR6 kainate receptor (GRIK2): molecular cloning, expression, polymorphism, and chromosomal assignment. Genomics 20:435–440

Paschen W, Schmitt J, Gissel C, Dux E (1997) Developmental changes of RNA editing of glutamate receptor subunits GluR5 and GluR6: in vivo versus in vitro. Brain Res Dev Brain Res 98:271–280

Paternain AV, Morales M, Lerma J (1995) Selective antagonism of AMPA receptors unmasks kainate receptor-mediated responses in hippocampal neurons. Neuron 14:185–189

Paternain AV, Rodríguez-Moreno A, Villarroel A, Lerma J (1998) Activation and desensitization properties of native and recombinant kainate receptors. Neuropharmacology 37:1249–1259

Paternain AV, Vicente MA, Nielsen EØ, Lerma J (1996) Comparative antagonism of kainate-activated AMPA and kainate receptors in hippocampal neurons. Eur. J. Neurosci 8:2129–2136

Patneau DK, Wright PW, Winters C, Mayer ML, Gallo V (1994) Glial cells of the oligodendrocyte lineage express both kainate- and AMPA-preferring subtypes of glutamate receptor. Neuron 12:357–371

Pemberton KE, Belcher SM, Ripellino JA, Howe JR (1998) High-affinity kainate-type ion channels in rat cerebellar granule cells. J Physiol (Lond) 510:401–420

Potier MC, Dutriaux A, Lambolez B, Bochet P, Rossier J (1993) Assignment of the human glutamate receptor gene GLUR5 to 21q22 by screening a chromosome 21 YAC library. Genomics 15:696–697

Price CJ, Raymond LA (1996) Evans blue antagonizes both alpha-amino-3-hydroxy-5-methyl-4-isoxazolepropionate and kainate receptors and modulates receptor desensitization. Mol Pharmacol 50:1665–1671

Raymond LA, Blackstone CD, Huganir RL (1993) Phosphorylation and modulation of recombinant GluR6 glutamate receptors by cAMP-dependent protein kinase. Nature 361:637–641

Rodriguez-Moreno A, Lerma J (1998) Kainate receptor modulation of GABA release involves a metabotropic function. Neuron 20:1211–1218

Rodríguez-Moreno A, Herreras O, Lerma J (1997) Kainate receptors presynaptically downregulate GABAergic inhibition in the rat hippocampus. Neuron 19:893–901

Rosenmund C, Stern-Bach Y, Stevens CF (1998) The tetrameric structure of a glutamate receptor channel. Science 280:1596–1599

Ruano D, Lambolez B, Rossier J, Paternain AV, Lerma J (1995) Kainate receptors subunits expressed in single cultured hippocampal neurons: molecular and functional variants by RNA editing. Neuron 14:1009–1017

Sahara Y, Noro N, Iida Y, Soma K, Nakamura Y (1997) Glutamate receptor subunits GluR5 and KA2 are coexpressed in rat trigeminal ganglion neurons. J Neurosci 17:6611–6620

Sakimura K, Morita T, Kushiya E, Mishina M (1992) Primary structure and expression of the γ2 subunit of the glutamate receptor channel selective for kainate. Neuron 8:267–274

Sander T, Hildmann T, Kretz R, Fürst R, Sailer U, Bauer G, Schmitz B, Beck-Mannagetta G, Wienker T, Janz D (1997) Allelic association of juvenile absence epilepsy with a GluR5 kainate receptor gene (GRIK1) polymorphism. Am. J. Med. Genet. 74:416–421

Sander T, Janz D, Ramel C, Ross CA, Paschen W, Hildmann T, Wienker TF, Bianchi A, Bauer G, Sailer U, et al (1995) Refinement of map position of the human GluR6

kainate receptor gene (GRIK2) and lack of association and linkage with idiopathic generalized epilepsies. Neurology 45:1713–1720
Schiffer HH, Swanson GT, Heinemann SF (1997) Rat GluR7 and a carboxy-terminal splice variant, GluR7b, are functional kainate receptor subunits with a low sensitivity to glutamate. Neuron 19:1141–1146
Seeburg PH (1996) The role of RNA editing in controlling glutamate receptor channel properties. J Neurochem 66:1–5
Simmons RM, Li DL, Hoo KH, Deverill M, Ornstein PL, Iyengar S (1998) Kainate GluR5 receptor subtype mediates the nociceptive response to formalin in the rat. Neuropharmacology 37:25–36
Sloviter RS, Damiano BP (1981) On the relationship between kainic acid-induced epileptiform activity and hippocampal neuronal damage. Neuropharmacology 20:1001–1011
Smirnov SV, Aaronson PI (1992) Ca^{2+} currents in single myocytes from human mesenteric arteries: evidence for a physiological role of L-type channels. J Physiol (Lond.) 457:455–475
Sommer B, Burnashev N, Verdoorn TA, Keinänen K, Sakmann B, Seeburg PH (1992) A glutamate receptor channel with high affinity for domoate and kainate. EMBO J 11:1651–1656
Sommer B, Seeburg PH (1992) Glutamate receptor channels: novel properties and new clones. Trends Pharmacol Sci 13:291–296
Swanson, GT, Heinemann SF (1998) Heterogeneity of homomeric GluR5 kainate receptor desensitization expressed in HEK293 cells. J Physiol (Lond) 15:639–646
Swanson GT, Feldmeyer D, Kaneda M, Cull-Candy SG (1996) Effect of RNA editing and subunit co-assembly single-channel properties of recombinant kainate receptors. J Physiol (Lond) 492:129–142
Swanson GT, Green T, Heinemann SF (1998) Kainate receptors exhibit differential sensitivities to (S)-5-iodowillardiine. Mol Pharmacol 53:942–949
Swanson GT, Kamboj SK, Cull-Candy SG (1997) Single-channel properties of recombinant AMPA receptors depend on RNA editing, splice variation, and subunit composition. J Neurosci 17:58–69
Swope SL, Moss SJ, Blackstone CD, Huganir RL (1992) Phosphorylation of ligand-gated ion channels: a possible mode of synaptic plasticity. FASEB J 6:2514–2523
Traynelis SF, Wahl P (1997) Control of rat GluR6 glutamate receptor open probability by protein kinase A and calcineurin. J Physiol (Lond) 503:513–531
Verdoorn TA, Johansen TH, Drejer J, Nielsen EØ (1994), Selective block of recombinant GluR6 receptors by NS-102, a novel non-NMDA receptor antagonist, Eur J Pharmacol (Mol Pharmacol Sect) 269:43–49
Vignes M, Bleakman D, Lodge D, Collingridge GL (1997) The synaptic activation of the GluR5 subtype of kainate receptor in area CA3 of the rat hippocampus. Neuropharmacology 36:1477–1481
Vignes M, Collingridge GL (1997) The synaptic activation of kainate receptors. Nature 388:179–182
Villmann C, Bull L, Hollmann M (1997) Kainate binding proteins possess functional ion channel domains. J Neurosci 17:7634–7643
Vizi ES, Mike A, Tarnawa I (1996) 2,3-Benzodiazepines (GYKI 52466 and analogs): Negative allosteric modulators of AMPA receptors. CNS Drug Rev. 2:91–126
Wada K, Dechesne CJ, Shimasaki S, King RG, Kusano K, Buonanno A, Hampson DR, Banner C, Wenthold RJ, Nakatani Y (1989) Sequence and expression of a frog brain complementary DNA encoding a kainate-binding protein. Nature 342: 684–689
Wang Y, Small DL, Stanimirovic DB, Morley P, Durkin JP (1997) AMPA receptor-mediated regulation of a G_i-protein in cortical neurons. Nature 389:502–504
Wang LY, Taverna FA, Huang XP, MacDonald JF, Hampson DR (1993) Phosphorylation and modulation of a kainate receptor (GluR6) by cAMP-dependent protein kinase. Science 259:1173–1175

Wenthold RJ, Hampson DR, Wada K, Hunter C, Oberdorfer MD, Dechesne CJ (1990) Isolation, localization, and cloning of a kainic acid binding protein from frog brain. J Histochem Cytochem 38:1717–1723

Werner P, Voigt M, Keinänen K, Wisden W, Seeburg PH (1991) Cloning of a putative high-affinity kainate receptor expressed predominantly in hippocampal CA3 cells. Nature 351:742–744

Westbrook GL, Lothman EW (1983) Cellular and synaptic basis of kainic acid-induced hippocampal epileptiform activity. Brain Res 273:97–109

Wilding TJ, Huettner JE (1995) Differential antagonism of α-amino-3-hydrosy-5-methyl-4-isoxazolepropionic acid-preferring and kainate-preferring receptors by 2,3-benzodiazepines. Mol Pharmacol 47:582–587

Wilding TJ, Huettner JE (1996) Antagonism pharmacology of kainate- and α-amino-3-hydroxy-5-methyl-4-isoxazolepropionic acid-preferring receptors. Mol Pharmacol 49:540–546

Wilding TJ, Huettner JE (1997) Activation and desensitization of hippocampal kainate receptors. J Neurosci 17:2713–2721

Williams SR, Toth TI, Turner JP, Hughes SW, Crunelli V (1997) The 'window' component of the low threshold Ca^{2+} current produces input signal amplification and bistability in cat and rat thalamocortical neurons. J Physiol (Lond) 505:689–705

Wo ZG, Oswald RE (1994) Transmembrane topology of two kainate receptor subunits revealed by *N*-glycosylation. Proc Natl Acad Sci USA 91:7154–7158

Wong LA, Mayer ML (1993) Differential modulation by cyclothiazide and concanavalin A of desensitization at native alpha-amino-3-hydroxy-5-methyl-4-isoxazolepropionic acid- and kainate-preferring glutamate receptors. Mol Pharmacol 44:504–510

Wong LA, Mayer ML, Jane DE, Watkins JC (1994) Willardiines differentiate agonist binding sites for kainate- versus AMPA-preferring glutamate receptors in DRG and hippocampal neurons. J Neurosci 14, 3881–3897

Young AB, Fagg GE (1990) Excitatory amino acid receptors in the brain: membrane binding and receptor autoradiographic approaches. Trends Pharmacol Sci 11:126–133

Zhou L-M, Gu Z-Q, Costa AM, Yamada KA, Mansson PE, Giordano T, Skolnick P, Jones KA (1997) (2S,4R)-4-methylglutamic acid (SYM 2081): a selective, high-affinity ligand for kainate receptors. J Pharmacol Exp Ther 280:422–427

CHAPTER 9

Molecular Determinants Controlling Functional Properties of AMPARs and NMDARs in the Mammalian CNS

H. MONYER, P. JONAS, and J. ROSSIER

Abbreviations

AMPARs	α-Amino-3-hydroxy-5-methyl-4-isoxazolepropionate receptors
Ba^{2+}	Barium
C-terminal	Carboxy-terminal
Ca1/3	Ca1/3 fields of the hippocampus
Ca^{2+}	Calcium
CNS	Central nervous system
EPSC	Excitatory postsynaptic current
GABAR	γ-Amino-butyric-acid receptor
GluR	Glutamate receptors
H	Histidine
HEK cells	Human embryonic kidney cells
I–V curve	Current–voltage relation
IC_{50}	Half-maximal inhibitory concentration
K_d	Dissociation constant
LTD	Long-term depression
LTP	Long-term potentiation
M2	Membrane segment 2
Mg^{2+}	Magnesium
mRNA	Messenger ribonucleic acid
N-terminal	Amino-terminal
N	Asparagine
NMDARs	*N*-methyl-D-aspartate receptors
Q	Glutamine
R	Arginine
RT-PCR	Reverse transcriptase–polymerase chain reaction

A. Introduction

L-α-amino-3-hydroxy-5-methyl-4-isoxazolepropionate receptors (AMPARs) and *N*-methyl-D-aspartate receptors (NMDARs) are the two major types of postsynaptic glutamate receptors (GluRs) that mediate excitatory synaptic transmission in the mammalian central nervous system (CNS). Both AMPARs and NMDARs are multimeric proteins, probably tetramers, formed by a variety of molecularly distinct subunits. AMPARs can be assembled from four types of subunits, termed GluR-A, -B, -C, and -D (or, in an alternative nomenclature, GluR1, GluR2, GluR3, and GluR4). Additional molecular diversity of AMPARs is generated by alternative splicing of the flip-flop module and RNA editing at the Q/R and R/G site. NMDARs are heteromers primarily assembled from NR1 subunits and NR2A, B, C, or D subunits. Various splice variants have been identified for the NR1 subunit, and a new NR3 subunit has been discovered recently. Considering all combinatorial possibilities, the molecular diversity of glutamate-receptor channels is considerable (HOLLMANN, this volume).

What is the functional significance of the molecular diversity of GluRs? In this chapter, we review the relation between structure and function for both recombinant and native GluRs. We suggest that GluR-subunit expression, alternative splicing, and RNA editing contribute to differential signalling at glutamatergic synapses in the mammalian CNS.

B. AMPA Receptors

I. Recombinant Receptors

Members of the AMPAR subunit family (GluR-A, -B, -C, and -D) have been recombinantly expressed in both mammalian cell lines, e.g. human embryonic kidney (HEK) cells, and *Xenopus* oocytes (HOLLMANN et al. 1989; NAKANISHI et al. 1990; VERDOORN et al. 1991; BURNASHEV et al. 1992a). Unlike many other ligand-gated ion channels, such as nicotinic acetylcholine receptors of skeletal muscle, glycine receptors, or γ-amino-butyric-acid-A receptors ($GABA_A$Rs), heterologously expressed AMPAR subunits can form both homomeric and heteromeric receptors.

1. Divalent Permeability

Homomeric AMPARs assembled from different subunits differ substantially in their permeability to Ca^{2+} ions (HOLLMANN et al. 1991; BURNASHEV et al. 1992a). Receptors formed from GluR-B subunits have a low Ca^{2+} permeability (P), with $P_{Ca}/P_{monovalent}$ values of <0.1 (BURNASHEV et al. 1992a), as described previously for native AMPARs in hippocampal and spinal neurons (MAYER and WESTBROOK 1987; IINO et al. 1990, JONAS and SAKMANN 1992). In contrast, receptors assembled from GluR-A, -C, and-D subunits are highly Ca^{2+} -permeable, with $P_{Ca}/P_{monovalent}$ values of approximately 2 (BURNASHEV et al. 1992a). In addition, homomeric GluR-A, -C, and -D channels are also permeable to Mg^{2+} (BURNASHEV et al. 1992a) and Ba^{2+} (DINGLEDINE et al. 1992).

A single amino acid difference in the pore forming segment M2 was identified, by mutational analysis, as the molecular determinant of the subunit-specific difference in divalent permeability (HUME et al. 1991; BURNASHEV et al. 1992a; KUNER et al., this volume). In this position, the GluR-B subunit contains a positively charged arginine (R), whereas GluR-A, -C, and -D subunits have a neutral glutamine residue (Q). When the R residue in GluR-B is replaced by Q, recombinant channels with high Ca^{2+} permeability are generated (BURNASHEV et al. 1992a). Similarly, when the Q residue in GluR-A, -C, or -D is replaced by R, channels with low Ca^{2+} permeability are formed (BURNASHEV et al. 1992a; DINGLEDINE et al. 1992). Thus, the residue at the Q/R site is the main molecular determinant of the Ca^{2+} permeability of recombinant AMPARs.

Very surprising was the finding that the genomic DNA sequences encoding M2s of all AMPAR subunits carry a Q codon, and that the R codon is selectively introduced into GluR-B mRNA by RNA editing (SOMMER et al. 1991). Editing at the Q/R site has been shown to be dependent on the presence, in the GluR-B gene, of an intronic sequence complementary to the editing site (HIGUCHI et al. 1993), and potential editases mediating the adenosine-to-inosine conversion have been identified (MELCHER et al. 1996). For details, the reader is referred to a recent review (SEEBURG et al. 1998).

In heteromeric combinations, GluR-B(R) subunits dominate the Ca^{2+} permeability of recombinant receptors (BURNASHEV et al. 1992a; WASHBURN et al. 1997). When Q-form and R-form subunits are expressed in equal amounts, heteromeric channels with low Ca^{2+} permeability are formed [similar to GluR-B(R) homomers]. Hence, the abundant expression of GluR-B subunit mRNA and protein in the mammalian CNS (KEINÄNEN et al. 1990; PETRALIA et al. 1997) and the high degree of editing of GluR-B at the Q/R site (virtually 100% in the adult brain; BURNASHEV et al. 1992a) explain the low Ca^{2+} permeability of native AMPARs in the majority of central neurons (MAYER and WESTBROOK 1987; IINO et al. 1990; JONAS and SAKMANN 1992). Coexpression of Q-form and R-form subunits at a ratio greater than 1 (for example, 10:1), however, resulted in the formation of recombinant AMPAR mosaics with intermediate average Ca^{2+} permeabilities ($P_{Ca}/P_{monovalent}$ = ~0.5; BURNASHEV et al. 1992a). Variation of the relative abundances of GluR-B transcripts resulted in average divalent permeabilities that extended over a wide range (BURNASHEV et al. 1992a; WASHBURN et al. 1997).

2. Blocking by Polyamines, and Single-Channel Conductance

In addition to the Ca^{2+} permeability, the GluR-B(R) subunit also determines current rectification and single-channel conductance of recombinant AMPARs. The current–voltage relation (I–V) of homomeric GluR-B(R) channels or heteromeric channels containing GluR-B(R) is linear or outwardly rectifying. In contrast, the I–V relation of AMPAR channels assembled from GluR-A, -C, and -D subunits has an inwardly or doubly rectifying shape (NAKANISHI et al. 1990; VERDOORN et al. 1991; HOLLMANN et al. 1991) with a region of low conductance between 0 mV and approximately +50 mV. Subsequent studies have shown that current rectification is due to voltage-dependent blocking by intracellular polyamines, such as spermine or spermidine, which block GluR-B(R)-free, but not GluR-B(R)-containing, AMPARs (BOWIE et al., this volume). Since polyamines are ubiquitous cytoplasmic factors (PEGG 1988; WATANABE et al. 1991), rectification generated via blocking by polyamines is generally thought to be a reliable marker of GluR-B(R)-free receptors. Furthermore, blocking by intracellular polyamines may be of physiological significance (ROZOV et al. 1998; see below).

In addition to controlling blocking by naturally occurring polyamines from the inside, the GluR-B(R) content of AMPARs determines the degree of blocking by external polyamines, polyamine toxins (argiotoxin, joro spider toxin, philanthotoxin) and synthetic analogues [*N*-(4-hydroxyphenylpropanoyl)-spermine]. As observed for internal blocking, externally applied polyamines have strong effects on GluR-B(R)-free AMPARs and negligible effects on GluR-B(R)-containing receptors (BLASCHKE et al. 1993; HERLITZE et al. 1993). Thus, polyamine toxins are useful tools for identifying the putative subunit composition of native receptors (IINO et al. 1996; MAHANTY and SAH 1998; TOTH and MCBAIN 1998).

High Ca^{2+} permeability and the doubly rectifying I–V relation of AMPAR-mediated currents appear to be often coupled, but the correlation is not absolute. First, I–V relations in the absence of intracellular polyamines are linear or outwardly rectifying independent of the subunit composition. Second, Ca^{2+} permeability and blocking by internal polyamines can be dissociated by mutation. When the Q residue at the Q/R site is replaced by a histidine (H) or an asparagine (N) or when aspartate residues in the inner mouth of the channel are replaced by neutral amino acids, the recombinant channels are Ca^{2+} permeable but show a linear I–V relation (CURUTCHET et al. 1992; BURNASHEV et al. 1992a; DINGLEDINE et al. 1992). Finally, the relation between $P_{divalent}/P_{monovalent}$ and the relative abundance of GluR-B(R) mRNA is very steep, whereas that between the K_d for blocking by internal polyamines and the relative abundance of GluR-B(R) mRNA is more gradual (WASHBURN et al. 1997). Possibly, a single GluR-B(R) subunit per channel is sufficient to suppress the Ca^{2+} permeability, whereas multiple GluR-B(R) subunits are necessary to abolish polyamine sensitivity. This further supports the idea that, within the constraints of stoichiometry, the number of a particular subunit can vary.

The single-channel conductance of recombinant AMPARs is also dependent on the subunit composition. Homomeric GluR-D channels have a substantially larger single-channel conductance (main conductance: 7–8 pS) than homomeric GluR-B(R) channels (300 fS; SWANSON et al. 1997); GluR-B(R)/GluR-D heteromeric channels showed intermediate channel conductances. Together with previous studies on recombinant kainate receptors, these results indicate that the residue at the Q/R site is a critical determinant of the single-channel conductance (SWANSON et al. 1996, 1997).

3. Gating Properties

Unlike the molecular determinants of Ca^{2+} permeability, blocking by polyamines, and single-channel conductance, the molecular factors shaping the gating properties of recombinant AMPARs are more complex. Subunit composition (MOSBACHER et al. 1994; PARTIN et al. 1994; SEKIGUCHI et al. 1997), alternative splicing of the flip-flop module (SOMMER et al. 1990; MOSBACHER et al. 1994), and editing of GluR-B, -C, and -D at the R/G site (LOMELI et al. 1994) have an influence on the time courses of deactivation, desensitization, and recovery from desensitization. Both the flip-flop module and the R/G site are, presumably, located in the extracellular loop between M3 and M4, which is thought to be part of the agonist-binding site; this suggests relationships between agonist binding and gating (STERN-BACH et al. 1994, 1998).

The desensitization time constants of recombinant AMPARs (obtained by fitting with a single exponential the decay of the current activated by a step application of glutamate) vary between 0.8 ms and 6.1 ms for recombinant AMPARs expressed in oocytes (at 22°C; MOSBACHER et al. 1994; MOSBACHER 1995) and between 3 ms and 34 ms for AMPARs expressed in HEK cells (BURNASHEV 1993; PARTIN et al. 1994; SEKIGUCHI et al. 1997). Subunit composition, alternative splicing, and editing control the kinetics of desensitization. Although dependent on the expression system, the desensitization time constants follow the sequence: GluR-A_i ≈ GluR-A_o ≈ GluR-C_i ≈ GluR-D_i >> GluR-C_o ≈ GluR-D_o, with "o" designating the flop and "i" the flip splice form (MOSBACHER et al. 1994; PARTIN et al. 1996; SEKIGUCHI et al. 1997). Homomeric GluR-B channels are difficult to measure due to the low current density, which is probably caused by the small single-channel conductance (SWANSON et al. 1997). BURNASHEV (1993) has reported a desensitization time constant of 32 ms for homomeric GluR-B$(R)_i$ channels, but the glutamate application in these experiments appears to be relatively slow. Heteromeric channels formed by GluR-B_i in combination with other subunits, however, have much slower desensitization than the respective homomers, whereas GluR-B_o has only a small effect (MOSBACHER et al. 1994; PARTIN et al. 1996; SEKIGUCHI et al. 1997). These results suggest that GluR-B_i is a critical determinant of slow AMPAR desensitization. The effect of editing at the R/G site is variable. Unedited (R) forms show faster desensitization than edited (G) forms for GluR-C_i and -D_i receptors, but opposite effects are observed for GluR-D_o receptors (LOMELI et al. 1994).

Whereas the desensitization time course is a robust parameter that is relatively easy to measure experimentally, the time courses of deactivation and recovery from desensitization may be more important for determining the shapes and amplitudes of the excitatory postsynaptic currents (EPSCs; COLQUHOUN et al. 1992; TRUSSELL et al. 1993). For the deactivation time constant, a marked correlation to the desensitization time constant has been noted (MOSBACHER 1995; PARTIN et al. 1996), suggesting that deactivation and desensitization are controlled by similar, if not identical, structural properties. For the recovery from desensitization, the molecular determinants are more complex. The time constant of recovery from desensitization of recombinant AMPARs varies between 6 and 61 ms (with 1-ms glutamate pulses; LOMELI et al. 1994), depending on subunit composition, splicing, and editing. Here too, no uniform picture has emerged; the slowest recovery is found for GluR-A, with little difference between flip and flop forms (MOSBACHER 1995), whereas faster recovery is observed for GluR-C and -D, with flop variants being slower than flip variants (LOMELI et al. 1994). A consistent effect on recovery from desensitization is exerted by R/G editing; AMPARs assembled from edited (G)-form subunits recover from desensitization more rapidly than those assembled from unedited (R)-form subunits (LOMELI et al. 1994).

II. Native AMPA Receptors

1. Differential Developmental and Regional Expression of AMPAR Subunits

The expression of all AMPAR subunits is regulated developmentally and differs substantially among neuron types. The GluR-B subunit is the first subunit that appears during development (the earliest time point that had been investigated was E16 in the rat; DURAND and ZUKIN 1993), whereas the expression of the other AMPAR subunits is upregulated at a slightly later stage during development (MONYER et al. 1991). Thus, in the embryonic brain, the GluR-B subunit mRNA is expressed more abundantly than the other subunits. In the adult brain, AMPAR-subunit mRNA and protein are expressed differentially in different circuits and in different types of neurons in the same circuit (KEINÄNEN et al. 1990; PETRALIA and WENTHOLD 1992). For example, GluR-A, -B, and -C are expressed abundantly in both the hippocampus and the cerebellum, with higher GluR-C mRNA levels in the cortex than in the hippocampus. In contrast, GluR-D mRNA is found at high levels in the cerebellum, thalamus, and brain stem (KEINÄNEN et al. 1990; Fig. 1A).

High expression levels of flip versions are found at birth, whereas the expression of flop versions gradually increases postnatally and shows a more restricted expression pattern (MONYER et al. 1991). Cell-type-specific differences in the expression of splice versions are remarkable. In pyramidal neurons of the hippocampal CA3 region, flip versions dominate whereas, in neurons of the CA1 subfield, flip and flop versions coexist (SOMMER et al. 1990).

Editing at both the Q/R and R/G sites is regulated developmentally, but the temporal profiles are very different. The extent of Q/R-site editing in the rat is 99% at E14 and rises to virtually 100% after birth in the rat (BURNASHEV et al. 1992a). In contrast, editing at the R/G site increases gradually from 20% at E14 to 70% at P21 in the rat (LOMELI et al. 1994). The differential developmental regulation of Q/R- and R/G-site editing is not surprising, since editing at the two sites appears to require different editases

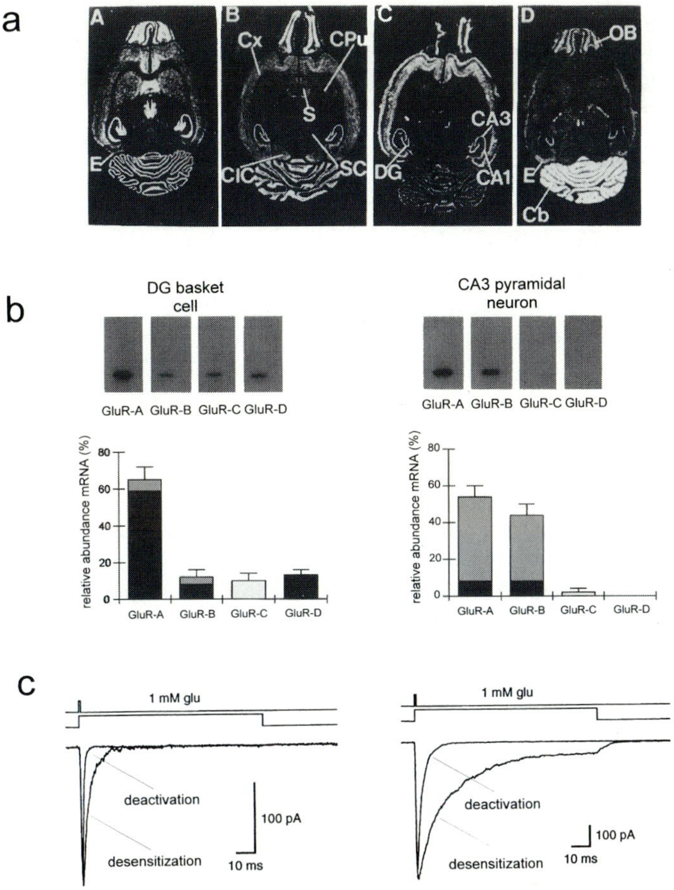

Fig. 1. **A** *In situ* hybridization of GluR-A, -B, -C, and -D messenger ribonucleic acids in adult rat brain horizontal sections using oligonucleotide-labelled probes (Keinänen et al. 1990). Abbreviations: *Cx*, cortex; *CPu*, caudate putamen; *S*, septal nuclei; *SC*, superior colliculus; *E*, entorhinal cortex; *DG*, dentate gyrus; *CA1/CA3*, CA1 and CA3 fields, respectively, of the hippocampus; *OB*, olfactory bulb; *Cb*, cerebellum. **B** Representative Southern blot of a polymerase chain reaction (PCR) product obtained from a basket cell or CA3 pyramdal neuron after probing with GluR-A to -D-specific oligonucleotide probes. Relative abundance of the GluR mRNAs after single-cell-PCR analysis in dentate gyrus basket cells and CA3 pyramidal neurons (*n*=10 cells in each group). White and black parts of the bars denote the flip and flop versions, respectively. **C** Deactivation and desensitization in a dentate gyrus basket cell and in a CA3 pyramidal neuron after a 1 ms and 100 ms pulse, respectively, of 1 mM glutamte (details for B and C in Geiger et al. vice 1995)

and is dependent on different intronic sequences complementary to editing sites (Seeburg et al. 1998). The developmental and cell type-specific differences were shown to occur both at the mRNA level when viewed by *in situ* hybridization (Keinänen et al. 1990; Sommer et al. 1990; Monyer et al. 1991) and at the protein level using subunit-specific antibodies (PETRALIA and WENTHOLD 1992, 1997). The levels of mRNA and protein were found to be in approximate agreement.

2. Relation Between Functional Properties of Native Receptors and Putative Subunit Composition

Analysis of structure–function relations of native AMPARs has concentrated on the Ca^{2+} permeability and deactivation and desensitization time courses. Early studies indicated that native AMPARs, unlike NMDARs, show low permeability to Ca^{2+} ions, linear or outwardly rectifying I–V relations, and low single-channel conductance (MAYER and WESTBROOK 1987; JONAS and SAKMANN 1992). The functional dominance of GluR-B(R) in determining the Ca^{2+} permeability, and the abundant expression of the GluR-B(R) in the CNS (KEINÄNEN et al. 1990; BURNASHEV et al. 1992a), provided a straightforward explanation of these functional properties.

The view that all native AMPARs are Ca^{2+} impermeable was challenged by two key findings. First, in 1990, IINO and coworkers found that, in a small, unidentified subset of cultured hippocampal neurons ("type II"), AMPARs were highly Ca^{2+} permeable and showed an inwardly rectifying I–V relation (IINO et al. 1990). In agreement with the data from recombinant AMPARs (see above), subsequent single-cell reverse transcriptase–polymerase chain reaction (RT-PCR) analysis showed that the GluR-B subunit was not expressed in "type-II" neurons (BOCHET et al. 1994). Second, Ca^{2+}-permeable AMPARs were identified in Bergmann glial cells in the cerebellum (BURNASHEV et al. 1992b; MÜLLER et al. 1992; GEIGER et al. 1995). As already suggested by previous *in situ* hybridization analysis (KEINÄNEN et al. 1990; BURNASHEV et al. 1992b), subsequent single-cell RT-PCR analysis showed unequivocally the lack of GluR-B mRNA in these cells (GEIGER et al. 1995). These results provided evidence for a strong link between the Ca^{2+} permeability of native AMPARs and the expression of the GluR-B subunit (JONAS and BURNASHEV 1995).

Analysis of the Ca^{2+} permeabilities of AMPARs in a variety of identified neurons in cortical slices led to the conclusion that Ca^{2+}-permeable AMPARs are primarily expressed in GABAergic interneurons in both the hippocampus (MCBAIN and DINGLEDINE 1993; KOH et al. 1995a; GEIGER et al. 1995; ISA et al. 1996) and the neocortex (JONAS et al. 1994; ITAZAWA et al. 1997). However, the $P_{Ca}/P_{monovalent}$ values were generally lower than in cultured type-II neurons or in Bergmann glial cells and varied over a wide range (approximately between 0.5 and 1.8). In all interneuron types examined, high Ca^{2+} permeability was correlated with both an inwardly rectifying shape for the I–V relation of the AMPAR-mediated currents in the presence of intracellular polyamines (KOH et al. 1995a,b; ISA et al. 1996; ITAZAWA et al. 1997) and a high single-channel conductance (HESTRIN 1993; KOH et al. 1995a; ANGULO et al. 1997; HAVERKAMPF et al. 1997).

Surprisingly, single-cell RT-PCR analysis demonstrated that GABAergic interneurons in the hippocampus and the neocortex expressed GluR-B(R) subunit mRNA; a complete lack of GluR-B mRNA appeared to be the exception. Quantitative analysis, however, revealed that the relative abundance of GluR-B mRNA in interneurons is much lower than in principal neurons (JONAS et al. 1994; GEIGER et al. 1995; LAMBOLEZ et al. 1996; Fig. 1B). The relation between the average $P_{Ca}/P_{monovalent}$ and the relative abundance of GluR-B is consistent with the functional dominance of GluR-B in determining the Ca^{2+} permeability of native AMPARs (GEIGER et al. 1995), in agreement with results from recombinant receptors (see above). Similarly, immunocytochemical analysis showed that GluR-B subunit protein is present in the majority of interneuron subtypes but that the relative levels are lower than in principal neurons. Double-labelling analysis revealed that the majority of parvalbumin- and calretinin-positive interneurons express the GluR-B subunit only at very low levels (LERANTH et al. 1996; RACCA et al. 1996; VISSAVAIJJHALA et al. 1996; KONDO et al. 1997; PETRALIA et al. 1997; CATANIA et al. 1998). Thus, Ca^{2+}-permeable AMPARs in interneurons are generated by the low relative abundance of GluR-B subunit mRNA and protein.

Analysis of neurons in other circuitries indicated that the expression of Ca^{2+}-permeable AMPARs is not restricted to GABAergic interneurons. Ca^{2+}-permeable AMPARs have been found in brain-stem neurons (in the auditory pathway, cochlear nucleus, and medial nucleus of the trapezoid body; OTIS et al. 1995; GEIGER et. al. 1995), dorsal-horn neurons (KYROZIS et al. 1995; GOLDSTEIN et al. 1995), septal neurons (SCHNEGGENBURGER et al. 1993), cholinergic interneurons of the striatum (GÖTZ et al. 1997), glutamatergic, subthalamic-nucleus neurons (GÖTZ et al. 1997), and retinal ganglion cells (GILBERTSON et al. 1991; TASCHENBERGER and GRANTYN 1998). Thus, it appears that the expression of Ca^{2+}-permeable AMPARs in neurons is widespread and is not strictly related to the transmitter phenotype.

Like the gating of recombinant AMPARs assembled from different subunit combinations, the desensitization time courses of native AMPARs differ by more than an order of magnitude. The desensitization time constants were as follows: auditory system (~1 ms; TRUSSELL et al. 1993; GEIGER et al. 1995)<hippocampal GABAergic interneurons (KOH et al. 1995a; GEIGER et al. 1995) ≈ neocortical GABAergic interneurons (HESTRIN 1993; JONAS et al. 1994; ANGULO et al. 1997) ≈ cerebellar granule cells (SILVER et al. 1996)<hippocampal granule cells (COLQUHOUN et al. 1992; GEIGER et al. 1995) ≈ neocortical pyramidal neurons (HESTRIN 1992a; JONAS et al. 1994) ≈ hippocampal pyramidal cell AMPARs (~15 ms; COLQUHOUN et al. 1992; GEIGER et al. 1995; Fig. 1C). Whilst the deactivation time constants followed the same sequence as the desensitization time constants, recovery from desensitization appeared to be controlled independently. AMPARs in auditory neurons desensitize and resensitize rapidly (TRUSSELL et al. 1993), interneuron AMPARs show rapid desensitization but slow recovery (HESTRIN 1993; JONAS et al. 1994; KOH et al. 1995a), CA3 pyramidal neuron AMPARs desensitize slowly but show rapid recovery (COLQUHOUN et al. 1992), and neocortical pyramidal neuron AMPARs show slow desensitization and resensitization (HESTRIN 1992a; JONAS et al. 1994). Analysis of fast spiking and regularly

spiking interneurons in layers 4–6 of the sensory motor cortex led to a similar conclusion (ANGULO et al. 1997).

Which subunits are primarily responsible for regulating the gating properties of native AMPARs? *In situ* hybridization, single-cell RT-PCR, and immunocytochemistry revealed that the various neurons that were examined expressed very different combinations of AMPAR subunits. Auditory AMPARs are mainly assembled from GluR-D_o subunits (GEIGER at al. 1995; RUBIO and WENTHOLD 1997). Cortical GABAergic interneurons express primarily GluR-A_o and GluR-D_o subunits (MARTIN et al. 1993; BAUDE et al. 1995; CATANIA et al. 1998; GEIGER et al. 1995; LAMBOLEZ et al. 1996; LERANTH et al. 1996; ANGULO et al. 1997). In contrast, cortical principal neurons express mainly GluR-A_i and GluR-B_i (KEINÄNEN et al. 1990; SOMMER et al. 1990; GEIGER et al. 1995; Fig. 1A, B).

Whereas the GluR-B(R) subunit is the main determinant of the Ca^{2+} permeability of native AMPARs, the molecular determinants of gating appear to be more complex. In a study by GEIGER et al. (1995), nine different cell types in the CNS were analyzed by a correlated functional and single-cell RT-PCR analysis. The highest correlation coefficients were found between the desensitization time constant and the relative abundances of GluR-B_i (positive correlation) and GluR-D (negative correlation) mRNA. These results are consistent with the data from recombinant AMPARs, suggesting that GluR-B_i and GluR-D_o are determinants of slow and fast gating, respectively (MOSBACHER et al. 1994; LOMELI et al. 1994; BURNASHEV 1993; PARTIN et al. 1994; SEKIGUCHI et al. 1997; see above). In a study by ANGULO et al. 1997, fast spiking and regularly spiking interneurons in layer 4–6 of the sensory motor cortex were examined. In this heterogeneous population of interneurons, a positive correlation between desensitization and the relative abundance of GluR-B_i was also detected, but the correlation coefficient was relatively low in this sample (ANGULO et al. 1997). No significant correlation was found between gating and the total percentage of flip or flop variants or the extent of editing at the R/G site (GEIGER et al. 1995; ANGULO et al. 1997). In conclusion, these results indicate that gating kinetics is not controlled by a single molecular determinant but is regulated by multiple factors that probably differ among cell types.

Two lines of evidence indicate that the Ca^{2+} permeability and gating properties of native AMPARs can be controlled independently. First, in regularly spiking nonpyramidal neurons in the neocortex, fast or slow desensitization could be associated with either rectifying (high Ca^{2+} permeability) or linear I–V curves (low Ca^{2+} permeability) (ANGULO et al. 1997). Second, AMPARs in GABAergic neurons in the substantia nigra, presumably assembled from GluR-A, -B/-C, and -D subunits (MARTIN et al. 1993), show low Ca^{2+} permeability but fast gating (GÖTZ et al. 1997).

When gating of native and recombinant AMPARs assembled from different subunits are compared, one problem remains. Whereas there is good quantitative agreement on the fast side of the range of desensitization time constants of native and recombinant AMPARs, there is less agreement on the slow side (depending on the type of host cell used for heterologous expression). For example, the desensitization of native GluR-A_i/GluR-B_i receptors is similar to that of recombinant receptors expressed in HEK cells (BURNASHEV 1993; PARTIN et al. 1994; SEKIGUCHI et al. 1997) but is considerably slower than that of receptors in *Xenopus* oocytes (MOSBACHER et al. 1994; LOMELI et al. 1994). The reasons for this discrepancy remain unclear.

The flexible number of a particular subunit within receptor assemblies as well as the large number of splice variants expressed in certain cell types (LAMBOLEZ et al. 1992) may suggest that AMPAR complexes with different subunit compositions coexist within single neurons and could be segregated to different portions of the somatodendritic domain. In cultured hippocampal neurons, differences in rectification have suggested the possibility that Ca^{2+}-permeable AMPARs are segregated to the dendrites, whereas Ca^{2+}-impermeable receptors are located in the soma (LERMA et al. 1994). However, direct recordings from dendrites of CA3 and CA1 pyramidal neurons (within 15–174 μm of the soma) have shown that dendritic AMPARs are functionally similar to somatic ones; both are impermeable to Ca^{2+} ions (SPRUSTON et al. 1995). In fusiform neurons in the dorsal cochlear nucleus, GluR-D subunit immunoreactivity is found in the basal dendrites (where the auditory synapses terminate) but not in the apical dendrites (where the parallel-fiber synapses are formed; RUBIO and WENTHOLD 1997). In interneurons of the stratum lucidum in the hippocampal CA3 region, differential sensitivity to external philanthotoxin suggests that Ca^{2+}-permeable AMPARs are present in the proximal dendrites (the mossy-fiber termination zone), whereas Ca^{2+}-impermeable receptors are segregated to distal dendrites (the region of recurrent collateral input; TÓTH and MCBAIN 1998). These examples point to the utmost importance of differential targeting of receptors (WENTHOLD et al. and SEEBURG et al., this volume).

The functional significance of AMPAR diversity for glutamatergic synaptic transmission in different types of target neurons is not entirely clear. Several possibilities have been proposed. First, Ca^{2+} influx through GluR-B-free subunit assemblies could trigger new forms of synaptic plasticity (GU et al. 1996; MAHANTY and SAH 1998; JIA et al. 1996). Second, blocking of GluR-B-free AMPARs by intracellular polyamines could contribute to the facilitation of synaptic currents in interneurons (ROZOV et al. 1998) and switch off AMPAR channels during action potentials, thus facilitating dendritic spike initiation (TRAUB and MILES 1995). Third, the larger single-channel conductance of GluR-B-free receptors will have a direct impact on the amplitude of the quantal postsynaptic conductance change (GEIGER et al. 1997). Finally, the faster gating of AMPARs in auditory neurons and interneurons could underlie coincident detection of synaptic events and the short latency of activation of the postsynaptic neuron (GEIGER et al. 1997), whereas the slow gating of principal neuron AMPARs facilitates integration of synaptic events. Analysis of transgenic animals may provide further information. Regarding synaptic transmission in the auditory system, the reader is referred to the chapter by TRUSSELL (this volume); regarding synaptic excitation of interneurons, the reader is referred to the chapter by GEIGER et al. (this volume).

C. NMDA Receptors

A number of functional properties distinguish the NMDAR from the AMPAR. These include, most importantly, the very high permeability for Ca^{2+} ions, the voltage-dependent blocking by extracellular Mg^{2+}, and the very slow gating (MCBAIN and MAYER 1994).

I. Recombinant Receptors

Members of the NMDAR-subunit family (NR1; NR2A, B, C, D; NR3; MORIYOSHI et al. 1991; ISHII et al. 1993; MEGURO et al. 1992; KUTSUWADA et al. 1992; MONYER et al. 1992; SUCHER et al. 1995; CIABARRA et al. 1995) have been recombinantly expressed in both mammalian cells and *Xenopus* oocytes (MORIYOSHI et al. 1991; MONYER et al. 1992). Evidence has accumulated that the NR2 subunit comprises the binding site for glutamate (LAUBE et al. 1997; ANSON et al. 1998), whereas the NR1 subunit carries the binding site for glycine (KURYATOV et al. 1994; WAFFORD et al. 1995; WILLIAMS et al. 1996). In mammalian cells, coexpression of NR1 and NR2 subunits is required to produce functional receptors. In *Xenopus* oocytes, however, NR1 subunits (but not NR2 subunits) form functional channels (MORIYOSHI et al. 1991). The molecular nature of the permissive factor in the oocyte system is unknown. Oocytes express an endogenous NMDAR subunit (termed XenU1; SOLOVIEV and BARNARD 1997) that may form heteromers with the heterologously expressed NR1 subunits. Independent of the question whether homomers can assemble from heterologously expressed subunits, however, there is good evidence that native receptors are heteromers (SHENG et al. 1994). For further details regarding NMDAR subunit assembly and stoichiometry, the reader should refer to the review by BEHE et al. (this volume).

Coexpression of NR1 and one of the four NR2 subunits results in the formation of NMDARs that differ in several of their functional properties. Since the formation of homomeric channels is less favorable for NMDARs than for AMPARs, analysis of the structure–function relations was more complicated (HOLLMANN and HEINEMANN 1994; MCBAIN and MAYER 1994).

1. Ion Permeation and Blocking

A key property of recombinant NMDARs is their very high permeability to Ca^{2+} ions. $P_{Ca}/P_{monovalent}$ values between 2 and approximately 15 have been reported (TSUZUKI et al. 1994; BURNASHEV et al. 1995; SHARMA and STEVENS 1996; IINO et al. 1990 1997). Unlike the Ca^{2+} permeability of AMPARs, that of recombinant NMDARs is largely independent of the subunit composition (MONYER et al. 1992; BURNASHEV et al. 1995).

Another key property of recombinant NMDARs is the voltage-dependent blocking by extracellular Mg^{2+} ions. At negative potentials, external Mg^{2+} induces rectification with a region of negative slope conductance that is approximately the mirror image of the rectification of GluR-B-free AMPAR combinations induced by internal polyamines (see above). The extent of blocking of NMDARs by Mg^{2+} is dependent on the NR2 subunit (MONYER et al. 1994; KUNER and SCHOEPFER 1996). For NR1–NR2A and 2B channels, the inward current is maximal at approximately –20 mV with 1 mM external Mg^{2+}, and the IC_{50} value at –100 mV is approximately 2 μM. In contrast, for NR1–NR2C and 2D channels, the inward current is maximal at –40 mV with 1 mM Mg^{2+}, and the IC_{50} at –100 mV is approximately 10 μM (MONYER et al. 1992 1994; KUTSUWADA et al. 1992; KUNER and SCHOEPFER 1996).

A third property of the NMDAR that is determined by the subunit composition is the single-channel conductance. Recombinant NMDARs can be classified into "high-conductance" receptors (NR1–NR2A and NR1–NR2B; ~50/40 pS) and "low-conductance" receptors (NR1–NR2C and NR1–NR2D; ~36/18 pS with 0.85 mM external Ca^{2+}; STERN et al. 1992; WYLLIE et al. 1996). Hence, the single-channel conductance is correlated with the sensitivity to external Mg^{2+}; high-conductance receptors are more sensitive than low-conductance receptors. Additionally, between 18-pS and 36-pS conductance levels, NR1–NR2D receptors show asymmetrical transitions that are diagnostic for this subunit combination (WYLLIE et al. 1996).

Mutational analysis revealed that the Ca^{2+} permeability of the recombinant NMDARs is controlled by a residue at a position equivalent to the Q/R site of AMPAR subunits. In this position, both NR1 and NR2 subunits contain an N residue. When the N is replaced by a Q residue in the NR1 subunit, $P_{Ca}/P_{monovalent}$ of the mutant channels is diminished (BURNASHEV et al. 1992c). When the N is replaced by an R residue in the NR1 subunit, $P_{Ca}/P_{monovalent}$ is reduced even more, to values of approximately 0.1 (BURNASHEV et al. 1992c). In contrast, replacement of the N by a Q residue in the NR2 subunits has a relatively small effect (BURNASHEV et al. 1992c; SHARMA and STEVENS 1996). This suggests that the residue at the Q/R site of the NR1 subunit is the main determinant of the Ca^{2+} permeability of recombinant NMDARs.

The molecular determinants of blocking by Mg^{2+} are more complex. Mutational analysis, surprisingly, revealed that the two residues downstream from the Q/R/N site of the NR2 subunit have a major role in controlling blocking by Mg^{2+}. The residue at the Q/R/N site of the NR2 subunit has a smaller influence on Mg^{2+} sensitivity, and that at the Q/R/N site of the NR1 subunit is even less important (MORI et al. 1992; BURNASHEV et al. 1992c; WOLLMUTH et al. 1998). Furthermore, analysis of NR2 chimeric subunits shows that the difference in Mg^{2+} sensitivity between NR1–NR2A and B and NR1–NR2C and D subunits is generated at multiple locations, including M1, the M1/M2 linker, and the M4 segment (KUNER and SCHOEPFER 1996). Regarding the properties of the NMDAR pore, the reader is referred to the review by KUNER et al. (this volume).

2. Gating Properties

Binary combinations formed from NR1 and NR2 subunits differ markedly in the time courses of both deactivation (following brief glutamate pulses) and desensitization (during long glutamate pulses). The deactivation time constants of recombinant NMDARs follow the sequence NR1–NR2A (~100 ms) < NR1–NR2B ≈ NR1–NR2C (500 ms) << NR1–NR2D (5 s at 22°C; MONYER et al. 1992 1994; WYLLIE et al. 1998; VICINI et al. 1998). This sequence is very similar to the sequence of affinities for glutamate (KUTSUWADA et al. 1992; IKEDA et al. 1992), consistent with the notion that the unbinding of glutamate is the rate-limiting step in the deactivation time course.

Desensitization of NMDARs is a complex issue, since at least three forms have been described for native NMDARs: glycine-independent desensitization (SATHER et al. 1990, 1992), glycine-dependent desensitization (MAYER et al. 1989), and calcium-dependent inactivation (LEGENDRE et al. 1993; MAYER et al. 1995). Marked subunit-specific differences have been reported for glycine-independent desensitization measured in the presence of saturating glycine concentrations. NR1–NR2A and NR1–NR2B receptors exhibit this form of desensitization, whereas NR1–NR2C and NR1–NR2D receptors do not show it (MONYER et al. 1994; KRUPP et al. 1998). Analysis of the functional properties of NR2A–2C chimeras revealed that two *N*-terminal regions of the NR2 subunit flanking the ligand-binding site (a leucine–isoleucine–valine binding protein-like domain and a domain in the pre-M1 region) are involved in subunit-specific, glycine-independent desensitization (KRUPP et al. 1998; VILLARROEL et al. 1998). Of note regarding this point is the difference from AMPARs, where the agonist-binding site itself is involved in desensitization (STERN-BACH et al. 1998).

Glycine-dependent desensitization is also modulated by the NR2 subunit. This form of desensitization is observed in the presence of low glycine concentrations and is interpreted as a negative allosteric interaction between glycine and glutamate binding at native (MAYER et al. 1989; BENVISTE et al. 1990) and recombinant NMDARs (KENDRICK et al. 1998). The ability of glycine to inhibit desensitization is correlated with the affinity of the receptors for glycine (KENDRICK et al. 1998), which follows the order NR2A < NR2B ≈ NR2C < NR2D (KUTSUWADA et al. 1992; IKEDA et al. 1992).

Finally, Ca^{2+}-dependent inactivation is NR2 subunit-dependent. Ca^{2+}-dependent inactivation of NMDARs (LEGENDRE et al. 1993; TONG and JAHR 1994; MEDINA et al. 1995) can not be considered a classical form of desensitization, since Ca^{2+} influx through NMDARs themselves, but also through voltage-gated Ca^{2+} channels or Ca^{2+}-permeable AMPAR induces this form of desensitization (LEGENDRE et al. 1993; KYROZIS et al. 1995). The underlying mechanism has been proposed to be Ca^{2+}-dependent binding of calmodulin to a low-affinity binding site at the NR1 *C*-terminus, releasing the receptor complex from the cytoskeleton and, thus, leading to inactivation (Rosenmund and Westbrook 1993; Ehlers et al. 1996; Rafiki et al. 1997; Zhang et al. 1998). Recombinant NR1–NR2A and NR1–NR2D receptors show marked Ca^{2+}-dependent inactivation, whereas no significant effect was apparent in NR2C-containing receptors (Krupp et al. 1996). The presence of Ca^{2+}-dependent inactivation in NR2B-containing receptors is controversial (Medina et al. 1995; Krupp et al. 1996).

Alternative splicing of the NR1 subunit may also regulate NMDAR gating. The NR1 subunit exists in eight splice variants generated by an *N*-terminally located exon insertion (–exon 5 = NR1a, +exon 5 = NR1b) and two *C*-terminal exon deletions, compared with the originally described sequence of NMDAR1 by Moriyoshi et al. (1991) (Zukin and Bennett 1995). Recombinant NR1b–NR2 channels differ from NR1a–NR2 channels regarding the faster deactivation time course (van Hooft et al. 1998) and the lower affinity for glutamate (Durand et al. 1992, 1993). Thus, the inclusion of exon 5 in the NR1 subunit appears to accelerate NMDAR deactivation by promoting the unbinding of glutamate. Regarding the details of NMDAR gating, the reader is referred to the review by Behe et al. (this volume).

3. Sensitivity to Protons, Zn^{2+} Ions, Endogenous Polyamines, and Selective Antagonists

In addition to the binding sites for the agonist glutamate and the co-agonist glycine, NMDARs comprise a variety of binding sites for modulatory substances and pharmacological agents on the extracellular surface of the protein. For various modulatory agents, the sensitivity is dependent on the subunit composition. Alternative splicing of the NR1 subunit at the *N*-terminus (–exon 5 versus +exon 5) appears to be a major factor that controls the modulatory properties of the NMDAR.

Similar to native receptors (Traynelis and Cull-Candy 1991), several recombinant NMDARs are blocked by extracellular protons in the physiological pH range. Homomeric NMDARs assembled from NR1b subunits show a much lower proton sensitivity (pKa = 7.3) than NR1a assemblies (pKa = 6.8). Similarly, heteromeric NMDARs built from NR1b and NR2A, NR2B, and NR2D subunits show a much lower proton sensitivity than NR1a-containing receptors (Traynelis et al. 1995). NR1–NR2C channels, by contrast, appear to be insensitive to proton regulation, independent of the NR1 splice variant (pKa = 6.2; Traynelis et al. 1995).

Extracellular polyamines have both potentiating and inhibiting effects on NMDARs. The mechanisms of these effects are complex, including glycine-independent potentiation, glycine-dependent potentiation, voltage-dependent channel blocking, and a decrease in affinity for glutamate (Williams et al. 1994; Williams 1995). Glycine-independent potentiation was found in NR1a homomeric channels but not in NR1b homomers (DURAND et al. 1993). Similarly, glycine-independent potentiation was found in NR1a–NR2B heteromers, but not in NR1a–NR2A, C, and D heteromers (WILLIAMS et al. 1994; WILLIAMS 1995). Thus, exon 5 and NR2A, C, and D suppress this form of potentiation. Glycine-dependent potentiation, however, was controlled differently by the NR2 subunit; this form of stimulation was observed in NR1a homomers and NR1a–NR2A and NR1a–NR2B heteromers but not in NR1a–NR2C and NR1a–NR2D heteromers (WILLIAMS 1995).

Extracellular Zn^{2+} ions also have diverse effects on NMDARs. Homomeric NR1a receptors expressed in oocytes are potentiated by low concentrations and are blocked by high concentrations of Zn^{2+} (HOLLMANN et al. 1993). In contrast, homomeric NR1b receptors and heteromeric receptors containing NR2 subunits show blocking but no potentiation by extracellular Zn^{2+} (HOLLMANN et al. 1993; TRAYNELIS et al. 1998), similar to native NMDARs (WESTBROOK and MAYER 1987; CHRISTINE and CHOI 1990). Blocking by Zn^{2+} can be further dissected into a voltage-independent component at low concentrations and a voltage-dependent component at high concentrations (CHRISTINE and CHOI 1990; PAOLETTI et al. 1997). In heteromeric NMDARs, voltage-independent blocking by Zn^{2+} has the following affinity sequence: NR1a–NR2A (~20 nM) > NR1a–NR2B (2 μM) >>NR1a–NR2C and NR1a–NR2D (20 μM; PAOLETTI et al. 1997; TRAYNELIS et al. 1998). Exon 5 in the NR1 subunit reduces Zn^{2+} sensitivity by an order of magnitude in heteromers with NR2A and NR2B but not in those with NR2C and NR2D (TRAYNELIS et al. 1998).

A model of structural convergence of the three modulators – protons, polyamines, and Zn^{2+} – has been proposed by TRAYNELIS and colleagues (1998). This model implies that polyamines and the intrinsic sequences of exon 5 act in a similar manner, attenuating channel blocking by both protons and Zn^{2+}. This hypothesis is also supported by the structural motifs of exon 5, which, like polyamines, contains clusters of positively charged amino acids (ZHENG et al. 1994).

Recombinant NMDARs also differ in their sensitivity to a variety of pharmacological agents. Recombinant NMDARs are differentially sensitive to the phenylethanolamine ifenprodil, which originally was developed as a vasodilator. Ifenprodil blocks NR1–NR2B receptors at much lower concentrations than it blocks NR1–NR2A, NR1–NR2C, and NR1–NR2D receptors (WILLIAMS 1993, 1995). Similar NR2B-specific effects were found for haloperidol (ILYIN et al. 1996; LYNCH and GALLAGHER 1996) and the ifenprodil derivative CP101,606 (CHENARD et al. 1995). Thus ifenprodil, haloperidol, and CP101,606 are potentially useful tools for identification of the putative subunit composition of native NMDARs (see below). Recent evidence indicates that phenylethanolamines inhibit NMDARs by enhancing proton inhibition (MOTT et al. 1998). These results imply that the effects of modulators (protons, Zn^{2+}, polyamines) and subunit-selective pharmacological tools (ifenprodil, CP101,606, haloperidol) partially result from common mechanisms of action.

II. Native NMDA Receptors

1. Differential Developmental and Regional Expression of NMDAR Subunits

The NR1 subunit displays ubiquitous expression in the brain, with little alteration during development (Moriyoshi et al. 1991; Monyer et al. 1994). NR1 subunit mRNA appears as early as E14, peaks around P20, and subsequently declines to adult levels (Moriyoshi et al. 1991; Laurie and Seeburg 1994; Monyer et al. 1994). The expression of the NR1 splice versions, however, shows considerable regional and developmental heterogeneity (Laurie and Seeburg 1994).

The expression of NR2B and NR2D begins at least as early as E14, whereas NR2A and NR2C are first detected perinatally. All NR2 transcripts are maximal around P20–P25 except NR2D, which peaks around P7 and subsequently declines to adult levels. In pyramidal neurons of the neocortex and the hippocampus, NR2B appears to be partly replaced by NR2A; in granule cells of the cerebellum, NR2B is replaced by NR2A and NR2C (Monyer et al. 1994). In the adult CNS, NR2A and NR2B are the major subunits expressed in the neocortex and hippocampus; NR2C expression is concentrated in the cerebellum, whereas NR2D expression is preponderant in midline structures, in the thalamus, and in putative GABAergic interneurons in the hippocampus, neocortex, and neostriatum (Monyer et al. 1994; Standaert et al. 1996; Fig. 2). The subunit changes were shown to occur both at the mRNA level when analyzed by *in situ* hybridization (Monyer et al. 1992, 1994; Watanabe et al. 1993; Watanabe et al. 1994) and at the protein level using subunit-specific antibodies (Sheng et al. 1994; Wenzel et al. 1997) and ligand-binding assays with rat brain membranes (Williams et al. 1993).

2. Relation Between Functional Properties of Native Receptors and Putative Subunit Composition

Analysis of the structure–function relations of native receptors has concentrated on three functional characteristics of NMDARs: gating, Mg^{2+} sensitivity, and single-channel conductance. In the cerebellum, substantial differences were found in kinetics, Mg^{2+} sensitivity, and single-channel conductance of native NMDARs at different developmental stages. In cerebellar granule cells, the time course of NMDAR-mediated currents becomes faster, whereas the Mg^{2+} sensitivity is reduced during development (Takahashi et al. 1996). Furthermore, analysis of single NMDAR channels in granule cells revealed that pre-migratory and migratory cells mainly expressed high-conductance channels, whereas a mixture of high- and low-conductance channels was present in mature granule cells (Farrant et al. 1994).

What is the molecular basis of these complex developmental changes? *In situ* hybridization studies had indicated that NR2A- and NR2C- subunit expression was upregulated, whereas NR2B was concomitantly downregulated in cerebellar granule cells during development (Monyer et al. 1992, 1994; Akazawe et al. 1994; Watanabe et al. 1994). Single-cell PCR analysis showed that comparable changes occur in cerebellar granule cells in culture, although their extent and time course do not exactly parallel those in the *in vivo* situation (Audinat et al. 1994; Vallano et al. 1996). To dissect the effects of the different subunits on the functional properties of native NMDARs, Takahashi et al. (1996) examined the properties of NMDARs in granule cells of NR2A knockout mice. After targeted disruption of the NR2A gene, the time course of the NMDAR-mediated EPSC component was slowed, indicating that NMDAR kinetics was controlled by the NR2A subunit (Takahashi et al. 1996). In contrast, the Mg^{2+} sensitivity and single-channel properties remained unaltered, suggesting that these characteristics were independent of the NR2A subunit (Takahashi et al. 1996). Taken together, these results show that a developmental switch from NR2B to NR2A and NR2C occurs in granule cells, with NR2A being responsible for the faster gating of native NMDARs and NR2C for the reduced sensitivity to Mg^{2+} and the lower single-channel conductance.

Functional properties of native NMDARs also differ significantly among cell types. This is exemplified by the expression of low-conductance channels in cerebellar granule cells and the expression of high-conductance channels in hippocampal neurons (Gibb and Colquhoun 1992; Spruston et al. 1995). Further analysis, however, revealed that low-conductance channels were also present in other cell types. Low-conductance NMDAR channels were identified in Purkinje cells, deep cerebellar nuclei, and dorsal horn (Momiyama et al. 1996; reviewed by Cull-Candy et al. 1998). Regarding NMDAR expression, Purkinje cells are probably unique in the entire CNS. First, NMDARs are expressed only transiently in these cells, during the first postnatal week (Dupont et al. 1987), and second, in these cells, NMDARs that show low conductance and weak Mg^{2+} sensitivity are expressed in isolation (Momiyama et al. 1996). Low-conductance channels were also found in neurons of deep-cerebellar nuclei and dorsal horn but, in these cells as in cerebellar granule cells, they coexisted with high-conductance channels (Momiyama et al. 1996). *In situ* hybridization studies suggested that low-conductance channels in the three types of neurons were NR1–NR2D receptors, unlike those in cerebellar granule cells, which were likely to be assembled from NR1–NR2C (Akazawa et al. 1994; Monyer et al. 1994). Furthermore, low-conductance NMDARs in Purkinje cells, neurons of deep cerebellar nuclei, and neurons in the dorsal horn showed asymmetry in the transitions between subconductance states, which is diagnostic for NR2D-containing receptors (Momiyama et al. 1996).

In the cortex, NMDAR gating changes substantially during development. Both the time course of the glutamate-activated current and the time course of the NMDAR-mediated EPSC become faster with age in pyramidal neurons of both the visual (Carmignoto and Vicini 1992; Nase et al. 1997) and somatosensory cortexes (Flint et al. 1997). Several lines of evidence indicate that a developmental change consisting of an increase of the NR2A subunit expression relative to the NR2B subunit expression determines the acceleration in NMDAR gating (Fig. 3C,D). First, pharmacological analysis indicates that the ifenprodil-sensitive component of glutamate-activated currents and NMDAR-mediated EPSC decreases substantially with age in cortical neurons (KIRSON and YAARI 1996; KEW et al. 1998). Second, single-cell RT-PCR analysis revealed an upregulation of NR2A and a concomitant downregulation of NR2B in both somatosensory-layer-2 pyramidal cells (FLINT et al. 1997) and visual-cortex-layer-4 pyramidal cells (NASE et al. 1997). Third, developmental shortening of the NMDAR-mediated EPSCs is correlated with an increase in NR2A mRNA levels (FLINT et al. 1997). Finally, the later maturation of the visual cortex compared with the somatosensory cortex is paralleled by a delayed change in NMDAR kinetics and a later increase in NR2A mRNA levels (NASE et al. 1997).

Whether the functional and molecular properties of NMDARs differ between the various types of cortical neurons is not known. A recent study suggests that spiny stellate cells in layer 4 of the mouse barrel cortex express NMDARs with reduced Mg^{2+} sensitivity (FLEIDERVISH et al. 1998). This may suggest that subsets of neocortical neurons express NR1–NR2C or NR1–NR2D receptors.

In conclusion, a developmental switch from NR2B to NR2A and NR2C in the cerebellum and a significant increase of NR2A expression relative to NR2B expression in the cortex regulate the functional properties of NMDAR-mediated synaptic transmission in these circuits. In both regions of the CNS, the developmental changes in NMDAR-subunit expression are regulated by neuronal activity. In cerebellar granule cells in culture, the switch from NR2B to NR2A that occurs during postnatal development depends on neuronal activity and activation of NMDARs; it is blocked by tetrodotoxin and enhanced by K^+ depolarization or NMDA (AUDINAT et al. 1994; BESSHO et al. 1994; VALLANO et al. 1996). In contrast, the upregulation of NR2C expression also requires the release of neuregulin-β from cerebellar mossy fibers (OZAKI et al. 1997). In the visual cortex, dark rearing delays the change in gating kinetics (CARMIGNOTO and VICINI 1992) and extends the time window in which the NR2B-rich "juvenile form" of the NMDAR is expressed (NASE et al. 1997). Finally, in hippocampal and neocortical neurons, the switch from NR2B to NR2A that occurs with increasing time in culture is dependent on exocytosis; it can be suppressed by tetanus toxin (HOFFMANN et al. 1997; LINDLBAUER et al. 1998).

Coexpression of multiple NMDAR subunits in a single cell raises the question of whether NMDARs are subject to differential, input-specific targeting. This issue has been addressed in CA3 pyramidal neurons, which receive three segregated glutamatergic inputs: the mossy-fiber input in the stratum lucidum, the commissural–associational input to the basal dendrites, and the fimbrial input to the distal, apical dendrites. FRITSCHY and colleagues (1998) found NR2A and NR2B immunoreactivity in stratum oriens and radiatum of CA1–CA3 fields but only NR2A immunoreactivity in stratum lucidum of the

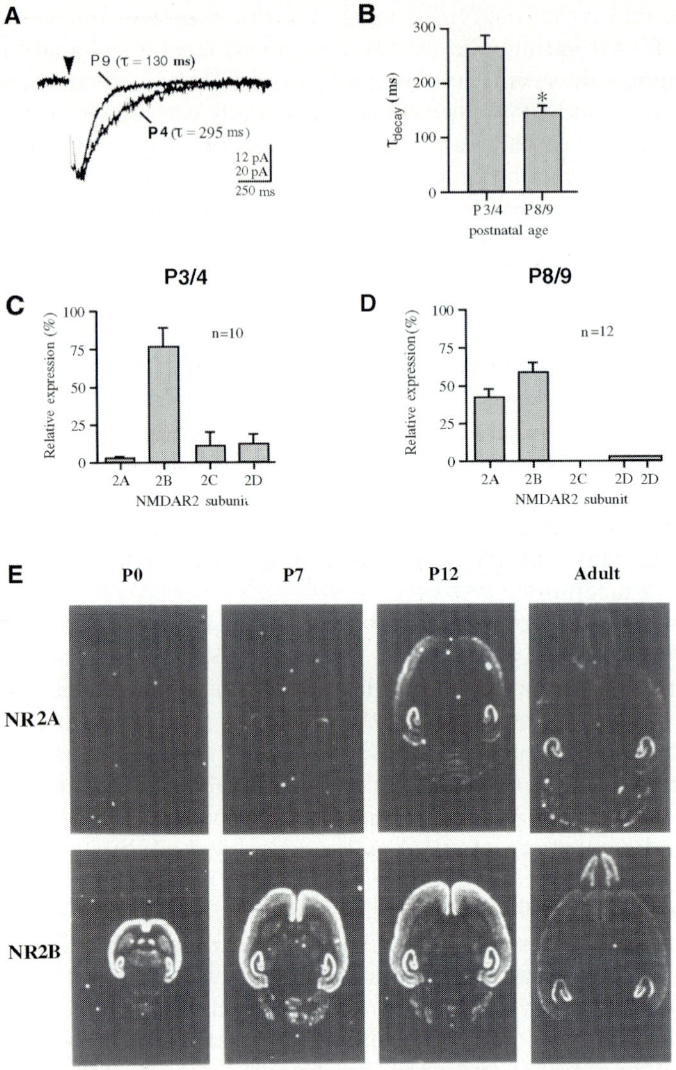

Fig. 2. **A** NMDAR EPSCs at P4 and P9 in rat somatosensory cortex. Each trace is an average of three responses. The amplitude of the P9 trace is normalized to that one of the P4 trace. *Scale bar*: 12 pA for the P4 trace, 20 pA for the P9 trace. **B** Decrease in the mean decay time constant from P4 ($n = 30$) to P9 ($n = 41$). **C, D** Expression profiles of the four NMDAR2 subunits in pyramidal somatosensory neurons at P3/4 and P8/9, respectively (Flint et al. 1997). **E** Developmental expression profile of NR2A and -B at P0, P7, P12 and in the adult rat brain using subunit-specific oligonucleotide labelled probes (Monyer et al. 1994)

CA3 subfield. Similarly, WATANABE and colleagues (1998) detected low expression of NR1 and NR2A and almost no expression of NR2B in stratum lucidum of the CA3 regions. The low abundance and distinct subunit composition of postsynaptic NMDARs at mossy-fiber synapses may explain why long-term-potentiation (LTP) induction at this synapse is NMDAR independent. Furthermore, targeted disruption of the NR2A subunit (homozygous mice) resulted in selective reduction of NMDA EPSCs and LTP at the commissural–associational input with little effect on the fimbrial input, whereas the opposite was the case in mice with disruption of the NR2B subunit (heterozygous mice were used; ITO et al. 1997). In conclusion, these results suggest that NR2A and NR2B subunits are distributed unevenly over the somatodendritic surface of CA3 pyramidal neurons.

The functional significance NMDAR diversity for glutamatergic synaptic transmission is even less well understood than that of AMPAR diversity. Several ideas have been suggested. NMDARs with low sensitivity to Mg^{2+} ions could contribute to the generation of excitatory synaptic potentials at rest, unlike NMDARs with high Mg^{2+} sensitivity. Differences in the deactivation time course and the Mg^{2+} sensitivity may set the temporal window and threshold for the induction of synaptic plasticity, as NMDARs have a key role in the induction of LTP or long-term depression (LTD) and operate as detectors of synchronous pre- and postsynaptic activity (MONYER et al. 1994). Finally, the Ca^{2+} influx mediated by synaptic activation of slow NMDARs could contribute to stabilization of synapses during development (CARMIGNOTO and VICINI 1992; HESTRIN 1992b).

D. Summary and Conclusions

A large number of subunits are available for the assembly of native GluRs. GluR-subunit expression differs substantially among different types of neurons in the mature brain and varies markedly during development. Subunit composition determines the Ca^{2+} permeability of AMPARs, the Mg^{2+} sensitivity of NMDARs, and the gating and single-channel properties of both types of receptors. This enables single neurons to obtain genetic control over the signalling properties of their glutamatergic synapses. Moreover, the cell-type-specific expression of GluR subunits is subject to activity-dependent regulation. This may contribute to the control of activity and plasticity in the CNS.

The information in this chapter provides ample evidence that tremendous progress has been made in recent years in the elucidation of the molecular determinants controlling the functional properties of AMPARs and NMDARs. The concluding remarks will not summarize what was said above – which was, itself, a condensed summary – but will name just a few questions that have remained unsolved and merit attention if we are to gain further insight into why this subunit diversity is required. (1) The elucidation of cell-type-specific expression profiles of GluRs needs to be extended to additional cell types. (2) Insights obtained from the molecular and functional characterization of GluRs need to be viewed in the context of neuronal assemblies, e.g. why are Ca^{2+}-permeable AMPA receptors in GABAergic interneurons important at the system level? (3) The coexistence of different GluR types within the same cell raises the question of differential synapse targeting and the identification of proteins involved in this targeting process and their regulation.

Acknowledgements. The authors thank Drs. J. R. P. Geiger and P. H. Seeburg for critically reading the manuscript. The studies were supported by the Deutsche Forschungsgemeinschaft, the German Israeli Foundation, and the Human Frontiers Science Program Organization.

References

Akazawa C, Shigemoto R, Bessho Y, Nakanishi S, Mizuno N (1994) Differential expression of five *N*-methyl-D-aspartate receptor subunit mRNAs in the cerebellum of developing and adult rats. J Comp Neurol 347:150–160

Angulo MC, Lambolez B, Audinat E, Hestrin S, Rossier J (1997) Subunit composition, kinetic, and permeation properties of AMPA receptors in single neocortical nonpyramidal cells. J Neurosci 17:6685–6696

Anson LC, Chen PE, Wyllie DJA, Colquhoun D, Schoepfer R (1998) Identification of amino acid residues of the NR2A subunit that control glutamate potency in recombinant NR1/NR2A NMDA receptors. J Neurosci 18:581–589

Audinat E, Lambolez B, Rossier J, Crépel F (1994) Activity-dependent regulation of *N*-methyl-D-aspartate receptor subunit expression in rat cerebellar granule cells. Eur J Neurosci 6:1792–1800

Baude A, Nusser Z, Molnár E, McIlhinney RAJ, Somogyi P (1995) High-resolution immunogold localization of AMPA type glutamate receptor subunits at synaptic and non-synaptic sites in rat hippocampus. Neurosci 69:1031–1055

Benveniste M, Clements J, Vyklicky L Jr, Mayer ML (1990) A kinetic analysis of the modulation of *N*-methyl-D-aspartic acid receptors by glycine in mouse cultured hippocampal neurones. J Physiol (Lond) 428:333–357

Bessho Y, Nawa H, Nakanishi S (1994) Selective up-regulation of an NMDA receptor subunit mRNA in cultured cerebellar granule cells by K^+-induced depolarization and NMDA treatment. Neuron 12:87–95

Blaschke M, Keller BU, Rivosecchi R, Hollmann M, Heinemann S, Konnerth A (1993) A single amino acid determines the subunit-specific spider toxin block of α-amino-3-hydroxy-5-methylisoxazole-4-propionate/kainate receptor channels. Proc Natl Acad Sci USA 90:6528–6532

Bochet P, Audinat E, Lambolez B, Crépel F, Rossier J, Iino M, Tsuzuki K, Ozawa S (1994) Subunit composition at the single-cell level explains functional properties of a glutamate-gated channel. Neuron 12:383–388

Burnashev N, Monyer H, Seeburg PH, Sakmann B (1992a) Divalent ion permeability of AMPA receptor channels is dominated by the edited form of a single subunit. Neuron 8:189–198

Burnashev N, Khodorova A, Jonas P, Helm PJ, Wisden W, Monyer H, Seeburg PH, Sakmann B (1992b) Calcium-permeable AMPA-kainate receptors in fusiform cerebellar glial cells. Science 256:1566–1570

Burnashev N, Schoepfer R, Monyer H, Ruppersberg JP, Günther W, Seeburg PH, Sakmann B (1992c) Control by asparagine residues of calcium permeability and magnesium blockade in the NMDA receptor. Science 257:1415–1419

Burnashev N (1993) Recombinant ionotropic glutamate receptors: functional distinctions imparted by different subunits. Cell Physiol Biochem 3:318–331

Burnashev N, Zhou Z, Neher E, Sakmann B (1995) Fractional calcium currents through recombinant GluR channels of the NMDA, AMPA and kainate receptor subtypes. J Physiol (Lond) 485:403–418

Carmignoto G, Vicini S (1992) Activity-dependent decrease in NMDA receptor responses during development of the visual cortex. Science 258:1007–1011

Catania MV, Bellomo M, Giuffrida R, Giuffrida R, Stella AMG, Albanese V (1998) AMPA receptor subunits are differentially expressed in parvalbumin- and calretinin-positive neurons of the rat. Eur J Neurosci 10:3479–3490

Chenard BL, Bordner J, Butler TW, Chambers LK, Collins MA, De Costa DL, Ducat MF, Dumont ML, Fox CB, Mena EE, et al. (1995) (1S,2S)-1-(4-hydroxyphenyl)-2-(4-hydroxy-4-phenylpiperidino)-1-propanol: a potent new neuroprotectant which blocks *N*-methyl-D-aspartate responses. J Med Chem 38:3138–3145

Christine CW, Choi DW (1990) Effect of zinc on NMDA receptor-mediated channel currents in cortical neurons. J Neurosci 10:108–116

Ciabarra AM, Sullivan JM, Gahn LG, Pecht G, Heinemann S, Sevarino KA (1995) Cloning and characterization of χ-1: a developmentally regulated member of a novel class of the ionotropic glutamate receptor family. J Neurosci 15:6498–6508

Colquhoun D, Jonas P, Sakmann B (1992) Action of brief pulses of glutamate on AMPA/kainate receptors in patches from different neurones of rat hippocampal slices. J Physiol (Lond) 458:261–287

Cull-Candy SG, Brickley SG, Misra C, Feldmeyer D, Momiyama A, Farrant M (1998) NMDA receptor diversity in the cerebellum: identification of subunits contributing to functional receptors. Neuropharmacol 37:1369–80

Curutchet P, Bochet P, Prado de Carvalho L, Lambolez B, Stinnakre J, Rossier J (1992) In the GluR1 glutamate receptor subunit a glutamine to histidine point mutation suppresses inward rectification but not calcium permeability. BBRC 182: 1089–1093

Dingledine R, Hume RI, Heinemann SF (1992) Structural determinants of barium permeation and rectification in non-NMDA glutamate receptor channels. J Neurosci 12:4080–4087

Dupont JL, Fournier R, Crepel F (1987) Postnatal development of chemosensitivity of rat cerebellar Purkinje cells to excitatory amino acids. An in vitro study. Develop Brain Res 34:59–68

Durand GM, Bennett MV, Zukin RS (1993) Splice variants of the *N*-methyl-D-aspartate receptor NR1 identify domains involved in regulation by polyamines and protein kinase C. Proc Natl Acad Sci USA 90:6731–6735

Durand GM, Gregor P, Zheng X, Bennett MV, Uhl GR, Zukin RS (1992) Cloning of an apparent splice variant of the rat *N*-methyl-D-aspartate receptor NMDAR1 with altered sensitivity to polyamines and activators of protein kinase C. Proc Natl Acad Sci USA 89:9359–9363

Durand GM, Zukin RS (1993) Developmental regulation of mRNAs encoding rat brain kainate/AMPA receptors: a northern analysis study. J Neurochem 61: 2239–2246

Ehlers MD, Zhang S, Bernhardt JP, Huganir RL (1996) Inactivation of NMDA receptors by direct interaction of calmodulin with the NR1 subunit. Cell 84:745–755

Farrant M, Feldmeyer D, Takahashi T, Cull-Candy SG (1994) NMDA-receptor channel diversity in the developing cerebellum. Nature 368:335–339

Fleck MW, Bähring R, Patneau DK, Mayer ML (1996) AMPA receptor heterogeneity in rat hippocampal neurons revealed by differential sensitivity to cyclothiazide. J Neurophysiol 75:2322–2333

Fleidervish IA, Binshtok AM, Gutnick MJ (1998) Functionally distinct NMDA receptors mediate horizontal connectivity within layer 4 of mouse barrel cortex. Neuron 21:1055–1065

Flint AC, Maisch US, Weishaupt JH, Kriegstein AR, Monyer H (1997) NR2A subunit expression shortens NMDA receptor synaptic currents in developing neocortex. J Neurosci 17:2469–2476
Freund TF, Buzsáki G (1996) Interneurons of the hippocampus. Hippocamp 6:347–470
Fritschy JM, Weinmann O, Wenzel A, Benke D (1998) Synapse-specific localization of NMDA and GABA(A) receptor subunits revealed by antigen-retrieval immunohistochemistry. J Comp Neurol 390:194–210
Geiger JRP, Lübke J, Roth A, Frotscher M, Jonas P (1997) Submillisecond AMPA receptor-mediated signaling at a principal neuron–interneuron synapse. Neuron 18:1009–1023
Geiger JRP, Melcher T, Koh D-S, Sakmann B, Seeburg PH, Jonas P, Monyer H (1995) Relative abundance of subunit mRNAs determines gating and Ca^{2+} permeability of AMPA receptors in principal neurons and interneurons in rat CNS. Neuron 15:193–204
Gibb AJ, Colquhoun D (1992) Activation of N-methyl-D-aspartate receptors by L-glutamate in cells dissociated from adult rat hippocampus. J Physiol (Lond) 456:143–179
Gilbertson TA, Scobey R, Wilson M (1991) Permeation of calcium ions through non-NMDA glutamate channels in retinal bipolar cells. Science 251:1613–1615
Goldstein PA, Lee CJ, MacDermott AB (1995) Variable distributions of Ca^{2+}-permeable and Ca^{2+}-impermeable AMPA receptors on embryonic rat dorsal horn neurons. J Neurophysiol 73:2522–2534
Götz T, Kraushaar U, Geiger J, Lübke J, Berger T, Jonas P (1997) Functional properties of AMPA and NMDA receptors expressed in identified types of basal ganglia neurons. J Neurosci 17:204–215
Gu JG, Albuquerque C, Lee CJ, MacDermott AB (1996) Synaptic strengthening through activation of Ca^{2+}-permeable AMPA receptors. Nature 381:793–796
Haverkampf K, Lübke J, Jonas P (1997) Single-channel properties of native AMPA receptors depend on the putative subunit composition. Soc for Neurosci Abstr 23:478.23
Herlitze S, Raditsch M, Ruppersberg JP, Jahn W, Monyer H, Schoepfer R (1993) Argiotoxin detects molecular differences in AMPA receptor channels. Neuron 10:1131–40
Hestrin S (1992a) Activation and desensitization of glutamate-activated channels mediating fast excitatory synaptic currents in the visual cortex. Neuron 9:991–999
Hestrin S (1992b) Developmental regulation of NMDA receptor-mediated synaptic currents at a central synapse. Nature 357:686–689
Hestrin S (1993) Different glutamate receptor channels mediate fast excitatory synaptic currents in inhibitory and excitatory cortical neurons. Neuron 11:1083–1091
Higuchi M, Single FN, Köhler M, Sommer B, Sprengel R, Seeburg PH (1993) RNA editing of AMPA receptor subunit GluR-B: a base-paired intron–exon structure determines position and efficiency. Cell 75:1361–1370
Hoffmann H, Hatt H, Gottmann K (1997) Presynaptic exocytosis regulates NR2A mRNA expression in cultured neocortical neurones. Neurorep 8:3731–3735
Hollmann M, Boulter J, Maron C, Beasley L, Sullivan J, Pecht G, Heinemann S (1993) Zinc potentiates agonist-induced currents at certain splice variants of the NMDA receptor. Neuron 10:943–954
Hollmann M, Hartley M, Heinemann S (1991) Ca^{2+} permeability of KA-AMPA-gated glutamate receptor channels depends on subunit composition. Science 252:851–853
Hollmann M, Heinemann S (1994) Cloned glutamate receptors. Ann Rev of Neurosci 17:31–108
Hollmann M, O'Shea-Greenfield A, Rogers SW, Heinemann S (1989) Cloning by functional expression of a member of the glutamate receptor family. Nature 342:643–648

Hume RI, Dingledine R, Heinemann SF (1991) Identification of a site in glutamate receptor subunits that controls calcium permeability. Science 253:1028–1031

Iino M, Koike M, Isa T, Ozawa S (1996) Voltage-dependent blockage of Ca^{2+}-permeable AMPA receptors by joro spider toxin in cultured rat hippocampal neurones. J Physiol (Lond) 496:431–437

Iino M, Ozawa S, Tsuzuki K (1990) Permeation of calcium through excitatory amino acid receptor channels in cultured rat hippocampal neurones. J Physiol (Lond) 424:151–165

Iino M, Ciani S, Tsuzuki K, Ozawa S, Kidokoro Y (1997) Permeation properties of Na^+ and Ca^{2+} ions through the mouse $\varepsilon 2/\zeta 1$ NMDA receptor channel expressed in *Xenopus* oocytes. J Membrane Biol 155:143–156

Ikeda K, Nagasawa M, Mori H, Araki K, Sakimura K, Watanabe M, Inoue Y, Mishina M (1992) Cloning and expression of the $\varepsilon 4$ subunit of the NMDA receptor channel. FEBS Letters 313:34–38

Ilyin VI, Whittemore ER, Guastella J, Weber E, Woodward RM (1996) Subtype-selective inhibition of *N*-methyl-D-aspartate receptors by haloperidol. Mol Pharmacol 50:1541–1550

Isa T, Itazawa S, Iino M, Tsuzuki K, Ozawa S (1996) Distribution of neurones expressing inwardly rectifying and Ca^{2+}-permeable AMPA receptors in rat hippocampal slices. J Physiol (Lond) 491:719–733

Ishii T, Moriyoshi K, Sugihara H, Sakurada K, Kadotani H, Yokoi M, Akazawa C, Shigemoto R, Mizuno N, Masu M, Nakanishi S (1993) Molecular characterization of the family of the *N*-methyl-D-aspartate receptor subunits. J Biol Chem 268:2836–2843

Itazawa SI, Isa T, Ozawa S (1997) Inwardly rectifying and Ca^{2+}-permeable AMPA-type glutamate receptor channels in rat neocortical neurons. J Neurophysiol 78: 2592–2605

Ito I, Futai K, Katagiri H, Watanabe M, Sakimura K, Mishina M, Sugiyama H (1997) Synapse-selective impairment of NMDA receptor functions in mice lacking NMDA receptor ε1 or ε2 subunit. J Physiol (Lond) 500:401–408

Jia Z, Agopyan N, Miu P, Xiong Z, Henderson J, Gerlai R, Taverna FA, Velumian A, MacDonald J, Carlen P, Abramow-Newerly W, Roder J (1996) Enhanced LTP in mice deficient in the AMPA receptor GluR2. Neuron 17:945–956

Jonas P, Burnashev N (1995) Molecular mechanisms controlling calcium entry through AMPA-type glutamate receptor channels. Neuron 15:987–990

Jonas P, Racca C, Sakmann B, Seeburg PH, Monyer H (1994) Differences in Ca^{2+} permeability of AMPA-type glutamate receptor channels in neocortical neurons caused by differential GluR-B subunit expression. Neuron 12:1281–1289

Jonas P, Sakmann B (1992) Glutamate receptor channels in isolated patches from CA1 and CA3 pyramidal cells of rat hippocampal slices. J Physiol (Lond) 455:143–171

Keinänen K, Wisden W, Sommer B, Werner P, Herb A, Verdoorn TA, Sakmann B, Seeburg PH (1990) A family of AMPA-selective glutamate receptors. Science 249:556–560

Kendrick SJ, Dichter MA, Wilcox KS (1998) Characterization of desensitization in recombinant *N*-methyl-D-aspartate receptors: comparison with native receptors in cultured hippocampal neurons. Brain Res. Mol Brain Res 57:10–20

Kew JN, Richards JG, Mutel V, Kemp JA (1998) Developmental changes in NMDA receptor glycine affinity and ifenprodil sensitivity reveal three distinct populations of NMDA receptors in individual rat cortical neurons. J Neurosci 18:1935–1943

Kirson ED, Yaari Y (1996) Synaptic NMDA receptors in developing mouse hippocampal neurones: functional properties and sensitivity to ifenprodil. J Physiol (Lond) 497:437–455

Koh D-S, Geiger JRP, Jonas P, Sakmann B (1995a) Ca^{2+}-permeable AMPA and NMDA receptor channels in basket cells of rat hippocampal dentate gyrus. J Physiol (Lond) 485:383–402

Koh D-S, Burnashev N, Jonas P (1995b) Block of native Ca^{2+}-permeable AMPA receptors in rat brain by intracellular polyamines generates double rectification. J Physiol (Lond) 486:305–312
Kondo M, Sumino R, Okado H (1997) Combinations of AMPA receptor subunit expression in individual cortical neurons correlate with expression of specific calcium-binding proteins. J Neurosci 17:1570–1581
Krupp JJ, Vissel B, Heinemann SF, Westbrook GL (1996) Calcium-dependent inactivation of recombinant *N*-methyl-D-aspartate receptors is NR2 subunit specific. Mol Pharmacol 50:1680–1688
Krupp JJ, Vissel B, Heinemann SF, Westbrook GL (1998) *N*-terminal domains in the NR2 subunit control desensitization of NMDA receptors. Neuron 20:317–327
Kuner T, Schoepfer R (1996) Multiple structural elements determine subunit specificity of Mg^{2+} block in NMDA receptor channels. J Neurosci 16:3549–3558
Kuryatov A, Laube B, Betz H, Kuhse J (1994) Mutational analysis of the glycine-binding site of the NMDA receptor: structural similarity with bacterial amino acid-binding proteins. Neuron 12:1291–1300
Kutsuwada T, Kashiwabuchi N, Mori H, Sakimura K, Kushiya E, Araki K, Meguro H, Masaki H, Kumanishi T, Arakawa M, Mishina M (1992) Molecular diversity of the NMDA receptor channel. Nature 358:36–41
Kyrozis A, Goldstein PA, Heath MJS, MacDermott AB (1995) Calcium entry through a subpopulation of AMPA receptors desensitized neighboring NMDA receptors in rat dorsal horn neurons. J Physiol (Lond) 485:373–381
Lambolez B, Audinat E, Bochet P, Crépel F, Rossier J (1992) AMPA receptor subunits expressed by single Purkinje cells. Neuron 9:247–258
Lambolez B, Ropert N, Perrais D, Rossier J, Hestrin S (1996) Correlation between kinetics and RNA splicing of α-amino-3-hydroxy-5-methylisoxazole-4-propionic acid receptors in neocortical neurons. Proc Natl Acad Sci USA 93:1797–1802
Laube B, Hirai H, Sturgess M, Betz H, Kuhse J (1997) Molecular determinants of agonist discrimination by NMDA receptor subunits: analysis of the glutamate binding site on the NR2B subunit. Neuron 18:493–503
Laurie DJ, Seeburg PH (1994) Regional and developmental heterogeneity in splicing of the rat brain NMDAR1 mRNA. J Neurosci 14:3180–3194
Legendre P, Rosenmund C, Westbrook GL (1993) Inactivation of NMDA channels in cultured hippocampal neurons by intracellular calcium. J Neurosci 13:674–684
Leranth C, Szeidemann Z, Hsu M, Buzsáki G (1996) AMPA receptors in rat and primate hippocampus: a possible absence of GluR2/3 subunits in most interneurons. Neurosci 70:631–652
Lerma J, Morales M, Ibarz JM, Somohano F (1994) Rectification properties and Ca^{2+} permeability of glutamate receptor channels in hippocampal cells. Eur J Neurosci 6:1080–1088
Lindlbauer R, Mohrmann R, Hatt H, Gottmann K (1998) Regulation of kinetic and pharmacological properties of synaptic NMDA receptors depends on presynaptic exocytosis in rat hippocampal neurons. J Physiol (Lond) 508:495–502
Lomeli H, Mosbacher J, Melcher T, Höger T, Geiger JRP, Kuner T, Monyer H, Higuchi M, Bach A, Seeburg PH (1994) Control of kinetic properties of AMPA receptor channels by nuclear RNA editing. Science 266:1709–1713
Lynch DR, Gallagher MJ (1996) Inhibition of *N*-methyl-D-aspartate receptors by haloperidol: developmental and pharmacological characterization in native and recombinant receptors. J Pharmacol Exp Therapeut 279:154–161
Mahanty NK, Sah P (1998) Calcium-permeable AMPA receptors mediate long-term potentiation in interneurons in the amygdala. Nature 394:683–687
Martin LJ, Blackstone CD, Levey AI, Huganir RL, Price DL (1993) AMPA glutamate receptor subunits are differentially distributed in rat brain. Neurosci 53:327–358
Mayer ML, Partin KM, Patneau DK, Wong LA, Vyklicki L Jr, Benveniste M, Bowie D (1995) Desensitization at AMPA, kainate, and NMDA receptors. In: Excitatory amino acids and synaptic transmission. H. Wheal and A. Thomson (eds), London, Academic

Mayer ML, Vyklicky L Jr, Clements J (1989) Regulation of NMDA receptor desensitization in mouse hippocampal neurons by glycine. Nature 338:425–427

Mayer ML, Westbrook GL (1987) Permeation and block of N-methyl-D-aspartic acid receptor channels by divalent cations in mouse cultured central neurones. J Physiol (Lond) 394:501–527

McBain CJ, Dingledine R (1993) Heterogeneity of synaptic glutamate receptors on CA3 stratum radiatum interneurons of rat hippocampus. J Physiol (Lond) 462:373–392

McBain CJ, Mayer ML (1994) N-methyl-D-aspartic acid receptor structure and function. Physiol Rev 74:723–760

Medina I, Filippova N, Charton G, Rougeole S, Ben-Ari Y, Khrestchatisky M, Bregestovski P (1995) Calcium-dependent inactivation of heteromeric NMDA receptor-channels expressed in human embryonic kidney cells. J Physiol (Lond) 482:567–573

Meguro H, Mori H, Araki K, Kushiya E, Kutsuwada T, Yamazaki M, Kumanishi T, Arakawa M, Sakimura K, Mishina M (1992). Functional characterization of a heteromeric NMDA receptor channel expressed from cloned cDNAs. Nature 357:70–74

Melcher T, Maas S, Herb A, Sprengel R, Seeburg PH, Higuchi M (1996) A mammalian RNA editing enzyme. Nature 379:460–464

Momiyama A, Feldmeyer D, Cull-Candy SG (1996) Identification of a native low-conductance NMDA channel with reduced sensitivity to Mg^{2+} in rat central neurones. J Physiol (Lond) 494:479–492

Monyer H, Burnashev N, Laurie DJ, Sakmann B, Seeburg PH (1994) Developmental and regional expression in the rat brain and functional properties of four NMDA receptors. Neuron 12:529–540

Monyer H, Seeburg PH, Wisden W (1991) Glutamate-operated channels: developmentally early and mature forms arise by alternative splicing. Neuron 6:799–810

Monyer H, Sprengel R, Schoepfer R, Herb A, Higuchi M, Lomeli H, Burnashev N, Sakmann B, Seeburg PH (1992) Heteromeric NMDA receptors: molecular and functional distinction of subtypes. Science 256:1217–1221

Mori H, Masaki H, Yamakura T, Mishina M (1992). Identification by mutagenesis of a Mg^{2+}-block site of the NMDA receptor channel. Nature 358:673–675

Moriyoshi K, Masu M, Ishii T, Shigemoto R, Mizuno N, Nakanishi S (1991) Molecular cloning and characterization of the rat NMDA receptor. Nature 354:31–37

Mosbacher J, Schoepfer R, Monyer H, Burnashev N, Seeburg PH, Ruppersberg, JP (1994) A molecular determinant for submillisecond desensitization in glutamate receptors. Science 266:1059–1062

Mosbacher JC (1995) Der Einfluss posttranskriptionaler Modifikationen auf die kinetischen Eigenschaften von AMPA-Rezeptoren. PhD Thesis, Heidelberg

Mott DD, Doherty JJ, Zhang S, Washburn MS, Frendley MJ, Lyuboslavsky P, Traynelis SF, Dingledine R (1998) Phenylethanolamines inhibit NMDA receptors by enhancing proton inhibition. Nature Neurosci 1:659–667

Müller T, Möller T, Berger T, Schnitzer J, Kettenmann H (1992) Calcium entry through kainate receptors and resulting potassium-channel blockade in Bergmann glial cells. Science 256:1563–1566

Nakanishi N, Shneider NA, Axel R (1990) A family of glutamate receptor genes: evidence for the formation of heteromultimeric receptors with distinct channel properties. Neuron 5:569–581

Nase G, Weishaupt J, Stern P, Singer W, Monyer H (1997) Activity-dependent regulation of NMDA receptor expression in the visual cortex. Soc Neurosci Abstr 23:371.5

Osten P, Srivastava S, Inman GJ, Vilim FS, Khatri L, Lee LM, States BA, Einheber S, Milner TA, Hanson PI, Ziff EB (1998) The AMPA receptor GluR2 *C* terminus can mediate a reversible, ATP-dependent interaction with NSF and α- and β-SNAPs. Neuron 21:99–110

Otis TS, Raman IM, Trussell LO (1995) AMPA receptors with high Ca^{2+} permeability mediate synaptic transmission in the avian auditory pathway. J Physiol (Lond) 482:309–315

Ozaki M, Sasner M, Yano R, Lu HS, Buonanno A (1997) Neuregulin-β induces expression of an NMDA-receptor subunit. Nature 390:691–694

Paoletti P, Ascher P, Neyton J (1997) High-affinity zinc inhibition of NMDA NR1-NR2A receptors. J Neurosci 17:5711–5725

Partin KM, Fleck MW, Mayer ML (1996) AMPA receptor flip/flop mutants affecting deactivation, desensitization, and modulation by cyclothiazide, aniracetam, and thiocyanate. J Neurosci 16:6634–6647

Partin KM, Patneau DK, Mayer ML (1994) Cyclothiazide differentially modulates desensitization of alpha-amino-3-hydroxy-5-methyl-4-isoxazolepropionic acid receptor splice variants. Mol Pharmacol 46:129–138

Pegg AE (1988) Polyamine metabolism and its importance in neoplastic growth and as a target for chemotherapy. Cancer Res 48:759–774

Petralia RS, Wang YX, Mayat E, Wenthold RJ (1997) Glutamate receptor subunit 2-selective antibody shows a differential distribution of calcium-impermeable AMPA receptors among populations of neurons. J Comp Neurol 385:456–476

Petralia RS, Wenthold RJ (1992) Light and electron immunocytochemical localization of AMPA-selective glutamate receptors in the rat brain. J Comp Neurol 318: 329–354

Racca C, Catania MV, Monyer H, Sakmann B (1996) Expression of AMPA-glutamate receptor B subunit in rat hippocampal GABAergic neurons. Eur J Neurosci 8:1580–1590

Rafiki A, Gozlan H, Ben-Ari Y, Khrestchatisky M, Medina I (1997) The calcium-dependent transient inactivation of recombinant NMDA receptor-channel does not involve the high affinity calmodulin binding site of the NR1 subunit. Neurosci Letters 223:137–139

Rosenmund C, Westbrook GL (1993) Calcium-induced actin depolymerization reduces NMDA channel activity. Neuron 10:805–814

Rozov A, Zilberter Y, Wollmuth LP, Burnashev N (1998) Facilitation of currents through rat Ca^{2+}-permeable AMPA receptor channels by activity-dependent relief from polyamine block. J Physiol (Lond) 511:361–377

Rubio ME, Wenthold RJ (1997) Glutamate receptors are selectively targeted to postsynaptic sites in neurons. Neuron 18:939–950

Sather W, Dieudonné S, MacDonald JF, Ascher P (1992) Activation and desensitization of *N*-methyl-D-aspartate receptors in nucleated outside-out patches from mouse neurones. J Physiol (Lond) 450:643–672

Sather W, Johnson JW, Henderson G, Ascher P (1990) Glycine-insensitive desensitization of NMDA responses in cultured mouse embryonic neurons. Neuron 4:725–731

Schneggenburger R, Zhou Z, Konnerth A, Neher E (1993) Fractional contribution of calcium to the cation current through glutamate receptor channels. Neuron 11:133–143

Seeburg PH, Higuchi M, Sprengel R (1998) RNA editing of brain glutamate receptor channels: mechanism and physiology Brain Res. Brain Res Rev 26:217–229

Sekiguchi M, Fleck MW, Mayer ML, Takeo J, Chiba Y, Yamashita S, Wada K (1997) A novel allosteric potentiator of AMPA receptors: 4–2-(phenylsulfonylamino)ethylthio-2,6-difluoro-phenoxyacetamide. J Neurosci 17:5760–5771

Sharma G, Stevens CF (1996). Interactions between two divalent ion binding sites in *N*-methyl-D-aspartate receptor channels. Proc Natl Acad Sci USA 93:14170–14175

Sheng M, Cummings J, Roldan LA, Jan YN, Jan LY (1994) Changing subunit composition of heteromeric NMDA receptors during development of rat cortex. Nature 368:144–147

Silver RA, Colquhoun D, Cull-Candy SG, Edmonds B (1996) Deactivation and desensitization of non-NMDA receptors in patches and the time course of EPSCs in rat cerebellar granule cells. J Physiol (Lond) 493:167–173

Soloviev MM, Barnard EA (1997) *Xenopus* oocytes express a unitary glutamate receptor endogenously. J Mol Biol 273:14–18
Sommer B, Keinänen K, Verdoorn TA, Wisden W, Burnashev N, Herb A, Köhler M, Takagi T, Sakmann B, Seeburg PH (1990) Flip and flop: a cell-specific functional switch in glutamate-operated channels of the CNS. Science 249:1580–1585
Sommer B, Köhler M, Sprengel R, Seeburg PH (1991) RNA editing in brain controls a determinant of ion flow in glutamate-gated channels. Cell 67:11–19
Spruston N, Jonas P, Sakmann B (1995) Dendritic glutamate receptor channels in rat hippocampal CA3 and CA1 pyramidal neurons. J Physiol (Lond) 482:325–352
Standaert DG, Landwehrmeyer GB, Kerner JA, Penney JB Jr, Young AB (1996) Expression of NMDAR2D glutamate receptor subunit mRNA in neurochemically identified interneurons in the rat neostriatum, neocortex and hippocampus. Brain Res. Mol Brain Res. 42:89–102
Stern P, Behe P, Schoepfer R, Colquhoun D (1992) Single-channel conductances of NMDA receptors expressed from cloned cDNAs: comparison with native receptors. Proc Roy Soc London B: Biol Sci 250:271–277
Stern-Bach Y, Bettler B, Hartley M, Sheppard PO, O'Hara PJ, Heinemann SF (1994) Agonist selectivity of glutamate receptors is specified by two domains structurally related to bacterial amino acid-binding proteins. Neuron 13:1345–1357
Stern-Bach Y, Russo S, Neuman M, Rosenmund C (1998) A point mutation in the glutamate binding site blocks desensitization of AMPA receptors. Neuron 21:907–918
Sucher NJ, Akbarian S, Chi CL, Leclerc CL, Awobuluyi M, Deitcher DL, Wu MK, Yuan JP, Jones EG, Lipton SA (1995) Developmental and regional expression pattern of a novel NMDA receptor-like subunit (NMDAR-L) in the rodent brain. J Neurosci 15:6509–6520
Swanson GT, Feldmeyer D, Kaneda M, Cull-Candy SG (1996) Effect of RNA editing and subunit co-assembly single-channel properties of recombinant kainate receptors. J Physiol (Lond) 492:129–142
Swanson GT, Kamboj SK, Cull-Candy SG (1997) Single-channel properties of recombinant AMPA receptors depend on RNA editing, splice variation, and subunit composition. J Neurosci 17:58–69
Takahashi T, Feldmeyer D, Suzuki N, Onodera K, Cull-Candy SG, Sakimura K, Mishina M (1996) Functional correlation of NMDA receptor epsilon subunits expression with the properties of single-channel and synaptic currents in the developing cerebellum. J Neurosci 16:4376–4382
Taschenberger H, Grantyn R (1998) Interaction of calcium-permeable non-*N*-methyl-D-aspartate receptor channels with voltage-activated potassium and calcium currents in rat retinal ganglion cells in vitro. Neurosci 84:877–896
Tong G, Jahr CE (1994) Regulation of glycine-insensitive desensitization of the NMDA receptor in outside-out patches. J Neurophysiol 72:754–761
Tóth K, McBain CJ (1998) Afferent-specific innervation of two distinct AMPA receptor subtypes on single hippocampal interneurons. Nature Neurosci 1:572–578
Traub RD, Miles R (1995) Pyramidal cell-to-inhibitory cell spike transduction explicable by active dendritic conductances in inhibitory cell. J Comput Neurosci 2:291–298
Traynelis SF, Burgess MF, Zheng F, Lyuboslavsky P, Powers JL (1998) Control of voltage-independent zinc inhibition of NMDA receptors by the NR1 subunit. J Neurosci 18:6163–6175
Traynelis SF, Cull-Candy SG (1991) Pharmacological properties and H^+ sensitivity of excitatory amino acid receptor channels in rat cerebellar granule neurones. J Physiol (Lond) 433:727–763
Traynelis SF, Hartley M, Heinemann SF (1995) Control of proton sensitivity of the NMDA receptor by RNA splicing and polyamines. Science 268:873–876
Trussell LO, Zhang S, Raman IM (1993) Desensitization of AMPA receptors upon multiquantal neurotransmitter release. Neuron 10:1185–1196

Tsuzuki K, Mochizuki S, Iino M, Mori H, Mishina M, Ozawa S (1994) Ion permeation properties of the cloned mouse $\varepsilon 2/\zeta 1$ NMDA receptor channel. Brain Res. Mol Brain Res 26:37–46

Vallano ML, Lambolez B, Audinat E, Rossier J (1996) Neuronal activity differentially regulates NMDA receptor subunit expression in cerebellar granule cells. J Neurosci 16:631–639

van Hooft JA, Haas TF, Fuchs EC, Monyer H (1998) Different functional properties and specific expression patterns of exon 5 splice variants of the NMDAR1 subunit. Eur J Neurosci 10:289

Verdoorn TA, Burnashev N, Monyer H, Seeburg PH, Sakmann B (1991) Structural determinants of ion flow through recombinant glutamate receptor channels. Science 252:1715–1718

Vicini S, Wang JF, Li JH, Zhu WJ, Wang YH, Luo JH, Wolfe BB, Grayson DR (1998) Functional and pharmacological differences between recombinant *N*-methyl-D-aspartate receptors. J Neurophysiol 79:555–566

Villarroel A, Regalado MP, Lerma J (1998) Glycine-independent NMDA receptor desensitization: localization of structural determinants. Neuron 20:329–339

Vissavajjhala P, Janssen WGM, Hu Y, Gazzaley AH, Moran T, Hof PR, Morrison JH (1996) Synaptic distribution of the AMPA-GluR2 subunit and its colocalization with calcium-binding proteins in rat cerebral cortex: an immunohistochemical study using a GluR2-specific monoclonal antibody. Exp Neurol 142:296–312

Wafford KA, Kathoria M, Bain CJ, Marshall G, Le Bourdelles B, Kemp JA, Whiting PJ (1995) Identification of amino acids in the *N*-methyl-D-aspartate receptor NR1 subunit that contribute to the glycine binding site. Mol Pharmacol 47:374–380

Washburn MS, Numberger M, Zhang S, Dingledine R (1997) Differential dependence of GluR2 expression of three characteristic features of AMPA receptors. J Neurosci 17:9393–9406

Watanabe M, Fukaya M, Sakimura K, Manabe T, Mishina M, Inoue Y (1998) Selective scarcity of NMDA receptor channel subunits in the stratum lucidum (mossy fibre-recipient layer) of the mouse hippocampal CA3 subfield. Eur J Neurosci 10:478–487

Watanabe M, Inoue Y, Sakimura K, Mishina M (1993) Distinct distributions of five *N*-methyl-D-aspartate receptor channel subunit mRNAs in the forebrain. J Comp Neurol 338:377–390

Watanabe M, Mishina M, Inoue Y (1994) Distinct spatiotemporal expressions of five NMDA receptor channel subunit mRNAs in the cerebellum. J Comp Neurol 343:513–519

Watanabe S, Kusama-Eguchi K, Kobayashi H, Igarashi K (1991) Estimation of polyamine binding to macromolecules and ATP in bovine lymphocytes and rat liver. J Biol Chem 266:20803–20809

Wenzel A, Fritschy JM, Mohler H, Benke D (1997) NMDA receptor heterogeneity during postnatal development of the rat brain: differential expression of the NR2A, NR2B, and NR2C subunit proteins. J Neurochem 68:469–478

Westbrook GL, Mayer ML (1987) Micromolar concentrations of Zn^{2+} antagonize NMDA and GABA responses of hippocampal neurons. Nature 328:640–643

Williams K (1993) Ifenprodil discriminates subtypes of the *N*-methyl-D-aspartate receptor: selectivity and mechanisms at recombinant heteromeric receptors. Mol Pharmacol 44:851–859

Williams K (1995) Pharmacological properties of recombinant *N*-methyl-D-aspartate (NMDA) receptors containing the $\varepsilon 4$ (NR2D) subunit. Neurosci Letters 184: 181–184

Williams K, Chao J, Kashiwagi K, Masuko T, Igarashi K (1996) Activation of *N*-methyl-D-aspartate receptors by glycine: role of an aspartate residue in the M3-M4 loop of the NR1 subunit. Mol Pharmacol 50:701–708

Williams K, Russell SL, Shen YM, Molinoff PB (1993) Developmental switch in the

expression of NMDA receptors occurs in vivo and in vitro. Neuron 10:267–278
Williams K, Zappia AM, Pritchett DB, Shen YM, Molinoff PB (1994) Sensitivity of the *N*-methyl-D-aspartate receptor to polyamines is controlled by NR2 subunits. Mol Pharmacol 45:803–809
Wollmuth LP, Kuner T, Sakmann B (1998) Adjacent asparagines in the NR2-subunit of the NMDA receptor channel control the voltage-dependent block by extracellular Mg^{2+}. J Physiol (Lond) 506:13–32
Wyllie DJ, Behe P, Colquhoun D (1998) Single-channel activations and concentration jumps: comparison of recombinant NR1a/NR2A and NR1a/NR2D NMDA receptors. J Physiol (Lond) 510:1–18
Wyllie DJ, Behe P, Nassar M, Schoepfer R, Colquhoun D (1996) Single-channel currents from recombinant NMDA NR1a/NR2D receptors expressed in *Xenopus* oocytes. Proc Roy Soc London B: Biol Sci 263:1079–1086
Zhang S, Ehlers MD, Bernhardt JP, Su CT, Huganir RL (1998) Calmodulin mediates calcium-dependent inactivation of *N*-methyl-D-aspartate receptors. Neuron 21: 443–453
Zheng X, Zhang L, Durand GM, Bennett MV, Zukin RS (1994) Mutagenesis rescues spermine and Zn^{2+} potentiation of recombinant NMDA receptors. Neuron 12:811–818
Zukin RS, Bennett MV (1995) Alternatively spliced isoforms of the NMDARI receptor subunit. Trends in Neurosci 18:306–313

Section III
Glutamatergic Synaptic Transmission

CHAPTER 10

Morphological Characteristics of Glutamatergic Synapses in the Hippocampus

M. FROTSCHER, K. MEWS, and G. ADELMANN

A. Introduction and Historical Background

When Sir Charles Sherrington created the term "synapse" as early as 1897 (FOSTER and SHERRINGTON 1897), he mainly used it to describe functional interactions between nerve cells. It was much later on, after the introduction of electron microscopy to the study of nervous tissue, that this term was also applied to specialized membrane contacts visible in thin sections. The first reports on the fine structure of synaptic contacts between individual neurons and their processes appeared in the middle of the 1950s (PALADE and PALAY 1954; PALAY 1956, 1958) and described all the essential components of a synaptic contact such as the synaptic cleft between the pre- and postsynaptic elements, the specialized membrane appositions, and synaptic vesicles. Traditionally, synapses in the central nervous system (CNS) were assumed to occur between an axon terminal and a dendrite. However, it later became obvious that not only dendritic shafts but also dendritic spines, cell bodies, and even axons may be postsynaptic elements. In addition, specialized dendrites may act as presynaptic elements, for instance in the dendrodendritic synapses of the olfactory bulb (SHEPHERD 1972).

From a functional point of view, it is of utmost interest to know the nature of the cells that establish synapses with each other. Thus, attempts were made to classify different types of synapses and to identify them as excitatory or inhibitory contacts. In the brain, the most common excitatory transmitter is glutamate, whereas GABA (γ-aminobutyric acid) is the main inhibitory transmitter. Based upon physiological data suggesting that inhibition is generated mainly at the cell body and excitation at the dendrite, electron microscopists attempted to define differences between axosomatic and axodendritic synaptic contacts (GRAY 1959; UCHIZONO 1965). In his classic paper, GRAY (1959) described two types of synaptic contacts, later on termed Gray type-I and Gray type-II synapses. Type-I synapses, often referred to as asymmetric synapses, are found on dendritic shafts and spines, whereas symmetric type-II synapses predominate on cell bodies. The terms "symmetric" and "asymmetric" referred to the thickness of the presynaptic and postsynaptic membrane specializations; while these membranes were found to be of equal thickness in the case of type-II synapses, type-I synapses displayed a thicker postsynaptic density. Based upon their preferential location on cell bodies, symmetric contacts were regarded as inhibitory synapses, whereas asymmetric contacts, mainly found on dendritic spines and shafts, were interpreted as the structural elements of excitatory synaptic transmission. It is remarkable that this early classification of synaptic contacts, solely based on structural criteria, was largely confirmed later on by the application of more sophisticated techniques.

With the advent of electron-microscopic immunocytochemistry, it was shown that presynaptic terminals of symmetric synapses often stained for the inhibitory transmitter GABA or its synthesizing enzyme glutamate decarboxylase (RIBAK and ROBERTS 1990). In contrast, asymmetric synapses were found to contain glutamate receptor (GluR) subunits in or near their postsynaptic membranes (PETRALIA and WENTHOLD 1992; BAUDE et al. 1993, 1995; MOLNÁR et al. 1993; NUSSER et al. 1994; LUJAN et al. 1996; MATSUBARA et al. 1996; PETRALIA et al. 1996; BERNARD et al. 1997; OTTERSEN and LANDSEND 1997), indicating that glutamate is the transmitter released by the presynaptic terminal.

In this chapter we will briefly summarize what is known about the structural components of glutamatergic synapses, taking the synapses of the trisy-

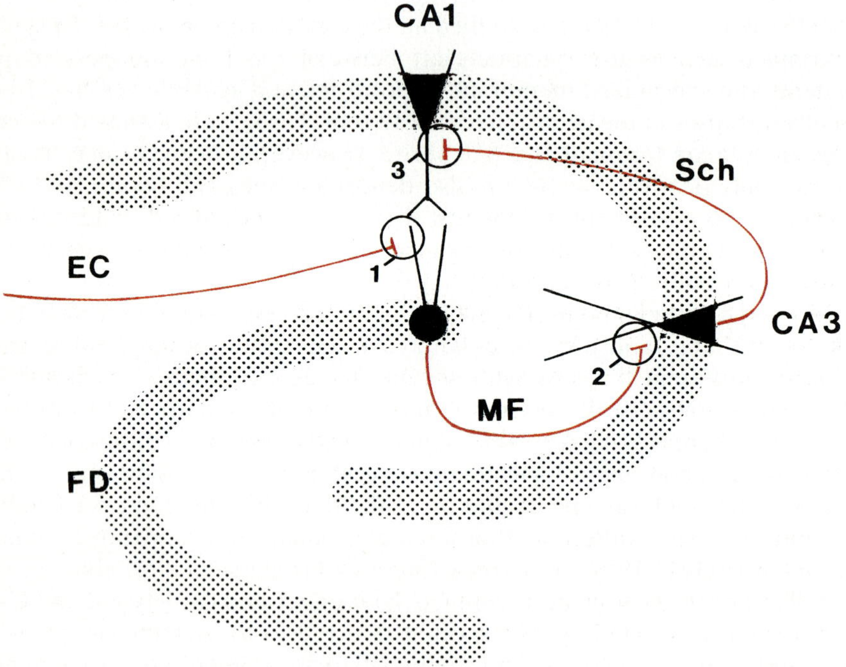

Fig. 1. The glutamatergic, excitatory, trisynaptic pathway of the hippocampus referred to in the present chapter. Glutamatergic fibers from the entorhinal cortex (*EC*) establish asymmetric synapses on distal dendrites of granule cells in the fascia dentata (*FD*, first synapse). Granule cell axons, the mossy fibers (*MF*), impinge on proximal dendrites of CA3 pyramidal cells (second synapse). The Schaffer collaterals (*Sch*) of CA3 pyramidal cell axons terminate on dendrites of CA1 pyramidal neurons (third synapse)

naptic excitatory hippocampal pathway (ANDERSEN et al. 1971; Fig. 1) as an example. By using electron-microscopic immunocytochemical techniques (cryosubstitution and immunogold labeling), we will demonstrate that different types of asymmetric synapses in the hippocampus assemble GluRs in their postsynaptic densities, whereas symmetric synapses lack these receptors but accumulate $GABA_A$-receptor subunits in their postsynaptic membranes.

B. Principal Components of a Glutamatergic Synapse

Several recent reviews have summarized our knowledge about the structural components of central synapses (CALVERLEY and JONES 1990; LISMAN and HARRIS 1993; HARRIS and KATER 1994; EDWARDS 1995; FROTSCHER 1996), and the reader is referred to these articles for more details. When approached from an electrophysiologist's viewpoint (EDWARDS 1995), there is a general need to quantify synaptic structures in order to understand function. However, data concerning the sizes of structures, e.g., the dimensions of the synaptic cleft, are difficult to obtain, because the individual tissue components are subject to shrinkage to a varying extent during processing for fine-structural analysis.

The majority of synapses in the cortex including the hippocampus are likely to be excitatory (84% in the cat visual cortex, according to BEAULIEU and COLONNIER 1985), most of them using glutamate as a transmitter. Principal components of supposedly glutamatergic synapses in the hippocampus are illustrated in Fig. 2. Glutamatergic synapses are formed by relatively small presynaptic boutons containing abundant round vesicles 35–50 nm in diameter and, regularly, one or several mitochondria. In contrast, GABAergic terminals establishing symmetric contacts on the cell body are usually larger and contain flattened vesicles (UCHIZONO 1965). However, the shape of the vesicle may be subject to alterations depending on the fixative used. Therefore, with the availability of antibodies against GABA and glutamate decarboxylase, the identification of GABAergic inhibitory synapses is done most reliably by immunocytochemical labeling. In contrast, immunostaining for glutamate does not allow one to identify glutamatergic neurons, because glutamate is a ubiquitous metabolite (SORIANO and FROTSCHER 1994). As will be discussed below, immunogold labeling of transmitter receptors may be an alternative way of characterizing these synaptic contacts.

Synaptic vesicles often accumulate at release sites between the regularly spaced dense projections that appear as triangular appositions of the presynaptic membrane in cross sections through a synapse (Fig. 2, inset). The dense projections are opposed to the postsynaptic density, which is, as mentioned before, thicker than the presynaptic membrane in supposedly glutamatergic synapses. As a rule, pre- and postsynaptic membranes run in parallel; often their course is curved (Fig. 2). The synaptic cleft between the two widens at the site of membrane specialization (15–20 nm; EDWARDS 1995) and contains some electron-dense material. Many asymmetric synapses, particularly in the

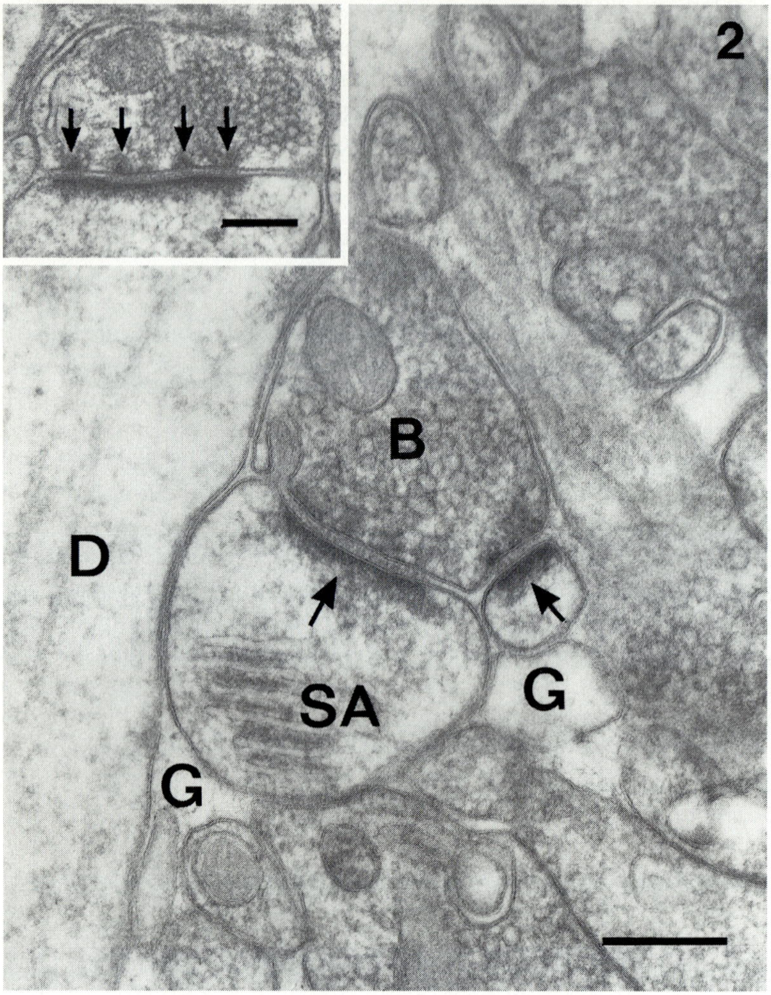

Fig. 2. Principal components of asymmetric, supposedly glutamatergic, excitatory synapses. The figure illustrates a spine synapse in the stratum radiatum of the mouse hippocampus. A relatively small bouton (*B*), probably a bouton of a Schaffer collateral, establishes asymmetric synaptic contacts (*arrows*) with two spines located next to an ascending CA1 pyramidal cell dendrite (*D*). The bouton contains numerous clear vesicles and a mitochondrion. The postsynaptic density of the spines is more pronounced than the presynaptic membrane specializations (asymmetric synapse). The large spine contains a prominent spine apparatus (*SA*). Glial processes (*G*) are seen in close proximity to the synaptic structures. *Inset*: An asymmetric synapse in the stratum radiatum of CA3. *Arrows* point to regularly spaced dense projections of the presynaptic bouton. *Scale bars*: 0.2 μm

molecular layer of the fascia dentata, are "perforated", i.e., two adjacent active zones are separated by a region lacking pre- and postsynaptic membrane specialization. This gap of nonspecialized membranes is often formed by a spinule protruding from the postsynaptic element into the presynaptic bouton at spine synapses. The number of spinules in synapses of the fascia dentata was found to change following long-term potentiation (SCHUSTER et al. 1990).

The prominent postsynaptic density of asymmetric synapses is not sharply delineated against the cytoplasm of the postsynaptic element. Rather, it continues by some blurred material extending deep into the postsynaptic cytoplasm (Fig. 2). It has been suggested that this subsynaptic web provides a structural matrix that clusters ion channels in the postsynaptic membrane and anchors signaling molecules, such as phosphatases and kinases, at the synapse (ZIFF 1997). Often, the postsynaptic element is a spine. According to BEAULIEU and COLONNIER (1985), 79% of all excitatory synapses in the cat visual cortex are on dendritic spines, light-microscopically visible dendritic appendages. The shape and size of the spine may vary considerably (SHEPHERD 1996), ranging from short, stubby spines to mushroom-shaped spines and filiform spines. Often, a spine apparatus is observed (Fig. 2), an ensemble of membrane-surrounded cisterns separated by electron-dense plates. In contrast to presynaptic boutons, spines often contain portions of smooth endoplasmic reticulum (SPACEK and HARRIS 1997). They also contain actin filaments that may cause rapid changes in spine shape (FISCHER et al. 1998).

As a rule, asymmetric synapses are found on the thickened spine head, but symmetric synapses, mainly on the spine neck connecting the spine head with the dendritic shaft, have been described as well (FREUND et al. 1984; FROTSCHER and LERANTH 1986; GOLDMAN-RAKIC et al. 1989). At the base of the spine, particularly in the case of the large thorns or excrescences postsynaptic to hippocampal mossy fibers, accumulations of ribosomes are often observed (STEWARD and LEVY 1982; Fig. 4). It has been hypothesized that these ribosomes are important for local protein synthesis in connection with plastic changes at the synapse (STEWARD and BANKER 1992).

C. Glial Wrapping is a General Feature of Glutamatergic Synapses

Astrocytes give rise to numerous fine processes intermingling with the processes of neurons. Their fine, ramified processes are particularly well seen when the immunolabeling of astrocytes is silver intensified (DEROUICHE and FROTSCHER 1991). In thin sections, one regularly finds that these fine astrocytic processes surround asymmetric, supposedly glutamatergic synapses (DEROUICHE and FROTSCHER 1991). As these processes stain for glutamine synthetase, which metabolizes glutamate from the transmitter pool, it has been assumed that these processes play an important role not only in limiting the lateral diffusion of transmitter, but also in glutamate degradation (DEROUICHE and Frotscher 1991). In fact, these astrocytic processes are highly active in glutamate uptake (Hösli and Hösli 1976; Drejer et al. 1982; Yu et al. 1982; Otis and Kavanaugh, this volume) and were found to contain glutamate transporters in their membranes (Chaudhry et al. 1995).

In Fig. 3, astrocytic processes are shown, identified as such by immunogold labeling for glial-fibrillary acidic protein (GFAP). The immunostaining is confined to the glial fibrils only, indicating specific immunolabeling as well as precise subcellular localization of immunobinding with this approach. When astrocytes are stained using diaminobenzidine (DAB) procedures, one finds a more or less homogeneous staining of the astrocytic processes. As seen in Fig. 3, portions of the astrocytic processes surround asymmetric, supposedly glutamatergic synapses. In the context of this chapter, we would like to emphasize that this glial wrapping is a regularly observed feature of glutamatergic synapses (Figs. 2–4). We conclude that the compartment of a glutamatergic synapse comprises the presynaptic terminal releasing the transmitter, the postsynaptic element containing the relevant transmitter receptors in its membrane, and the astrocytic wrapping that may influence the time course of glutamate clearance from the synaptic cleft.

D. Different Types of Glutamatergic Synapses in the Hippocampus

In our description of glutamatergic synapses in the hippocampus, we will focus on the synapses involved in the main excitatory pathway (Andersen et al. 1971; Fig. 1). This pathway has often been called "trisynaptic", because the excitatory signal flow from the entorhinal cortex involves excitatory entorhinal synapses on the dendrites of the granule cells (first station), the large synapses of the granule cell axons (the mossy fibers) on CA3 pyramidal neurons (second station), and the synapses of the Schaffer collaterals of CA3 neurons on CA1 pyramidal cell dendrites (third station). Numerous studies have provided evidence that these three synapses are excitatory and use glutamate as a transmitter (White et al. 1977; Storm-Mathisen and Ottersen 1984; Bramham et al. 1990). It should be pointed out, however, that our view of the main excitatory pathway of the hippocampal formation as involving only three synapses is certainly an oversimplification (Frotscher et al. 1994). For example, the axons of the granule cells, the mossy fibers, give rise to numerous collaterals in the hilus, thereby activating, for instance, the mossy cells, supposedly glutamatergic neurons (Soriano and Frotscher 1994) that project back to the fascia dentata both ipsi- and contralaterally (Ribak et al. 1985; Frotscher et al. 1991). The mossy fibers not only contact spines of excitatory neurons but also establish synapses with a variety of GABAergic interneurons in the hilar region and in CA3 (Frotscher et al. 1994; Geiger et al., this volume). It has recently been pointed out that the majority of the mossy-fiber synapses are probably established with interneurons (Acsády

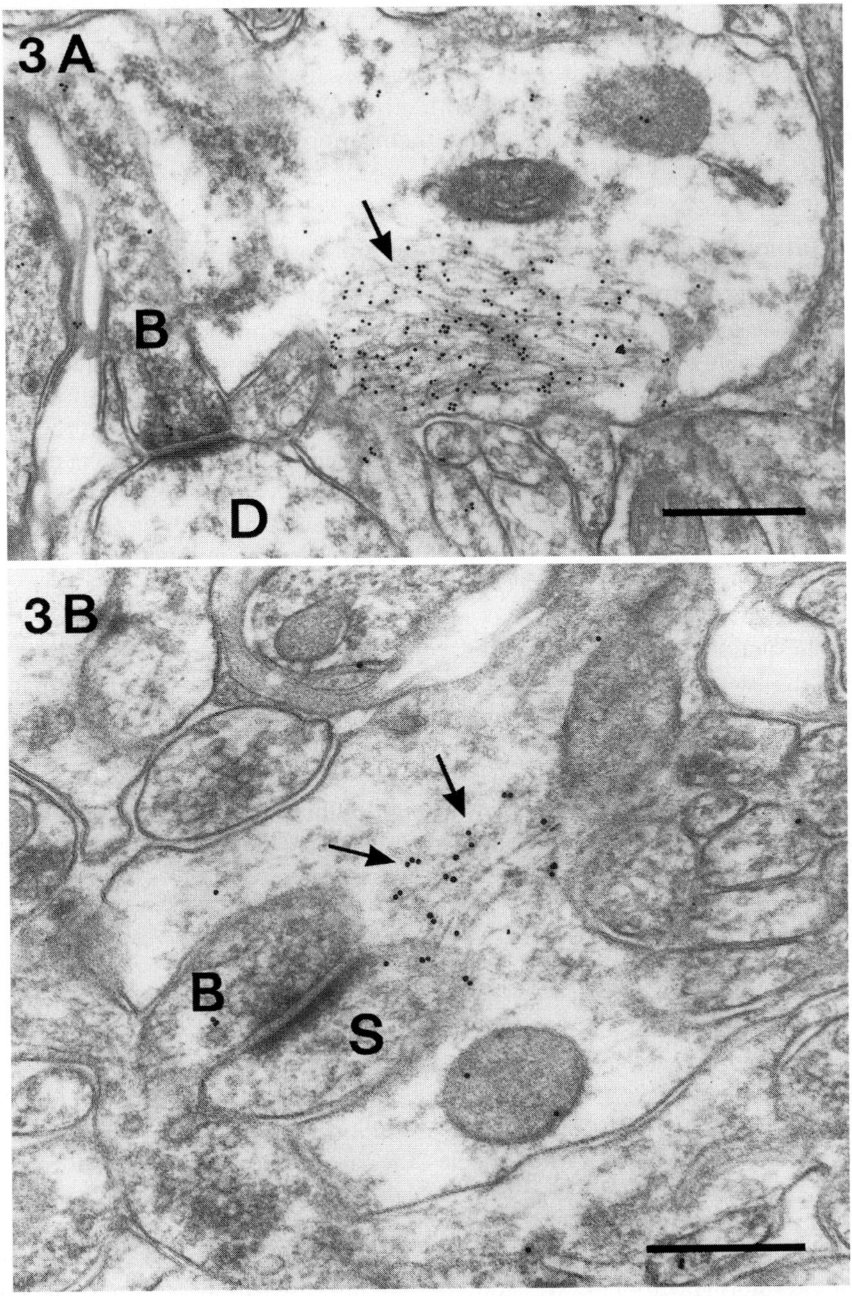

Fig. 3A,B. Glial wrapping of asymmetric, supposedly glutamatergic synapses. Two presynaptic boutons (*B*) of the stratum radiatum in CA3 of the rat hippocampus are shown, one contacting a dendritic shaft (*D*; **A**), the other one contacting a spine (*S*; **B**). These synaptic structures are surrounded by astrocytic processes identified by immunogold labeling for glial fibrillary acidic protein (GFAP). Note the specificity of the immunogold labeling, which is only seen over glial fibrils (*arrows*). *Scale bars*: 0.5 μm (**A**); 0.2 μm (**B**)

et al. 1998). Irrespective of these modifications, the concept of the trisynaptic excitatory pathway has been of great heuristic value and has led to numerous discoveries regarding hippocampal organization, and the function of glutamatergic synapses in particular.

I. Glutamatergic Entorhinal Synapses on Granule Cells

Although the entorhino-hippocampal projection has a GABAergic component (Germroth et al. 1989), the majority of entorhinal fibers are certainly excitatory, using glutamate as a transmitter. Indeed, entorhinal axons labeled by anterogradely transported tracers or by anterograde degeneration were regularly found to establish characteristic, asymmetric synapses (Nafstad 1967; Deller et al. 1996). The majority of entorhinal fibers to the fascia dentata terminate in the outer two thirds of the molecular layer, but a few fibers invading the inner molecular layer, granule cell layer, and hilus have also been described (Deller et al. 1996). Besides innervating the granule cells, entorhinal axons also establish contacts with GABAergic interneurons, as revealed in double-labeling studies (Zipp et al. 1989; Deller et al. 1996).

The majority of the entorhinal contacts are established with spines on distal granule cell dendrites. The morphology of these synapses is similar to that of the synapse shown in Fig. 2, which, however, is from a thin section of the CA1 stratum radiatum. Thus, the relatively small boutons of entorhinal axons establish typical asymmetric contacts with these spines, which, in the dentate molecular layer, often show a spinule separating two adjacent contact zones. The morphology of the spines varies with the distance from the granule cell soma (Desmond and Levy 1985).

II. Mossy-Fiber Synapses

Mossy fibers are thin axons that are often impregnated in Golgi-stained material, due to the lack of a myelin sheath. Therefore, the course and termination of the mossy fibers have been known for a long time (Golgi 1886; Koelliker 1896; Ramón and Cajal 1911). After giving off numerous collaterals to hilar neurons, the mossy fibers enter the CA3 region, where they are concentrated in the stratum lucidum. In most species, the mossy fibers stop at the border to the CA1 region. In CA3, they establish unique, large synapses with characteristic complex spines (excrescences) originating from the proximal dendrites of the CA3 pyramidal cells (Blackstad and Kjaerheim 1961; Hamlyn 1962; Chicurel and Harris 1992 for a detailed quantitative analysis from serial thin sections). The giant mossy-fiber expansions are densely filled with clear, round synaptic vesicles intermingled with a few dense-core vesicles. The vesicles are accumulated at the asymmetric contacts established with the large, branched spines of the pyramidal cells (Fig. 4). The spines often appear embedded in the giant bouton; one spine originating from the parent dendrite may give rise to as many as 16 branches (Chicurel and Harris 1992). The postsynaptic

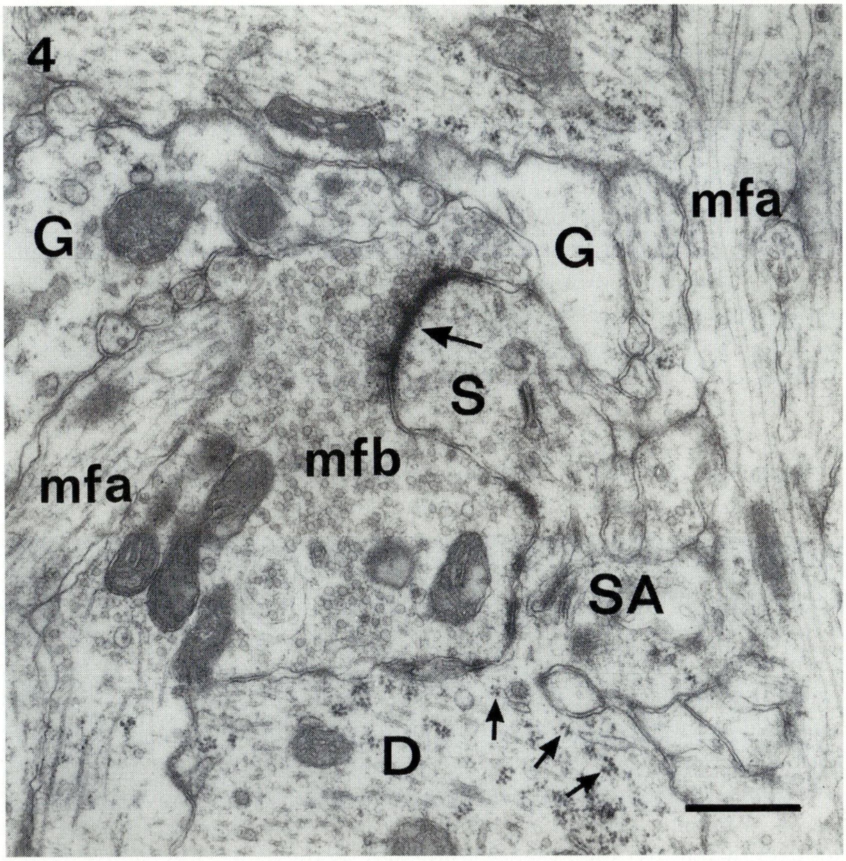

Fig. 4. Structural characteristics of a giant mossy-fiber synapse in the stratum lucidum of the rat hippocampus. Unmyelinated mossy-fiber axons (*mfa*) give rise to large mossy-fiber boutons (*mfb*) that establish asymmetric synapses (*arrow*) with large, often branched, complex spines (*S*) arising from proximal CA3 pyramidal cell dendrites (*D*). These spines regularly contain a spine apparatus (*SA*) and show ribosomes (*small arrows*) at their origins at the dendrite. Astrocytic processes (*G*) surround the synaptic compartment. *Scale bar*: 0.5 μm

densities occupy about 10–15% of the spine-head membrane, a value that is consistent with values for the spines of other cells (Chicurel and Harris 1992).

The mossy fibers use glutamate as a transmitter (Storm-Mathisen et al. 1983) but also contain dynorphin, as revealed by immunocytochemical studies (Gall 1988; Gall et al. 1990). The mossy fibers are known for their high zinc content and, thus, are heavily stained by the Timm method for heavy metals (Timm 1958; Danscher 1981). As mentioned, the mossy fibers not only contact the spines of CA3 pyramidal neurons but form structurally different asymmetric synapses with a variety of other cell types in the hilus and in CA3, including mossy cells and various interneurons (FROTSCHER 1985, 1989; RIBAK et al. 1985; FROTSCHER et al. 1991; 1994; SORIANO and FROTSCHER 1993; DELLER et al. 1994; ACSÁDY et al. 1998).

III. Schaffer Collateral Synapses

The asymmetric, supposedly glutamatergic synapse illustrated in Fig. 2 is located in the stratum radiatum of hippocampal region CA1. Here, the majority of the Schaffer collaterals, associational collaterals of CA3 pyramidal cell axons, are known to terminate. The density of this projection was underestimated in the past, as revealed by recent intracellular-labeling studies (ISHIZUKA et al. 1990; LI et al. 1994) and tracer studies with anterogradely transported *Phaseolus vulgaris Leucoagglutinin* (DELLER et al. 1994). Therefore, the contacts shown in Fig. 2, representing the most common type of synapse in CA1, are likely to be Schaffer collateral synapses, although this has not been proven. Other boutons forming asymmetric synapses in this region, such as the commissural fibers from the contralateral hippocampus, establish very similar types of spine synapses. This is not surprising, as the commissural fibers originate from the same cells as the Schaffer collaterals, i.e., the CA3 pyramidal neurons (SWANSON et al. 1981). The bouton illustrated in Fig. 2 appears to establish contacts with two spines, the larger one containing a prominent spine apparatus. It should be pointed out, however, that an analysis of serial sections would be required to prove that the two contacts were, indeed, on separate postsynaptic elements. In a quantitative analysis of serial thin sections, SORRA and HARRIS (1993) studied postsynaptic densities of spine synapses in the stratum radiatum of CA1 and found considerable size differences even when two spines were contacted by the same presynaptic terminal (range in size of the postsynaptic density: 0.02–0.26 μm^2). The conclusion to be drawn from these data is that it is not the presynaptic axon alone that determines the size and shape of the postsynaptic element.

E. Receptor Labeling Identifies Glutamatergic Synapses

We mentioned already that immunostaining for glutamate does not ensure transmitter identification, because glutamate is a ubiquitous metabolite. As one would expect, because the relevant transmitter receptors are localized to the postsynaptic membrane, an alternative method of synapse identification would be to stain these receptors by using adequate immunocytochemical labeling techniques. As DAB is a diffusible reaction product, immunogold techniques in combination with cryosubstitution to preserve as much antigenicity as possible are the methods of choice. In Fig. 5, two dendritic profiles from the stratum radiatum of the CA3 region show that both are contacted by two boutons. One bouton in each figure forms an asymmetric contact while the other establishes a symmetric synapse. Immunogold labeling for the

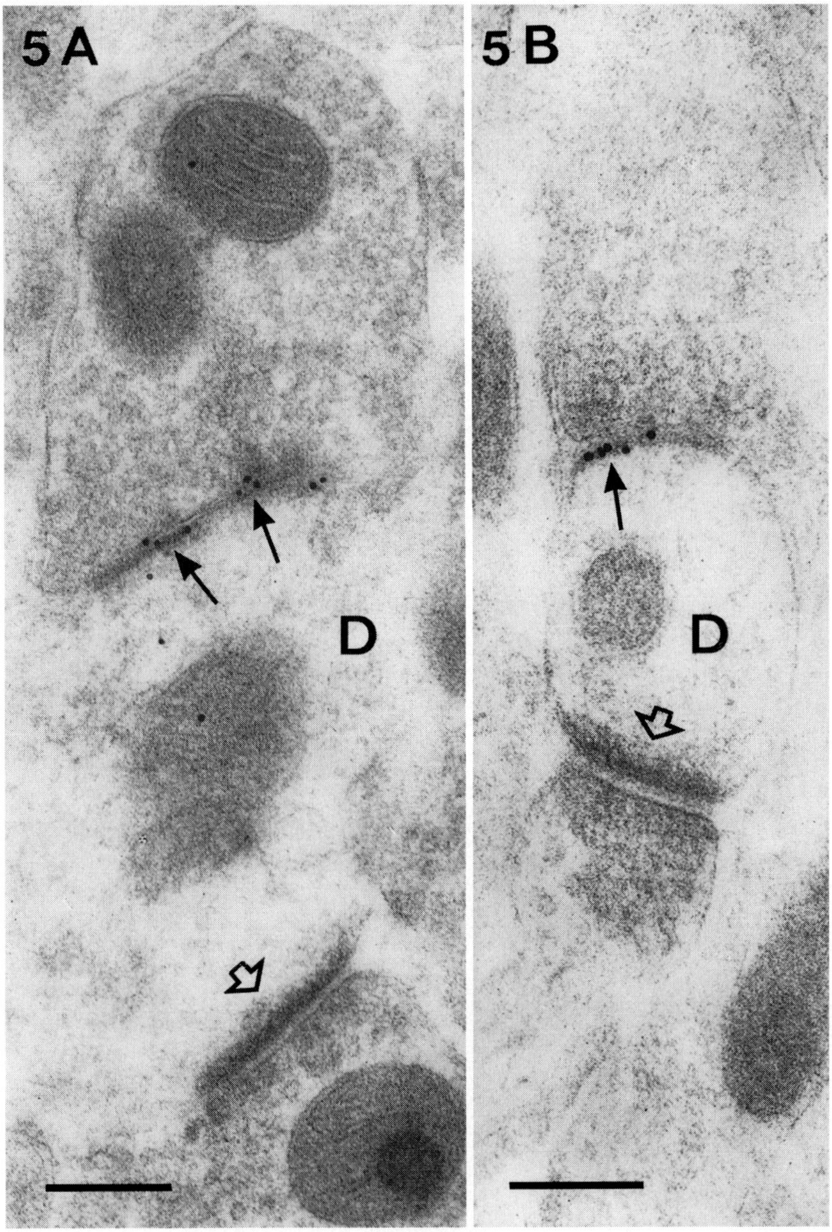

Fig. 5A,B. Immunogold staining of the α1-subunit of the γ-amino-butyric-acid-A ($GABA_A$) receptor labels symmetric synapses (*arrows*) but leaves asymmetric synapses unstained (*open arrows*). Two dendritic shafts (*D*) in CA3 of the rat hippocampus are shown, each of them establishing both symmetric and asymmetric synaptic contacts with presynaptic boutons. *Scale bars*: 0.2 μm (**A**); 0.1 μm (**B**). Method: Samples from perfusion-fixed (in 4% paraformaldehyde, 0.1% glutaraldehyde, 0.2% picric acid in 0.1 M phosphate buffer) rat hippocampus were cryoprotected and then plunged rapidly into liquid propane cooled to –180°C. Following immersion in 0.5% uranyl acetate, they were embedded in Lowicryl HM20 resin. For postembedding immunocytochemistry, ultrathin sections were treated with a saturated NaOH solution (for 1–2 s) and incubated with primary antibodies to the α1 subunit of the $GABA_A$ receptor, glutamate-receptor subunit 1 (GluR1), GluR2, and GluR4, respectively, and then in secondary gold-coupled antibodies

α1-subunit of the $GABA_A$ receptor (Nusser et al. 1996, 1997; Fletcher et al. 1998) accordingly occurrs in the membranes of the symmetric synapses but not in those of the asymmetric contacts. This is consistent with the concept that symmetric synapses are GABAergic and asymmetric synapses are exci-

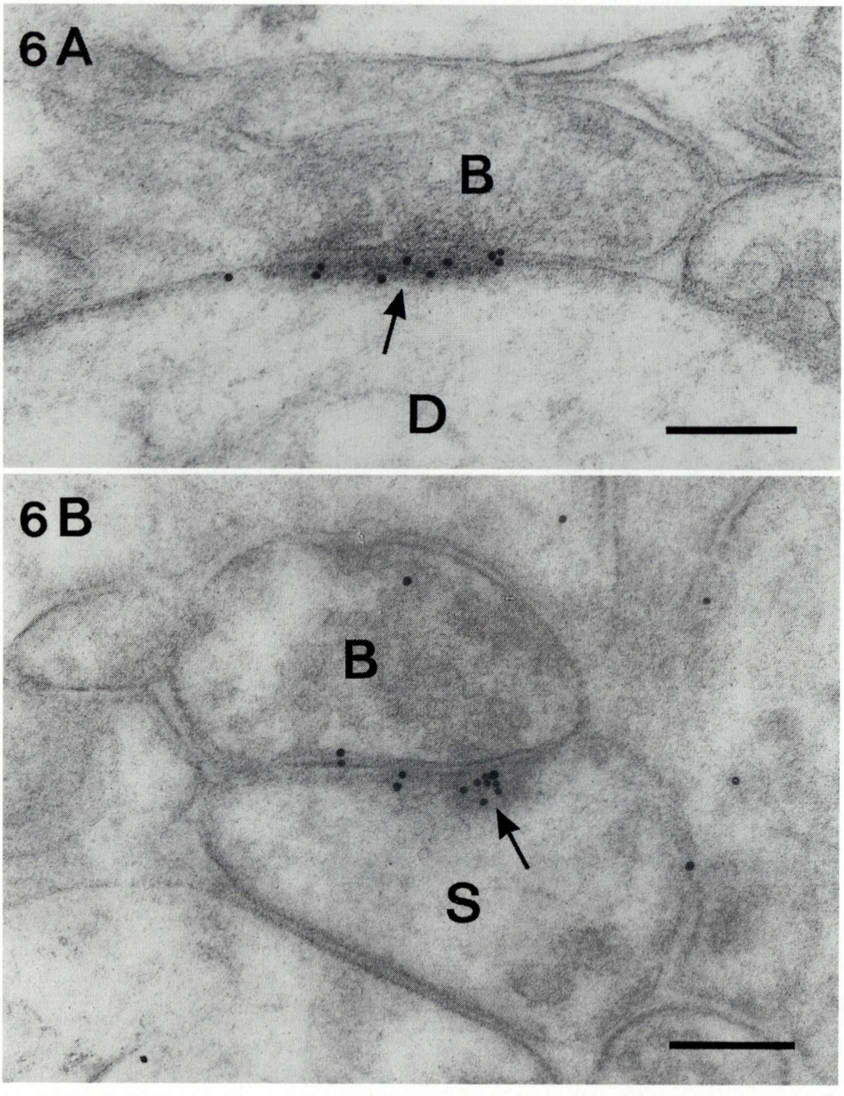

Fig. 6A,B. Immunogold labeling for glutamate-receptor subunit 1. Presynaptic boutons (*B*) in the stratum radiatum of CA3 (rat hippocampus) establish asymmetric synapses with (**A**) a dendritic shaft (*D*) and (**B**) a spine (*S*). *Arrows* point to gold grains concentrated at the postsynaptic membrane specialization. *Scale bars*: 0.1 μm. The method was the same as that described in Fig. 5

tatory, most likely glutamatergic. In fact, when antibodies against ionotropic GluRs were employed, the reverse picture was seen: symmetric synapses remained unlabeled, whereas asymmetric spine and shaft synapses displayed gold grains in their postsynaptic membranes (Figs. 6–8). These findings are in agreement with a variety of previous studies with different antibodies against subunits of ionotropic GluRs (NUSSER et al. 1994; BAUDE et al. 1995; MATSUBARA et al. 1996; POPRATILOFF et al. 1996; BERNARD et al. 1997; LANDSEND et al. 1997; OTTERSEN and LANDSEND 1997; HE et al. 1998; SOMOGYI et al. 1998).

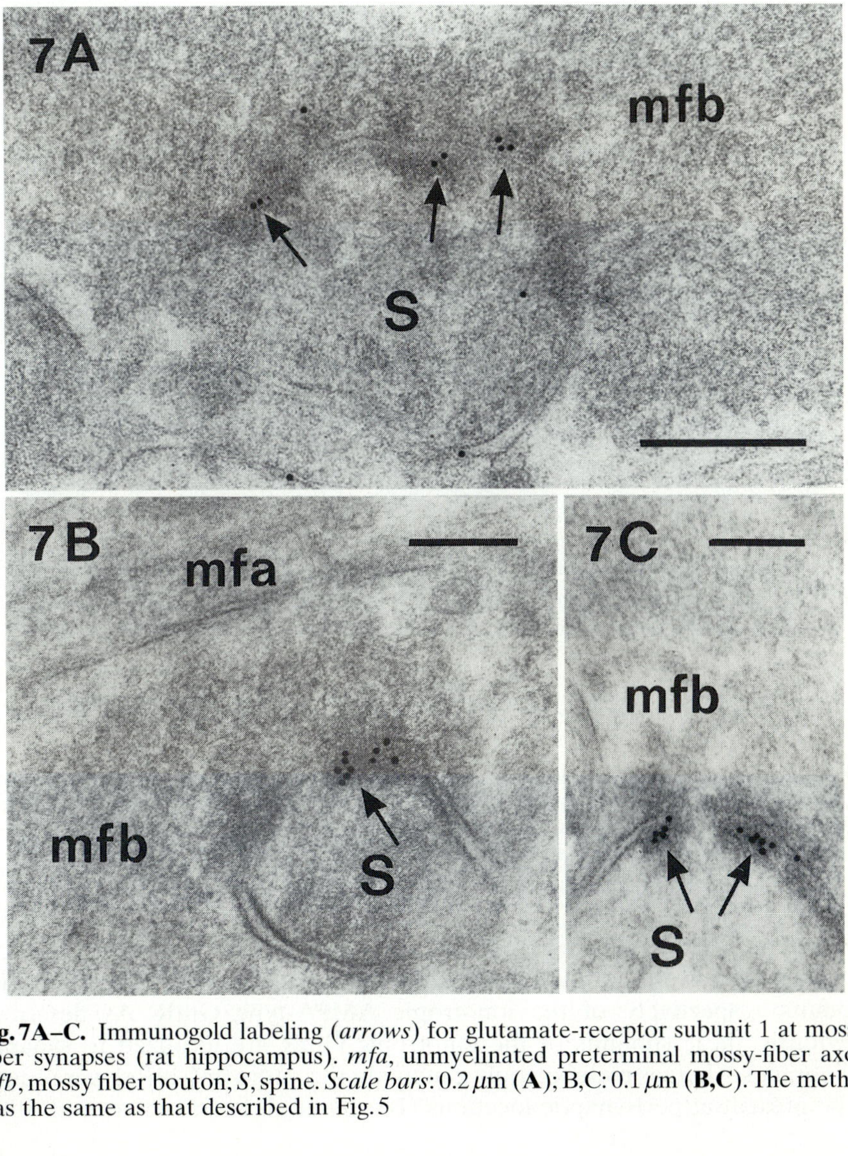

Fig. 7A–C. Immunogold labeling (*arrows*) for glutamate-receptor subunit 1 at mossy-fiber synapses (rat hippocampus). *mfa*, unmyelinated preterminal mossy-fiber axon; *mfb*, mossy fiber bouton; *S*, spine. *Scale bars*: 0.2 μm (**A**); B,C: 0.1 μm (**B,C**). The method was the same as that described in Fig. 5

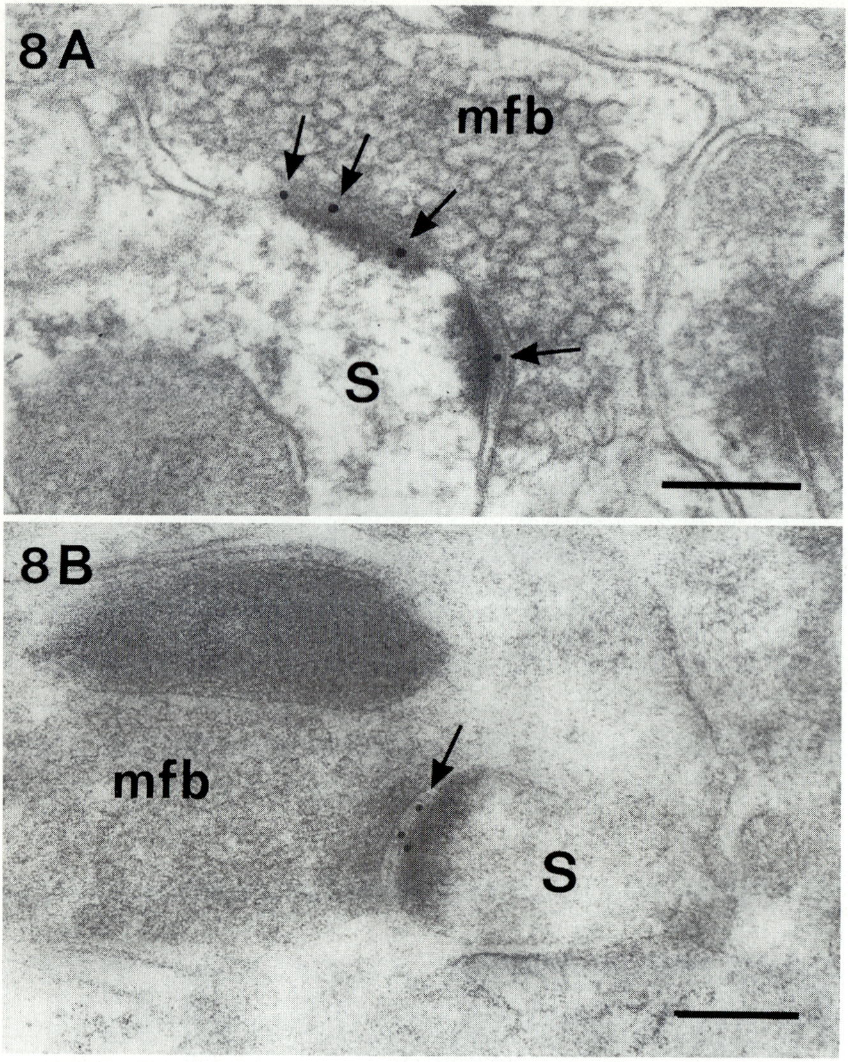

Fig. 8A,B. Immunogold labeling of rat mossy-fiber synapses for (A) glutamate-receptor subunit 4 (GluR4) and (B) GluR2. *Arrows* point to gold grains located near the postsynaptic membrane specialization. *mfb*, mossy-fiber bouton; *S*, spine. *Scale bars*: 0.1 μm. The method was the same as that described in Fig. 5

In Figs. 6–8 are illustrated different types of asymmetric hippocampal synapses that were immunostained for the GluR1, GluR2, and GluR4 subunits, respectively, of the ionotropic AMPA-type GluR. As described previously, these subunits of the ionotropic GluR are localized to synaptic membranes, whereas metabotropic glutamate receptors (mGluRs) were found to be present at perisynaptic locations (BAUDE et al. 1993; LUJAN et al. 1996,

1997; SOMOGYI et al. 1998). Recent studies have shown that immunogold labeling for GluR2 is lower at glutamatergic synapses on GABAergic cells than on pyramidal cells (HE et al. 1998), suggesting an increased calcium influx at glutamatergic synapses on interneurons (HOLLMANN et al. 1991; HUME et al. 1991; BURNASHEV et al. 1992; JONAS et al. 1994; GEIGER et al. 1995; JONAS and BURNASHEV 1995; BURNASHEV 1996). Using a pan-AMPA-receptor antibody, NUSSER et al. (1998) were recently able to demonstrate that there are considerable differences in the densities of AMPA receptors among different types of hippocampal synapses. Moreover, even within one population of synapses, the Schaffer collateral synapses in CA1, they observed a large variability in AMPA-receptor number. This variability in the number of AMPA receptors may reflect different functional states of these synapses.

Taken together, the results summarized here show that immunogold labeling of GluRs provides a reliable tool for the identification of glutamatergic synapses in the absence of a definitive marker for the transmitter itself. Immunogold labeling of GluR subunits may also allow for the description of some of the functional properties of these contacts.

F. Summary and Conclusions

By focusing on the hippocampus, we have summarized in this chapter some structural characteristics of glutamatergic synapses. We regard it as a main result that glutamatergic synapses can be identified reliably by immunogold labeling of subunits of ionotropic GluRs in the postsynaptic membrane. Data in the literature indicate that there is even a quantitative relationship between immunogold labeling and the number of receptors present (BAUDE et al. 1995; HE et al. 1998; NUSSER et al. 1998; SOMOGYI et al. 1998). No GluR labeling was found in symmetric GABAergic synapses, which, in turn, could be labeled by antibodies against the α1-subunit of the $GABA_A$ receptor. It is remarkable that this identification of glutamatergic and GABAergic synapses is in perfect agreement with the traditional distinction between asymmetric excitatory and symmetric inhibitory synapses.

A previously underestimated feature of glutamatergic synapses is their glial wrapping. As suggested in this review, the synaptic compartment should be considered as consisting of four components: the presynaptic bouton, the postsynaptic structure, the synaptic cleft between the two, and the surrounding astrocytic processes.

Finally, there is a structural diversity of glutamatergic synapses, which is poorly understood at present. In the hippocampus, such different types of glutamatergic synapses are found as the small contacts of Schaffer collaterals with single release sites and the giant mossy-fiber synapses with their numerous contacts on the complex spines of CA3 pyramidal cells. A better understanding of the diversity of glutamatergic synapses requires correlative physiological and morphological analyses, including paired recordings from the pre- and postsynaptic neurons and the *post hoc* morphological identification of the contacts involved (Miles and Poncer 1996; Geiger et al. 1997).

Acknowledgements. The authors wish to thank Drs. O.P. Ottersen and H. Schwarz for their help with the establishment of the cryosubstitution technique and Drs. R.J. Wenthold and J.-M. Fritschy for supplying us with antibodies against the GluR2 and GluR4 subunits of the AMPA receptor and the α1-subunit of the $GABA_A$ receptor, respectively. This work was supported by the Deutsche Forschungsgemeinschaft (SFB 505, TP C6).

References

Acsády L, Kamondi A, Sík A, Freund T, Buzsáki G (1998) GABAergic cells are the major postsynaptic targets of mossy fibers in the rat hippocampus. J Neurosci 18:3386–3403

Andersen P, Bliss TVP, Skrede KK (1971) Lamellar organization of hippocampal excitatory pathways. Exp Brain Res 13:222–238

Baude A, Nusser Z, Roberts JDB, Mulvihill E, McIlhinney RAJ, Somogyi P (1993) The metabotropic glutamate receptor (mGluR1α) is concentrated at perisynaptic membrane of neuronal subpopulations as detected by immunogold reaction. Neuron 11:771–787

Baude A, Nusser Z, Molnár E, McIlhinney RAJ, Somogyi P (1995) High-resolution immunogold localization of AMPA type glutamate receptor subunits at synaptic and non-synaptic sites in rat hippocampus. Neuroscience 69:1031–1055

Beaulieu C, Colonnier M (1985) A laminar analysis of the number of round-asymmetrical and flat-symmetrical synapses on spines, dendritic trunks, and cell bodies in area 17 of the cat. J Comp Neurol 231:180–189

Bernard V, Somogyi P, Bolam JP (1997) Cellular, subcellular, and subsynaptic distribution of AMPA-type glutamate receptor subunits in the neostriatum of the rat. J Neurosci 17:819–823

Blackstad TW, Kjaerheim A (1961) Special axo-dendritic synapses in the hippocampal cortex: electron and light microscopic studies on the layer of mossy fibres. J Comp Neurol 117:113–159

Bramham CR, Torp R, Zhang N, Storm-Mathisen J, Ottersen OP (1990) Distribution of glutamate-like immunoreactivity in excitatory hippocampal pathways: a semi-quantitative electron microscopic study in rats. Neuroscience 39:405–417

Burnashev N (1996) Calcium permeability of glutamate-gated channels in the central nervous system. Curr Opin Neurobiol 6:311–317

Burnashev N, Monyer H, Seeburg PH, Sakmann B (1992) Divalent ion permeability of AMPA receptor channels is dominated by the edited form of a single subunit. Neuron 8:189–198

Calverley RKS, Jones DG (1990) Contributions of dendritic spines and perforated synapses to synaptic plasticity. Brain Res Rev 15:215–249

Chaudhry FA, Lehre KP, van Lookeren Campagne M, Ottersen OP, Danbolt NC, Storm-Mathisen J (1995) Glutamate transporters in glial plasma membranes: highly differentiated localizations revealed by quantitative ultrastructural immunocytochemistry. Neuron 15:711–720

Chicurel ME, Harris KM (1992) Three-dimensional analysis of the structure and composition of CA3 branched dendritic spines and their synaptic relationships with mossy fiber boutons in the rat hippocampus. J Comp Neurol 325:169–182

Danscher G (1981) Histochemical demonstration of heavy metals: a revised version of the sulphide silver method suitable for both light and electron microscopy. Histochemistry 71:1–16

Deller T, Nitsch R, Frotscher M (1994) Associational and commissural afferents of parvalbumin-immunoreactive neurons in the rat hippocampus: a combined immunocytochemical and PHA-L study. J Comp Neurol 350:612–622
Deller T, Martinez A, Nitsch R, Frotscher M (1996) A novel entorhinal projection to the rat dentate gyrus: direct innervation of proximal dendrites and cell bodies of granule cells and GABAergic neurons. J Neurosci 16:3322–3333
Derouiche A, Frotscher M (1991) Astroglial processes around identified glutamatergic synapses contain glutamine synthetase: evidence for transmitter degradation. Brain Res 552:346–350
Desmond NL, Levy WB (1985) Granule cell dendritic spine density in the rat hippocampus varies with spine shape and location. Neurosci Lett 54:219–224
Drejer J, Larsson OM, Schousboe A (1982) Characterization of L-glutamate uptake into and release from astrocytes and neurons cultured from different brain regions. Exp Brain Res 47:259–269
Edwards FA (1995) Anatomy and electrophysiology of fast central synapses lead to a structural model for long-term potentiation. Physiol Rev 75:759–787
Fischer M, Kaech S, Knutti D, Matus A (1998) Rapid actin-based plasticity in dendritic spines. Neuron 20:847–854
Fletcher EL, Koulen P, Wässle H (1998) GABAA and GABAC receptors on mammalian rod bipolar cells. J Comp Neurol 396:351–365
Foster M, Sherrington CS (1897) A text book of physiology, part III: The central nervous system, 7th edn. Macmillan, London
Freund TF, Powell JF, Smith AD (1984) Tyrosine hydroxylase-immunoreactive boutons in synaptic contact with identified striatonigral neurons, with particular reference to dendritic spines. Neuroscience 13:1189–1215
Frotscher M (1985) Mossy fibres form synapses with identified pyramidal basket cells in the CA3 region of the guinea pig hippocampus: a combined Golgi-electron microscope study. J Neurocytol 14:245–259
Frotscher M (1989) Mossy fiber synapses on glutamate decarboxylase-immunoreactive neurons: evidence for feed-forward inhibition in the CA3 region of the hippocampus. Exp Brain Res 75:441–445
Frotscher M (1996) Synaptic transmission. In: Greger R, Windhorst U (eds) Comprehensive Human Physiology, vol 1. Springer Verlag, Berlin, Heidelberg, pp 321–334
Frotscher M, Leranth C (1986) The cholinergic innervation of the rat fascia dentata: identification of target structures on granule cells by combining choline acetyltransferase immunocytochemistry and Golgi impregnation. J Comp Neurol 243:58–70
Frotscher M, Seress L, Schwerdtfeger WK, Buhl E (1991) The mossy cells of the fascia dentata: a comparative study of their fine structure and synaptic connections in rodents and primates. J Comp Neurol 312:145–163
Frotscher M, Soriano E, Misgeld U (1994) Divergence of hippocampal mossy fibers. Synapse 16:148–160
Gall C (1988) Seizures induce dramatic and distinctly different changes in enkephalin, dynorphin, and cholecystokinin immunoreactivities in mouse hippocampal mossy fibers. J Neurosci 8:1852–1862
Gall C, Lauterborn J, Isackson P, White J (1990) Seizures, neuropeptide regulation, and mRNA expression in the hippocampus. Prog Brain Res 83:371–390
Geiger JR, Melcher T, Koh DS, Sakmann B, Seeburg PH, Jonas P, Monyer H (1995) Relative abundance of subunit mRNAs determines gating and Ca2+ permeability of AMPA receptors in principal neurons and interneurons in rat CNS. Neuron 15:193–204
Geiger JRP, Lübke J, Roth A, Frotscher M, Jonas P (1997) Submillisecond AMPA receptor-mediated signaling at a principal neuron-interneuron synapse. Neuron 18:1009–1023
Germroth P, Schwerdtfeger WK, Buhl EH (1989) GABAergic neurons in the entorhinal cortex project to the hippocampus. Brain Res 494:187–192

Goldman-Rakic PS, Leranth C, Williams SM, Mons N, Geffard M (1989) Dopamine synaptic complex with pyramidal neurons in primate cerebral cortex. Proc Natl Acad Sci USA 86:9015–9019
Golgi C (1886) Sulla fina anatomia degli organi centrali del sistema nervoso. Hoepli, Milan
Gray EG (1959) Axo-somatic and axo-dendritic synapses of the cerebral cortex: an electron microscope study. J Anat 93:420–433
Hamlyn LH (1962) The fine structure of the mossy fibre endings in the hippocampus of the rabbit. J Anat 97:112–120
Harris KM, Kater SB (1994) Dendritic spines: cellular specializations imparting both stability and flexibility to synaptic function. Annu Rev Neurosci 17:341–371
He Y, Janssen WGM, Vissavajjhala P, Morrison JH (1998) Synaptic distribution of GluR2 in hippocampal GABAergic interneurons and pyramidal cells: a double-label immunogold analysis. Exp Neurol 150:1–13
Hollmann M, Hartley M, Heinemann S (1991) Ca2+ permeability of KA-AMPA-gated glutamate receptor channels depends on subunit composition. Science 252: 851–853
Hösli E, Hösli L (1976) Uptake of L-glutamate and L-aspartate in neurons and glial cells of cultured human and rat spinal cord. Experientia 32:219–222
Hume RI, Dingledine R, Heinemann SF (1991) Identification of a site in glutamate receptor subunits that controls calcium permeability. Science 253:1028–1031
Ishizuka N, Weber J, Amaral DG (1990) Organization of intrahippocampal projections originating from CA3 pyramidal cells in the rat. J Comp Neurol 295:580–623
Jonas P, Burnashev N (1995) Molecular mechanisms controlling calcium entry through AMPA-type glutamate receptor channels. Neuron 15:987–990
Jonas P, Racca C, Sakmann B, Seeburg PH, Monyer H (1994) Difference in Ca2+ permeability of AMPA-type glutamate receptor channels in neocortical neurons caused by differential GluR-B subunit expression. Neuron 12:1281–1289
Koelliker A (1896) Handbuch der Gewebelehre des Menschen. Zweiter Band: Nervensystem des Menschen und der Thiere. Wilhelm Engelmann, Leipzig
Landsend AS, Amiry-Moghaddam M, Matsubara A, Bergersen L, Usami S, Wenthold RJ, Ottersen OP (1997) Differential localization of delta glutamate receptors in the rat cerebellum: coexpression with AMPA receptors in parallel fiber-spine synapses and absence from climbing fiber-spine synapses. J Neurosci 17:834–842
Li XG, Somogyi P, Ylinen A, Buzsáki G (1994) The hippocampal CA3 network; an in vivo intracellular labeling study. J Comp Neurol 339:181–208
Lisman JE, Harris KM (1993) Quantal analysis and synaptic anatomy: integrating two views of hippocampal plasticity. TINS 16:141–147
Lujan R, Nusser Z, Roberts JDB, Shigemoto R, Somogyi P (1996) Perisynaptic location of metabotropic glutamate receptors mGluR1 and mGluR5 on dendrites and dendritic spines in the rat hippocampus. Eur J Neurosci 8:1488–1500
Lujan R, Roberts JDB, Shigemoto R, Ohishi H, Somogyi P (1997) Differential plasma membrane distribution of metabotropic glutamate receptors mGluR1α, mGluR2 and mGluR5, relative to neurotransmitter release sites. J Chem Neuroanat 13: 219–241
Matsubara A, Laake JH, Davanger S, Usami S-I, Ottersen OP (1996) Organization of AMPA receptor subunits at a glutamate synapse: a quantitative immunogold analysis of hair cell synapses in the rat organ of Corti. J Neurosci 16:4457–4467
Miles R, Poncer JC (1996) Paired recordings from neurones. Curr Opin Neurobiol 6:387–394
Molnár E, Baude A, Richmond SA, Patel PB, Somogyi P, McIlhinney RAJ (1993) Biochemical and immunocytochemical characterization of antipeptide antibodies to a cloned GluR1 glutamate receptor subunit: cellular and subcellular distribution in the rat forebrain. Neuroscience 53:307–326
Nafstad PH (1967) An electron microscope study on the termination of the perforant path fibres in the hippocampus and the fascia dentata. Z Zellforsch Mikrosk Anat 76:532–542

Nusser Z, Mulvihill E, Streit P, Somogyi P (1994) Subsynaptic segregation of metabotropic and ionotropic glutamate receptors as revealed by immunogold localization. Neuroscience 61:421–427
Nusser Z, Sieghart W, Benke D, Fritschy J-M, Somogyi P (1996) Differential synaptic localization of two major gamma-aminobutyric acid type A receptor α subunits on hippocampal pyramidal cells. Proc Natl Acad Sci USA 93:11939–11944
Nusser Z, Cull-Candy S, Farrant M (1997) Differences in synaptic GABAA receptor number underlie variation in GABA mini amplitude. Neuron 19:697–709
Nusser Z, Lujan R, Laube G, Roberts JDB, Molnár E, Somogyi P (1998) Cell type and pathway dependence of synaptic AMPA receptor number and variability in the hippocampus. Neuron 21:545–559
Ottersen OP, Landsend AS (1997) Organization of glutamate receptors at the synapse. Eur J Neurosci 9:2219–2224
Palade GE, Palay SL (1954) Electron microscope observations of interneuronal and neuromuscular synapses. Anat Rec 118:335–336
Palay SL (1956) Synapses in the central nervous system. J Biophys Biochem Cytol 2:193–206
Palay SL (1958) The morphology of synapses of the central nervous system. Exp Cell Res 5:275–293
Petralia RS, Wenthold RJ (1992) Light and electron immunocytochemical localization of AMPA-selective glutamate receptors in the rat brain. J Comp Neurol 318:329–354
Petralia RS, Wang Y-X, Zhao H-M, Wenthold RJ (1996) Ionotropic and metabotropic glutamate receptors show unique postsynaptic, presynaptic, and glial localizations in the dorsal cochlear nucleus. J Comp Neurol 372:356–383
Popratiloff A, Weinberg RJ, Rustioni A (1996) AMPA receptor subunits underlying terminals of fine-caliber primary afferent fibers. J Neurosci 16:3363–3372
Ramón y Cajal SR (1911) Histologie du Système Nerveux de l'Homme et des Vertébrés, vol II. Maloine, Paris
Ribak CE, Roberts RC (1990) GABAergic synapses in the brain identified with antisera to GABA and its synthesizing enzyme, glutamate decarboxylase. J Electron Microsc Technique 15:34–48
Ribak CE, Seress L, Amaral DG (1985) The development, ultrastructure and synaptic connections of the mossy cells of the dentate gyrus. J Neurocytol 14:835–857
Shepherd GM (1972) Synaptic organization of the mammalian olfactory bulb. Physiol Rev 52:864–917
Schuster T, Krug M, Wenzel J (1990) Spinules in axospinous synapses of the rat dentate gyrus: changes in density following long-term potentiation. Brain Res 523:171–174
Shepherd GM (1996) The dendritic spine: a multifunctional integrative unit. J Neurophysiol 75:2197–2210
Somogyi P, Nusser Z, Roberts JDB, Lujan R (1998) Precision and variability in the placement of pre- and postsynaptic receptors in relation to transmitter release sites. In: Faber DS, Korn H, Redman SJ, Thompson SM, Altman JS (eds) Central Synapses: Quantal Mechanisms and Plasticity. Human Frontier Science Program, Strasbourg, pp 82–95
Soriano E, Frotscher M (1993) Spiny nonpyramidal neurons in the CA3 region of the rat hippocampus are glutamate-like immunoreactive and receive convergent mossy fiber input. J Comp Neurol 332:435–448
Soriano E, Frotscher M (1994) Mossy cells of the rat fascia dentata are glutamate-immunoreactive. Hippocampus 4:65–70
Sorra KE, Harris KM (1993) Occurrence and three-dimensional structure of multiple synapses between individual radiatum axons and their target pyramidal cells in hippocampal area CA1. J Neurosci 13:3736–3748
Spacek J, Harris KM (1997) Three-dimensional organization of smooth endoplasmic reticulum in hippocampal CA1 dendrites and dendritic spines of the immature and mature rat. J Neurosci 17:190–203

Steward O, Banker GA (1992) Getting the message from the gene to the synapse: sorting and intracellular transport of RNA in neurons. TINS 15:180–186
Steward O, Levy WB (1982) Preferential localization of polyribosomes under the base of dendritic spines in granule cells of the dentate gyrus. J Neurosci 2:284–291
Storm-Mathisen J, Ottersen OP (1984) Neurotransmitters in the hippocampal formation. In: Reinoso-Suárez F, Ajmone-Marsan C (eds) Cortical integration. Raven, New York, pp 105–130
Storm-Mathisen J, Leknes AK, Bore AT, Vaaland JL, Edminson P, Haug FMS, Ottersen OP (1983) First visualization of glutamate and GABA in neurones by immunocytochemistry. Nature 301:517–520
Swanson LW, Sawchenko PE, Cowan WM (1981) Evidence for collateral projections by neurons in Ammon's horn, the dentate gyrus, and the subiculum: a multiple retrograde labeling study in the rat. J Neurosci 1:548–559
Timm F (1958) Zur Histochemie der Schwermetalle, das Sulfid-Silber-Verfahren. Dtsch Z Gesamte Gerichtl Med 46:706–711
Uchizono K (1965) Characteristics of excitatory and inhibitory synapses in the central nervous system of the cat. Nature 207:642–643
White WF, Nadler JV, Hamberger A, Cotman CW, Cummins JT (1977) Glutamate as transmitter of the hippocampal perforant path. Nature 270:356–357
Yu ACH, Schousboe A, Hertz L (1982) Metabolic fate of 14C-labeled glutamate in astrocytes. J Neurochem 39:954–966
Ziff EB (1997) Enlightening the postsynaptic density. Neuron 19:1163–1174
Zipp F, Nitsch R, Soriano E, Frotscher M (1989) Entorhinal fibers form synaptic contacts on parvalbumin-immunoreactive neurons in the rat fascia dentata. Brain Res 495:161–166

CHAPTER 11

Glutamate-Mediated Synaptic Excitation of Cortical Interneurons

J.R.P. GEIGER, A. ROTH, B. TASKIN, and P. JONAS

Abbreviations

ACPD	*trans*-1-aminocyclopentane-1,3-dicarboxylic acid
AMPARs	α-Amino-3-hydroxy-5-methyl-4-isoxazolepropionate receptors
ATPA	2-Amino-3-(3-hydroxy-5-tert-butylisoxazole-4-yl)propanoate
CNS	Central nervous system
D	Diffusion coefficient for glutamate
EPSC	Excitatory postsynaptic current
EPSP	Excitatory postsynaptic potential
GABA	γ-aminobutyric acid
HICAP	Hilar–commissural-association pathway
HIPP	Hilar–perforant path
KARs	Kainate receptors
LTD	Long-term depression
LTP	Long-term potentiation
mGluRs	Metabotropic glutamate receptors
MCPG	α-methyl-4-carboxyphenyl glycine
MOPP	Molecular layer–perforant path
NMDARs	*N*-methyl-D-aspartate receptors
OALM	Oriens-alveus–lacunosum moleculare
OML	Outer molecular layer
PPD	Paired-pulse depression
PPF	Paired-pulse facilitation

A. Introduction

Principal neurons and interneurons are the two main classes of cells in cortical neuronal networks. Principal neurons (granule cells or pyramidal neurons) have transregional axonal projections and release glutamate onto their postsynaptic target cells. In contrast, interneurons have local, but often extensive, axonal arborizations and use γ-aminobutyric acid (GABA) as a transmitter. Although interneurons represent only approximately 10% of the neuronal population, they control the electrical activity of the entire network (FREUND and BUZSÁKI 1996). Interneurons forming inhibitory synapses on the somata or axon initial segments of their postsynaptic target cells are thought to set the threshold of action potential initiation (MILES et al. 1996) and can synchronize the collective activities of large principal neuron ensembles (COBB et al. 1995). In contrast, interneurons establishing inhibitory synapses mainly on dendrites could suppress dendritic Na^+ or Ca^{2+} spikes (BUZSÁKI et al. 1996; MILES et al. 1996) and, thus, regulate plasticity at glutamatergic synapses in the cortex (DAVIES et al. 1991).

If interneurons control activity and plasticity in neuronal networks, a central question is: how are interneurons excited? How do they sense the activity level of the principal neuron ensemble? In this chapter, we review the functional properties of glutamatergic principal neuron–interneuron synapses, with focus on the hippocampus. These synapses, established either by local recurrent collaterals or by long-range projection components of principal neuron axons, form the main source of interneuron excitation. Very surprisingly, it turns out that their signaling properties differ substantially from those of synapses between principal neurons, implying a high degree of synaptic specialization within neuronal microcircuits.

B. Unitary Glutamate-Mediated Synaptic Potentials and Currents Generated at Principal Neuron–Interneuron Synapses

I. Sources of Glutamatergic Synaptic Input

GABAergic interneurons mediate feedforward or feedback inhibition (FREUND and BUZSÁKI 1996). In the feedforward circuitry, an excitatory afferent input activates the interneuron directly. In the feedback system, the excitatory afferent input evokes action potentials in principal neurons, which in turn activate the interneuron via recurrent collaterals. Whether an interneuron operates as a feedforward or feedback element is defined by the source of its glutamatergic synaptic input. In the hippocampus, where the neuronal circuitry shows a laminar organization, the location of the somatodendritic domain of the interneuron is decisive (Table 1).

Many hippocampal interneurons have dendrites on both sides of the principal neuron layer, consistent with both feedforward and feedback activation. These include the classical interneuron subtypes, the basket cells and axo-axonic cells in the dentate gyrus and the CA1 region (Fig. 1A; Table 1; FREUND and BUZSÁKI 1996). Hilar–commissural-association pathway (HICAP) interneurons in the dentate gyrus (Table 1; HAN et al. 1993) and bistratified and trilamellar interneurons in the CA1 region also fall into this class (Table 1; SIK et al. 1995; for a nomenclature of interneuron subtypes see Table 1 and FREUND and BUZSÁKI 1996). In contrast, molecular layer–perforant path (MOPP) interneurons and outer molecular layer (OML) interneurons in the dentate gyrus (HAN et al. 1993; CERANIK et al. 1997), and a subset of interneurons in

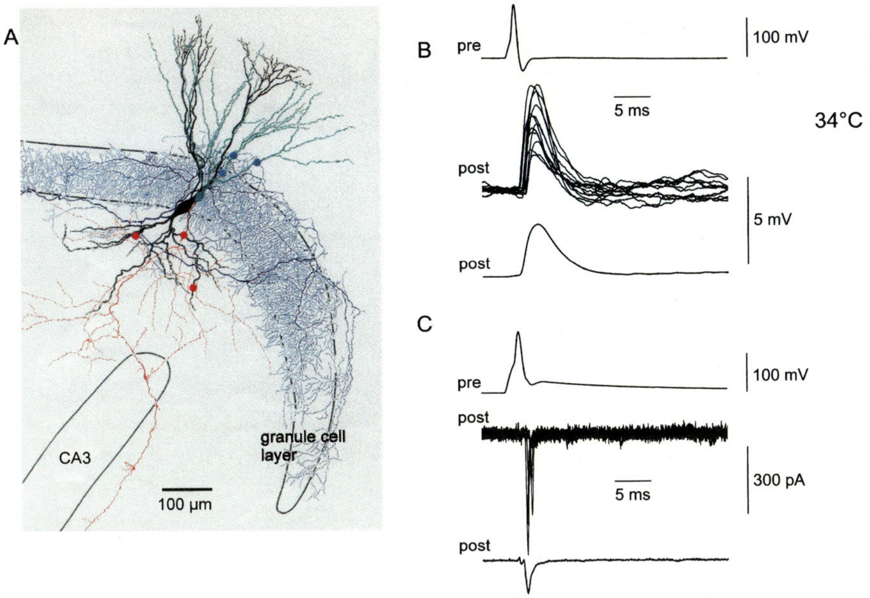

Fig. 1A–C. Rapid time course excitatory postsynaptic potentials (EPSPs) generated at anatomically defined principal neuron–interneuron synapses. **A** *Camera lucida* reconstruction of a synaptically connected granule cell–basket cell pair in the dentate gyrus of a rat hippocampal slice. *Green*, somatodendritic domain of the granule cell; *red*, axon of the granule cell; *black*, somatodendritic domain of the basket cell; *blue*, axon of the basket cell. Three putative excitatory (*red dots*) and three putative inhibitory synaptic contacts (*green dots*, one contact equivocal) were identified. **B** α-Amino-3-hydroxy-5-methyl-4-isoxazolepropionate receptor (AMPAR)-mediated EPSPs generated at the granule cell-basket cell synapse recorded in the current-clamp configuration. The action potential in the presynaptic granule cell is shown on *top*, individual EPSPs recorded in the basket cell are shown in the *center*, and the average EPSP is shown at the *bottom*. The resting potential was –62 mV. **C** AMPAR-mediated excitatory postsynaptic currents (EPSCs) recorded in the voltage-clamp configuration. The action potential in the presynaptic granule cell is shown on *top*, individual EPSCs recorded at –70 mV in the basket cell are shown in the *center*, and the average EPSC is shown at the *bottom*. Different pairs are shown in B and C. Bath solution with 2 mM Ca^{2+}/1 mM Mg^{2+}. The recording temperature was 34°C. Modified from GEIGER et al. 1997, Copyright by Cell Press

stratum radiatum and stratum lacunosum-moleculare, receive predominantly feedforward excitatory input (Table 1; MACCAFERRI and MCBAIN 1996; MCMAHON and KAUER 1997; VIDA et al. 1998). Finally, both hilar–perforant path (HIPP) interneurons in the dentate gyrus (HAN et al. 1993) and oriens-alveus–lacunosum-moleculare (OALM) interneurons in the CA1 subfield receive predominantly feedback excitatory input (Table 1; BLASCO-IBÁÑEZ and FREUND 1995; MACCAFERRI and MCBAIN 1996). Thus, interneurons are positioned strategically relative to the location of the major glutamatergic pathways and, thus, receive feedback, feedforward, or dual excitation. The source of the excitatory input, in turn, defines their functional role within the circuitry.

Table 1. Different types of hippocampal interneurons and their sources of glutamatergic synaptic input

Interneuron	Somatodendritic domain	Axonal domain	Activation mode	AP frequency	Neurochemical markers	Reference
Interneurons receiving dual (feedback and feedforward) excitation						
Basket interneuron, axo-axonic interneuron (DG, CA1)	Soma close to principal neuron layer, dendrites perpendicular through all layers	Principal cell layer (soma, proximal dendrites vs axon initial segment)	Dual	Very fast	Parvalbumin	HAN et al. 1993; SIK et al. 1995; VIDA et al. 1998
HICAP interneuron (DG)	Soma in hilus, dendrites perpendicular through all layers	Inner molecular layer coaligned with commissural-association pathway	Dual	Very fast		HAN et al. 1993; FREUND and BUZSÁKI 1996
Bistratified interneuron (CA1)	Soma in stratum oriens-alveus, dendrites in stratum oriens-alveus and radiatum	Stratum oriens-alveus, stratum radiatum	Dual	Very fast	Calbindin (all?)	SIK et al. 1995; FREUND and BUZSÁKI 1996

Interneurons receiving feedforward excitation						
MOPP interneuron (DG)	Molecular layer	OML coaligned with perforant path	Feedforward	Fast		Han et al. 1993; Freund and Buzsáki 1996
OML interneuron (DG)	Molecular layer	Local: molecular layer; long range: subiculum	Feedforward	Fast		Ceranik et al. 1997
Stratum radiatum/lacunosum-moleculare interneurons (CA1)	Stratum radiatum/lacunosum-moleculare	Stratum pyramidale/radiatum/lacunosum-moleculare	Mainly feedforward	Slow		Maccaferri and McBain 1996; McMahon and Kauer 1997; Vida et al. 1998
Interneurons receiving feedback excitation						
HIPP interneuron (DG)	Hilus	OML coaligned with perforant path	Feedback	Fast	Somatostatin, NP-Y (most), mGluR1	Han et al. 1993; Freund and Buzsáki 1996
OALM[a] interneuron (CA1)	Stratum oriens-alveus	Stratum lacunosum-moleculare coaligned with perforant path	Feedback	Fast	Somatostatin, NP-Y (most), mGluR1	Maccaferri and McBain 1996; Blasco-Ibáñez and Freund 1995; Freund and Buzsáki 1996

AP, action potential; CA, cornu ammonis; DG, dentate gyrus; HICAP, hilar–commissural-association pathway; HIPP, hilar–perforant path; MOPP, molecular layer–perforant path; NP-Y, neuropeptide Y; OALM, oriens-alveus–lacunosum-moleculare; OML, outer molecular layer
Interneuron nomenclature is derived from the location of the soma and that of the axonal arborization, relative to major glutamatergic pathways

[a] Stratum oriens interneurons with horizontal dendrites are likely to be identical to OALM interneurons, whereas stratum oriens interneurons with vertical dendrites represent basket cells with projections to the pyramidal cell layer (Maccaferri and McBain 1996)

II. Fast Excitatory Postsynaptic Potentials at Principal Neuron–Interneuron Synapses

In general, glutamatergic synaptic transmission involves both ionotropic and metabotropic glutamate receptors. Ionotropic receptors are further subdivided into L-α-amino-3-hydroxy-5-methyl-4-isoxazolepropionate receptors (AMPARs), kainate receptors (KARs), and *N*-methyl-D-aspartate receptors (NMDARs) depending on their pharmacological properties (HOLLMANN, this volume). The subthreshold excitatory postsynaptic potential (EPSP) at principal neuron–interneuron synapses is predominantly mediated by AMPARs, although the other types of glutamate receptors also participate (GEIGER et al. 1997; THOMSON 1997; COSSART et al. 1998; FRERKING et al. 1998; MILES and PONCER 1993).

Surprisingly, it turns out that glutamate-mediated EPSPs at principal neuron–interneuron synapses are much briefer than those generated at synapses between principal neurons in the same circuitry (Fig. 1B; Table 2). The half-duration of the subthreshold EPSP in interneurons (recorded at the soma) is between 3.7 ms and 9.8 ms at near-physiological temperatures (32–37°C; Table 2; MILES 1990; SCHARFMAN et al. 1990; DEBANNE et al. 1995; GEIGER et al. 1997; BUHL et al. 1997; ALI and THOMSON 1998; ALI et al. 1998). In contrast, the half-duration of the EPSP in principal cells is substantially longer (17–38 ms; Table 2; MILES and WONG 1986; DEUCHARS and THOMSON 1996; MARKRAM et al. 1997). The time course of the fastest principal neuron–interneuron EPSPs is comparable to that of EPSPs generated at auditory synapses (half-duration approximately 2 ms at 29–32°C and at the resting potential; ZHANG and TRUSSELL 1994). Furthermore, EPSPs generated at principal neuron–interneuron synapses tend to have relatively large amplitudes; the average EPSP amplitude varies between 0.9 mV and 3.4 mV in different interneurons (Table 2). In certain interneuron subtypes, single EPSPs are sufficiently large to evoke action potentials with high reliability (MILES 1990; TRAUB and MILES 1995).

III. Factors Involved in Generating Fast EPSPs

Multiple factors, including the electrotonic properties of the postsynaptic neuron, the location of the synaptic contacts on the somatodendritic domain of the target cell, and the time course of the GluR-mediated postsynaptic conductance change, could be involved in generating EPSPs with rapid time courses in interneurons (RALL 1967; JACK and REDMAN 1971). The simulations depicted in Fig. 2 illustrate the importance of each of these factors. The fastest EPSPs are generated if the membrane time constant (τ_m) of the postsynaptic neuron is fast, if the postsynaptic conductance change is brief, and if the synapse is located close to the soma. The slowest EPSPs are generated if τ_m is slow, if the postsynaptic conductance change is long lasting, and if the synapse is located distally. For proximal synapses, the postsynaptic conductance change

Table 2. Unitary excitatory postsynaptic potentials (EPSPs) generated at principal neuron–interneuron synapses

Synapse	EPSP peak amplitude	EPSP half-duration	τ_m	Method	Reference
Principal neuron–interneuron synapses					
Granule cell–basket interneuron (DG)	2.1 mV	3.7 ms	8.4 ms	wc	Geiger et al. 1997
Pyramidal–OALM interneuron (CA1)	0.93 mV	7.5 ms	12.8 ms	sh	Ali and Thomson 1998
Pyramidal–basket/bistratified interneuron (CA1)	1.4/3.4 mV	5.4/7.6 ms	4.8–11.4 ms	sh	Ali et al. 1998
Pyramidal–interneuron (CA3)	1.9 mV	9.8 ms	9 ms	sh	Miles 1990
Pyramidal–interneuron (neocortex layer 2/3)	1.0 mV	4.7 ms	9.2 ms	sh	Buhl et al. 1997
Principal neuron–principal neuron synapses					
Pyramidal–pyramidal (CA3)	1.4 mV	27 ms	~20 ms[a]	sh	Miles and Wong 1986
			66 ms	pp	Spruston and Johnston 1992
			93 ms (22°C)	wc	Major et al. 1994
Pyramidal–pyramidal (CA1)	0.7 mV	16.8 ms	7.5 ms	sh	Deuchars and Thomson 1996
			28 ms	wc	Spruston and Johnston 1992
Pyramidal–pyramidal (neocortex layer 5)	1.3 mV	38 ms[b]	51 ms[c]	wc	Markram et al. 1997
			12.4 ms	wc	Stuart and Spruston 1998

Shown is a comparison with EPSPs at synapses between principal neurons in the same circuitry. All data are mean values at near-physiological temperature unless specified differently. Note that sharp microelectrode recording is thought to induce a somatic shunt between intracellular and extracellular space around the electrode, which reduces the apparent value of τ_m

pp, tight-seal perforated-patch configuration; sh, sharp microelectrode recording; wc, tight-seal whole-cell configuration of the patch-clamp technique

[a] τ_m was measured from Fig. 3E in Miles and Wong 1986

[b] Half-duration of the EPSP was estimated from the values of 20–80% rise time and decay time constant given in Markram et al. 1997.

[c] τ_m was estimated from the simulations of EPSPs (Roth, unpublished)

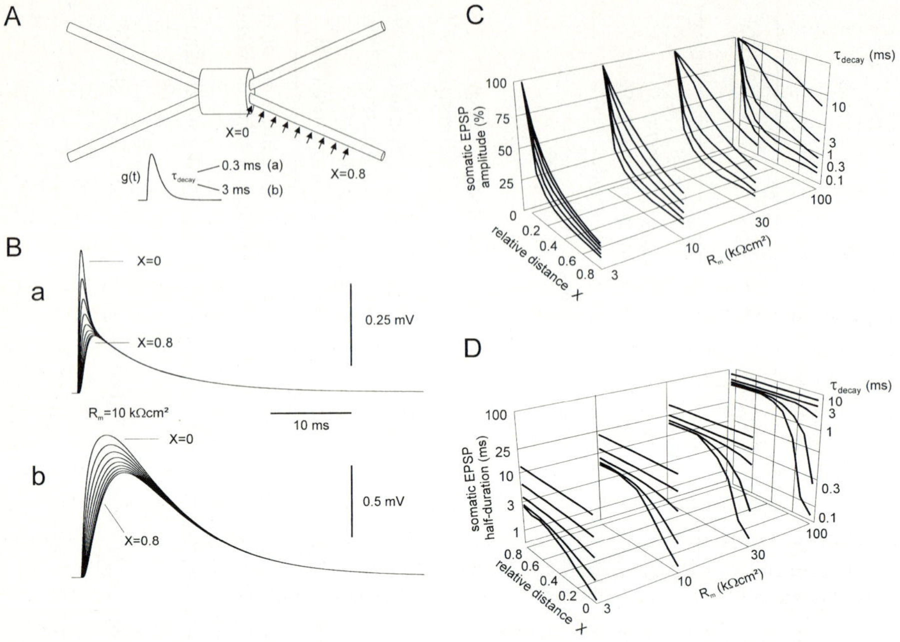

Fig. 2A–D. Dependence of the time course of simulated excitatory postsynaptic potentials (EPSPs) on synapse location, time course of postsynaptic conductance change, and specific membrane resistance, R_m. **A** Schematic representation of the neuron used for simulations. The neuron has a cylindrical soma (height and diameter 20 μm) and four processes attached to it (physical diameter $d = 2.6\,\mu$m; physical length $l = 565\,\mu$m, corresponding to an electrotonic length L of 0.7 for $R_m = 10\,\text{k}\Omega\text{cm}^2$). C_m was assumed to be $0.8\,\mu\text{Fcm}^{-2}$ and R_i was assumed to be $100\,\Omega$cm (uniform for all segments). Voltage changes were simulated using NEURON version 3.2.2 (HINES and CARNEVALE 1997), running on a Sparcstation 5 (Sun Microsystems). The simulated postsynaptic conductance change was proportional to the sum of two exponentials [one for the rise and one for the decay; $y = -\exp(-t/\tau_{rise}) + \exp(-t/\tau_{decay})$]. Further simulation parameters were: the number of segments (10 for the soma and 100 for each process), the time step (5 μs), τ_{rise} (100 μs, except for the simulated conductance change with the fastest decay, where τ_{rise} was 50 μs), peak conductance change (1 nS), resting potential (–70 mV), and reversal potential of the simulated conductance change (0 mV). **B** Somatic EPSPs for conductance changes simulated on one process, at relative physical locations X, which were integer multiples of 10% of the entire process length. $\tau_{decay} = 0.3$ ms (*a*) and 3 ms (*b*). R_m was assumed to be $10\,\text{k}\Omega\text{cm}^2$, which gave a membrane time constant $\tau_m = R_m C_m = 8$ ms and an input resistance of 49 MΩ. The passive properties of the schematic neuron were thus comparable to those of basket cells in the dentate gyrus reported previously (GEIGER et al. 1997). **C, D** Attenuation of EPSPs (ratio of somatic and dendritic EPSP peak amplitude; **C**) and half-duration of EPSPs (**D**) plotted against relative location (X) and R_m (3, 10, 30, and $100\,\text{k}\Omega\text{cm}^2$) for different values of τ_{decay} (0.1, 0.3, 1, 3, and 10 ms, respectively). The half-duration of the EPSP is shown on a logarithmic scale

has a significant "shaping effect" on the somatic postsynaptic potential (Fig. 2; GEIGER et al. 1997).

Combined electrophysiological and morphological analysis reveals that interneurons use all possible options to generate fast somatic EPSPs. First, τ_m is faster in interneurons than in principal neurons (Table 2). The difference in τ_m between the two types of cells is related to a difference in the specific membrane resistance, R_m ($\tau_m = R_m C_m$, and the specific membrane capacitance, C_m is approximately constant, ~1 μFcm^{-2}). Second, the synaptic contacts of principal neuron–interneuron synapses are located mainly in the perisomatic region of the postsynaptic target cells (Table 3; Fig. 1A). Finally, excitatory postsynaptic currents (EPSCs) generated at principal neuron–interneuron synapses rise and decay very rapidly; the decay time constant of the average EPSC is less than a millisecond at 34°C, significantly faster than at excitatory synapses between principal neurons (Table 3).

The implications of the fast EPSP time course for the operation of interneurons in the network are twofold. First, rapid EPSPs may enable interneurons to detect synchronized principal neuron activity and, thus, to operate as coincidence detectors (GEIGER et al. 1997; BUHL et al. 1997). Second, rapid EPSPs will minimize the delay of feedback and feedforward inhibition and increase its temporal precision, in analogy to the auditory system (TRUSSELL 1997).

In conclusion, EPSPs generated at glutamatergic principal neuron–interneuron synapses are substantially faster than those generated at synapses between principal neurons. This rapid time course is generated by the synergistic effects of three factors: the low specific resistance of the interneuron membrane, the perisomatic location of the synaptic contacts, and the fast time course of the postsynaptic conductance change.

C. Submillisecond AMPAR-Mediated Signaling at Principal Neuron–Interneuron Synapses

Which factors contribute to the rapid time course of the postsynaptic conductance change at principal neuron–interneuron synapses? In general, its average kinetics are determined by the time course of quantal transmitter release (i.e. the synchrony or asynchrony of synaptic vesicle fusion) and the time course of the quantal postsynaptic conductance change (i.e. the conductance change evoked by the release of the contents of a single synaptic vesicle). The time course of the quantal conductance change, in turn, is shaped by the duration of the glutamate pulse in the synaptic cleft and the gating properties of the postsynaptic receptors. If the quantal synaptic glutamate pulse is brief, the EPSC is expected to decay rapidly due to AMPAR deactivation following transmitter removal. If the glutamate pulse is very long, the EPSC would decay more slowly due to AMPAR desensitization while the transmitter is present (JONAS and SPRUSTON 1994; BÉHÉ et al., this volume). Interactions between

Table 3. Location of synaptic contacts and time course of the postsynaptic conductance change at principal neuron–interneuron synapses

Synapse	Location of synaptic contacts on postsynaptic target cell[a]	Unitary EPSC decay τ (~22°C)	Unitary EPSC decay τ (~34°C)	Reference
Principal neuron–interneuron synapses				
Granule cell–basket cell synapse (dentate gyrus, hippocampus)	7–147 µm	av 2.5 ms; qu 1.3 ms	av 772 µs; qu 367 µs	Geiger et al. 1997
Pyramidal neuron–interneuron synapses (neocortex layer 2/3)	0–250 µm	sp 2.5 ms		Hestrin 1993; Buhl et al. 1997
Pyramidal neuron–interneuron synapses (neocortex layer 4)		av 2.4 ms		Stern et al. 1992
Principal neuron–principal neuron synapses				
Commissural-association/perforant path–granule cell synapse (dentate gyrus, hippocampus)		av 3–9 ms		Keller et al. 1991
Granule cell–CA3 pyramidal cell synapse (hippocampus)	10–120 µm	av 5.5 ms		Jonas et al. 1993
CA3–CA1 pyramidal cell synapse (hippocampus)	9–370 µm	av 4–8 ms		Hestrin et al. 1990; Sorra and Harris 1993
Pyramidal neuron autapses (hippocampal culture)		av 6.3 ms; mi 4.2 ms	av 3.9 ms; mi 2.3 ms	Tong and Jahr, 1994
Pyramidal neuron–pyramidal neuron synapse (neocortex layer 2/3)		sp 4.6 ms		Hestrin 1993
Pyramidal neuron–pyramidal neuron synapse (neocortex layer 5)	80–585 µm[b]			Markram et al. 1997
			sp 3.3 ms	Stuart and Sakmann 1995

Shown is a comparison with synapses between principal neurons in the same circuitry.
av, average evoked excitatory postsynaptic current (EPSC); mi, miniature EPSCs (in the presence of tetrodotoxin); qu, quantal evoked EPSC; sp, spontaneous EPSCs

[a] Measured from the contacts to the soma of the postsynaptic neuron

[b] Mean geometric distances of contacts of one connection

quantal glutamate pulses originating at different release sites, referred to as "cross-talk" (BARBOUR and HÄUSSER 1997) or "spillover" (KULLMANN and ASZTELY 1998), could further complicate the picture (TRUSSELL, this volume). All factors relevant to the signaling time course have been examined systematically at the granule cell–basket cell synapse in the dentate gyrus, a prototypic principal neuron–interneuron synapse (GEIGER et al. 1995, 1997; KOH et al. 1995a).

I. Estimates of Quantal Content, Quantal Size, and Number of Anatomical Release Sites

The mean peak amplitude of the average AMPAR-mediated EPSC generated at the granule cell–basket cell synapse (excluding failures) is 315 pA, and that of the quantal EPSC, examined under conditions of reduced Ca^{2+}/Mg^{2+} concentration ratio, is 160 pA (at 34°C, −70 mV; GEIGER et al. 1997). This suggests a quantal content of approximately 2 at this synapse. As the average single-channel conductance of AMPARs expressed in basket cells is 23 pS (at 22°C; KOH et al. 1995a; Table 4), this may indicate that a quantal EPSC is generated by the opening of approximately 70 AMPAR channels at the peak (assuming a Q_{10} value of 1.3 for the single-channel conductance without correction for voltage-clamp errors).

The estimate of the quantal content at the granule cell–basket cell synapse is consistent with the small number of anatomical release sites. Morphological analysis reveals that the number of putative synaptic contacts per connection is 2–4 for this synapse (GEIGER et al. 1997). Electron-microscopic analysis shows that the synaptic contacts are small and probably comprise a single active zone (ACSÁDY et al. 1998; GEIGER et al., unpublished; GULYÁS et al. 1993; BUHL et al. 1994). Thus, principal neuron–interneuron connections involve a relatively small total number of transmitter release sites. This could minimize intersite variability in the timing of release (GEIGER et al. 1997).

II. Synchrony of Transmitter Release and Time Course of the Quantal Conductance Change

Quantal transmitter release at the granule cell–basket cell synapse is highly synchronized. The release period decays with a time constant of 300 μs at 34°C (GEIGER et al. 1997), comparable to that at the neuromuscular junction and at auditory synapses (at >30°C; DATYNER and GAGE 1980; ISAACSON and WALMSLEY 1995). The time course of the quantal EPSC at the granule cell–basket cell synapse is extremely fast, substantially faster than that at the majority of other glutamatergic synapses. Quantal EPSCs in basket cells decay with an average time constant of 367 μs at 34°C and 1.3 ms at 22°C; this is substantially faster than in principal neurons of the same region (Table 3; GEIGER et al. 1997).

The mechanisms underlying the quantal EPSC time course can be addressed using the technique of fast application of glutamate to isolated

Table 4. Comparison of gating properties of interneuron α-amino-3-hydroxy-5-methyl-4-isoxazolepropionate receptors (AMPARs; dentate gyrus basket cells) and principal neuron AMPARs (CA3 pyramidal neurons)

Parameter	Dentate gyrus basket cell	Reference	CA3 pyramidal neuron	Reference
Gating properties				
20–80% rise time (1 mM glu)	~0.2 ms	Koh et al. 1995a	~0.2 ms	Colquhoun et al. 1992
$\tau_{deactivation}$ (1 mM glu)	1.2 ms	Koh et al. 1995a	2.5 ms	Colquhoun et al. 1992
$\tau_{desensitization}$ (1 mM glu)	3.7 ms	Koh et al. 1995a	11.3 ms	Colquhoun et al. 1992
Steady state/peak current (1 mM glu)	<1%	Koh et al. 1995a	5.3%	Geiger et al. 1995
$\tau_{desensitization}$ (0.1 mM glu)	11.9 ms	Geiger et al. 1997	~12 ms	Jonas and Sakmann 1992
$\tau_{desensitization}$ (10 mM glu)	3.2 ms	Geiger et al. 1997	~8 ms	Jonas and Sakmann 1992
Peak open probability (3 mM glu)	0.54	Koh et al. 1995a	0.71	Jonas et al. 1993
Activation glutamate EC_{50}	813 μM	Koh et al. 1995a	342 μM	Jonas and Sakmann 1992
respective Hill coefficient	1.3	Koh et al. 1995a	1.3	Jonas and Sakmann 1992
Resensitization τ_1 (amplitude contribution)[a]	34 ms (40%)	Koh et al. 1995a	48 ms (60%)	Colquhoun et al. 1992
Resensitization τ_2 (amplitude contribution)	387 ms (26%)	Koh et al. 1995a		
Equilibrium desensitization IC_{50}[b]	2.2 μM	Koh and Jonas, unpublished	9.6 μM	Colquhoun et al. 1992
respective Hill coefficient	1.5	Koh and Jonas, unpublished	1.2	Colquhoun et al. 1992
Conductance properties				
P_{Ca}/P_{Na}	1.79	Koh et al. 1995a	0.05–0.1	Jonas and Sakmann 1992; Colquhoun et al. 1992; Geiger et al. 1995
Rectification index[c] (25 μM intracellular spermidine)	0.5	Koh et al. 1995b	1.12	Koh et al. 1995b
Spermidine sensitivity IC_{50} (intracellular)	9.9 μM	Koh et al. 1995b	>100 μM	Koh et al. 1995b
Single-channel conductance	22.6 pS[d]	Koh et al. 1995a	8.5 pS	Jonas et al. 1993
	11, 23, 38 pS[e]	Haverkampf et al. 1997		

All measurements were made at approximately 22°C

glu, glutamate; EC_{50}, half-maximal activating concentration; IC_{50}, half-maximal inhibitory concentration

[a] Recovery from desensitization was investigated by two 1-ms pulses of 1 mM glutamate separated by intervals of variable duration

[b] Equilibrium desensitization was measured using long prepulses of low glutamate concentration followed by a 1-ms pulse of 1 mM glutamate

[c] Determined from chord conductances at +40 mV and –80 mV

[d] Estimated by nonstationary fluctuation analysis

[e] Obtained by direct single-channel recording

outside-out patches. The fast application system and the patch may together be regarded as a surrogate synapse where the time course of transmitter concentration can be controlled precisely (Franke et al. 1987; Dudel et al. 1990; Colquhoun et al. 1992; Hestrin 1993; Koh et al. 1995a; Spruston et al. 1995). In basket cells in the dentate gyrus, the mean deactivation time constant is 1.2 ms, and the desensitization time constant is 3.7 ms at 22°C (Fig. 3; Koh et al. 1995a). A comparison of quantal synaptic currents and glutamate-activated currents in patches indicates that the decay of the EPSC at the granule cell–basket cell synapse is determined by AMPAR deactivation rather than desensitization (Geiger et al. 1997; Koh et al. 1995a; Geiger et al. 1995). If AMPARs in postsynaptic densities are functionally identical to those in outside-out patches, these results imply that glutamate is cleared from the synaptic cleft very rapidly. Furthermore, a comparison of AMPARs in differ-

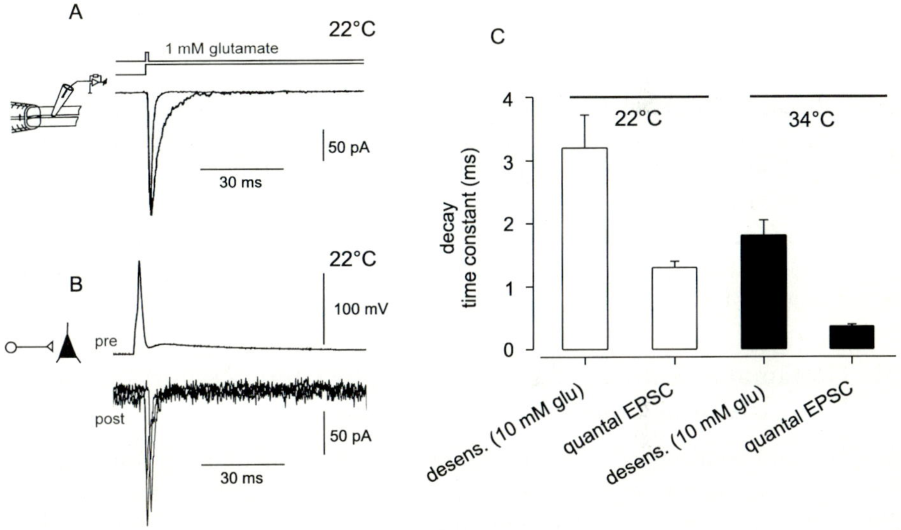

Fig. 3A–C. Time course of quantal conductance change is determined by α-amino-3-hydroxy-5-methyl-4-isoxazolepropionate receptor (AMPAR) deactivation. **A** Traces of currents activated by a 1 ms and a 100 ms pulse of 1 mM glutamate in an outside-out patch isolated from the soma of a basket cell. The holding potential was –60 mV and the recording temperature was 22°C. **B** Quantal excitatory postsynaptic currents (EPSCs) recorded in a granule cell–basket cell pair. The percentage of failures was 82.5%, suggesting that the majority of events were quantal, caused by the release of the contents of a single synaptic vesicle. The action potential evoked in the presynaptic granule cell is shown on *top*, and three single EPSCs are shown superimposed at the *bottom*. The holding potential was –70 mV. The recording temperature was 22°C. The time scale is the same as for the traces shown in **A**. **C** Comparison of the desensitization time constant for 10 mM glutamate and the decay time constant of the quantal EPSC at 22°C and 34°C. As the desensitization time constant of basket cell AMPARs decreases with concentration (Geiger et al. 1997), the difference between the EPSC decay time constant and desensitization time constant would become even larger at lower glutamate concentrations. Modified from Geiger et al. 1997, Copyright by Cell Press

ent types of neurons reveals that the receptors expressed in basket cells are similar to those in other types of interneurons but are very different from those in principal neurons, particularly with respect to their gating kinetics but also in their specific conductance properties (Table 4; COLQUHOUN et al. 1992; HESTRIN 1993; LIVSEY et al. 1993; JONAS et al. 1994; GEIGER et al. 1995; ISA et al. 1996; ANGULO et al. 1997).

What are the molecular determinants that confer the specific functional properties of the interneuron receptor? AMPARs are multimeric proteins assembled from four types of subunits (GluR-A, -B, -C, and -D or, in an alternative nomenclature, GluR1, GluR2, GluR3, and GluR4), each expressed in two alternatively spliced versions (flip and flop; HOLLMANN, this volume). *In situ* hybridization, single-cell reverse transcription–polymerase chain reaction, and immunocytochemical analysis have revealed three main differences between AMPARs in interneurons and those in principal neurons. First, interneurons express GluR-B subunit mRNA and protein at markedly lower levels (BOCHET et al. 1994; JONAS et al. 1994; GEIGER et al. 1995; ANGULO et al. 1997). Second, interneurons preferentially express AMPAR subunits in the flop splice version, whereas principal neurons mainly express flip-version subunits (BOCHET et al. 1994; GEIGER et al. 1995; ANGULO et al. 1997). Finally, some interneuron subtypes strongly express GluR-D-subunit mRNA and protein (GEIGER et al. 1995; BAUDE et al. 1995). The low GluR-B_{flip} content and the high GluR-D content of interneuron AMPARs are thought to be primarily responsible for their rapid deactivation (MONYER et al., this volume; HE et al. 1998; BAUDE et al. 1995), which in turn is the main determinant of the decay of the quantal EPSC.

III. Glutamate Clearance from the Synaptic Cleft: Role of Diffusion, Ultrastructural Constraints, and Uptake

How is rapid transmitter clearance at principal neuron–interneuron synapses achieved? In general, rapid diffusion and buffering appear to be the main possible elimination mechanisms at glutamatergic synapses. Transmitter degradation is not involved, because a high-turnover enzyme analogous to the acetylcholine esterase at the neuromuscular junction is not present. Thus, the three main parameters that shape the spatiotemporal profile of transmitter concentration in the synaptic cleft are: the diffusion coefficient (D), the geometry of the synaptic cleft, and the concentration of glutamate transporters. Rapid binding of glutamate to transporters is probably more important than the subsequent slow translocation step (DIAMOND and JAHR 1997; OTIS and KAVANAUGH, this volume).

To address the contribution of transmitter clearance and receptor gating to the time course of the postsynaptic conductance change quantitatively, we have simulated the instantaneous release of glutamate, the diffusion in the synaptic cleft, and the subsequent activation of postsynaptic AMPARs using Monte Carlo methods (BARTOL et al. 1991; STILES et al. 1998; FABER et al. 1992;

WAHL et al. 1996). The cyclic reaction scheme of AMPAR gating used is shown in Fig. 4, and the results of the simulation are depicted in Fig. 5. With plausible assumptions of the value of D and the synaptic contact size, the time course of the quantal EPSC can be well reproduced (Fig. 5). D was set to 3×10^{-6} cm^2s^{-1} [approximately half of that of glutamine in aqueous solution (LONGSWORTH 1953)], which takes into account macromolecular material in the synaptic cleft (RUSAKOV and KULLMANN 1998a, 1998b). The synaptic contact area was assumed to be $2\,\mu m \times 2\,\mu m$, which is consistent with electron-microscopic data of these synapses (ACSÁDY et al. 1998). Approximately 250 μs after instantaneous release, postsynaptic AMPARs have maximal open probability (Figs. 5A, C) and 2 ms after release, the majority of channels have closed again (a large fraction of them being in desensitized states; Figs. 5A, D).

Simulations in which the values of D, the synaptic contact area, and the transporter density were varied systematically indicate that the value of D is the main determinant of the time course of transmitter clearance (Fig. 6). Decreasing the value of D slows the decay and increases the amplitude of

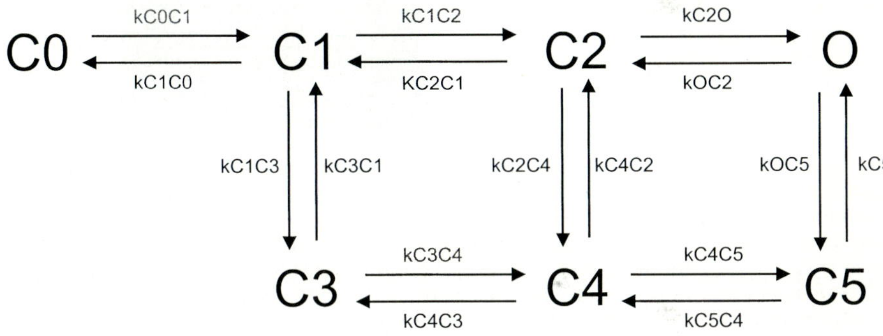

Fig. 4. The gating scheme of interneuron α-amino-3-hydroxy-5-methyl-4-isoxazolepropionate receptors at 22°C. The model includes 3 closed available states (*C0–C2*), 3 closed desensitized states (*C3–C5*), and a single open state (*O*). Initial estimates of the rates were determined by trial and error. Estimates were refined using an iterative algorithm (programmed in Mathematica 2.2; Wolfram Research) running on a Pentium-II computer. The algorithm minimized the deviations between observed and predicted values; the weights were determined using the experimental standard errors. The model is certainly not unique but predicts the experimental observations very closely (Table 4). The estimates of the rate constants were as follows: $k_{C0C1} = 17.1 \times 10^6\,M^{-1}s^{-1}$, $k_{C1C0} = 0.157 \times 10^3\,s^{-1}$, $k_{C1C2} = 3.24 \times 10^6\,M^{-1}s^{-1}$, $k_{C2C1} = 3.76 \times 10^3\,s^{-1}$, $k_{C2O} = 14.9 \times 10^3\,s^{-1}$, $k_{OC2} = 4.00 \times 10^3\,s^{-1}$, $k_{C3C4} = 0.611 \times 10^6\,M^{-1}s^{-1}$, $k_{C4C3} = 0.00200 \times 10^3\,s^{-1}$, $k_{C4C5} = 1.59 \times 10^3\,s^{-1}$, $k_{C5C4} = 899 \times 10^3\,s^{-1}$, $k_{C1C3} = 1.53 \times 10^3\,s^{-1}$, $k_{C3C1} = 0.408 \times 10^3\,s^{-1}$, $k_{C2C4} = 0.502 \times 10^3\,s^{-1}$, $k_{C4C2} = 0.000377 \times 10^3\,s^{-1}$, $k_{OC5} = 0.121 \times 10^3\,s^{-1}$, $k_{C5O} = 0.191 \times 10^3\,s^{-1}$. The predictions of the model were: 20–80% rise time [1 mM glutamate (glu)] = 0.25 ms, $\tau_{deactivation}$ (1 mM glu) = 1.2 ms, $\tau_{desensitization}$ (1 mM glu) = 3.6 ms, steady state/peak (1 mM glu) = 0.8%, $\tau_{desensitization}$ (0.1 mM glu) = 13.9 ms, $\tau_{desensitization}$ (10 mM glu) = 4.4 ms, peak open probability (3 mM glu) = 0.56, activation glutamate EC_{50} = 806 μM, respective Hill coefficient = 1.2, resensitization τ_1 (amplitude contribution) = 33.6 ms (42%), resensitization τ_2 (amplitude contribution) = 395 ms (32%), equilibrium desensitization IC_{50} = 2.3 μM, respective Hill coefficient = 1.1

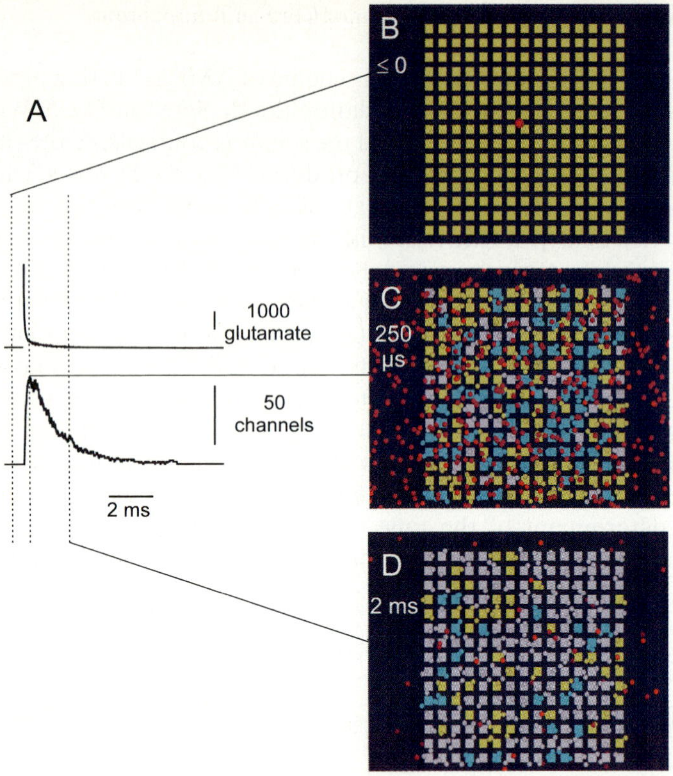

Fig. 5A–D. Monte Carlo simulation of transmitter diffusion and postsynaptic receptor activation at a principal neuron–interneuron synapse. **A** Simulated number of glutamate molecules in the center of the synaptic cleft (*upper trace*) and number of open α-amino-3-hydroxy-5-methyl-4-isoxazolepropionate receptor (AMPAR) channels (*lower trace*). Free glutamate molecules were counted in the cleft subvolume adjacent to the postsynaptic density. Simulations were performed using MCell version 1.21, written by Tom Bartol (Computational Neurobiology Laboratory, Salk Institute) and Joel Stiles (Section of Neurobiology and Behaviour, Cornell University), running on a Sparcstation 5 (Sun Microsystems) or on an Origin 2000 server (Silicon Graphics). The diffusion coefficient for glutamate was assumed to be $3 \times 10^{-6}\,cm^2 s^{-1}$, approximately half of that of glutamine in aqueous solution (LONGSWORTH 1953). The number of glutamate molecules per vesicle was assumed to be 5000, consistent with biochemical estimates (RIVEROS et al. 1986). The morphological assumptions were: width of the synaptic cleft = 20 nm, postsynaptic density area = 250 nm × 250 nm, and synaptic contact area = 2 μm × 2 μm, in agreement with morphological data (ACSÁDY et al. 1998; FROTSCHER et al., this volume). The AMPAR density was assumed to be 4000 receptors per μm^2 (restricted to the postsynaptic density, evenly spaced). This gives a number of postsynaptic AMPARs of 225, consistent with the large quantal size (Sect. C.I) and the strong AMPAR subunit immunoreactivity of the postsynaptic densities of principal neuron–interneuron synapses (BAUDE et al. 1995; NUSSER et al. 1998). Glutamate was assumed to be released instantaneously from a site with a diameter of 5 nm. The time step was 100 ns in all simulations. Glutamate transporters were not incorporated here. **B–D** View from the synaptic cleft onto the postsynaptic density. Snapshots were taken at $t = 0$ (**B**), 250 μs (**C**), and 2 ms (**D**) after transmitter release. Glutamate molecules are *red*, unliganded receptors are *yellow*, open receptors are *cyan*, and desensitized receptors are *gray*. Note that a large fraction of AMPARs is in desensitized states (*gray*) at $t = 2$ ms

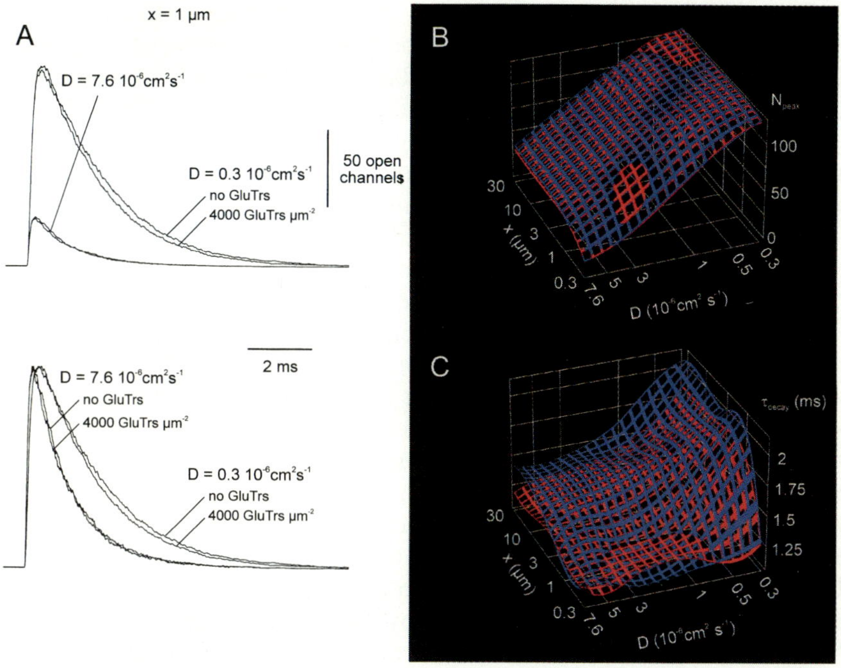

Fig. 6A–C. Determinants of the time courses and amplitudes of simulated excitatory postsynaptic currents at a principal neuron–interneuron synapse. **A** Simulated synaptic events for diffusion coefficient (D) = $7.6 \times 10^{-6}\,cm^2s^{-1}$ and $0.3 \times 10^{-6}\,cm^2s^{-1}$ and for transporter densities of $0\,\mu m^{-2}$ and $4000\,\mu m^{-2}$. *Upper traces* show the number of open α-amino-3-hydroxy-5-methyl-4-isoxazolepropionate receptors; *lower traces* were normalized to illustrate differences in the time course. **B** Number of open channels at the peak EPSC and **C** decay time constant of the simulated EPSC plotted against contact size (0.3, 1, 3, 10, and $30\,\mu m$ half side length) and D (7.6, 5, 3, 1, 0.5, and $0.3 \times 10^{-6}\,cm^2s^{-1}$) for different densities of glutamate transporters (*blue* = $0\,\mu m^{-2}$ and *red* = $4000\,\mu m^{-2}$). The surfaces were generated by bicubic spline interpolation of the data points (which are shown superimposed). Assumptions for the simulations were identical to those in Fig. 5, unless specified differently. For each parameter set, 25 traces (50 in the case of D = $7.6 \times 10^{-6}\,cm^2s^{-1}$) were simulated with different sequences of random numbers. Traces were averaged and analyzed in Mathematica. Glutamate transporters were incorporated into the postsynaptic membrane surrounding the postsynaptic density. Transporter kinetics were modeled by a three-state model, with states *T*1 (unliganded), *T*2 (glutamate bound outside), and *T*3 (glutamate bound inside). The rate constants were assumed to be $k_{T1T2} = 18 \times 10^6\,M^{-1}s^{-1}$, $K_{T2T1} = 180\,s^{-1}$, $k_{T2T3} = 180\,s^{-1}$, and $k_{T3T1} = 25.7\,s^{-1}$. This model reproduced the major binding, unbinding, and translocation characteristics of native glutamate transporters (Otis and Jahr 1998)

the simulated EPSC (Fig. 6). A lower limit for D of approximately $0.5 \times 10^{-6}\,cm^2s^{-1}$ would be consistent with the rapid decay of the quantal EPSC at the granule cell–basket cell synapse. Increasing the synaptic contact area slightly prolongs the simulated EPSC, whereas incorporation of glutamate transporters accelerates the decay in most conditions except for intermediate values of D and small contact sizes (Fig. 6).

Fast signaling is likely to be a general characteristic of principal neuron–interneuron synapses but may be generated by different mechanisms in different circuits. Fast EPSCs have been found in all GABAergic interneurons studied to date, including those in the hilar region of the hippocampus (LIVSEY and VICINI 1992), in the neocortex (HESTRIN 1993), and in the cerebellum (BARBOUR et al. 1994). In the hilus and the neocortex, the differences between EPSCs in interneurons and principal neurons are mainly conferred by AMPAR gating (LIVSEY et al. 1993; HESTRIN 1993; GEIGER et al. 1995). In the cerebellum, however, AMPAR gating is indistinguishable between interneurons and principal neurons, and the main difference is the absence and presence of spines, respectively (leading to differences in the kinetics of transmitter clearance; BARBOUR et al. 1994).

The fast time course of signaling at the granule cell–basket cell synapse is comparable to that at auditory synapses, which operate at the absolute temporal limits of synaptic transmission (TRUSSELL, this volume). The synchrony of quantal release and the time course and amplitude of quantal EPSCs are very similar in the two types of synapses (ISAACSON and WALMSLEY 1995). Transmitter clearance may be slower at auditory synapses (due to the substantially larger synaptic contact area of calyx and endbulb synapses), but this appears to be compensated by the faster gating of the auditory AMPAR (TRUSSELL 1997; GEIGER et al. 1995). Accordingly, the subunit composition of the auditory AMPAR is an extreme version of that of the interneuron AMPAR, being predominantly composed of GluR-D_{flop} subunits. In conclusion, pre- and postsynaptic factors contribute to the fast time course of AMPAR-mediated signaling at the granule cell–basket cell synapse, including the high synchrony of quantal release (perhaps related to a small number of release sites), the rapid clearance of transmitter from the synaptic cleft (probably related to a high value of D, a small synaptic contact area, and a high concentration of glutamate transporters), and the rapid deactivation of postsynaptic AMPARs (caused by their specific subunit composition).

D. Slow Interneuron Excitation Mediated by KARs, NMDARs, and Metabotropic GluRs

Although EPSPs and EPSCs at principal neuron–interneuron synapses are primarily generated by AMPAR activation, KARs, NMDARs, and mGluRs also contribute to various degrees. Collectively, these receptors mediate a second form of interneuron excitation that is distinguished from AMPAR-mediated interneuron excitation by a slow time course (FRERKING et al. 1998; COSSART et al. 1998; STERN et al. 1992; PEROUANSKY and YAARI 1993; MCBAIN and DINGLEDINE 1993; ISA et al. 1996; GEIGER et al. 1997; MILES and PONCER 1993).

I. Postsynaptic KARs in Interneurons

A KAR-mediated component of EPSPs and EPSCs was found in OALM interneurons and in interneurons of the stratum radiatum of the CA1 region of the hippocampus (FRERKING et al. 1998; COSSART et al. 1998). The KAR-mediated EPSC component, isolated pharmacologically using the selective AMPAR antagonist GYKI53655, shows a slow rise (approximately 5 ms rise time) and decay (approximately 100 ms decay time constant at 22°C). Although the contribution to the peak EPSC is minor (approximately 10% or less), the fraction of the postsynaptic charge mediated by KARs may be substantial (FRERKING et al. 1998).

KARs are assembled from GluR5, 6, and 7 subunits and KA-1, -2 subunits (HOLLMANN, this volume). The subunit composition of KARs in interneurons is not precisely known and may differ between interneuron subtypes. Those in OALM interneurons are putative GluR5-containing receptors (as they are selectively activated by 2-amino-3-(3-hydroxy-5-tert-butylisoxazol-4-yl) propanoate .(ATPA) and selectively antagonized by LY293558; COSSART et al. 1998; LERMA, this volume). Those expressed in stratum radiatum interneurons are less sensitive to ATPA and LY293558, suggesting that they are assembled from other subunits. These results are consistent with the preferential expression of GluR5 mRNA in scattered cells in stratum oriens-alveus (BAHN et al. 1994). The abundant expression of KA-2 subunit mRNA in the central nervous system (CNS) further suggests the possibility of KAR heteromer formation in interneurons (LERMA, this volume).

The slow time course of the KAR-mediated EPSC component remains unexplained. Recombinant and native KARs activated by brief glutamate pulses or glutamate steps show rapid rise and decay (HECKMANN et al. 1996), unlike the KAR-mediated synaptic-current components. This could be due to a difference in subunit composition between postsynaptic and extrasynaptic receptors or, alternatively, the remote location of the receptors relative to the release site and their preferential activation by glutamate spillover.

II. Postsynaptic NMDARs in Interneurons

Postsynaptic NMDARs are also activated by synaptically released glutamate at principal neuron–interneuron synapses. Their contribution to subthreshold EPSPs is minor (GEIGER et al. 1997); postsynaptic NMDARs are blocked by Mg^{2+} ions at the resting potential, and the local dendritic voltage changes mediated by AMPAR or KAR activation are either too short or too small to relieve the block. By contrast, the contribution of NMDARs to the EPSCs at positive membrane potentials or in the absence of extracellular Mg^{2+} is considerable (STERN et al. 1992; PEROUANSKY and YAARI 1993; MCBAIN and DINGLEDINE 1993; ISA et al. 1996; GEIGER et al. 1997; but see MAHANTY and SAH 1998 for a principal neuron–interneuron synapse in the amygdala that lacks NMDARs).

While the slow time course of NMDAR-mediated EPSCs is a general characteristic throughout the CNS, NMDAR-mediated EPSC components with exceptionally slow rises were observed in a subset of stratum oriens-alveus interneurons, probably OALM cells (10–90% rise time 31 ms at 19–22°C; Perouansky and Yaari 1993). In other interneurons, however, the time courses of both NMDAR-mediated EPSCs and NMDAR-mediated currents activated by glutamate in outside-out patches are similar to those in principal neurons; this has been reported in both the hippocampus and the neocortex (EPSCs: Stern et al. 1992; Perouansky and Yaari 1993; McBain and Dingledine 1993; Isa et al. 1996; Geiger et al. 1997; glutamate-activated currents: Koh et al. 1995a; Spruston et al. 1995).

NMDARs are heteromultimeric proteins assembled from NR1 and NR2A, NR2B, NR2C, or NR2D subunits (Hollmann, this volume). *In situ* hybridization analysis indicates that NR2D subunits are selectively expressed in interneurons in the hilar region, stratum oriens-alveus, and stratum radiatum, probably including basket cells and OALM interneurons; (Monyer et al. 1994; Standaert et al. 1996). Recombinant NR1/NR2D channels expressed in host cells, unlike heteromeric receptors assembled from any other subunit combination, activate slowly, deactivate within seconds, and do not desensitize in the maintained presence of glutamate (Monyer et al. 1994; Wyllie et al. 1998; Vicini et al. 1998). In addition, NR1/NR2D receptors show very high affinity for both the agonist glutamate and the co-agonist glycine (Ikeda et al. 1992) and are relatively insensitive to blocking by external Mg^{2+} ions (Monyer et al. 1994). Whether functional NR1/NR2D channels are assembled in interneurons is not known; their kinetic properties would be compatible with the idea that they mediate the NMDAR-mediated EPSCs in a subset of oriens-alveus interneurons (Perouansky and Yaari 1993).

III. Postsynaptic Metabotropic GluRs in Interneurons

High-frequency stimulation (100 Hz) of afferent axons results in the synaptic activation of metabotropic glutamate receptors (mGluRs) in interneurons (Miles and Poncer 1993; Whittington et al. 1995). This form of interneuron excitation occurs in the presence of antagonists of AMPARs, KARs, and NMDARs, is occluded by the mGluR agonist *trans*-1-aminocyclopentane-1,3-dicarboxylic acid (ACPD), and is blocked by the mGluR antagonist L-α-methyl-4-carboxyphenyl glycine (MCPG, Miles and Poncer 1993; Whittington et al. 1995). The requirement for repetitive stimulation suggests that mGluRs are located at a substantial diffusional distance from the release site, consistent with the perisynaptic location of mGluR1 immunoreactivity in a subset of principal neuron–interneuron synapses (Baude et al. 1993). In OALM interneurons, 100 μM ACPD induces large inward currents in the voltage-clamp mode (McBain et al. 1994). In interneurons in the CA3 region, low concentrations (5–20 μM) of ACPD abolish the afterhyperpolarization following a spike, and higher concentrations (50–100 μM) induce a depolarization (MILES and PONCER 1993). This may indicate both the expression of different mGluRs in interneurons and their coupling to diverse molecular targets.

MGluRs are G-protein-coupled receptors; several subtypes, which fall into three classes according to structural similarities and coupling to different intracellular signaling cascades, have been identified by molecular cloning (PIN and DUVOISIN 1995). Interneurons express group-1 mGluR (mGluR1 and mGluR5) mRNA and protein at high levels (BAUDE et al. 1993; ROMANO et al. 1995). Thus, it appears likely that these mGluRs mediate the currents induced by ACPD and high-frequency synaptic stimulation. MGluR7 is also found in hippocampal and neocortical interneurons (OHISHI et al. 1995), and mGluR2 is expressed in neocortical interneurons (OHISHI et al. 1993a). In contrast, mGluR3 and mGluR4 have not been detected in interneurons (OHISHI et al. 1993b, 1995). MGluR1 is expressed differentially, with high mRNA and protein levels in somatostatin-positive feedback interneurons, such as HIPP and OALM cells (Table 1), and at low levels in principal neurons and other interneuron subtypes.

In conclusion, interneurons receive fast glutamatergic synaptic excitation mediated by AMPARs and slow glutamatergic synaptic excitation mediated by KARs, NMDARs, and mGluRs. Slow excitation may integrate synaptic activity and may help to adjust the membrane potential of the interneuron close to the threshold for action potential initiation. Slow synaptic interneuron excitation appears to be important for the operation of neuronal networks, since threshold adjustments in interneurons are essential for associative pattern completion in model circuitries (McNAUGHTON and BARNES 1990).

How is coordination between fast and slow interneuron excitation achieved? Fast glutamate release presumably coactivates AMPARs, KARs, NMDARs, and mGluRs; thus, factors that regulate release are expected to affect fast and slow components synergistically. In contrast, glutamate spillover and ambient glutamate will have opposite effects on fast and slow interneuron excitation by desensitizing postsynaptic AMPARs and KARs and activating NMDARs and mGluRs; the half-maximal effective glutamate concentrations for these effects are comparable [0.4 μM for activation of NR1/NR2D channels (IKEDA et al. 1992); 10 μM for glutamate binding to group-1 mGluRs (PIN and DUVOISIN 1995); 2.2 μM for equilibrium desensitization of basket cell AMPARs (Table 4); 0.3 μM for equilibrium desensitization of recombinant GluR6 KARs (HECKMANN et al. 1996)].

E. Short-Term Dynamics at Glutamatergic Principal Neuron–Interneuron Synapses

Principal neuron–interneuron synapses show substantial short-term changes in synaptic strength; they can either exhibit paired-pulse depression (PPD) or paired-pulse facilitation (PPF) following repeated stimulation, depending on the type of principal neuron–interneuron synapse.

I. Paired-Pulse Depression

PPD occurs at synapses between principal neurons and classical interneuron subtypes that innervate the perisomatic regions of their postsynaptic target cells (Table 1). These include granule cell–basket cell synapses in the dentate gyrus (Fig. 7; GEIGER et al. 1996, 1997), CA1 pyramidal neuron–basket cell and –bistratified cell synapses in the CA1 region of the hippocampus, and pyramidal neuron–basket cell synapses in the neocortex (Table 5; ALI et al. 1998;

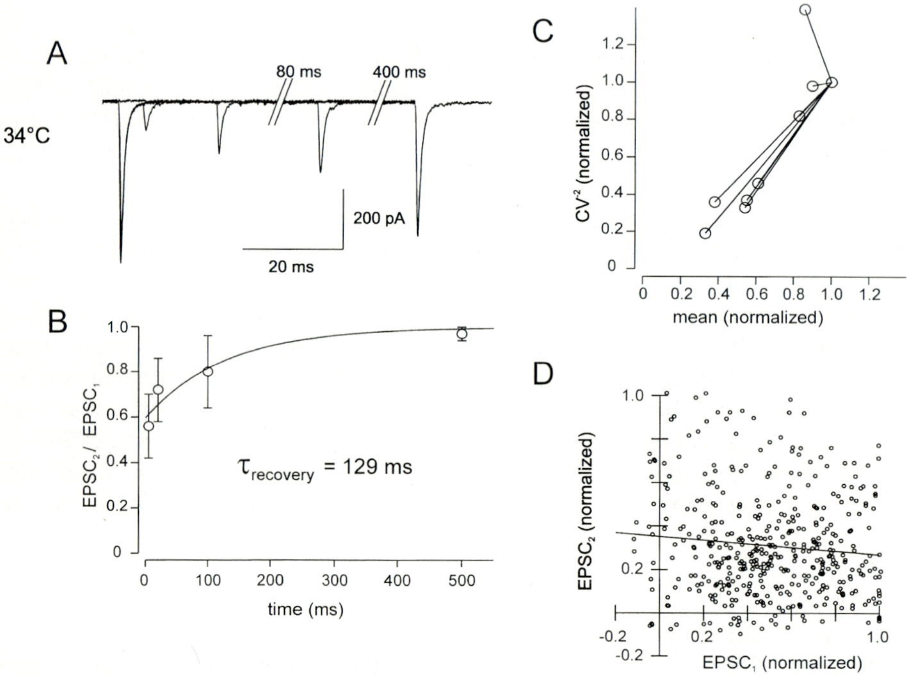

Fig. 7A–D. Paired-pulse depression at the granule cell–basket cell synapse. **A** Average excitatory postsynaptic currents (EPSCs) evoked by pairs of action potentials in the presynaptic granule cell, separated by intervals of variable duration. **B** Paired-pulse depression ratio (peak amplitude of the second EPSC divided by that of the first EPSC) plotted against interpulse interval. The time constant of recovery from paired-pulse depression was 129 ms, and the corresponding exponential function is shown superimposed. **C** The inverse of the square of the coefficient of variation (CV^{-2}) of the EPSC peak amplitude of the second EPSC was plotted against the mean amplitude; data were normalized by the CV^{-2} and the mean, respectively, of the first EPSC. The interval between the two EPSCs was 20 ms. The CV and mean were calculated from 50–60 evoked EPSCs. The location of the data points suggests that the PPD was mainly presynaptic in origin (MALINOW and TSIEN 1990). **D** Plot of the peak amplitude of the second EPSC against that of the first EPSC, both of which were normalized to the maximal amplitude of the first EPSC; the line shows the result of linear regression analysis. The data in **B** are from 4 granule cell–basket cell pairs; the data in **C** and **D** are from 8 pairs. The holding potential was –70 mV in all cases. Pairs of action potentials were evoked at a frequency of $0.5 s^{-1}$. From GEIGER and JONAS, unpublished data

Table 5. Paired-pulse modulation at principal neuron–interneuron synapses

Synapse	PPD/PPF	Interval between APs	$\tau_{recovery}$	Method	Reference
Granule cell–basket interneuron (DG)	44%	20 ms	129 ms	wc, EPSC	Geiger et al. 1996; Fig. 7
Pyramidal–OALM interneuron (CA1)	253%	<20 ms		sh, EPSP	Ali and Thomson 1998
Pyramidal–basket/bistratified interneuron (CA1)	59%/72%	<15 ms		sh, EPSP	Ali et al. 1998
Pyramidal–basket interneuron (layer 2/3)	80%	10–50 ms		sh, EPSP	Buhl et al. 1997
Pyramidal–parvalbumin+/somatostatin+ interneuron (layer 2/3)	70%/191%	100 ms		wc, EPSP	Reyes et al. 1998

All data are mean values at near-physiological temperature and physiological Ca^{2+}/Mg^{2+} concentration ratios
AP, action potential; EPSC, excitatory postsynaptic current; EPSP, excitatory postsynaptic potential; PPD, paired-pulse depression; OALM, oriens-alveus–lacunosum-moleculare; PPF, paired-pulse facilitation; sh, sharp microelectrode recording, with recording of either EPSCs or EPSPs; wc, tight-seal, whole-cell configuration of the patch-clamp technique

BUHL et al. 1997). Analyses of the coefficient of variation and the number of failures indicate that the PPD is mainly presynaptic in origin (Fig. 7C). Correlation plots of the amplitudes of the first and second synaptic events further suggest that the PPD is largely independent of prior release in granule cell–basket cell synapses of the hippocampus (Fig. 7D) and neocortical pyramidal neuron–basket cell synapses of the cat visual cortex (BUHL et al. 1997) but not in synapses on multipolar, parvalbumin-positive interneuron synapses of the rat somatosensory cortex (REYES et al. 1998). This may imply that the molecular process generating PPD is located upstream of the release machinery at some but not all principal neuron–interneuron synapses; one potential mechanism is inactivation of presynaptic Ca^{2+} channels. In a subset of neocortical pyramidal neuron–interneuron synapses, however, PPD is reduced by cyclothiazide, an inhibitor of AMPAR desensitization (ROZOV and BURNASHEV, personal communication), suggesting that desensitization of postsynaptic interneuron AMPARs contributes to PPD. This is not unexpected, since postsynaptic AMPAR channels enter frequently desensitized states following quantal release, and recovery from desensitization is slow (Fig. 5; Table 4).

II. Paired-Pulse Facilitation

PPF mainly occurs at synapses between principal neurons and interneuron subtypes that innervate preferentially distal dendrites of their postsynaptic target cells (Table 1). These include pyramidal–OALM interneuron synapses in CA1 (ALI and THOMSON 1998) and synapses between pyramidal neurons and bitufted, somatostatin-positive interneurons in the neocortex (Table 5; REYES et al. 1998). As observed for PPD, PPF appears to be mainly presynaptic in origin. Furthermore, PPF may be largely independent of prior release; this could be due to cumulative activation of Ca^{2+} channels or temporal summation of Ca^{2+} transients in presynaptic elements. In addition, postsynaptic factors – for example, relief from block by polyamines following AMPAR activation – could contribute to facilitation (ROZOV et al. 1998; BOWIE et al., this volume).

A single presynaptic pyramidal neuron in the neocortex can establish depressing synapses on multipolar, parvalbumin-positive interneurons and facilitating synapses on bitufted, somatostatin-positive interneurons (REYES et al. 1998). Moreover, a single presynaptic pyramidal neuron forms synapses with PPD on other pyramidal neurons and synapses showing PPF on interneurons (THOMSON 1997; MARKRAM et al. 1998). It is not known, however, whether all glutamatergic synapses on a single interneuron are homogeneous in their paired-pulse behaviour, or whether mixtures of facilitating and depressing synapses can occur.

The differences in paired-pulse behaviour among principal neuron–interneuron synapses have substantial implications for the operation of interneurons in the network. PPD in combination with the fast EPSP time course may enable interneurons to detect selectively coincident single-spike activity in several presynaptic principal cells (GEIGER et al. 1997; BUHL et al.

1997). In contrast, PPF in combination with the fast EPSP time course would allow an interneuron to detect high-frequency burst activity in a single principal neuron (in addition to coincident single-spike activity in many presynaptic cells). Thus, different activity patterns of the principal neuron ensemble will activate different interneuron types and trigger different forms of inhibition (perisomatic vs distal-dendritic inhibition).

In conclusion, paired-pulse modulation at principal neuron–interneuron synapses is determined by a variety of pre- and postsynaptic factors. Inactivation or cumulative activation of presynaptic Ca^{2+} channels, summation of presynaptic Ca^{2+} transients, vesicle depletion and mobilization, AMPAR desensitization, and relief from block by polyamines are potentially involved, but their precise contribution at different synapses remains to be addressed (TRUSSELL, this volume, regarding the cellular mechanisms of paired-pulse depression at auditory synapses).

F. Presynaptic Modulation of Glutamate Release at Principal Neuron–Interneuron Synapses

Several pieces of evidence suggest that neuromodulators influence the efficacy of principal neuron–interneuron synapses. First, interneurons and synapses established on their somatodendritic domains are preferential targets of neuromodulatory inputs (FREUND and BUZSÁKI 1996). Second, amplitudes of EPSPs at principal neuron–interneuron synapses in *in vitro* slice preparations vary over a wide range (Table 2). Finally, recent *in vivo* recordings indicate that the impact of principal neuron–interneuron synapses is dependent on the behavioral state of the animal (CSICSVARI et al. 1998).

I. Presynaptic mGluRs and Adenosine Receptors at Principal Neuron–Interneuron Synapses

The primary candidate for a modulator is glutamate itself acting on presynaptic mGluRs at principal neuron–interneuron synapses. Activation of mGluRs has very heterogeneous effects. In OALM interneurons of the CA1 region, the nonselective mGluR agonist ACPD enhances the amplitude of evoked EPSCs; miniature EPSCs are unchanged, consistent with a presynaptic nature of the effect (MCBAIN et al. 1994). In contrast, in interneurons of the dentate gyrus (probably basket cells) and interneurons of the stratum radiatum of the CA1 subfield, ACPD reduces the amplitude of evoked EPSCs; an increase in the percentage of failures suggests that the effect is also mainly presynaptic in origin (DOHERTY and DINGLEDINE 1997; DESAI et al. 1994).

Independent lines of evidence suggest that principal neuron–interneuron synapses contain highly variable sets of mGluRs in their presynaptic elements. First, selective group-2 mGluR agonists suppress EPSCs in dentate gyrus interneurons evoked by stimulation of granule cells but not CA3 pyramidal neurons (DOHERTY and DINGLEDINE 1998). Second, immunocytochemical analysis reveals that mGluR7-positive boutons are preferentially established on mGluR1α-positive interneurons (probably including HIPP and OALM interneurons; Table 1) but are largely absent on boutons established on principal neurons and other interneuron subtypes (SHIGEMOTO et al. 1996, 1997).

Adenosine is another candidate molecule for neuromodulation of interneuron excitation (FORTUNATO et al. 1996; DOHERTY and DINGLEDINE 1997). It inhibits evoked EPSCs in dentate gyrus interneurons (probably basket cells). Furthermore, the adenosine A_1 receptor antagonists 8-cyclopentyl-1,3-dipropylxanthine and *N*-cyclopentyl-9-methyladenine augment EPSCs, suggesting that the ambient adenosine concentration is sufficiently high to activate presynaptic A_1 receptors (DOHERTY and DINGLEDINE 1997). The increase in the percentage of failures indicates that the effect is presynaptic in origin (DOHERTY and DINGLEDINE 1997).

II. Activation of Presynaptic Receptors

Presynaptic mGluRs and adenosine A_1 receptors could be activated in both physiological and pathophysiological conditions. During acute hypoxia in slices, synaptic interneuron excitation is depressed (CONGAR et al. 1995; DOHERTY and DINGLEDINE 1997; KHAZIPOV et al. 1995), probably via a rise in the extracellular glutamate and adenosine concentrations and a subsequent activation of presynaptic receptors. This leads to a reversible form of "dormancy" of interneurons (SLOVITER 1991). MGluRs appear to be primarily involved in interneurons of the dentate gyrus (DOHERTY and DINGLEDINE 1997), whereas adenosine A_1 receptors are relevant in stratum radiatum interneurons (KHAZIPOV et al. 1995). Principal neuron–interneuron synapses are particularly sensitive to hypoxia; the depression of glutamatergic synaptic input develops more rapidly and is more complete than at synapses between principal neurons (CONGAR et al. 1995). Thus, hypoxia induces a selective reduction of excitation of GABAergic interneurons, which in turn may lead to epileptiform activity.

In conclusion, principal neuron–interneuron synapses are modulated by presynaptic mGluRs and adenosine A_1 receptors. These receptors are activated under physiological and pathophysiological conditions and mainly suppress interneuron excitation. The effect of other neuromodulators remains to be investigated.

G. Long-Term Changes in Efficacy at Glutamatergic Principal Neuron–Interneuron Synapses?

I. Principal Neuron–Interneuron Synapses Show Less Plasticity than Those Between Principal Neurons

Several principal neuron–interneuron synapses appear to have more static transmission properties than synapses between principal neurons (BLISS and

COLLINGRIDGE 1993). Low-frequency (1 Hz) synaptic stimulation, a paradigm that induces robust long-term depression (LTD) at glutamatergic synapses on principal neurons, causes little change in the amplitude of monosynaptic EPSPs in OALM interneurons (evoked by antidromic stimulation of CA1 pyramidal cell axons in the alveus; MACCAFERRI and MCBAIN 1995). Pairing synaptic stimulation with depolarization of the postsynaptic neuron, a paradigm that induces robust long-term potentiation (LTP) at glutamatergic synapses between principal neurons, also has little effect on both stratum oriens-alveus and stratum radiatum interneurons of the CA1 region (MACCAFERRI and MCBAIN 1996).

The effect of high-frequency stimulation, an alternative paradigm for LTP induction in pyramidal neurons, is less clear (Table 6). Consistent LTD was found in stratum radiatum interneurons in the CA1 subfield (both basket and bistratified); the peak amplitude of the EPSP or EPSC was reduced to approximately 50% of the control value (measured 30 min after high-frequency stimulation; MCMAHON and KAUER 1997). No change was reported in stratum-lucidum interneurons of the CA3 region (MACCAFERRI et al. 1998) and in stratum lacunosum-moleculare interneurons in the CA1 subfield (OUARDOUZ and LACAILLE 1995). Plastic changes of variable extent and direction were detected in stratum pyramidale interneurons (putative basket, axo-axonic, or bistratified cells) of the CA1 subfield (COWAN et al. 1998). Finally, consistent LTP was observed in stratum oriens-alveus interneurons; the peak amplitude of the EPSC was potentiated to approximately 130% of the control value (OUARDOUZ and LACAILLE 1995). A main difference between interneuron LTP or LTD and the respective changes in principal neurons is the lack of synapse specificity (MCMAHON and KAUER 1997; COWAN et al. 1998). This may be related to the absence of spines in interneurons, which may enable the spillover of transmitters and retrograde messengers (BLISS and COLLINGRIDGE 1993).

II. Distinct Mechanisms of Interneuron Plasticity

Why are principal neuron–interneuron synapses less plastic, and why is the direction of change of synaptic efficacy more variable? First, it appears that key molecules that mediate synaptic plasticity in principal neurons are absent in interneurons. Both Ca^{2+}/calmodulin-dependent protein kinase II, a kinase involved in LTP induction (BLISS and COLLINGRIDGE 1993), and calcineurin, a Ca^{2+}-dependent serine/threonine phosphatase that suppresses the intermediate phase of LTP in CA1 pyramidal neurons (WINDER et al. 1998), are absent in interneurons (JONES et al. 1994; SÍK et al. 1998). Second, it is possible that Ca^{2+} influx through GluR-B-free AMPARs could trigger new forms of plasticity at principal neuron–interneuron synapses. In the amygdala, Ca^{2+}-permeable AMPARs mediate a form of LTP that is dependent on a rise in postsynaptic Ca^{2+} concentration, independent of NMDAR activation, and

Table 6. Long-term changes in synaptic efficacy at principal neuron–interneuron synapses induced by high-frequency stimulation

Interneuron	Paradigm[a]	Change	Properties	Recording configuration	Recording temperature	Reference
Stratum radiatum interneurons: basket, bistratified (CA1)	HFS: 2 × 100 Hz, 1 s	LTD	No synapse specificity	wc	29–31°C	McMahon and Kauer 1997
Stratum lucidum interneuron (CA3)	HFS	No change/LTD		wc	24°C	Maccaferri et al. 1998
Stratum lacunosum-moleculare interneuron (CA1)	HFS: 100 Hz, 1 s	No change		wc	22–24°C	Ouardouz and Lacaille 1995
Stratum pyramidale interneurons: basket, axo-axonic, bistratified (CA1)	HFS: 4 × 100 Hz, 0.4 s	LTP, LTD	No synapse specificity; blocked by BAPTA	wc	30 ± 1°C	Cowan et al. 1998
Horizontal/vertical stratum oriens-alveus, stratum radiatum interneuron (CA1)	HFS: 4 × 100 Hz, 1 s	Polysynaptic LTP	Monosynaptic LTP (pairing) only in glut stratum radiatum neurons[b]	wc, pp	22–26°C	Maccaferri and McBain 1996
Stratum oriens-alveus interneuron (CA1)	HFS: 100 Hz, 1 s	LTP	blocked by AP-5, BAPTA, MCPG	wc	22–24°C	Ouardouz and Lacaille 1995

BAPTA, 1,2-bis (2-aminophenoxy)-ethane-N,N,N',N'-tetraacetic acid; HFS, high-frequency stimulation; LTD, long-term depression; LTP, long-term potentiation; MCPG, L-α-methyl-4-carboxyphenyl glycine; pp, perforated patch configuration of the patch-clamp technique; wc, whole-cell recording configuration

[a] Excitatory synaptic events were evoked by extracellular stimulation of presynaptic axonal tracts.

[b] These giant stratum radiatum cells have been identified as glutamatergic and probably represent ectopic pyramidal neurons (Gulyás et al. 1997)

inducible by high-frequency stimulation, but not inducible by pairing of low-frequency stimulation with postsynaptic depolarization (MAHANTY and SAH 1998). Whether the Ca^{2+} inflow through postsynaptic AMPARs contributes to plastic changes at glutamatergic synapses in cortical interneurons, however, remains to be investigated.

In conclusion, the available evidence may suggest that glutamatergic principal neuron–interneuron synapses are less plastic than principal neuron–principal neuron synapses. However, even if principal neuron–interneuron synapses lack any form of plasticity, polysynaptic forms of plasticity (also referred to as "passively propagated"; MACCAFERRI and MCBAIN 1995, 1996) could occur. It is difficult to exclude a contribution of polysynaptic forms of LTP or LTD induced by high-frequency stimulation in the majority of studies mentioned above because, often, several presynaptic axons were stimulated (Table 6). Thus, rigorous analysis of interneuron plasticity will require paired recording experiments, probably in combination with stimulation of axonal tracts.

H. Summary and Perspectives

Interneurons receive two fundamentally different forms of glutamate-mediated synaptic input: fast excitation mediated by AMPARs and slow excitation mediated by KARs, NMDARs, and mGluRs. Fast AMPAR-mediated signaling may enable coincidence detection of EPSPs, high temporal precision, and short latency of action potential initiation. Slow interneuron excitation may contribute to threshold adjustments. The cellular and molecular mechanisms of fast AMPAR-mediated excitation are well understood; perisomatic location of synaptic contacts, high synchrony of quantal glutamate release, rapid transmitter clearance, and rapid gating of the postsynaptic AMPARs appear to be the most important factors. In contrast, the mechanisms of the slow interneuron excitation mediated by KARs, NMDARs, and mGluRs in interneurons are less clear. Principal neuron–interneuron synapses are diverse in their short-term dynamics, neuromodulatory properties, and long-term plasticity, and we are just beginning to understand the underlying mechanisms. The final aim should be to obtain a cellular and molecular picture of interneuron excitation and its heterogeneity. This will provide the basis for a better understanding of the dynamic operation of interneurons in complex neuronal circuits.

Acknowledgements. The authors thank Drs. J. Bischofberger and M. Martina for critically reading the manuscript, and J. Stiles and T. Bartol for providing MCell. A. R. thanks Dr. B. Sakmann for support. The studies were supported by the Deutsche Forschungsgemeinschaft, the German Israeli Foundation, and the Human Frontiers Science Program Organization.

References

Acsády L, Kamondi A, Sík A, Freund T, Buzsáki G (1998) GABAergic cells are the major postsynaptic targets of mossy fibers in the rat hippocampus. J Neurosci 18: 3386–3403

Ali AB, Thomson AM (1998) Facilitating pyramid to horizontal oriens-alveus interneurone inputs: dual intracellular recordings in slices of rat hippocampus. J Physiol (Lond) 507:185–199

Ali AB, Deuchars J, Pawelzik H, Thomson AM (1998) CA1 pyramidal to basket and bistratified cell EPSPs: dual intracellular recordings in rat hippocampal slices. J Physiol (Lond) 507:201–217

Angulo MC, Lambolez B, Audinat E, Hestrin S, Rossier J (1997) Subunit composition, kinetic, and permeation properties of AMPA receptors in single neocortical nonpyramidal cells. J Neurosci 17:6685–6696

Bahn S, Volk B, Wisden W (1994) Kainate receptor gene expression in the developing rat brain. J Neurosci 14:5525–5547

Barbour B, Keller BU, Llano I, Marty A (1994) Prolonged presence of glutamate during excitatory synaptic transmission to cerebellar Purkinje cells. Neuron 12:1331–1343

Barbour B, Häusser M (1997) Intersynaptic diffusion of neurotransmitter. Trends in Neurosci 20:377–384

Bartol TM, Land BR, Salpeter EE, Salpeter MM (1991) Monte Carlo simulation of miniature endplate current generation in the vertebrate neuromuscular junction. Biophys J 59:1290–1307

Baude A, Nusser Z, Roberts JDB, Mulvihill E, McIlhinney RAJ, Somogyi P (1993) The metabotropic glutamate receptor (mGluR1α) is concentrated at perisynaptic membrane of neuronal subpopulations as detected by immunogold reaction. Neuron 11:771–787

Baude A, Nusser Z, Molnár E, McIlhinney RAJ, Somogyi P (1995) High-resolution immunogold localization of AMPA type glutamate receptor subunits at synaptic and non-synaptic sites in rat hippocampus. Neuroscience 69:1031–1055

Blasco-Ibáñez JM, Freund TF (1995) Synaptic input of horizontal interneurons in stratum oriens of the hippocampal CA1 subfield: Structural basis of feed-back activation. Eur J Neurosci 7:2170–2180

Bliss TVP, Collingridge GL (1993) A synaptic model of memory: long-term potentiation in the hippocampus. Nature 361:31–39

Bochet P, Audinat E, Lambolez B, Crépel F, Rossier J, Iino M, Tsuzuki K, Ozawa S (1994) Subunit composition at the single-cell level explains functional properties of a glutamate-gated channel. Neuron 12:383–388

Buhl EH, Halasy K, Somogyi P (1994) Diverse sources of hippocampal unitary inhibitory postsynaptic potentials and the number of synaptic release sites. Nature 368:823–828

Buhl EH, Tamás G, Szilágyi T, Stricker C, Paulsen O, Somogyi P (1997) Effect, number and location of synapses made by single pyramidal cells onto aspiny interneurones of cat visual cortex. J Physiol (Lond) 500:689–713

Buzsáki G, Penttonen M, Nádasdy Z, Bragin A (1996) Pattern and inhibition-dependent invasion of pyramidal cell dendrites by fast spikes in the hippocampus *in vivo*. Proc Natl Acad Sci USA 93:9921–9925

Ceranik K, Bender R, Geiger JRP, Monyer H, Jonas P, Frotscher M, Lübke J (1997) A novel type of GABAergic interneuron connecting the input and the output regions of the hippocampus. J Neurosci 17:5380–5394

Cobb SR, Buhl EH, Halasy K, Paulsen O, Somogyi P (1995) Synchronization of neuronal activity in hippocampus by individual GABAergic interneurons. Nature 378:75–78

Colquhoun D, Jonas P, Sakmann B (1992) Action of brief pulses of glutamate on AMPA/kainate receptors in patches from different neurones of rat hippocampal slices. J Physiol (Lond) 458:261–287

Congar P, Khazipov R, Ben-Ari Y (1995) Direct demonstration of functional disconnection by anoxia of inhibitory interneurons from excitatory inputs in rat hippocampus. J Neurophysiol 73:421–426

Cossart R, Esclapez M, Hirsch JC, Bernard C, Ben-Ari Y (1998) GluR5 kainate receptor activation in interneurons increases tonic inhibition of pyramidal cells. Nature Neurosci 1:470–478

Cowan AI, Stricker C, Reece LJ, Redman SJ (1998) Long-term plasticity at excitatory synapses on aspinous interneurons in area CA1 lacks synaptic specificity. J Neurophysiol 79:13–20

Csicsvari J, Hirase H, Czurko A, Buzsáki G (1998) Reliability and state dependence of pyramidal cell-interneuron synapses in the hippocampus: an ensemble approach in the behaving rat. Neuron 21:179–189

Datyner NB, Gage PW (1980) Phasic secretion of acetylcholine at a mammalian neuromuscular junction. J Physiol (Lond) 303:299–314

Davies CH, Starkey SJ, Pozza MF, Collingridge GL (1991) $GABA_B$ autoreceptors regulate the induction of LTP. Nature 349:609–611

Debanne D, Guérineau NC, Gähwiler BH, Thompson SM (1995) Physiology and pharmacology of unitary synaptic connections between pairs of cells in areas CA3 and CA1 of rat hippocampal slice cultures. J Neurophysiol 73:1282–1294

Desai MA, McBain CJ, Kauer JA, Conn PJ (1994) Metabotropic glutamate receptor-induced disinhibition is mediated by reduced transmission at excitatory synapses onto interneurons and inhibitory synapses onto pyramidal cells. Neurosci Lett 181:78–82

Deuchars J, Thomson AM (1996) CA1 pyramid-pyramid connections in rat hippocampus *in vitro*: Dual intracellular recordings with biocytin filling. Neuroscience 74:1009–1018

Diamond JS, Jahr CE (1997) Transporters buffer synaptically released glutamate on a submillisecond time scale. J Neurosci 17:4672–4687

Doherty J, Dingledine R (1997) Regulation of excitatory input to inhibitory interneurons of the dentate gyrus during hypoxia. J Neurophysiol 77:393–404

Doherty J, Dingledine R (1998) Differential regulation of synaptic inputs to dentate hilar border interneurons by metabotropic glutamate receptors. J Neurophysiol 79:2903–2910

Dudel J, Franke C, Hatt H (1990) Rapid activation, desensitization, and resensitization of synaptic channels of crayfish muscle after glutamate pulses. Biophys J 57: 533–545

Faber DS, Young WS, Legendre P, Korn H (1992) Intrinsic quantal variability due to stochastic properties of receptor-transmitter interactions. Science 258:1494–1498

Fortunato C, Debanne D, Scanziani M, Gähwiler BH, Thompson SM (1996) Functional characterization and modulation of feedback inhibitory circuits in area CA3 of rat hippocampal slice cultures. Eur J Neurosci 8:1758–1768

Franke C, Hatt H, Dudel J (1987) Liquid filament switch for ultra-fast exchanges of solutions at excised patches of synaptic membrane of crayfish muscle. Neurosci Lett 77:199–204

Frerking M, Malenka RC, Nicoll RA (1998) Synaptic activation of kainate receptors on hippocampal interneurons. Nature Neurosci 1:479–486

Freund TF, Buzsáki G (1996) Interneurons of the hippocampus. Hippocampus 6: 347–470

Geiger JRP, Melcher T, Koh D-S, Sakmann B, Seeburg PH, Jonas P, Monyer H (1995) Relative abundance of subunit mRNAs determines gating and Ca^{2+} permeability of AMPA receptors in principal neurons and interneurons in rat CNS. Neuron 15:193–204

Geiger JRP, Lübke J, Frotscher M, Jonas P (1996) Rapid kinetics of AMPA receptor-mediated excitatory synaptic transmission onto GABAergic interneurons. Pflügers Arch 431:R19

Geiger JRP, Lübke J, Roth A, Frotscher M, Jonas P (1997) Submillisecond AMPA receptor-mediated signaling at a principal neuron-interneuron synapse. Neuron 18:1009–1023

Gulyás AI, Miles R, Sík A, Tóth K, Tamamaki N, Freund TF (1993) Hippocampal pyramidal cells excite inhibitory neurons through a single release site. Nature 366:683–687

Gulyás AI, Tóth K, McBain CJ, Freund TF (1997) Anatomical and physiological characterization of stratum radiatum giant cells: a unique type of principal cell in the CA1 area of the rat hippocampus. Society for Neuroscience Abstracts 23:188.13

Han Z-S, Buhl EH, Lörinczi Z, Somogyi P (1993) A high degree of spatial selectivity in the axonal and dendritic domains of physiologically identified local-circuit neurons in the dentate gyrus of the rat hippocampus. Eur J Neurosci 5:395–410

Haverkampf K, Lübke J, Jonas P (1997) Single-channel properties of native AMPA receptors depend on the putative subunit composition. Society for Neuroscience Abstracts 23:478.23

He Y, Janssen WGM, Vissavajjhala P, Morrison JH (1998) Synaptic distribution of GluR2 in hippocampal GABAergic interneurons and pyramidal cells: a double-label immunogold analysis. Exp Neurol 150:1–13

Heckmann M, Bufler J, Franke C, Dudel J (1996) Kinetics of homomeric GluR6 glutamate receptor channels. Biophys J 71:1743–1750.

Hestrin S, Nicoll RA, Perkel DJ, Sah P (1990) Analysis of excitatory synaptic action in pyramidal cells using whole-cell recording from rat hippocampal slices. J Physiol (Lond) 422:203–225

Hestrin S (1993) Different glutamate receptor channels mediate fast excitatory synaptic currents in inhibitory and excitatory cortical neurons. Neuron 11:1083–1091

Hines ML, Carnevale NT (1997) The NEURON simulation environment. Neural Comput 9:1179–1209

Ikeda K, Nagasawa M, Mori H, Araki K, Sakimura K, Watanabe M, Inoue Y, Mishina M (1992) Cloning and expression of the ε4 subunit of the NMDA receptor channel. FEBS Letters 313:34–38

Isa T, Itazawa S, Iino M, Tsuzuki K, Ozawa S (1996) Distribution of neurones expressing inwardly rectifying and Ca^{2+}-permeable AMPA receptors in rat hippocampal slices. J Physiol (Lond) 491:719–733

Isaacson JS, Walmsley B (1995) Counting quanta: direct measurements of transmitter release at a central synapse. Neuron 15:875–884

Jack JJB, Redman SJ (1971) The propagation of transient potentials in some linear cable structures. J Physiol (Lond) 215:283–320

Jonas P, Sakmann B (1992) Glutamate receptor channels in isolated patches from CA1 and CA3 pyramidal cells of rat hippocampal slices. J Physiol (Lond) 455:143–171

Jonas P, Major G, Sakmann B (1993) Quantal components of unitary EPSCs at the mossy fibre synapse on CA3 pyramidal cells of rat hippocampus. J Physiol (Lond) 472:615–663

Jonas P, Spruston N (1994) Mechanisms shaping glutamate-mediated excitatory postsynaptic currents in the CNS. Curr Opin Neurobiol 4:366–372

Jonas P, Racca C, Sakmann B, Seeburg PH, Monyer H (1994) Differences in Ca^{2+} permeability of AMPA-type glutamate receptor channels in neocortical neurons caused by differential GluR-B subunit expression. Neuron 12:1281–1289

Jones EG, Huntley GW, Benson DL (1994) Alpha calcium/calmodulin-dependent protein kinase II selectively expressed in a subpopulation of excitatory neurons in monkey sensory-motor cortex: comparison with GAD-67 expression. J Neurosci 14:611–629

Keller BU, Konnerth A, Yaari Y (1991) Patch clamp analysis of excitatory synaptic currents in granule cells of rat hippocampus. J Physiol (Lond) 435:275–293

Khazipov R, Congar P, Ben-Ari Y (1995) Hippocampal CA1 lacunosum-moleculare interneurons: comparison of effects of anoxia on excitatory and inhibitory postsynaptic currents. J Neurophysiol 74:2138–2149
Koh D-S, Geiger JRP, Jonas P, Sakmann B (1995a) Ca^{2+}-permeable AMPA and NMDA receptor channels in basket cells of rat hippocampal dentate gyrus. J Physiol (Lond) 485:383–402
Koh D-S, Burnashev N, Jonas P (1995b) Block of native Ca^{2+}-permeable AMPA receptors in rat brain by intracellular polyamines generates double rectification. J Physiol (Lond) 486:305–312
Kullmann DM, Asztely F (1998) Extrasynaptic glutamate spillover in the hippocampus: evidence and implications. Trends in Neurosci 21:8–14
Livsey CT, Vicini S (1992) Slower spontaneous excitatory postsynaptic currents in spiny versus aspiny hilar neurons. Neuron 8:745–755
Livsey CT, Costa E, Vicini S (1993) Glutamate-activated currents in outside-out patches from spiny versus aspiny hilar neurons of rat hippocampal slices. J Neurosci 13:5324–5333
Longsworth LG (1953) Diffusion measurements, at 25°, of aqueous solutions of amino acids, peptides and sugars. J Am Chem Soc 75:5705–5709
Maccaferri G, McBain CJ (1995) Passive propagation of LTD to stratum oriens-alveus inhibitory neurons modulates the temporoammonic input to the hippocampal CA1 region. Neuron 15:137–145
Maccaferri G, McBain CJ (1996) Long-term potentiation in distinct subtypes of hippocampal nonpyramidal neurons. J Neurosci 16:5334–5343
Maccaferri G, Tóth K, McBain CJ (1998) Target-specific expression of presynaptic mossy fiber plasticity. Science 279:1368–1370
Mahanty NK, Sah P (1998) Calcium-permeable AMPA receptors mediate long-term potentiation in interneurons in the amygdala. Nature 394:683–687
Major G, Larkman AU, Jonas P, Sakmann B, Jack JJB (1994) Detailed passive cable models of whole-cell recorded CA3 pyramidal neurons in rat hippocampal slices. J Neurosci 14:4613–4638
Malinow R, Tsien RW (1990) Presynaptic enhancement shown by whole-cell recordings of long-term potentiation in hippocampal slices. Nature 346:177–180
Markram H, Lübke J, Frotscher M, Roth A, Sakmann B (1997) Physiology and anatomy of synaptic connections between thick tufted pyramidal neurons in the developing rat neocortex. J Physiol (Lond) 500:409–440
Markram H, Wang Y, Tsodyks M (1998) Differential signaling via the same axon of neocortical pyramidal neurons. Proc Natl Acad Sci USA 95:5323–5328
McBain CJ, Dingledine R (1993) Heterogeneity of synaptic glutamate receptors on CA3 stratum radiatum interneurones of rat hippocampus. J Physiol (Lond) 462: 373–392
McBain CJ, DiChiara TJ, Kauer JA (1994) Activation of metabotropic glutamate receptors differentially affects two classes of hippocampal interneurons and potentiates excitatory synaptic transmission. J Neurosci 14:4433–4445
McMahon LL, Kauer JA (1997) Hippocampal interneurons express a novel form of synaptic plasticity. Neuron 18:295–305
McNaughton BL, Barnes CA (1990) From cooperative synaptic enhancement to associative memory: bridging the abyss. Seminars in Neurosci 2:403–416
Miles R, Wong RKS (1986) Excitatory synaptic interactions between CA3 neurons in the guinea-pig hippocampus. J Physiol (Lond) 373:397–418
Miles R (1990) Synaptic excitation of inhibitory cells by single CA3 hippocampal pyramidal cells of the guinea-pig *in vitro*. J Physiol (Lond) 428:61–77
Miles R, Poncer J-C (1993) Metabotropic glutamate receptors mediate a post-tetanic excitation of guinea-pig hippocampal inhibitory neurones. J Physiol (Lond) 463: 461–473
Miles R, Tóth K, Gulyás AI, Hájos N, Freund TF (1996) Differences between somatic and dendritic inhibition in the hippocampus. Neuron 16:815–823

Monyer H, Burnashev N, Laurie DJ, Sakmann B, Seeburg PH (1994) Developmental and regional expression in the rat brain and functional properties of four NMDA receptors. Neuron 12:529–540

Nusser Z, Lujan R, Laube G, Roberts JDB, Molnar E, Somogyi P (1998) Cell type and pathway dependence of synaptic AMPA receptor number and variability in the hippocampus. Neuron 21:545–559

Ohishi H, Shigemoto R, Nakanishi S, Mizuno N (1993a) Distribution of the messenger RNA for a metabotropic glutamate receptor, mGluR2, in the central nervous system of the rat. Neuroscience 53:1009–1018

Ohishi H, Shigemoto R, Nakanishi S, Mizuno N (1993b) Distribution of the mRNA for a metabotropic glutamate receptor (mGluR3) in the rat brain: an in situ hybridization study. J Comp Neurol 335:252–266

Ohishi H, Akazawa C, Shigemoto R, Nakanishi S, Mizuno N (1995) Distributions of the mRNAs for L-2-amino-4-phosphonobutyrate-sensitive metabotropic glutamate receptors, mGluR4 and mGluR7, in the rat brain. J Comp Neurol 360:555–570

Otis TS, Jahr CE (1998) Anion currents and predicted glutamate flux through a neuronal glutamate transporter. J Neurosci 18:7099–7110

Ouardouz M, Lacaille J-C (1995) Mechanisms of selective long-term potentiation of excitatory synapses in stratum oriens/alveus interneurons of rat hippocampal slices. J Neurophysiol 73:810–819

Perouansky M, Yaari Y (1993) Kinetic properties of NMDA receptor-mediated synaptic currents in rat hippocampal pyramidal cells *versus* interneurons. J Physiol (Lond) 465:223–244

Pin J-P, Duvoisin R (1995) The metabotropic glutamate receptors: structure and functions. Neuropharmacology 34:1–26

Rall W (1967) Distinguishing theoretical synaptic potentials computed for different soma-dendritic distributions of synaptic input. J Neurophysiol 30:1138–1168

Reyes A, Lujan R, Rozov A, Burnashev N, Somogyi P, Sakmann B (1998) Target-cell-specific facilitation and depression in neocortical circuits. Nature Neurosci 1: 279–285

Riveros N, Fiedler J, Lagos N, Muñoz C, Orrego F (1986) Glutamate in rat brain cortex synaptic vesicles: influence of the vesicle isolation procedure. Brain Res 386:405–408

Romano C, Sesma MA, McDonald CT, O'Malley K, van den Pol AN, Olney JW (1995) Distribution of metabotropic glutamate receptor mGluR5 immunoreactivity in rat brain. J Comp Neurol 355:455–469

Rozov A, Zilberter Y, Wollmuth LP, Burnashev N (1998) Facilitation of currents through rat Ca^{2+}-permeable AMPA receptor channels by activity-dependent relief from polyamine block. J Physiol (Lond) 511:361–377

Rusakov DA, Kullmann DM (1998a) Geometric and viscous components of the tortuosity of the extracellular space in the brain. Proc Natl Acad Sci USA 95: 8975–8980

Rusakov DA, Kullmann DM (1998b) Extrasynaptic glutamate diffusion in the hippocampus: ultrastructural constraints, uptake, and receptor activation. J Neurosci 18:3158–3170

Scharfman HE, Kunkel DD, Schwartzkroin PA (1990) Synaptic connections of dentate granule cells and hilar neurons: results of paired intracellular recordings and intracellular horseradish peroxidase injections. Neuroscience 37:693–707

Shigemoto R, Kulik A, Roberts JDB, Ohishi H, Nusser Z, Kaneko T, Somogyi P (1996) Target-cell-specific concentration of a metabotropic glutamate receptor in the presynaptic active zone. Nature 381:523–525

Shigemoto R, Kinoshita A, Wada E, Nomura S, Ohishi H, Takada M, Flor PJ, Neki A, Abe T, Nakanishi S, Mizuno N (1997) Differential presynaptic localization of metabotropic glutamate receptor subtypes in the rat hippocampus. J Neurosci 17:7503–7522

Sik A, Penttonen M, Ylinen A, Buzsáki G (1995) Hippocampal CA1 interneurons: An *in vivo* intracellular labeling study. J Neurosci 15:6651–6665

Sík A, Hájos N, Gulácsi A, Mody I, Freund TF (1998) The absence of a major Ca^{2+} signaling pathway in GABAergic neurons of the hippocampus. Proc Natl Acad Sci USA 95:3245–3250

Sloviter RS (1991) Permanently altered hippocampal structure, excitability, and inhibition after experimental status epilepticus in the rat: the "dormant basket cell" hypothesis and its possible relevance to temporal lobe epilepsy. Hippocampus 1:41–66

Sorra KE, Harris KM (1993) Occurrence and three-dimensional structure of multiple synapses between individual radiatum axons and their target pyramidal cells in hippocampal area CA1. J Neurosci 13:3736–3748

Spruston N, Johnston D (1992) Perforated patch-clamp analysis of the passive membrane properties of three classes of hippocampal neurons. J Neurophysiol 67: 508–529

Spruston N, Jonas P, Sakmann B (1995) Dendritic glutamate receptor channels in rat hippocampal CA3 and CA1 pyramidal neurons. J Physiol (Lond) 482:325–352

Standaert DG, Landwehrmeyer GB, Kerner JA, Penney JB, Young AB (1996) Expression of NMDAR2D glutamate receptor subunit mRNA in neurochemically identified interneurons in the rat neostriatum, neocortex and hippocampus. Mol Brain Res 42:89–102

Stern P, Edwards FA, Sakmann B (1992) Fast and slow components of unitary EPSCs on stellate cells elicited by focal stimulation in slices of rat visual cortex. J Physiol (Lond) 449:247–278

Stiles JR, Bartol TM, Salpeter EE, Salpeter MM (1998) Monte Carlo simulation of neurotransmitter release using MCell, a general simulator of cellular physiological processes. In: Bower J (ed) Computational Neuroscience. Plenum press, New York, pp. 279–284

Stuart G, Sakmann B (1995) Amplification of EPSPs by axosomatic sodium channels in neocortical pyramidal neurons. Neuron 15:1065–1076

Stuart G, Spruston N (1998) Determinants of voltage attenuation in neocortical pyramidal neuron dendrites. J Neurosci 18:3501–3510

Thomson AM (1997) Activity-dependent properties of synaptic transmission at two classes of connections made by rat neocortical pyramidal axons *in vitro*. J Physiol (Lond) 502:131–147

Tong G, Jahr CE (1994) Block of glutamate transporters potentiates postsynaptic excitation. Neuron 13:1195–1203

Traub RD, Miles R (1995) Pyramidal cell-to-inhibitory cell spike transduction explicable by active dendritic conductances in inhibitory cell. J Comput Neurosci 2:291–298

Trussell LO (1997) Cellular mechanisms for preservation of timing in central auditory pathways. Curr Opin Neurobiol 7:487–492

Vicini S, Wang JF, Li JH, Zhu WJ, Wang YH, Luo JH, Wolfe BB, Grayson DR (1998) Functional and pharmacological differences between recombinant *N*-methyl-D-aspartate receptors. J Neurophysiol 79:555–566

Vida I, Halasy K, Szinyei C, Somogyi P, Buhl EH (1998) Unitary IPSPs evoked by interneurons at the stratum radiatum–stratum lacunosum-moleculare border in the CA1 area of the rat hippocampus *in vitro*. J Physiol (Lond) 506:755–773

Wahl LM, Pouzat C, Stratford KJ (1996) Monte Carlo simulation of fast excitatory synaptic transmission at a hippocampal synapse. J Neurophysiol 75:597–608

Whittington MA, Traub RD, Jefferys JGR (1995) Synchronized oscillations in interneuron networks driven by metabotropic glutamate receptor activation. Nature 373:612–615

Winder DG, Mansuy IM, Osman M, Moallem TM, Kandel ER (1998) Genetic and pharmacological evidence for a novel, intermediate phase of long-term potentiation suppressed by calcineurin. Cell 92:25–37

Wyllie DJA, Béhé P, Colquhoun D (1998) Single-channel activations and concentration jumps: comparison of recombinant NR1a/NR2A and NR1a/NR2D NMDA receptors. J Physiol (Lond) 510:1–18

Zhang S, Trussell LO (1994) A characterization of excitatory postsynaptic potentials in the avian nucleus magnocellularis. J Neurophysiol 72:705–718

CHAPTER 12

Physiology of Glutamatergic Transmission at Calyceal and Endbulb Synapses of the Central Auditory Pathway

L.O. TRUSSELL

A. Introduction

The structure and function of fast excitatory synapses in the central nervous system often represent compromises between several physiological imperatives. To achieve a short delay between the pre- and postsynaptic signal, the transmitter glutamate directly gates ion channels, leading to rapid depolarization and initiation of action potentials. However, the need to integrate signals from thousands of different inputs and to perform miniature computations on the signals from small groups of inputs requires that synapses be distributed over a complex dendritic tree, which necessarily slows the onset of excitation. In the auditory system, the need for speed outweighs all others, and it is there that giant, somatic synapses are found. In recent years, these *calyceal* and *endbulb* synapses have been rediscovered by biophysicists because of technical advantages the synapses offer for the study of basic features of glutamatergic transmission in the central nervous system, such as better voltage control for measuring the synaptic current and electrical and physical access to the presynaptic terminal. In the course of these studies, we have learned that the demands of acoustic processing have not only selected for the unique morphology of the giant synapses but also for biophysical specializations which compromise (or make more interesting) their service as "model synapses". In this chapter, we will look at the control of synaptic strength at calyceal/endbulb synapses, with particular, but not exclusive, emphasis on postsynaptic issues.

B. Giant Glutamatergic Synapses are Key Elements of Auditory Processing

Large endbulbs of Held are terminations of auditory nerve fibers on spherical bushy cells (SBCs) of the mammalian anteroventral cochlear nucleus (AVCN) and on neurons of the nucleus magnocellularis (nMAG) of birds and reptiles (CANT 1992; CANT and MOREST 1979; CARR and CODE 1998; LENN and REESE 1966; PARKS 1981). The SBC relays signals to the medial-superior olive, a pathway homologous to the nMAG-to-nucleus-laminaris path of birds and reptiles (CARR and CODE 1998; OERTEL 1999; RHODE and GREENBERG 1992). Thus, SBC and nMAG neurons probably have a common function. Both neurons fire in a pattern termed "primary-like", mimicking the poststimulus time histograms of the primary afferent input (CARR and CODE 1998; OERTEL 1999; RHODE and GREENBERG 1992). It is generally believed that the matching of the input and output in terms of these period histogram reflects the faithful suprathreshold transmission of the occurrence and timing of presynaptic signals, and that this relay behavior is essential to the role of these cells in the localization of low-frequency sound. Later, we will discuss the degree to which such precision is maintained in the context of short-term synaptic plasticity.

Principal neurons of the medial nucleus of the trapezoid body (MNTB) receive glutamatergic innervation from contralateral globular bushy cells of the AVCN (OERTEL 1999), through the well-known calyces of Held. Unlike SBCs and nMAG cells, MNTB neurons are sign inverting, providing feedforward, glycinergic inhibitory innervation to the ipsilateral-lateral and medial-superior olives (GROTHE and SANES 1994; WU and KELLY 1995; WU and KELLY 1994). Because of their connectivity and structure, MNTB neurons provide a rapid relay of the bushy cell input so that the arrival of excitation from the cochlear nucleus to the olive can be matched with inhibition driven by acoustic signals from the contralateral ear (JORIS and YIN 1998; SMITH et al. 1998). The excitation in the olive is, therefore, monosynaptic and the inhibition disynaptic with respect to the cochlear nucleus. The large size of MNTB synapses, their somatic termination, and the large size of their axons would appear to be features needed to minimize the extra synaptic delay and conduction times in the disynaptic circuit. JORIS and YIN (1998) determined that the disynaptic inhibitory signal arrives at the lateral superior olive only 200 μs after the mono-synaptic excitatory signal in the cat. In contrast to SBC/nMAG cells, globular bushy cells and MNTB cells respond mainly to high-frequency sound (i.e., they have a high characteristic frequency or CF), firing reliably and with little jitter early, but not late, in an acoustic stimulus (GUINAN and LI 1990; JORIS and YIN 1998; OERTEL 1999; RHODE and GREENBERG 1992; SMITH et al. 1998). However, for those cells that do have a low CF, the responses show good phase locking, consistent with the fact that they have synaptic features similar to those in SBC/nMAG cells (JORIS and YIN 1998; SMITH et al. 1998).

C. Form and Function in Giant Synapses

I. Morphology of the Mature Synapse

Giant auditory synapses vary considerably in morphology. At maturity, the auditory nerve terminals on SBC/nMAG cells range in shape from dense clusters of large boutons to thick fingers studded with short branchlets (CANT and MOREST 1979; JHAVERI and MOREST 1982a; PARKS 1981; RYUGO and SENTO 1991; RYUGO et al. 1996). Figure 1A illustrates this complexity in a horseradish-peroxidase-filled endbulb on a cat SBC. Each postsynaptic cell is innervated by

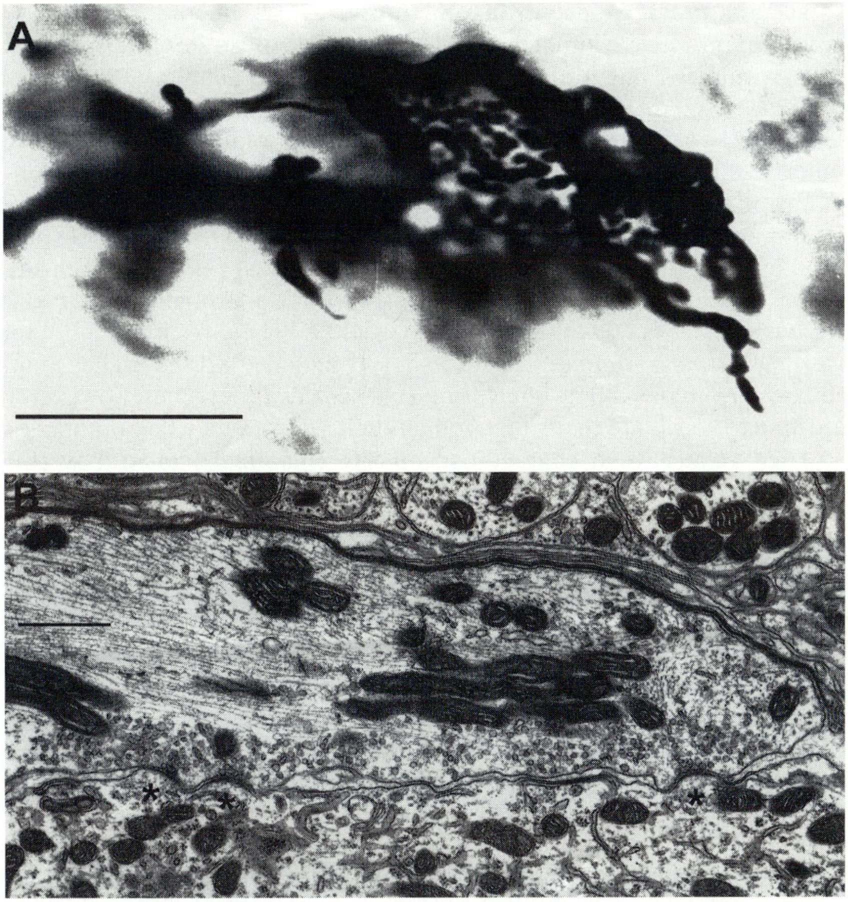

Fig. 1A,B. Endbulb of Held on a cat spherical bushy cell. **A** Horseradish-peroxidase-labeled endbulb shows a complex structure, with thick major branches giving rise to strings of smaller, closely spaced boutons. *Scale bar* 10 μm. **B** Electron micrograph of a single endbulb finger. This profile shows three release sites (*). *Scale bar* 1 μm. Adapted from a micrograph originally published in CANT (1992)

1–3 auditory-nerve axons (LIBERMAN 1991; RYUGO and SENTO 1991). By contrast, the globular bushy cells in cats receive a large number of auditory axonal terminations (20–40/cell) which form smaller, "modified" endbulbs (LIBERMAN 1991). The calyx of Held in the MNTB is an extreme case of concentrated glutamatergic innervation, as each cell body receives only a single large input that occupies 25–50% of the cell-body surface (CASEY and FELDMAN 1988; LENN and REESE 1966; SMITH et al. 1998). The calyceal synapse tends to have a simpler morphology than the endbulb and forms a claw-like structure around the postsynaptic cell (FORSYTHE 1994; KUWABARA et al. 1991; SMITH et al. 1991; SPIROU et al. 1990).

Each mature terminal probably forms several hundred release sites, although quantitative estimates are rare. Figure 1B illustrates multiple release sites (*) within one finger of a cat SBC endbulb. RYUGO et al. (1996) estimated that auditory-nerve terminals on cat SBCs contain 400–1700 release sites, based upon the number of sites in electron-micrographic profiles and the total terminal size measured in the light microscope. In the smaller nMAG cells of chicks, peak quantal content (an indicator of the minimal number of release sites) has been estimated to be 113 (ZHANG and TRUSSELL 1994a). Given dispersion of release over time, as well as a probability of release at each site of less than one, the number of available release sites is probably closer to 200. In rat bushy cells, ISAACSON and WALMSLEY (1995) show response amplitudes suggesting a similar number of active sites. It may be that the higher number counted morphologically in the cat reflects species differences; however, it is interesting to speculate that some or many release sites may be nonfunctional. Bushy-like cells have a characteristic bushy (mammals) or stubby (birds/reptiles) dendrite that receives little innervation (CANT and MOREST 1979; ISAACSON and WALMSLEY 1995; JHAVERI and MOREST 1982a; PARKS 1981). RYUGO and SENTO (1991) found, however, that the distal dendrites of SBCs receive input from fine branches of endbulb terminals on adjacent SBCs. While such innervation might contribute to the output of the SBC, it is clear the majority of the excitatory drive originates at the cell body. Like endbulbs, MNTB calyces feature many closely spaced, synaptic release sites (CASEY and FELDMAN 1988; KIL et al. 1995; LENN and REESE 1966) but, again, detailed counts are not available. BORST and SAKMANN (1996) estimated a quantal content of 210, similar to that seen in the cochlear nucleus.

Despite the overall similarity among bushy-like cells, structural differences may reflect subtle differences in their functions. Auditory-nerve fibers vary in their rates of firing in the absence of sound, which presumably reflects the resting physiology of the hair cells and the densities of release sites on each cochlear ganglion cell dendrite (GEISLER 1998). Afferent fibers with different spontaneous firing rates form endbulbs of characteristic morphology (RYUGO and SENTO 1991; RYUGO et al. 1996). High-rate fibers (>18 Hz) form larger synapses, with more vesicle release sites than low-rate fibers (<18 Hz). Curiously, the synaptic sites formed by these high-rate fibers are also more rounded (RYUGO et al. 1996), perhaps due to differential expression of cytoskeletal elements in the postsynaptic densities in SBCs (GULLEY and REESE 1981). In a study of the barn owl, KÖPPL (1994) examined the morphology of nMAG synapses across the tonotopic axis, finding that low-CF cells were innervated by more highly branched axons and terminals, while high-CF cells had more fat endbulb terminations from unbranched axons. Although physiological specializations for maintaining tonotopic representation are generally considered the domain of the periphery (FETTIPLACE and FUCHS 1999), the observations of Köppl and of Ryugo and colleagues raise the intriguing possibility of frequency-dependent synaptic specializations in the cochlear nucleus.

II. Development of the Synapse

Giant synapses in AVCN, nMAG and MNTB at ages most ideal for in vitro biophysical studies (due to the reduced myelination) tend to be web like, and less fenestrated than adult synapses, radiating fine filopodial extensions (JHAVERI and MOREST 1982a, 1982b; KANDLER and FRIAUF 1995; KIL et al. 1995; MOREST 1968; RYUGO and FEKETE 1982). In fact, these features are themselves vestiges of the earlier morphological transformation endbulbs undergo during the approach to maturity, in which young postsynaptic cells (just after birth in rats or on day 12–13 in chick embryos) extend multiple dendrites that become entangled in the fine processes of immature auditory-nerve fibers (JHAVERI and MOREST 1982a, 1982b; MOREST 1969; NEISES et al. 1982). Over the next 5–7 days (to postnatal day 16 in rats and to embryonic day 16 in chicks) the dendrites are withdrawn, pulling the presynaptic structures onto the cell body, where they coalesce into giant somatic synapses. Thus, the giant synapse develops from more conventional, distributed synapses. PARKS and colleagues have shown that the characteristic morphology of the synapse is a result of specific pre- and postsynaptic interactions, as innervation of nMAG neurons by aberrant contralateral nMAG axons results in somatic synapses having very different morphologies than the auditory-nerve terminals (JACKSON and PARKS 1988; PARKS et al. 1990).

Several recent studies have examined physiological events that occur during calyx/endbulb formation. Presynaptically, calcium currents of the calyx of Held begin as a mixture of the N, P/Q, and R subtypes but, after postnatal day 10, they are predominantly of the P/Q type (IWASAKI and TAKAHASHI 1998; WU et al. 1998). In parallel, there is a consolidation of the release process, such that quanta are released over a more restricted time course (CHUHMA and OHMORI 1998). Similarly, in SBCs, endbulbs increase their quantal content during this period (BELLINGHAM et al. 1998). Postsynaptically, in the SBC, the amplitude of the NMDA component of EPSCs falls dramatically, while there is an increase in the amplitude of the α-amino-3-hydroxy-5-methylisoxazole-4-propionic-acid (AMPA)-mediated quantal conductance (BELLINGHAM et al. 1998); this change in quantal size was not seen in MNTB (CHUHMA and OHMORI 1998). In nMAG, AMPA receptors also change in their kinetic properties following innervation such that the time course of desensitization accelerates over time, reaching the mature level just before hatching (LAWRENCE and TRUSSELL 1998).

D. Giant EPSPs

Excitatory postsynaptic potentials (EPSPs) in giant auditory synapses tend to be quite large and decay within 1–1.5 ms (OERTEL 1983; WU and KELLY 1991). Overall, these features promote reliability of transmission, both in the sense that an incoming signal will lead quickly to a suprathreshold output, and in that there is little history dependence in the timing or probability of firing, as might occur with temporal summation or refractory periods. Under voltage clamp, excitatory postsynaptic currents (EPSCs) in these somatic synapses are quite large relative to those of other glutamatergic synapses. Assuming a reversal potential of zero, synaptic conductances of 66 (BORST and SAKMANN 1996), 45 (ISAACSON and WALMSLEY 1995), and 211 nS (ZHANG and TRUSSELL 1994a) have been reported, which produce currents many times those needed to drive neurons to threshold. Clearly, the large size of terminals and the abundant release sites are major determinants of EPSC amplitude. However, the fact that so many of the release sites are active must also reflect an intrinsically high release probability at each site. While no true single-site estimates of probability are yet available at these terminals under normal ionic conditions, estimates of variance (ZHANG and TRUSSELL 1994a) and the tendency of the synapses to show paired-pulse depression (see below) are indicators that, for many of the release sites, arrival of a presynaptic action potential is very likely to result in exocytosis, at least for low-frequency stimuli.

An additional factor that is not often considered in explaining the large auditory EPSPs is the amplitude of the quantal response. Measurement of quantal size is fraught with the perils of filtering introduced by dendrites and by the patch-electrode access resistance. Nevertheless, it is informative to contrast measurements made across preparations. Table 1 shows glutamatergic miniature EPSCs (mEPSCs) or evoked single-quantal currents in different preparations recorded under conditions in which AMPA receptors are likely to generate most or all of the quantal current. Included in the table are examples that have some hope of reliability. Those with mEPSC rise times less than 0.5 ms are likely to be electrotonically close to the soma. Some were chemically evoked by applying depolarizing or hypertonic solutions near the cell body, while others likely arose from more distant dendritic sites but were "corrected" to estimate the unfiltered amplitude. Note the variety of temperatures. The data suggest that the quantal sizes of currents mediated by AMPA receptors vary widely among brain regions, and that the endbulb/calyx synapses are among those having the larger quanta. Several mechanisms could generate a large quantal size. Single AMPA receptor/channel conductance in nMAG has been estimated at about 20 pS (RAMAN and TRUSSELL 1995), roughly twice that measured in cortical pyramidal cells (HESTRIN 1992); assuming equivalent numbers of receptors, the quanta would be expected to be larger in the auditory cells. However, the auditory cells may have more receptors than some other cell types, since cerebellar granule cells have a similar single-channel conductance but much smaller mEPSCs (SILVER et al. 1996; TRAYNELIS et al. 1993). Most interestingly, there may be presynaptic factors that contribute to the larger quantal size. Several reports suggest that the diameter of synaptic vesicles at endbulb or auditory-nerve synapses are larger than at bouton synapses by about 20% (IBATA and PAPPAS 1976; KANE 1974; LENN and REESE 1966; RYUGO 1992). While comparisons of vesicle size are complicated by the variable effects of fixatives, such differences in size seem to be observed repeatedly in the cochlear nucleus. A 20% difference in diameter corresponds

Table 1. Amplitude of quantal currents mediated by α-amino-3-hydroxy-5-methylisoxazole-4-propionic acid receptors. Values have been converted to conductance, assuming a reversal potential of zero and a linear conductance–voltage relation[a]

Preparation	Quantal size (nS)	Temperature (°C)
nMAG[a]	2.0	31
nMAG[b]	0.98 ± 0.31	24
Bushy cell[c]	0.9–1	23
Bushy cell[d]	1.0 ± 0.66	34
MNTB[e]	0.54 ± 0.06	23
Dentate-gyrus basket cell[f]	2.29 ± 0.44	34
Hippocampal culture[g]	0.46 ± 0.19	23
Hippocampal culture[h]	0.8	23
Hippocampal culture[i]	0.44 ± 0.14	23
Hippocampal culture[j]	0.2–0.6	24
CA3 pyramidal neuron[k]	0.12 (best case); 0.55 worst case	23
Retinal ganglion cell[l]	0.89 ± 0.30	23
Purkinje neuron[m]	0.77	23
Spinal-cord culture[n]	0.69	23
Cerebellar granule cell[o]	0.14 ± 0.02	23
Dentate-gyrus granule[p]	0.1	23
Nucleus tractus solitarius[q]	0.34 ± 0.18	23

nMAG, nucleus magnocellularis; MNTB, medial nucleus of the trapezoid body

[a] Miniature EPSCs (mEPSCs; Zhang and Trussell 1994a)

[b] Otis and Trussell, unpublished analysis of mEPSCs from cells included in Otis et al. (1996a)

[c] Bellingham et al. (1998) reported 0.72 ± 0.09 nS for mEPSCs, but indicated in a personal communication that, at maturity, the value is higher, as indicated

[d] Unpublished measurements of mEPSCs from the study of Gardner et al. (1998)

[e] Borst and Sakmann (1996); from the study of Bellingham et al. (1998), these values are about 60% of the mature level, assuming a similar developmental time course in MNTB and spherical bushy cells. Sect. C.II

[f] Geiger et al. (1997); evoked quantal current in low-Ca^{2+} solutions

[g] Diamond and Jahr (1997); chemically evoked mEPSCs at proximal locations

[h] Bekkers and Stevens (1996); "corrected" with consideration of the cable properties of the neuron

[i] Abdul-Ghani et al. (1996); Sr^{++}-induced mEPSCs with fast rise times

[j] Liu and Tsien (1995); chemically evoked mEPSCs at proximal locations

[k] Jonas et al. (1993); mEPSCs with "best"- and "worst"-case estimations of the magnitudes of effects of cable filtering on quantal size

[l] Taschenberger et al. (1995); mEPSCs in electrically compact neurons

[m] Silver et al. (1998); Sr^{++}-induced mEPSCs with a worst-case estimate of cable filtering

[n] Ulrich and Luscher (1993); mEPSCs with corrections for cable properties

[o] Silver et al. (1996); mEPSCs in electrically compact neurons

[p] Bekkers (1998); Sr^{++}-induced mEPSCs in electrically compact cells

[q] Titz and Keller (1997); mEPSCs in electrically compact neurons

to a 72% larger volume, a difference that is potentially significant for quantal size, assuming that the contents of a vesicle do not saturate the receptors at a given release site. Perhaps, however, a larger glutamate concentration is needed to compensate for the faster kinetics and lower affinity of the glutamate receptors at the giant synapses (see below).

E. Control of EPSP Duration

I. Receptors

The short EPSPs of auditory cells are a consequence of the duration of the synaptic current and of the membrane time constant. The intrinsic membrane properties that generate the fast membrane time constant in these auditory neurons, including the high expression levels of several potassium-channel subtypes, have been previously reviewed (TRUSSELL 1999; TRUSSELL 1997). The duration of the EPSC itself is a reflection of the structure and biophysics of the postsynaptic transmitter receptors and of the rate of clearance of transmitters from the cleft, which will be discussed below.

Transmission from the auditory nerve to the cochlear nucleus, or from the globular bushy cell to the MNTB, is mediated by glutamate and the AMPA subtype of glutamate receptors. Glutamate is localized to the endbulbs and calyces of Held (GRANDES and STREIT 1989; HACKNEY et al. 1996). Moreover, application of antagonists of AMPA receptors blocks transmission in both the cochlear nucleus and MNTB (FORSYTHE and BARNES-DAVIES 1993; ISAACSON and WALMSLEY 1995; WU and KELLY 1992; ZHANG and TRUSSELL 1994a; ZHOU and PARKS 1992). However, these features alone cannot account for the speed of synaptic currents in the auditory system, which are briefer than at any other glutamatergic synapse, with decay times of quantal currents at room temperature of less than $500\,\mu s$ (ISAACSON and WALMSLEY 1995; OTIS et al. 1996a). Direct comparison of the kinetics of the response to glutamate in "auditory AMPA receptors" in nMAG with those in other brain pathways revealed that the speed of the EPSC may be attributed to the intrinsic response rate of the receptor to the transmitter (RAMAN and TRUSSELL 1992; RAMAN et al. 1994). For example, the decay of the glutamate response after application of glutamate for 1 ms ("deactivation") matches precisely the decay time of miniature synaptic currents (OTIS et al. 1996a). Desensitization time constants measured during prolonged application of glutamate in nMAG, nucleus angularis, nucleus laminaris and cochlear ganglion of the chick and MNTB of the rat are all uniformly rapid (GEIGER et al. 1995; RAMAN et al. 1994).

What structural hallmarks underlie this kinetic phenotype of the AMPA receptor in the auditory system? As discussed elsewhere in this volume, native AMPA receptors are believed to be multimeric complexes whose varied subunit composition reflects variation in gene expression, alternative splicing, and RNA editing. Expression studies suggest that receptors composed of the flop slice variant, especially from the GluR-D gene, have kinetics of desensitization and deactivation similar to those of nMAG and MNTB (MOSBACHER et al. 1994). Accordingly, single-cell RT-PCR (i.e., amplification, by reverse transcriptase polymerase chain reaction, of mRNA from single neurons) of MNTB neurons revealed a high level of GluR-D flop (GEIGER et al. 1995). Immunocytochemical labeling and *in situ* hybridization also confirm the presence of high levels of GluR-D in the cochlear nuclei of rat and owl (HUNTER et al. 1993; LEVIN et al. 1997; PETRALIA and WENTHOLD 1992; SATO et al. 1993; WANG et al. 1998). However, rapid-response kinetics have also been associated with reduced levels of GluR-B flip subunits (ANGULO et al. 1997; GEIGER et al. 1995). Given that the presence of GluR-B reduces calcium flux through AMPA receptors/channels, and that the channels in nMAG and MNTB have a high calcium permeability, the receptor kinetics in auditory synapses may also be influenced by reduced GluR-B expression.

II. Release and Clearance of Glutamate

In addition to the intrinsic kinetics of the receptors, other factors contribute to controlling the time course of the EPSC. Upon arrival of an action potential at the giant axon terminal, the several hundred synaptic sites do not release simultaneously but, rather, over a period of hundreds of microseconds. ISAACSON and WALMSLEY (1995) and BORST and SAKMANN (1996) have shown that the EPSC duration in some giant auditory synapses may be described almost entirely by the convolution of the release time course and the duration of the single-quantal current. This important result indicates that, as is the case at the neuromuscular junction (BARRETT and STEVENS 1972), the kinetics of the transmitter-release machinery is crucial to the temporal aspects of signaling in the brain. As the measurements were made under conditions of low quantal content (low bath calcium or addition of Cd^{2+}), the argument requires that the duration of the EPSC is independent of the quantal content. Otherwise, the measured release profile could not be assumed to be the same as under normal ionic conditions.

While this condition is clearly met in the experiments described above, some large synapses show different behaviors, raising the possibility that additional factors besides release time course and quantal duration contribute to the shaping of the EPSC. For example, in nMAG, reduction of release by baclofen, low Ca^{2+}, or Cd^{2+} reduces the amplitude but also narrows the EPSC (OTIS and TRUSSELL 1996; TRUSSELL et al. 1998; TRUSSELL et al. 1993). Similar observations have been made at other, nonauditory synapses (SILVER et al. 1996). These results suggest a relationship between the amount of transmitter released and the lifetime of its action. Such would not be the case if each release site/postsynaptic receptor complex acted as an independent unit of transmission. Rather, transmitters from adjacent sites may exit the cleft by common pathways; thus, when the number of active sites is increased, clearance would be impeded and the period of transmitter action would be prolonged (OTIS et al. 1996a; ROSSI et al. 1995; SILVER et al. 1996; TRUSSELL et al. 1993). The common-pathway idea is supported by consideration of the morphology of calyceal/endbulb synapses, in which multiple synaptic release sites line a common synaptic cleft. OTIS et al. (1996a) explored the consequences of adjacent, active release sites in a simple, analytical solution to the diffusion equation for a point source and a cleft-like diffusion path, assuming zero uptake. Summation of diffusing molecules from regularly spaced sites over time and space predicts a transmitter profile that initially falls rapidly at each site but, after about 0.2 ms (for a 0.71-μm site spacing), falls much more slowly, declining to submicromolar levels only after many tens of milliseconds. The consequences of such biphasic diffusion were estimated in Otis et al. (1996a) by creating a chemical-kinetic model of the AMPA receptor with the predicted diffusion profile, showing that the time course of activation of each AMPA receptor is longer with multiquantal release, even at early times in the synaptic event, because receptors have a greater opportunity for rebinding transmitter. Another prediction of the model was a slow tail current after the initial rapid fall of the simulated EPSC. This tail was a result of the delayed clearance of transmitter.

Confirming these predicted consequences of shared cleft space during a real EPSC is, therefore, critical to the notion that synaptic morphology can dictate glutamate clearance rates and the EPSC time course. The fact that enhancing release widens the EPSC is partial support. However, Otis et al (1996a) also found a small and very slow tail in the nMAG EPSC, whose pharmacology and voltage sensitivity indicated that it was due to AMPA receptors. Comparison of the decay kinetics of glutamate responses in patches with that of the slow EPSC indicated that the latter must have resulted from rebinding of transmitter, as suggested by the model. Therefore, glutamate must persist in the synaptic cleft during the period of the tail current. Most interestingly, the amplitude and duration of the tail current were prolonged by blockers of glutamate uptake, with little effect on the amplitude or duration of the fast components of the EPSC. These data show that transporters do not turn over fast enough to shape the early profile of glutamate at these synapses but can markedly accelerate transmitter removal over the time scale of tens of milliseconds (Otis and Kavanaugh, this volume).

Given the relative contributions of the quantal time course, release time course and glutamate clearance, we can form a generalizable picture of the events during the EPSC at large auditory synapses activated at low frequency. Transmitter release results in a rapidly rising EPSC due to the near-synchronous release of about 200 vesicles. This release time course at physiological temperatures is probably complete within a few hundred microseconds. From the standpoint of the calyceal/endbulb synaptic cleft, this time is probably short, as passive diffusion in the face of massive release will quickly result in pooling of transmitter. Why, then, do MNTB and SBC EPSCs not speed up when release is reduced? Moreover, why is the slow phase of the EPSC observed in nMAG not reported in these other large synapses? We suggest that these differences reflect variation in the density of active release sites in the chick and rat synapses. Modeling indicates that the pooling effect should be enhanced when release sites are closer together (Otis et al. 1996a). This situation is likely to be the case in nMAG. Unlike MNTB, three axon terminals share roughly 45% of the cell-body surface in nMAG (Jackson and Parks 1982; Parks et al. 1990) and yet, despite their smaller size, they produce EPSCs that are several times larger (see above). Assuming similar single-channel properties between species, more glutamate must be released over a smaller area in the chick, which probably explains the tendency toward delayed clearance. In summary, the duration of the AMPA EPSC at SBC/nMAG/MNTB synapses is determined first by the duration of the release process and of the quantal currents, and then by a transmitter-clearance process, which is itself shaped by the release probability, the spacing of release sites and the avidity of glutamate transporters.

F. Synaptic Depression

Depression of glutamatergic EPSCs during repetitive synaptic stimulation is characteristic of patch-clamp studies of endbulb and calyceal synapses. Activation at rates of 10–20 Hz resulted in nearly 50% attenuation, while rates of 100–300 Hz reduced the EPSC to only a few percent (BORST et al. 1995; ZHANG and TRUSSELL 1994a). We will consider here whether or not depression occurs in vivo and what its cellular mechanisms are, particularly as they relate to the issues of transmitter clearance discussed above.

I. Depression In Vivo

Given the ubiquity of depression, and of studies of depression, it is of more than passing interest to ask if depression occurs in the auditory system in vivo. Recent studies of cortical neurons suggest that depression at the glutamatergic synapse may serve a key computational role in balancing the strength of different synaptic inputs dependent upon activity level (ABBOTT et al. 1997; MARKRAM and TSODYKS 1996). However, the large auditory synapses are often portrayed in the literature as archetypal, "reliable" synapses, where plasticity should be avoided. Indeed, earlier studies of calyx and endbulb synapses using microelectrodes and performed on mature animals did not report any depression during high-frequency stimulation (OERTEL 1985; WU and KELLY 1991). However, virtually all recent studies of giant auditory synapses in brain slices report significant depression of EPSCs (BARNES-DAVIES and FORSYTHE 1995; BORST et al. 1995; BRENOWITZ et al. 1998; ISAACSON and WALMSLEY 1996; von GERSDORFF et al. 1997; WANG and KACZMAREK 1998; ZHANG and TRUSSELL 1994a, 1994b). Indeed, one might expect that some depression is inevitable, given the high release probability and poor transmitter clearance described above. Several lines of evidence suggest that, in vivo, depression to subthreshold levels during acoustic stimuli probably is not common, but that EPSP depression may occur just enough to delay significantly the approach to spike threshold.

For example, the spontaneous rate of firing of globular bushy cells and their postsynaptic targets in the MNTB are very similar, suggesting that spiking may be maintained during ongoing synaptic activity at around 10 Hz (SMITH et al. 1998). Extracellular recordings of endbulb and MNTB activity characteristically feature a "prepotential" thought to reflect active currents in the calyx, followed by a larger response thought to reflect the action potential in the postsynaptic cell. GUINAN and LI (1990) found that prepotentials were always followed by postsynaptic spikes in MNTB during trains of shocks delivered to the trapezoid body at up to 500 Hz. By careful, event-triggered data collection, RUBSAMEN et al. (1998) found that, during acoustically driven activity, most prepotentials in AVCN or MNTB were followed by postsynaptic spikes, so it would seem unlikely that EPSP depression to subthreshold levels was common in response to sound. In favor of depression, for SBCs, the quality of phase locking during these stimuli is progressively reduced during the course of a tone, such that peaks in poststimulus time histograms widen (SMITH et al. 1993). This result is consistent with the more variable timing of threshold crossing due to reduced EPSC size and other factors, such as accumulated inactivation of sodium channels. Depression with maintained spiking is consistent with in vitro studies in nMAG as well. There, orthodromic spike trains may remain suprathreshold for tens of stimuli while showing a progressive increase in jitter in the onset time of the spike, consistent with profiles in the period histograms (ZHANG and TRUSSELL 1994b). The in vitro studies may overemphasize the extent of depression because of the younger animals used or because of the low recording temperatures. BRENOWITZ et al. (1998) observed that the magnitude of depression is highly temperature sensitive, and recordings made close to physiological temperatures reveal less depression and more reliable spike generation. Thus, it would seem that depression may occur physiologically in the auditory system, but it probably does not block suprathreshold transmission as much as it introduces temporal jitter in the spike onset. In pathways where information is encoded in spike timing as opposed to rate, this enhanced variability may, indeed, reduce the information content of the spikes.

II. Mechanisms of Depression

The cellular mechanisms of depression have been addressed in a variety of studies in the cochlear nucleus and MNTB. Comparison of the magnitudes of depression of AMPA and NMDA components in MNTB indicated that depression was due to a stimulus-dependent reduction in glutamate release (VON GERSDORFF et al. 1997). Reduction in release is also suggested by the progressive increase in the coefficient of variation in EPSC amplitude during stimulus trains (TRUSSELL et al. 1993; ZHANG and TRUSSELL 1994a). Moreover, while blockade of AMPA-receptor desensitization may reduce depression (see below), it does not eliminate it (ISAACSON and WALMSLEY 1996; TRUSSELL et al. 1993; ZHANG and TRUSSELL 1994a), indicating that reduced release makes an ubiquitous contribution to depression at endbulb/calyx synapses.

What molecular mechanism is responsible for reduced transmitter release? One possibility is feedback of glutamate onto inhibitory autoreceptors. However, even though presynaptic, metabotropic receptors can exert a powerful inhibition of release and of presynaptic calcium current (BARNES-DAVIES and FORSYTHE 1995; TAKAHASHI et al. 1996), these receptors are probably not activated by a neurally-released transmitter in MNTB, at least with stimulus rates of up to 10 Hz (VON GERSDORFF et al. 1997). Possibly, Ca^{2+} channels might become inactive and, indeed, prolonged stimulation produces significant inactivation of calcium current in MNTB calyces (FORSYTHE et al. 1998). While this effect undoubtedly influences depression in prolonged stimulus trains, depression proceeds before significant reduction in the calcium-channel current (CUTTLE et al. 1998; FORSYTHE et al. 1998). Acute presynaptic depression in the calyx must result from a process downstream of calcium influx, such as depletion of vesicles from a readily releasable pool, depletion of some other factor, refractoriness of the release apparatus, or accumulation of some cytosolic inhibitory factor. Simple models of depression suggest that the onset of depression proceeds to an equilibrium determined by the probability of release and the mean recovery time of the release-ready state (ABBOTT et al. 1997). Recent studies have, therefore, focused on the factors determining the recovery rate. WANG and KACZMAREK (1998) have shown that intraterminal calcium may accelerate recovery in MNTB calyces, consistent with other studies indicating that resupply of vesicles to the releasable pool is regulated by calcium ions. In the cerebellum, this effect of calcium ions was correlated with a reduction in the extent of depression, as expected from consideration of a depression model incorporating calcium-dependent recovery (DITTMAN and REGEHR 1998); thus, calcium and calcium-dependent regulatory processes may regulate synaptic strength during ongoing, acoustically driven synaptic activity.

Depression through reductions in AMPA receptor sensitivity or desensitization has been studied in the rat and chick cochlear nucleus and in the MNTB. From an examination of the response of AMPA receptors to repetitive pulses of glutamate, desensitization seems to be inevitable, as even the briefest exposures to glutamate lead to a reduced responsiveness (COLQUHOUN et al. 1992; HESTRIN 1992; RAMAN and TRUSSELL 1995). However, examination of the synaptic currents themselves has yielded more diverse conclusions. As noted, desensitization in MNTB is not thought to account for depression at rates of 10 Hz or less (VON GERSDORFF et al. 1997). WANG and KACZMAREK (1998), using the desensitization blocker cyclothiazide, found no effect on depression in MNTB. In the SBC, ISAACSON and WALMSLEY (1996) found that cyclothiazide reduced depression, suggesting that the transmitter persisted long enough to desensitize receptors. In nMAG, cyclothiazide reduced depression and broadened the EPSC in a calcium-dependent manner, indicating that persistent glutamate and the desensitization process contribute to depression (TRUSSELL et al. 1993). How may we account for these different conclusions? Following the arguments presented earlier, synapses with a lower density of active release sites will show less transmitter pooling and, thus, less rebinding of transmitter to receptors, a condition that would minimize desensitization. Indeed, all effects ascribed to transmitter pooling are lessened when release is lowered (OTIS et al. 1996a; OTIS and TRUSSELL 1996; TRUSSELL et al. 1993). Moreover, cyclothiazide, by prolonging EPSCs, accentuates the problem of receptor saturation in measuring overlapping events (TANG et al. 1994), which could, in some instances, lead to an overestimation of depression. Finally, cyclothiazide has been shown to have presynaptic effects in some preparations (DIAMOND and JAHR 1995) but not others (ISAACSON and WALMSLEY 1996); such actions would confound any conclusions about the extent of postsynaptic depression (DITTMAN and REGEHR 1998).

Direct evidence for postsynaptic depression in nMAG suggests that roughly 35–50% of the receptors may desensitize after a single EPSC elicited at low frequencies. OTIS et al. (1996) showed that the sensitivity of receptors assayed by a pulse of rapidly applied glutamate or by the amplitude of the quantal currents was reduced immediately after an EPSC. Additionally, the EPSC was able to inhibit the response to the quasi-nondesensitizing agonist kainate. Finally, paired-pulse depression was found to be reduced upon depolarization of the postsynaptic cell, a condition known to reduce desensitization. However, the extent to which desensitization might contribute to depression at the other auditory synapses or the extent to which postsynaptic depression in nMAG is maintained throughout long trains of stimuli remains unclear.

The results described above show depression to be a consequence of transmitter release, i.e., release of a large bolus of transmitter results in depletion and, at some synapses, desensitization. In this light, depression may be an inevitable consequence of a system used to drive a neuron with a very low input resistance (BREW and FORSYTHE 1995; OERTEL 1983; ZHANG and TRUSSELL 1994b) to threshold as quickly as possible. The machinery of release is probably standard; the receptors are only a little different, but the whole is arranged in such an unusual way, with dense clustering of active release sites in a small area, that EPSC depression inevitably occurs. However, depression may be a regulated phenomenon. As described above, calcium may modulate the recovery rate (WANG and KACZMAREK 1998) and, perhaps, second messenger pathways might do so as well. BRENOWITZ et al. (1998) found that depression could be minimized when release is partially inhibited by the $GABA_B$ agonist baclofen. Remarkably, baclofen rendered the EPSC resistant to stimulus rate, over a range of frequencies that would normally depress the EPSC nearly to zero. Though slowing the onset to spike threshold, the GABAergic receptors made the synapse more reliable during high-frequency stimulation, suggesting that GABA might serve to regulate responsiveness to high-intensity activity.

G. Conclusion

Attention has recently focused on the giant, somatic terminals of the auditory system because of the advantages they offer for biophysical measurements. We have learned that these synapses retain many of the basic mechanisms believed to be employed at "conventional"synapses and remind us of many of the unresolved problems as well. What determines synaptic strength at a given synapse? What happens during depression? How are the pre- and postsynaptic components of transmission assembled in a coordinated fashion during development? It is hoped that the pairing of biophysical and molecular approaches to these synapses may provide the answers. However, studies of giant synapses have also served as object lessons in biological design. The adaptations used for synaptic speed and strength come with a cost: enhancement of synapse area, release site number, and release probability lead to limitations in sustained transmission and the breakdown of precise temporal coding.

Acknowledgments. I wish to thank S. Brenowitz, D. Oertel, I. Raman and T. Yin for helpful comments, and N. Cant, P. Smith, and W. Rhode for the photomicrographs in Fig. 1. Support for this work was provided by the National Institutes of Health.

References

Abbott LF, Varela JA, Sen K, Nelson SB (1997) Synaptic depression and cortical gain control. Science 275:220–224

Abdul-Ghani MA, Valiante TA, Pennefather PS (1996) Sr^{2+} and quantal events at excitatory synapses between mouse hippocampal neurons in culture. J Physiol 495: 113–125

Angulo MC, Lambolez B, Audinat E, Hestrin S, Rossier J (1997) Subunit composition, kinetic, and permeation properties of AMPA receptors in single neocortical nonpyramidal cells. J Neurosci 17:6685–6696

Barnes-Davies M, Forsythe ID (1995) Pre- and postsynaptic glutamate receptors at a giant excitatory synapse in rat auditory brainstem slices. J Physiol 488:387–406

Barrett EF, Stevens CF (1972) The kinetics of transmitter release at the frog neuromuscular junction. J Physiol 227:691–708

Bekkers JM (1998) Strontium-evoked mEPSCs reveal large quantal variance at central glutamate synapses. In: Faber DS, Korn H, Redman SJ, Thompson SM, Altmann JS (eds) Central synapses: quantal mechanisms and plasticity. HFSP, Strasbourg, pp. 115–123

Bekkers JM, Stevens CF (1996) Cable properties of cultured hippocampal neurons determined from sucrose-evoked miniature EPSCs. J Neurophysiol 75:1250–1255

Bellingham MC, Lim R, Walmsley B (1998) Developmental changes in EPSC quantal size and quantal content at a central glutamatergic synapse. J. Physiol 511:861–869

Borst JG, Helmchen F, Sakmann B (1995) Pre- and postsynaptic whole-cell recordings in the medial nucleus of the trapezoid body of the rat. J Physiol 489:825–840

Borst JG, Sakmann B (1996) Calcium influx and transmitter release in a fast CNS synapse. Nature 383:431–434

Brenowitz S, David J, Trussell L (1998) Enhancement of synaptic efficacy by presynaptic $GABA_B$ receptors. Neuron 20:135–141

Brew HM, Forsythe ID (1995) Two voltage-dependent K^+ conductances with complementary functions in postsynaptic integration at a central auditory synapse. J Neurosci 15:8011–8022

Brownell WE (1975) Organization of the cat trapezoid body and the discharge characteristics of its fibers. Brain Res 94:413–433

Cant NB (1992) The cochlear nucleus: neuronal types and their synaptic organization. In: Webster DB, Popper AN, Fay RR (eds) The mammalian auditory pathway: neuroanatomy. Springer-Verlag, New York, pp. 66–116

Cant, NB, Morest DK (1979) The bushy cells in the anteroventral cochlear nucleus of the cat. A study with the electron microscope. Neuroscience 4:1925–1945

Carr CD, Code RA (1998) The central auditory system of reptiles and birds. In: Dooling ANPR, Fay RR (eds) The Auditory System of Birds and Reptiles. Springer-Verlag, New York, in press

Casey MA, Feldman ML (1988) Age-related loss of synaptic terminals in the rat medial nucleus of the trapezoid body. Neuroscience 24:189–194

Chuhma N, Ohmori H (1998) Postnatal development of phase-locked high-fidelity synaptic transmission in the medial nucleus of the trapezoid body of the rat. J Neurosci 18:512–520

Colquhoun D, Jonas P, Sakmann B (1992) Action of brief pulses of glutamate on AMPA/kainate receptors in patches from different neurons of rat hippocampal slices. J Physiol 458:261–287

Cuttle MF, Tsujimoto T, Forsythe ID, Takahashi T (1998) Facilitation of the presynaptic calcium current at an auditory synapse in rat brainstem. J Physiol, in press

Diamond JS, Jahr CE (1995) Asynchronous release of synaptic vesicles determines the time course of the AMPA receptor-mediated EPSC. Neuron 15:1097–1107

Diamond JS, Jahr CE (1997) Transporters buffer synaptically released glutamate on a submillisecond time scale. J Neurosci 17:4672–4687

Dittman JS, Regehr WG (1998) Calcium dependence and recovery kinetics of presynaptic depression at the climbing fiber to purkinje cell synapse. J Neurosci 18:6147–6162

Fettiplace R, Fuchs P (1999) Mechanisms of hair cell tuning. Annual Rev. Physiol. 61: in press

Forsythe ID (1994) Direct patch recording from identified presynaptic terminals mediating glutamatergic EPSCs in the rat CNS, in vitro. J Physiol 479:381–387

Forsythe ID, Barnes-Davies M (1993) The binaural auditory pathway: excitatory amino acid receptors mediate dual timecourse excitatory postsynaptic currents in the rat medial nucleus of the trapezoid body. Proc R Soc Lond B 251:151–157

Forsythe ID, Tsujimoto T, Barnes-Davies M, Cuttle MF, Takahashi T (1998) Inactivation of presynaptic calcium current contributes to synaptic depression at a fast central synapse. Neuron 20:797–807

Gardner SM, Trussell LO, Oertel D (1998) Time course of AMPA receptor-mediated miniature EPSCs in the mouse cochlear nuclei. Soc. Neurosci. Abstracts 24:1635

Geiger JR, Lübke J, Roth A, Frotscher M, Jonas P (1997) Submillisecond AMPA receptor-mediated signaling at a principal neuron- interneuron synapse. Neuron 18: 1009–1023

Geiger JR, Melcher T, Koh DS, Sakmann B, Seeburg PH, Jonas P, Monyer H (1995) Relative abundance of subunit mRNAs determines gating and Ca^{2+} permeability of AMPA receptors in principal neurons and interneurons in rat CNS. Neuron 15:193–204

Geisler CD (1998) From Sound to Synapse. Physiology of the Mammalian Ear. Oxford University Press, New York

Grandes P, Streit P (1989) Glutamate-like immunoreactivity in calyces of Held. J Neurocytol 18:685–693

Grothe B, Sanes, DH (1994) Synaptic inhibition influences the temporal coding properties of medial superior olivary neurons: an in vitro study. J Neurosci 14: 1701–1709

Guinan JJ, Jr, Li RY (1990) Signal processing in brainstem auditory neurons which receive giant endings (calyces of Held) in the medial nucleus of the trapezoid body of the cat. Hear Res 49:321–334

Gulley RL, Reese TS (1981) Cytoskeletal organization at the postsynaptic complex. J Cell Biol 91:298–302

Hackney CM, Osen KK, Ottersen OP, Storm-Mathisen J, and Manjaly G (1996) Immunocytochemical evidence that glutamate is a neurotransmitter in the cochlear nerve: a quantitative study in the guinea-pig anteroventral cochlear nucleus. Eur J Neurosci 8:79–91

Hestrin S (1992) Activation and desensitization of glutamate-activated channels mediating fast excitatory synaptic currents in the visual cortex. Neuron 9:991–999

Hunter C, Petralia RS, Vu T, Wenthold RJ (1993) Expression of AMPA-selective glutamate receptor subunits in morphologically defined neurons of the mammalian cochlear nucleus. J Neurosci 13:1932–1946

Ibata Y, Pappas GD (1976) The fine structure of synapses in relation to the large spherical neurons in the anterior ventral cochlear of the cat. J Neurocytol 5:395–406

Isaacson JS, Walmsley B (1995) Counting quanta: direct measurements of transmitter release at a central synapse. Neuron 15:875–884

Isaacson JS, Walmsley B (1995) Receptors underlying excitatory synaptic transmission in slices of the rat anteroventral cochlear nucleus. J Neurophysiol 73:964–973

Isaacson JS, Walmsley B (1996) Amplitude and time course of spontaneous and evoked excitatory postsynaptic currents in bushy cells of the anteroventral cochlear nucleus. J Neurophysiol 76:1566–1571

Iwasaki S, Takahashi T (1998) Developmental changes in calcium channel types mediating synaptic transmission in rat auditory brainstem. J Physiol (Lond) 509: 419–423

Jackson H, Parks TN (1982) Functional synapse elimination in the developing avian cochlear nucleus with simultaneous reduction in cochlear nerve axon branching. J Neurosci 2:1736–1743

Jackson H, Parks TN (1988) Induction of aberrant functional afferents to the chick cochlear nucleus. J Comp Neurol 271:106–114

Jhaveri S, Morest DK (1982a) Neuronal architecture in nucleus magnocellularis of the chicken auditory system with observations on nucleus laminaris: a light and electron microscope study. Neuroscience 7:809–836

Jhaveri S, Morest DK (1982b) Sequential alterations of neuronal architecture in nucleus magnocellularis of the developing chicken: a Golgi study. Neuroscience 7:837–853

Jhaveri S, Morest DK (1982c) Sequential alterations of neuronal architecture in nucleus magnocellularis of the developing chicken: an electron microscope study. Neuroscience 7:855–870

Jonas P, Major G, Sakmann B (1993) Quantal components of unitary EPSCs at the mossy fibre synapse on CA3 pyramidal cells of rat hippocampus. J Physiol 472: 615–663

Joris PX, Yin TCT (1998) Envelope coding in the lateral superior olive. III. Comparison with afferent pathways. J Neurophysiol 79:253–269

Kandler K, Friauf E (1995) Development of glycinergic and glutamatergic synaptic transmission in the auditory brainstem of perinatal rats. J Neurosci 15:6890–6904

Kane EC (1974) Synaptic organization in the dorsal cochlear nucleus of the cat: a light and electron microscopic study. J Comp Neurol 155:301–329

Kil J, Kageyama GH, Semple MN, Kitzes LM (1995) Development of ventral cochlear nucleus projections to the superior olivary complex in gerbil. J Comp Neurol 353:317–340

Köppl C (1994) Auditory nerve terminals in the cochlear nucleus magnocellularis: differences between low and high frequencies. J Comp Neurol 339:438–446

Kuwabara N, DiCaprio RA, Zook JM (1991) Afferents to the medial nucleus of the trapezoid body and their collateral projections. J Comp Neurol 314:684–706

Lawrence JJ, Trussell LO (1998) Developmental regulation of AMPA receptor kinetics in the avian cochlear nucleus. Society for Neuroscience Abstracts 24:846

Lenn NJ, Reese TS (1966) The fine structure of nerve endings in the nucleus of the trapezoid body and the ventral cochlear nucleus. Am J Anat 118:375–389

Levin MD, Kubke MF, Schneider M, Wenthold R, Carr CE (1997) Localization of AMPA-selective glutamate receptors in the auditory brainstem of the barn owl. J Comp Neurol 378:239–253

Liberman MC (1991) Central projections of auditory-nerve fibers of differing spontaneous rate. I. Anteroventral cochlear nucleus. J Comp Neurol 313:240–258
Liu G, Tsien RW (1995) Properties of synaptic transmission at single hippocampal synaptic boutons. Nature 375:404–408
Markram H, Tsodyks M (1996) Redistribution of synaptic efficacy between neocortical pyramidal neurons. Nature 382:807–810
Morest DK (1969) The differentiation of cerebral dendrites: A study of the post- migratory neuroblast in the medial nucleus of the trapezoid body. Z Anat Entwicklungsgesch 128:271–289
Morest DK (1968) The growth of synaptic endings in the mammalian brain: a study of the calyces of the trapezoid body. Z Anat Entwicklungsgesch 127:201–220
Mosbacher J, Schoepfer R, Monyer H, Burnashev N, Seeburg PH, Ruppersberg JP (1994) A molecular determinant for submillisecond desensitization in glutamate receptors. Science 266:1059–1062
Neises GR, Mattox DE, Gulley RL (1982) The maturation of the end bulb of Held in the rat anteroventral cochlear nucleus. Anat Rec 204:271–279
Oertel D (1999) The role of timing in the auditory brainstem nuclei of vertebrates. Ann. Rev. Physiol 61: in press
Oertel D (1983) Synaptic responses and electrical properties of cells in brain slices of the mouse anteroventral cochlear nucleus. J Neurosci 3:2043–2053
Oertel D (1985) Use of brain slices in the study of the auditory system: spatial and temporal summation of synaptic inputs in cells in the anteroventral cochlear nucleus of the mouse. J Acoust Soc Am 78:328–333
Otis TS, Trussell LO (1996) Inhibition of transmitter release shortens the duration of the excitatory synaptic current at a calyceal synapse. J Neurophysiol 76:3584–3588
Otis TS, Wu YC, Trussell LO (1996a) Delayed clearance of transmitter and the role of glutamate transporters at synapses with multiple release sites. J Neurosci 16: 1634–1644
Otis T, Zhang S, Trussell LO (1996b) Direct measurement of AMPA receptor desensitization induced by glutamatergic synaptic transmission. J Neurosci 16:7496–7504
Parks TN (1981) Morphology of axosomatic endings in an avian cochlear nucleus: nucleus magnocellularis of the chicken. J Comp Neurol 203:425–440
Parks TN, Taylor DA, Jackson H (1990) Adaptations of synaptic form in an aberrant projection to the avian cochlear nucleus. J Neurosci 10:975–984
Petralia RS, Wenthold RJ (1992) Light and electron immunocytochemical localization of AMPA-selective glutamate receptors in the rat brain. J Comp Neurol 318: 329–354
Raman IM, Trussell LO (1992) The kinetics of the response to glutamate and kainate in neurons of the avian cochlear nucleus. Neuron 9:173–186
Raman IM, Trussell LO (1995) The mechanism of alpha-amino-3-hydroxy-5-methyl-4-isoxazolepropionate receptor desensitization after removal of glutamate. Biophys J 68:137–146
Raman IM, Zhang S, Trussell LO (1994) Pathway-specific variants of AMPA receptors and their contribution to neuronal signaling. J Neurosci 14:4998–5010
Rhode WS, Greenberg S (1992) Physiology of the cochlear nuclei. In: Popper AN, Fay RR (eds) The mammalian auditory pathway: neurophysiology. Springer, New York, pp 94–152
Rossi DJ, Alford S, Mugnaini E, Slater NT (1995) Properties of transmission at a giant glutamatergic synapse in cerebellum: the mossy fiber-unipolar brush cell synapse. J Neurophysiol 74:24–42
Rubsamen R, Kopp C, Dorrscheidt GJ (1998) Principal component analysis applied to action potentials reveals neuronal interaction in auditory brainstem nuclei. In: Palmer AR, Rees A, Summerfield AQ, Meddis R, (eds) Psychophysical and physiological advances in hearing. Whurr, London, pp 352–358
Ryugo DK (1992) The auditory nerve: peripheral innervation, cell body morphology, and central projections. In: Webster DB, Popper AN, Fay RR (eds) The mammalian auditory pathway: neuroanatomy. Springer, New York, pp 23–65

Ryugo DK, Fekete DM (1982) Morphology of primary axosomatic endings in the anteroventral cochlear nucleus of the cat: a study of the endbulbs of Held. J Comp Neurol 210:239–257

Ryugo DK, Sento S (1991) Synaptic connections of the auditory nerve in cats: relationship between endbulbs of held and spherical bushy cells. J Comp Neurol 305: 35–48

Ryugo DK, Wu MM, Pongstaporn T (1996) Activity-related features of synapse morphology: a study of endbulbs of held. J Comp Neurol 365:141–158

Sato K, Kiyama H, Tohyama M (1993) The differential expression patterns of messenger RNAs encoding non-N- methyl-D-aspartate glutamate receptor subunits (GluR1–4) in the rat brain. Neuroscience 52:515–539

Silver RA, Colquhoun D, Cull-Candy SG, Edmonds B (1996) Deactivation and desensitization of non-NMDA receptors in patches and the time course of EPSCs in rat cerebellar granule cells . J Physiol 493:167–173

Silver RA, Cull-Candy SG, Takahashi T (1996) Non-NMDA glutamate receptor occupancy and open probability at a rat cerebellar synapse with single and multiple release sites. J Physiol 494:231–250

Silver RA, Momiyama A, Cull-Candy SG (1998) Locus of frequency-dependent depression identified with multiple- probability fluctuation analysis at rat climbing fibre-Purkinje cell synapses. J Physiol 510:881–902

Smith PH, Joris PX, Carney LH, Yin TC (1991) Projections of physiologically characterized globular bushy cell axons from the cochlear nucleus of the cat. J Comp Neurol 304:387–407

Smith PH, Joris PX, Yin TC (1998) Anatomy and physiology of principal cells of the medial nucleus of the trapezoid body (MNTB) of the cat. J Neurophysiol 79: 3127–3142

Smith PH, Joris PX, Yin TC (1993) Projections of physiologically characterized spherical bushy cell axons from the cochlear nucleus of the cat: evidence for delay lines to the medial superior olive. J Comp Neurol 331:245–260

Spirou GA, Brownell WE, Zidanic M (1990) Recordings from cat trapezoid body and HRP labeling of globular bushy cell axons. J Neurophysiol 63:1169–1190

Takahashi T, Forsythe ID, Tsujimoto T, Barnes-Davies M, Onodera K (1996) Presynaptic calcium current modulation by a metabotropic glutamate receptor. Science 274:594–597

Tang CM, Margulis M, Shi QY, Fielding A (1994) Saturation of postsynaptic glutamate receptors after quantal release of transmitter. Neuron 13:1385–1393

Taschenberger H, Engert F, Grantyn R (1995) Synaptic current kinetics in a solely AMPA-receptor-operated glutamatergic synapse formed by rat retinal ganglion neurons. J Neurophysiol 74:1123–1136

Titz S, Keller BU (1997) Rapidly deactivating AMPA receptors determine excitatory synaptic transmission to interneurons in the nucleus tractus solitarius from rat. J Neurophysiol 78:82–91

Traynelis SF, Silver RA, Cull-Candy SG (1993) Estimated conductance of glutamate receptor channels activated during EPSCs at the cerebellar mossy fiber-granule cell synapse. Neuron 11:279–289

Trussell L (1999) Synaptic mechanisms for coding timing in auditory neurons. Annual reviews of physiology 61: in press

Trussell L, Brenowitz S, Otis T (1998) Postsynaptic mechanisms underlying synaptic depression. In: Faber DS, Korn H, Redman SJ, Thompson SM, Altmann JS (eds) Central synapses: quantal mechanisms and plasticity. HFSP, Strasbourg, pp 149–158

Trussell LO (1997) Cellular mechanisms for preservation of timing in central auditory pathways. Curr Opin Neurobiol 7:487–492

Trussell LO, Zhang S, Raman IM (1993) Desensitization of AMPA receptors upon multiquantal neurotransmitter release. Neuron 10:1185–1196

Ulrich D, Lüscher HR (1993) Miniature excitatory synaptic currents corrected for dendritic cable properties reveal quantal size and variance. J Neurophysiol 69:1769–1773 von Gersdorff H, Schneggenburger R, Weis S, Neher E (1997) Presynaptic depression at a calyx synapse: the small contribution of metabotropic glutamate receptors. J Neurosci 17:8137–8146

Wang LY, Kaczmarek LK (1998) High-frequency firing helps replenish the readily releasable pool of synaptic vesicles. Nature 394:384–388

Wang YX, Wenthold RJ, Ottersen OP, Petralia RS (1998) Endbulb synapses in the anteroventral cochlear nucleus express a specific subset of AMPA-type glutamate receptor subunits. J Neurosci 18:1148–1160

Wu LG, Borst JG, Sakmann B (1998) R-type Ca2+ currents evoke transmitter release at a rat central synapse. Proc Natl Acad Sci USA 95 4720–4725

Wu SH, Kelly JB (1995) Inhibition in the superior olivary complex: pharmacological evidence from mouse brain slice. J Neurophysiol 73:256–269

Wu SH, Kelly JB (1994) Physiological evidence for ipsilateral inhibition in the lateral superior olive: synaptic responses in mouse brain slice. Hear Res 73:57–64

Wu SH, Kelly JB (1991) Physiological properties of neurons in the mouse superior olive: membrane characteristics and postsynaptic responses studied in vitro. J Neurophysiol 65:230–246

Wu SH, Kelly JB (1992) Synaptic pharmacology of the superior olivary complex studied in mouse brain slice. J Neurosci 12:3084–3097

Zhang S, Trussell LO (1994a) Voltage clamp analysis of excitatory synaptic transmission in the avian nucleus magnocellularis. J Physiol 480:123–136

Zhang S, Trussell LO (1994b) A characterization of excitatory postsynaptic potentials in the avian nucleus magnocellularis. J Neurophysiol 72:705–718

Zhou N, Parks TN (1992) Developmental changes in the effects of drugs acting at NMDA or non-NMDA receptors on synaptic transmission in the chick cochlear nucleus (nuc. magnocellularis). Dev Brain Res 67:145–152

CHAPTER 13

Glutamate Transporters and Their Contributions to Excitatory Synaptic Transmission

T.S. OTIS and M.P. KAVANAUGH

A. Introduction

Following exocytosis, the excitatory neurotransmitter L-glutamate diffuses freely within a complicated network of extracellular spaces in the brain, where it can interact with various target receptors (some of which are the topic of the rest of this book). Released transmitter molecules will eventually be removed from the space surrounding cells by excitatory amino acid transporters (EAATs), proteins localized in the plasma membrane of neuronal and glial cells. It is an interplay among the processes of release, diffusion, and uptake by EAATs that determines the temporal and spatial dynamics of L-glutamate concentration. As has been detailed in several of the preceding chapters (COLQUHOUN, this volume; GEIGER, this volume; TRUSSELL, this volume), the precise dynamics in glutamate concentration can have important consequences for the behavior of target receptors. For example, ionotropic glutamate receptors are believed to have at least two binding sites (ROSENMUND et al. 1998; CLEMENTS et al. 1998). As a result of this, the time course of changes in glutamate concentration might influence whether an α-amino-3-hydroxy-5-methylisoxazole-4-propionic-acid (AMPA)-type glutamate receptor remains unbound, is incompletely bound (in which case it may desensitize before opening) or is fully bound and opens.

The spatial distributions of potential targets for glutamate are very different. Both AMPA- and *N*-methyl-D-aspartate (NMDA)-type glutamate receptors are localized in a patchy distribution in the postsynaptic membrane in register with presynaptic sites of transmitter release (KORNAU, this volume; WENTHOLD, this volume). These neighboring patches of receptors are separated by distances of less than 1 μm in certain brain regions (FROTSCHER, this volume). By contrast, metabotropic glutamate receptors are typically located at much greater distances from release sites.

Given such temporal and spatial complexities, EAAT function might be expected to play key roles in synaptic signaling. Can EAATs influence the strength of excitatory connections by modulating glutamate-concentration transients at individual synapses? Do transporters help to maintain synapse specificity by limiting the spillover of synaptically released glutamate? What is the ambient concentration of glutamate at the various target receptors? Under what conditions does glutamate reach distant receptor types, and what concentrations are achieved? Furthermore, because excessive activation of glutamate receptors is thought to underlie several neurological diseases (Choi, this volume; Rothstein, this volume), different questions arise regarding the role of transport in the pathophysiology of glutamatergic signaling. Do uptake mechanisms fail, or are they somehow impaired or overwhelmed under pathological conditions?

In an effort to eventually answer such questions, much of the recent research on EAATs has concentrated on three key questions: (1) Where does transport occur within the microstructure of the synapse? (2) How rapidly is glutamate removed from the extracellular space by EAATs? and (3) What is the maximum achievable concentration gradient of glutamate? Initial answers have come from careful examination of the density and localization of the various EAAT subtypes, from the study of energetic and kinetic properties of transporters, and from experiments in which EAAT function is either directly measured or impaired and then indirectly monitored at intact excitatory synapses.

The first part of this chapter will present what is currently known about the EAATs, with special attention to those characteristics that might be important for their operation within synapses. Sections in the first part discuss the molecular identities of glutamate transporters, their structural and functional properties, and their relative localization in the ultrastructure of the synapse. The second part will review our current understanding of EAAT function at three different, extensively studied excitatory synapses.

B. Transporter Properties

I. Molecular Structure

Molecular-cloning techniques have identified a family of EAATs that, to date, consists of at least five members. In 1992, three glutamate-transporter complementary DNAs (cDNAs) were isolated and exogenously expressed; glutamate–aspartate transporter (GLAST; Storck et al. 1992; Tanaka 1993) and glutamate transporter (GLT)-1 (Pines et al. 1992) were isolated from rat brain, while excitatory amino acid carrier (EAAC1) was isolated from rabbit intestine (Kanai and Hediger 1992) but was also shown to be expressed in neurons. Soon after, the corresponding sequences were cloned from the human motor cortex (Arriza et al. 1994) and termed EAAT1–EAAT3 (EAAT1 = GLAST, EAAT2 = GLT-1, EAAT3 = EAAC1). Homology-based screening led to the identification of EAAT4, which is expressed in the human cerebellum (Fairman et al. 1995). EAAT5 was similarly identified from salamander (Eliasof et al. 1998) and human (Arriza et al. 1997) retinal libraries. The cDNAs predict amino acid sequences of 524–574 residues in length (Table 1), with between 30% and 65% identity when compared to one another and, typically, greater than 75% identity when comparing a particular isoform across

Table 1. Comparative properties of the five molecular isoforms of glutamate transporters

Predicted size in rodents (humans)	GLAST/(EAAT1) 543 (542) aa	GLT1/(EAAT2) 573 (574) aa	EAAC1/(EAAT3) 524 (525) aa	EAAT4 (564) aa	EAAT5 (560) aa
Expression pattern	Bergmann glia, less in other astrocytes	Most astrocytes in the CNS	Neurons, peripheral epithelia	Cerebellar Purkinje neurons	Retinal glial cells and neurons
Subcellular localization	Cell bodies and processes opposed to neuropil	Cell bodies and processes opposed to neuropil	Dendritic membrane, presynaptic terminals?	Perisynaptic, surrounding PF and CF PSDs	Müller cells, outer plexiform layer
Steady-state EC_{50} (L-Glu)	20 μM	18 μM	28 μM	3 μM	64 μM
Turnover rate at 22–24°C	16 s^{-1}	15 s^{-1}; 60 s^{-1}?		12 s^{-1}	

The predicted size is the number of amino acid residues predicted from the mRNA sequence for rodent; the prediction for human sequence is in parentheses. The steady-state EC_{50} is the L-glutamate concentration yielding a half maximal current in voltage-clamped *Xenopus* oocytes (Arriza et al. 1994, 1997; Fairman et al. 1995). See text for the sources for estimates of the turnover rates

aa, amino acids; CF, climbing fiber; CNS, central nervous system; EAAC, excitatory amino acid carrier; EAAT, excitatory amino acid transporter; GLAST, glutamate–aspartate transporter; GLT glutamate transporter; PF, parallel fiber; PSDs, postsynaptic densities

species (salamander EAAT5 with human EAAT5). Together with two neutral amino acid transporters (Kanai 1997), these molecules are members of a gene family distinct from the Na^+/Cl^--dependent carrier family responsible for the uptake of most other neurotransmitters, including γ-aminobutyric acid (GABA), glycine, serotonin, and catecholamines (Amara and Arriza 1993). The functional properties of the EAATs further suggest that they are likely to be molecularly distinct from the proton/glutamate antiporters expressed in vesicular membranes (Naito and Ueda 1985), which have not yet been identified at the molecular level.

The transmembrane topology of the EAATs is controversial. There is general agreement on the disposition of the first six transmembrane segments, beginning with an intracellular *N*-terminus. However, in the *C*-terminal ~150 residues, which exhibit a strongly conserved primary sequence and hydrophobic moment, the assignment of transmembrane segments is uncertain, having initially led to proposals of zero (Storck et al. 1992), two (Pines et al. 1992; Tanaka 1993) or four (Kanai and Hediger 1992) additional transmembrane domains. Recent studies have used two strategies to elucidate how the protein is situated within the membrane: truncated constructs fused with tagged sequences, and cysteine-accessibility mutagenesis. Slotboom et al. (1996), relying upon a series of alkaline-phosphatase-reporter-fusion constructs with a bacterial glutamate transporter, proposed a structure with four additional α-helical transmembrane segments. By contrast, an approach employing a reporter peptide containing *N*-glycosylation sites fused to GLAST1 sequences yielded an alternate model with four β-sheet transmembrane domains located in different regions of the *C*-terminal sequence (Wahle and Stoffel 1996). Still different conclusions were reached based on the study of a series of functional GLT1 transporters containing engineered cysteine residues accessible to chemical modification from the intracellular or extracellular compartments (Grunewald et al. 1998). They proposed a structure in which the seventh and eighth α-helical transmembrane segments flank a 22-amino-acid-long re-entrant sequence reminiscent of a P-loop structure like that found in voltage-gated channels, such as Shaker K^+ channels. Between this re-entrant sequence and TM8, they suggest that an extremely exposed hydrophobic sequence lies in close association with the membrane. Resolution of these disagreements is important, because this *C*-terminal region of EAAT structure is highly conserved and appears to be a part of the molecule involved in key functions. Experiments on a point mutant have identified a residue (E404 in GLT1) involved in substrate selectivity and required for K^+ coupling (Kavanaugh et al. 1997), while chimeras have suggested that sequence variations among EAATs in this region confer differences in substrate discrimination (Vandenberg et al. 1995) and chloride permeability (Mitrovic et al. 1998). Further experiments on functional transporters using site-directed mutagenesis should help to further refine the structural models.

Immunoprecipitation and chemical-crosslinking studies have raised the possibility that transporters may associate as homomultimers. Sodiumdodecyl-sulfate (SDS) gel electrophoresis of membrane preparations from rat brain exposed to oxidative crosslinking agents showed evidence that, under such conditions, GLT-1 and GLAST form trimers and dimers, and that EAAC1 and EAAT4 can form dimers (HAUGETO et al. 1996; DEHNES et al. 1998). Immunoprecipitation following crosslinking further demonstrated that homomultimers are specifically preferred, as GLT-1 and GLAST did not form complexes with one another. These findings are in agreement with functional studies suggesting that, in oocytes, coexpression of EAAT1 and EAAT2 or EAAT2 and EAAT3 does not result in the appearance of functional properties different from the summed properties of the individual homomers (ZERANGUE and KAVANAUGH, manuscript in preparation). Conservatively interpreted, these data imply that EAATs exist in high-density clusters of identical subunits, in which crosslinking agents can promote disulfide-bond formation. However, the evidence that multimer formation is required for transporter function is limited. Radiation-inactivation experiments on brain membranes suggest a functional unit of 160kD, which is approximately 2.5 times the monomer size (64kDa; HAUGETO et al. 1996). This result highlights the possible involvement of ancillary proteins associated with EAATs. It is worth noting that, for EAAT5, a domain at the *C*-terminus has been identified (ARRIZA et al. 1997) that might bind to a motif termed a PDZ-binding domain, which is involved in the localization of several other ion-channel types. This raises the possibility that EAATs may associate with localization proteins. Definitive answers to the question of whether the EAATs function as homomultimeric complexes or with other unidentified protein subunits awaits further biochemical and molecular studies.

II. Energetics

A hallmark of EAAT function is that translocation of glutamate is coupled to the cotransport of sodium ions, while reorientation of the substrate-binding site is linked to the countertransport of potassium ions (KANNER and SHARON 1978; STALLCUP et al. 1979; KANNER and BENDAHAN 1982). Coupled flux of a pH-changing ion during a transport cycle also occurs, with intracellular acidification accompanying uptake (NELSON et al. 1983; ZERANGUE and KAVANAUGH 1996a; BILLUPS et al. 1996). Net positive charge accompanies glutamate entry so that transport is electrogenic (Sect. B.III). The free energy required to concentrate glutamate against its electrochemical gradient is not derived from adenosine triphosphate hydrolysis but rather from stoichiometric movement of the other ions (Na^+, H^+, and K^+) down their transmembrane electrochemical gradients. With ion gradients similar to those found in the brain, *Xenopus* oocytes expressing the neuronal transporter EAAT3 maintain nanomolar extracellular concentrations of glutamate in the face of millimolar intracellular concentrations (ZERANGUE and KAVANAUGH 1996a). The ability to maintain such large concentration gradients of glutamate prevents excitotoxicity, ensures that receptors are available for synaptic stimulation, and may be an important factor in preserving sufficient cytoplasmic glutamate levels for vesicle loading by the intracellular vesicular transporter.

The precise stoichiometry of a transport cycle determines how large a concentration gradient of glutamate can theoretically be generated at steady state. For this reason, the dependencies of transport on K^+, pH and Na^+ gradients have been studied intensively. Early work on synaptosomes isolated from rat brain demonstrated that Na^+ is cotransported and K^+ is countertransported during a transport cycle (KANNER and SHARON 1978). Glutamate transport was also shown to be coupled to the pH gradient (NELSON et al. 1983). The observation that anions are transported out of the cell during uptake led to the hypothesis that OH^- or HCO_3^- was likely to be a countertransported pH-changing ion (BOUVIER et al. 1992). However, in light of the subsequent discovery of an uncoupled anion conductance and the evidence that a proton can be cotransported as an amino acid ion pair (ZERANGUE and KAVANAUGH 1996a, 1996b), proton cotransport with glutamate appears to be more likely.

A lack of consensus regarding sodium stoichiometry has existed, which may stem from differences in the approaches that have been used to estimate this parameter. Stoichiometries of 2 or 3 have been suggested from Hill coefficients for $[Na^+]_o$-response relationships for radiolabelled-substrate flux (ERECINSKA et al. 1983) or current (BREW and ATTWELL 1987; BARBOUR et al. 1991; KLOCKNER et al. 1993; KANAI et al. 1995). Comparisons of radiolabelled Na^+ and glutamate fluxes have yielded estimates of approximately two Na^+ ions per glutamate molecule (STALLCUP et al. 1979; KANAI et al. 1995). The ambiguity in the number of Na^+ ions entering per cycle may result from several confounding factors: reliance on the Hill slope, which can be a misleading indicator of the coupling number, imperfect isolation of the Na^+ and glutamate fluxes through the transporter, lack of control of membrane potential, and efflux of the label in the radiotracer studies. These issues are largely avoided by another approach, that of measuring reversal potentials of the coupled charge movements (isolated by using a specific antagonist) in different $[Na^+]_o$, $[K^+]_o$, and $[H^+]_o$ and fitting the data to the zero-flux equation. With this strategy, recent studies have concluded that three Na^+ ions, one H^+, and one glutamate are cotransported, while one K^+ is countertransported in each cycle (ZERANGUE and KAVANAUGH 1996a; LEVY et al. 1998).

What are the theoretical glutamate gradients that may be reached, given this stoichiometry? Predictions can be made following rearrangement of the zero-flux equation (ZERANGUE and KAVANAUGH 1996a):

$$(Glu_o/Glu_i)^{nGlu} = (Na_i/Na_o)^{nNa}(H_i/H_o)^{nH}(K_o/K_i)^{nK} \exp[(n_{Na} + n_H - n_{Glu} - n_K)VF/RT], \quad (1)$$

where n_{Ion} is the number of given ions coupled to a cycle, and Ion_o (Ion_i) is the extracellular (intracellular) concentration for a given ion. V, R, T and F are the resting membrane potential, gas constant, temperature, and Faraday constant, respectively. Assuming that three Na^+ and one glutamate (Glu) are transported per cycle, $V = -90\,mV$ for a typical glial cell and, with ionic gradients

(Ion_o/Ion_i) of 146 mM/20 mM for Na^+, 4 mM/140 mM for K^+, and a pH gradient of 7.6 (out)/7.3 (in), $Glu_o/Glu_i = 1.7 \times 10^{-7}$ at 37°C. By comparison, with two Na^+ per cycle, the gradient is 3.6×10^{-5}. Thus, despite millimolar intracellular glutamate concentrations, the theoretical limit for extracellular glutamate levels is in the nanomolar or subnanomolar range. In line with these theoretical predictions, direct measurements have shown that oocytes expressing the neuronal transporter EAAT3 are capable of generating gradients near this limit (ZERANGUE and KAVANAUGH 1996a). This is an important issue, because maintenance of glutamate concentrations below micromolar levels would ensure maximal availability of AMPA receptors and minimal activation of NMDA receptors. However, it should be recognized that the ability of EAATs to reach this limit in the vicinity of the synapse will depend on many factors, including rates of glutamate release, binding rates and affinities of transporters (see below), and the localization and density of the transporters.

III. Additional Conductances

1. Anion Conductance

The fluxes of stoichiometrically coupled ions through the transporter are not the only charge movements mediated by EAATs. This was first suggested on the basis of a chloride permeability associated with activation of the native glutamate transporter in photoreceptors (ELIASOF and WERBLIN 1993). Anion fluxes of varying relative magnitudes were later measured through each of the five EAAT isoforms subsequent to their molecular identification (WADICHE et al. 1995a; FAIRMAN et al. 1995; ARRIZA et al. 1997).

A central question is whether the anion conductance arises from an integral part of the transporter structure or from an unidentified accessory subunit. To date, a large body of evidence is consistent with the former idea. (1) Where there is glutamate transport, there is an anion conductance. Anion current is observed in EAATs exogenously expressed in different expression systems, including *Xenopus* oocytes, CHO cells (LEVY et al. 1998), and HEK 293 cells (OTIS and KAVANAUGH in preparation). Likewise, native transporters from glial cells (BILLUPS et al. 1996; ELIASOF and JAHR 1996; BERGLES and JAHR 1997; CLARK and BARBOUR 1997; BERGLES et al. 1997) and neurons (GRANT and DOWLING 1995; PICAUD et al. 1995a; TAKAHASHI et al. 1996; OTIS et al. 1997) exhibit anion currents with a pharmacology identical to that of the radiotracer flux. (2) The relative magnitude of anion conductance is characteristic for each transporter isoform and for each substrate. In Cl^--containing Ringer solution, the substrate-induced current of each EAAT isoform has a distinct reversal potential, suggesting a distinct Cl^- current magnitude relative to the stoichiometrically coupled current (WADICHE et al. 1995a). The relative magnitude of anion conductance follows the sequence: EAAT4~EAAT5 > EAAT1 > EAAT3 > EAAT2 (WADICHE et al. 1995a; FAIRMAN et al. 1995; ARRIZA et al. 1997). Furthermore, the size of the anion conductance through a particular

EAAT is substrate specific; D-aspartate activates a relatively larger anion conductance through EAAT1 and EAAT4 than does L-glutamate, resulting in different reversal potentials for these substrates (WADICHE et al. 1995a; FAIRMAN et al. 1995). (3) Mutations in the EAAT structure can increase or decrease the anion conductance. Chimeric studies indicate that swapping a *C*-terminal region of the transporter confers differences in anion conductance (MITROVIC et al. 1998). (4) The anion conductance does not appear to be due to a previously recognized chloride channel. The anion conductance is not antagonized by relevant concentrations of the usual chloride-channel blockers, such as 4-acetamido-4′-isothiocyanatostilbene-2,2′-disulfonic acid (SITS), 4,4′-diisothiocyanatostilbene-2,2′-disulfonic acid (DIDS), and niflumic acid (WADICHE et al. 1995a; KAVANAUGH, unpublished observations). In addition, while the conductance displays a relatively common lyotropic selectivity sequence, it exhibits a uniquely wide range of anion permeabilities. For example, with EAAT1, SCN^- is approximately 70-fold more permeant than Cl^-, a ratio not seen for any previously described Cl^- channel (WADICHE and KAVANAUGH 1998). (5) Rapidly activating anion conductances are observed in cell-free patches in response to transporter substrates. Anion conductances activate in <1 ms in response to jumps into saturating concentrations of the transporter substrates L-glutamate, D-aspartate, D,L-threo-β-hydroxyaspartic acid (THA) and L-*trans*-2,4-pyrrolidine dicarboxcylic acid (PDC) (BERGLES and JAHR 1997; BERGLES et al. 1997; OTIS et al. 1997; WADICHE and KAVANAUGH 1998; OTIS, unpublished observations). For the hippocampal astrocyte transporter, the kinetics of the anion conductance are very similar to the kinetics of the transporter current measured at the anion-reversal potential (BERGLES and JAHR 1997, and this is also true for EAAT2 expressed in HEK cells (OTIS and KAVANAUGH, in preparation). The dynamics of anion currents activated by glutamate jumps are affected by intracellular glutamate and Na^+, as expected for a transporter (OTIS and JAHR 1998). Finally, (6) the dynamics of the anion currents can be described with cyclical, Markov-style kinetic schemes based on traditional models of transport (OTIS and JAHR 1998; WADICHE and KAVANAUGH 1998). While these data cannot rule out the possibility that a ubiquitously expressed accessory subunit accounts for the anion conductance, this seems very unlikely.

In the previous section it was argued that cation fluxes are obligatorily coupled to the transport cycle. In contrast, the anion-conducting pathway of the transporter seems to behave as a channel. A working hypothesis is that part of the transporter structure forms an anion channel that opens more frequently while substrate is bound. While the channel is open, anions flow down their electrochemical gradient, and current flow through this hypothetical channel has no influence on the energetics of transport. Several results support this view. The magnitude of anion flux, the direction of anion flux, and the type of anion do not affect the rate of substrate transport (WADICHE et al. 1995a; WADICHE and KAVANAUGH 1998). In fact, transport occurs at the same rate in the absence of permeant anions (FAIRMAN et al. 1995; WADICHE et al. 1995a). Under exchange conditions where no net transport is expected, anion current is still observed (BILLUPS et al. 1996; KAVANAUGH et al. 1997; OTIS and JAHR 1998). Lastly, transport blockers that are not transported (kainate and dihydrokainate) reduce a tonic anion conductance seen in the absence of substrate (BERGLES and JAHR 1997; OTIS and JAHR 1998; WADICHE and KAVANAUGH 1998), as if binding of these constrained analogues of glutamate blocks a constitutive open state of the channel.

Unfortunately, no single-channel events have been recorded, and evidence suggests that if a channel exists it has a very small unitary conductance. Noise analyses of currents from oocyte patches containing EAAT1 suggest a unitary conductance of 0.6–1 fS in chloride (WADICHE and KAVANAUGH 1998). However, LARSSON et al. (1996), based on noise analysis from photoreceptors, have proposed a single-channel current of 0.7 pS in physiological chloride gradients. This difference could be due to the presence in photoreceptors of EAAT5, which has a large anion conductance (ELIASOF et al. 1998). The large size of these unitary currents raises the possibility that single-channel events may yet be resolved in EAAT5 in the presence of highly permeant anions, allowing critical tests of whether the anions pass through a channel and how the channel's microscopic gating is controlled.

What physiological significance, if any, does this anion conductance have? It seems unlikely that the conductance significantly affects the excitability of many neurons where the EAATs are expressed, including cortical pyramidal cells, hippocampal neurons, and Purkinje neurons. Even for those neuronal EAAT subtypes in which the transport current is dominated by anion conductance (EAAT4 in Purkinje neurons), this conductance is almost certainly dwarfed by conductances carried by ligand-gated channels. Climbing-fiber (CF)-mediated synaptic activation of the glutamate transporter generates an anion conductance that can be measured with NO_3^--filled pipettes but that is undetectable (<200 pS) in whole-cell recordings in physiological chloride gradients (OTIS et al. 1997). This small transporter-anion current is superimposed on an excitatory postsynaptic current, mediated by AMPA receptors, that is typically greater than 200 nS. Furthermore, Purkinje neurons experience a high frequency barrage of very large (up to 20 nS) GABAergic synaptic currents. Under these circumstances, transporter currents will exert negligible effects on the membrane potential. However, this conclusion cannot be generalized to all neurons. An important electrical role for an EAAT-associated chloride conductance seems reasonable in at least one case – the synapse formed by cone photoreceptors and ON bipolar cells in the retina. Data in support of this idea will be discussed in the final section. Regardless of whether the chloride conductance has evolved to serve any signaling role, it seems to have an ancient relationship with this transporter family, as it is also found in the related neutral amino acid transporter alanine, serine, and cysteine transporter 1 (ZERANGUE and KAVANAUGH 1996c).

2. Proton Conductance

Two recent reports have described an uncoupled proton flux through the EAAT4 transporter that is activated by arachidonic acid (Fairman et al. 1998; Tzingounis et al. 1998). The proton conductance does not alter the rate of substrate transport and appears to be additive to the anion conductance. Further study will be required before it is clear whether this modulation influences cellular properties by changing intracellular pH but, in Purkinje neurons, depolarization-stimulated arachidonic-acid release is capable of significantly increasing a transporter current presumably carried by protons (Kataoka et al. 1997).

IV. Kinetic Properties

1. Binding Rate and Affinity

Diffusion of transmitter within the synaptic cleft should (theoretically) be rapid (Ogston 1955) and, thus, the binding rate of a glutamate transporter is likely to be an important determinant of its effectiveness at capturing glutamate following synaptic release. Given the simple assumption that the activation of the anion conductance occurs only after the transporter has bound substrate, a lower limit of the binding rates can be experimentally measured. Rapid application of different concentrations of glutamate to outside-out patches has shown that the rate of onset of the anion current is dose dependent. Based on these measurements, modeling of the binding and unbinding rates for recombinant EAAT1 (Wadiche and Kavanaugh 1998) and native Purkinje neuron transporter (Otis and Jahr 1998) yielded estimates of $6.8 \times 10^6 M^{-1}s^{-1}$ and $1.8 \times 10^7 M^{-1}s^{-1}$, respectively. In comparison, AMPA receptors are thought to have similar binding rates, typically ranging between $5 \times 10^6 M^{-1}s^{-1}$ and $3 \times 10^7 M^{-1}s^{-1}$ (Raman and Trussell 1996; Jonas et al. 1993; Häusser and Roth 1997). Binding rates for other native transporters expressed in hippocampal astrocytes (Bergles and Jahr 1997) and Bergmann glial cells (Bergles et al. 1997) might be slightly faster, based on the dose dependence of the rise times measured in patches. Such rapid binding rates raise the possibility that, if transporters are localized in sufficient numbers near release sites, on a rapid time scale they may "compete" with postsynaptic receptors for transmitter, as has been recently suggested (Tong and Jahr 1994; Diamond and Jahr 1997).

Apparent affinities for glutamate of the various transporters have been reported based on the dose dependence of either steady-state currents or rates of radiolabelled substrate uptake. A comparison of EC_{50} values for steady-state currents carried by the five identified human EAATs is shown in Table 1. EAAT1–EAAT3 have similar affinities of $20 \mu M$, EAAT4 has approximately a tenfold-higher and EAAT5 a lower affinity than the other transporters. In general, the affinities for radiotracer uptake are slightly lower (48, 97, 62 and $2.5 \mu M$ for EAATs 1–4, respectively). The significance of affinity differences among isoforms and between current and radiolabelled uptake is unclear, as these assays reflect a combination of the microscopic binding affinity and the turnover rate. The binding rates and micromolar-range affinities presented above provide some information about the microscopic kinetics, but important questions remain regarding the capture efficiency and the probability that a bound glutamate molecule will be transported rather than becoming unbound.

2. Turnover Rate

A fundamental property of a transporter is the rate at which it cycles. Two very different methods have been used to measure this rate for glutamate transporters. WADICHE et al. (1995b) used nonlinear capacitance measurements to estimate the number of EAAT2 transporters expressed by a *Xenopus* oocyte. Estimating the charge stoichiometry per glutamate molecule allowed the conversion of currents activated by glutamate to flux of glutamate per oocyte. The slope of the correlation between glutamate flux and the number of transporters plotted for many individual oocytes gave a voltage-dependent turnover rate that was $14.6\,s^{-1}$ at $-80\,mV$. The same approach was used recently to estimate an EAAT1 turnover rate of $16\,s^{-1}$ at $-80\,mV$ (WADICHE and KAVANAUGH 1998). These values are displayed in Table 1 for comparison with estimates for other transporters.

Another method for assessing the turnover rate relies upon measurements of the anion conductance made from outside-out patches. In response to prolonged (>10 ms) pulses of glutamate, both the anion current (BERGLES and JAHR 1997; BERGLES et al. 1997; OTIS et al. 1997) and the stoichiometrically coupled current (BERGLES and JAHR 1997; OTIS and KAVANAUGH, unpublished observations) exhibit a transient followed by a sustained current, which is typically 10–30% of the peak current. Given this depression, delivery of pairs of pulses of substrate to the transporters in a patch can be used to monitor the time course of recovery from depression. Cyclical, Markov-style kinetic models of different transporters support the contention that this depression and recovery of the anion current reflects the disappearance and reappearance, respectively, of the extracellular glutamate-binding site during the transporter cycle (OTIS and JAHR 1998; WADICHE and KAVANAUGH 1998).

For patches removed from Purkinje neurons, the recovery time constant indicates a turnover rate of $12\,s^{-1}$ (OTIS and JAHR 1998). Based on immunocytochemical localization studies (Sect. C below), this measurement probably represents mostly EAAT4 transporter and has been included as such in Table 1. The recovery rate determined from hippocampal astrocyte patches (BERGLES and JAHR 1997; BERGLES and JAHR 1998), which express mostly GLT1 (EAAT2), matches that from patches excised from HEK 293 cells expressing EAAT2 (OTIS and KAVANAUGH, unpublished observations) and suggests a cycling rate of about $60\,s^{-1}$.

Why does the first method yield a slower turnover rate than the second for EAAT2? Formal possibilities include differences in post-translational processing in the different expression systems or differences in the recovery of the anion current relative to the transporter-binding site. Another factor that may contribute to the faster recovery time in patches is the possibility that a glutamate molecule can, in principle, unbind from either membrane face after activation of the anion current. However, we favor an alternative to these explanations: oocytes contain millimolar intracellular concentrations of the transporter substrates L-glutamate and L-aspartate as well as 10–20 mM Na^+ (Zerangue and Kavanaugh 1996b). These intracellular species, if present at similar concentrations in mammalian cells, are presumably dialyzed in the outside-out patches. Patch data from Purkinje-neuron transporters are consistent with the idea that similar concentrations of Na^+ and substrate significantly slow the cycling rate of the transporter, and a kinetic model also supports this hypothesis (Otis and Jahr 1998). Clearly, a critical test of this proposal will be measurement of recovery of currents in patches removed from oocytes expressing EAAT2, both with and without internal Na–glutamate.

C. Transporter Localization

Immunocytochemical studies and *in situ* hybridization data have led to a general consensus that EAAT1 and EAAT2 are expressed in glial astrocytes but not oligodendrocytes, while EAAT3 and EAAT4 are exclusively neuronal transporters. EAAT5 is expressed in both glial cells and neurons in the retina. The following describes the cellular and subcellular localization of these molecules, and a summary is presented in Table 1.

I. Glial EAATs

1. Excitatory Amino Acid Transporter 1 (GLAST)

Antibodies to GLAST have shown that this transporter is localized primarily in the cerebellar Bergmann glial cell but is also present, albeit at greater than fivefold lower levels, in astrocytes in the olfactory bulb, hippocampus, neocortex, and striatum (Lehre et al. 1995). Non-Bergmann glial astrocytes in the cerebellar granule cell layer also stain positive for GLAST, but astrocytes in the white-matter tracts had low levels of staining. At the ultrastructural level, analysis with gold-particle-coupled antibodies uncovered higher levels of GLAST protein in astrocytic processes opposed to neuropil-containing excitatory synapses versus those opposed to endothelial cells surrounding capillaries, the pial surface, or Purkinje cell bodies (Chaudry et al. 1995). Interestingly, processes surrounding pre- and postsynaptic elements of excitatory and inhibitory synapses had similarly high particle densities. Furthermore, parallel-fiber (PF) synapses made onto dendritic spines on putative Purkinje neurons had significantly more GLAST-associated particles located nearby (within 1 μm), compared to PF synapses made onto presumed interneuron-dendritic shafts. This may explain the reported differences in the effects of transport blockers at the two types of synapse (Barbour et al. 1994).

2. Excitatory Amino Acid Transporter 2 (GLT-1)

EAAT2/GLT-1 is presumed to be the most abundant transporter in the nervous system. Astrocytes in nearly all areas of the brain and spinal cord localize GLT-1 to some degree (Rothstein et al. 1994; Lehre et al. 1995). GLT-1 staining at the ultrastructural level is similar to that reported for GLAST, with the signal localized to areas of high densities of excitatory synaptic contacts. Fibrous astrocytes, but not oligodendrocytes in white-matter areas, also express GLT-1 (Lehre et al. 1995). It has been reported that GLT-1 is expressed transiently in neurons on migrating growth cones at embryonic ages (Yamada et al. 1998) and, at low levels, in mature neuronal-axon terminals (Lehre et al. 1995). Cultures of hippocampal neurons also seem to express functional GLT-1 (Mennerick et al. 1998). The significance of this neuronal localization is unclear and may be related to culturing conditions.

II. Neuronal EAATs

1. Excitatory Amino Acid Transporter 3 (EAAC1)

EAAT3/EAAC1 is expressed in a wide range of glutamatergic and GABAergic neurons in the central nervous system, including neurons of the dorsal horn of the spinal cord, medium-sized neurons in striatum, CA1 and CA3 pyramidal cells, cerebellar granule neurons, and neocortical pyramidal neurons from layers II/III and V (Rothstein et al. 1994). Localization in cerebellar Purkinje neurons is controversial; three groups find expression in the cells (Rothstein et al. 1994; Kanai et al. 1995; Dehnes et al. 1998) while one does not (Shibata et al. 1996). Surprisingly, the transporter has an almost exclusive localization to dendrites and perikarya. However, if the localization of EAAT3 in Purkinje neurons is confirmed, it will serve as a rare example of a transporter abundantly localized in an axon terminal (Rothstein et al. 1994; Furuta et al. 1997). Examination of EAAT3 localization with antibodies made to different parts of the sequence, in addition to settling this controversy, may uncover further neuronal localization in presynaptic terminals. Such findings could account for the clear demonstrations that nerve terminals possess avid glutamate-uptake mechanisms (Gundersen et al. 1993).

2. Excitatory Amino Acid Transporter 4

Dendrites of Purkinje neurons in the cerebellum are heavily labeled by antibodies directed to EAAT4 (Yamada et al. 1996; Furuta et al. 1997; Nagao et al. 1997; Dehnes et al. 1998), while very faint expression has been reported in other neurons in the striatum, hippocampus, and neocortex. Labeling in the cerebellum appears in a striking parasagittal banding pattern, colocalizing (at the level of individual Purkinje cells) with zebrin II, another protein expressed in a compartmental manner in the cerebellum (Nagao et al. 1997; Dehnes et al. 1998). At the ultrastructural level, pre-embedding electron microscopy (EM) suggests that the transporter is excluded from the postsynaptic density but is concentrated just outside, in a perisynaptic profile surrounding (PF) and (CF) synapses (Tanaka et al. 1997). Postembedding techniques have provided more details; EAAT4 is present in postsynaptic densities, but is concentrated at its highest levels in postsynaptic spine membranes opposed to the astrocytic membranes that envelop the spines (Dehnes et al. 1998). This specific localization, typically within 1 μm of transmitter release sites, coupled with the high density of EAAT4 molecules (roughly estimated by Dehnes et al. as more than 3600 μm^{-2}), may help to ensure that signals at postsynaptic spines remain compartmentalized (Sect D). Finally, these studies also concluded that a substantial fraction of the EAAT4 protein is localized intracellularly in smooth endoplasmic reticulum and multivesicular bodies, raising the possibility that intracellular trafficking actively regulates the number of transporters at the plasma membrane.

III. EAAT5, a Retinal Transporter

Northern-blot analysis of human tissues indicates that EAAT5 is specifically expressed in the retina (Arriza et al. 1997). Antibodies generated to the homologous protein in salamanders gave patterns of localization consistent with expression in Müller glial cells and the outer plexiform layer, possibly in association with photoreceptor terminals and/or bipolar cell dendrites (Eliasof et al. 1998). Two distinct isoforms of EAAT5 (a and b), are present in salamander retina, with 77% and 58% sequence identity to human EAAT5, respectively. While these two forms were extensively colocalized, only EAAT5a cRNA was functional when injected into *Xenopus* oocytes. No ultrastructural data are available at this time to indicate where the transporter is located within the various retinal synapses, though a domain at the *C*-terminus of the predicted sequence suggests the possibility that the transporter may interact with PDZ-domain-containing proteins, such as postsynaptic-density protein 95 (Arriza et al. 1997).

D. Transporter Function at Several Well-Studied Synapses

I. Parallel-Fiber and Climbing-Fiber Synapses in the Cerebellum

At both major types of excitatory synaptic input to cerebellar Purkinje neurons, the PF synapses and the CF synapses, transporters play a clear role in speeding the clearance of transmitter from synaptic receptors. This was first described by BARBOUR et al. (1994), who showed that the decay of the AMPA-receptor-mediated postsynaptic currents (EPSCs) in Purkinje neurons were markedly slowed by the transporter substrates (i.e. competitive antagonists) D-aspartate, THA, and PDC. In contrast, PF inputs to inhibitory interneurons were not altered by uptake blockers. The authors suggested a basis for these differences based on the ultrastructural geometries of the different synapses. PF synapses and CF synapses are formed on dendritic spines of Purkinje cell dendrites, which are often completely enveloped by Bergmann glial cell membranes, thereby creating a more restricted diffusional exit. PF contacts to interneurons, however, are made on dendritic shafts with increased access to the larger extracellular volume. Analytical simulations of diffusion in hypothetical synaptic spaces supported these interpretations, showing that the contents of a single vesicle released onto a "covered" spine head were cleared more slowly than those released onto an infinite plane. However, these simulations did not include transport, so it is not known whether transporters with realistic properties and localizations could speed clearance enough to account for the faster EPSC decays in controls.

TAKAHASHI et al. (1995) found similar effects on the EPSC time course and further extended the examination of these synapses, considering whether the amount of transmitter released affects the rate of decay of the EPSC. This is important because, depending on the density of active release sites, impairment of transport might slow the clearance of transmitter not only by slowing clearance at individual synapses but also by allowing the contents of single vesicles to overlap more in space, a sort of "pooling" effect (TRUSSELL, this volume). The data showed that, under control conditions, release probability may be sufficient for crosstalk to occur, but the effects of blocking uptake at low release probability, a condition where the pooling effect would be minimal, were not examined.

Recently, transporter currents have been recorded in Purkinje neurons (OTIS et al. 1997) and Bergmann glial cells (BERGLES et al. 1997; CLARK and BARBOUR 1997) in response to PF and CF activity. By calibrating the charge transferred during a CF-elicited transporter current, we estimated the relative amount of glutamate transported into the Purkinje neuron (as opposed to the Bergmann glial cell or the CF terminal; OTIS et al. 1997). The calibration was performed by measuring ratios of charge to radiotracer flux in *Xenopus* oocytes expressing EAAT3 or EAAT4. These experiments suggested that, on average, greater than 22% of the released glutamate is cleared by postsynaptic transporters. Providing support for this proposal, the selective blockade of postsynaptic transporters by inclusion of D-aspartate solutions has been reported to slow the CF EPSC (TAKAHASHI et al. 1996), although the possibility that heteroexchange may liberate D-aspartate from the neuron in these experiments could temper this conclusion.

How can we interpret these findings in light of the function and localization of EAATs at these synapses? Estimated turnover times for all of the

EAATs are probably too slow for multiple cycles to occur during the EPSC decay (Table 1). Therefore, the density of transporters is critical if sequestration of glutamate by binding (presumably with later transport) is to play a role in speeding clearance. Modeling with realistic localizations, estimates of densities, and transporter properties will be required to determine whether EAATs can indeed speed the clearance of the contents of single vesicles of glutamate from "covered" spines. However, given their location, forming a sort of ring around the necks or bases of spines, GLAST and EAAT4 transporters may serve more to isolate release sites from one another than to speed the concentration transient at a single site. Supporting this idea, the synaptically elicited transporter currents in Bergmann glial cells (BERGLES et al. 1997; CLARK and BARBOUR 1997) and Purkinje neurons (OTIS et al. 1997) have slow rise times (1–3 ms) compared to the intrinsic rise times that the transporters exhibit at high glutamate concentrations (BERGLES et al. 1997; OTIS et al. 1997). Other synapses that show large effects of transporter blockers on the EPSC decay also have a very high density of active release sites (OTIS et al. 1996; KINNEY et al. 1997).

II. Schaffer Collateral/Commissural Synapses in the Hippocampus

Unlike the effect on cerebellar synapses, blocking EAAT function has surprisingly little effect on the function of the excitatory connection between CA3 and CA1 pyramidal neurons. In brain slices, AMPA-receptor-mediated EPSCs elicited in CA1 pyramidal cells by stimulation of Schaffer collateral/commissural inputs are not affected by the transporter antagonist/substrate PDC (ISAACSON and NICOLL 1993; SARANTIS et al. 1993). An EAAT2/GLT-1-specific antagonist, dihydrokainate, also did not affect the AMPA component of the EPSC but caused the NMDA component to increase, with no change in the decay rate (HESTRIN et al. 1990). The results of uptake blockade in hippocampal cultures are less clear, with a slowing of the glutamate-concentration transient being suggested in some experiments (MENNERICK and ZORUMSKI 1994) but not others (TONG and JAHR 1994; DIAMOND and JAHR 1996). Rather, in these experiments, an increase in receptor occupancy was seen without any evidence of slowing of the concentration transient. These differences in the culture studies are likely to be explained by technical differences regarding the extent to which glial cells cover synaptic elements and the degree of perfusion of the extracellular space. Glial processes in the CA1 region are closely associated with many excitatory synapses (SORRA and HARRIS 1993), but they do not seem to ensheath excitatory contacts to the degree seen for synapses in the cerebellar molecular layer (PALAY and CHAN-PALAY 1974). A less extensive diffusional restriction at hippocampal synapses may explain why impairment of glutamate uptake does not slow the clearance of glutamate at individual contacts. The possibility should be considered, however, that glutamate transport in the hippocampal region might be of such high capacity that it is impossible to completely block it with substrate antagonists (Bergles and Jahr 1997). It may also be that the density of active release sites is too low, at least for AMPA receptors, for a pooling or crosstalk effect to slow the EPSC. These conclusions would need to be specific for the receptor type; for the higher-affinity NMDA receptors, evidence indicates that crosstalk may be significant under some conditions (Asztely et al. 1997). Similarly, activation of presynaptic metabotropic receptors during repetitive stimulation of mossy-fiber inputs to CA3 pyramidal cells is, apparently, limited by glutamate transport (Scanziani et al. 1997).

Synaptically evoked transporter currents in hippocampal astrocytes are likely to provide key evidence to eventually solve this puzzle. An examination of the kinetics of synaptic responses in comparison to responses with glutamate pulses in outside-out patches led Bergles and Jahr (1997) to propose that glial membranes outside the synaptic cleft sense glutamate within 1 ms after release. They also suggested that glutamate may persist, in the vicinity of these processes, at effective concentrations for more than 10 ms, longer than is believed to be the case in the synaptic cleft (Clements et al. 1992). These results are consistent with the EM localization data for GLT-1, placing the transporters at ~1-μm distances from synaptic specializations. The density estimates for transporters in outside-out patches were on the order of 18,000 μm^{-2}; such high densities are in line with the idea that the transport capacity in the CA1 neuropil is extremely high. A comparison of the physiological data with realistic simulations of diffusion and transporter activation should prove informative and will, hopefully, provide insight into why blocking transport at these types of synapses has little effect on the excitatory signal.

III. ON-Bipolar Cell Synapse in the Teleost Retina

ON-bipolar cells in the retina depolarize in response to light inputs despite the fact that their presynaptic partners, the photoreceptors, respond to light by releasing less glutamate. This sign inversion seems to occur via two distinct ionic mechanisms in the postsynaptic bipolar cells (Nawy and Copenhagen 1990; Grant and Dowling 1996). The first involves a metabotropic glutamate receptor coupled to a phosphodiesterase that degrades cyclic guanosine monophosphate (cGMP), thus allowing cGMP-activated cation channels to close (Nawy and Jahr 1990; Shiells and Falk 1990). The other glutamate-elicited hyperpolarization results from activation of a glutamate-gated chloride conductance (Grant and Dowling 1995). Pharmacological experiments in the Grant and Dowling study showed that the chloride current is mediated by a glutamate transporter with a large anion conductance, which may correspond to EAAT5 (Eliasof et al. 1998). This is the best evidence to date for a physiological role for the accessory anion conductance of the transporter. EAAT5 or another transporter may also serve as an inhibitory autoreceptor on photoreceptor terminals (Picaud et al. 1995b). Alternatively, a transporter on the terminal may play a novel role linking photoreceptor membrane potential to glutamate concentration in the cleft (Gaal et al. 1998). As mentioned earlier, the *C*-terminal sequence of EAAT5 implies the possibility of clustering via pre- and/or postsynaptic anchoring proteins, and the ultrastructural localization of transporters may turn out to be very interesting at these synapses. Will there be differences in spatial distributions of transporters depending on their involvement in clearance, as compared to the hypothesized additional-signaling roles?

It seems fitting that this final example of transporter function includes the irony that the glutamate transporter, rather than terminating a signal, functions to transmit a signal as an inhibitory receptor. The glimpse, presented in this review, of the structure, functional characteristics, localization, and synaptic physiology of EAATs leaves the impression that there remain many functions that are yet to be identified for transporters. We should probably prepare ourselves for surprising future discoveries of novel roles for EAATs in synaptic transmission.

References

Amara SG, Arriza JL (1993) Neurotransmitter transporters: three distinct gene families. Curr Opin Neurobio 3:337–344

Arriza JL, Fairman WA, Wadiche JL, Murdoch GH, Kavanaugh MP, Amara SG (1994) Functional comparisons of three glutamate transporter subtypes cloned from human motor cortex. J Neurosci 14:5559–5569

Arriza JL, Eliasof S, Kavanaugh MP, Amara SG (1997) EAAT5, a retinal glutamate transporter coupled to a chloride conductance. Proc Natl Acad Sci USA 95:4155–4160

Asztely F, Erdemli G, Kullmann DM (1997) Extrasynaptic glutamate spillover in the hippocampus: dependence on temperature and the role of active glutamate uptake. Neuron 18:281–293

Barbour B, Brew H, Attwell D (1991) Electrogenic uptake of glutamate and aspartate into glial cells isolated from the salamander (*Ambystoma*) retina. J Physiol 436: 169–193

Barbour B, Keller BU, Llano I, Marty A (1994) Prolonged presence of glutamate during excitatory synaptic transmission to cerebellar Purkinje cells. Neuron 12: 1331–1343

Bergles DE, Jahr CE (1997) Synaptic activation of glutamate transporters in hippocampal astrocytes. Neuron 19:1297–1308

Bergles DE, Dzubay JA, Jahr CE (1997) Glutamate transporter currents in Bergmann glial cells follow the time course of extrasynaptic glutamate. Proc Natl Acad Sci USA 94:14821–14825

Bergles DE, Jahr CE (1998) Glial contributions to glutamate uptake at schaffer collateral-commissural synapses in the hippocampus. J Neurosci 18:7709–7716

Billups B, Rossi D, Attwell D (1996) Anion conductance behavior of the glutamate uptake carrier in salamander retinal cells. J Neurosci 16:6722–6731

Bouvier M, Szatkowski M, Amato A, Attwell D (1992) The glial cell glutamate uptake carrier countertransports pH-changing anions. Nature 360:471–474

Brew H, Attwell D (1987) Electrogenic glutamate uptake is a major current carrier in the membrane of axolotl retinal glial cells. Nature 327:707–709

Chaudry FA, Lehre KP, van Lookeren Campagne M, Ottersen OP, Danbolt NC, Storm-Mathisen (1995) Glutamate transporters in glial plasma membranes: highly differentiated localizations revealed by quantitative ultrastructural immunocytochemistry. Neuron 15:711–720

Clark BA, Barbour B (1997) Currents evoked in Bergmann glial cells by parallel fibre stimulation in rat cerebellar slices. J Physiol 502:335–350
Clements JD, Lester RA, Tong G, Jahr CE, Westbrook GL (1992) The time course of glutamate in the synaptic cleft. Science 258:1498–1501
Clements JD, Feltz A, Sahara Y, Westbrook GL (1998) Activation kinetics of AMPA receptor channels reveal the number of functional agonist binding sites. J Neurosci 18:119–127
Dehnes Y, Chaudhry FA, Ullensvang K, Lehre KP, Storm-Mathisen J, Danbolt NC (1998) The glutamate transporter EAAT4 in rat cerebellar Purkinje cells: A glutamate gated chloride channel concentrated near the synapse in parts of the dendritic membrane facing astroglia. J Neurosci 18:3606–3619
Diamond JS, Jahr CE (1997) Transporters buffer synaptically released glutamate on a millisecond time scale. J Neurosci 17: 4672–4687
Eliasof S, Jahr CE (1996) Retinal glial cell glutamate transporter is coupled to an anionic conductance. Proc Natl Acad Sci USA 93:4153–4158
Eliasof S, Werblin F (1993) Characterization of the glutamate transporter in retinal cones of the tiger salamander. J Neurosci 13:402–411
Eliasof S, Arriza JL, Kavanaugh MP, Amara SG (1998) Excitatory amino-acid transporters of the salamander retina: identification, localization and function. J Neurosci 18:698–712
Erecinska M, Wantorsky D, Wison DF (1983) Aspartate transport in synaptosomes from rat brain. J Biol Chem 258:9069–9077
Fairman WA, Vandenberg RJ, Arriza JL, Kavanaugh MP, Amara SG (1995) An excitatory amino-acid transporter with properties of a ligand gated chloride channel. Nature 375:599–603
Fairman WA, Sonders MS, Murdoch GH, Amara SG (1998) Arachidonic acid elicits a substrate-gated proton current associated with the glutamate transporter EAAT4. Nature Neuroscience 1:105–113
Furuta A, Martin LJ, Lin C-LG, Dykes-Hoberg M, Rothstein JD (1997) Cellular and synaptic localization of the neuronal glutamate transporters excitatory amino acid transporter 3 and 4. Neurosci 81:1031–1042
Gaal L, Roska B, Picaud SA, Wu SM, Marc R, Werblin FS (1998) Postsynaptic response kinetics are controlled by a glutamate transporter at cone photoreceptors. J Neurophys 79:190–196
Grant GB, Dowling JE (1995) A glutamate-activated chloride current in cone-driven ON bipolar cells of the white perch retina. J Neurosci 15:3852–3862
Grant GB, Dowling JE (1996) ON bipolar cell responses in the teleost retina are generated by two distinct mechanisms. J Neurophys 76:3842–3849
Grunewald, M., Bendahan, A., and Kanner, B.I. (1998) Biotinylation of single cysteine mutants of the glutamate transporter GLT-1 from rat brain reveals its unusual topology. Neuron 21:623–632
Gundersen V, Danbolt NC, Ottersen OP, Storm-Mathisen J (1993) Demonstration of glutamate/aspartate uptake activity in nerve endings by use of antibodies recognizing exogenous D-aspartate. Neurosci 57:97–111
Haugeto O, Ullensvang K, Levy LM, Chaudry FA, Honore T, Nielsen M, Lehre KP, Danbolt NC (1996) Brain glutamate transporters form homomultimers. J Biol Chem 271:27715–27722
Häusser M, Roth A (1997) Dendritic and somatic glutamate receptor channels in rat cerebellar Purkinje cells. J Physiol 501:77–95
Hestrin S, Sah P, Nicoll RA (1990) Mechanisms generating the time course of dual component excitatory synaptic currents recorded in hippocampal slices. Neuron 5: 247–253
Isaacson JS, Nicoll RA (1993) The uptake inhibitor L-*trans*-PDC enhances responses to glutamate but fails to alter the kinetics of excitatory postsynaptic currents. J Neurophys 70:2187–2191
Jonas P, Major G, Sakmann B (1993) Quantal components of unitary EPSCs at the

mossy-fiber synapse on CA3 pyramidal cells of rat hippocampus. J Physiol 472: 615–663
Kanai Y Hediger MA (1992) Primary structure and functional characterization of a high-affinity glutamate transporter. Nature 360:467–471
Kanai Y, Bhide PG, DiFiglia M, Hediger MA (1995) Neuronal high-affinity glutamate transport in the rat central nervous system. NeuroReport 6:2357–2362
Kanai Y (1997) Family of neutral and acidic amino acid transporters: molecular biology, physiology and medical implications. Curr Opin Cell Biol 9:565–572
Kanner BI, Sharon I (1978) Active transport of glutamate by membrane vesicles isolated from rat brain. Biochemistry 17:3949–3953
Kanner BI, Bendahan A (1982) Binding order of substrates to the sodium and potassium ion coupled glutamic acid transporter from rat brain. Biochemistry 21: 6327–6330
Kataoka Y, Morii H, Watanabe Y, and Ohmori H (1997) A postsynaptic excitatory amino acid transporter with chloride conductance functionally regulated by neuronal activity in cerebellar Purkinje cells. J Neurosci 17:7017–24
Kavanaugh MP, Bendahan A, Zerangue N, Zhang Y, Kanner BI (1997) Mutation of an amino acid residue influencing potassium coupling in the glutamate transporter GLT-1 induces obligate exchange. J Biol Chem 272:1703–1708
Kinney GA, Overstreet LS, Slater NT (1997) Prolonged physiological entrapment of glutamate in the synaptic cleft of cerebellar unipolar brush border cells. J Neurophys 78:1320–1333
Klockner U, Storck T, Conradt M, Stoffel W (1993) Electrogenic L-glutamate uptake in *Xenopus laevis* oocytes expressing a cloned rat brain L-glutamate/L-aspartate transporter (GLAST-1). J Biol Chem 268:14594–14596
Larsson HP, Picaud SA, Werblin FS, Lecar H (1996) Noise analysis of the glutamate-activated current in photoreceptors. Biophys J 70:733–742
Lehre KP, Levy LM, Otterson OP, Storm-Mathisen J, Danbolt NC (1995) Differential expression of two glial glutamate transporters in the rat brain: quantitative and immunocytochemical observations. J Neurosci 15:1835–1853
Levy LM, Warr O, Attwell D (1998) Stoichiometry of the glial glutamate transporter GLT-1 expressed inducibly in a CHO cell line selected for low endogenous Na^+-dependent glutamate uptake. J Neurosci, in press
Mennerick S, Zorumski CF (1994) Glial contribution to excitatory transmission in cultured hippocampal cells. Nature 368:59–62
Mennerick S, Dhond RP, Benz A, Xu W, Rothstein JD, Danbolt NC, Isenberg KE, Zorumski CF (1998) Neuronal expression of the glutamate transporter GLT-1 in hippocampal microcultures. J Neurosci 18:4490–4499
Mitrovic AD, Amara SG, Johnston GAR, Vandenberg RJ (1998) Identification of functional domains of the human glutamate transporters EAAT1 and EAAT2. J Biol Chem 273:14698–14706
Nagao S, Kwak S, Kanazawa I (1997) EAAT4, a glutamate transporter with properties of a chloride channel, is predominantly localized in Purkinje cell dendrites, and forms parasagittal compartments in rat cerebellum. Neurosci 78:929–933
Naito S, Ueda T (1985) Characterization of glutamate uptake into synaptic vesicles. J Neurochem 44:99–109
Nawy S, Copenhagen DR (1990) Intracellular cesium separates two glutamate conductances in retinal bipolar cells of goldfish. Vision Res 30: 967–972
Nawy S, Jahr C (1990) Suppression by glutamate of cGMP-activated conductance in retinal bipolar cells. Nature 346:269–271
Nelson, PJ, Dean, GE, Aronson, PS, and Rudnick, G (1983) Hydrogen ion cotransport by the renal brush border glutamate transporter. Biochemistry 22: 5459–5463
Ogston AG (1955) Removal of acetylcholine from a limited volume by diffusion. J Physiol 128:222–223
Otis TS, Kavanaugh MP, Jahr CE (1997) Postsynaptic glutamate transport at the climbing fiber-Purkinje cell synapse. Science 277:1515–1518

Otis TS, Jahr CE (1998) Anion currents and predicted glutamate flux through a neuronal glutamate transporter. J Neurosci 18:7099–7110
Palay SL, Chan-Palay V (1974) Cerebellar cortex. Springer Verlag. New York
Picaud S, Larsson HP, Grant GB, Lecar H, Werblin FS (1995a) Glutamate-gated chloride channel with glutamate-transporter-like properties in cone photoreceptors of the tiger salamander. J Neurophysiol 74:1760–1771
Picaud, SA, Larsson, HP, Wellis, DP, Lecar, H, Werblin, FS (1995b) Cone photoreceptors respond to their own glutamate release in the tiger salamander Proc Natl Acad Sci USA 92:9417–9421
Pines G, Danbolt NC, Bjoras M, Zhang Y, Bendahan A, Eide L, Koespell H, Storm-Mathisen J, Seeberg E, Kanner BI (1992) Cloning and expression of a rat brain L-glutamate transporter. Nature 360:464–467
Raman IM, Trussell LO (1996) The mechanism of α-amino-3-hydroxy-5-methyl-4-isoxazolepropionate receptor desensitization after removal of glutamate. Biophys J 68:137–146
Rosenmund C, Stern-Bach Y, Stevens CF (1998) The tetrameric structure of a glutamate receptor channel. Science 280:1596–1599
Rothstein JD, Martin L, Levey AI, Dykes-Hoberg M, Jin L, Wu D, Nash N, Kuncl RW (1994) Localization of neuronal and glial transporters. Neuron 13:713–725
Sarantis M, Ballerini L, Miller B, Silver RA, Edwards M, Attwell D (1993) Glutamate uptake from the synaptic cleft does not shape the decay of the non-NMDA component of the synaptic current. Neuron 11:541–549
Scanziani M, Salin PA, Vogt KE, Malenka RC, Nicoll RA (1997) Use-dependent increases in glutamate concentration activate presynaptic metabotropic glutamate receptors. Nature 385:630–634
Shibata T, Watanabe M, Tanaka K, Wada K, Inoue Y (1996) Dynamics changes in expression of glutamate transporter mRNAs in developing brain. NeuroReport 7:705–709
Sheills RA, Falk G (1990) Glutamate receptors of rod bipolar cells are linked to a cyclic GMP cascade via a G-protein. Proc R Soc Lond B Biol Sci 242:91–94
Slotboom DJ, Lolkema JS, Konings WN (1996) Membrane topology of the *C*-terminal half of the neuronal, glial and bacterial glutamate transporter family. J Biol Chem 271:31317–31321
Sorra KE, Harris KM (1993) Occurrence and three dimensional structure of multiple synapses between individual radiatum axons and their target pyramidal cells in hippocampal area CA1. J Neurosci 13:3736–3748
Stallcup WB, Bulloch K, Baetge EE (1979) Coupled transport of glutamate and sodium in a cerebellar nerve cell line. J Neurochem 32:57–65
Storck T, Schulte S, Hofmann K, Stoffel W (1992) Structure, expression, and functional analysis of a Na^+-dependent glutamate/aspartate transporter from rat brain. Proc Natl Acad Sci USA 89:10955–10959
Takahashi M, Kovalchuk Y, Attwell D (1995) Pre- and postsynaptic determinants of EPSC waveform at cerebellar climbing fiber and parallel fiber to Purkinje cell synapses. J Neurosci 15:5693–5702
Takahashi M, Sarantis M, Attwell D (1996) Postsynaptic glutamate uptake in rat cerebellar Purkinje cells. J Physiol 497:523–530
Tanaka J, Ichikawa R, Watanabe M, Tanaka K, Inoue Y (1997) Extra-junctional localization of glutamate transporter EAAT4 at excitatory Purkinje cell synapses. NeuroReport 8:2461–2464
Tanaka K (1993) Expression cloning of a rat glutamate transporter. Neurosci Res 16:149–153
Tong G, Jahr CE (1994) Block of glutamate transporters potentiates postsynaptic excitation. Neuron 13:1195–1203
Tzingounis AV, Lin CL, Rothstein JD, Kavanaugh MP (1998) Arachidonic acid activates a proton current in the rat glutamate transporter EAAT4. J Biol Chem 273: 17315–17317

Vandenberg RJ, Arriza JL, Amara SG, Kavanaugh MP (1995) Constitutive ion fluxes and substrate binding domains of human glutamate transporters. J Biol Chem 270:17668–17671
Wadiche JI, Amara SG, Kavanaugh MP (1995a) Ion fluxes associated with excitatory amino acid transport. Neuron 15:721–728
Wadiche JI, Arriza JL, Amara SG, Kavanaugh MP (1995b) Kinetics of a human glutamate transporter. Neuron 14:1019–1027
Wadiche JI, Kavanaugh MP (1998) Macroscopic and microscopic properties of a cloned glutamate transporter/chloride channel. J Neurosci 18:7650–7661
Wahle S, Stoffel W (1996) Membrane topology of the high-affinity L-glutamate transporter (GLAST-1) of the central nervous system. J Cell Biol 135:1867–1877
Yamada K, Watanabe M, Shibata T, Nagashima M, Tanaka K, Inoue Y (1998) Glutamate transporter GLT-1 is transiently localized on growing axons of the mouse spinal cord before establishing astrocytic expression. J Neurosci 18:5706–5713
Zerangue N, Kavanaugh MP (1996a) Flux coupling in a neuronal glutamate transporter. Nature 383:634–637
Zerangue N, Kavanaugh MP (1996b) Interaction of L-cysteine with a human excitatory amino acid transporter. J Physiol 493:419–423
Zerangue N, Kavanaugh MP (1996c) ASCT-1 is a neutral amino acid exchanger with chloride channel activity. J Biol Chem 271:27991–27994

Section IV
Involvement of Glutamate Receptors and Transporters in Neurological Diseases

CHAPTER 14

Glutamate-Mediated Excitotoxicity

G.A. KERCHNER, A.H. KIM, and D.W. CHOI

A. Introduction

Excitotoxicity – the ability of glutamate receptor activation to trigger neuronal cell death – has been recognized for more than four decades (LUCAS and NEWHOUSE 1957; OLNEY 1969; CHOI 1988b). Over the last 10–15 years, there has been an accumulation of evidence that suggests that glutamate toxicity contributes to brain or spinal cord tissue damage in a variety of disease settings, such as brain ischemia, seizures, traumatic brain injury, and hypoglycemia. The concept of excitotoxicity has undergone substantial evolution. The earlier notion that glutamate receptors must be overactivated to induce cell death has yielded to a newer view that physiological levels of activation may be lethal to neurons rendered vulnerable by energy depletion or other derangements. Recent evidence suggesting that oligodendrocytes may be as susceptible to excitotoxicity as many neurons has further broadened the original definition of this process. In addition, as distinctions between necrosis and programmed cell death have become better defined, increased scrutiny has been accorded to the modes of cell death induced by glutamate-receptor activation. Furthermore, with an explosion of new knowledge regarding the molecular composition and modulation of ionotropic glutamate receptors, an increased appreciation of how these factors may affect excitotoxicity is developing. Finally, the late downstream effectors of excitotoxic cell death are beginning to be recognized.

B. Glutamate Receptors and Excitotoxicity

Ionotropic glutamate receptors play a central role in the induction of excitotoxic cell death. A healthy neuron can tolerate substantial levels of glutamate-receptor activation in the course of normal synaptic transmission, but abnormally intense or prolonged activation can result in cell death. Alternatively, normal ambient concentrations of glutamate may be lethal to a neuron whose function is already compromised for other reasons, a principle established by the demonstration that glucose deprivation potentiated cerebellar granule cell vulnerability to glutamate-induced death (NOVELLI et al. 1988). Based on this notion, excitotoxicity has been identified as a pathophysiological factor not only in the acute neurological disorders mentioned above, but also (and more speculatively) in chronic neurodegenerative conditions, such as amyotrophic lateral sclerosis, Parkinson's disease, Huntington's disease, and Alzheimer's disease, where it is possible that even physiological levels of glutamate-receptor stimulation may induce neuronal death (BEAL et al. 1993).

The link between ionotropic glutamate-receptor activation and consequent excitotoxic cell death is, predictably enough, a disruption of ionic gradients across the plasma membrane. While Na^+ influx through *N*-methyl-D-aspartate (NMDA)- or α-amino-3-hydroxy-5-methylisoxazole-4-propionic-acid (AMPA)/kainate-receptor-gated channels (with accompanying Cl^- and water influx) accounts for most of the cell-body swelling that occurs acutely during glutamate overexposure, Ca^{2+} influx through NMDA receptors and other routes is the predominant factor in the neurodegeneration that occurs over subsequent hours (CHOI 1988a, 1995; see below). More recently, the possibility has been raised that shifts in other ions may contribute to excitotoxic neuronal loss; Zn^{2+}, released from glutamatergic presynaptic nerve terminals may enter postsynaptic neurons in toxic quantities via routes facilitated by AMPA/kainate- and NMDA-receptor activation (CHOI and KOH 1998), and K^+, by exiting neurons via voltage-gated K^+ channels or NMDA- or AMPA/kainate-receptor-gated channels, may promote programmed cell death (YU et al. 1997b).

I. NMDA and AMPA/Kainate Receptors in Neuronal Excitotoxicity

1. Rapidly Versus Slowly Triggered Excitotoxicity

Intense exposure to glutamate for only 2–3 min is sufficient to trigger the delayed death of cultured cortical neurons over the next 24 h, a phenomenon we have called "rapidly triggered excitotoxicity" (CHOI et al. 1987; CHOI 1987, 1992). This injury is characterized by two components distinguishable both in time course and ionic dependence: (1) immediate (within minutes) swelling of neuronal cell bodies and dendrites, which is dependent on the presence of extracellular Na^+ and Cl^- and is mediated by both NMDA and AMPA/kainate receptors; and (2) delayed degeneration of neurons over the next hours, dependent on the presence of extracellular Ca^{2+} and mediated predominantly by NMDA receptors. The dominant participation of NMDA receptors in the neuronal death induced by brief, intense glutamate exposure likely reflects their ability to mediate high levels of Ca^{2+} influx directly.

In contrast, cortical cultures exposed to saturating concentrations of kainate for up to 60 min develop pronounced, reversible dendrosomatotoxic swelling but little neurodegeneration. A similar exposure to AMPA, a powerful excitant, does not even produce swelling (KOH et al. 1990). However, AMPA/kainate-receptor activation is not benign, as low micromolar concentrations of AMPA or kainate alone can kill cultured cortical neurons if the exposure time is extended to 24 h. Like rapidly triggered excitotoxicity, this "slowly triggered excitotoxicity" is likely mediated by excessive Ca^{2+} entry

(Garthwaite and Garthwaite 1986; Rothman et al. 1987) occurring via several routes, including voltage-gated Ca^{2+} channels and reverse operation of neuronal Na^+–Ca^{2+} exchangers (due to membrane depolarization and elevated intracellular Na^+; Yu and Choi 1997).

The observation that NMDA-receptor activation can kill cortical neurons more rapidly than AMPA/kainate-receptor activation does not imply that the former will always mask the latter. As described below, several factors, including, among others, extracellular acidity, the co-release of Zn^{2+} with glutamate, and activity dependent NMDA-receptor desensitization, may reduce the contribution of NMDA receptors relative to AMPA/kainate receptors in inducing excitotoxic death. In addition, the expression of high levels of Ca^{2+}-permeable AMPA receptors in some neurons may enhance their susceptibility to AMPA-receptor-mediated excitotoxicity (see below). The therapeutic potential of AMPA/kainate-receptor antagonist drugs has been highlighted in some studies of transient global ischemia in rats (Sheardown et al. 1990; but also Colbourne et al. 1997) and various animal models of spinal cord injury (Wrathall et al. 1992; Xu et al. 1993).

2. NMDA-Receptor Subtypes

The impressive ability of even brief NMDA-receptor activation to induce neuronal death encourages a detailed characterization of the excitotoxic potential of various NMDA-receptor subtypes. Further incentive for bringing an analysis of excitotoxicity to the subtype level is provided by recent, disappointing results obtained in clinical-stroke trials with non-subtype-selective NMDA-receptor antagonist drugs (Chap. 4), a strategy which may be limited by psychotomimetic and motor side effects (Olney et al. 1991). In addition, the anti-excitotoxic benefits of such broad-spectrum blockade may be counterbalanced by an enhancement of Ca^{2+}-starvation-induced apoptosis (see below). Several groups have now hypothesized that a better therapeutic index might be obtainable with subtype-selective NMDA-receptor blockade.

Each of the four NR2 subtypes (NR2A-D) confers a unique set of characteristics upon the resultant NMDA receptor, such as sensitivity to Mg^{2+} block, glycine and glutamate affinity, and single-channel conductance, making extrapolations to excitotoxicity challenging (Monyer et al. 1992; Kutsuwada et al. 1992; Stern et al. 1992). Results obtained from heteromeric NR1–NR2-expression systems, however, have provided insight into the relative excitotoxic potentials of individual NR2 subunits. Several groups have reported that expression of certain NR1 and NR2 subunit combinations in human embryonic kidney (HEK) cells in the presence of ambient levels of glutamate in the culture medium causes the cells to die (Cik et al. 1993; Chazot et al. 1994; Anegawa et al. 1995). These studies indicated that, in the absence of NMDA antagonists, co-transfection of NR1 and NR2A resulted in a greater frequency of cell death than transfection of NR1 with NR2B, while co-transfection of NR1 and NR2C yielded no death despite comparable amplitudes of NMDA-induced whole-cell currents with each cell line (Chazot et al. 1994; Anegawa et al. 1995). Measurements of intracellular Ca^{2+} concentration in these cell lines have shown that the magnitudes of peak agonist-induced Ca^{2+} responses correlate well with excitotoxic vulnerability (Grimwood et al. 1996; Grant et al. 1997).

Similarly, the electrophysiological and pharmacological properties of recombinant NMDA receptors can be substantially altered by the inclusion of different NR1 splice variants (Nakanishi and Masu 1994; Zukin and Bennett 1995). The implications of differential NR1-splice-variant expression for excitotoxic vulnerability are difficult to predict given the diverse characteristics of individual isoforms. For instance, homomeric channels comprised of NR1 splice variants harboring the amino-terminal cassette, N1, exhibit up to threefold-larger current amplitudes (Hollmann et al. 1993; Zheng et al. 1994) but also demonstrate a fivefold reduction in agonist affinity (Durand et al. 1993). Inclusion of carboxy-terminal NR1 cassettes, C1, C2, and C2', may also modify NR1 characteristics and, therefore, excitotoxic potential in a complex fashion.

Individual NR2 subunits and NR1 splice variants exhibit different spatial distributions. While NR2A is expressed throughout the adult brain, other NR2 subunits are found in more region-specific patterns: NR2B expression is restricted to the forebrain, NR2C to the cerebellum, and NR2D to the thalamus, midbrain, and brainstem (Watanabe et al. 1993; Monyer et al. 1994; Wenzel et al. 1995). The spatial distribution of NR2B renders this particular subunit a potentially important therapeutic target, since many forms of stroke and certain traumatic brain injuries predominantly involve the forebrain. As discussed in Chap. 4, novel NR2B-specific antagonists have shown promise in reducing damage resulting from middle-cerebral-artery occlusion in rodents. With the development of cassette-specific probes and antibodies, the spatial and neuron-type-specific distribution of NR1 splice variants is just beginning to be known (Standaert et al. 1994; Laurie and Seeburg 1994; Weiss et al. 1998).

3. Non-NMDA-Receptor Subtypes

The notion that AMPA and kainate receptors are potentially important mediators of excitotoxic cell death is supported by the ability of recombinant-receptor-subunit expression to confer sensitivity to agonist-induced cell death upon normally insensitive, non-neuronal cells. Transfection into cultured fibroblasts (Gahring et al. 1996) or HEK cells (Raymond et al. 1996) of the AMPA-receptor subunit GluR-A (which, in the absence of GluR-B, would lead to the formation of Ca^{2+}-permeable channels) or expression of the Ca^{2+}-permeable kainate-receptor subunit GluR6 in HEK cells (Raymond et al. 1996; Carver et al. 1996) renders these cells vulnerable to kainate-induced toxicity.

Neurons that express Ca^{2+}-permeable AMPA-receptor-gated channels lacking the edited form of the GluR-B subunit (Hollmann et al. 1991; Bur-

NASHEV et al. 1992) are present in many regions of the brain and can be identified by kainate-induced Co^{2+} uptake (PRUSS et al. 1991). Consistent with the view that the contribution of a given type of glutamate receptor to excitotoxicity correlates with its permeability to Ca^{2+} (or Zn^{2+}), kainate exposure as brief as 10–60 min induces preferential rises in intracellular Ca^{2+} and selective Ca^{2+}-dependent death in a GluR-B-less subpopulation of cultured cortical neurons, even when membrane depolarization and consequent activation of voltage-gated Ca^{2+} channels are prevented by removing extracellular Na^{+} (TURETSKY et al. 1994). These same neurons, when exposed to extracellular Zn^{2+} and kainate, show enhanced elevations in intracellular Zn^{2+} and increased Zn^{2+}-induced death (YIN and WEISS 1995).

While mice lacking a functional GluR-B gene do not exhibit any decrease in neuronal survival despite increased AMPA-receptor-mediated Ca^{2+} flux in CA1 neurons (JIA et al. 1996), mutant mice expressing GluR-B subunits deficient in glutamine/arginine editing (due to a disruption of the editing-site complementary sequence in intron 11 of the GluR-B gene) form increased numbers of Ca^{2+}-permeable AMPA receptors in hippocampal pyramidal neurons and die early in life after recurrent seizures and hippocampal neuronal cell loss (BRUSA et al. 1995). Further studies will be needed to delineate the bases of these superficially discordant results, but the most likely explanation is that compensatory mechanisms occurring during development – for example, increased intracellular expression of Ca^{2+}-binding proteins, such as parvalbumin or calbindin – are, for some reason, better able to handle the former perturbation than the latter.

Certain subpopulations of neurons are naturally deficient in GluR-B expression, including the γ-aminobutyric-acid (GABA)ergic interneurons of the neocortex (JONAS et al. 1994) and hippocampus (IINO et al. 1990; BOCHET et al. 1994) and reduced nicotinamide adenine dinucleotide phosphate (NADPH)-diaphorase/neuronal nitric oxide synthase (nNOS)-positive neurons in the neocortex, striatum, and hippocampus (CATANIA et al. 1995). Cultured cortical GABAergic neurons also exhibit unusually prominent sensitivity to death induced by NMDA (TECOMA and CHOI 1989) or kainate (YIN et al. 1994). Thus, it is surprising that GABAergic interneurons in the hippocampal CA1 region are resistant to death induced by transient global cerebral ischemia (JOHANSEN et al. 1983; INGLEFIELD et al. 1997); presumably the explanation lies in other cellular characteristics or cell-extrinsic factors, such as synaptic-input patterns. Likewise, although NADPH-diaphorase/nNOS-positive neurons in the striatum may be highly sensitive to kainate toxicity in vitro (KOH et al. 1986), they appear to be selectively resistant to ischemic injury (UEMURA et al. 1990).

The vulnerability of hippocampal CA1 neurons to death after transient global ischemia may be amplified by dynamic changes in AMPA-receptor subunit composition consequent to an ischemic insult, specifically by the enhanced formation of Ca^{2+}-permeable AMPA receptors (PELLEGRINI-GIAMPIETRO et al. 1997). Following transient global cerebral ischemia, GluR-B mRNA levels are reduced both absolutely and relative to the levels of GluR-A and -C in CA1 hippocampal pyramidal neurons (PELLIGRINI-GIAMPIETRO et al. 1992) – precisely the neurons that, ultimately, go on to die. Both the reduction in GluR-B expression in these selectively vulnerable neurons and the accompanying enhancement of AMPA-induced rises in intracellular Ca^{2+} (GORTER et al. 1997) occur before there is any evidence of neurodegeneration, suggesting that the formation of Ca^{2+}-permeable AMPA-receptor-gated channels precedes – and could, therefore, play a role in – neuronal loss. Ca^{2+}-permeable AMPA receptors could be formed not only by the decreased expression of GluR-B but also by the increased expression of other AMPA-receptor subunits; after sublethal oxygen–glucose deprivation, cultured hippocampal neurons demonstrated increased AMPA-receptor-mediated Ca^{2+} influx, a concurrent increase in GluR-D expression, and no significant change in GluR-B expression, as detected by single-cell polymerase chain reaction (YING et al. 1997).

It seems peculiar that neurons vulnerable to excitotoxic death would, in response to ischemia, set in motion a means to enhance Ca^{2+} influx. Perhaps a post-ischemic enhancement of AMPA-receptor-mediated Ca^{2+} influx is truly maladaptive and contributes to the degeneration of CA1 pyramidal neurons following transient global ischemia. However, it could be that this increased Ca^{2+} influx is instead beneficial, as at least some delayed hippocampal CA1-neuronal death may occur through programmed cell death (see below), and neuronal apoptosis can be associated with and, indeed, promoted by low levels of intracellular free Ca^{2+}. This alternative speculation is supported by demonstrations that increasing Ca^{2+} influx can prevent growth-factor-deprivation-induced neuronal apoptosis (FRANKLIN et al. 1995), along with recent fura-2 measurements revealing low basal intracellular free-Ca^{2+} levels in post-ischemic CA1 neurons (CONNOR, personal communication).

In addition to site-selective glutamine/arginine editing of GluR-B, two other classes of AMPA-receptor-subunit variant exist: Three subunits, GluR-B, -C, and -D, are subject to an arginine/glycine switch at another site (LOMELI et al. 1994), and all four AMPA-receptor mRNAs contain one of two exons, termed "flip" and "flop," introduced by alternative splicing (SOMMER et al. 1990). These variants, instead of altering the permeability sequence of the channel, influence channel kinetics. Arginine/glycine editing confers a faster recovery rate from desensitization (LOMELI et al. 1994), and flip channels desensitize about four times more slowly than flop channels (MOSBACHER et al. 1994). As with NMDA receptors, these AMPA-receptor-subunit variants – by differential expression either spatially or temporally – may confer susceptibility or resistance to excitotoxic neurodegeneration on specific neurons. Mice overexpressing the GluR-B flip subunit driven by an α-Ca^{2+}-calmodulin-dependent kinase II (CaM-KII) promoter sustain larger infarctions after permanent middle-cerebral-artery occlusion, and cultured cortical neurons from these mice are more sensitive to glutamate toxicity than controls (LE et al. 1997), observations which could be explained by altered kinetics of AMPA-receptor-mediated responses due to an increased flip/flop ratio or, alternatively, by a net increase in the total number of AMPA receptors per neuron. Potentially important changes in the GluR-B flip/flop ratio have also been demonstrated in hippocampal neurons following kainate-induced seizures (Pollard et al. 1993).

II. Non-NMDA Receptors in Excitotoxic White-Matter Injury

While excitotoxicity has traditionally been considered to involve only neurons, recent evidence suggests that glutamate-receptor activation may also kill oligodendroglia. Oka et al. (1993) first described glutamate toxicity in cultured oligodendrocytes, but cell death in their system was due to inhibition of cysteine uptake and consequent glutathione depletion, not to glutamate-receptor activation. Primary oligodendrocytes and immortalized oligodendroglial lineage cells in culture express several AMPA and kainate-receptor subunits, and exposure to non-desensitizing AMPA/kainate-receptor agonists kills these cells in a Ca^{2+}-dependent manner (Yoshioka et al. 1995; Matute et al. 1997; McDonald et al. 1998a). Furthermore, injection of AMPA/kainate-receptor agonists into mouse subcortical white matter or rabbit optic nerve was shown to kill oligodendroglia in the vicinity of the injection site (Matute et al. 1997), an effect inhibited by non-NMDA-receptor antagonists (McDonald et al. 1998a). The prominent vulnerability of oligodendroglia to excitotoxic death may, in part, reflect their expression of Ca^{2+}-permeable AMPA or kainate receptors (Holzwarth et al. 1994; Puchalski et al. 1994).

III. Metabotropic Glutamate Receptors

To date, a total of eight glutamate receptors linked to G-proteins rather than ion channels have been identified (Nakanishi and Masu 1994; Conn and Pin 1997). These metabotropic glutamate receptors (mGluRs) are divided into three groups – designated groups I, II, and III – based on sequence homology and mechanism of signal transduction. Unfortunately, many of the currently available mGluR agonists and antagonists cross-react at multiple mGluR subtypes or even at NMDA receptors (Contractor et al. 1998), making it difficult to resolve the functions of individual receptor subtypes. However, it appears that manipulating mGluR activation can affect the excitotoxic death mediated by ion-channel-linked glutamate receptors.

Some effects of mGluR activation may be anti-excitotoxic. Activation of group-II mGluRs can reduce currents through high-voltage-activated Ca^{2+} channels (Stefani et al. 1996), including presynaptic N-type [ω-conotoxin GVIA-sensitive] channels, and activation of group-I or group-II mGluRs may enhance presynaptic, voltage-gated, 4-aminopyridine-sensitive K^+ currents (Sladeczek et al. 1993), thus inhibiting Ca^{2+}-dependent vesicle release. Inhibition of voltage-gated Ca^{2+} channels could also attenuate excitotoxicity by acting postsynaptically to reduce toxic Ca^{2+} (as well as Zn^{2+}) influx.

Other effects of mGluR activation, particularly of group-I subtypes, may facilitate excitotoxicity by increasing neuronal excitability (Gereau and Conn 1995b; Davies et al. 1995). In hippocampal neurons, group-I-mGluR activation appears to inhibit several K^+ conductances, including a resting K^+ current (Guerineau et al. 1994) and various voltage-sensitive K^+ currents (Charpak et al. 1990; Desai and Conn 1991; Luthi et al. 1996), and, in cerebellar granule cells, group-I mGluRs trigger an enhancement of L-type Ca^{2+}-channel current (Chavis et al. 1995). In addition, presynaptic mGluRs downregulate GABA release in several brain areas, including hippocampal CA1 neurons, where group-I receptors again play the dominant role (Gereau and Conn 1995a). Finally, group-I-mGluR agonists may selectively enhance NMDA-receptor-mediated responses in hippocampal neurons (Aniksztejn et al. 1991; Fitzjohn et al. 1996), although a membrane-delimited reduction in NMDA-receptor-mediated currents on cortical neurons has also been described (Yu et al. 1997a).

Given the low pharmacological specificity and broad and incompletely understood range of mGluR actions, it has been challenging to establish an effective neuroprotective strategy for excitotoxicity; nevertheless, a few preliminary principles have emerged. The NMDA-induced death of cultured neurons has been shown to be potentiated by agonists of group-I mGluRs (Bruno et al. 1995b) and attenuated by selective group-II and -III agonists (Bruno et al. 1995a; Buisson et al. 1996; Pizzi et al. 1996). Likewise, group-I agonists amplify, and group-III agonists decrease, neuronal death induced by traumatic injury in vitro and in vivo (Gong et al. 1995; Mukhin et al. 1997; Faden et al. 1997). Therefore, a combination of drugs or multifunctional agents that act as agonists at group-II and -III mGluRs and antagonists at group-I mGluRs appears, at present, an attractive therapeutic strategy. One cautionary note when designing mGluR-based strategies is that the influence of a certain mGluR subtype on neuronal survival may depend on the predominant mode of death – apoptotic or necrotic – induced by the insult (see below); notably, cerebellar granule cells undergoing low K^+-induced apoptosis are protected by group-I activation (Copani et al. 1995).

IV. Anchoring Proteins for Glutamate Receptors

A group of anchoring molecules, PDZ proteins, interact directly with glutamate-receptor subunits and may influence post-synaptic responses through their ability to aggregate receptors, ion channels, and signaling molecules at specific subcellular locations (Craven and Bredt 1998; Chap 1). All four NR2 subunits (Kornau et al. 1995; Niethammer et al. 1996), certain splice variants of NR1 (Lin et al. 1998), AMPA receptor subunits GluR-B and -C (same refs.) (Dong et al. 1997; Srivastava et al. 1998), and group-I mGluRs (Brakeman et al. 1997; Kato et al. 1998) bind to the PDZ domains of a growing family of anchoring proteins that are thought either to aggregate or retain receptor subunits at synapses (Kim et al. 1996; Rao and Craig 1997). Furthermore, PDZ proteins package NMDA-receptor subunits together with signal-transduction proteins, such as the novel guanosine triphosphate (GTP)ase-activating protein SynGAP (Kim et al. 1998; Chen et al. 1998), and nNOS (Brenman et al. 1996) and, thus, may create a signaling module in which downstream signal-transducing enzymes are favorably positioned to respond to NMDA-receptor-mediated Ca^{2+} influx. Anchoring proteins may be essential for glutamate-receptor function in vivo, as mice with targeted disruptions of NR2-subtype carboxy termini, which contain PDZ interaction motifs, exhibit a phenotype as severe as that resulting from gene deletion of the corresponding subunit (Sprengel et al. 1998).

Taken together, these observations raise the interesting possibility that dynamic changes in anchoring interactions may influence the course of excitotoxic death. Following an excitotoxic insult, Ca^{2+}-activated proteases might digest PDZ proteins, leading to dispersal of NMDA or AMPA receptors from post-synaptic sites and the uncoupling of nNOS from NMDA receptors, thus acting as a brake on excitotoxic damage.

C. Modulators of Excitotoxicity

The function of ionotropic glutamate receptors and possibly their relative contributions to excitotoxic cell death are modulated by several endogenous intra- and extracellular factors discussed in preceding chapters. The influence of these modulatory factors can be expected to change dynamically following acute insults and the initiation of excitotoxic cascades, resulting in a multitude of complex, interlocking and, at times, opposing effects.

I. Magnesium

One of the most prominent endogenous modulators of NMDA-receptor current is Mg^{2+}, the ion which confers voltage dependence on the channel by blocking its pore at resting membrane potential (Nowak et al. 1984; Mayer et al. 1984). As might be predicted, reducing extracellular Mg^{2+} induces NMDA-receptor-mediated neuronal death in vitro (Abele et al. 1990; Rose et al. 1990). NMDA receptors exhibit a relative loss of voltage dependence 48 h after transient forebrain ischemia, probably due to altered Mg^{2+} sensitivity (Hori and Carpenter 1994), an effect which may result from the action of protein kinase C (PKC) on the receptor (Chen and Huang 1992). Raising brain Mg^{2+} concentrations through the infusion of Mg^{2+} salts is neuroprotective after brain ischemia in animals and possibly also humans (Izumi et al. 1991; Tsuda et al. 1991; Muir 1998). Interestingly, in the absence of extracellular Na^{+} and Ca^{2+}, Mg^{2+} is able to permeate NMDA-receptor-gated channels (Stout et al. 1996) and may itself contribute to excitotoxic neuronal death (Hartnett et al. 1997).

II. Protons

During seizures and brain ischemia, extracellular pH in the brain decreases due to the buildup of CO_2 and lactic acid. Currents through NMDA- (Tang et al. 1990; TRAYNELIS and CULL-CANDY 1990) and non-NMDA- (GIFFARD et al. 1990b; TANG et al. 1990; CHRISTENSEN and HIDA 1990) receptor-gated channels are inhibited by the binding of protons. Indeed, moderate decreases in extracellular pH decrease the intracellular Ca^{2+} accumulation and death associated with glutamate toxicity and oxygen–glucose deprivation in vitro (GIFFARD et al. 1990b; TOMBAUGH and SAPOLSKY 1990). However, extracellular acidity has been shown to enhance the slowly triggered, AMPA-receptor-mediated death of cultured neurons, perhaps by interfering with the restoration of intracellular Ca^{2+} homeostasis (MCDONALD et al. 1998b). Other potential injury-enhancing effects of extracellular protons might include the gating of acid-sensitive ion channels (ASICs), which are amiloride-sensitive, Na^{+}-, Ca^{2+}-, and K^{+}-permeable channels widely distributed throughout the brain (WALDMANN et al. 1997), and damage to glial cells (KRAIG et al. 1987; NORENBERG et al. 1987; GIFFARD et al. 1990a).

III. Zinc

As mentioned above, Zn^{2+} is released, along with glutamate, at excitatory synaptic terminals at concentrations that may reach high micromolar levels in the extracellular space during intense activity (ASSAF and CHUNG 1984). Zn^{2+} modulates the activity of a broad array of ion channels (SMART et al. 1994; HARRISON and GIBBONS 1994), specifically reducing currents through NMDA-receptor-gated channels (PETERS et al. 1987; WESTBROOK and MAYER 1987; CHRISTINE and CHOI 1990) and voltage-gated Ca^{2+} channels (WINEGAR and LANSMAN 1990). However, chelation of extracellular Zn^{2+} with CaEDTA (ethylenediaminetetraacetic acid saturated with equimolar Ca^{2+}) attenuates, rather than increases, hippocampal CA1-neuronal death after transient global ischemia (KOH et al. 1996). We have hypothesized that the beneficial effects of extracellular Zn^{2+} in reducing NMDA-receptor-mediated toxicity is more than counterbalanced by its direct neurotoxic effects. Since Zn^{2+}-induced neuronal death appears to be associated with its entry through several routes facilitated by glutamate-receptor activation, including voltage-gated Ca^{2+} channels, the Na^{+}–Ca^{2+} exchanger, and NMDA- and Ca^{2+}-permeable AMPA-receptor-gated channels, Zn^{2+} may be considered an alternative to Ca^{2+} in the mediation of excitotoxicity (WEISS et al. 1993; SENSI et al. 1997; CHOI and KOH 1998). Furthermore, there is some suggestion that Zn^{2+} can potentiate AMPA-receptor current (PETERS et al. 1987; RASSENDREN et al. 1990), an effect that might enhance the Ca^{2+}-dependent toxicity mediated by these channels.

IV. Free Radicals

Both NMDA- and AMPA-receptor activation increases neuronal production of oxygen and nitrogen free radicals, which appear to be important mediators of excitotoxic cell damage (see below). However, as with increases in extracellular Zn^{2+} and H^+ concentrations, buildup of free radicals also selectively diminishes NMDA-receptor current by direct oxidation of redox-sensitive site(s) (AIZENMAN et al. 1989). Redox modulation of NMDA-receptor currents might be especially prominent on channels containing NR2A subunits versus NR2B, NR2C, or NR2D subunits (KOHR et al. 1994).

V. Neurotrophins

Following cerebral ischemia or hypoglycemia in rats, levels of growth-factor transcripts, including brain-derived neurotrophic factor (BDNF), nerve-growth factor (NGF), and neurotrophin 3 (NT-3), were elevated (LINDVALL et al. 1992), a result reproduced upon direct activation of NMDA receptors in vivo (LINDEFORS et al. 1992). During development, neurotrophin molecules are necessary for the survival of central neurons, and their withdrawal leads to neuronal apoptosis (OPPENHEIM 1991). However, neurotrophins may act oppositely on neurons suffering from an intense excitotoxic insult. BDNF, NT-3, and NT-4/5 all potentiated excitotoxic cortical-neuronal death in vitro (KOH et al. 1995; PREHN 1996). Neurotrophins induce a selective increase in NR2A-receptor expression (relative to NR1 or NR2B; YING et al. 1995) as well as an increase in neuronal vulnerability to free-radical-mediated injury (GWAG et al. 1995). Furthermore, BDNF and NGF have been shown to potentiate excitatory neurotransmission, an effect that may be accounted for by an increase in the probability of transmitter release (CARMIGNOTO et al. 1997) or enhanced current through NMDA-receptor-gated channels (JARVIS et al. 1997).

VI. Phosphorylation

The activity of all ionotropic glutamate receptors is potentiated by phosphorylation and reduced by the action of phosphatases (Chap 1). Synaptic activity may, indirectly, regulate the phosphorylation state of receptors. Glutamate release, by triggering Ca^{2+} entry through NMDA receptors, can lead rapidly to the phosphorylation of AMPA receptors by CaM-KII (TAN et al. 1994), and, thus, may promote AMPA-receptor-mediated toxicity. The effect of activity on NMDA-receptor phosphorylation may be mixed; although group-I-mGluR activation can direct the phosphorylation of these receptors by PKC, thereby enhancing current amplitude and reducing Mg^{2+} sensitivity (see above), prolonged synaptic activity is associated with a Ca^{2+}-dependent desensitization of NMDA responses (ZORUMSKI et al. 1989) mediated by calcineurin (TONG et al. 1995), a Ca^{2+}-activated phosphatase highly enriched in the post-synaptic density (GOTO et al. 1986). More generally, as excitotoxic cascades progress in an injured neuron and cellular energy metabolism becomes stressed (see below), limitations in the availability of high-energy phosphates would be expected to lead to receptor dephosphorylation and the consequent loss of receptor activity.

VII. Other Modulatory Influences

Various potentiators of NMDA-receptor current may enhance the contribution of these receptors to glutamate-induced damage. Glycine enhances NMDA-induced neuronal death in vitro, possibly by acting at sites in addition to the NMDA receptor (MCNAMARA and DINGLEDINE 1990). Endogenous polyamines, including putrescine in particular, selectively enhance NMDA-receptor-gated currents (MCGURK et al. 1990) and may be released from injured neurons following transient brain ischemia (PASCHEN et al. 1992). In addition, arachidonic acid, which is released from membranes following NMDA-receptor activation (see below), potentiates NMDA-receptor-mediated currents while reducing current through non-NMDA-receptor-gated channels (MILLER et al. 1992; KOVALCHUK et al. 1994).

Two kynurenine metabolites of tryptophan degradation include quinolinate and kynurenate, the former acting as a weak agonist at the NMDA receptor, and the latter acting as a weak pan-antagonist at all ionotropic glutamate receptors (STONE 1993). Kynurenate levels in the brain fall during neuronal activity (SCHWARCZ et al. 1992), a change that might sensitize neurons to excitotoxic damage if baseline concentrations reach the high micromolar to low millimolar concentrations generally required to affect excitatory neurotransmission (GANONG et al. 1983).

D. Consequences of Excitotoxic Glutamate-Receptor Activation

I. Apoptosis Versus Necrosis

Controversy currently exists regarding whether excitotoxic glutamate-receptor stimulation leads to neuronal necrosis or apoptosis (CHOI 1996; LEIST and NICOTERA 1998; MARTIN et al. 1998). One might predict that the opening of large plasma-membrane conductances and the consequent massive influx of cations and water, events tantamount to plasma-membrane failure, should induce necrosis. Indeed, many groups have observed that the ultrastructural, pharmacological, and biochemical features of excitotoxic death in vitro are consistent with necrosis, including membrane and organelle disruption and insensitivity to inhibitors of protein synthesis or caspase activity (DESSI et al. 1993; REGAN et al. 1995; GWAG et al. 1997; GOTTRON et al. 1997).

However, other observations have complicated this picture, demonstrating that the excitotoxic degeneration of cultured neurons is sometimes marked by features of programmed cell death. In particular, positive TUNEL [terminal transferase-mediated deoxyuridine triphosphate (dUTP)-digoxigenin nick end labeling] stain, internucleosomal DNA fragmentation (DNA laddering), and chromatin condensation detected by propidium-iodide staining have been used as evidence of apoptosis in neurons exposed to excitotoxins in vitro

(Kure et al. 1991; Ankarcrona et al. 1995; Bonfoco et al. 1995; Simonian et al. 1996). Even in excitotoxic paradigms with a predominance of necrotic features, dying neurons exhibited TUNEL-positive nuclei and transient DNA laddering (Didier et al. 1996; Gwag et al. 1997). These observations are consistent with the idea that excitotoxic insults can trigger both apoptotic and necrotic cascades in parallel, with ultimate predominance dependent upon several factors (see below). Indeed, in some in vitro models, anti-apoptotic interventions, including Bax-gene deletion (Xiang et al. 1998), inhibition of protein synthesis (Dreyer et al. 1995; Finiels et al. 1995), and caspase inhibition (Tenneti et al. 1998), did reduce excitotoxic death.

Mixed features of necrosis and apoptosis are similarly apparent in models of excitotoxic cell death in vivo. Although neurons in adult rat brains may die a morphologically necrotic death after intrastriatal injection of glutamate-receptor agonists (Ferrer et al. 1995; Portera-Cailliau et al. 1995), these neurons may exhibit TUNEL positivity and transient DNA laddering (Qin et al. 1996). Non-NMDA agonists also induce chromatin clumping, implying a greater tendency for activation of these receptors to initiate programmed-cell-death pathways (Portera-Cailliau et al. 1997). Likewise, following focal and global ischemia, evidence of DNA laddering (Linnik et al. 1993; Nitatori et al. 1995) and *in situ* double-strand breaks (MacManus et al. 1993; Li et al. 1995a) in degenerating neurons suggests that these insults can also mobilize a cell's apoptotic machinery. As with agonist injections, ultrastructural evidence suggests that these cells may, ultimately, die either by necrosis (Deshpande et al. 1995; van Lookeren Campagne and Gill 1996) or apoptosis (Nitatori et al. 1995; Li et al. 1995b; Charriaut-Marlangue et al. 1996).

The presence in a single cell of both necrotic and apoptotic characteristics suggests that cell death may be best interpreted not in binary terms, but rather on the basis of an apoptosis–necrosis continuum. Cascades leading to both forms of death may occur simultaneously, with the final death phenotype determined by the relative contribution and speed of a particular process. Factors favoring apoptosis may include milder insult intensity (less likely to trigger fulminant necrosis), cell immaturity, low intracellular Ca^{2+}, and low trophic-factor availability. Furthermore, especially in vivo, initial excitotoxic insults may be compounded by subsequent loss of trophic influences, for example, due to damage to processes or loss of innervating or target cells.

The concept of a death continuum introduces a higher level of complexity to anti-excitotoxic strategies, since interventions that ameliorate one form of death may have no effect or even an opposite influence on the other form. For example, in focal cerebral ischemia, glutamate-receptor antagonists may attenuate damage due to excitotoxic necrosis, especially in the core, but may promote Ca^{2+}-starvation and, therefore, hasten cell death in penumbral cells dying predominantly by apoptotic means. Future neuroprotective approaches to excitotoxic death may need to include both anti-necrosis and anti-apoptosis strategies.

II. Downstream Mediators of Injury

Regardless of the final form of cell death, it remains clear that Ca^{2+} overload represents the predominant mediator of excitotoxicity. As mentioned above, Ca^{2+} permeates NMDA receptors, specific subtypes of AMPA/kainate receptors, voltage-gated Ca^{2+} channels, and Na^+/Ca^{2+} exchangers. Additionally, given the ability of dantrolene, a ryanodine-receptor antagonist, to decrease excitotoxic death, the magnitude of cytosolic Ca^{2+} elevation may be augmented by the release of Ca^{2+} from intracellular stores (FRANDSEN and SHOUSBOE 1991).

A broad range of endonucleases, proteases, phospholipases, phosphatases, and kinases are activated by increases in intracellular Ca^{2+}. Although these proteins carry out important physiological functions under normal conditions, excessive activation secondary to cellular Ca^{2+} loading may convert them into executioners. The calpain group of cytoskeleton-remodeling proteases may be especially prominent in mediating excitotoxicity (SIMAN and NOSZEK 1988); calpain inhibition protected neurons from agonist-induced death in vitro (BRORSON et al. 1994) as well as from delayed neurodegeneration subsequent to transient global ischemia (LEE et al. 1991). Cytoplasmic phospholipase A_2 ($cPLA_2$), another Ca^{2+}-dependent enzyme implicated in the expression of excitotoxicity, can break down membrane lipids to yield arachidonic acid (DUMUIS et al. 1988). In addition to altering membrane composition, release of arachidonic acid leads to the formation of inflammatory eicosanoids (such as prostaglandins and leukotrienes), a decrease in glutamate re-uptake (YU et al. 1986), a facilitation of glutamate release (FREEMAN et al. 1990), and generation of free radicals (see below). Interestingly, the $cPLA_2$ knockout mouse exhibited decreased brain injury after transient focal cerebral ischemia compared to littermates, although it is unclear which downstream effects of $cPLA_2$ elimination account for this resistance (BONVENTRE et al. 1997). Ca^{2+}-activated endonucleases have been implicated as part of the excitotoxic death machinery as well, but a causal link between endonuclease activation and cell death has not been demonstrated in any system, given current pharmacology (KURE et al. 1991; ROBERTS-LEWIS et al. 1993; but also POSNER et al. 1995).

Glutamate-receptor activation and subsequent Ca^{2+} overload also activate a destructive free-radical cascade that can peroxidize lipids, oxidize proteins, and damage DNA. Application of antioxidants decreased glutamate neurotoxicity in vitro (DYKENS et al. 1987; MONYER et al. 1990). Additionally, glutamate excitotoxicity in hippocampal cultures was reduced under hypoxic conditions, which limit oxygen free-radical formation (DUBINSKY et al. 1995). Excess Ca^{2+} influx and free-radical generation are coupled through several mechanisms. One link is the Ca^{2+}-induced conversion of xanthine dehydrogenase to xanthine oxidase, a source of superoxide radicals ($\cdot O_2^-$; DYKENS et al. 1987), but it is questionable whether this oxidant pathway is relevant in the human brain (SARNESTO et al. 1996). After the Ca^{2+}-activated release of arachidonic acid by $cPLA_2$, metabolism of arachidonic acid by cyclooxygenase can also lead to the production of $\cdot O_2^-$ (CHAN et al. 1985; WEI et al. 1986). A third link between Ca^{2+} overload and free-radical formation is the Ca^{2+}-dependent activation of nNOS (DAWSON and SNYDER 1994). Nitric oxide alone has weak oxidizing activity but can react with $\cdot O_2^-$ to form the more powerful oxidant peroxynitrite (BECKMAN and KOPPENOL 1996). Pharmacological-inhibition and gene-deletion studies have demonstrated that nNOS plays a prominent role in some in vitro models of NMDA toxicity (DAWSON et al. 1991; DAWSON et al. 1996). Importantly, nNOS knockout mice demonstrated increased resistance to focal cerebral ischemia and intrastriatal injection of NMDA compared to controls (HUANG et al. 1994; AYATA et al. 1997).

Increasing evidence has pointed to the critical role of mitochondria in determining whether and how a cell dies after glutamate-receptor activation. First, mitochondria represent an additional critical link between intracellular Ca^{2+} handling and the production of free radicals. Although a number of groups have demonstrated that mitochondria have a prodigious ability to buffer large increases in cytosolic Ca^{2+} concentrations (GUNTER and PFEIFFER 1990; WANG and THAYER 1996), excessive mitochondrial Ca^{2+} loading may uncouple electron transport from adenosine triphosphate synthesis, causing increased production of mitochondrial reactive oxygen species and a derangement of cellular energy metabolism (REYNOLDS and HASTINGS 1995; DUGAN et al. 1995; WHITE and REYNOLDS 1996; SCHINDER et al. 1996). Additional destructive processes might be initiated if impaired mitochondria inadequately buffer further increases in cytosolic Ca^{2+} or release their Ca^{2+} stores into the cytoplasm (WHITE and REYNOLDS 1996). Second, mitochondrial dysfunction within either neurons or glia may lead to enhanced excitotoxicity, in the latter case perhaps by interfering with protective glutamate uptake. Third, recent work has demonstrated that mitochondria may play an active role in the apoptotic cascade. The Bcl-2 family of proteins, for instance, may produce pro- and anti-apoptotic effects by interacting with mitochondria (GREEN and REED 1998).

Recent studies have shown that the DNA-repair enzyme, poly[adenosine diphosphate (ADP)-ribose] synthetase (PARS), also known as poly(ADP-ribose) polymerase (PARP), is involved in mediating excitotoxic death (SZABO and DAWSON 1998). PARS inhibitors reduced injury induced by glutamate-receptor agonists in culture (ZHANG et al. 1994), and cortical cultures derived from PARS knockout mice demonstrated decreased excitotoxic cell death compared with controls (ELIASSON et al. 1997), results consistent with the idea that the metabolic cost associated with PARS activation in the setting of excitotoxicity has lethal consequences. In addition, PARS knockout mice exhibited reduced infarction after transient focal cerebral ischemia (ELIASSON et al. 1997; ENDRES et al. 1997). By providing more specific links between glutamate-receptor activation and consequent cell loss, such new knowledge reveals potential targets of anti-excitotoxic therapy.

References

Abele AE, Scholz KP, Scholz WK, Miller RJ (1990) Excitotoxicity induced by enhanced excitatory neurotransmission in cultured hippocampal pyramidal neurons. Neuron 4:413–419

Aizenman E, Lipton SA, Loring RH (1989) Selective modulation of NMDA responses by reduction and oxidation. Neuron 2:1257–1263

Anegawa NJ, Lynch DR, Verdoorn TA, Pritchett DB (1995) Transfection of *N*-methyl-D-aspartate receptors in a nonneuronal cell line leads to cell death. J Neurochem 64:2004–2012

Aniksztejn L, Bregestovski P, Ben-Ari Y (1991) Selective activation of quisqualate metabotropic receptor potentiates NMDA but not AMPA responses. Eur J Pharmacol 205:327–328

Ankarcrona M, Dypbukt JM, Bonfoco E, Zhivotovsky B, Orrenius S, Lipton SA, Nicotera P (1995) Glutamate-induced neuronal death: a succession of necrosis or apoptosis depending on mitochondrial function. Neuron 15:961–973

Assaf SY, Chung SH (1984) Release of endogenous Zn^{2+} from brain tissue during activity. Nature 308:734–736

Ayata C, Ayata G, Hara H, Matthews RT, Beal MF, Ferrante RJ, Endres M, Kim A, Christie RH, Waeber C, Huang PL, Hyman BT, Moskowitz MA (1997) Mechanisms of reduced striatal NMDA excitotoxicity in type I nitric oxide synthase knock-out mice. J Neurosci 17:6908–6917

Beal MF, Hyman BT, Koroshetz W (1993) Do defects in mitochondrial energy metabolism underlie the pathology of neurodegenerative diseases? Trends Neurosci 16:125–131

Beckman JS, Koppenol WH (1996) Nitric oxide, superoxide, and peroxynitrite: the good, the bad, and ugly. Am J Physiol 271:C1424–1437

Bochet P, Audinat E, Lambolez B, Crepel F, Rossier J, Iino M, Tsuzuki K, Ozawa S (1994) Subunit composition at the single-cell level explains functional properties of a glutamate-gated channel. Neuron 12:383–388

Bonfoco E, Krainc D, Ankarcrona M, Nicotera P, Lipton SA (1995) Apoptosis and necrosis: two distinct events induced, respectively, by mild and intense insults with *N*-methyl-D-aspartate or nitric oxide/superoxide in cortical cell cultures. Proc Natl Acad Sci USA 92:7162–7166

Bonventre JV, Huang Z, Taheri MR, O'Leary E, Li E, Moskowitz MA, Sapirstein A (1997) Reduced fertility and postischaemic brain injury in mice deficient in cytosolic phospholipase A2. Nature 390:622–625

Brakeman PR, Lanahan AA, O'Brien R, Roche K, Barnes CA, Huganir RL, Worley PF (1997) Homer: a protein that selectively binds metabotropic glutamate receptors. Nature 386:284–288

Brenman JE, Chao DS, Gee SH, McGee AW, Craven SE, Santillano DR, Wu Z, Huang F, Xia H, Peters MF, Froehner SC, Bredt DS (1996) Interaction of nitric oxide synthase with the postsynaptic density protein PDS-95 and α1-syntrophin mediated by PDZ domains. Cell 84:757–767

Brorson JR, Manzolillo PA, Miller RJ (1994) Ca^{2+} entry via AMPA/KA receptors and excitotoxicity in cultured cerebellar Purkinje cells. J Neurosci 14:187–197

Bruno V, Battaglia G, Copani A, Giffard RG, Raciti G, Raffaele R, Shinozaki H, Nicoletti F (1995a) Activation of class II or III metabotropic glutamate receptors protects cultured cortical neurons against excitotoxic degeneration. Eur J Neurosci 7:1906–1913

Bruno V, Copani A, Knopfel T, Kuhn R, Casabona G, Dell'Albani P, Condorelli DF, Nicoletti F (1995b) Activation of metabotropic glutamate receptors coupled to inositol phospholipid hydrolysis amplifies NMDA-induced neuronal degeneration in cultured cortical cells. Neuropharmacology 34:1089–1098

Brusa R, Zimmerman F, Koh DS, Feldmeyer D, Gass P, Seeburg PH, Sprengel R (1995) Early-onset epilepsy and postnatal lethality associated with an editing-deficient GluR-B allele in mice. Science 270:1677–1680

Buisson A, Yu SP, Choi DW (1996) DCG-IV selectively attenuates rapidly triggered NMDA-induced neurotoxicity in cortical neurons. Eur J Neurosci 8:138–143

Burnashev N, Monyer H, Seeburg PH, Sakmann B (1992) Divalent ion permeability of AMPA receptor channels is dominated by the edited form of a single subunit. Neuron 8:189–198

Carmignoto G, Pizzorusso T, Tia S, Vicini S (1997) Brain-derived neurotrophic factor and nerve growth factor potentiate excitatory synaptic transmission in the rat visual cortex. J Physiol (Lond) 498:153–164

Carver JM, Mansson PE, Cortes-Burgos L, Shu J, Zhou LM, Howe JR, Giordano T (1996) Cytotoxic effects of kainate ligands on HEK cell lines expressing recombinant kainate receptors. Brain Res 720:69–74

Catania MV, Tolle TR, Monyer H (1995) Differential expression of AMPA receptor subunits in NOS-positive neurons of cortex, striatum, and hippocampus. J Neurosci 15:7046–7061

Chan PH, Fishman RA, Longar S, Chen S, Yu A (1985) Cellular and molecular effects of polyunsaturated fatty acids in brain ischemia and injury. Prog Brain Res 63:227–235

Charpak S, Gahwiler BH, Do KQ, Knopfel T (1990) Potassium conductances in hippocampal neurons blocked by excitatory amino-acid transmitters. Nature 347:765–767

Charriaut-Marlangue C, Margaill I, Represa A, Popovici T, Plotkine M, Ben-Ari Y (1996) Apoptosis and necrosis after reversible focal ischemia: an in situ DNA fragmentation analysis. J Cereb Blood Flow Metab 16:186–194

Chavis P, Fagni L, Bockaert J, Lansman JB (1995) Modulation of calcium channels by metabotropic glutamate receptors in cerebellar granule cells. Neuropharmacology 34:929–937

Chazot PL, Coleman SK, Cik M, Stephenson FA (1994) Molecular characterization of *N*-methyl-D-aspartate receptors expressed in mammalian cells yields evidence for the coexistence of three subunit types within a discrete receptor molecule. J Biol Chem 269:24403–24409

Chen L, Huang LY (1992) Protein kinase C reduces Mg^{2+} block of NMDA-receptor channels as a mechanism of modulation. Nature 356:521–523

Chen HJ, Rojas-Soto M, Oguni A, Kennedy MB (1998) A synaptic Ras-GTPase activating protein (p135 SynGAP) inhibited by CaM kinase II. Neuron 20:895–904

Choi DW (1987) Ionic dependence of glutamate neurotoxicity. J Neurosci 7:369–379

Choi DW (1988a) Calcium-mediated neurotoxicity: relationship to specific channel types and role in ischemic damage. Trends Neurosci 11:465–469

Choi DW (1988b) Glutamate neurotoxicity and diseases of the nervous system. Neuron 1:623–634

Choi DW (1992) Excitotoxic cell death. J Neurobiol 23:1261–1276

Choi DW (1995) Calcium: still center-stage in hypoxic-ischemic neuronal death. Trends Neurosci 18:58–60

Choi DW (1996) Ischemia-induced neuronal apoptosis. Curr Opin Neurobiol 6:667–672

Choi DW, Koh JY (1998) Zinc and brain injury. Annu Rev Neurosci 21:347–375

Choi DW, Maulucci-Gedde M, Kriegstein AR (1987) Glutamate neurotoxicity in cortical cell culture. J Neurosci 7:357–368

Christensen BN, Hida E (1990) Protonation of histidine groups inhibits gating of the quisqualate/kainate channel protein in isolated catfish cone horizontal cells. Neuron 5:471–478

Christine CW, Choi DW (1990) Effect of zinc on NMDA receptor-mediated channel currents in cortical neurons. J Neurosci 10:108–116

Cik M, Chazot PL, Stephenson FA (1993) Optimal expression of cloned NMDAR1/NMDAR2A heteromeric glutamate receptors: a biochemical characterization. Biochem J 296:877–883

Colbourne F, Sutherland G, Corbett D (1997) Postischemic hypothermia. A critical appraisal with implications for clinical treatment. Mol Neurobiol 14:171–201

Conn PJ, Pin JP (1997) Pharmacology and functions of metabotropic glutamate receptors. Annu Rev Pharmacol Toxicol 37:205–237

Contractor A, Gereau RW 4th, Green T, Heinemann SF (1998) Direct effects of metabotropic glutamate receptor compounds on native and recombinant *N*-methyl-D-aspartate receptors. Proc Natl Acad Sci USA 95:8969–8974

Copani A, Bruno VM, Barresi V, Battaglia G, Condorelli DF, Nicoletti F (1995) Activation of metabotropic glutamate receptors prevents neuronal apoptosis in culture. J Neurochem 64:101–108

Craven SE, Bredt DS (1998) PDZ proteins organize synaptic signaling pathways. Cell 93:495–498

Crepel V, Aniksztejn L, Ben-Ari Y, Hammond C (1994) Glutamate metabotropic receptors increase a Ca^{2+}-activated nonspecific cationic current in CA1 hippocampal neurons. J Neurophysiol 72:1561–1569

Davies CH, Clarke VR, Jane DE, Collingridge GL (1995) Pharmacology of postsynaptic metabotropic glutamate receptors in rat hippocampal CA1 pyramidal neurons. Br J Pharmacol 116:1859–1869

Dawson TM, Snyder SH (1994) Gases as biological messengers: nitric oxide and carbon monoxide in the brain. J Neurosci 14:5147–5159

Dawson VL, Dawson TM, London ED, Bredt DS, Snyder SH (1991) Nitric oxide mediates glutamate neurotoxicity in primary cortical cultures. Proc Natl Acad Sci USA 88:6368–6371

Dawson VL, Kizushi VM, Huang PL, Snyder SH, Dawson TM (1996) Resistance to neurotoxicity in cortical cultures from neuronal nitric oxide synthase-deficient mice. J Neurosci 1996 16:2479–2487

Desai MA, Conn PJ (1991) Excitatory effects of ACPD receptor activation in the hippocampus are mediated by direct effects on pyramidal cells and blockade of synaptic inhibition. J Neurophysiol 66:40–52

Deshpande J, Bergstedt K, Linden T, Kalimo H, Wieloch T (1992) Ultrastructural changes in the hippocampal CA1 region following transient cerebral ischemia: evidence against programmed cell death. Exp Brain Res 88:91–105

Dessi F, Charriaut-Marlangue C, Khrestchatisky M, Ben-Ari Y (1993) Glutamate-induced neuronal death is not a programmed cell death in cerebellar culture. J Neurochem 60:1953–1955

Didier M, Bursztajn S, Adamec E, Passani L, Nixon RA, Coyle JT, Wei JY, Berman SA (1996) DNA strand breaks induced by sustained glutamate excitotoxicity in primary neuronal cultures. J Neurosci 16:2238–2250

Dong H, O'Brien RJ, Fung ET, Lanahan AA, Worley PF, Huganir RL (1997) GRIP: a synaptic PDZ domain-containing protein that interacts with AMPA receptors. Nature 386:279–284

Dreyer EB, Zhang D, Lipton SA (1995) Transcriptional or translational inhibition blocks low dose NMDA-mediated cell death. Neuroreport 6:942–944

Dubinsky JM, Kristal BS, Elizondo-Fournier M (1995) An obligate role for oxygen in the early stages of glutamate-induced, delayed neuronal death. J Neurosci 15:7071–7078

Dugan LL, Sensi SL, Canzoniero LM, Handran SD, Rothman SM, Lin TS, Goldberg MP, Choi DW (1995) Mitochondrial production of reactive oxygen species in cortical neurons following exposure to *N*-methyl-D-aspartate. J Neurosci 15:6377–6388

Dumuis A, Sebben M, Haynes L, Pin JP, Bockaert J (1988) NMDA receptors activate the arachidonic acid cascade system in striatal neurons. Nature 336:68–70

Durand GM, Bennett MV, Zukin RS (1993) Splice variants of the *N*-methyl-D-aspartate receptor NR1 identify domains involved in regulation by polyamines and protein kinase C. Proc Natl Acad Sci USA 90:6731–6735

Dykens JA, Stern A, Trenkner E (1987) Mechanism of kainate toxicity to cerebellar neurons in vitro is analogous to reperfusion tissue injury. J Neurochem 49: 1222–1228

Eliasson MJ, Sampei K, Mandir AS, Hurn PD, Traystman RJ, Bao J, Pieper A, Wang ZQ, Dawson TM, Snyder SH, Dawson VL (1997) Poly(ADP-ribose) polymerase gene disruption renders mice resistant to cerebral ischemia. Nat Med 3:1089–1095

Endres M, Wang ZQ, Namura S, Waeber C, Moskowitz MA (1997) Ischemic brain injury is mediated by the activation of poly(ADP-ribose) polymerase. J Cereb Blood Flow Metab 17:1143–1151

Faden AI, Ivanova SA, Yakovlev AG, Mukhin AG (1997) Neuroprotective effects of group III mGluR in traumatic neuronal injury. J Neurotrauma 14:885–895

Ferrer I, Martin F, Serrano T, Reiriz J, Perez-Navarro E, Alberch J, Macaya A, Planas AM (1995) Both apoptosis and necrosis occur following intrastriatal administration of excitotoxins. Acta Neuropathol (Berl) 90:504–510

Finiels F, Robert JJ, Samolyk ML, Privat A, Mallet J, Revah F (1995) Induction of neuronal apoptosis by excitotoxins associated with long-lasting increase of 12-O-tetradecanoylphorbol-13-acetate-responsive element-binding activity. J Neurochem 65:1027–1034

Fitzjohn SM, Irving AJ, Palmer MJ, Harvey J, Lodge D, Collingridge GL (1996) Activation of group I mGluRs potentiates NMDA responses in rat hippocampal slices. Neurosci Lett 203:211–213

Frandsen A, Schousboe A (1991) Dantrolene prevents glutamate cytotoxicity and Ca^{2+} release from intracellular stores in cultured cerebral cortical neurons. J Neurochem 56:1075–1078

Franklin JL, Sanz-Rodriguez C, Juhasz A, Deckwerth TL, Johnson EM Jr (1995) Chronic depolarization prevents programmed death of sympathetic neurons in vitro but does not support growth: requirement for Ca^{2+} influx but not Trk activation. J Neurosci 15:643–664

Freeman EJ, Terrian DM, Dorman RV (1990) Presynaptic facilitation of glutamate release from isolated hippocampal mossy fiber nerve endings by arachidonic acid. Neurochem Res 15:743–750

Gahring LC, Cauley K, Rogers SW (1996) Kainic acid induced excitotoxicity and cfos expression in fibroblasts transfected with glutamate receptor subunit, GluR1. J Neurobiol 31:56–66

Ganong AH, Lanthorn TH, Cotman CW (1983) Kynurenic acid inhibits synaptic and acidic amino acid-induced responses in the rat hippocampus and spinal cord. Brain Res 273:170–174

Garthwaite G, Garthwaite J (1986) Neurotoxicity of excitatory amino acid receptor agonists in rat cerebellar slices: dependence on calcium concentration. Neurosci Lett 66:193–198

Gereau RW 4th, Conn PJ (1995a) Multiple presynaptic metabotropic glutamate receptors modulate excitatory and inhibitory synaptic transmission in hippocampal area CA1. J Neurosci 15:6879–6889

Gereau RW 4th, Conn PJ (1995b) Roles of specific metabotropic glutamate receptor subtypes in regulation of hippocampal CA1 pyramidal cell excitability. J Neurophysiol 74:122–129

Giffard RG, Monyer H, Choi DW (1990a) Selective vulnerability of cultured cortical glia to injury by extracellular acidosis. Brain Res 530:138–141

Giffard RG, Monyer H, Christine CW, Choi DW (1990b) Acidosis reduces NMDA receptor activation, glutamate neurotoxicity, and oxygen-glucose deprivation neuronal injury in cortical cultures. Brain Res 506:339–342

Gong QZ, Delahunty TM, Hamm RJ, Lyeth BG (1995) Metabotropic glutamate antagonist, MCPG, treatment of traumatic brain injury in rats. Brain Res 700:299–302

Gorter JA, Petrozzino JJ, Aronica EM, Rosenbaum DM, Opitz T, Bennett MV, Connor JA, Zukin RS (1997) Global ischemia induces downregulation of Glur2 mRNA and increases AMPA receptor-mediated Ca^{2+} influx in hippocampal CA1 neurons of gerbil. J Neurosci 17:6179–6188

Goto S, Matsukado Y, Mihara Y, Inoue N, Miyamoto E (1986) The distribution of calcineurin in rat brain by light and electron microscopic immunohistochemistry and enzyme-immunoassay. Brain Res 397:161–172

Gottron FJ, Ying HS, Choi DW (1997) Caspase inhibition selectively reduces the apoptotic component of oxygen-glucose deprivation-induced cortical neuronal cell death. Mol Cell Neurosci 9:159–169

Grant ER, Bacskai BJ, Pleasure DE, Pritchett DB, Gallagher MJ, Kendrick SJ, Kricka LJ, Lynch DR (1997) *N*-methyl-D-aspartate receptors expressed in a nonneuronal cell line mediate subunit-specific increases in free intracellular calcium. J Biol Chem 272:647–656

Green DR, Reed JC (1998) Mitochondria and apoptosis. Science 281:1309–1312

Grimwood S, Gilbert E, Ragan CI, Hutson PH (1996) Modulation of $^{45}Ca^{2+}$ influx into cells stably expressing recombinant human NMDA receptors by ligands acting at distinct recognition sites. N Neurochem 66:2589–2595

Guerineau NC, Bossu JL, Gahwiler BH, Gerber U (1995) Activation of a nonselective cationic conductance by metabotropic glutamatergic and muscarinic agonists in CA3 pyramidal neurons of the rat hippocampus. J Neurosci 15:4395–4407

Guerineau NC, Gahwiler BH, Gerber U (1994) Reduction of resting K^+ current by metabotropic glutamate and muscarinic receptors in rat CA3 cells: mediation by G-proteins. J Physiol (Lond) 474:27–33

Gunter TE, Pfeiffer DR (1990) Mechanisms by which mitochondria transport calcium. Am J Physiol 258:C755–786

Gwag BJ, Koh JY, DeMaro JA, Ying HS, Jacquin M, Choi DW (1997) Slowly triggered excitotoxicity occurs by necrosis in cortical cultures. Neuroscience 77:393–401

Harrison NL, Gibbons SJ (1994) Zn^{2+}: an endogenous modulator of ligand- and voltage-gated ion channels. Neuropharmacology 33:935–952

Hartnett KA, Stout AK, Rajdev S, Rosenberg PA, Reynolds IJ, Aizenman E (1997) NMDA receptor-mediated neurotoxicity: a paradoxical requirement for extracellular Mg^{2+} in Na^+/Ca^{2+}-free solutions in rat cortical neurons in vitro. J Neurochem 68:1836–1845

Hollmann M, Boulter J, Maron C, Beasley L, Sullivan J, Pecht G, Heinemann S (1993) Zinc potentiates agonist-induced currents at certain splice variants of the NMDA receptor. Neuron 10:943–954

Hollmann M, Hartley M, Heinemann S (1991) Ca^{2+} permeability of KA-AMPA-gated glutamate receptor channels depends on subunit composition. Science 252:851–853

Holzwarth JA, Gibbons SJ, Brorson JR, Philipson LH, Miller RJ (1994) Glutamate receptor agonists stimulate diverse calcium responses in different types of cultured rat cortical glial cells. J Neurosci 14:1879–1891

Hori N, Carpenter DO (1994) Transient ischemia causes a reduction of Mg^{2+} blockade of NMDA receptors. Neurosci Lett 173:75–78

Huang Z, Huang PL, Panahian N, Dalkara T, Fishman MC, Moskowitz MA (1994) Effects of cerebral ischemia in mice deficient in neuronal nitric oxide synthase. Science 265:1883–1885

Iino M, Ozawa S, Tsuzuki K (1990) Permeation of calcium through excitatory amino acid receptor channels in cultured rat hippocampal neurones. J Physiol (Lond) 424:151–165

Inglefield JR, Wilson CA, Schwartz-Bloom RD (1997) Effect of transient cerebral ischemia on γ-aminobutyric-acid-A receptor α1-subunit-immunoreactive interneurons in the gerbil CA1 hippocampus. Hippocampus 7:511–523

Izumi Y, Roussel S, Pinard E, Seylaz J (1991) Reduction of infarct volume by magnesium after middle cerebral artery occlusion in rats. J Cereb Blood Flow Metab 11:1025–1030

Jarvis CR, Xiong ZG, Plant JR, Churchill D, Lu WY, MacVicar BA, MacDonald JF (1997) Neurotrophin modulation of NMDA receptors in cultured murine and isolated rat neurons. J Neurophysiol 78:2363–2371

Jia Z, Agopyan N, Miu P, Xiong Z, Henderson J, Gerlai R, Taverna FA, Velumian A, MacDonald J, Carlen P, Abramow-Newerly W, Roder J (1996) Enhanced LTP in mice deficient in the AMPA receptor GluR2. Neuron 17:945–956

Johansen FF, Jorgensen MB, Diemer NH (1983) Resistance of hippocampal CA-1 interneurons to 20 min of transient cerebral ischemia in the rat. Acta Neuropathol (Berl) 61:135–140

Jonas P, Racca C, Sakmann B, Seeburg PH, Monyer H (1994) Differences in Ca^{2+} permeability of AMPA-type glutamate receptor channels in neocortical neurons caused by differential GluR-B subunit expression. Neuron 12:1281–1289

Kato A, Ozawa F, Saitoh Y, Fukazawa Y, Sugiyama H, Inokuchi K (1998) Novel members of the Vesl/Homer family of PDZ proteins that bind metabotropic glutamate receptors. J Biol Chem 273:23969–23975

Kim E, Cho KO, Rothschild A, Sheng M (1996) Heteromultimerization and NMDA receptor-clustering activity of Chapsyn-110, a member of the PSD-95 family of proteins. Neuron 17:103–113

Kim JH, Liao D, Lau LF, Huganir RL (1998) SynGAP: a synaptic RasGAP that associates with the PSD-95/SAP90 protein family. Neuron 20:683–691

Koh JY, Goldberg MP, Hartley DM, Choi DW (1990) Non-NMDA receptor-mediated neurotoxicity in cortical culture. J Neurosci 10:693–705

Koh JY, Gwag BJ, Lobner D, Choi DW (1995) Potentiated necrosis of cultured cortical neurons by neurotrophins. Science 268:573–575

Koh JY, Peters S, Choi DW (1986) Neurons containing NADPH-diaphorase are selectively resistant to quinolinate toxicity. Science 234:73–76

Koh JY, Suh SW, Gwag BJ, He YY, Hsu CY, Choi DW (1996) The role of zinc in selective neuronal death after transient global cerebral ischemia. Science 272: 1013–1016

Kohr G, Eckardt S, Luddens H, Monyer H, Seeburg PH (1994) NMDA receptor channels: subunit-specific potentiation by reducing agents. Neuron 12:1031–1040

Kornau HC, Schenker LT, Kennedy MB, Seeburg PH (1995) Domain interaction between NMDA receptor subunits and the postsynaptic density protein PSD-95. Science 269:1737–1740

Kovalchuk Y, Miller B, Sarantis M, Attwell D (1994) Arachidonic acid depresses non-NMDA receptor currents. Brain Res 643:287–295

Kraig RP, Petito CK, Plum F, Pulsinelli WA (1987) Hydrogen ions kill brain at concentrations reached in ischemia. J Cereb Blood Flow Metab 7:379–386

Kure S, Tominaga T, Yoshimoto T, Tada K, Narisawa K (1991) Glutamate triggers internucleosomal DNA cleavage in neuronal cells. Biochem Biophys Res Commun 179:39–45

Kutsuwada T, Kashiwabuchi N, Mori H, Sakimura K, Kushiya E, Araki K, Meguro H, Masaki H, Kumanishi T, Arakawa M, Michina M (1992) Molecular diversity of the NMDA receptor channel. Nature 358:36–41

Laurie DJ, Seeburg PH (1994) Regional and developmental heterogeneity in splicing of the rat brain NMDAR1 mRNA. J Neurosci 14:3180–3194

Le D, Das S, Wang YF, Yoshizawa T, Sasaki YF, Takasu M, Nemes A, Mendelsohn M, Dikkes P, Lipton SA, Nakanishi N (1997) Enhanced neuronal death from focal ischemia in AMPA-receptor transgenic mice. Mol Brain Res 52:235–241

Leist M, Nicotera P (1998) Apoptosis, excitotoxicity, and neuropathology. Exp Cell Res 239:183–201

Li Y, Chopp M, Jiang N, Yao F, Zaloga C (1995a) Temporal profile of in situ DNA fragmentation after transient middle cerebral artery occlusion in the rat. J Cereb Blood Flow Metab 15:389–397

Li Y, Sharov VG, Jiang N, Zaloga C, Sabbah HN, Chopp M (1995b) Ultrastructural and light microscopic evidence of apoptosis after middle cerebral artery occlusion in the rat. Am J Pathol 146:1045–1051

Lin JW, Wyszynski M, Madhavan R, Sealock R, Kim JU, Sheng M (1998) Yotiao, a novel protein of neuromuscular junction and brain that interacts with specific splice variants of NMDA receptor subunit NR1. J Neurosci 18:2017–2027

Lindefors N, Ballarin M, Ernfors P, Falkenberg T, Persson H (1992) Stimulation of glutamate receptors increases expression of brain-derived neurotrophic factor mRNA in rat hippocampus. Ann NY Acad Sci 648:296–299

Lindvall O, Ernfors P, Bengzon J, Kokaia Z, Smith ML, Siesjo BK, Persson H (1992) Differential regulation of mRNAs for nerve growth factor, brain-derived neurotrophic factor, and neurotrophin 3 in the adult rat brain following cerebral ischemia and hypoglycemic coma. Proc Natl Acad Sci USA 89:648–652

Linnik MD, Zobrist RH, Hatfield MD (1993) Evidence supporting a role for programmed cell death in focal cerebral ischemia in rats. Stroke 24:2002–2008

Lomeli H, Mosbacher J, Melcher T, Hoger T, Geiger JR, Kuner T, Monyer H, Higuchi M, Bach A, Seeburg PH (1994) Control of kinetic properties of AMPA receptor channels by nuclear RNA editing. Science 266:1709–1713

Lucas DR, Newhouse JP (1957) The toxic effects of sodium L-glutamate on the inner layers of the retina. Arch Ophthalmol 58:193–201

Luthi A, Gahwiler BH, Gerber U (1996) A slowly inactivating potassium current in CA3 pyramidal cells of rat hippocampus in vitro. J Neurosci 16:586–594

MacManus JP, Buchan AM, Hill IE, Rasquinha I, Preston E (1993) Global ischemia can cause DNA fragmentation indicative of apoptosis in rat brain. Neurosci Lett 164:89–92

Martin LJ, Al-Abdulla NA, Brambrink AM, Kirsch JR, Sieber FE, Portera-Cailliau C (1998) Neurodegeneration in excitotoxicity, global cerebral ischemia, and target deprivation: a perspective on the contributions of apoptosis and necrosis. Brain Res Bull 46:281–309

Matute C, Sanchez-Gomez MV, Martinez-Millan L, Miledi R (1997) Glutamate receptor-mediated toxicity in optic nerve oligodendrocytes. Proc Natl Acad Sci USA 94:8830–8835

Mayer ML, Westbrook GL, Guthrie PB (1984) Voltage-dependent block by Mg^{2+} of NMDA responses in spinal cord neurons. Nature 309:261–263

McDonald JW, Althomsons SP, Hyrc KL, Choi DW, Goldberg MP (1998a) Oligodendrocytes from forebrain are highly vulnerable to AMPA/kainate receptor-mediated excitotoxicity. Nat Med 4:291–297

McDonald JW, Bhattacharyya T, Sensi SL, Lobner D, Ying HS, Canzoniero LM, Choi DW (1998b) Extracellular acidity potentiates AMPA receptor-mediated cortical neuronal death. J Neurosci 18:6290–6299

McGurk JF, Bennett MV, Zukin RS (1990) Polyamines potentiate responses of *N*-methyl-D-aspartate receptors expressed in *Xenopus* oocytes. Proc Natl Acad Sci USA 87:9971–9974

McNamara D, Dingledine R (1990) Dual effect of glycine on NMDA-induced neurotoxicity in rat cortical cultures. J Neurosci 10:3970–3976

Miller B, Sarantis M, Traynelis SF, Attwell D (1992) Potentiation of NMDA receptor currents by arachidonic acid. Nature 355:722–725

Monyer H, Burnashev N, Laurie DJ, Sakmann B, Seeburg PH (1994) Developmental and regional expression in the rat brain and functional properties of four NMDA receptors. Neuron 12:529–540

Monyer H, Hartley DM, Choi DW (1990) 21-Aminosteroids attenuate excitotoxic neuronal injury in cortical cell cultures. Neuron 5:121–126

Monyer H, Sprengel R, Schoepfer R, Herb A, Higuchi M, Lomeli H, Burnashev N, Sakmann B, Seeburg PH (1992) Heteromeric NMDA receptors: molecular and functional distinction of subtypes. Science 256:1217–1221

Mosbacher J, Schoepfer R, Monyer H, Burnashev N, Seeburg PH, Ruppersberg JP (1994) A molecular determinant for submillisecond desensitization in glutamate receptors. Science 266:1059–1062

Muir KW (1998) New experimental and clinical data on the efficacy of pharmacological magnesium infusions in cerebral infarcts. Magnes Res 11:43–56

Mukhin AG, Ivanova SA, Faden AI (1997) mGluR modulation of post-traumatic neuronal death: role of NMDA receptors. Neuroreport 8:2561–2566

Nakanishi S, Masu M (1994) Molecular diversity and functions of glutamate receptors. Annu Rev Biophys Biomol Struct 23:619–348

Niethammer M, Kim E, Sheng M (1996) Interaction between the C terminus of NMDA receptor subunits and multiple members of the PSD-95 family of membrane-associated guanylate kinases. J Neurosci 16:2157–2163

Nitatori T, Sato N, Waguri S, Karasawa Y, Araki H, Shibanai K, Kominami E, Uchiyama Y (1995) Delayed neuronal death in the CA1 pyramidal cell layer of the gerbil hippocampus following transient ischemia is apoptosis. J Neurosci 15:1001–1011

Norenberg MD, Mozes LW, Gregorios JB, Norenberg LO (1987) Effects of lactic acid on astrocytes in primary culture. J Neuropathol Exp Neurol 46:154–166

Novelli A (1988) Glutamate becomes neurotoxic via the *N*-methyl-D-aspartate receptor when intracellular energy levels are reduced. Brain Res 451:205–212

Nowak L, Bregestovski P, Ascher P, Herbet A, Prochiantz A (1984) Magnesium gates glutamate-activated channels in mouse central neurons. Nature 307:462–465

Oka A, Belliveau MJ, Rosenberg PA, Volpe JJ (1993) Vulnerability of oligodendroglia to glutamate: pharmacology, mechanisms, and prevention. J Neurosci 13:1441–1453

Olney JW (1969) Brain lesion, obesity and other disturbances in mice treated with monosodium glutamate. Science 164:719–721

Olney JW, Labruyere J, Wang G, Wozniak DF, Price MT, Sesma MA (1991) NMDA antagonist neurotoxicity: mechanism and prevention. Science 254:1515–1518

Oppenheim RW (1991) Cell death during development of the nervous system. Annu Rev Neurosci 13:453–501

Paschen W, Widmann R, Weber C (1992) Changes in regional polyamine profiles in rat brains after transient cerebral ischemia (single versus repetitive ischemia): evidence for release of polyamines from injured neurons. Neurosci Lett 135:121–124

Pellegrini-Giampietro DE, Gorter JA, Bennett MV, Zukin RS (1997) The GluR2 (GluR-B) hypothesis: Ca^{2+}-permeable AMPA receptors in neurological disorders. Trends Neurosci 20:464–470

Pellegrini-Giampietro DE, Zukin RS, Bennett MV, Cho S, Pulsinelli WA (1992) Switch in glutamate receptor subunit gene expression in CA1 subfield of hippocampus following global ischemia in rats. Proc Natl Acad Sci USA 89:10499–10503

Peters S, Koh J, Choi DW (1987) Zinc selectively blocks the action of *N*-methyl-D-aspartate on cortical neurons. Science 236:589–593

Pizzi M, Consolandi O, Memo M, Spano PF (1996) Activation of multiple metabotropic glutamate receptor subtypes prevents NMDA-induced excitotoxicity in rat hippocampal slices. Eur J Neurosci 8:1516–1521

Pollard H, Charriaut-Marlangue C, Cantagrel S, Represa A, Robain O, Moreau J, Ben-Ari Y (1994) Kainate-induced apoptotic cell death in hippocampal neurons. Neuroscience 63:7–18

Pollard H, Heron A, Moreau J, Ben-Ari Y, Khrestchatisky M (1993) Alterations of the GluR-B AMPA receptor subunit flip/flop expression in kainate-induced epilepsy and ischemia. Neuroscience 57:545–554

Portera-Cailliau C, Hedreen JC, Price DL, Koliatsos VE (1995) Evidence for apoptotic cell death in Huntington disease and excitotoxic animal models. J Neurosci 15: 3775–3787

Portera-Cailliau C, Price DL, Martin LJ (1997) Non-NMDA and NMDA receptor-mediated excitotoxic neuronal deaths in adult brain are morphologically distinct: further evidence for an apoptosis-necrosis continuum. J Comp Neurol 378:88–104

Posner A, Raser KJ, Hajimohammadreza I, Yuen PW, Wang KK (1995) Aurintricarboxylic acid is an inhibitor of μ- and m-calpain. Biochem Mol Biol Int 36:291–299

Prehn JH (1996) Marked diversity in the action of growth factors on *N*-methyl-D-aspartate-induced neuronal degeneration. Eur J Pharmacol 306:81–88

Pruss RM, Akeson RL, Racke MM, Wilburn JL (1991) Agonist-activated cobalt uptake identifies divalent cation-permeable kainate receptors on neurons and glial cells. Neuron 7:509–518

Puchalski RB, Louis JC, Brose N, Traynelis SF, Egebjerg J, Kukekov V, Wenthold RJ, Rogers SW, Lin F, Moran T, Morrison, JH, Heinemann, SF (1994) Selective RNA editing and subunit assembly of native glutamate receptors. Neuron 13:131–147

Qin ZH, Wang Y, Chase TN (1996) Stimulation of *N*-methyl-D-aspartate receptors induces apoptosis in rat brain. Brain Res 725:166–176

Rao A, Craig AM (1997) Activity regulates the synaptic localization of the NMDA receptor in hippocampal neurons. Neuron 19:801–812

Rassendren FA, Lory P, Pin JP, Nargeot J (1990) Zinc has opposite effects on NMDA and non-NMDA receptors expressed in *Xenopus* oocytes. Neuron 4:733–740

Raymond LA, Moshaver A, Tingley WG, Huganir RL (1996) Glutamate receptor ion channel properties predict vulnerability to cytotoxicity in a transfected nonneuronal cell line. Mol Cell Neurosci 7:102–115

Regan RF, Panter SS, Witz A, Tilly JL, Giffard RG (1995) Ultrastructure of excitotoxic neuronal death in murine cortical culture. Brain Res 705:188–198

Reynolds IJ, Hastings TG (1995) Glutamate induces the production of reactive oxygen species in cultured forebrain neurons following NMDA receptor activation. J Neurosci 15:3318–3327

Roberts-Lewis JM, Marcy VR, Zhao Y, Vaught JL, Siman R, Lewis ME (1993) Aurintricarboxylic acid protects hippocampal neurons from NMDA- and ischemia-induced toxicity in vivo. J Neurochem 61:378–381

Rose K, Christine CW, Choi DW (1990) Magnesium removal induces paroxysmal neuronal firing and NMDA receptor-mediated neuronal degeneration in cortical cultures. Neurosci Lett 115:313–317

Rothman SM, Thurston JH, Hauhart RE (1987) Delayed neurotoxicity of excitatory amino acids in vitro. Neuroscience 22:471–480

Sarnesto A, Linder N, Raivio KO (1996) Organ distribution and molecular forms of human xanthine dehydrogenase/xanthine oxidase protein. Lab Invest 74:48–56

Schinder AF, Olson EC, Spitzer NC, Montal M (1996) Mitochondrial dysfunction is a primary event in glutamate neurotoxicity. J Neurosci 16:6125–6133

Schwarcz R, Du F, Schmidt W, Turski WA, Gramsbergen JB, Okuno E, Roberts RC (1992) Kynurenic acid: a potential pathogen in brain disorders. Ann NY Acad Sci 648:140–153

Sensi SL, Canzoniero LM, Yu SP, Ying HS, Koh JY, Kerchner GA, Choi DW (1997) Measurement of intracellular free zinc in living cortical neurons: routes of entry. J Neurosci 17:9554–9564

Sheardown MJ, Nielsen EO, Hansen AJ, Jacobsen P, Honore T (1990) 2,3-Dihydroxy-6-nitro-7-sulfamoyl-benzo(F)quinoxaline: a neuroprotectant for cerebral ischemia. Science 247:571–574

Siman R, Noszek JC (1988) Excitatory amino acids activate calpain I and induce structural protein breakdown in vivo. Neuron 1:279–287

Simonian NA, Getz RL, Leveque JC, Konradi C, Coyle JT (1996) Kainic acid induces apoptosis in neurons. Neuroscience 75:1047–1055

Sladeczek F, Momiyama A, Takahashi T (1993) Presynaptic inhibitory action of a metabotropic glutamate-receptor agonist on excitatory transmission in visual cortical neurons. Proc R Soc Lond B Biol Sci 253:297–303

Smart TG, Xie X, Krishek BJ (1994) Modulation of inhibitory and excitatory amino acid receptor ion channels by zinc. Prog Neurobiol 42:393–341

Sommer B, Keinanen K, Verdoorn TA, Wisden W, Burnashev N, Herb A, Kohler M, Takagi T, Sakmann B, Seeburg PH (1990) Flip and flop: a cell-specific functional switch in glutamate-operated channels of the CNS. Science 249:1580–1585

Sprengel R, Suchanek B, Amico C, Brusa R, Burnashev N, Rozov A, Hvalby O, Jensen V, Paulsen O, Andersen P, Kim JJ, Thompson RF, Sun W, Webster LC, Grant SG, Eilers J, Konnerth A, Li J, McNamara JO, Seeburg PH (1998) Importance of the intracellular domain of NR2 subunits for NMDA receptor function in vivo. Cell 92:279–289

Srivastava S, Osten P, Vilim FS, Khatri L, Inman G, States B, Daly C, DeSouza S, Abagyan R, Valtschanoff JG, Weinberg RJ, Ziff EB (1998) Novel anchorage of GluR2/3 to the postsynaptic density by the AMPA receptor-binding protein ABP. Neuron 21:581–591

Standaert DG, Testa CM, Young AB, Penney JB Jr (1994) Organization of *N*-methyl-D-aspartate glutamate receptor gene expression in the basal ganglia of the rat. J Comp Neurol 343:1–16

Stefani A, Pisani A, Mercuri NB, Calabresi P (1996) The modulation of calcium currents by the activation of mGluRs: functional implications. Mol Neurobiol 13: 81–95

Stern P, Behe P, Schoepfer R, Colquhoun D (1992) Single-channel conductances of NMDA receptors expressed from cloned cDNAs: comparison with native receptors. Proc R Soc Lond Biol Sci 250:271–277

Stone TW (1993) Neuropharmacology of quinolinic and kynurenic acids. Pharmacol Rev 45:309–379

Stout AK, Li-Smerin Y, Johnson JW, Reynolds IJ (1996) Mechanisms of glutamate-stimulated Mg^{2+} influx and subsequent Mg^{2+} efflux in rat forebrain neurons in culture. J Physiol (Lond) 492:641–657

Szabo C, Dawson VL (1998) Role of poly(ADP-ribose) synthetase in inflammation and ischaemia-reperfusion. Trends Pharmacol Sci 19:287–298

Tan SE, Wenthold RJ, Soderling TR (1994) Phosphorylation of AMPA-type glutamate receptors by calcium/calmodulin-dependent protein kinase II and protein kinase C in cultured hippocampal neurons. J Neurosci 14:1123–1129

Tang CM, Dichter M, Morad M (1990) Modulation of the *N*-methyl-D-aspartate channel by extracellular H^+. Proc Natl Acad Sci USA 87:6445–6449

Tecoma ES, Choi DW (1989) GABAergic neocortical neurons are resistant to NMDA receptor-mediated injury. Neurology 39:676–682

Tenneti L, D'Emilia DM, Troy CM, Lipton SA (1998) Role of caspases in *N*-methyl-D-aspartate-induced apoptosis in cerebrocortical neurons. J Neurochem 71: 946–959

Tombaugh GC, Sapolsky RM (1990) Mild acidosis protects hippocampal neurons from injury induced by oxygen and glucose deprivation. Brain Res 506:343–345

Tong G, Shepherd D, Jahr CE (1995) Synaptic desensitization of NMDA receptors by calcineurin. Science 267:1510–1512

Traynelis SF, Cull-Candy SG (1990) Proton inhibition of *N*-methyl-D-aspartate receptors in cerebellar neurons. Nature 345:347–350

Tsuda T, Kogure K, Nishioka K, Watanabe T (1991) Mg^{2+} administered up to twenty-four hours following reperfusion prevents ischemic damage of the Ca1 neurons in the rat hippocampus. Neuroscience 44:335–341

Turetsky DM, Canzoniero LM, Sensi SL, Weiss JH, Goldberg MP, Choi DW (1994) Cortical neurons exhibiting kainate-activated Co^{2+} uptake are selectively vulnerable to AMPA/kainate receptor-mediated toxicity. Neurobiol Dis 1:101–110

Uemura Y, Kowall NW, Beal MF (1990) Selective sparing of NADPH-diaphorase–somatostatin–neuropeptide Y neurons in ischemic gerbil striatum. Ann Neurol 27:620–625

van Lookeren Campagne M, Gill R (1996) Ultrastructural morphological changes are not characteristic of apoptotic cell death following focal cerebral ischaemia in the rat. Neurosci Lett 213:111–114

Waldmann R, Champigny G, Bassilana F, Heurteaux C, Lazdunski M (1997) A proton-gated cation channel involved in acid-sensing. Nature 386:173–177

Wang GJ, Thayer SA (1996) Sequestration of glutamate-induced Ca^{2+} loads by mitochondria in cultured rat hippocampal neurons. J Neurophysiol 76:1611–1621

Watanabe M, Inoue Y, Sakimura K, Mishina M (1993) Distinct distributions of five *N*-methyl-D-aspartate receptor channel subunit mRNAs in the forebrain. J Comp Neurol 338:377–390

Wei EP, Ellison MD, Kontos HA, Povlishock JT (1986) O_2 radicals in arachidonate-induced increased blood-brain barrier permeability to proteins. Am J Physiol 251:H693–699

Weiss JH, Hartley DM, Koh JY, Choi DW (1993) AMPA receptor activation potentiates zinc neurotoxicity. Neuron 10:43–49

Weiss SW, Albers DS, Iadarola MJ, Dawson TM, Dawson VL, Standaert DG (1998) NMDAR1 glutamate receptor subunit isoforms in neostriatal, neocortical, and hippocampal nitric oxide synthase neurons. J Neurosci 18:1725–34

Wenzel A, Scheurer L, Kunzi R, Fritschy JM, Mohler H, Benke D (1995) Distribution of NMDA-receptor-subunit proteins NR2A, 2B, 2C and 2D in rat brain. Neuroreport 7:45–48

Westbrook GL, Mayer ML (1987) Micromolar concentrations of Zn^{2+} antagonize NMDA and GABA responses of hippocampal neurons. Nature 328:640–643

White RJ, Reynolds IJ (1996) Mitochondrial depolarization in glutamate-stimulated neurons: an early signal specific to excitotoxin exposure. J Neurosci 16:5688–5697

Winegar BD, Lansman JB (1990) Voltage-dependent block by zinc of single calcium channels in mouse myotubes. J Physiol (Lond) 425:563–578

Wrathall JR, Teng YD, Choiniere D, Mundt DJ (1992) Evidence that local non-NMDA receptors contribute to functional deficits in contusive spinal cord injury. Brain Res 586:140–143

Xiang H, Kinoshita Y, Knudson CM, Korsmeyer SJ, Schwartzkroin PA, Morrison RS (1998) Bax involvement in p53-mediated neuronal cell death. J Neurosci 18:1363–1373

Xu XJ, Hao JX, Seiger A, Wiesenfeld-Hallin Z (1993) Systemic excitatory amino acid receptor antagonists for the α-amino-3-hydroxy-5-methyl-4-isoxazoleproprionic acid (AMPA) receptor and of the *N*-methyl-D-aspartate (NMDA) receptor relieve mechanical hypersensitivity after transient spinal cord ischemia in rats. J Pharmacol Exp Ther 267:140–144

Yin HZ, Turetsky D, Choi DW, Weiss JH (1994) Cortical neurones with Ca^{2+} permeable AMPA/kainate channels display distinct receptor immunoreactivity and are GABAergic. Neurobiol Dis 1:43–49

Yin HZ, Weiss JH (1995) Zn^{2+} permeates Ca^{2+} permeable AMPA/kainate channels and triggers selective neural injury. Neuroreport 6:2553–2556

Ying HS, Gwag BJ, Behrens MM, Koh J, Lobner D, Choi DW (1995) Neurotrophins induce NMDA receptor expression in cultured rat neocortical neurons. Soc Neurosci Abstr 21:1031

Ying HS, Weishaupt JH, Grabb M, Canzoniero LMT, Sensi SL, Sheline CT, Monyer H, Choi DW (1997) Sublethal oxygen-glucose deprivation alters hippocampal neuronal AMPA receptor expression and vulnerability to kainate-induced death. J Neurosci 17:9536–9544

Yoshioka A, Hardy M, Younkin DP, Grinspan JB, Stern JL, Pleasure D (1995) α-Amino-3-hydroxy-5-methyl-4-isoxazolepropionate (AMPA) receptors mediate excitotoxicity in the oligodendroglial lineage. J Neurochem 64:2442–2448

Yu AC, Chan PH, Fishman RA (1986) Effects of arachidonic acid on glutamate and gamma-aminobutyric acid uptake in primary cultures of rat cerebral cortical astrocytes and neurons. J Neurochem 47:1181–1189

Yu SP, Choi DW (1997) Na^+-Ca^{2+} exchange currents in cortical neurons: concomitant forward and reverse operation and effect of glutamate. Eur J Neurosci 9:1273–1281

Yu SP, Sensi SL, Canzoniero LM, Buisson A, Choi DW (1997a) Membrane-delimited modulation of NMDA currents by metabotropic glutamate receptor subtypes 1/5 in cultured mouse cortical neurons. J Physiol (Lond) 499:721–732

Yu SP, Yeh CH, Sensi SL, Gwag BJ, Canzoniero LM, Farhangrazi ZS, Ying HS, Tian M, Dugan LL, Choi DW (1997b) Mediation of neuronal apoptosis by enhancement of outward potassium current. Science 278:114–117
Zhang J, Dawson VL, Dawson TM, Snyder SH (1994) Nitric oxide activation of poly(ADP-ribose) synthetase in neurotoxicity. Science 263:687–689
Zheng X, Zhang L, Durand GM, Bennett MV, Zukin RS (1994) Mutagenesis rescues spermine and Zn^{2+} potentiation of recombinant NMDA receptors. Neuron 12:811–818
Zorumski CF, Yang J, Fischbach GD (1989) Calcium-dependent, slow desensitization distinguishes different types of glutamate receptors. Cell Mol Neurobiol 9:95–104
Zukin RS, Bennett MV (1995) Alternatively spliced isoforms of the NMDAR1 receptor subunit. Trends Neursci 18:306–313

CHAPTER 15

Glutamate Transporter Dysfunction and Neuronal Death

R. GANEL and J.D. ROTHSTEIN

A. Introduction

This chapter summarizes the normal function of the four major glutamate transporters in the mammalian central nervous system (CNS), as well as their structure, distribution, and cellular localization. We will provide an overview of the regulatory processes involved in normal expression and function of glutamate transporters, identifying critical steps where a point failure might lead to transporter dysfunction resulting in pathological conditions. We will also review the cellular events leading from glutamate transport dysfunction to neuronal death. In particular, we will describe how a loss of a certain glutamate transporter could cause/contribute to neuronal degeneration in the neurodegenerative disease amyotrophic lateral sclerosis (ALS). Finally, we will review the various experimental models being used in these studies.

B. Glutamate Transporters – Function

Glutamate is an excitatory amino acid neurotransmitter. When released from presynaptic terminals, it can activate *N*-methyl-D-aspartate (NMDA) and non-NMDA ionotropic receptors, as well as metabotropic glutamate receptors. Excitatory synaptic transmission is terminated by transporter proteins located on surrounding astroglia and neurons. Four distinct, high-affinity, sodium-dependent glutamate transporters have been identified in animal and human CNS. Glutamate/aspartate transporter (GLAST), glutamate transporter 1 (GLT-1), excitatory amino acid carrier 1 (EAAC1) [excitatory amino acid transporter 1 (EAAT1), EAAT2 and EAAT3, respectively, in human] and EAAT4 differ in structure, pharmacological properties and tissue distribution (KANAI and HEDIGER 1992; PINES et al. 1992; STOCK et al. 1992; ARRIZA et al. 1994; FAIRMAN et al. 1995). Another member of the family has been recently cloned. EAAT5, whose expression is limited to the retina (ARRIZA et al. 1997), will not be discussed here. Glutamate transporters share over 50% amino acid sequence identity with each other and display almost identical hydrophobic profiles, with six prominent hydrophobic peaks followed by a small hydrophobic peak and a long hydrophobic stretch, suggesting a great deal of structural similarity (KANAI et al. 1993). These proteins transport L-glutamate, D-aspartate and L-aspartate and some other acidic amino acids, such as threo-β-

hydroxyaspartate (THA) and cysteate (KANAI et al. 1997). However, the transporters display distinct properties in substrate or inhibitor selectivity, e.g. dihydrokainate is a specific inhibitor of GLT-1 (ARRIZA et al. 1994), and EAAC1 transports cysteine with much higher affinity than the other transporters (ZERANGUE and KAVANAUGH 1996b). These differences suggest that small but significant differences do exist between the transporters' substrate-binding sites. HAUGETO et al. (1996) have shown, using chemical cross-linking, that GLAST, GLT-1 and EAAC1 form homomultimers. They hypothesized, based on results from radiation-inactivation experiments, that the oligomeric structure is required for transport activity. In a more recent study, DEHNES et al. (1998) have shown that EAAT4 forms dimers as well. However, GRUNEWALD et al. (1998), using biotinylation studies of GLT-1 wild-type and mutants, could not detect any biotinylated dimer forms of GLT-1 and concluded that the dimerization is a property of immature transporters rather than functional ones. Thus, although transporters may form dimers, this process may be an artifact of experimental conditions or may occur subcellularly; physiological transport may only require monomers of the protein.

Glutamate transport is a sodium- and potassium-coupled process capable of concentrating intracellular glutamate up to 10,000-fold compared with the extracellular environment. There is a discrepancy in the literature regarding the stoichiometry of the process. Experiments with salamander retina glial cells suggest that the transport process is coupled to the co-transport of two Na^+ ions and the counter-transport of one K^+ and one OH^- ion (BOUVIER et al. 1992). However, ZERANGUE and KAVANAUGH (1996a) have demonstrated, in EAAC1, that one glutamate is co-transported with three Na^+ and one H^+ ion, with the counter-transport of one K^+. In contrast, the comparison of the direct influx of ^{14}C-glutamate and $^{22}Na^+$ through rabbit EAAC1 suggests that two Na^+ ions, rather than three, are co-transported with one glutamate molecule (KANAI et al. 1995). This discrepancy regarding the exact stoichiometry of the transport process is as yet unresolved.

Glutamate transporters can also possess channel-like properties. They conduct Cl^- flux not thermodynamically coupled to substrate transport, although a transportable substrate is required for the Cl^- conductance (WADICHE et al. 1995; SONDERS and AMARA 1996; WADICHE and KAVANAUGH 1998). The current hypothesis is that, by binding glutamate, the transporter changes its conformational state to form the Cl^- channel (WADICHE et al. 1995; SONDERS and AMARA 1996). The physiological relevance of the chloride fluxes associated with glutamate transport is not yet fully understood.

Immunohistochemical studies have showed that GLAST and GLT-1 are localized primarily in astrocytes (CHAUDHRY et al. 1995; LEHRE et al. 1995; MILTON et al. 1997). In the adult CNS, GLT-1 is widely distributed throughout the brain and spinal cord in astroglial cell bodies and processes, while GLAST protein is localized in glial cells of cerebellar molecular and granule cell layers and in some astroglia throughout the brain (ROTHSTEIN et al. 1994b). Double-labelling post-embedding electron-microscopic immunocytochemistry showed the two glial transporters GLT-1 and GLAST expressed in the same cell membrane. Each protein forms oligomeric complexes, but GLT-1 and GLAST do not complex with each other (Haugeto et al. 1996). Antisense knock-down studies have shown that these two glial transporters are responsible for over 80% of glutamate uptake in the brain (Rothstein et al. 1996), an observation later confirmed in GLT-1 null mice (Tanaka et al. 1997). Based on immunoprecipitation of solubilized glutamate transporters, Haugeto et al. have concluded that GLT-1 and GLAST account for about 93% and 7%, respectively, of the total transporter activity in adult rat forebrain (Haugeto et al. 1996). Lehre and Danbolt (1998) have shown, using quantitative immunoblotting that the glial transporters are quite abundant. GLAST and GLT-1, respectively, are found at levels of 2300 and 8500 molecules per μm^2 in CA1 hippocampus membrane and 4700 and 740 molecules per μm^2 in the cerebellar molecular layer.

Developmental studies reveal differential expression and cellular and regional distribution of GLT-1 and GLAST mRNA and protein. In embryonic mouse brain, GLAST and GLT-1 mRNAs are prominently expressed in the proliferative ventricular zone from embryonic day 13 to day 18 (Sutherland et al. 1996; Shibata et al. 1996). Thereafter, the levels of expression increase significantly in the mantle zone and peak at postnatal days 7–14. GLAST expression then shifts mainly to the cerebellum, whereas GLT-1 expression remains throughout most of the CNS. The dramatic up-regulation of GLT-1 gene expression at postnatal day 14 coincides with the postnatal development of glutamatergic transmission in the cortex (Shibata et al. 1996; Sutherland et al. 1996). Similar observations were made based on immunoreactivity of the actual expressed proteins in the developing CNS (Bar-Peled et al. 1997; Furuta et al. 1997; Ullensvang et al. 1997). In rat cerebellar Bergmann glia, expression of GLAST protein, which is low prenatally, becomes enriched in the early postnatal period and is also present in the forebrain in the later postnatal period (Furuta et al. 1997). Similarly, GLAST immunoreactivity was not detectable in human cerebellum during several gestation stages studied (Bar-Peled et al. 1997). Another interesting observation in developing human tissue was an intense GLAST immunoreactivity in the periventricular zone, similar to that observed in the developing brain of the mouse (Bar-Peled et al. 1997; Shibata et al. 1996; Sutherland et al. 1996). GLT-1 immunoreactivity was observed in fetal rat brain and spinal cord, and its expression increases to adult levels throughout the neuroaxis by postnatal day 26 (Furuta et al. 1997). In human, GLT-1 is the most prominent transporter throughout CNS development, and its expression appears to be low in mid-gestation, increasing dramatically postnatally (Bar-Peled et al. 1997).

GLT-1 mRNA (Brooks-Kayal et al. 1998) and protein (Mennerick et al. 1998) were also found in hippocampal neurons in culture. A recent study in rat brain demonstrates, using differential double hybridization, that the presence of GLT-1 mRNA in neurons is more widespread than previously thought, encompassing the majority of neurons in the neocortex, neurons in the olfactory bulb, neurons in the thalamus, CA3 pyramidal neurons in the hippocampus, and neurons in the inferior olive (BERGER and HEDIGER 1998). While several researchers (ULLENSVANG et al. 1997) have not been able to detect GLAST or GLT-1 proteins in neurons of rat brain in vivo, YAMADA et al. (1998) report detection of transiently localized GLT-1 on growing axons of mouse spinal cord before astrocytic expression is established. In fact, we previously observed transient expression of both GLT-1 and GLAST proteins along axonal pathways in prenatal rat brain (FURUTA et al. 1997) as well as in CA1 and dentate gyrus of human fetus (BAR-PELED et al. 1997). Since the latter study was done in frozen tissue, high-power microscopy determination of the glial transporters' cellular distribution in the developing hippocampus was not possible. However, a comparison of GLT-1 and GLAST distribution with that of a glial marker suggests that the glial transporters were localized to non-glial cells (BAR-PELED et al. 1997). A possible explanation for neuronal localization of GLT-1 and GLAST could be that, during development, prior to their glial expression, these transporters, expressed on neurons, function to prevent toxic extracellular accumulation of glutamate. Additional studies in models of ischemic brain injury (MARTIN et al. 1997) and in fetal ovine brain (NORTHINGTON et al. 1998) suggest neuronal expression of GLT-1 as well.

EAAC1 and EAAT4 are neuronal transporters. EAAC1 immunoreactivity is particularly high in regions such as the hippocampus, cerebellum and basal ganglia (ROTHSTEIN et al. 1994b). It is widely distributed in neurons, such as large cortical pyramidal neurons, and is also present in non-glutamatergic neurons, including γ-aminobutyric acid (GABA)ergic cerebellar Purkinje cells. Ultrastructural studies suggest that EAAC1 is not a presynaptic transporter of glutamatergic neurons. In fact, EAAC1 appears to be localized in the somatodendritic compartment (ROTHSTEIN et al. 1994b; SHASHIDHARAN et al. 1997) and is already expressed at stages preceding synaptic contact formation (COCO et al. 1997). EAAC1 is also widely expressed outside the CNS (KANAI and HEDIGER 1992; FURUTA et al. 1997), so it may serve metabolic functions in neurons. For example, it may provide glutamate for re-synthesis of GABA in GABAergic terminals where the protein has been localized (ROTHSTEIN et al. 1994b). In fact, studies using antisense oligonucleotides to inhibit EAAC1 suggest that this transporter may, in part, regulate GABA synthesis (SEPKUTY et al. 1997).

EAAT4 is largely expressed in the cerebellum, with very faint levels of expression in hippocampus, neocortex, striatum, brain stem and thalamus in both the adult human and rat CNS (FURUTA et al. 1997). DEHNES et al. (1998) have shown that EAAT4 is present at low concentrations in the synaptic membrane but is highly enriched in the parts of the dendritic and spine membranes facing astrocytes. Several groups have hypothesized that a functional relationship may exist between EAAT4 and the glial transporters and that EAAT4, having a prominent Cl^- channel property (FAIRMAN et al. 1995; VANDENBERG et al. 1995), may function as a combined transporter and inhibitory glutamate receptor. The average density of EAAT4 protein in the

Purkinje cell membrane has been calculated to be 1800 molecules per μm^2. Immunohistochemistry and immunoblot analysis demonstrated that, during development, EAAT4 protein is expressed in the human cerebellum both pre- and postnatally, while its expression in the frontal cortex is restricted to fetal stages (Bar-Peled et al. 1997). In the cerebellum, Purkinje cells show faint EAAT4 immunoreactivity at gestation week 17. However, EAAT4 expression becomes increasingly intense from gestation week 23 to the infantile period. After the late infantile period, EAAT4 immunoreactivity shows the same pattern as in adults (Itoh et al. 1997). The intracellular localization of EAAT4 also changes with development. In the early embryonic period, its immunoreactivity is demonstrated in the short processes of the Purkinje cells. In the late fetal to early infantile periods, EAAT4 immunoreactivity is found in the cell bodies and dendrites and, in the late infantile period, it is found in the spines (Itoh et al. 1997).

In adult human and rat CNS, both neuronal glutamate transporters are localized at somatodendritic compartments of Purkinje cells and are found at postsynaptic locations (Furuta et al. 1997). However, in regions of dense glutamatergic innervation, e.g. striatum or hippocampus, no presynaptic membranes have been found to contain either neuronal transporter (Rothstein et al. 1994b; Furuta et al. 1997). Furthermore, lesions of glutamate pathways indicate that the bulk of all glutamate transport is either postsynaptic or astroglial (Ginsberg et al. 1995). This observation suggests that presynaptic glutamatergic terminals do not transport glutamate or that other neuronal transporters exist and have yet to be found. However, molecular and functional studies suggest that EAAC1, GLT1 and GLAST can account for about 100% of all glutamate-transport activity in forebrain. The accepted model of high-affinity glutamate transporters was that they served to transport glutamate from the extracellular space and that they were localized to both surrounding glia and *presynaptic* terminals of glutamatergic neurons (Hertz 1979; Nicholls and Attwell 1990). The accumulated data on the cloned transporters as described above suggests that neuronal glutamate transport is largely postsynaptic, requiring a revision of the model.

C. Excitotoxicity

Functional glutamate transport is essential for the normal function of the brain. Excessive uptake of glutamate may result in glutamatergic hypofunction, a proposed mechanism for the development of schizophrenia and other psychoses (Tamminga 1998). However, maintenance of a low extracellular concentration of glutamate in the CNS is important to ensure a high signal-to-noise ratio during synaptic activation and to prevent neuronal damage from excessive activation of glutamate receptors. The first reports on the neurotoxic properties of glutamate were made by Lucas and Newhouse, who showed that systemic administration of glutamate to infant mice caused retinal degeneration (LUCAS and NEWHOUSE 1957). Olney demonstrated that oral administration of glutamate to infant rodents and primates resulted in neuronal degeneration in brain areas that lack the blood–brain barrier (OLNEY 1969, 1971; OLNEY and HO 1970; OLNEY et al. 1972). Further work demonstrated a direct correlation between neuroexcitatory and neurotoxic properties of glutamate and linked glutamate-induced neurotoxicity to activation of excitatory amino acid receptors (OLNEY 1969, 1971, 1982; OLNEY and HO 1970). The term "excitotoxicity" was proposed to describe this process. A number of excellent reviews have been published on glutamate-mediated toxicity in the last decade (CHOI and ROTHMAN 1990; CHOI 1988; MELDRUM and GARTHWAITE 1990), including a chapter in this book. Various mechanisms have been proposed to explain how acute or chronic glutamate exposure could be neurotoxic. Most models involve enzyme activation by an increase in cytosolic calcium following glutamate-receptor stimulation. These toxic pathways include activation of calpain and other proteases, phospholipase A_2 and other lipases, protein kinase C (PKC), endonucleases and nitric oxide synthase (BITTIGAU and IKONOMIDOU 1997). There has also been recent interest in the increased cellular oxidative stress following glutamate-receptor activation (COYLE and PUTTFARCKEN 1993). Specifically, recent studies suggest that oxidative stress, such as oxygen-radical excess, can be toxic to motor neurons and that glutamate may participate in this toxicity (MICHIKAWA et al. 1994; ROTHSTEIN et al. 1994a; ZEMAN et al. 1994).

Morphologically, acute and chronic excitotoxicity cause somatodendritic swelling, chromatin condensation into irregular clumps, and organelle damage in many neurons. These are features typical of cellular necrosis. However, in other neurons, excitotoxicity causes cytological features more reminiscent of apoptosis (MARTIN et al. 1998). Thus, it has been suggested that the morphological phenotype of excitotoxic neuronal death exists as an intermediate form of cell death that lies along a structural continuum with apoptosis and necrosis at its extremes. The resulting phenotype is influenced by the subtype of glutamate receptors activated and, therefore, excitotoxic neuronal death is not identical in every neuron (MARTIN et al. 1998).

D. Glutamate Transporter Dysfunction

The effective removal of excessive glutamate is a crucial mechanism, and failure or dysfunction of glutamate transporters may aggravate neurotoxic damage. For example, if glutamate uptake is blocked, as little as $1\,\mu M$ exogenous glutamate is sufficient to induce excitotoxic death in cortical neurons (FRANDSEN and SCHOUSBOE 1990). Furthermore, under conditions of energy failure, such as ischemia, hypoglycemia and hypoxia, when ion-gradient rundown occurs, glutamate transporters may operate in the reverse direction (NICHOLLS and ATTWELL 1990). In support of this proposal, Szatkowski and co-workers have shown in vitro that changing the forces driving the transporter can cause it to work in reverse (SZATKOWSKI et al. 1990), leading to nonvesicular release of glutamate rather than removal of glutamate from the extracellular space. Other studies suggest that reversed transport can occur under conditions of physiological depolarization as well (HERON et al. 1995).

In a study of glaucoma in vertebrate retina, MAGUIRE et al. (1998) show a calcium-independent release of glutamate under ischemic conditions and speculate that a possible source of the elevated glutamate surrounding retinal ganglion cells in glaucoma is release through reversed operation of a glial glutamate transporter. In audiogenic seizure-susceptible mice, marked differences were observed in glutamate transport kinetics and pharmacology when compared with control mice (CORDERO et al. 1994; ORTIZ et al. 1996).

A significant and selective decrease in glutamate uptake was also observed in feline astrocytes infected with FIV, the feline analog of human immunodeficiency virus (HIV) (YU et al. 1998). Furthermore, the pro-inflammatory cytokine, tumor necrosis factor α (TNF-α, produced by HIV-1-infected macrophages and microglia) induces a decrease in glutamate transport in primary human fetal astrocytes (FINE et al. 1996). Since previous reports demonstrate the ability of glutamate-receptor antagonists to block the neurotoxic effects of several HIV-1 neurotoxin candidates (GELBARD et al. 1993, 1994; LIPTON and GENDELMAN 1995; MAGNUSON et al. 1995), these findings may indicate that a selective loss of normal glutamate-transporter function may contribute to the neurodegeneration and dementia caused by HIV-1 virus.

Numerous studies have shown glutamate-transporter dysfunction in association with neurodegenerative pathologies in humans as well as in animal models. A significant decrease in ^{3}H-D-aspartate binding, a marker of glutamate-transporter sites (MASLIAH et al. 1996), as well as a reduction in GLT-1 (EAAT2) protein (LI et al. 1997) have been reported in frontal brain tissue from patients with Alzheimer's disease. In a perinatal model of hypoxia–ischemia, extracellular glutamate levels increase transiently in striatum (HAGBERG et al. 1987; GORDON et al. 1991), possibly by vesicular exocytosis of glutamate (ROTHMAN 1984), reversed glutamate-transporter function (SZATKOWSKI and ATTWELL 1994), defective uptake of glutamate (SILVERSTEIN et al. 1986) or astrocyte swelling (KIMELBERG et al. 1990). Astrocytes are damaged early, and a loss of 85% of GLT-1 occurs (MARTIN et al. 1997). This initial astrocytic alteration may contribute to delayed neuronal damage, since in vitro studies show that neurons in an astrocyte-poor environment are more vulnerable to excitotoxicity (ROSENBERG et al. 1992). Abnormal decreases in glutamate-transport function and specifically in GLT-1 protein expression were also observed in hepatic encephalopathy, acute liver failure (KNECHT et al. 1997) and ALS (see below).

E. Amyotrophic Lateral Sclerosis

ALS is an adult-onset chronic neuromuscular disease characterized pathologically by the relatively selective progressive degeneration of cortical motor neurons (upper motor neurons) and motor neurons in the brainstem and spinal cord (lower motor neurons). Clinically, the disease is characterized by muscle wasting, weakness and spasticity that begins typically as an asymmetric weakness in two or more limbs, progressing to complete paralysis (KUNCL et al. 1992). The disease was first described over 130 years ago (ARAN 1850) and occurs in both sporadic and familial forms. Familial ALS (FALS) is inherited as an autosomal dominant trait and accounts for 5–10% of the cases, of which ~20% (i.e. ~1% of all ALS cases) have mutations in the enzyme Cu/Zn superoxide dismutase (SOD) (ROSEN et al. 1993). These mutations are not present in sporadic ALS, and there is no environmental, geographical, occupational, nutritional or any specific life-style risk factor known to account for the disease (KUNCL et al. 1992).

Currently, there are four major hypotheses addressing putative causal factors for motor-neuron degeneration in ALS. (1) The oxidative stress hypothesis, based in part on the Cu/Zn SOD mutations mentioned above, which have been shown to reproduce the disease faithfully in transgenic animals (ROSEN et al. 1993; GURNEY et al. 1994). (2) The autoimmune hypothesis, suggesting that antibodies to L-type calcium channels could cause or contribute to the sporadic form of the disorder (SMITH et al. 1992). (3) The cytoskeletal hypothesis, suggesting that cytoskeleton abnormalities, such as excessive accumulation of neurofilaments, cause motor-neuron degeneration (LEE et al. 1994; WONG et al. 1995; BRUIJN and CLEVELAND 1996). (4) The excitotoxicity hypothesis, suggesting that excessive synaptic glutamate, due in part to glutamate-transporter dysfunction and/or dysfunction of other metabolic enzymes, could contribute to motor-neuron loss (PLAITAKIS 1990; ROTHSTEIN et al. 1992; SHAW et al. 1994). It is likely that multiple primary insults could result in a common ALS phenotype. The effect of the FALS-linked SOD mutant proteins has been studied in transgenic mice expressing various SOD mutations. These mice develop clinical symptoms similar to those seen in ALS patients, i.e. weakness of the limbs, muscle wasting, motor-neuron degeneration and death (LEE et al. 1996), and are a widely accepted model of the familial form of the disease. Recently, studies in two different lines of such transgenic mice have shown a dramatic drop in glutamate uptake (V_{max}) in spinal-cord synaptosomes, along with a marked decrease in GLT-1 protein level (BRUIJN et al. 1997; CANTON et al. 1998). These observations suggest that excitotoxicity resulting from glutamate-transporter dysfunction could be a common link contributing to the chronic motor-neuron degeneration seen in the disease.

What other evidence links motor-neuron degeneration and ALS with glutamate-transport dysfunction? Several studies suggest that serum or cerebrospinal fluid (CSF) glutamate levels may be elevated in sporadic ALS (PLAITAKIS and CAROSCIO 1987; IWASAKI et al. 1992; ROTHSTEIN et al. 1992; CAMU et al. 1993). Furthermore, studies have demonstrated significant decrease in brain or spinal-cord glutamate levels in ALS (PERRY et al. 1987; PLAITAKIS et al. 1988; TSAI et al. 1991). Motor-cortex and spinal-cord glutamate levels are decreased 30–45%, and these changes appear to be selective for gray matter (Tsai et al. 1991). This general alteration in glutamate led Plaitakis (1990) and others to suggest that there might be a generalized defect in the metabolism of glutamate, such as an alteration in glutamate transport or astroglial glutamate metabolism, release or membrane permeability. The two anabolic enzymes responsible for the major synthesis of glutamate are glutamine synthase and glutamate dehydrogenase. Our lab examined motor-cortex activities of glutamate dehydrogenase and glutamine synthase in ALS and found no difference from control enzyme activities (Rothstein 1996). Thus, the abnormalities in tissue glutamate or extracellular levels cannot be attributed to altered synthetic enzyme activity. Using synaptosomal membrane preparations of postmortem tissue, a significant loss of high-affinity sodium-dependent glutamate transport was found in ALS tissue when compared to control (Rothstein et al. 1992). A 70% and 59% decrease in glutamate uptake was observed in the affected motor cortex and spinal cord, respectively, whereas other areas of ALS CNS not affected by the disease, such as the hippocampus or striatum, showed normal glutamate-transport function. Furthermore, the effect is selective, since other sodium-dependent transporters, such as transporters for GABA and phenylalanine, were unaffected. In addition, Shaw et al. (1994) found that the abnormalities in glutamate transport were not present in the purely lower motor-neuron disease progressive muscular atrophy.

What could be the reason for a loss of glutamate transport in ALS tissue? Could it merely be due to the deaths of neurons? To answer this question, tissue from various brain regions of ALS patients and controls were examined by immunoblot or immunocytochemical methods using transporter-subtype-specific antibodies developed in our lab. Up to 60–70% of sporadic ALS patients have a 30–95% loss of EAAT2 (human GLT-1) protein both in motor cortex and in spinal cord. The loss in EAAT2 appears to be specific to these regions in most but not all patients (Rothstein et al. 1995). A modest loss (20% decrease compared with control) in EAAT3 (human EAAC1) immunoreactivity was observed in ALS motor cortex as well. This minor loss in EAAT3 protein could be secondary to loss of cortical motor neurons and other small cortical neurons known to degenerate in ALS. Interestingly, the dramatic loss in EAAT2 occurred in spite of the complete preservation of astroglia, as assessed by both immunoblots for glial fibrillary acidic protein and histological observation. Therefore, the loss of EAAT2 could represent either the selective loss of the protein from astroglia or the loss of an EAAT2-positive astroglia subset only in neuropathologically affected regions. Since studies in rat and human suggest, as described earlier, that EAAT2 is ubiquitously localized to astroglia, we hypothesized that the change in EAAT2 protein in ALS represents a selective loss or down-regulation of this protein. In support of this, the other astroglial transporter, EAAT1 (human GLAST), was unaffected in ALS. Subsequent studies utilized Northern-blot analysis in an attempt to examine the mRNA level for each glutamate-transporter subtype in ALS-affected motor cortex. There was no quantitative change, compared with control, in mRNA for any of the transporters, even in patients with large loss of EAAT2 and decreased tissue glutamate-transport activity (BRISTOL and ROTHSTEIN 1996).

What could account for the selective loss of one astroglial glutamate transporter? Recent findings in our lab suggest that the loss of EAAT2 in ALS is due to truncated EAAT2 mRNA, which could result from mRNA processing errors (LIN et al. 1998a). Multiple abnormal EAAT2 mRNAs, including intron retention and exon skipping, have been identified from the affected regions of ALS CNS. The aberrant mRNAs were specific to ALS-affected regions and were found in 65% of sporadic ALS patients. The aberrant species were quite abundant, representing more than 70% of total EAAT2 RNA. They were not generally found in normal or other disease controls. Rarely, a single truncated species was found in very low abundance in some control specimens, reflecting either alternate splice species or illegitimate RNA. They were also found in the CSF of living ALS patients at early stages of the disease.

Are these truncated EAAT2 RNA species relevant to the loss of EAAT2 in ALS patients? In vitro expression studies suggest that proteins translated from these aberrant mRNA species may undergo rapid degeneration and/or produce a dominant negative effect on normal GLT-1, resulting in loss of protein and activity. Germline mutations in the glutamate transporter GLT-1 gene have been found but are infrequent and do not explain the presence of variant mRNA transcripts in many ALS patients (AOKI et al. 1998). The available evidence suggests that the truncated RNA species may result from abnormal RNA metabolism. Interestingly, the childhood motor-neuron disease spinal muscular atrophy is due to mutations in the survival motor neuron (SMN) gene (CRAWFORD and PARDO 1996). SMN protein was recently found to be an essential component for assembly of the spliceosomal complex responsible for cellular RNA processing (LIU and DREYFUSS 1996; FISCHER and DREYFUSS 1997). Identification of the events that lead to aberrant RNA processing could provide an opportunity for the development of specific therapies designed to prevent the loss of glutamate transport in ALS. It may also further our understanding of the pathogenesis of diseases that are not yet fully understood.

F. Causes of Glutamate Transporter Dysfunction

Glutamate transporters are highly regulated proteins (for review see GEGELASHVILI and SCHOUSBOE 1997) with various mechanisms modulating different levels of gene expression and protein activity. Theoretically, any step along the process of producing a functional protein can represent a potential target to errors, resulting in inappropriate up- or down-regulation of the glutamate-uptake process.

Neurons regulate the expression of glial transporters in a complex manner. Lesions that destroy excitatory (cortical) input to the striatum result in decreased GLT-1 and GLAST protein levels (GINSBERG et al. 1995). However, excitotoxic insult to co-cultures of astrocytes and neurons, which kills the neurons, causes a decrease in GLT-1 expression by the astrocytes without affecting the level of GLAST (SCHLAG et al. 1998). The neuronal regulation is probably carried by a diffusible factor that activates a signaling pathway mediated by cyclic adenosine monophosphate (cAMP) production (GEGELASHVILI et al. 1997; SWANSON et al. 1997; SCHLAG et al. 1998). In studies with purified cortical astrocytes, treatment with cAMP analogs induced the protein levels of both glial transporters (SCHLAG et al. 1998). However, kinetic and pharmacological data suggest that the induced GLT-1 was not functional. Furthermore, the authors have observed, in astrocyte cultures, a single EAAT4 immunoreactive band which is smaller than that observed in cerebellar homogenate and which is identical in size to the deglycosylated band of EAAT4 observed in cerebellar tissue (SCHLAG et al. 1998). It appears that, in these studies, treatment with cAMP analogs induced the expression of GLT-1 and EAAT4 proteins. However, important components that are required for the functional assembly of the transporter protein in the plasma membrane in an active form may have been missing or not induced in this system. The existence of specific transporter-interactive proteins with regulation or trafficking roles has been proposed. A family of such proteins has been identified for the NMDA and non-NMDA glutamate receptors (KIM et al. 1996; BRAKEMAN et al. 1997; DONG et al. 1997). Recently, we have identified several unique proteins which function to regulate both membrane localization and activity of neuronal (EAAC1 and EAAT4) glutamate transporters (LIN et al. 1998b; ORLOV et al. 1998). In addition, certain interacting proteins may serve to communicate metabolic signalling following transporter activation.

Studies describing regulation of glutamate-transporter expression during development suggest a close connection between synapse formation and mature expression of the glial transporters. GLAST and GLT-1 levels increase with maturation, while EAAC1 levels peak in neonatal rat brain and then decrease to adult levels (SUTHERLAND et al. 1996; FURUTA et al. 1997). Recent studies suggest that glutamate may regulate GLAST expression through activation of ionotropic glutamate receptors (GEGELASHVILI et al. 1996).

As part of the normal regulation of the biological clock, EAAC1 mRNA levels vary with the circadian rhythm in the suprachiasmatic nuclei (CAGAMPANG et al. 1996). In a renal epithelial cell line, EAAC1 mRNA levels decrease in response to amino acid deprivation (PLAKIDOU-DYMOCK and MCGIVAN 1993).

In pathological conditions, GLAST mRNA expression altered after photochemically induced focal ischemia in rat brain (YIN et al. 1998). Lowered levels of GLT-1 mRNA are observed in post-ischemic rat hippocampus, suggesting a possible mechanism for the decreased clearance of glutamate in ischemia models (TORP et al. 1995).

The expression level of two glutamate transporters altered in amygdala-kindled rats. GLAST protein was down-regulated in the piriform cortex–amygdala region of kindled rats as early as 24h after one stage-3 seizure and persisting through multiple stage-5 seizures. In contrast, kindling induced an increase in EAAC1 levels in piriform cortex–amygdala and hippocampus once the animals reached the stage-5 level (Miller et al. 1997).

Various drugs can affect the expression of glutamate transporters. Chronic neuroleptic treatment alters expression of glial glutamate transporter GLT-1 mRNA in the striatum (Schneider et al. 1998). Lithium inhibits glutamate uptake in the mouse cortex by decreasing the V_{max}, without affecting the K_m, while chronic treatment with therapeutic levels of lithium causes up-regulation (Dixon and Hokin 1998).

Numerous studies suggest that regulation of activity occurs on the level of post-translational modifications of glutamate transporters. Differential modulation of glutamate-transporter subtypes has been demonstrated with arachidonic acid. EAAT1 (GLAST) activity was inhibited via a decrease in V_{max}, while EAAT2 (GLT-1) was potentiated via an increase in apparent affinity. There was almost no effect on the activity of EAAT3 (EAAC1) by arachidonic-acid treatment (Zerangue et al. 1995). Conflicting evidence suggests that PKC activation can increase (Casado et al. 1993), decrease (Ganel and Crosson 1998) or have no effect (Tan et al. 1999) on GLT-1 activity in different systems. Chronic exposure to ethanol caused an increase in glutamate transport V_{max} in astrocytes via a post-translational modification. The ethanol-induced increase in transport V_{max} was reversed upon withdrawal of ethanol from the medium and was mediated by PKC activation (Smith 1997).

The ability to traffic transporters rapidly may represent an endogenous protective mechanism to limit excitotoxicity damage by increasing the cell surface expression of transporters to promote clearance of extracellular glutamate or by removing transporters from the cell surface and, thus, reducing the reversal activity of the transporter thought to contribute to the rise in extracellular glutamate observed during excitotoxicity insult. The trafficking of EAAC1 is regulated by two independent signalling pathways: activation of PKC and PI3K (Davis et al. 1998).

A processing mechanism for normal editing of transporter mRNA may exist. As mentioned above, an error in such a process may be the cause of production of the aberrant mRNA species observed in sporadic ALS. In a similar manner, any of the regulatory processes described above (or other mechanisms similar to these) may be the source of defective production of a dysfunctional transporter.

G. Experimental Models

Many studies, some discussed above, have shown a close association between glutamate-transporter dysfunction and neuronal cell death. However, the hypothesis that glutamate-transport dysfunction can actually lead to neurodegeneration has been studied through various experimental approaches.

The organotypic spinal-cord culture model was developed to study whether a chronic loss in glutamate uptake could produce elevations in extracellular glutamate and lead to a slow loss of motor neurons, such as that seen in ALS. In these cultures, prepared from 8-day-old rat pups, thin slices of lumbar spinal cord are maintained morphologically intact, including ventral motor-neuron pools, for up to 3 months (Rothstein et al. 1993). Organotypic cultures offer several advantages relevant to the study of the above hypothesis: they are prepared from postnatal spinal cord and, thus, represent relatively mature motor neurons; spinal synaptic connectivity is largely maintained; and normal astroglia–neuron interactions are preserved. To mimic glutamate-transport dysfunction or loss, a non-selective blocker of glutamate uptake, THA (Bridges et al. 1991), was used chronically. Chronic blockage of glutamate transport in this in vitro system produced persistent elevations of extracellular glutamate to levels of 3–25 μM depending on THA concentration (Rothstein et al. 1993). The physiological significance of this result is supported by a very similar result obtained by similar blockage of glutamate transport in vivo as measured through the use of microdialysis (Bargett et al. 1994; Rothstein et al. 1996). Furthermore, chronic inhibition of glutamate transport in organotypic spinal-cord cultures produced a slow, selective loss of motor neurons (Rothstein et al. 1993). From these studies, it appears that a chronic dysfunction of glutamate transport can lead to elevated extracellular glutamate and to slow motor-neuron degeneration.

To date, no selective inhibitors of the glutamate-transporter subtypes are available. Therefore, other models were developed to study the role of individual transporter subtypes in neuronal degeneration. Antisense oligodeoxynucleotides (ODN) are a particularly powerful and highly specific method to knock out the synthesis of targeted proteins. They act primarily through hybridization of the sense mRNA to prevent translation and protein synthesis. They can be administered in vitro by adding to the cell-culture medium or in vivo through indwelling intraventricular cannulas. The ODN offer several advantages over the use of nonselective pharmacological inhibitors or the development of transgenic knock-out mice: they are highly selective, cause a reversible effect, can be used in vitro and in vivo, are relatively inexpensive, do not require specialized equipment and allow a short-term study, thereby avoiding the development of compensatory mechanisms. Experiments with antisense ODN in organotypic spinal-cord cultures have shown that knock-out of the glial transporters GLT-1 and GLAST caused a gradual motor-neuron degeneration (Rothstein et al. 1996). The knock-out of the neuronal transporter EAAC1 in this system, however, did not induce neurodegeneration. When antisense ODN to individual glutamate transporters were administered intraventricularly to rats, the loss of either astroglial glutamate transporter produced a progressive paralysis beginning first in the hind limbs. Furthermore, extracellular glutamate, measured by microdialysis, was markedly increased (ROTHSTEIN et al. 1996). In vivo knock-out of the neuronal transporter EAAC1 did not affect extracellular glutamate levels and did not cause paralysis. Instead, EAAC1 knock-out rats developed a motor-neuron syndrome with slow hind-limb movement, atactic gait, hind-limb paresis and epilepsy with clonic seizures (ROTHSTEIN et al. 1996).

In another approach to study the role of individual transporters, TANAKA et al. (1997) used homologous recombination to inactivate GLT-1. The levels of the other transporter subtype mRNAs were not appreciably affected by the mutation, and the glutamate uptake in cortical crude synaptosomes of homozygous mutant mice was decreased to 5.8% of that in synaptosomes from wild-type mice. This observation suggests that GLT-1 is responsible for more then 90% of cerebral glutamate transport. No significant differences in the time course of NMDA- and non-NMDA-receptor-mediated excitatory post-synaptic currents (EPSCs) were found in hippocampus slices between the mutant and wild-type mice, indicating that GLT-1 does not determine the decay time of EPSCs in glutamatergic synapses. However, the peak concentration of synaptically released glutamate was increased in the mutant mice, and the glutamate remained elevated in the synaptic cleft for longer periods, indicating that GLT-1 is an important determinant of the clearance of free glutamate from the synaptic cleft, recently confirmed in hippocampal tissue (BERGLES and JAHR 1998). Furthermore, selective neuronal degeneration was found in mutant mouse hippocampus. The mutant mice experienced spontaneous epileptic seizures with a pattern similar to those of NMDA-induced seizures, which in most cases resulted in the death of the mice within a few seconds of seizure onset. In addition, the mutant mice were significantly more susceptible to the development of acute glutamate neurotoxicity after trauma. This study demonstrates that GLT-1 function is important in the maintenance of low extracellular glutamate levels and in the prevention of acute excitotoxicity after brain trauma and that, without GLT-1 function, glutamate levels rise enough to cause epilepsy and cell death.

The same approach was used by PEGHINI et al. (1997) to develop a mutant mouse line deficient in EAAC1. EAAC1 is not brain specific but is strongly expressed in kidney and small intestine. In the kidney, this transporter is responsible for tubular reabsorption of glutamate and aspartate from the glomerular filtrate. EAAC1 deficiency leads to excessive glutamate and aspartate excretion by the kidney, resulting in dicarboxylic aminoaciduria. However, no neurological or cognitive abnormalities were observed in EAAC1-deficient mice. The lack of EAAC1 did not produce any histological or neurological changes associated with neurodegeneration. Furthermore, no spontaneous epileptic seizures have been observed in EAAC1-lacking mice, and their susceptibility to certain pharmacologically induced seizures did not differ from that of wild-type mice. These results imply that either EAAC1 does not play an important role in the prevention of long-term neurotoxicity or that compensatory mechanisms have occurred during the development of these mice;

however, no enhanced expression of GLAST or GLT-1 mRNA or protein were observed in EAAC1-deficient mice.

Transgenic mice often make important models for various genetic diseases. The development of mice carrying different SOD mutations provides the best animal model for studying FALS. These mice, described in great detail in the review by LEE et al. (1996), develop a progressive motor-neuron degeneration characteristic of ALS. Interestingly, the mutant-SOD transgenic mice exhibit reduced glutamate uptake as well as the loss of GLT-1 (BRUIJN et al. 1997; CANTON et al. 1998). This observation suggests that, even in the subset of FALS where the origin of the disease is clearly a mutation in SOD enzyme, dysfunction of glutamate transporter and excitotoxicity are important secondary processes in the development of motor-neuron neurodegeneration.

The various experimental models described have additional importance; they can be used in screening putative therapeutic agents against neurodegeneration. We have used the organotypic spinal-cord cultures to evaluate putative neuroprotective factors. These studies revealed that agents that can inhibit glutamate release (riluzole) or synthesis (methionine sulfoximine; gabapentin) or block non-NMDA receptors (6-cyano-5-nitroquinoxaline-2,3-dione, 6-nitro-7-sulphamoylbenzo-[f]-quinoxaline-2,3-dione, GYKI-52466) are the most potent neuroprotectants (ROTHSTEIN and KUNCL 1995). Interestingly, some of these agents (riluzole, gabapentin) were subsequently found to extend survival in mice with mutant SOD (GURNEY et al. 1994). Several double-blind, placebo-controlled trials have shown riluzole to be efficacious at extending survival in ALS (BENSIMON et al. 1994; LACOMBLEZ et al. 1996); it is the first Food and Drug Administration approved therapy for ALS. Similarly, recent studies with gabapentin also suggest an effect at slowing decline in muscle strength (MILLER et al. 1996). The organotypic culture model has been also used to screen various trophic factors. Selected members of the transforming growth factor $\beta 1$ and glial-derived neurotrophic factor families, as well as the non-immunosuppressive FK506 analog GPI-1046, were found to protect motor neurons from chronic excitotoxicity (HO et al. 1998; STEINER et al. 1998), at least partially by inducing an increase in GLT-1 expression (GANEL et al. 1998).

Studies of glutamate transport began first in the early 1970s. As this review demonstrates, over the last 4 years, there has been dramatic progress in understanding the normal synaptic and molecular biologies of these proteins and their role in pathophysiologic conditions.

References

Aoki M, Lin CLG, Rothstein JD, Geller BA, Hosler BA, Munsat TL, Horvitz HR, Brown RH (1998) Mutations in the glutamate transporter EAAT2 gene do not cause abnormal EAAT2 transcripts in amyotrophic lateral sclerosis. Ann Neurol 43(5):645–653

Aran F (1850) Studies of an as yet undescribed disease of the muscular system. Arch Gen Med (Paris) 24:4–35

Arriza JL, Fairman WA, Wadiche JI, Murdoch GH, Kavanaugh MP, Amara SG (1994) Functional comparisons of three glutamate transporter subtypes cloned from human motor cortex. J Neurosci 14(9):5559–5569

Arriza JL, Eliasof S, Kavanaugh MP, Amara SG (1997) Excitatory amino acid transporter 5, a retinal glutamate transporter coupled to a chloride conductance. Proc Natl Acad Sci USA 94(8):4155–4160

Bar-Peled O, BenHur H, Biegon A, Groner Y, Dewhurst S, Furuta A, Rothstein JD (1997) Distribution of glutamate transporter subtypes during human brain development. J Neurochem 69(6):2571–2580

Bargett ME, Heritage RL, Harding J, Brenner DM (1994) Accumulation of extracellular excitatory amino acids after glutamate uptake inhibition: an in vivo microdialysis study of the ventral striatum. Soc Neurosci Abs 20:926

Bensimon G, Lacomblez L, Meininger V, the ALS/Riluzole Study Group (1994) A controlled trial of riluzole in amyotrophic lateral sclerosis. N Engl J Med 330:585–591

Berger UV, Hediger MA (1998) Comparative analysis of glutamate transporter expression in rat brain using differential double in situ hybridization. Anat Embryol (Berl) 198(1):13–30

Bergles DE, Jahr CE (1998) Glial contribution to glutamate uptake at Schaffer collateral-commissural synapses in the hippocampus, J Neurosci 18(19):7709–7716

Bittigau P, Ikonomidou C (1997) Glutamate in neurologic diseases. J Child Neurol 12(8):471–485

Bouvier M, Szatkowski M, Amato A, Attwell D (1992) The glial cell glutamate uptake carrier countertransports pH-changing anions. Nature 360:471–474

Brakeman PR, Lanahan AA, O'Brien R, Roche K, Barnes CA, Huganir RL, Worley PF (1997) Homer: a protein that selectively binds metabotropic glutamate receptors, Nature 386(6622):284–288

Bridges RJ, Stanley MS, Anderson MW, Cotman CW, Chamberlin AR (1991) Conformationally defined neurotransmitter analogues. Selective inhibition of glutamate uptake by one pyrrolidine-2,4-dicarboxylate diastereomer. J Med Chem 34:717–725

Bristol LA, Rothstein JD (1996) Glutamate transporter gene expression in amyotrophic lateral sclerosis motor cortex, Ann Neurol 39:676–679

Brooks-Kayal AR, Munir M, Jin H, Robinson MB (1998) The glutamate transporter, GLT-1, is expressed in cultured hippocampal neurons. Neurochem Int 33(2):95–100

Bruijn LI, Cleveland DW (1996) Mechanisms of selective motor neuron death in ALS: Insights from transgenic mouse models of motor neuron disease. Neuropathol Appl Neurobiol 22(5):373–387

Bruijn LI, Becher MW, Lee MK, Anderson KL, Jenkins NA, Copeland NG, Sisodia SS, Rothstein JD, Borchelt DR, Price DL, Cleveland DW (1997) ALS-linked SOD1 mutant G85R mediates damage to astrocytes and promotes rapidly progressive disease with SOD1-containing inclusions. Neuron 18(2):327–338

Cagampang FR, Rattray M, Powell JF, Chong NW, Campbell IC, Coen CW (1996) Circadian variation of EAAC1 glutamate transporter messenger RNA in the rat suprachiasmatic nuclei. Brain Res Mol Brain Res 35:190–196

Camu W, Billiard M, Baldy-Moulinier M (1993) Fasting plasma and CSF amino acid levels in amyotrophic lateral sclerosis: a subtype analysis. Acta Neurol Scand 88(1):51–55

Canton T, Pratt J, Stutzmann JM, Imperato A, Boireau A (1998) Glutamate uptake is decreased tardively in the spinal cord of FALS mice, Neuroreport 9(5):775–778

Casado M, Bendahan A, Zafra F, Danbolt NC, Aragon C, Gimenez C, Kanner BI (1993) Phosphorylation and modulation of brain glutamate transporters by protein kinase C. J Biol Chem 268(36):27313–27317

Chaudhry FA, Lehre KP, van Lookeren Campagne M, Ottersen OP, Danbolt NC, Storm-Mathisen J (1995) Glutamate transporters in glial plasma membranes:

highly differentiated localizations revealed by quantitative ultrastructural immunocytochemistry. Neuron 15(3):711–720

Choi DW (1988) Glutamate neurotoxicity and diseases of the nervous system. Neuron 1:623–634

Choi DW, Rothman SM (1990) The role of glutamate neurotoxicity in hypoxic–ischemic neuronal death. Annu Rev Neurosci 13:171–182

Coco S, Verderio C, Trotti D, Rothstein JD, Volterra A, Matteoli M (1997) Non-synaptic localization of the glutamate transporter EAAC1 in cultured hippocampal neurons. Eur J Neurosci 9(9):1902–1910

Cordero ML, Ortiz JG, Santiago G (1994) High-affinity [^{3}H]glutamate-uptake systems in normal and audiogenic seizure-susceptible mice. Brain Res Dev Brain Res 78(1):44–48

Coyle JT, Puttfarcken P (1993) Oxidative stress, glutamate, and neurodegenerative disorders. Science 262:689–695

Crawford TO, Pardo CA (1996) The neurobiology of childhood spinal muscular atrophy, Neurobiol Disease 3(2):97–110

Davis KE, Straff DJ, Weinstein EA, Bannerman PG, Correale DM, Rothstein JD, Robinson MB (1998) Multiple signaling pathways regulate cell surface expression and activity of the excitatory amino acid carrier 1 subtype of Glu transporter in C6 glioma, J Neurosci 18(7):2475–2485

Dehnes Y, Chaudhry FA, Ullensvang K, Lehre KP, Storm-Mathisen J, Danbolt NC (1998) The glutamate transporter EAAT4 in rat cerebellar Purkinje cells: a glutamate-gated chloride channel concentrated near the synapse in parts of the dendritic membrane facing astroglia. J Neurosci 18(10):3606–3619

Dixon JF, Hokin LE (1998) Lithium acutely inhibits and chronically up-regulates and stabilizes glutamate uptake by presynaptic nerve endings in mouse cerebral cortex. Proc Natl Acad Sci USA 95(14):8363–8368

Dong H, O'Brien RJ, Fung ET, Lanahan AA, Worley PF, Huganir RL (1997) GRIP: a synaptic PDZ domain-containing protein that interacts with AMPA receptors. Nature 386(6622):279–284

Fairman WA, Vandenberg RJ, Arriza JL, Kavanaugh MP, Amara SG (1995) An excitatory amino-acid transporter with properties of a ligand-gated chloride channel, Nature 375(6532):599–603

Fine SM, Angel RA, Perry SW, Epstein LG, Rothstein JD, Dewhurst S, Gelbard HA (1996) Tumor necrosis factor α inhibits glutamate uptake by primary human astrocytes. Implications for pathogenesis of HIV-1 dementia, J Biol Chem 271(26): 15303–15306

Fischer U, Dreyfuss G (1997) The SMN-SIP1 complex has an essential role in spliceosomal snRNP biogenesis. Cell 90:1023–1029

Frandsen A, Schousboe A (1990) Development of excitatory amino acid induced cytotoxicity in cultured neurons. Int J Dev Neurosci 8(2):209–216

Furuta A, Rothstein JD, Martin LJ (1997) Glutamate transporter protein subtypes are expressed differentially during rat CNS development. J Neurosci 17(21):8363–8375

Ganel R, Crosson CE (1998) Modulation of human glutamate transporter activity by phorbol ester. J Neurochem 70(3):993–1000

Ganel R, Ho T, Coccia C, Sakal C, Steiner J, Dykes Hoberg M, Robinson MB, Rothstein JD (1998) Excitotoxicity and neurodegeneration – a novel therapeutic approach. Soc Neurosci Abs 24:2069

Gegelashvili G, Schousboe A (1997) High-affinity glutamate transporters: regulation of expression and activity. Mol Pharmacol 52(1):6–15

Gegelashvili G, Civenni G, Racagni G, Danbolt NC, Schousboe I, Schousboe A (1996) Glutamate receptor agonists up-regulate glutamate transporter GLAST in astrocytes. Neuroreport 8(1):261–265

Gegelashvili G, Danbolt NC, Schousboe A (1997) Neuronal soluble factors differentially regulate the expression of the GLT1 and GLAST glutamate transporters in cultured astroglia. J Neurochem 69(6):2612–2615

Gelbard HA, Dzenko KA, DiLoreto D, del Cerro C, del Cerro M, Epstein LG (1993) Neurotoxic effects of tumor necrosis factor α in primary human neuronal cultures are mediated by activation of the glutamate AMPA-receptor subtype: implications for AIDS neuropathogenesis. Dev Neurosci 15:417–422
Gelbard HA, Nottet HS, Swindells S, Jett M, Dzenko KA, Genis P, White R, Wang L, Choi YB, Zhang D, et al (1994) Platelet-activating factor: a candidate human immunodeficiency virus type 1-induced neurotoxin. J Virol 68:4628–4635
Ginsberg SD, Martin LJ, Rothstein JD (1995) Regional de-afferentation down-regulates subtypes of glutamate transporter proteins. J Neurochem 65:2800–2803
Gordon KE, Simpson J, Statman D, Silverstein FS (1991) Effects of perinatal stroke on striatal amino acid efflux in rats studied with in vivo microdialysis. Stroke 22(7):928–932
Grunewald M, Bendahan A, Kanner BI (1998) Biotinylation of single cysteine mutants of the glutamate transporter GLT-1 from rat brain reveals its unusual topology. Neuron 21(3):623–632
Gurney ME, Pu H, Chiu AY, Dal Canto MC, Polchow CY, Alexander DD, Caliendo J, Hentati A, Kwon YW, Deng H-X, Chen W, Zhai P, Sufit RL, Siddique T (1994) Motor neuron degeneration in mice that express a human Cu, Zn superoxide dismutase mutation. Science 264:1772–1775
Hagberg H, Andersson P, Kjellmer I, Thiringer K, Thordstein M (1987) Extracellular overflow of glutamate, aspartate, GABA and taurine in the cortex and basal ganglia of fetal lambs during hypoxia–ischemia. Neurosci Lett 78(3):311–317
Haugeto O, Ullensvang K, Levy LM, Chaudhry FA, Honore T, Nielsen M, Lehre KP, Danbolt NC (1996) Brain glutamate transporter proteins form homomultimers. J Biol Chem 271(44):27715–27722
Heron A, Springhetti V, Seylaz J, Lasbennes F (1995) Effects of a glutamate uptake inhibitor on glutamate release induced by veratridine and ischemia. Neurochem Int 26(6):593–599
Hertz L (1979) Functional interactions between neurons and astrocytes I. Turnover and metabolism of putative amino acid transmitters. Prog Neurobiol 13(3):277–323
Ho T, Li Y, Coccia C, Milbrant H, Johnson E, Phillips J, Griffin J, Rothstein JD (1998) GDNF and neurturin induce motor axon outgrowth: a novel organotypic spinal cord culture method. Soc Neurosci Abs 24:555
Itoh M, Watanabe Y, Watanabe M, Tanaka K, Wada K, Takashima S (1997) Expression of a glutamate transporter subtype, EAAT4, in the developing human cerebellum. Brain Res 767(2):265–271
Iwasaki Y, Ikeda K, Kinoshita M (1992) Plasma amino acid levels in patients with amyotrophic lateral sclerosis. J Neurol Sci 107(2):219–222
Kanai Y, Hediger MA (1992) Primary structure and functional characterization of a high-affinity glutamate transporter. Nature 360:467–471
Kanai Y, Smith CP, Hediger MA (1993) A new family of neurotransmitter transporters: the high-affinity glutamate transporters. FASEB J 7(15):1450–1459
Kanai Y, Nussberger S, Romero MF, Boron WF, Hebert SC, Hediger MA (1995) Electrogenic properties of the epithelial and neuronal high affinity glutamate transporter. J Biol Chem 270(28):16561–16568
Kanai Y, Trotti D, Nussberger S, Hediger MA (1997) The high-affinity glutamate transporter family – structure, function, and physiological relevance. In: Reith MEA (ed) Neurotransmitter transporters. Humana, Totowa, pp 171–213
Kim E, Cho KO, Rothschild A, Sheng M (1996) Heteromultimerization and NMDA receptor-clustering activity of Chapsyn- 110, a member of the PSD-95 family of proteins. Neuron 17(1):103–113
Kimelberg HK, Goderie SK, Higman S, Pang S, Waniewski RA (1990) Swelling-induced release of glutamate, aspartate, and taurine from astrocyte cultures. J Neurosci 10(5):1583–1591
Knecht K, Michalak A, Rose C, Rothstein JD, Butterworth RF (1997) Decreased glutamate transporter (GLT-1) expression in frontal cortex of rats with acute liver failure. Neurosci Lett 229(3):201–203

Kuncl RW, Crawford TO, Rothstein JD, Drachman DB (1992) Motor neuron diseases. In: Asbury AK, McKhann GM, McDonald WI (eds) Diseases of the nervous system, vol II. Saunders, Philadelphia, pp 1179–1208

Lacomblez L, Bensimon G, Leigh PN, Guillet P, Meininger V (1996) Dose-ranging study of riluzole in amyotrophic lateral sclerosis. Lancet 347:1425–1431

Lee MK, Marszalek JR, Cleveland DW (1994) A mutant neurofilament subunit causes massive, selective motor neuron death: implications for the pathogenesis of human motor neuron disease. Neuron 13:975–988

Lee MK, Borchelt DR, Wong PC, Sisodia SS, Price DL (1996) Transgenic models of neurodegenerative diseases. Curr Opin Neurobiol 6:651–660

Lehre KP, Danbolt NC (1998) The number of glutamate transporter subtype molecules at glutamatergic synapses: Chemical and stereological quantification in young adult rat brain. J Neurosci 18(21):8751–8757

Lehre KP, Levy LM, Ottersen OP, Storm-Mathisen J, Danbolt NC (1995) Differential expression of two glial glutamate transporters in the rat brain: quantitative and immunocytochemical observations. J Neurosci 15:1835–1853

Li S, Mallory M, Alford M, Tanaka S, Masliah E (1997) Glutamate transporter alterations in Alzheimer disease are possibly associated with abnormal APP expression. J Neuropathol Exp Neurol 56(8):901–911

Lin CLG, Bristol LA, Jin L, Dykes-Hoberg M, Crawford T, Clawson L, Rothstein JD (1998a) Aberrant RNA processing in a neurodegenerative disease: the cause for absent EAAT2, a glutamate transporter, in amyotrophic lateral sclerosis. Neuron 20(3):589–602

Lin CLG, Jackson M, Jin L, Song W, Orlov I, Dykes-Hoberg M, Rothstein JD (1998b) Identification and characterization of Purkinje cell-specific glutamate transporter EAAT4 associated proteins (GTRAP4). Soc Neurosci Abs 24:2068

Lipton SA, Gendelman HE (1995) Seminars in medicine of the Beth Israel Hospital, Boston: dementia associated with the acquired immunodeficiency syndrome. N Engl J Med 332(14):934–940

Liu G, Dreyfuss G (1996) A novel nuclear structure containing the survival of motor neuron protein. EMBO J 15(14):3555–3565

Lucas DR, Newhouse JP (1957) The toxic effect of sodium l-glutamate on the inner layers of the retina. Arch Ophthalmol 58:193–201

Magnuson DS, Knudsen BE, Geiger JD, Brownstone RM, Nath A (1995) Human immunodeficiency virus type 1 tat activates non-*N*-methyl-D- aspartate excitatory amino acid receptors and causes neurotoxicity. Ann Neurol 37(3):373–380

Maguire G, Simko H, Weinreb RN, Ayoub G (1998) Transport-mediated release of endogenous glutamate in the vertebrate retina. Pflugers Arch 436(3):481–484

Martin LJ, Brambrink AM, Lehmann C, Portera-Cailliau C, Koehler R, Rothstein J, Traystman RJ (1997) Hypoxia-ischemia causes abnormalities in glutamate transporters and death of astroglia and neurons in newborn striatum. Ann Neurol 42(3):335–348

Martin LJ, AlAbdulla NA, Brambrink AM, Kirsch JR, Sieber FE, Portera Cailliau C (1998) Neurodegeneration in excitotoxicity, global cerebral ischemia, and target deprivation: A perspective on the contributions of apoptosis and necrosis. Brain Res Bull 46(4):281–309

Masliah E, Alford M, DeTeresa R, Mallory M, Hansen L (1996) Deficient glutamate transport is associated with neurodegeneration in Alzheimer's disease. Ann Neurol 40(5):759–766

Meldrum B, Garthwaite J (1990) Excitatory amino acid neurotoxicity and neurodegenerative disease. Trends Pharmacol Sci 11(9):379–387

Mennerick S, Dhond RP, Benz A, Xu W, Rothstein JD, Danbolt NC, Isenberg KE, Zorumski CF (1998) Neuronal expression of the glutamate transporter GLT-1 in hippocampal microcultures. J Neurosci 18(12):4490–4499

Michikawa M, Lim KT, McLarnon JG, Kim SU (1994) Oxygen radical-induced neurotoxicity in spinal cord cultures. J Neurosci Res 37:62–70

Miller HP, Levey AI, Rothstein JD, Tzingounis AV, Conn PJ (1997) Alterations in glutamate transporter protein levels in kindling-induced epilepsy. J Neurochem 68(4):1564–1570
Miller RG, Gelinas D, Moore D, Mendoza M, Quien A, Young L, Armon C, Linda L, Barohn R, Bryan W, Bromberg M, Petajan J, Neville H, Ringel S, Parry G, Ravits J, Ross M, Leiderman D (1996) A placebo-controlled trial of gabapentin in amyotrophic lateral sclerosis. Neurology 46:A469
Milton ID, Banner SJ, Ince PG, Piggott NH, Fray AE, Thatcher N, Horne CHW, Shaw PJ (1997) Expression of the glial glutamate transporter EAAT2 in the human CNS: an immunohistochemical study. Brain Res Mol Brain Res 52(1):17–31
Nicholls DG, Attwell D (1990) The release and uptake of excitatory amino acids. Trends Pharmacol Sci 11:462–468
Northington FJ, Traystman RJ, Koehler RC, Rothstein JD, Martin LJ (1998) Regional and cellular expression of glial (GLT1) and neuronal (EAAC1) glutamate transporter proteins in ovine fetal. Neuroscience 85(4):1183–1193
Olney JW (1969) Brain lesions, obesity, and other disturbances in mice treated with monosodium glutamate. Science 164(880):719–721
Olney JW (1971) Glutamate-induced neuronal necrosis in the infant mouse hypothalamus. An electron microscopic study. J Neuropathol Exp Neurol 30(1):75–90
Olney JW (1982) The toxic effects of glutamate and related compounds in the retina and the brain. Retina 2(4):341–359
Olney JW, Ho OL (1970) Brain damage in infant mice following oral intake of glutamate, aspartate or cysteine. Nature 227(258):609–611
Olney JW, Sharpe LG, Feigin RD (1972) Glutamate-induced brain damage in infant primates. J Neuropathol Exp Neurol 31(3):464–688
Orlov I, Lin CLG, Song W, Jin L, Dykes-Hoberg M, Rothstein JD (1998) GTRAP3: Identification and characterization of a neuronal transporter EAAC1-associated protein. Soc Neurosci Abs 24:2068
Ortiz JG, Claudio O, Santiago G, Cordero ML, Nieves J (1996) Possible regulation of high-affinity glutamate uptake in synaptosomes of normal and epileptic mice. Mol Chem Neuropathol 28(1–3):127–133
Peghini P, Janzen J, Stoffel W (1997) Glutamate transporter EAAC-1-deficient mice develop dicarboxylic aminoaciduria and behavioral abnormalities but no neurodegeneration. EMBO J 16(13):3822–3832
Perry TL, Hansen S, Jones K (1987) Brain glutamate deficiency in amyotrophic lateral sclerosis. Neurology 37(12):1845–1848
Pines G, Danbolt NC, Bjoras M, Zhang Y, Bendahan A, Eide L, Koepsell H, Storm-Mathisen J, Seeberg E, Kanner BI (1992) Cloning and expression of a rat brain L-glutamate transporter. Nature 360(6403):464–467
Plaitakis A (1990) Glutamate dysfunction and selective motor neuron degeneration in amyotrophic lateral sclerosis: a hypothesis. Ann Neurol 28:3–8
Plaitakis A, Caroscio JT (1987) Abnormal glutamate metabolism in amyotrophic lateral sclerosis. Ann Neurol 22(5):575–579
Plaitakis A, Constantakakis E, Smith J (1988) The neuroexcitotoxic amino acids glutamate and aspartate are altered in the spinal cord and brain in amyotrophic lateral sclerosis. Ann Neurol 24(3):446–449
Plakidou-Dymock S, McGivan JD (1993) Regulation of the glutamate transporter by amino acid deprivation and associated effects on the level of EAAC1 mRNA in the renal epithelial cell line NBL-I. Biochem J 295:749–755
Rosen DR, Siddique T, Patterson D, Figlewicz DA, Sapp P, Hentati A, Donaldson D, Goto J, O'Regan JP, Deng H-X, Rahmani Z, Krizus A, McKenna-Yasek D, Cayabyab A, Gaston SM, Berger R, Tanzi RE, Halperin JJ, Herzfeldt B, Van den Bergh R, Hung W-Y, Bird T, Deng G, Mulder DW, Smyth C, Laing NG, Soriano E, Pericak-Vance MA, Haines J, Rouleau GA, Gusella JS, Horvitz HR, Brown RH Jr (1993) Mutations in Cu/Zn superoxide dismutase gene are associated with familial amyotrophic lateral sclerosis. Nature 362:59–62

Rosenberg PA, Amin S, Leitner M (1992) Glutamate uptake disguises neurotoxic potency of glutamate agonists in cerebral cortex in dissociated cell culture. J Neurosci 12(1):56–61
Rothman S (1984) Synaptic release of excitatory amino acid neurotransmitter mediates anoxic neuronal death. J Neurosci 4(7):1884–1891
Rothstein JD (1996) Excitotoxicity and neurodegeneration in amyotrophic lateral sclerosis. Clinical Neurosci 3:348–359
Rothstein JD, Kuncl RW (1995) Neuroprotective strategies in a model of chronic glutamate- mediated motor neuron toxicity. J Neurochem 65:643–651
Rothstein JD, Martin LJ, Kuncl RW (1992) Decreased glutamate transport by the brain and spinal cord in amyotrophic lateral sclerosis. N Engl J Med 326(22):1464–1468
Rothstein JD, Jin L, Dykes-Hoberg M, Kuncl RW (1993) Chronic inhibition of glutamate uptake produces a model of slow neurotoxicity. Proc Natl Acad Sci USA 90(14):6591–6595
Rothstein JD, Bristol LA, Hosler B, Brown RH Jr, Kuncl RW (1994a) Chronic inhibition of superoxide dismutase produces apoptotic death of spinal neurons. Proc Natl Acad Sci USA 91(10):4155–4159
Rothstein JD, Martin L, Levey AI, Dykes-Hoberg M, Jin L, Wu D, Nash N, Kuncl RW (1994b) Localization of neuronal and glial glutamate transporters. Neuron 13:713–725
Rothstein JD, Van Kammen M, Levey AI, Martin LJ, Kuncl RW (1995) Selective loss of glial glutamate transporter GLT-1 in amyotrophic lateral sclerosis. Ann Neurol 38:73–84
Rothstein JD, Dykes-Hoberg M, Pardo CA, Bristol LA, Jin L, Kuncl RW, Kanai Y, Hediger MA, Wang Y, Schielke J, Welty DF (1996) Knockout of glutamate transporters reveals a major role for astroglial transport in excitotoxicity and clearance of glutamate. Neuron 16:675–686
Schlag BD, Vondrasek JR, Munir M, Kalandadze A, Zelenaia OA, Rothstein JD, Robinson MB (1998) Regulation of the glial Na^+-dependent glutamate transporters by cyclic AMP analogs and neurons. Mol Pharmacol 53(3):355–369
Schneider JS, Wade T, Lidsky TI (1998) Chronic neuroleptic treatment alters expression of glial glutamate transporter GLT-1 mRNA in the striatum. Neuroreport 9(1):133–136
Sepkuty J, Eccles CU, Lesser RP, Dykes-Hoberg M, Rothstein JD (1997) Molecular Knockdown of neuronal glutamate transporter EAAT3 produces epilepsy and dysfunction of GABA metabolism. Soc Neurosci Abs 23:1484
Shashidharan P, Huntley GW, Murray JM, Buku A, Moran T, Walsh MJ, Morrison JH, Plaitakis A (1997) Immunohistochemical localization of the neuron-specific glutamate transporter EAAC1 (EAAT3) in rat brain and spinal cord revealed by a novel monoclonal antibody. Brain Res 773(1–2):139–148
Shaw PJ, Chinnery RM, Ince PG (1994) [^{3}H]D-aspartate binding sites in the normal human spinal cord and changes in motor neuron disease: a quantitative autoradiographic study. Brain Res 655(1–2):195–201
Shibata T, Watanabe M, Tanaka K, Wada K, Inoue Y (1996) Dynamic changes in expression of glutamate transporter mRNAs in developing brain. Neuroreport 7(3):705–709
Silverstein FS, Buchanan K, Johnston MV (1986) Perinatal hypoxia-ischemia disrupts striatal high-affinity [^{3}H]glutamate uptake into synaptosomes. J Neurochem 47(5):1614–1619
Smith RG, Hamilton S, Hoffman F, Schneider T, Nastainczyk WCS, Birnbaumer L, Stafani E, Appel SH (1992) Serum antibodies to L-type calcium channels in patients with amyotrophic lateral sclerosis. N Engl J Med 327:1721–1728
Smith TL (1997) Regulation of glutamate uptake in astrocytes continuously exposed to ethanol. Life Sci 61(25):2499–2505
Sonders MS, Amara SG (1996) Channels in transporters. Curr Opin Neurobiol 6(3):294–302

Steiner J, Ho T, Lai M, Coccia C, Griffin J, Snyder S, Rothstein JD (1998) The neuroimmunophilin ligand GPI-1046 protects motor neurons from chronic excitotoxicity. Soc Neurosci Abs 24:297

Stock T, Schulte S, Hofmann K, Stoffel W (1992) Structure, expression, and functional analysis of a Na^+-dependent glutamate/aspartate transporter from rat brain. Proc Natl Acad Sci USA 89(22):10955–10959

Sutherland ML, Delaney TA, Noebels JL (1996) Glutamate transporter mRNA expression in proliferative zones of the developing and adult murine CNS. J Neurosci 16(7):2191–2207

Swanson RA, Miller JW, Rothstein JD, Farrell K, Stein BA, Longuemare MC (1997) Neuronal regulation of glutamate transporter subtype expression in astrocytes. J Neurosci 17(3):932–940

Szatkowski M, Attwell D (1994) Triggering and execution of neuronal death in brain ischaemia: two phases of glutamate release by different mechanisms. Trends Neurosci 17(9):359–365

Szatkowski M, Barbour B, Attwell D (1990) Non-vesicular release of glutamate from glial cells by reversed electrogenic glutamate uptake. Nature 348:443–446

Tamminga CA (1998) Schizophrenia and glutamatergic transmission. Crit Rev Neurobiol 12(1–2):21–36

Tan J, Zelenaia O, Rothstein JD, Robinson MB (1999) Expression of the GLT-1 subtype of the Na^+-dependent glutamate transporter: pharmacological characterization and lack of regulation by protein kinase C activation, submitted to J Pharmacol and Experiment Therapeut

Tanaka K, Watase K, Manabe T, Yamada K, Watanabe M, Takahashi K, Iwama H, Nishikawa T, Ichihara N, Hori S, Takimoto M, Wada K (1997) Epilepsy and exacerbation of brain injury in mice lacking the glutamate transporter GLT-1. Science 276(5319):1699–1702

Torp R, Lekieffre D, Levy LM, Haug FM, Danbolt NC, Meldrum BS, Ottersen OP (1995) Reduced postischemic expression of a glial glutamate transporter, GLT1, in the rat hippocampus. Exp Brain Res 103(1):51–58

Tsai GC, Stauch-Slusher B, Sim L, Hedreen JC, Rothstein JD, Kuncl R, Coyle JT (1991) Reductions in acidic amino acids and *N*-acetylaspartylglutamate in amyotrophic lateral sclerosis CNS. Brain Res 556:151–156

Ullensvang K, Lehre KP, Storm-Mathisen J, Danbolt NC (1997) Differential developmental expression of the two rat brain glutamate transporter proteins GLAST and GLT. Eur J Neurosci 9(8):1646–1655

Vandenberg RJ, Arriza JL, Amara SG, Kavanaugh MP (1995) Constitutive ion fluxes and substrate binding domains of human glutamate transporters. J Biol Chem 270(30):17668–17671

Wadiche JI, Kavanaugh MP (1998) Macroscopic and microscopic properties of a cloned glutamate transporter/chloride channel. J Neurosci 18(19):7650–7661

Wadiche JI, Amara SG, Kavanaugh MP (1995) Ion fluxes associated with excitatory amino acid transport. Neuron 15(3):721–728

Wong PC, Pardo CA, Borchelt DR, Lee MK, Copeland NG, Jenkins NA, Sisodia SS, Cleveland DW, Price DL (1995) An adverse property of a familial ALS-linked SOD1 mutation causes motor neuron disease characterized by vacuolar degeneration of mitochondria. Neuron 14:1105–1116

Yamada K, Watanabe M, Shibata T, Nagashima M, Tanaka K, Inoue Y (1998) Glutamate transporter GLT-1 is transiently localized on growing axons of the mouse spinal cord before establishing astrocytic expression. J Neurosci 18(15):5706–5713

Yin KJ, Yan YP, Sun FY (1998) Altered expression of glutamate transporter GLAST mRNA in rat brain after photochemically induced focal ischemia. Anat Rec 251(1):9–14

Yu N, Billaud JN, Phillips TR (1998) Effects of feline immunodeficiency virus on astrocyte glutamate uptake: Implications for lentivirus-induced central nervous system diseases. Proc Natl Acad Sci USA 95(5):2624–2629

Zeman S, Lloyd C, Meldrum B, Leigh PN (1994) Excitatory amino acids, free radicals and the pathogenesis of motor neuron disease. Neuropathol Appl Neurobiol 20(3):219–231

Zerangue N, Kavanaugh MP (1996a) Flux coupling in a neuronal glutamate transporter. Nature 383(6601):634–637

Zerangue N, Kavanaugh MP (1996b) Interaction of L-cysteine with a human excitatory amino acid transporter. J Physiol (Lond) 493(Pt 2):419–423

Zerangue N, Arriza JL, Amara SG, Kavanaugh MP (1995) Differential modulation of human glutamate transporter subtypes by arachidonic acid. J Biol Chem 270(12):6433–6435

CHAPTER 16

NMDA Receptor Antagonists and Their Potential as Neuroprotective Agents

J.A. KEMP, J.N.C. KEW, and R. GILL

A. Introduction

L-Glutamate is the major excitatory transmitter in the mammalian central nervous system (CNS) and it mediates its effects by actions on a variety of ionotropic (ligand-gated cation channels) and metabotropic (G-protein coupled) receptors. *N*-methyl-D-aspartate (NMDA) receptors are a subtype of ionotropic receptors and play key roles not only in several important physiological functions, particularly synaptic plasticity (BLISS and COLLINGRIDGE 1993), but also in neuropathological states, such as epilepsy and acute neurodegeneration (MELDRUM and GARTHWAITE 1990). These latter findings prompted extensive research into the development of selective NMDA-receptor antagonists for the treatment of these indications.

The concept of "excitotoxic" neuronal cell damage emerged following the finding that the neurotoxic potency of excitatory amino acids (EAAs) appeared to parallel their excitatory effects on neurones (OLNEY et al. 1971; OLNEY 1978). The pattern of selective neuronal vulnerability following injections of EAAs into the brain and the "axon-sparing" nature of the damage was reminiscent of ischaemic neuronal injury (JOHANSEN et al. 1984). Furthermore, microdialysis studies showed that there was a massive release of glutamate and aspartate into the extracellular space during and following periods of brain ischaemia (BENVENISTE et al. 1984; HAGBERG et al. 1985). These findings led to the suggestion that the neuronal death caused by periods of cerebral ischaemia was due to the over-activation of specific post-synaptic receptors by the excessive release of the endogenous EAAs L-glutamate and L-aspartate [which is also an NMDA, but not α-amino-3-hydroxy-5-methyl-4-isoxazole propionic acid (AMPA)/kainate, receptor agonist] (CHOI 1988; MELDRUM 1990; DIEMER et al. 1993; Chap. 14). The demonstration that NMDA-receptor antagonists were neuroprotective in a wide variety of neurodegenerative models in a number of different species confirmed the role played by NMDA receptors in mediating acute ischaemic and traumatic brain injury. The marked neurotoxic potential of the NMDA receptor appears to result from: its high permeability to calcium (MACDERMOTT et al. 1986), a known mediator of cell damage (SIESJÖ 1981); its high affinity for glutamate compared to AMPA/kainate receptors; and its relative lack of desensitisation during prolonged activation.

B. Molecular Structure of NMDA Receptors

Molecular cloning studies have identified five NMDA-receptor subunits: NMDAR1 and four related NR2 subunits. NMDAR1 has been cloned from rat (Moriyoshi et al. 1991), mouse (where it was named ζ1; Yamazaki et al. 1992) and human brain (Planells-Cases et al. 1993; Foldes et al. 1993; Karp et al. 1993). Eight functional splice variants and one non-functional truncated splice variant of NMDAR1 have been described (reviewed in McBain and Mayer 1994). The NR2 subunits share 38–53% amino acid sequence identity and exhibit about 20% homology with NMDAR1 (Ikeda et al. 1992; Kutsuwada et al. 1992; Meguro et al. 1992; Monyer et al. 1992; Ishii et al. 1993). The NR2 subunits cloned from rat and mouse are termed NR2A-D and ε1–4, respectively (Ikeda et al. 1992; Kutsuwada et al. 1992; Meguro et al. 1992; Monyer et al. 1992; Ishii et al. 1993), and splice variants have only been reported for NR2D (Ishii et al. 1993). The mature functional NMDAR1 polypeptides have a size similar to that of AMPA and kainate receptor subunits, whilst the NR2 subunits are distinguished by their much larger carboxy termini (reviewed in McBain and Mayer 1994). Both NMDAR1 and NR2 polypeptides possess five common hydrophobic domains; these are thought to represent an amino-terminal signal domain and four membrane-residing domains (reviewed in McBain and Mayer 1994).

The ionotropic glutamate receptor family subunits are distinguished from the subunits of other members of the ligand-gated ion-channel superfamily by virtue of their relatively high molecular mass (97–163 kDa), which is approximately twice as large as that of the subunits of the acetylcholine, γ-aminobutyric acid, glycine and 5-HT3 receptors. The glutamate-receptor family is also distinguished by its transmembrane topology. A series of elegant experiments with AMPA and kainate receptor subunits has demonstrated that the transmembrane topology of the glutamate-receptor family appears to differ from the classic four-transmembrane-domain model of the nicotinic acetylcholine receptor, with an extracellular amino- and carboxy-terminus and a large intracellular loop between transmembrane domains 3 and 4. The current model for the glutamate receptor consists of an extracellular amino terminus followed by a transmembrane domain and then a second membrane-residing domain that does not cross the membrane but forms a re-entrant loop entering from and exiting to the cytoplasm. Thus, the fourth membrane domain is followed by an intracellular carboxy terminus (reviewed in Kemp and Kew 1998). The recently described crystal structure of the GluR2 ligand-binding domain, which comprises polypeptides in both the amino terminus and the loop between transmembrane domains 3 and 4, lends further support to this topological model for the glutamate-receptor family (Armstrong et al. 1998). Notably, these topology studies have not been extended to all members of the ionotropic glutamate-receptor family, and particularly not to the NMDA-receptor subunits. However, point mutations of amino acids within the putative extracellular loop between transmembrane domains 3 and 4 suggest that these regions form part of the ligand-binding pocket in both NMDAR1 (Hirai et al. 1996) and NR2 (Laube et al. 1997; Anson et al. 1998), in agreement with the recent model.

The identification of membrane-residing domain 2 as a non-transmembrane domain is of particular interest, as this domain contains the asparagine residue in NMDA receptors and the corresponding arginine/glutamine-editing site in AMPA and kainate receptors, which controls the ion selectivity and rectification properties of the receptors. These residues are thought to be located within the channel pore. Thus, the second membrane-residing domain is likely to form at least a portion of the channel pore.

The glutamate-receptor family appears to differ further from the other members of the ligand-gated ion-channel superfamily in their receptor-subunit stoichiometry. Recent evidence suggests that both AMPA (Rosenmund et al. 1998; Mano and Teichberg 1998) and NMDA (Laube et al. 1998) receptors are tetramers. In the case of the NMDA receptor, these tetramers are believed to be composed of two NMDAR1 and two NR2 subunits (Behe et al. 1995; Laube et al. 1998), which is compatible with earlier electrophysiological evidence demonstrating that NMDA-receptor activation requires occupation of two independent glycine-binding sites and two independent glutamate sites (Benveniste and Mayer 1991; Clements and Westbrook 1991). These are believed to be located on the NMDAR1 (Kuryatov et al. 1994; Wafford et al. 1995; Hirai et al. 1996) and NR2 (Laube et al. 1997; Anson et al. 1998) subunits, respectively. However, controversy remains, as evidence still exists for a pentameric structure (Ferrier-Montiel and Montal 1996; Premkumar and Auerbach 1997). If the tetrameric structure is upheld, then the glutamate-receptor stoichiometry would be distinct from the pentameric structure of other members of the ligand-gated ion-channel superfamily, such as the well-studied nicotinic receptor (Changeux et al. 1992). In fact, they would better resemble the tetrameric voltage-gated potassium channels (Mackinnon 1991; Liman et al. 1992) and cyclic nucleotide-gated channels (Root and Mackinnon 1993), which also possess re-entrant membrane-residing domains that form the channel pore (Mackinnon 1995).

NMDAR1 is believed to be an obligate component of functional receptors, as NR2 subunits are unable to form functional NMDA receptors when expressed alone. Expression of NMDAR1 alone in *Xenopus* oocytes yields functional receptors (albeit at low levels) that have electrophysiological and pharmacological properties characteristic of the native NMDA receptor. However, *Xenopus* oocytes may express an endogenous glutamate-receptor subunit that can combine with NR1 to form functional receptors (Soloviev and Barnard 1997). Furthermore, a novel endogenous NR2-like subunit has recently been identified in *Xenopus* oocytes (Teichert et al., personal communication); this subunit may also assemble with NMDAR1 to form functional receptors. Interestingly, the apparent homomeric NR1-subunit-containing receptors expressed in *Xenopus* oocytes have properties like NR2-subunit-containing receptors, in that they are potently inhibited by ifenprodil

(Williams et al. 1993) and potentiated by polyamines (Durand et al. 1992). Expression of NMDAR1 alone in mammalian cell lines fails to generate functional channels, although specific binding of both the channel blocker MK-801 and glycine-site antagonists has been reported (Chazot et al. 1992; Grimwood et al. 1995). Interestingly, however, cell-surface expression of human NMDAR1a in a mouse cell line appears to require the co-expression of NR2A (McIlhinney et al. 1996).

A number of other proteins have been proposed as prospective NMDA receptor subunits, including the 71-kDa glutamate-binding protein (Kumar et al. 1991), GR33 (Smirnova et al. 1993) and NR3A/NMDAR-L/χ-1 (Ciabarra et al. 1995; Sucher et al. 1995; Das et al. 1998). Whilst the physiological roles of the 71-kDa glutamate-binding protein and GR33 and their involvement, if any, in NMDA-receptor structure and function remain unclear, NR3A has recently been shown to co-immunoprecipitate with NMDAR1 and NR2B, and expression of NR3A together with NMDAR1 and NR2A results in the appearance of smaller unitary conductances than observed with NMDAR1 and NR2A alone (Das et al. 1998). Furthermore, NR3A knockout mice exhibit enhanced NMDA responses and increased numbers of dendritic spines in early postnatal cerebrocortical neurons and, accordingly, NR3A has been suggested to regulate synaptic development by modulating NMDA-receptor activity (Das et al. 1998).

Co-expression of NMDAR1 with one or more of the NR2 subunits yields receptors with distinct functional and pharmacological properties that appear to best resemble native receptors (Kutsuwada et al. 1992; Monyer et al.1992). The NR2 subunits are expressed in mammalian brain in a distinct spatiotemporal manner (Watanabe et al. 1992; 1993; Monyer et al. 1994; Sheng et al. 1994; Zhong et al. 1995; Wenzel et al. 1997; Kew et al. 1998a). NMDAR1 exhibits a widespread distribution, with the splice variants exhibiting both distinct expression patterns and functional properties (Laurie and Seeburg 1994; Laurie et al. 1995; Zukin and Bennett 1995). Thus, native NMDA receptors are likely to be composed of a variety of different subunit combinations, and electrophysiological studies have demonstrated the heterogeneity of native NMDA receptors from different brain regions (Priestley and Kemp 1993). In the adult rodent and human, the predominant NR2 subunits in the forebrain are NR2A and NR2B, with NR2C expressed largely in the cerebellum and various select nuclei, and NR2D expression confined to the diencephalon and midbrain (Kutsuwada et al. 1992; Monyer et al. 1992, 1994, Ishii et al. 1993; Rigby et al. 1996). Immunoprecipitation and electrophysiological studies have suggested the existence of three-subunit-containing NMDAR1/NR2A/NR2B receptors (Sheng et al. 1994; Brimecombe et al. 1997; Chazot and Stephenson 1997; Luo et al. 1997; Kew et al. 1998a) and NMDAR1/NR2A/NR2C receptors (Wafford et al. 1993; Chazot et al. 1994). Thus, in the adult forebrain, the most abundant heteromeric combinations are likely to be NMDAR1/NR2A, NMDAR1/NR2B and the three-subunit-containing NMDAR1/NR2A/NR2B.

C. Types of NMDA Receptor Antagonists

The NMDA receptor is unique amongst ligand-gated ion channels in its requirement of two co-agonists (glutamate and glycine) for channel activation (JOHNSON and ASCHER 1987; KLECKNER and DINGLEDINE 1988). Interestingly, site-directed mutagenesis showed that amino acids in the amino-terminal and L3 regions of NMDAR1, homologous to the bacterial amino-acid-binding proteins, appear to be critical for binding of the co-agonist glycine rather than glutamate, as in the AMPA and kainate receptors (KURYATOV et al. 1994; STERNBACH et al. 1994; WAFFORD et al. 1995; HIRAI et al. 1996). It was subsequently shown that, on the NR2 subunits, these regions have remained the binding site for glutamate (LAUBE et al. 1997; ANSON et al. 1998). An allosteric interaction between the glutamate- and glycine-binding sites of the NMDA receptor has been demonstrated, indicating that ligand binding at either site can affect the affinity of the other site for its agonist (BENVENISTE et al. 1990; KEMP and PRIESTLEY 1991; LESTER et al. 1993; PRIESTLEY and KEMP 1994). The NMDA receptor has a number of regulatory sites subject to modulation by both endogenous and exogenous compounds. These include a site within the channel pore, where Mg^{2+} binds to confer the well-described voltage-dependence of receptor activation (MAYER et al. 1984; NOWAK et al. 1984). NMDA receptors are also subject to modulation by Zn^{2+} (PETERS et al. 1987; WESTBROOK and MAYER 1987; CHRISTINE and CHOI 1990), redox state (GOZLAN and BEN-ARI 1995), protons (TANG et al. 1990; TRAYNELIS and CULL-CANDY 1990; TRAYNELIS et al. 1995) and polyamines (MCGURK et al. 1990; LERMA 1992; BENVENISTE and MAYER 1993; TRAYNELIS et al. 1995). Mg^{2+} can also enhance NMDA-receptor activation by binding at sites distinct from that within the channel pore (PAOLETTI et al. 1995; WANG and MACDONALD 1995) and which appear to be the same as the polyamine-binding sites (PAOLETTI et al. 1995; KEW and KEMP 1998). Thus, it seems that Mg^{2+} may be the physiological ligand for the polyamine site. Many of these regulatory sites have attracted attention as possible pharmacological targets to prevent the over-activation of NMDA receptors and associated neurotoxicity.

I. Ion-Channel Blockers

The ion-channel blockers, which include such compounds as dizocilpine (MK-801), aptiganel (cerestat, CNS 1102), phencyclidine and ketamine (IVERSEN and KEMP 1994), are activity-dependent, i.e. they require channel opening in order to bind to and block the receptor (WONG et al. 1986; HUETTNER and BEAN 1988; KEMP et al. 1991) and are, therefore, uncompetitive antagonists. Such activity-dependency can be seen as a potential desirable feature of a therapeutic receptor blocker. One attractive strategy for neuroprotection with minimal side effects would be to block channels reversibly during periods of particularly high activity whilst leaving resting channels relatively unaffected. In practice, however, this does not seem to occur, and such compounds also inhibit physiologically activated NMDA receptors, possibly through a build-up of block due to their slow dissociation rates and ability to become trapped in the closed state of the channel. Thus, whilst the neuroprotective abilities of high-affinity compounds in this class are well established (Sect. D.II), a number of associated side effects have become evident, including behavioural (TRICKLEBANK et al. 1989), cardiovascular (LEWIS et al. 1989) and potentially cytotoxic (OLNEY et al. 1989; ALLEN and IVERSEN 1990) effects. It has been suggested that lower-affinity channel blockers with faster blocking and unblocking kinetics have better therapeutic profiles than the high-affinity compounds (PALMER et al. 1995; SUBRAMANIAM et al. 1996; DANYSZ et al. 1997; DAVIES 1997). However, in our experience, low-affinity compounds were not effective neuroprotective agents in a rat middle cerebral artery (MCA) occlusion model, even at doses that produced marked behavioural side effects (FISCHER et al. 1995).

II. Glutamate Recognition-Site Antagonists

A large number of competitive antagonists of the glutamate-recognition site of the NMDA receptor have been synthesised, and the vast majority are conformationally constrained α-amino carboxylic acids with an ω-phosphonic acid group an appropriate distance away (JANE et al. 1994). Pharmacophores for both agonists and antagonists have been proposed based upon the large number of structure–activity studies that have been performed (JANE et al. 1994). Glutamate-recognition-site antagonists tend to penetrate the blood–brain barrier poorly because of their highly polar nature. Attempts to overcome this by using less polar substituents met with limited success, as high polarity appears to be a prerequisite for high-affinity binding to the receptor (JANE et al. 1994). Thus, whilst neuroprotective in vivo, they are most effective when administered prior to the ischaemic insult and have a limited therapeutic time window.

Competitive compounds bear an additional potential disadvantage as therapeutic compounds in that, theoretically, they will inhibit NMDA receptors subjected to weaker levels of agonist stimulation more effectively than those subjected to excessive, potentially neurotoxic levels of glutamate. Thus, such compounds are more likely to target the normal glutamatergic function of the brain than regions of non-physiological receptor overactivation resulting, for example, from the high levels of extracellular glutamate generated following ischaemia. The competitive antagonists produced dose-limiting side effects in man in clinical trials for stroke and epilepsy (Sect. E).

III. Glycine Recognition-Site Antagonists

Compounds acting at the glycine-recognition site of the NMDA receptor have attracted considerable attention as potentially effective therapeutic agents. The reason for this is that, whilst both glutamate and glycine are co-agonists at the NMDA receptor, it is glutamate which plays the neurotransmitter role, being released from presynaptic terminals in an activity-dependent manner. Glycine, however, plays a more modulatory role, apparently continuously present in the extracellular fluid at more constant levels. Thus, a partial blockade of the glycine site would limit the level of NMDA-receptor activation, thus providing some neuroprotection whilst still permitting a certain level of physiological activation to take place. An attractive way of achieving this would be with partial agonists, which cannot produce a complete block even when dosed at high levels.

HA-966, one of the first compounds identified to be an NMDA antagonist, was shown to act at the glycine site shortly following its discovery (FLETCHER and LODGE 1988; FOSTER and KEMP 1989) and was characterised as a partial agonist (FOSTER and KEMP 1989; KEMP and PRIESTLEY 1991). Chemical modification of HA-966 led to higher-affinity compounds (KEMP and LEESON 1993; LEESON and IVERSEN 1994), with the best being L-687,414, the R-(+)-*cis*-4-methyl analogue. The advantage of these compounds is that not only are they partial agonists, they also have relatively good CNS bioavailability and are active in vivo following systemic administration. Thus, L-687,414 was demonstrated to be neuroprotective without adverse cardiovascular or CNS side effects in a rat model of stroke (Sect. D.II).

Another compound shown to be active at the glycine site was the broad-spectrum EAA-receptor antagonist, kynurenic acid, which has no partial agonist activity. Substitutions at the 5 and 7 positions improved both affinity and selectivity for the glycine site (KEMP et al. 1988, BARON et al. 1990; FOSTER et al. 1992). A subsequent medicinal chemistry programme led to the development of low-nanomolar-affinity compounds with in vivo activities of below 1 mg/kg, exemplified by L-701,324 (PRIESTLEY et al. 1996) with an increase in affinity of more than 10,000-fold over kynurenic acid (KEMP and LEESON 1993; LEESON and IVERSEN 1994). Another glycine-site antagonist in this chemical class, gavestinel, or GV 150526A (BORDI et al. 1997), is currently in phase-III clinical trials in stroke (Sect. E).

IV. NR2B Subunit-Selective NMDA Antagonists

One approach to the problem of preventing NMDA-receptor overactivation while permitting sufficient normal glutamatergic function to avoid unacceptable side effects has been the development of NMDA-receptor subunit-selective compounds. As already outlined, in the adult mammalian forebrain, the predominant NR2 subunits are NR2A and NR2B. Thus, an attractive strategy for neuroprotection with reduced side effects might be to selectively block NMDA receptors containing NR2B whilst leaving those containing NR2A relatively unaffected. Several NR2B-subunit-selective antagonists have now been identified, including ifenprodil, Ro 25-6981 (FISCHER et al. 1997), CP 101-606 (CHENARD et al. 1995), Ro 8-4304 (KEW et al. 1998b) and Ro 63-1908 (GILL et al. 1998). Ifenprodil is the prototypic NR2B-subunit-selective NMDA-receptor antagonist; it exhibits a markedly higher affinity for both recombinant receptors containing NR2B compared with those containing NR2A (WILLIAMS 1993), NR2C or NR2D (WILLIAMS 1995) and a subset of presumptive NR2B-containing native NMDA receptors (WILLIAMS 1993; PRIESTLEY et al. 1994). Interestingly, recent evidence suggests that ifenprodil binds with high affinity to and inhibits NMDA receptors containing three different subunits, i.e. NMDAR1/NR2A/NR2B, as well as those composed of just NMDAR1/NR2B (KEW et al. 1998a). However, the proportion of such three-subunit-containing receptors in the mammalian brain remains controversial (LUO et al. 1997; CHAZOT and STEPHENSON 1997). Ifenprodil was originally believed to act as a competitive antagonist at a polyamine-binding site of the NMDA receptor (CARTER et al. 1990), although various other studies have suggested that ifenprodil binds to a distinct site (REYNOLDS and MILLER 1989; GALLAGHER et al. 1996). Recently, we demonstrated a non-competitive allosteric interaction between spermine and both ifenprodil and the related NR2B-selective antagonist Ro 8-4304 in cultured rat cortical neurones (KEW and KEMP 1998).

Ifenprodil, Ro 25-6981, Ro 8-4304, Ro 63-1908 and probably also CP 101-606 act via a novel state-dependent mechanism of action (KEW et al. 1996, 1998b; FISCHER et al. 1997; GILL et al. 1998) such that these antagonists bind with a higher affinity to activated and desensitised states of the receptor than to the unliganded resting state and, thus, display an activity-dependent mode of action. It is predicted that such compounds will preferentially block NMDA receptors, which are continuously activated by sustained high glutamate levels in ischaemic brain areas whilst leaving those that are physiologically activated in normal brain areas by fast synaptic transmission relatively unaffected. Together with their subunit selectivity, this state-dependent mechanism of action is likely to underlie the desirable neuropharmacological profile of this class of compounds (Sect. F).

Recently, it has been suggested that ifenprodil and similar phenylethanolamines inhibit the NMDA receptor by increasing its sensitivity to proton inhibition (MOTT et al. 1998) and, similarly, the potency of ifenprodil and (to a lesser extent) CP101-606 was increased at lower pH. Since brain pH levels fall during ischaemia (SILVER and ERECINSKA 1992), this observation indicates that it may be possible to identify similar compounds that might be relatively inactive at physiological pH and exhibit a significant increase in potency at lower pHs (as encountered during ischaemia), perhaps further improving tolerability.

D. Neuroprotective Effects of NMDA Antagonists in Cerebral Ischaemia

The first studies demonstrating an excitatory and neurotoxic role for glutamate (HAYASHI 1952; ROTHMAN and OLNEY 1987; CHOI 1988) led to the realisation that this could have important therapeutic implications in epilepsy and in acute and chronic neurodegeneration. Glutamate receptors have been implicated in many neurodegenerative diseases, including stroke, Alzheimer's and Huntington's disease and amyotropic lateral sclerosis. However, the most compelling evidence for the involvement of glutamate-mediated excitotoxicity comes from cerebral ischaemia.

The early in vivo microdialysis studies demonstrated that there was an increase in the levels of glutamate after a period of transient forebrain ischaemia (BENVENISTE et al. 1984; HAGBERG et al. 1985). This was subsequently also shown to occur in a model of permanent MCA occlusion, which models the pathophysiological changes seen after a stroke (BUTCHER et al. 1990). Increases in glutamate have also been reported after stroke in humans and appear to be related to the severity of the stroke (BULLOCK et al. 1995; CASTILLOE et al. 1997; DAVALOS et al. 1997). Glutamate can act on NMDA, AMPA/kainate and metabotropic receptors to produce an increase in cytosolic free Ca^{2+}. High levels of cytosolic Ca^{2+} can activate enzymes such as protein kinase C, calpains, endonucleases and phospholipases, which can directly or indirectly (through free-radical formation) destroy cellular structure. Nitric oxide synthase is also activated via high levels of intracellular Ca^{2+}, resulting in the formation of free radicals, which may contribute to the consequent cell damage. The cytosolic Ca^{2+} levels are normally maintained via buffering by mitochondria and the Na^{+}–Ca^{2+} exchanger. Therefore, the high levels of glutamate in the extracellular space trigger a cascade of excitotoxic events, resulting in an uncontrolled influx of Ca^{2+} into the post-synaptic neurone and, through various mechanisms, this ultimately results in cell death (SIESJÖ 1981; CHOI 1988; SIESJÖ and BENGTSSON 1989). In the clinical setting, cerebral ischaemia can result from a stroke, cardiac arrest or head trauma or can follow cardiac bypass surgery.

Animal models of cerebral ischaemia have been developed to try and understand the pathological and neurochemical changes and to evaluate the effects of pharmacological intervention. Animal models of cerebral ischaemia fall into two main categories: transient forebrain ischaemia, which tries to mimic pathological changes seen following cardiac arrest, and focal cerebral ischaemia, which mimics the pathological changes seen following stroke. The first models developed were models of transient forebrain ischaemia in the rat, one of which involved occlusion of both carotid arteries and induced hypotension for 10 min (2-VO; SMITH et al. 1984). An alternative model involved occlusion of both vertebral arteries and, 24 h later, occlusion of both carotid arteries for 10–20 min (4-VO; PULSINELLI et al. 1982). Both of these models result in quite severe forebrain ischaemia and delayed neuronal degeneration of hippocampal CA1 neurones and some cortical and striatal damage. There is also a gerbil model of transient forebrain ischaemia which involves bilateral carotid occlusion for 5 min and results in delayed neuronal degeneration of hippocampal CA1 neurones (KIRINO 1982). The competitive NMDA antagonist D-2-amino-5-phosphonoheptanoate (D-AP7) was the first compound used to demonstrate neuroprotective effects of NMDA antagonists in vivo but, due to its physiochemical characteristics, it had to be administered intracerebro-ventricularly (Simon et al. 1984; Wieloch 1985). These studies reported that AP7 was able to attenuate neuronal degeneration produced by either transient forebrain ischaemia or hypoglycaemia.

Following these early studies and the evolving evidence demonstrating that NMDA receptors are highly permeable to calcium (MacDermott et al. 1986), which is a key component of ischaemia-induced neuronal degeneration, a search began for potent NMDA antagonists that could be administered systemically. MK-801 was discovered in 1986 (Wong et al. 1986); it was the most potent NMDA antagonist known and had a non-competitive mechanism of action (Kemp et al. 1987). In initial studies, it was demonstrated to be neuroprotective against NMDA-induced lesions in the hippocampus (Foster et al. 1987). It was also shown to be neuroprotective in the gerbil model of bilateral carotid occlusion for 5 min, which results in selective CA1 damage; in this model, MK-801 was also neuroprotective when administered post-ischaemia (Gill et al. 1987a, 1988). There was quite some controversy with regard to the neuroprotective effect of MK-801 in the gerbil model, as some authors reported that the protection was due to MK-801-induced hypothermia, rather than blockade of NMDA receptors (Buchan and Pulsinelli 1990; Colbourne et al. 1997). However, other evidence demonstrated that MK-801 was still neuroprotective in the gerbil model when brain and core temperature were maintained at 37°C (Gill and Woodruff; 1990; Warner et al. 1991; Hoffmann and Boast 1995). Parallel studies with the competitive NMDA antagonist CGS 19755 (or selfotel) demonstrated that this compound was also neuroprotective in gerbils (Boast et al. 1988), as was the non-competitive NMDA antagonist ketamine (Marcoux et al. 1988). Selfotel was the first of a series of brain-permeable competitive NMDA antagonists (Bennett et al. 1990).

I. NMDA Antagonists in Models of Transient Forebrain Ischaemia in Rats

In a rat 2-VO model of transient forebrain ischaemia, pretreatment with MK-801 significantly reduced the early ischaemic cell change seen in hippocampal CA1 neurones following 2-VO for 30 min and 2 h of reperfusion (Gill et al. 1987b). However, in this study, the animals were not allowed to survive for 7 days, as would be crucial in order to examine delayed neuronal degeneration. In further studies, MK-801 administered 1 h prior to 10 min of ischaemia in the 2-VO model with a 7 day recovery period resulted in partial protection of hippocampal CA1 neurones (Gill et al. 1989). This study also failed to show any neuroprotection when MK-801 was administered post-ischaemia. Others also reported a neuroprotective effect of MK-801 in the 2-VO model (Church et al. 1988; Rod and Auer 1989). However, studies by Wieloch's group (Nellgard et al. 1991) using the 2-VO model failed to show any protection using various treatment regimens starting prior to or after ischaemia (BUCHAN 1990). The 2-VO model has also been used to demonstrate the neuroprotective actions of other ion-channel blockers, such as ketamine, phencyclidine (SAUER et al. 1988) and dextrorphan (SWAN and MELDRUM 1990).

The 4-VO model of transient forebrain ischaemia has also produced negative results with ion-channel blockers. The early studies, however, used a 30 min period of ischaemia and 7-day recovery, which produces very dense ischaemic damage (BLOCK and PULSINELLI 1987), and it is likely that this will be difficult to reverse with any therapeutic intervention. The same group have since used shorter ischaemic periods of 5 and 15 min but were still unable to detect any protection with doses of MK-801 of 1–5 mg/kg (BUCHAN et al. 1991). However, TORTELLA et al. (1989) reported that dextromethorphan improved post-ischaemic hypoperfusion and electroencephalogram in the 4-VO model when administered prior to or after ischaemia, although they did not report any histological data in this study.

Ion-channel blockers and competitive NMDA antagonists have provided equivocal results in models of transient and complete forebrain ischaemia (BUCHAN 1990; SMALL and BUCHAN 1997). The emerging picture seems to be that these sorts of models produce a dense ischaemia that is not very amenable to therapeutic amelioration (SIESJÖ and BENGTSSON 1989). MK-801 did appear to produce some protection when given prior to or shortly after the ischaemic insult, but it was not as impressive as the results seen in the gerbil model and was variable among groups (GILL et al. 1987b; 1989; CHURCH et al. 1988; ROD and AUER 1989; SWAN and MELDRUM 1990). A major contributory factor of the controversial effects of MK-801 and other NMDA antagonists in these models is the density of ischaemia. These models produce variable outcome, which is dependent on many factors, such as species of rat, diet, physiological variables, anaesthesia and body temperature; thus, enormous variability in the density of ischaemia produced exists among different researchers (GINSBERG and BUSTO 1989; MELDRUM 1990). It seems that pretreatment with NMDA antagonists in the rat forebrain ischaemia models gives some partial protection of hippocampal CA1 neurones, but it is quite clear there is no post-ischaemic efficacy, and they do not completely block the delayed neuronal degeneration (GROTTA et al. 1990; NELLGARD et al. 1991). It was speculated that other mechanisms or receptor systems may be involved, but the major remaining argument in favour of EAA involvement in delayed neuronal degeneration came from lesion studies.

These lesion studies concentrated on the hippocampus and delayed neuronal degeneration, and a wide variety of techniques and means of removing the glutamatergic inputs to the hippocampus were utilised (JOHANSEN et al. 1986; JØRGENSEN et al. 1987; DIEMER et al. 1993). These various methods were demonstrated to ameliorate ischaemia-induced neuronal degeneration of hippocamal CA1 cells. BENVENISTE et al. (1989), used the 4-VO for 10 min plus hypotension (50–60 mmHg) model combined with CA3 lesions and reported that this resulted in preservation of CA1 neurones and an attenuation in the post-ischaemic increase in extracellular levels of glutamate and aspartate normally seen in this model.

Thus, it was demonstrated that post-ischaemic release of glutamate is at least partially dependent on intact glutamatergic innervation of CA1 hippocampal neurones and that glutamatergic pathways are important in delayed neuronal degeneration following forebrain ischaemia. Following the disappointing results with NMDA antagonists, attention was focused on AMPA antagonists, such as 2,3-dihydroxy-6-nitro-7-sulfamoyl-benzo(F) quinoxaline, which appeared to give more robust neuroprotective effects in these models of transient forebrain ischaemia (GILL 1994; GILL and LODGE 1997).

II. NMDA Antagonists in Models of Focal Ischaemia

In total contrast to the controversial results of NMDA antagonists in global ischaemia models, animal models of focal ischaemia have proved to be more reproducibly amenable to therapeutic intervention with NMDA antagonists. The majority of these models employ permanent or temporary occlusion of a major cerebral artery, usually the MCA, as this is widely regarded as being the most relevant to stroke in humans (GRAHAM 1988; MCAULEY 1995). Permanent MCA occlusion results in a densely ischaemic core of infarcted tissue that has no or very little blood flow, this tissue is irreversibly damaged and is not amenable to therapeutic intervention. Surrounding this core is an area of low flow, where there is some collateral circulation; this area is described as the "penumbra" (ASTRUP et al. 1981). It represents an area in which the cells are energetically compromised and are at risk of becoming incorporated into the infarct. The penumbral area has a compromised blood flow, which is 20% of the normal level. This is the threshold level at which synaptic transmission is abolished, but normal ionic homeostasis can be maintained. However, as energy demand is increased, e.g. during cortical spreading depression (CSD), or energy levels drop further, this area then becomes susceptible to infarction (ASTRUP et al. 1981; MCCULLOCH et al. 1991). The evolving lesion after permanent MCA occlusion in animals can be monitored using diffusion-weighted magnetic resonance imaging (MRI); after 4h, 80% of the ischaemic damage has occurred, and the volume of damage is at its maximum by 24h (GILL et al. 1995a). It has been demonstrated that CSD appears to contribute to evolution of the ischaemic damage over this 24-h period (NEDERGAARD and ASTRUP; 1986). MK-801 and other EAA antagonists can reduce the number and intensity of these CSDs in the penumbral area and, thus, prevent the spread of the lesion (GILL et al. 1992); such phenomenon have also been measured using MRI (GYNGELL et al. 1994; LATOUR et al. 1994; TAKANO et al. 1996). Studies using human stroke patients and various MRI techniques have demonstrated that the lesion develops heterogeneously in man over a period of 24–53 h (BAIRD and WARACH 1998); thus, a longer therapeutic window exists in man. These MRI studies have extended the initial therapeutic window of 6–8h identified using positron emission tomography studies (Baron et al. 1995; Furlan et al. 1996).

MK-801 was demonstrated to be neuroprotective in a cat model of focal ischaemia when administered prior to or 2h after MCA occlusion (Ozyurt et al. 1988; Park et al. 1988a). These were the first studies demonstrating a 50% reduction in the volume of cortical infarction with an NMDA antagonist and demonstrating post-ischaemic efficacy. These were acute studies; therefore, all physiological parameters, such as blood pressure, body temperature, blood gases, glucose and pH, were monitored throughout the experimental period, negating the concerns seen with the global ischaemia models. These were very important findings showing that therapy could actually be delayed following a stroke, thus making it possible for such compounds to have clinical application. In the rat, MK-801 [0.5mg/kg, intravenously (i.v.)] given 30min prior to or after MCA occlusion gave 50% reduction in the volume of cortical damage (Park et al. 1988b; Gill et al. 1991). The therapeutic window for treatment with MK-801 in the rat permanent focal ischaemia model was reported to be up to 2h post-ischaemia (Hatfield et al. 1992; Fig. 1), and 3h post-ischaemia was reported for treatment with phencyclidine (Bielenberg and Beck 1991).

The consensus of opinion on the neuroprotective effect of NMDA antagonists in these focal ischaemia models is that they protect against cortical ischaemic damage in the penumbra (Gill et al. 1991, 1992; Mcculloch et al. 1991). However, no protection is seen against caudate damage, because this represents the core of the lesion, where blood flow remains below threshold levels, resulting in irreversible damage.

MK-801 was also tested in a model of multiple focal infarcts induced by injection of microspheres into the internal carotid artery in rabbits. Using this technique, MK-801 improved function and survival in the animals when administered during or shortly after microsphere injection (Kochhar et al. 1988). This model of embolic stroke has important applications, because it can produce multiple small infarcts, which have been reported to occur following cardiac by-pass surgery; therefore, these compounds may have therapeutic use as a premedication to this type of surgery (Smith et al. 1986; Pugsly et al. 1994).

The main NMDA antagonists to emerge from these studies were MK-801, selfotel and aptiganel. These compounds, whilst demonstrating neuroprotective effects in animal models of focal ischaemia, were confounded by cardiovascular side effects (Lewis et al. 1989; Hargreaves et al. 1993a, 1994) and psychotomimetic effects and produced vacuolisation in neurones of the cingulate cortex (Olney et al. 1989). This side-effect profile is more commonly associated with high-affinity NMDA antagonists, whereas the low-affinity antagonists, such as memantine and remacimide, do not have this behavioural profile (Chen et al. 1992; 1998; Grant et al. 1996; Subramaniam et al. 1996; Danysz et al. 1997).

The glycine-site partial agonist L-687,414 was demonstrated to be neuroprotective in a rat model of stroke (Gill et al. 1995b) at doses that caused no

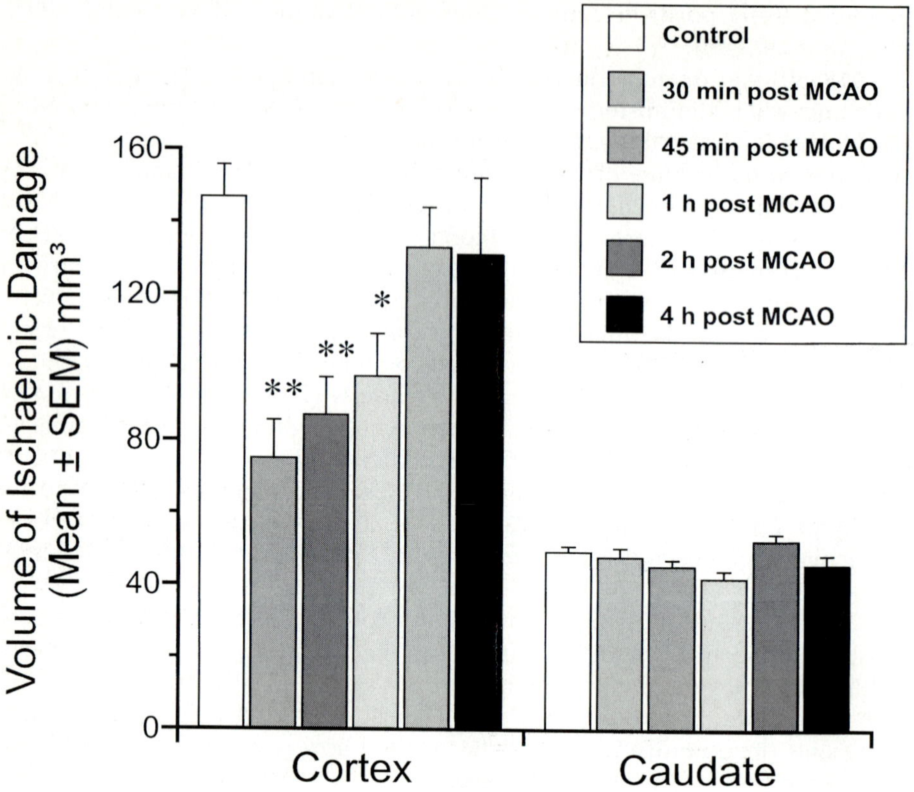

Fig. 1. The therapeutic window for neuroprotection with MK-801 in a model of focal ischaemia produced by permanent occlusion of the left middle cerebral artery (*MCA*) in rats. MK-801, at a dose of 3 mg/kg intraperitoneally (i.p.), was administered at various times up to 4 h post MCA occlusion. The control group of animals (*open bars*) was administered saline (1 ml/kg, i.p.). Significant protection against cortical infarction was seen when MK-801 was administered at 30, 45 and 60 min post MCA occlusion (n = 10–23). The data are expressed as means ± SEM for n animals (HATFIELD et al. 1992). Reproduced with kind permission of Elsevier Science Ltd., The Boulevard, Langford Lane, Kidlington, OXS IGB, UK.

cardiovascular side effects and did not cause vacuolisation in neurones of the cingulate and retrosplenial cortex (HARGREAVES et al. 1993b). Furthermore, L-687,414 produces fewer behavioural effects than the open-channel blockers and does not block long-term potentiation at neuroprotective doses (TRICKLEBANK et al. 1994; PRIESTLEY et al. 1998). Thus, it appears to prevent receptor overactivation and is neuroprotective whilst permitting a 'maintenance level' of normal glutamatergic neurotransmission. The glycine-site full antagonist gavestinel has been reported to be neuroprotective at a dose of 3 mg/kg i.v. administered up to 6 h after ischaemia, and the maximally protective dose caused no adverse cardiovascular or CNS side effects (BORDI et al. 1997).

E. Clinical Trials of the First-Generation NMDA Antagonists

Selfotel and aptiganel (cerestat) were the two main NMDA antagonists to emerge from this period and go into clinical trials for the treatment of stroke. Selfotel is a competitive NMDA antagonist which, in animal models, showed better separation between neuroprotective effects and behavioural side effects than ion-channel blockers (BENNETT et al. 1990; SAUER et al. 1993). However, in clinical studies, dose-limiting side effects, such as sedation, psychosis and hallucinations, were seen (MUIR and LEES 1995a; LEES 1998). A dose of 1.5 mg/kg was used for the phase-III efficacy study; however, this was not really optimal for neuroprotection, producing only around 30% protection in the permanent MCA-occlusion model (Fig. 2). Indeed, in other neuroprotection studies in cat and rabbit models, doses of around 30 mg/kg were used to obtain significant neuroprotection (PEREZ-PINZON et al. 1995; OKADA et al. 1997). The plasma concentration of selfotel predicted to be neuroprotective in animals was reported to be 40 μg/ml, but the target plasma concentration in the phase-III efficacy trial was 21 μg/ml (LEES 1996). Therefore, dosing in man was limited by the side effects and was below levels required to see significant protection in the animal studies (GROTTA et al. 1995). The result of the stroke trial showed a trend (which was not statistically significant) to increased mortality with the active dosage, and the trial was abandoned due to these worries and a low likelihood for demonstrating efficacy for stroke patients (DAVIS et al. 1997). SDZ EAA 494 (D-CPPene), another potent competitive NMDA antagonist, also produced CNS "side effects" in human volunteers and, in epilepsy patients, it was ineffective at doses that produced adverse events severe enough to stop the trial (LOWE et al. 1994).

Aptiganel is an ion-channel blocker acting in a manner similar to that of MK-801, with an in vitro potency approximately ten times less than that of MK-801 (KEMP et al. 1987; MCBURNEY 1997). Aptiganel was able to reduce ischaemic damage by 40–70% in animal models of focal ischaemia and was also effective when administered 1 h post-occlusion (MINEMATSU et al. 1993; MEADOWS et al. 1994). In rats, the lowest neuroprotective plasma concentration was around 10 ng/ml and was achieved with a dosing regime of 0.25 mg/kg followed by a 6 h infusion of 0.17 mg/kg/h. This dosing regime resulted in 24% cortical protection and was associated with some ataxia and increases in mean arterial blood pressure (MCBURNEY 1997). The dosing regime used for the phase-III efficacy studies of aptiganel was 4.5 mg/kg bolus followed by 0.75 mg/h for 12 h, resulting in a plasma concentration of 8–12 ng/ml (MCBURNEY 1997). The dose-limiting side effects seen in patients taking aptiganel were hypertension, sedation, agitation and hallucinations; the trial of aptiganel was suspended in July 1997, but the results have not yet been published (LEES 1997, 1998). Once again, intolerance to the CNS side effects was dose-limiting for the phase-III efficacy study.

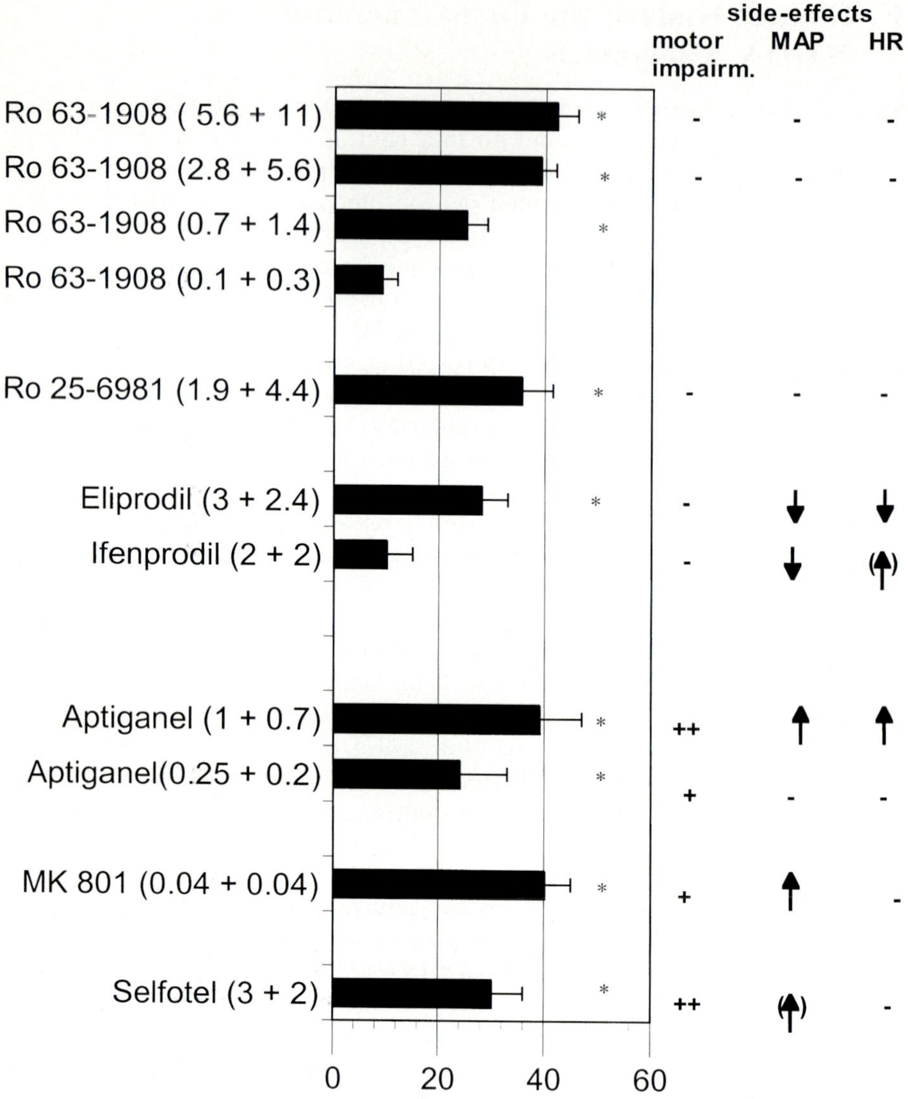

Fig. 2. Comparison of the neuroprotective effects of various *N*-methyl-D-aspartate antagonists in a permanent middle cerebral artery (MCA) occlusion model of focal ischaemia in the rat. The compounds were administered as a bolus dose (shown as the *first number* in the bracket, in mg/kg) intravenously (i.v.) 5 min after MCA occlusion followed by an i.v. infusion (*second number* in the bracket, mg/kg/h) for 5 h. The doses shown were the ones giving significant ($P < 0.05$) protection compared with control animals. The data are shown as the percent reduction in cortical infarct volume compared with the control group of animals. The side effect profile provides a semi-quantitative analysis of the severity of motor impairment, mean arterial blood pressure (*MAP*) and heart rate (*HR*). The *crosses* are a grading of the motor impairment or ataxia seen in these animals. The *arrows* indicate a significant increase or decrease in the MAP or HR, whereas the *arrows in the brackets* suggest a trend that did not reach significance (LOREZ et al. unpublished results)

Dextrorphan is also a non-competitive NMDA antagonist but may interact with the channel in a way slightly different than MK-801 and aptiganel (WONG and KEMP 1991). Dextrorphan was demonstrated to be neuroprotective in a rat model of temporary focal ischaemia (STEINBERG et al. 1989). In a phase-I tolerability study, dextrorphan administered intravenously was associated with hallucinations, agitation, and sedation (ALBERS et al. 1995), side effects common to both competitive and non-competitive NMDA antagonists. Dextrorphan also produced a hypotensive effect at higher doses, which is contraindicative in stroke and could be potentially problematic; thus, the development of dextrorphan was stopped after phase I.

Remacemide appears to be an NMDA ion-channel blocker "prodrug" and has recently been reported to be effective in improving the neuropsychological impairment, which may follow coronary-artery bypass surgery due to microemboli formed during the operation (ARROWSMITH et al. 1998). In this study, remacemide was administered prior to surgical intervention, and the side effects seen were dizziness, ataxia and drowsiness. In animal studies of focal ischaemia, maximal protection with remacemide was seen only when the drug was administered prior to focal ischaemia (BANNAN et al. 1994). Early phase-I trials demonstrated that remacemide could be well tolerated, apparently due to the fact that it is a prodrug and/or due to its lower affinity as an NMDA antagonist (MUIR and LEES 1995b).

Current stroke trials with a glycine-site antagonist of the NMDA receptor GV 150526A are ongoing (gavestinel, DI FABIO et al. 1997). This compound, which is currently in phase-III trials, demonstrated good tolerability without any adverse CNS side effects in phase-II trials (LEES 1996; LEES et al. 1998). However, at face value, it is difficult to understand why adequate blockade of NMDA receptors with this compound should fail to produce the side effects seen with other non-selective NMDA antagonists, and there is some question about its ability to readily penetrate the brain. It has been shown that this is a problem with other glycine-site antagonists of similar chemical class, largely due to their extremely tight binding to plasma proteins (ROWLEY et al. 1997). There also appear to be some increases in bilirubin levels, which have proved to be dose-limiting (LEES et al. 1998). Gavestinel was shown to be neuroprotective at a dose of 3 mg/kg i.v. after permanent MCA occlusion (BORDI et al. 1997); thus, in this phase-III study, the human dosing regime of 800 mg followed by 200 mg b.i.d. + 5 days is apparently well above the dose required to achieve neuroprotection in this animal study.

In conclusion, the clinical trials for the high-affinity NMDA antagonists in stroke have thus far been disappointing due to dose-limiting side effects, such as hypertension, sedation, agitation and hallucinations, and we await the phase-III results for the glycine-site antagonist gavestinel. The recent results with remacemide – demonstrating the first neuroprotective effect of an NMDA antagonist in man – are very encouraging.

F. Future Approaches to the Treatment of Stroke with NMDA Antagonists

The atypical NMDA antagonist ifenprodil was demonstrated to be neuroprotective in animal models without the adverse side effects of compounds like MK-801, aptiganel and selfotel (GOTTI et al. 1988). As already outlined in Sect. C.IV, ifenprodil is selective for the NR2B subunit of the NMDA receptor, with an activity-dependent mechanism of action (Sect. C.IV; WILLIAMS 1997). Eliprodil, which is from the same class of compounds, entered clinical trails for stroke, but these were abandoned in 1996 due to a lack of efficacy, which was probably due to underdosing (LEES 1996). However, the important finding to emerge from this trial was that side effects, such as hallucinations, agitation or sedation, were not seen in these patients (GIROUX et al. 1994). There were some problems (which were dose limiting) related to QT prolongation (GIROUX et al. 1994), hence the low dose (3 mg/kg) used in the efficacy trial. However, in addition to being a selective NR2B antagonist, eliprodil also acts on calcium, potassium and sodium channels, which may contribute to the cardiovascular side effects (GIROUX et al. 1994; CARTER et al. 1997).

CP-101,606 is also selective for NMDA NR2B subunits and is a structural analogue of ifenprodil but without the α1-adrenoceptor activity (CHENARD et al. 1995; MENNITI et al. 1997). CP-101,606 was demonstrated to protect against glutamate-induced toxicity in neuronal cultures, with a potency similar to that of MK-801 (MENNITI et al. 1997). It was also found to be neuroprotective in an acute model of focal ischaemia in the cat (DI et al. 1997). This compound has been reported to be in phase-II clinical trials for brain trauma.

We have identified state-dependent NR2B subtype-selective NMDA antagonists as potential neuroprotective agents. Ro 25–6981 was the first of this class of potent and selective NMDA antagonists (FISCHER et al. 1997), and it produced significant protection of 40% of the cortical infarct without any adverse cardiovascular or CNS side effects (FISCHER et al. 1996; Fig. 2). The selective NR2B antagonists are similarly protective in an animal model of stroke (permanent occlusion of the MCA in rats) to the non-selective NMDA antagonists but are much better tolerated. Ro 25–6981 does not block learning and memory in the Morris water-maze test or long-term potentiation in a hippocampal-slice model (BOURSON, unpublished results; KEMP, unpublished results).

Ro 63-1908 is the latest of the high-affinity NR2B subtype-selective NMDA antagonists. Ro 63-1908 produced a dose-related decrease in the volume of cortical damage, giving a maximum protection of 42% after a dosing regimen of a 5.6 mg/kg bolus followed by an infusion of 11 mg/kg/h. In a second experiment, a dosing regimen of 2.8 + 5.6 mg/kg/h gave a similar level of protection of 39%, suggesting that the extent of protection achieved reaches a plateau level (Fig. 3; GILL et al. 1998). There was no protection seen in the caudate, because the lenticulostriate artery, which is a branch of the MCA

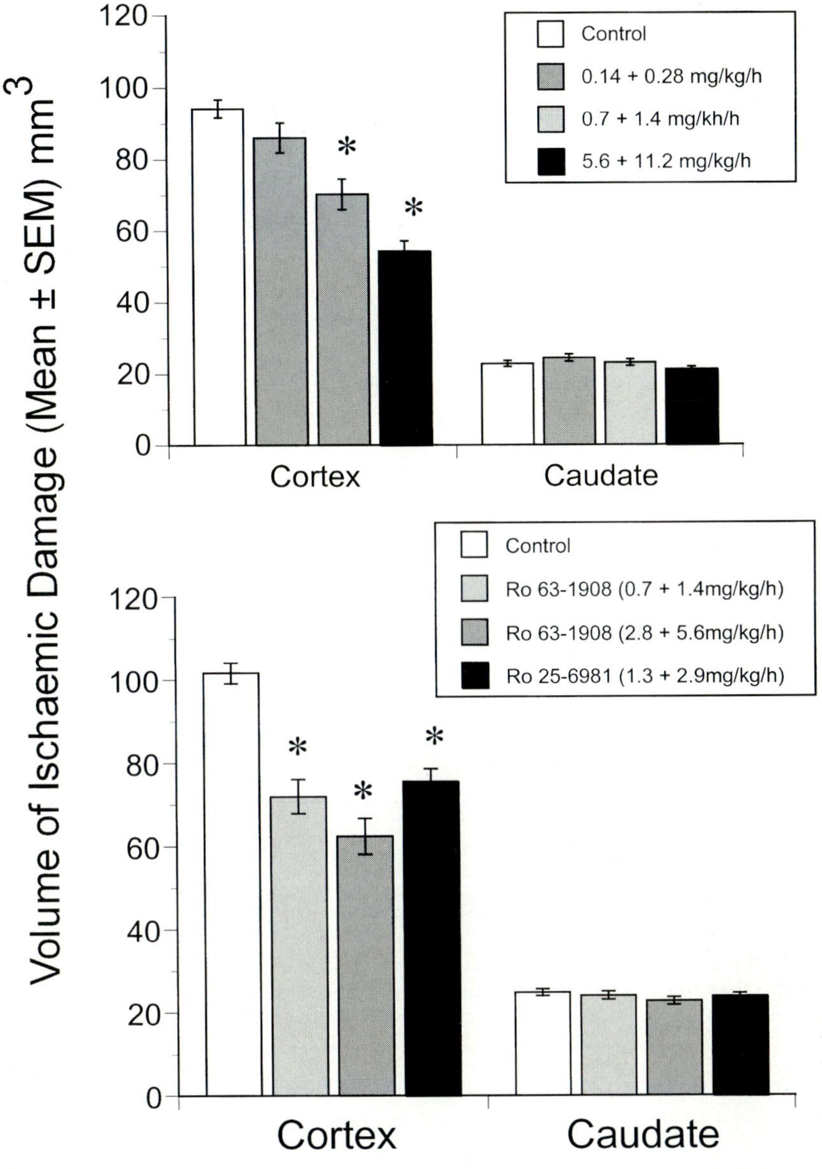

Fig. 3. Neuroprotective effects of Ro 63–1908 in the permanent middle cerebral artery (MCA) occlusion model of focal ischaemia. The compound was administered as an intravenous (i.v.) bolus dose 5 min after MCA occlusion followed by infusion for 5 h. The first *bar graph* shows a dose-related neuroprotective effect of Ro 63-1908 against the volume of cortical damage, with a maximum protection of 42% being seen with a dosing regime of a 5.6 mg/kg bolus dose followed by infusion of 11 mg/kg/h for 5 h. In a second experiment, a dose of 2.8 mg/kg bolus followed by infusion of 5.6 mg/kg/h for 5 h resulted in 39% protection. In this experiment, Ro 63-1908 gave a protective effect similar to that of Ro 25-6981 administered at a dose of 1.3 mg/kg i.v. bolus followed by infusion of 2.9 mg/kg/h for 5 h (Gill et al. 1998)

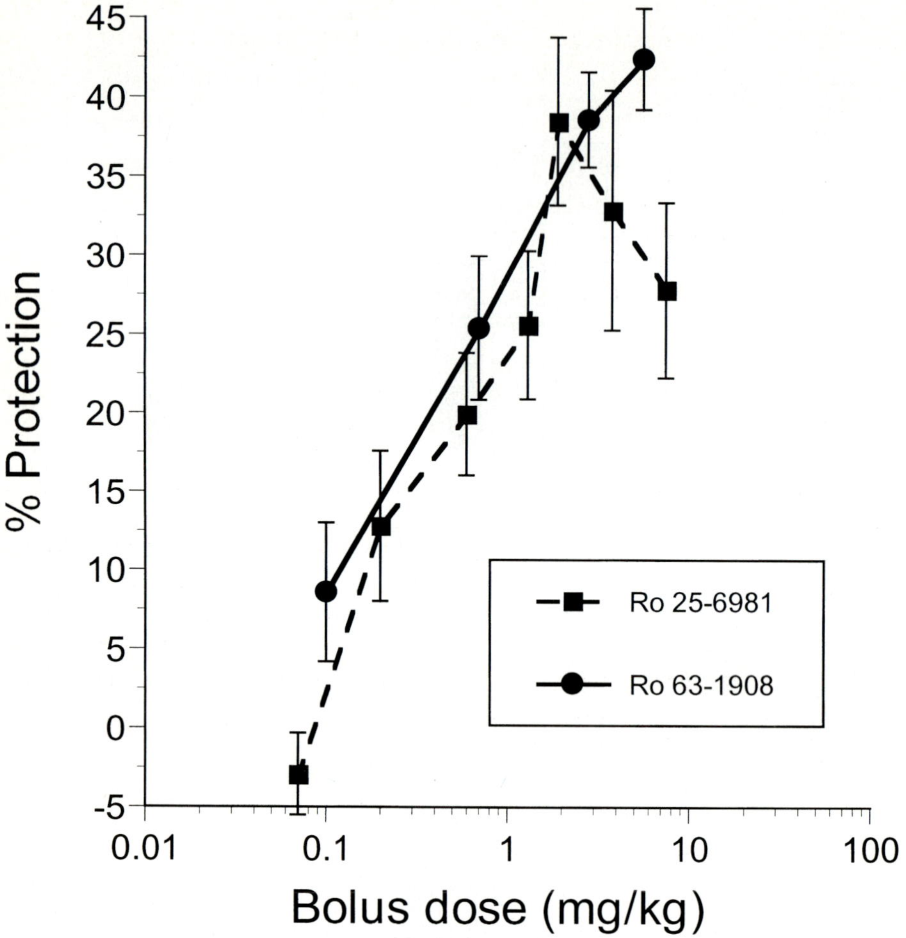

Fig. 4. Comparison of the neuroprotective effect of Ro 63-1908 and Ro 25-6981 in the rat focal ischaemia model. For comparative purposes, the percent protection has been plotted against the bolus dose of the compound. Both compounds gave a dose-related neuroprotective effect in this model. Significant protection was achieved with Ro 63-1908 and Ro 25-6981 without adverse cardiovascular or central nervous system side effects (FISCHER et al. 1996; GILL et al. 1998)

serving the caudate nucleus, is permanently occluded in this model of focal ischaemia.

The extent of protection seen with Ro 63-1908 is comparable to that seen with Ro 25-6981. For comparative purposes, the percentage protection after Ro 63-1908 and Ro 25-6981 has been plotted against the bolus dosage on a log scale (Fig. 4). The extent of protection achieved with the activity-dependent NR2B-selective blocker Ro 63-1908 is similar to that achieved with non-selective open-channel blockers of the NMDA receptor, such as aptiganel. The optimum neuroprotective effect of 39% protection was seen at a mean plasma concentration ranging from 465–534 ng/ml, with little variability between animals. A further doubling of the dose (and hence the plasma concentration) did not significantly increase the percentage of cortical protection seen with Ro 63-1908.

These dosing regimens of Ro 63-1908 were without effect on the CNS or on cardiovascular parameters (mean arterial pressure and heart rate) in these animals when measured at 0.5, 2 and 5 h following the start of dosing. This good tolerability is in marked contrast to earlier non-selective NMDA blockers, such as aptiganel and MK-801, which caused marked ataxia and increased blood pressure and heart rate at maximally protective doses in the permanent MCA-occlusion model.

This class of compounds is also without effect on neuronal vacuolisation (Olney et al. 1989) at the optimally neuroprotective doses. Assuming this improved tolerability in animals is predictive for humans, then maximum protective doses should be attained in patients without being limited by adverse CNS and cardiovascular side effects.

In conclusion, it appears that these NR2B subtype-selective NMDA antagonists show potential in the treatment of stroke and brain trauma. They have advantages over competitive and ion-channel-blocking NMDA antagonists in that they have a much improved tolerability in animals and, from initial reports, in man (Di et al. 1997). Therefore, it is conceivable that maximally neuroprotective doses of these compounds could be achieved in man without cardiovascular or CNS side effects.

References

Albers GW, Atkinson RP, Kelley RE, Rosenbaum DM (1995) Safety, tolerability, and pharmacokinetics of the *N*-methyl-D-aspartate antagonist dextrorphan in patients with acute stroke. Dextrorphan Study Group. Stroke 26:254–258

Allen HL, Iversen LL (1990) Phencyclidine, dizocilpine, and cerebrocortical neurons [comment]. Science 247:221

Anson LC, Chen PE, Wyllie DA, Colquhoun D, Schoepfer R (1998) Identification of amino acid residues of the NR2A subunit that control glutamate potency in recombinant NR1/NR2A NMDA receptors. J Neurosci 18:581–589

Armstrong N, Sun Y, Chen GQ, Gouaux E (1998) Structure of a glutamate-receptor ligand-binding core in complex with kainate. Nature 395:913–917

Arrowsmith JE, Harrison MJ, Newman SP, Stygall J, Timberlake N, Pugsley WB (1998) Neuroprotection of the brain during cardiopulmonary bypass: a randomized trial of remacemide during coronary artery bypass in 171 patients. Stroke 29:2357–2362

Astrup J, Siesjo BK, Symon L (1981) Thresholds in cerebral ischemia – the ischemic penumbra. Stroke 12:723–725

Baird AE, Warach S (1998) Magnetic resonance imaging of acute stroke. J Cereb Blood Flow Metab 18:583–609

Bannan PE, Graham DI, Lees KR, McCulloch J (1994) Neuroprotective effect of remacemide hydrochloride in focal cerebral ischemia in the cat. Brain Res 664: 271–275

Baron BM, Harrison BL, Miller FP, McDonald IA, Salituro FG, Schmidt CJ, Sorensen SM, White HS, Palfreyman MG (1990) Activity of 5,7-dichlorokynurenic acid, a potent antagonist at the N-methyl-D-aspartate receptor-associated glycine binding site. Mol Pharmacol 38:554–561

Baron JC, von Kummer R, del Zoppo GJ (1995) Treatment of acute ischemic stroke. Challenging the concept of a rigid and universal time window [editorial]. Stroke 26:2219–2221

Behe P, Stern P, Wyllie DJ, Nassar M, Schoepfer R, Colquhoun D (1995) Determination of NMDA NR1 subunit copy number in recombinant NMDA receptors. Proc R Soc Lond B Biol Sci 262:205–213

Bennett DA, Lehmann J, Bernard PS, Liebman JM, Williams M, Wood PL, Boast CA, Hutchison AJ (1990) CGS 19755: a novel competitive N-methyl-D-aspartate (NMDA) receptor antagonist with anticonvulsant, anxiolytic and anti-ischemic properties. Prog Clin Biol Res 361:519–524

Benveniste H, Drejer J, Schousboe A, Diemer NH (1984) Elevation of the extracellular concentrations of glutamate and aspartate in rat hippocampus during transient cerebral ischemia monitored by intracerebral microdialysis. J Neurochem 43:1369–1374

Benveniste H, Jorgensen MB, Sandberg M, Christensen T, Hagberg H, Diemer NH (1989) Ischemic damage in hippocampal CA1 is dependent on glutamate release and intact innervation from CA3. J Cereb Blood Flow Metab 9:629–639

Benveniste M, Mayer ML (1991) Kinetic analysis of antagonist action at N-methyl-D-aspartic acid receptors. Two binding sites each for glutamate and glycine. Biophys J 59:560–573

Benveniste M, Mayer ML (1993) Multiple effects of spermine on N-methyl-D-aspartic acid receptor responses of rat cultured hippocampal neurones. J Physiol (Lond) 464:131–163

Benveniste M, Clements J, Vyklicky LJ, Mayer ML (1990) A kinetic analysis of the modulation of N-methyl-D-aspartic acid receptors by glycine in mouse cultured hippocampal neurones. J Physiol Lond 428:333–357

Bielenberg GW, Beck T (1991) The effects of dizocilpine (MK-801), phencyclidine, and nimodipine on infarct size 48h after middle cerebral artery occlusion in the rat. Brain Res 552:338–342

Bliss TVP, Collingridge GL (1993) A synaptic model of memory – long-term potentiation in the hippocampus. Nature 361:31–39

Block, GA, and Pulsinelli, WA. Excitatory amino acid receptor antagonists: failure to prevent ischemic neuronal damage. J.Cereb.Blood Flow Metab 7, S149. 1987

Boast CA, Gerhardt SC, Pastor G, Lehmann J, Etienne PE, Liebman JM (1988) The N-methyl-D-aspartate antagonists CGS 19755 and CPP reduce ischemic brain damage in gerbils. Brain Res 442:345–348

Bordi F, Pietra C, Ziviani L, Reggiani A (1997) The glycine antagonist GV150526 protects somatosensory evoked potentials and reduces the infarct area in the MCAo model of focal ischemia in the rat. Exp Neurol 145:425–433

Brimecombe JC, Boeckman FA, Aizenman E (1997) Functional consequences of NR2 subunit composition in single recombinant N-methyl-D-aspartate receptors. Proc Natl Acad Sci USA 94:11019–11024

Buchan A (1990) Do NMDA antagonists protect against cerebral ischemia: are clinical trials warranted? Cerebrovasc Brain Metab Rev 2:1–26

Buchan A, Pulsinelli WA (1990) Hypothermia but not the N-methyl-D-aspartate antagonist, MK-801, attenuates neuronal damage in gerbils subjected to transient global ischemia. J Neurosci 10:311–316

Buchan A, Li H, Pulsinelli WA (1991) The N-methyl-D-aspartate antagonist, MK-801, fails to protect against neuronal damage caused by transient, severe forebrain ischemia in adult rats. J Neurosci 11:1049–1056

Bullock R, Zauner A, Woodward J, Young HF (1995) Massive persistent release of excitatory amino acids following human occlusive stroke. Stroke 26:2187–2189

Butcher SP, Bullock R, Graham DI, McCulloch J (1990) Correlation between amino acid release and neuropathologic outcome in rat brain following middle cerebral artery occlusion. Stroke 21:1727–1733

Carter C, Lloyd KG, Zivkovic B, Scatton B (1990) Ifenprodil and SL 82.0715 as cerebral antiischemic agents. III. Evidence for antagonistic effects at the polyamine modulatory site within the *N*-methyl-D-aspartate receptor complex. J Pharmacol Exp Ther 253:475–482

Carter C, Avenet P, Benavides J, Besnard F, Biton B, Cudennec A, Duverger D, Frost J, Giroux C, Graham D, Langer SZ, Nowicki JP, Oblin A, Perrault G, Pigasse S, Rosen P., Sanger DJ, Schoemaker H, Thenot JP, Scatton B (1997) Ifenprodil and eliprodil: neuroprotective NMDA receptor antagonists and calcium channel blockers. In: Herling P (ed) Excitatory amino acids-clinical results with antagonists. Academic, New York, p 57

Castillo J, Davalos A, Noya M (1997) Progression of ischaemic stroke and excitotoxic amino acids. Lancet 349:79–83

Changeux JP, Galzi JL, Devillers-Thiery A, Bertrand D (1992) The functional architecture of the acetylcholine nicotinic receptor explored by affinity labelling and site-directed mutagenesis. Q Rev Biophys 25:395–432

Chazot PL, Stephenson FA (1997) Molecular dissection of native mammalian forebrain NMDA receptors containing the NR1 C2 exon: direct demonstration of NMDA receptors comprising NR1, NR2A, and NR2B subunits within the same complex. J Neurochem 69:2138–2144

Chazot PL, Cik M, Stephenson FA (1992) Immunological Detection of the NMDAR1 glutamate receptor subunit expressed in human embryonic kidney-293 cells and in rat brain. J Neurochem 59:1176–1178

Chazot PL, Coleman SK, Cik M, Stephenson FA (1994) Molecular characterization of *N*-methyl-D-aspartate receptors expressed in mammalian cells yields evidence for the coexistence of three subunit types within a discrete receptor molecule. J Biol Chem 269:24403–24409

Chen HS, Pellegrini JW, Aggarwal SK, Lei SZ, Warach S, Jensen FE, Lipton SA (1992) Open-channel block of *N*-methyl-D-aspartate (NMDA) responses by memantine: therapeutic advantage against NMDA receptor-mediated neurotoxicity. J Neurosci 12:4427–4436

Chen HS, Wang YF, Rayudu PV, Edgecomb P, Neill JC, Segal MM, Lipton SA, Jensen FE (1998) Neuroprotective concentrations of the *N*-methyl-D-aspartate open-channel blocker memantine are effective without cytoplasmic vacuolation following post-ischemic administration and do not block maze learning or long-term potentiation. Neuroscience 86:1121–1132

Chenarp BL, Bordner J, Butler TW, Chambers LK, Collins MA, De Costa DL, Ducat MF, Dumont ML, Fox CB, Mena EE, et al. (1995) (1S,2S)-1-(4-hydroxyphenyl)-2-(4-hydroxy-4-phenylpiperidino)-1-propanol: a potent new neuroprotectant which blocks *N*-methyl-D-aspartate responses. J Med Chem 38:3138–3145

Choi DW (1988) Glutamate neurotoxicity and diseases of the nervous system. Neuron 1:623–634

Christine CW, Choi DW (1990) Effect of zinc on NMDA receptor-mediated channel currents in cortical neurons. J Neurosci 10:108–116

Church J, Zeman S, Lodge D (1988) The neuroprotective action of ketamine and MK-801 after transient cerebral ischemia in rats. Anesthesiology 69:702–709

Ciabarra AM, Sullivan JM, Gahn LG, Pecht G, Heinemann S, Sevarino KA (1995) Cloning and characterization of chi-1: a developmentally regulated member of a novel class of the ionotropic glutamate receptor family. J Neurosci 15:6498–6508

Clements JD, Westbrook GL (1991) Activation kinetics reveal the number of glutamate and glycine binding sites on the *N*-methyl-D-aspartate receptor. Neuron 7:605–613

Colbourne F, Sutherland G, Corbett D (1997) Postischemic hypothermia. A critical appraisal with implications for clinical treatment. Mol Neurobiol 14:171–201

Danysz W, Parsons CG, Kornhuber J, Schmidt WJ, Quack G (1997) Aminoadamantanes as NMDA receptor antagonists and antiparkinsonian agents–preclinical studies. Neurosci Biobehav Rev 21:455–468

Das S, Sasaki YF, Rothe T, Premkumar LS, Takasu M, Crandall JE, Dikkes P, Conner DA, Rayudu PV, Cheung W, Chen HS, Lipton SA, Nakanishi N (1998) Increased NMDA current and spine density in mice lacking the NMDA receptor subunit NR3A. Nature 393:377–381

Davalos A, Castillo J, Serena J, Noya M (1997) Duration of glutamate release after acute ischemic stroke. Stroke 28:708–710

Davies JA (1997) Remacemide hydrochloride: a novel antiepileptic agent. Gen Pharmacol 28:499–502

Davis SM, Albers GW, Diener HC, Lees KR, Norris J (1997) Termination of acute stroke studies involving selfotel treatment. ASSIST Steering Committee. Lancet 349:32

Di X, Bullock R, Watson J, Fatouros P, Chenard B, White F, Corwin F (1997) Effect of CP101,606, a novel NR2B subunit antagonist of the *N*-methyl-D-aspartate receptor, on the volume of ischemic brain damage off cytotoxic brain edema after middle cerebral artery occlusion in the feline brain. Stroke 28:2244–2251

Di Fabio R, Capelli AM, Conti N, Cugola A, Donati D, Feriani A, Gastaldi P, Gaviraghi G, Hewkin CT, Micheli F, Missio A, Mugnaini M, Pecunioso A, Quaglia AM, Ratti E, Rossi L, Tedesco G, Trist DG, Reggiani A (1997) Substituted indole-2-carboxylates as in vivo potent antagonists acting as the strychnine-insensitive glycine binding site. J Med Chem 40:841–850

Diemer NH, Valente E, Bruhn T, Berg M, Jorgensen MB, Johansen FF (1993) Glutamate receptor transmission and ischemic nerve cell damage: evidence for involvement of excitotoxic mechanisms. Prog Brain Res 96:105–123

Durand GM, Gregor P, Zheng X, Bennett MVL, Uhl GR, Zukin S (1992) Cloning of an apparent splice variant of the rat *N*-methyl-D-aspartate receptor NMDAR1 with altered sensitivity to polyamines and activators of protein kinase C. Proc Natl Acad Sci USA 89:9359:9363

Ferrer-Montiel AV, Montal M (1996) Pentameric subunit stoichiometry of a neuronal glutamate receptor. Proc Natl Acad Sci USA 93:2741–2744

Fischer G, Bourson A, Kemp JA, Lorez H-P, Mutel V, Trube G (1995) Characterization of morphanins highly protective in permanent middle cerebral artery occlusion (MCAO) in rats. Soc Neurosci Abstr 392.7

Fischer G, Bourson A, Kemp JA, Lorez HP (1996) The neuroprotective activity of Ro 25–6981, a NMDA receptor NR2B subtype selective blocker. Soc Neurosci Abstract 693.5

Fischer G, Mutel V, Trube G, Malherbe P, Kew JN, Mohacsi E, Heitz MP, Kemp JA (1997) Ro 25–6981, a highly potent and selective blocker of *N*-methyl-D-aspartate receptors containing the NR2B subunit. Characterization in vitro. J Pharmacol Exp Ther 283:1285–1292

Fletcher EJ, Lodge D (1988) Glycine reverses antagonism of *N*-methyl-D-aspartate (NMDA) by 1-hydroxy-3-aminopyrrolidone-2 (HA-966) but not by d-2-amino-5-phosphonovalerate (D-AP5) on rat cortical slices. Eur J Pharmacol 151:161–162

Foldes RL, Rampersad V, Kamboj RK (1993) Cloning and sequence analysis of cDNAs encoding human hippocampus *N*-methyl-D-aspartate receptor subunits – evidence for alternative RNA splicing. Gene 131:293–298

Foster AC, Kemp JA (1989) HA-966 antagonizes *N*-methyl-D-aspartate receptors through a selective interaction with the glycine modulatory site. J Neurosci 9: 2191–2196

Foster AC, Gill R, Kemp JA, Woodruff GN (1987) Systemic administration of MK-801 prevents *N*-methyl-D-aspartate-induced neuronal degeneration in rat brain. Neurosci Lett 76:307–311

Foster AC, Kemp JA, Leeson PD, Grimwood S, Donald AE, Marshall GR, Priestley T, Smith JD, Carling RW (1992) Kynurenic acid analogues with improved affinity and selectivity for the glycine site on the *N*-methyl-D-aspartate receptor from rat brain. Mol Pharmacol 41:914–922

Furlan M, Marchal G, Viader F, Derlon JM, Baron JC (1996) Spontaneous neurological recovery after stroke and the fate of the ischemic penumbra. Ann Neurol 40:216–226

Gallagher MJ, Huang H, Pritchett DB, Lynch DR (1996) Interactions between ifenprodil and the NR2B subunit of the *N*-methyl-D-aspartate receptor. J Biol Chem 271:9603–9611

Gallo V, Upson LM, Hayes WP, Vyklicky LJ, Winters CA, Buonanno A (1992) Molecular cloning and development analysis of a new glutamate receptor subunit isoform in cerebellum. J Neurosci 12:1010–1023

Gill R (1994) The pharmacology of alpha-amino-3-hydroxy-5-methyl-4-isoxazole propionate (AMPA)/kainate antagonists and their role in cerebral ischaemia. Cerebrovasc Brain Metab Rev 6:225–256

Gill R, Lodge D (1997) Pharmacology of AMPA antagonists and their role in neuroprotection. Int Rev Neurobiol 40:197–232

Gill R, Woodruff GN (1990) The neuroprotective actions of kynurenic acid and MK-801 in gerbils are synergistic and not related to hypothermia. Eur J Pharmacol 176:143–149

Gill R, Foster AC, Woodruff GN (1987a) Systemic administration of MK-801 protects against ischemia-induced hippocampal neurodegeneration in the gerbil. J Neurosci 7:3343–3349

Gill R, Foster AC, Woodruff GN (1987b) Systemic administration of MK-801 protects against ischaemic neuropathology in rats. Br J Pharmac Proc (Suppl.) 91:311P

Gill R, Foster AC, Woodruff GN (1988) MK-801 is neuroprotective in gerbils when administered during the post-ischaemic period. Neuroscience 25:847–855

Gill R, Foster AC, Woodruff GN (1989) Neuroprotective actions of MK-801 in rat global ischaemia model. J Cereb Blood Flow Metab (suppl 1) 9:S629

Gill R, Brazell C, Woodruff GN, Kemp JA (1991) The neuroprotective action of dizocilpine (MK-801) in the rat middle cerebral artery occlusion model of focal ischaemia. Br J Pharmacol 103:2030–2036

Gill R, Andine P, Hillered L, Persson L, Hagberg H (1992) The Effect of MK-801 on cortical spreading depression in the penumbral zone following focal ischaemia in the rat. J Cereb Blood Flow Metab 12:371–379

Gill R, Sibson NR, Hatfield RH, Burdett NG, Carpenter TA, Hall LD, Pickard JD (1995a) A comparison of the early development of ischaemic damage following permanent middle cerebral artery occlusion in rats as assessed using magnetic resonance imaging and histology. J Cereb Blood Flow Metab 15:1–11

Gill R, Hargreaves RJ, Kemp JA (1995b) The neuroprotective effect of the glycine site antagonist 3R-(+)-*cis*-4-methyl-HA966 (L-687,414) in a rat model of focal ischaemia. J Cereb Blood Flow Metab 15:197–204

Gill R, Lorez H-P, Wybrecht R, Miss M-T, Bourson A, Fischer G, Kew JN, Mutel V, Trube G, Roever S, Heitz MP, Kemp JA (1998) The neuroprotective effects of Ro 63–1908 a selective antagonist of NMDA receptors containing theNR2B subunit. Pharmacology of cerebral ischaemia. Marburg Drugl

Ginsberg MD, Busto R (1989) Rodent models of cerebral ischemia. Stroke 20: 1627–1642

Giroux C, Rosen P, Scatton B (1994) Preclinical pharmacology and clinical safety profile of eliprodil, an atypical NMDA receptor antagonist. In: Krieglstein J, Oberpichler-Schwenk H. (eds) Pharmacology of cerebral ischaemia 1996. Wissenschaftliche Verlagsgesellschaft mbH, Stuttgart, p643

Gotti B, Duverger D, Bertin J, Carter C, Dupont R, Frost J, Gaudilliere B, MacKenzie ET, Rousseau J, Scatton B, et al (1988) Ifenprodil and SL 82.0715 as cerebral anti-ischemic agents. I. Evidence for efficacy in models of focal cerebral ischemia. J Pharmacol Exp Ther 247:1211–1221

Gozlan H, Ben-Ari Y (1995) NMDA receptor redox sites: are they targets for selective neuronal protection? Trends Pharmacol Sci 16:368–374

Graham DI (1988) Focal cerebral infarction. J Cereb Blood Flow Metab 8:769–773

Grant KA, Colombo G, Grant J, Rogawski MA (1996) Dizocilpine-like discriminative stimulus effects of low-affinity uncompetitive NMDA antagonists. Neuropharmacology 35:1709–1719
Grimwood S, Kulagowski JJ, Mawer IM, Rowley M, Leeson PD, Foster AC (1995) Allosteric modulation of the glutamate site on the NMDA receptor by four novel glycine site antagonists. Eur J Pharmacol 290:221–226
Grotta J, Picone CM, Ostrow PT, Strong RA, Earls RM, Yao LP, Rhoades HM, Dedman JR (1990) CGS-19755, a competitive NMDA receptor antagonist, reduces calcium-calmodulin binding and improves outcome after global cerebral ischemia. Ann Neurol 27:612–619
Grotta J, Clark W, Coull B, Pettigrew LC, Mackay B, Goldstein LB, Meissner I, Murphy D, LaRue L (1995) Safety and tolerability of the glutamate antagonist CGS 19755 (Selfotel) in patients with acute ischemic stroke. Results of a phase-IIa randomized trial. Stroke 26:602–605
Gyngell ML, Back T, Hoehn-Berlage M, Kohno K, Hossmann KA (1994) Transient cell depolarization after permanent middle cerebral artery occlusion: an observation by diffusion-weighted MRI and localized 1H-MRS. Magn Reson Med 31:337–341
Hagberg H, Lehmann A, Sandberg M, Nystrom B, Jacobson I, Hamberger A (1985) Ischemia-induced shift of inhibitory and excitatory amino acids from intra- to extracellular compartments. J Cereb Blood Flow Metab 5:413–419
Hargreaves RJ, Rigby M, Smith D, Hill RG, Iversen LL (1993a) Competitive as well as uncompetitive *N*-methyl-D-aspartate receptor antagonists affect cortical neuronal morphology and cerebral glucose metabolism. Neurochem Res 18:1263–1269
Hargreaves RJ, Rigby M, Smith D, Hill RG (1993b) Lack of effect of L-687,414 ((+)-*cis*-4-methyl-HA-966), an NMDA receptor antagonist acting at the glycine site, on cerebral glucose metabolism and cortical neuronal morphology. Br J Pharmacol 110:36–42
Hargreaves RJ, Hill RG, Iversen LL (1994) Neuroprotective NMDA antagonists – the controversy over their potential for adverse effects on cortical neuronal morphology. Brain Edema IX 60:15–19
Hatfield RH, Gill R, Brazell C (1992) The dose-response relationship and therapeutic window for dizocilpine (MK-801) in a rat focal ischaemia model. Eur J Pharmacol 216:1–7
Hayashi T (1952) A physiological study of epileptic seizures following cortical stimulation in animals and its application to human clinics. Japn J Physiol 3:46–64
Hirai H, Kirsch J, Laube B, Betz H, Kuhse J (1996) The glycine binding site of the *N*-methyl-D-aspartate receptor subunit NR1: identification of novel determinants of co-agonist potentiation in the extracellular M3-M4 loop region. Proc Natl Acad Sci USA 93:6031–6036
Hoffman CA, Boast CA (1995) Neuroprotection by MK-801 in temperature maintained gerbils. Brain Res Bull 38:405–409
Huettner JE, Bean BP (1988) Block of *N*-methyl-D-aspartate-activated current by the anticonvulsant MK-801: selective binding to open channels. Proc Natl Acad Sci USA 85:1307–1311
Ikeda K, Nagasawa M, Mori H, Araki K, Sakimura K, Watanabe M, Inoue Y, Mishina M (1992) Cloning and expression of the epsilon 4 subunit of the NMDA receptor channel. FEBS Lett 313:34–38
Ishii T, Moriyoshi K, Sugihara H, Sakurada K, Kadotani H, Yokoi M, Akazawa C, Shigemoto R, Mizuno N, Masu M (1993) Molecular characterization of the family of the *N*-methyl-D-aspartate receptor subunits. J Biol Chem 268:2836–2843
Iversen LL, Kemp JA (1994) Non-competitive NMDA antagonists as drugs. In: Collingridge GL, Watkins JC (eds) The NMDA receptor. Oxford, New York, pp 469–486
Jane DE, Olverman HJ, Watkins JC (1994) Agonists and competitive antagonists: structure-activity and molecular modelling studies. In: Collingridge GL, Watkins JC (eds) The NMDA receptor. Oxford, New York, p 31

Johansen FF, Jorgensen MB, Ekstrom vLD, Diemer NH (1984) Selective dendrite damage in hippocampal CA1 stratum radiatum with unchanged axon ultrastructure and glutamate uptake after transient cerebral ischaemia in the rat. Brain Res 291:373–377
Johansen FF, Jorgensen MB, Diemer NH (1986) Ischemic CA-1 pyramidal cell loss is prevented by preischemic colchicine destruction of dentate gyrus granule cells. Brain Res 377:344–347
Johnson JW, Ascher P (1987) Glycine potentiates the NMDA response in cultured mouse brain neurons. Nature 325:529–531
Jorgensen MB, Johansen FF, Diemer NH (1987) Removal of the entorhinal cortex protects hippocampal CA-1 neurons from ischemic damage. Acta Neuropathol Berl 73:189–194
Karp SJ, Masu M, Eki T, Ozawa K, Nakanishi S (1993) Molecular cloning and chromosomal localization of the key subunit of the human *N*-methyl-D-aspartate receptor. J Biol Chem 268:3728–3733
Kemp JA, Kew JN (1998) NMDA Receptor antagonists. In: Design Leff P (ed) Receptor-based drug. Marcel Dekker, New York, pp297–321
Kemp JA, Leeson PD (1993) The glycine site of the NMDA receptor–five years on. Trends Pharmacol Sci 14:20–25
Kemp JA, Priestley T (1991) Effects of (+)-HA-966 and 7-chlorokynurenic acid on the kinetics of *N*-methyl-D-asparatate receptor agonist responses in rat cultured cortical neurons. Mol Pharmacol 39:666–670
Kemp JA, Foster AC, Wong EHF (1987) Non-competitive antagonists of excitatory amino acid receptors. Trends Neurosci 10:294–298
Kemp JA, Foster AC, Leeson PD, Priestley T, Tridgett R, Iversen LL, Woodruff GN (1988) 7-Chlorokynurenic acid is a selective antagonist at the glycine modulatory site of the *N*-methyl-D-aspartate receptor complex. Proc Natl Acad Sci USA 85: 6547–6550
Kemp JA, Marshall GR, Priestley T (1991) A comparison of the agonist-dependency of the block produced by uncompetitive NMDA receptor antagonists on rat cortical slices. Mol Neuropharmacol 1:65–70
Kew JN, Kemp JA (1998) An allosteric interaction between the NMDA receptor polyamine and ifenprodil sites in rat cultured cortical neurones. J Physiol (Lond) 512:17–28
Kew JN, Trube G, Kemp JA (1996) A novel mechanism of activity-dependent NMDA receptor antagonism describes the effect of ifenprodil in rat cultured cortical neurones. J Physiol (Lond) 497:761–772
Kew JN, Richards JG, Mutel V, Kemp JA (1998a) Developmental changes in NMDA receptor glycine affinity and ifenprodil sensitivity reveal three distinct populations of NMDA receptors in individual rat cortical neurons. J Neurosci 18:1935–1943
Kew JN, Trube G, Kemp JA (1998b) State-dependent NMDA receptor antagonism by Ro 8–4304, a novel NR2B selective, non-competitive, voltage-independent antagonist. Br J Pharmacol 123:463–472
Kirino T (1982) Delayed neuronal death in the gerbil hippocampus following ischemia. Brain Res 239:57–69
Kleckner NW, Dingledine R (1988) Requirement for glycine in activation of NMDA-receptors expressed in Xenopus oocytes. Science 241:835–837
Kochhar A, Zivin JA, Lyden PD, Mazzarella V (1988) Glutamate antagonist therapy reduces neurologic deficits produced by focal central nervous system ischemia. Arch Neurol 45:148–153
Kumar KN, Tilakaratne N, Johnson PS, Allen AE, Michaelis EK (1991) Cloning of cDNA for the glutamate-binding subunit of an NMDA receptor complex. Nature 354:70–73
Kuryatov A, Laube B, Betz H, Kuhse J (1994) Mutational analysis of the glycine-binding site of the NMDA receptor: structural similarity with bacterial amino acid-binding proteins. Neuron 12:1291–1300

Kutsuwada T, Kashiwabuchi N, Mori H, Sakimura K, Kushiya E, Araki K, Meguro H, Masaki H, Kumanishi T, Arakawa M (1992) Molecular diversity of the NMDA receptor channel. Nature 358:36–41

Latour LL, Hasegawa Y, Formato JE, Fisher M, Sotak CH (1994) Spreading waves of decreased diffusion coefficient after cortical stimulation in the rat brain. Magn Reson Med 32:189–198

Laube B, Hirai H, Sturgess M, Betz H, Kuhse J (1997) Molecular determinants of agonist discrimination by NMDA receptor subunits: analysis of the glutamate binding site on the NR2B subunit. Neuron 18:493–503

Laube B, Kuhse J, Betz H (1998) Evidence for a tetrameric structure of recombinant NMDA receptors. J Neurosci 18:2954–2961

Laurie DJ, Seeburg PH (1994) Regional and developmental heterogeneity in splicing of the rat brain NMDAR1 mRNA. J Neurosci 14:3180–3194

Laurie DJ, Putzke J, Zieglgansberger W, Seeburg PH, Tolle TR (1995) The distribution of splice variants of the NMDAR1 subunit mRNA in adult rat brain. Brain Res Mol Brain Res 32:94–108

Lees KR (1996) Clinical trials for acute cerebral ischaemia. In: Krieglstein J (ed) Pharmacology of cerebral ischaemia 1996. Medpharm Scientific, Stuttgart, p691

Lees KR (1997) Cerestat and other NMDA antagonists in ischemic stroke. Neurology 49:S66-S69

Lees KR (1998) Does neuroprotection improve stroke outcome? Lancet 351: 1447–1448

Lees KR, Lavelle JF, Hobbiger SF (1998) Safety and tolerability of GV150526 in acute stroke (GLYB2002). Cerebrovasc Dis 8:66–66

Leeson PD, Iversen LL (1994) The glycine site on the NMDA receptor: structure-activity relationships and therapeutic potential. J Med Chem 37:4053–4067

Lerma J (1992) Spermine regulates *N*-methyl-D-aspartate receptor desensitization. Neuron 8:343–352

Lester RA, Tong G, Jahr CE (1993) Interactions between the glycine and glutamate binding sites of the NMDA receptor. J Neurosci 13:1088–1096

Lewis SJ, Barres C, Jacob HJ, Ohta H, Brody MJ (1989) Cardiovascular effects of the *N*-methyl-D-aspartate receptor antagonist MK-801 in conscious rats. Hypertension 13:759–765

Liman ER, Tytgat J, Hess P (1992) Subunit stoichiometry of a mammalian K+ channel determined by construction of multimeric cDNAs. Neuron 9:861–871

Lowe DA, Emre M, Frey P, Kelly PH, Malanowski J, McAllister KH, Neijt HC, Rudeberg C, Urwyler S, White TG (1994) The pharmacology of SDZ EAA 494, a competitive NMDA antagonist. Neurochem Int 25:583–600

Luo J, Wang Y, Yasuda RP, Dunah AW, Wolfe BB (1997) The majority of *N*-methyl-D-aspartate receptor complexes in adult rat cerebral cortex contain at least three different subunits (NR1/NR2A/NR2B). Mol Pharmacol 51:79–86

Macdermott AB, Mayer ML, Westbrook GL, Smith SJ, Barker JL (1986) NMDA-receptor activation increases cytoplasmic calcium concentration in cultured spinal cord neurones. Nature 321:519–522

MacKinnon R (1991) New insights into the structure and function of potassium channels. Curr Opin Neurobiol 1:14–19

MacKinnon R (1995) Pore loops: an emerging theme in ion channel structure. Neuron 14:889–892

Mano I, Teichberg VI (1998) A tetrameric subunit stoichiometry for a glutamate receptor-channel complex. Neuroreport 9:327–331

Marcoux FW, Goodrich JE, Dominick MA (1988) Ketamine prevents ischemic neuronal injury. Brain Res 452:329–335

Mayer ML, Westbrook GL, Guthrie PB (1984) Voltage-dependent block by Mg2+ of NMDA responses in spinal cord neurones. Nature 309:261–263

McAuley MA (1995) Rodent models of focal ischemia. Cerebrovasc Brain Metab Rev 7:153–180

McBain CJ, Mayer ML (1994) *N*-methyl-D-aspartic acid receptor structure and function. Physiol Rev 74:723–760

Mcburney RN (1997) Development of the NMDA ion-channel blocker, aptiganel hydrochloride, as a neuroprotective agent for acute CNS injury. Int Rev Neurobiol 40:173–95:173–195

McCulloch J, Bullock R, Teasdale GM (1991) Excitatory amino acid antagonists: opportunities for the treatment of ischaemic brain damage in man. In: Meldrum BS (ed) Excitatory amino acid antagonists. Blackwell, London, pp 287–326

McGurk JF, Bennett MV, Zukin RS (1990) Polyamines potentiate responses of *N*-methyl-D-aspartate receptors expressed in xenopus oocytes. Proc Natl Acad Sci USA 87:9971–9974

McIlhinney RA, Molnar E, Atack JR, Whiting PJ (1996) Cell surface expression of the human *N*-methyl-D-aspartate receptor subunit 1a requires the co-expression of the NR2A subunit in transfected cells. Neuroscience 70:989–997

Meadows ME, Fisher M, Minematsu K (1994) Delayed treatment with a noncompetitive NMDA antagonist, CNS-1102, reduces infarct size in rats. Cerebrovasc Dis 4:26–31

Meguro H, Mori H, Araki K, Kushiya E, Kutsuwada T, Yamazaki M, Kumanishi T, Arakawa M, Sakimura K, Mishina M (1992) Functional characterization of a heteromeric NMDA receptor channel expressed from cloned cDNAs. Nature 357:70–74

Meldrum B (1990) Protection against ischaemic neuronal damage by drugs acting on excitatory neurotransmission. Cerebrovasc Brain Metab Rev 2:27–57

Meldrum B, Garthwaite J (1990) Excitatory amino acid neurotoxicity and neurodegenerative disease. Trends Pharmacol Sci 11:379–387

Menniti F, Chenard B, Collins M, Ducat M, Shalaby I, White F (1997) CP-101,606, a potent neuroprotectant selective for forebrain neurons. Eur J Pharmacol 331:117–126

Minematsu K, Fisher M, Li L, Davis MA, Knapp AG, Cotter RE, Mcburney RN, Sotak CH (1993) Effects of a novel NMDA antagonist on experimental stroke rapidly and quantitatively assessed by diffusion-weighted MRI. Neurology 43:397–403

Monyer H, Sprengel R, Schoepfer R, Herb A, Higuchi M, Lomeli H, Burnashev N, Sakmann B, Seeburg PH (1992) Heteromeric NMDA receptors: molecular and functional distinction of subtypes. Science 256:1217–1221

Monyer H, Burnashev N, Laurie DJ, Sakmann B, Seeburg PH (1994) Developmental and regional expression in the rat brain and functional properties of four NMDA receptors. Neuron 12:529–540

Moriyoshi K, Masu M, Ishii T, Shigemoto R, Mizuno N, Nakanishi S (1991) Molecular cloning and characterization of the rat NMDA receptor. Nature 354:31–37

Mott DD, Doherty JJ, Zhang S, Washburn MS, Fendley MJ, Lyuboslavsky P, Traynelis SF, Dingledine R, (1998) Phenylethanolamines inhibit NMDA receptors by enhancing proton inhibition. Nature Neuroscience 1:659–667

Muir KW, Lees KR (1995a) Clinical experience with excitatory amino acid antagonist drugs. Stroke 26:503–513

Muir KW, Lees KR (1995b) Initial experience with remacemide hydrochloride in patients with acute ischemic stroke. Ann N Y Acad Sci 765:322–323

Nedergaard M, Astrup J (1986) Infarct rim: effect of hyperglycemia on direct current potential and2-deoxyglucose phosphorylation. J Cereb Blood Flow Metab 6: 607–615

Nellgard B, Gustafson I, Wieloch T (1991) Lack of protection by the *N*-methyl-D-aspartate receptor blocker dizocilpine (MK-801) after transient severe cerebral ischemia in the rat. Anesthesiology 75:279–287

Nowak L, Bregestovski P, Ascher P, Herbet A, Prochiantz A (1984) Magnesium gates glutamate-activated channels in mouse central neurones. Nature 307:462–465

Okada M, Ueda H, Kometani M, Nakao K (1997) Effect of d-(E)-2-amino-4-methyl-5-phosphono-3-pentenoic acid on focal cerebral ischemia in cat. Arzneimittelforschung 47:703–705

Olney JW (1978) Neurotoxicity of excitatory amino acids. In:McGeer EG, Olney JW, McGeer PJ (eds) Kainic acid as a tool in neurobiology. Raven, New York, p95

Olney JW, Ho OL, Rhee V (1971) Cytotoxic effects of acidic and sulphur containing amino acids on the infant mouse central nervous system. Exp Brain Res 14:61–76

Olney JW, Labruyere J, Price MT (1989) Pathological changes induced in cerebrocortical neurons by phencyclidine and related drugs. Science 244:1360–1362

Ozyurt E, Graham DI, Woodruff GN, McCulloch J (1988) Protective effect of the glutamate antagonist, MK-801 in focal cerebral ischemia in the cat. J Cereb Blood Flow Metab 8:138–143

Palmer GC, Cregan EF, Borrelli AR, Willett F (1995) Neuroprotective properties of the uncompetitive NMDA receptor antagonist remacemide hydrochloride. Ann N Y Acad Sci 765:236–47; discussion 248:236–247

Paoletti P, Neyton J, Ascher P (1995) Glycine-independent and subunit-specific potentiation of NMDA responses by extracellular Mg2+. Neuron 15:1109–1120

Park CK, Nehls DG, Graham DI, Teasdale GM, McCulloch J (1988a) Focal cerebral ischaemia in the cat: treatment with the glutamate antagonist MK-801 after induction of ischaemia. J Cereb Blood Flow Metab 8:757–762

Park CK, Nehls DG, Graham DI, Teasdale GM, McCulloch J (1988b) The glutamate antagonist MK-801 reduces focal ischemic brain damage in the rat. Ann Neurol 24:543–551

Perez-Pinzon MA, Maier CM, Yoon EJ, Sun GH, Giffard RG, Steinberg GK (1995) Correlation of CGS 19755 neuroprotection against in vitro excitotoxicity and focal cerebral ischemia. J Cereb Blood Flow Metab 15:865–876

Peters S, Koh J, Choi DW (1987) Zinc selectively blocks the action of *N*-methyl-D-aspartate on cortical neurons. Science 236:589–593

Planells-Cases R, Sun W, Ferrer-Montiel AV, Montal M (1993) Molecular cloning, functional expression, and pharmacological characterization of an *N*-methyl-D-aspartate receptor subunit from human brain. Proc Natl Acad Sci USA 90:5057–5061

Premkumar LS, Auerbach A (1997) Stoichiometry of recombinant *N*-Methyl-D-Aspartate receptor channels inferred from single-channel current patterns. J. Gen. Physiol. 110: 485–502.

Priestley T, Kemp JA (1993) Agonist response kinetics of *N*-methyl-D-aspartate receptors in neurons cultured from rat cerebral cortex and cerebellum: evidence for receptor heterogeneity. Mol Pharmacol 44:1252–1257

Priestley T, Kemp JA (1994) Kinetic study of the interactions between the glutamate and glycine recognition sites on the *N*-methyl-D-aspartic acid receptor complex. Mol Pharmacol 46:1191–1196

Priestley T, Ochu E, Kemp JA (1994) Subtypes of NMDA receptor in neurones cultured from rat brain. Neuroreport 5:1763–1765

Priestley T, Laughton P, Macaulay AJ, Hill RG, Kemp JA (1996) Electrophysiological characterisation of the antagonist properties of two novel NMDA receptor glycine site antagonists, L-695,902 and L-701,324. Neuropharmacology 35:1573–1581

Priestley T, Marshall GR, Hill RG, Kemp JA (1998) L-687,414, a low efficacy NMDA receptor glycine site partial agonist in vitro, does not prevent hippocampal LTP in vivo at plasma levels known to be neuroprotective. Br J Pharmacol 124:1767–1773

Pugsley W, Klinger L, Paschalis C, Treasure T, Harrison M, Newman S (1994) The impact of microemboli during cardiopulmonary bypass on neuropsychological functioning. Stroke 25:1393–1399

Pulsinelli WA, Brierley JB, Plum F (1982) Temporal profile of neuronal damage in a model of transient forebrain ischemia. Ann Neurol 11:491–498

Reynolds IJ, Miller RJ (1989) Ifenprodil is a novel type of *N*-methyl-D-aspartate receptor antagonist: interaction with polyamines. Mol Pharmacol 36:758–765

Rigby M, Le Bourdelles B, Heavens RP, Kelly S, Smith D, Butler A, Hammans R, Hills R, Xuereb JH, Hill RG, Whiting PJ, Sirinathsinghji DJ (1996) The messenger RNAs for the *N*-methyl-D-aspartate receptor subunits show region-specific expression of different subunit composition in the human brain. Neuroscience 73:429–447

Rod MR, Auer RN (1989) Pre- and post-ischemic administration of dizocilpine (MK-801) reduces cerebral necrosis in the rat. Can J Neurol Sci 16:340–344

Root MJ, MacKinnon R (1993) Identification of an external divalent cation-binding site in the pore of a cGMP-activated channel. Neuron 11:459–466

Rosenmund C, Stern-Bach Y, Stevens CF (1998) The tetrameric structure of a glutamate receptor channel. Science 280:1596–1599

Rothman SM, Olney JW (1987) Excitotoxicity and the NMDA receptor. Trends Neurosci 10:299–302

Rowley M, Kulagowski JJ, Watt AP, Rathbone D, Stevenson GI, Carling RW, Baker R, Marshall GR, Kemp JA, Foster AC, Grimwood S, Hargreaves R, Hurley C, Saywell KL, Tricklebank MD, Leeson PD (1997) Effect of plasma protein binding on in vivo activity and brain penetration of glycine/NMDA receptor antagonists. J Med Chem 40:4053–4068

Sauer D, Nuglisch J, Rossberg C, Mennel HD, Beck T, Bielenberg GW, Krieglstein J (1988) Phencyclidine reduces postischemic neuronal necrosis in rat hippocampus without changing blood flow. Neurosci Lett 91:327–332

Sauer D, Allegrini PR, Cosenti A, Pataki A, Amacker H, Fagg GE (1993) Characterization of the cerebroprotective efficacy of the competitive NMDA receptor antagonist CGP40116 in a rat model of focal cerebral ischemia: an in vivo magnetic resonance imaging study. J Cereb Blood Flow Metab 13:595–602

Sheng M, Cummings J, Roldan LA, Jan YN, Jan LY (1994) Changing subunit composition of heteromeric NMDA receptors during development of rat cortex. Nature 368:144–147

Siesjo BK (1981) Cell damage in the brain: a speculative synthesis. J Cereb Blood Flow Metab 1:155–185

Siesjo BK, Bengtsson F (1989) Calcium fluxes, calcium antagonists, and calcium-related pathology in brain ischemia, hypoglycemia, and spreading depression: a unifying hypothesis. J Cereb Blood Flow Metab 9:127–140

Silver IA, Erecinska M (1992) Ion homeostasis in rat brain in vivo: intra- and extracellular [Ca2+] and [H+] in the hippocampus during recovery from short-term, transient ischemia. J Cereb Blood Flow Metab 12:759–772

Simon RP, Swan JH, Griffiths T, Meldrum BS (1984) Blockade of *N*-methyl-D-aspartate receptors may protect against ischemic damage in the brain. Science 226:850–852

Small DL, Buchan AM (1997) NMDA antagonists: their role in neuroprotection. Int Rev Neurobiol 40:137–71:137–171

Smirnova T, Stinnakre J, Mallet J (1993) Characterization of a presynaptic glutamate receptor. SCIENCE 262:430–433

Smith ML, Bendek G, Dahlgren N, Rosen I, Wieloch T, Siesjo BK (1984) Models for studying long-term recovery following forebrain ischemia in the rat. 2. A 2-vessel occlusion model. Acta Neurol Scand 69:385–401

Smith PL, Treasure T, Newman SP, Joseph P, Ell PJ, Schneidau A, Harrison MJ (1986) Cerebral consequences of cardiopulmonary bypass. Lancet 1:823–825

Steinberg GK, Saleh J, DeLaPaz R, Kunis D, Zarnegar SR (1989) Pretreatment with the NMDA antagonist dextrorphan reduces cerebral injury following transient focal ischemia in rabbits. Brain Res 497:382–386

Stern-Bach Y, Bettler B, Hartley M, Sheppard PO, O'Hara PJ, Heinemann SF (1994) Agonist selectivity of glutamate receptors is specified by two domains structurally related to bacterial amino acid-binding proteins. Neuron 13:1345–1357

Subramaniam S, Donevan SD, Rogawski MA (1996) Block of the *N*-methyl-D-aspartate receptor by remacemide and its des-glycine metabolite. J Pharmacol Exp Ther 276:161–168

Sucher NJ, Akbarian S, Chi CL, Leclerc CL, Awobuluyi M, Deitcher DL, Wu MK, Yuan JP, Jones EG, Lipton SA (1995) Developmental and regional expression pattern of a novel NMDA receptor-like subunit (NMDAR-L) in the rodent brain. J Neurosci 15:6509–6520
Swan JH, Meldrum BS (1990) Protection by NMDA antagonists against selective cell loss following transient ischaemia. J Cereb Blood Flow Metab 10:343–351
Takano K, Latour LL, Formato JE, Carano RA, Helmer KG, Hasegawa Y, Sotak CH, Fisher M (1996) The role of spreading depression in focal ischemia evaluated by diffusion mapping. Ann Neurol 39:308–318
Tang CM, Dichter M, Morad M (1990) Modulation of the *N*-methyl-D-aspartate channel by extracellular H+. Proc Natl Acad Sci USA 87:6445–6449
Tortella FC, Martin DA, Allot CP, Steel JA, Blackburn TP, Loveday BE, Russell NJ (1989) Dextromethorphan attenuates post-ischemic hypoperfusion following incomplete global ischemia in the anesthetized rat. Brain Res 482:179–183
Traynelis SF, Cull Candy SG (1990) Proton inhibition of *N*-methyl-D-aspartate receptors in cerebellar neurons. Nature 345:347–350
Traynelis SF, Hartley M, Heinemann SF (1995) Control of proton sensitivity of the NMDA receptor by RNA splicing and polyamines. Science 268:873–876
Tricklebank MD, Singh L, Oles RJ, Preston C, Iversen SD (1989) The behavioural effects of MK-801: a comparison with antagonists acting non-competitively and competitively at the NMDA receptor. Eur J Pharmacol 167:127–135
Tricklebank MD, Bristow LJ, Hutson PH, Leeson PD, Rowley M, Saywell K, Singh L, Tattersall FD, Thorn L, Williams BJ (1994) The anticonvulsant and behavioural profile of L-687,414, a partial agonist acting at the glycine modulatory site on the *N*-methyl-D-aspartate (NMDA) receptor complex. Br J Pharmacol 113:729–736
Wafford KA, Bain CJ, Le Bourdelles B, Whiting PJ, Kemp JA (1993) Preferential co-assembly of recombinant NMDA receptors composed of three different subunits. Neuroreport 4:1347–1349
Wafford KA, Kathoria M, Bain CJ, Marshall G, Le Bourdelles B, Kemp JA, Whiting PJ (1995) Identification of amino acids in the *N*-methyl-D-aspartate receptor NR1 subunit that contribute to the glycine binding site. Mol Pharmacol 47:374–380
Wang LY, MacDonald JF (1995) Modulation by magnesium of the affinity of NMDA receptors for glycine in murine hippocampal neurones. J Physiol (Lond) 486:83–95
Warner MA, Neill KH, Nadler JV, Crain BJ (1991) Regionally selective effects of NMDA receptor antagonists against ischemic brain damage in the gerbil. J Cereb Blood Flow Metab 11:600–610
Watanabe M, Inoue Y, Sakimura K, Mishina M (1992) Developmental changes in distribution of NMDA receptor channel subunit mRNAs. Neuroreport 3:1138–1140
Watanabe M, Inoue Y, Sakimura K, Mishina M (1993) Distinct distributions of five *N*-methyl-D-aspartate receptor channel subunit messenger RNAs in the forebrain. J Comp Neurol 338:377–390
Wenzel A, Fritschy JM, Mohler H, Benke D (1997) NMDA receptor heterogeneity during postnatal development of the rat brain: differential expression of the NR2A, NR2B, and NR2C subunit proteins. J Neurochem 68:469–478
Westbrook GL, Mayer ML (1987) Micromolar concentrations of Zn2+ antagonize NMDA and GABA responses of hippocampal neurons. Nature 328:640–643
Wieloch T (1985) Hypoglycemia-induced neuronal damage prevented by an *N*-methyl-D-aspartate antagonist. Science 230:681–683
Williams K (1993) Ifenprodil discriminates subtypes of the *N*-methyl-D-aspartate receptor: selectivity and mechanisms at recombinant heteromeric receptors. Mol Pharmacol 44:851–859
Williams K (1995) Pharmacological properties of recombinant *N*-methyl-D-aspartate (NMDA) receptors containing the epsilon 4 (NR2D) subunit. Neurosci Lett 184:181–184
Williams K (1997) Modulation and block of ion channels: a new biology of polyamines. Cell Signal 9:1–13

Williams K, Russel SL, Shen YM, Molinoff PB (1993) Developmental switch in the expression of NMDA receptors occurs in vivo and in vitro. Neuron 10:267–278

Wong EH, Kemp JA (1991) Sites for antagonism on the *N*-methyl-D-aspartate receptor channel complex. Annu Rev Pharmacol Toxicol 31:401–425

Wong EH, Kemp JA, Priestley T, Knight AR, Woodruff GN, Iversen LL (1986) The anticonvulsant MK-801 is a potent *N*-methyl-D-aspartate antagonist. Proc Natl Acad Sci USA 83:7104–7108

Yamazaki M, Mori H, Araki K, Mori KJ, Mishina M (1992) Cloning, expression and modulation of a mouse NMDA receptor subunit. FEBS Lett 300:39–45

Zhong J, Carrozza DP, Williams K, Pritchett DB, Molinoff PB (1995) Expression of mRNAs encoding subunits of the NMDA receptor in developing rat brain. J Neurochem 64:531–539

Zukin RS, Bennett MV (1995) Alternatively spliced isoforms of the NMDARI receptor subunit. Trends Neurosci 18:306–313

Subject Index

Printing: Saladruck, Berlin
Binding: Buchbinderei Lüderitz & Bauer, Berlin